ADVANCED THERMODYNAMICS

ADVANCED THERMODYNAMICS

Fundamentals - Mathematics - Applications

Mehrzad Tabatabaian, PhD, PEng
and
Er. R.K. Rajput

MERCURY LEARNING AND INFORMATION
Dulles, Virginia
Boston, Massachusetts
New Delhi

Publisher: David Pallai
MERCURY LEARNING AND INFORMATION
22841 Quicksilver Drive
Dulles, VA 20166
info@merclearning.com
www.merclearning.com
(800) 232-0223

M. Tabatabaian and Er. R.K. Rajput. *Advanced Thermodynamics: Fundamentals, Mathematics,
Applications.*
ISBN: 978-1-936420-27-8

Library of Congress Control Number: 2017934667

171819321   This book is printed on acid-free paper in the United States of America.

*To the precious memories of my mother
and to my two dear daughters and son
for their gift of love and support
Mehrzad Tabatabaian*

CONTENTS

PREFACE

Advanced thermodynamics is becoming a subject of renewed interest for its practical applications to related topics. The requirement to solve complex engineering systems along with the availability of computer power and software resources has rendered this interest more complex. Therefore, new textbooks with modern tools are in demand. This textbook is an attempt to help with this demand and also introduce some available software tools and Apps.

Simple language and consistency in presentation of the topics along with visual aids are used as a format for explaining thermodynamic fundamentals. This encourages readers to follow topics in sequence and/or choose specific ones. Mathematics is utilized as required, serving as a tool to formulate the concepts, solve problems and applications. Furthermore, numerous examples are provided to demonstrate the applications of thermodynamics for engineering problems and to enhance the use of concepts.

The broad scope and applications of thermodynamics has rendered a challenge to authors who seek to cover related topics both in textbooks and relevant publications. In general, thermodynamic textbooks have been two-sided, focused primarily on chemistry and chemical engineering, or power generation and mechanical engineering applications. Naturally, texts on statistical thermodynamics have been more linked to subjects related to physics.

This textbook focuses on the mechanical and power engineering applications of thermodynamics. It will also include statistical thermodynamic examples when relevant and pertinent. These examples will be shown either conceptually or numerically, as a means to elaborate on macroscopic thermodynamics related topics. Nevertheless, advanced topics are the central focus of this textbook, thus geared to readers versed on the fundamentals of thermodynamic at a university/college level introductory course, as well as multi-variable functions calculus.

In Chapter 1, a comprehensive and generalist view of thermodynamics is discussed, along with some historical developments in the field. The four laws of thermodynamics are also discussed.

In Chapter 2, the first and second laws are explained in detail, and concepts like entropy function and entropy generation are discussed.

In Chapter 3, statistical thermodynamics are concisely discussed, yet comprehensively explained for engineers and readers with limited previous knowledge of the topic. The chapter begins with a simple example and builds to more complex topics, such as, the Boltzmann factor, partition functions, as well as connections between microscopic and macroscopic entropies. Examples are presented to explain the physical meaning of partition functions and particle distributions, as well as calculations of average thermodynamics properties, such as, energy, entropy, and pressure.

In Chapter 4, mathematical tools are presented, such as, the Legendre transformation, the Euler chain rule, and the Jacobian methodology and applications for thermodynamic derivatives. A complete list of 336 derivatives related to 168 combinations of 56 thermodynamic functions are provided. In this Chapter, the derivations for Enthalpy, Helmholtz, and Gibbs energy functions using Legendre transformations are presented, as well as Maxwell relations and alternative relations for some thermodynamic functions. This Chapter includes explanations and formulations of the Joule-Thomson experiment and coefficient, along with the inversion curve.

In Chapter 5, material behavior including phase change (Clapeyron equation) and mathematical models (or Equations of State) are discussed. For real gases, several models and empirical approach based solutions are presented, including the compressibility Z-factor and departure functions for thermodynamic properties. The following models are also presented; the virial equation, van der Waals, Dieterici, Redlich-Kwong, Berthelot, Peng-Robinson, Beattie-Bridgeman, and lastly, Benedict-Webb-Rubin. Several worked-out examples are presented along with application of the Joule-Thompson inversion-curve measure for comparing some of these models performances. The Maxwell construct method and its application for multi-phase region are also discussed.

In Chapter 6, non-reactive and reactive gas mixtures are discussed. Mixture relations for several models, like Dalton's and Amagat's, Z-factor, and Kay's mixing rules are elaborated upon. The combustion phenomenon is the primary focus for reactive mixtures. Several relevant and worked-out numerical examples are provided for different combustion chambers, fuels, for products thermodynamic conditions with demonstrations of applying the first and second thermodynamic laws. Concepts like air-fuel ratio and adiabatic combustion temperature are also explained.

In Chapters 7 to 13, thermodynamic system applications like steam power cycles, compound steam engines, steam turbines, gas power cycles, gas turbines, jet propulsion, and refrigeration cycles are presented. Several relevant and worked-out examples are provided, in order to demonstrate the detail calculations related to these specific systems.

In Chapter 14, derivation of relations for several specific kinds of finite-time thermodynamic systems, or endo-reversible engines, such as Curzon-Ahlborn, Novikov, and Müse engines are explained along with their maximum-power efficiency relations. This is then compared with Carnot efficiency and practical efficiencies of several types of physical power plants.

Mehrzad Tabatabaian, PhD, PEng
Vancouver, BC
October 2017

ABOUT THE AUTHOR

Dr. Mehrzad Tabatabaian is a faculty member—instructor and program head—at the Mechanical Engineering Department, School of Energy at BCIT with several years of teaching experience. In addition to teaching courses in the fields of thermodynamics, energy systems management, and modeling, he also does research on renewable energy systems and modeling. Dr. Tabatabaian is also School of Energy Research Committee Chair and actively involved in energy-initiative activities. He has published several papers in various scientific journals and conferences, textbooks on multiphysics and turbulent flow modeling, and also holds several patents in the energy field.

Recently, Mehrzad volunteered to help establish a new division at Engineers and Geoscientists British Columbia (EGBC), Division for Energy Efficiency and Renewable Energy. Mehrzad offers several PD seminars for the EGBC members on the subjects of wind power, solar power, renewable energy, and multiphysics modeling methods for engineers.

Mehrzad Tabatabaian got his BEng from Sharif University of Technology, and graduated from McGill University (MEng and PhD). He has been an active academic, professor, and engineer in leading alternative energy, oil, and gas industries. Mehrzad also has a Leadership Certificate from the University of Alberta, holds an EGBC PEng license, and is an active member of ASME.

Acknowledgments

I would like to thank David Pallai and Jennifer Blaney of MERCURY LEARNING AND INFORMATION for their ongoing support and encouragement during the publication of this book.

CHAPTER 1

A HOLISTIC VIEW OF THERMODYNAMICS

In This Chapter

- Overview
- Some Historical Development
- Laws of Thermodynamics
- Some Basic Definitions
- Exercises

1.1 Overview

Thermodynamics is one of the major pillars of science and engineering with a very wide scope of applications in natural phenomena, the environment, and industry. Both its theoretical and practical outcomes have been serving us to better understand and interpret natural phenomena as well as the efficiency limit of power generation plants and heat engines, chemical processing facilities, and so forth, and to advance our state of practice. The first traces of historical development in recent history go back to the late seventeenth century when G. W. Leibnitz hinted at the 1st law in terms of conservation of kinetic and potential energy for a closed mechanical system [1], [2]. Major developments in thermodynamics occurred in the late eighteenth until the beginning of the twentieth century; highlighted is the work of Carnot, Joule, Kelvin, Clausius, Maxwell, Boltzmann, Gibbs, Planck, Einstein, and Nerst, among others [3], [4]. These developments consist of two distinguished paths or approaches which complement each other—the so-called *classical thermodynamics* and *statistical thermodynamics*.

The former takes a macroscopic viewpoint approach (i.e., very larger than the atomic scale of constituents), whereas the latter considers the atomic/particulate nature of things with a statistical analysis treatment. Classical thermodynamics (sometimes referred to as engineering thermodynamics) is successful in analyzing and predicting behavior of matter with a relatively small number of parameters (like temperature, pressure, volume—these will be defined later on) while ignoring or not needing to consider microscopic physics of matter [5], [6]. However, the microscopic viewpoint has important contributions in both the understanding of macroscopic variables and functions (like entropy and temperature) as well as building a foundation to link the whole science of thermodynamics to quantum mechanics [7], [8]. The vast scope and applications of thermodynamics have brought the challenge of covering the related topics as a whole in textbooks and related publications to the interested authors. In general, thermodynamics-related textbooks have focused on either chemistry and chemical engineering or power generation and mechanical engineering applications, with of course a common share of statistical thermodynamics, which is more linked in physics-related subjects.

For this book we focus on the engineering approach, including a summary of statistical thermodynamics for references where we find it useful, either conceptually or by numerical examples, to explain the related topics in macroscopic thermodynamics. The emphasis is, however, on advanced topics, and as mentioned in the preface we assume that readers are familiar with thermodynamics fundamentals at the level of a university/college introductory course. For a summary of basic thermodynamics concepts and definitions, see the relevant section at the end of this chapter.

1.2. Some Historical Development

Thermodynamics development was a cyclical process rather than a serial one. That is, the development of concepts and mathematical formulation came after some of its previously defined laws, which were expressed sporadically, mainly based on axioms, observations, and experimental studies [9]. The serious development of thermodynamics started [10] with the necessity of answering practical questions related to the engineering and performance of heat engines at the time of Sadi Carnot (named after the Persian poet Sadi of Shiraz [1210], [11]) in the early nineteenth century. Heat engines, or machines which convert heat to mechanical work, were very inefficient at the time, since a small percentage of input heat was converted to useful work and mostly not using a thermo-dynamic cycle. It was not clear to the industry and designers whether the inefficiency was due to poor design of the system, improper working fluid (e.g., water), or a natural limitation in the whole conversion process that prevented the improvement and maximization of the efficiency. Carnot's insightful thought experiment and analysis of an ideal heat engine not only answered these questions adequately, but also formed the founda-tion of thermodynamics, more precisely the 2nd law of thermodynamics [12]. He solved the problem of heat engine efficiency with his analysis considering the best possible, although ideal, engine and with just using the principle of energy conservation (cited as "fall of heat") and a postu-late compatible with observations in nature (i.e., heat cannot move from a relatively cold reservoir to a hot one, solely, without any external input/ work to the system). He concluded that the efficiency of his ideal revers-ible engine (consisting of a cycle with two isothermal expansion/heating and compression/cooling connected with two adiabatic expansion and compression paths) *only* depends on the temperatures of the heating and cooling reservoirs. He also showed that his engine had the highest effi-ciency of all heat machines that could possibly be made connected to the same heat sources. The huge breakthrough of Carnot's insight and analysis (although not recognized/realized fully at his time, [13]) guided the steam engine industry to build more efficient machines, knowing the limitations expressed by Carnot's ideal engine. But there was more to it! Carnot, through his analysis of thermal engines, opened a door to the formal science of thermodynamics. Later on, Rudolf Clausius (hinted at by Clapeyron and William Thomson, see Chapter 5) looked at Carnot's results and developed the concept of a new state function, so-called entropy S, which is of such central importance to the 2nd law that entropy and the 2nd law are nowadays usually considered synonymous. The 2nd law of thermodynamics is given by Clausius's inequality, Equation 1.1:

$$dS \geq \frac{\delta Q}{T}, \; \oint \frac{\delta Q}{T} \leq 0 \qquad (1.1)$$

where dS is the incremental change in entropy, δQ is the heat exchange, and T is the absolute temperature along a given thermodynamic path between two state points. The line-integral (symbolized by $\oint$) indi-cates integration around a closed loop or a thermodynamic cycle. Clausius (1865) also gave very complete definitions of the classical thermodynamic

1st and 2nd laws with his famous statements applicable to isolated systems [14]:

The Energy of the universe is constant.
The Entropy of the universe tends to a maximum.[1]

According to these statements the imbalances of energy in an isolated system tend to equilibrium by minimization of the total energy differences, and through this process the entropy maximizes while total energy is conserved. In other words, the total sum of the energy remains constant, but the usefulness of energy diminishes, which manifests itself through a measure of lost useful work, which is in turn measured by entropy. The main mechanism for the entropy increase is irreversibility, which exists in real processes like contact friction, losses in a pipe flow due to pipe friction, plastic deformation in solids, and so on.

Any phenomenon is allowed to happen, in nature or industry, if satisfying the 1st and 2nd laws of thermodynamics. It is possible though to conceptualize machines that satisfy the 1st law, but not the second law (so-called perpetual machines), hence the 1st law is necessary but not sufficient. It is the 2nd law that checks possible versus not-possible phenomena in our physical universe. Calculation of the entropy of a system is the key guide for engineers and scientists to guarantee that their suggested design, process, and so forth work, but from a practical point of view they are not usually concerned about what this entropy actually means physically. Questions like:

- what is entropy?

- what is entropy good for?

- what is entropy generation?

- what is the relation between entropy and efficiency?

are usually asked, and their answers come, at least more clearly, from the microscopic point of view of thermodynamics using statistical mechanics. We will be engaged with entropy and its calculations in Chapter 2 and in further chapters in this book. We believe that understanding entropy and entropy generation is useful for practitioners and engineers and can help them in the industrial design of machines.

---

[1] *Die Energie der Welt ist Konstant. Die Entropie der Welt strebt einem Maximum.*

Maxwell and Boltzmann made major contributions to the field of statistical thermodynamics [8]. To understand why we seek help from statistical methods with useful applications in mechanics and thermodynamics, let's consider a relatively simple system, say a container full of a monatomic gas like helium (*He*). Four grams of He (or one mole) contains ~6 × 10^{23} of helium atoms (Avogadro's number). Following the Newtonian mechanics approach to solve the motion of these particles that move around, colliding with one another and the container's walls, we have the following scenario. Each atom has six degrees of freedom (three positional and three for the momentum, ignoring other possible motions). Hence we have ~36 × 10^{23} unknowns and the same number of simultaneous equations at hand to solve at any given time during the motion of these atoms. Unless the atoms start from stationary initial positions and velocities (i.e., absolute zero temperature—not possible according to the 3rd law of thermodynamics), we also need to include the non-zero initial boundary/initial conditions for solving these equations. This analysis of a dynamical system of moving particles is an insurmountable task even for modern/upcoming exa-flop computers (however, it may be possible using quantum computers[2]). In addition, in general, we are not interested to know the logistics of every particle during its motion in the container; rather we are interested to know, for example, the volume, pressure, temperature, and other such properties of the whole system of particles. In any event, the particle-level dynamical approach is not practical, at least at the present time, for engineering problems, and hence a statistical averaging treatment of the motion of particles is needed. It is shown [6], [8] that the statistical ensemble type of sampling of particles/atoms/molecules represents the behavior of these particles. Maxwell-Boltzmann velocity distribution (experimentally verified by Miller and Kusch, [15]) shows the validity of this approach. Using the statistical thermodynamics approach, Boltzmann found the relationship between the macroscopic entropy and positional/momentum possible microscopic configurations of atoms at equilibrium [16]. This realization can be expressed as follows: in our example the helium gas at a given temperature and pressure occupies the container volume. From the naked-eye point of view nothing changes, and these variables (i.e., pressure, volume, temperature, etc.) are constant, as long as the experiment boundary conditions remain stationary/intact. But at the particle/atom level there is a lot of movement

---

[2] *A 5-qubit computer announced by IBM (http://www.research.ibm.com/quantum/, May 2016)*

and a huge number of particles changing their positions and velocities, which is referred to as *configuration*. However, at the macroscopic level these configurations have the same *complexion* to us. Entropy is proportional to the maximum number of particle configurations possible (the so-called configuration weight), W (through Boltzmann constant $k_B = 1.38 \times 10^{-23}$ *J/mol.K*) given as Equation 1.2

$$S = k_B \ln W \tag{1.2}$$

Entropy could be interpreted as our lack of information (or intentional ignorance!) at the macroscopic level from what is actually happening at the microscopic level. The "beauty" of classical thermodynamics is that the Carnot/Clausius formulation and interpretation give the same results for entropy as that of Boltzmann. These interpretations are related to each other through k_B (which could be interpreted as entropy per particle). Therefore, in practical applications of thermodynamics for engineering problems (except in some areas like chemistry and physical chemistry), we don't need, in general, to consider or even be aware of the fact that matter is made of atoms/molecules. In further chapters we will discuss statistical thermodynamics in more detail.

1.3. Laws of Thermodynamics

There are four fundamental concepts or laws that form the foundation of thermodynamics. These laws were discovered and developed based mainly on observations and experimental research, that is, a phenomenological approach [17], with the addition of their mathematical derivations and analyses and some geometrical interpretation of thermodynamics state functions ([8], [18]). The discovery and conceptual development of these laws was not chronological, but are complete as a whole. Each law defines a thermodynamic *concept* and a related *quantity*. These laws are:

- Zeroth law (*equilibrium* and *temperature*): defines the temperature as a property of a system based on the thermal equilibrium concept. In brief, if we have systems A and B in equilibrium with each other, and similarly systems B and C, then we can conclude that systems A and C are in equilibrium as well. This allows us to measure the temperature of systems that are not in physical contact (i.e., systems A and C) using a thermometer (i.e., system B).

- 1st law (*conservation* and *energy*): defines the state of energy of a system, usually focusing on internal energy, based on the principle of conservation of total energy. In thermodynamics, we consider the balance of heat and work in relation to internal energy (or other energy functions like enthalpy) for both closed and open systems.

- 2nd law (*irreversibility/reversibility* and *entropy*): defines entropy change as a state property of a system based on some principles, that is, those of Carnot, Clausius, and Kelvin-Plank [19]. For reversible cycles the entropy change is balanced with temperature-scaled heat exchanges, whereas in irreversible cycles it doesn't balance out. Clausius's inequality relation summarizes the 2nd law (see Equation 1.1).

- 3rd law (*absolute temperature* and *entropy*): defines a reference for entropy at absolute zero temperature. As temperature decreases the entropy change also decreases. Hence, we can set the entropy reference value at zero. Nevertheless, zero absolute temperature is not achievable in practice.

Among these laws, we mostly use the 1st and 2nd laws for calculations and analysis in engineering problems, which mainly involve the study of thermodynamic equilibrium states. We will give a review of these two important laws of thermodynamics in the next chapter of this book, as well as for future reference for the related terminologies and notations.

Modern thermodynamics (see, e.g., [20]) is more concerned with the non-equilibrium or close-to-equilibrium transformations. This point of view is a continuation of what Clausius [21] called "uncompensated transformation," a positive quantity with the dimensions of entropy which is the result of irreversibility that may exist in a system/process. Some authors refer to this quantity as entropy generation, $S_{Gen} = \Delta S - \int \frac{\delta Q}{T} > 0$. The treatment leads to the definition of *thermodynamic forces*, F which drive *thermodynamics flows*, dX to form the entropy of the system, $dS = FdX$. For example, temperature gradient is the thermodynamic force driving an irreversible thermodynamic flow of heat, or concentration gradient forces the flow of matter. A great contribution to the area of non-equilibrium thermodynamics is the work of Lars Onsager, who introduced and defined Onsager reciprocal relations [22], sometimes referred to as the 4th law of thermodynamics.

1.4. Some Basic Definitions

- Thermodynamics is an axiomatic science which deals with the relations among heat, work, and properties of systems which are in equilibrium. It basically entails four laws or axioms known as the Zeroth, 1st, 2nd, and 3rd laws of thermodynamics.

- A system (or object) is a finite quantity of matter for a prescribed region of space. An object can be separated from the rest of its surroundings.

- A system may be a closed, open, or isolated system.

- The boundary of a system is the physical interface between the system and its surroundings.

- A system is defined as being in isolation when no instantaneous changes occur within the system. This is an idealized assumption, but it is rather useful for the study of a system.

- Equilibrium for a system is defined when an isolated system is undergoing no change at all.

- A system in thermodynamic equilibrium can change from one state of equilibrium to another, hence equilibrium thermodynamics is the defined study of such changes.

- A property of a system is a characteristic of the system which depends upon its state, but not upon how the state is reached.

- Interaction between two systems, each originally at equilibrium, will impose new conditions to the equilibrium of the combined systems.

- State is the condition of the system at an instant of time as described or measured by its properties. Each unique condition of a system is called a state.

- A process occurs when the system undergoes a change in state or an energy transfer takes place at a steady state.

- Any process or series of processes whose end states are identical is termed a cycle.

- The pressure of a system is the force exerted by the system on a unit area of boundaries. A vacuum is defined as the absence of pressure.

- A reversible process is one which can be stopped at any stage and reversed so that the system and surroundings are exactly restored to their initial states.

- An irreversible process is one in which heat is transferred through a finite temperature.

- The Zeroth law of thermodynamics states that if two systems are each equal in temperature to a third system, they are equal in temperature to each other.

- The 3rd law of thermodynamics states that the entropy of all perfect crystalline solids is zero at absolute zero temperature.

- A phase is a quantity of matter which is homogeneous throughout in chemical composition and physical structure.

- A homogeneous system is one which consists of a single phase.

- A heterogeneous system is one which consists of two or more phases.

- A pure substance is one that has a homogeneous and invariable chemical composition even though there is a change of phase.

1.5. Exercises

1. Explain the differences and relations between classical and statistical thermodynamics.

2. Explain the four laws of thermodynamics and list the related concept and quantity associated with each one.

3. Explain entropy as defined in classical and statistical thermodynamics.

4. Explain the concepts of reversible and irreversible processes with some daily-life examples.

5. Explain the meaning of this statement: "Thermodynamics is an axiomatic science."

6. A cup of fresh coffee is left intact in a room with normal temperature. The next day, we reach the coffee and find out that: a) the coffee feels colder or b) the coffee feels hotter! Discuss each case with respect to the 1st and 2nd laws of thermodynamics. (Hint: Consider the cup of coffee as the system and the room as the surroundings, i.e., the room is insulated from the outside world.)

CHAPTER 2

REVIEW OF THE 1ST AND 2ND LAWS OF THERMODYNAMICS

In This Chapter

- State and Path Functions
- 1st Law of Thermodynamics
- 2nd Law of Thermodynamics
- Microscopic Boltzmann Entropy
- Summary of Relations for Common Thermodynamic Processes
- Exercises

2.1 State and Path Functions

In this section we review two laws of thermodynamics, the 1st and 2nd laws. These laws are major tools for engineering calculations of thermodynamic systems. In general, the laws of thermodynamics are relations among thermodynamic properties or variables of matter. These variables are usually divided into two groups: *state-variables* or functions and *path-variables*. A state is defined at a point in thermodynamic space coordinates, for example pressure-volume-temperature. Thermodynamic state functions depend only on variables related to an equilibrium point in a path or cycle. For example, internal energy, enthalpy, and entropy are state functions which could be functions of temperature, pressure, or volume. Obviously, the value of a state function does not change when we go around a thermodynamic cycle, provided we start and end the journey from/to the same point. This simple fact forms the foundation of cyclic work-producing machines, like internal combustion engines. In contrast, path-variables depend on the path we choose to move from one state point to another. For example, work and heat are path-dependent variables. If we start from a point in a cycle and return to the same point, then we cannot say that we have the same amount of heat or work exchanges, since we can travel along infinite number of paths and arrive at the same point that we started from. For this simple fact we cannot define heat or work for a given matter at a state point in a cycle. Figure 2.1 shows a sketch for state and path functions.

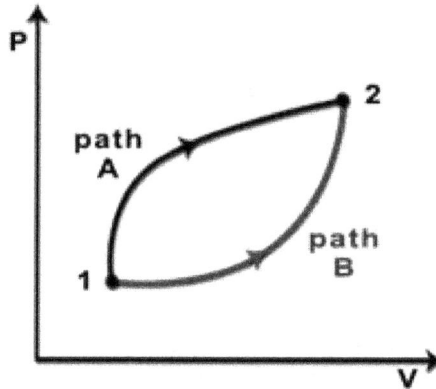

FIGURE 2.1 A sketch for thermodynamic state and path functions.

It is worth mentioning that the work-to-heat conversion process can occur with 100% efficiency without the need of having a cycle, whereas the heat-to-work conversion process has an efficiency of <100% and requires a cycle [23].

2.2. 1st Law of Thermodynamics

The concept of energy for matter and its conservation is the core of the 1st law. It was initiate, in its mechanical form and coined by Leibnitz in 1695 [4], and further developed for thermodynamics by Helmholtz and Clausius [24], [21]. Energy can be in different forms, for example, kinetic, potential, chemical, thermal, and mechanical work. For a closed system (i.e., only heat and/or work exchanges happen through the boundaries) we write the 1st law as the balance of internal energy, U and the amount of heat, δQ and work, δW exchanges or $dU = \delta Q - \delta W$, if we consider the addition of heat to the system to be positive (hence increasing the internal energy) and delivering work by the system to the surroundings to be positive, since it decreases the internal energy. Therefore, we can write $\delta Q = Q_{in} - Q_{out}$ and $\delta W = -W_{in} + W_{out}$. In other words, the sign of the heat exchange is positive when heat is input to the system and positive when work is delivered by the system. This is the traditional sign convention in thermodynamics, more suitable for engineers, consistent with its history in relation to steam engines. Some authors consider the opposite sign for work exchange, that is, work input is also considered as a positive quantity. In that case we have $dU = \delta Q + \delta W$. Therefore, we can write $\delta Q = Q_{in} - Q_{out}$ and $\delta W = W_{in} - W_{out}$. This is a relatively the modern sign convention (IUPAC, www.iupac.org) more suitable for chemistry and physics. According to both conventions, the final result is the same, having the work given as $W = -\int p dV$, where p is pressure and V is the volume of the matter. Careful considerations should be given to the definition of p. This is actually the external pressure acting on a moving wall of a system to produce work, for example a piston-cylinder filled with a gas. It could be equal to the internal pressure of the gas in the cylinder, if and only if the process is reversible, that is, the wall is moving very slowly among intermediate equilibrium states. Otherwise, the pressure inside the container cannot be defined properly as a variable at a point and hence is not equal to the external pressure everywhere inside the gas. It should be clear that in practice we can pull/push a piston in a cylinder very fast, but we cannot define a path between these two state-point positions in a state diagram, for example on a $p - V$ diagram. But if we give the system some time to reach its equilibrium state, we can consider internal pressure as a constant. Some authors prefer to define work as $W = -\int P_{ext} dV$ to emphasize on this subtle point. Table 2.1 shows a summary of both sign conventions and the resulting volume change and the 1st law. Both sign conventions result in the same final form of the 1st law or

$$dU = \delta Q - p dV \qquad (2.1)$$

Energy balance	Sign, δQ	Sign, δW	Sketch	1ˢᵗ law
$dU = \delta Q - \delta W$ (engineering sign convention)	heat taken <0	work taken >0 (U decreases)	$W = \int pdV, \Delta V > 0$	
	heat added >0	work added <0 (U inreases)	$W = \int pdV, \Delta V < 0$	$dU = \delta Q - pdV$
$dU = \delta Q + \delta W$ (modern sign convention)	heat taken <0	work taken <0 (U decreases)	$W = -\int pdV, \Delta V > 0$	
	heat added >0	work added >0 (U increases)	$W = -\int pdV, \Delta V < 0$	

TABLE 2.1 Sign Conventions for Heat and Work w.r.t. Gas in the System and Resulting Volume Change for the 1st Law Dashed lines show piston initial position.

For open systems (i.e., where heat, work, and/or matter exchanges happen through the boundaries) the energy transfer by exchanged mass, dE_{mass} should also be added to the energy balance. The total energy relation reads as follows:

$$dU = \delta Q + \delta W + dE_{mass}$$

Considering the kinetic and potential energies, in addition to the internal energy for substance mass flow rate $\dot{m}$, we can write the energy balance for an open system as

$$\dot{Q}_{in} + \dot{W}_{in} + \underbrace{\sum_{in} \dot{m}\left(h + \frac{v^2}{2} + gz\right)}_{for\ each\ inlet} = \dot{Q}_{out} + \dot{W}_{out} + \underbrace{\sum_{out} \dot{m}\left(h + \frac{v^2}{2} + gz\right)}_{for\ each\ outlet}$$

where v is velocity and $h = u + pV$ is specific enthalpy, g gravitational acceleration, and Z elevation. This equation is applicable to any medium in any steady flow.

2.3. 2nd Law of Thermodynamics

The 2nd law is considered one of the most important discoveries in physics (see Einstein's quote[3]). It is universally applicable to all natural phenomena and when satisfied we can determine if our human-made machines are feasible to work. This law was conceptualized by Carnot, although not explicitly mentioned at the time (see videos by Adam Shulman, "the The Discovery of Entropy" [25], and "An Introduction to Chemical Change and Thermodynamics" by George Porter [26]).

At the time of Carnot the heat was considered as a type of massless fluid or *caloric*—that is, heat as a "substance" is never destroyed but moves among bodies (Cleghorn [27], Lavoisier [28]). Modern physics [29], rejects the caloric theory, and considers heat as a form of energy that can be consumed and transferred to other forms, like mechanical work. Carnot believed in the caloric nature of heat [12], but the rejection of caloric heat later on by modern physics didn't jeopardize his major contribution and insight for laying down the foundation of the 2nd law. Carnot's insight came from his analogy of the heat engine with water wheels. He made the analogy of difference in heights with that of temperatures and the water with caloric heat. The same way that water wheels generates work by water falling through its the paddles, the rotation of heat driven by a temperature difference can generate mechanical work. The reverse operation is also possible, or we can lift heat by adding work to the system the same way as we can lift water by rotating a water wheel in the opposite direction, for example.

There are several statements for the 2nd law, those of Kelvin-Planck, Clausius, and Carnot, [19]. In brief, these statements are based on

---

[3]"A theory is the more impressive the greater the simplicity of its premises is, the more different kinds of things it relates, and the more extended is its area of applicability. Therefore, the deep impression which classical thermodynamics made upon me. It is the only physical theory of universal content concerning which I am convinced that within the framework of the applicability of its basic concepts, it will never be overthrown." *Autobiographical Notes* (1946), 33. Quoted in Gerald Holton and Yehuda Elkana, *Albert Einstein: Historical and Cultural Perspectives* (1997), 227.

observations and experiments that are consistent with our everyday life experiences and common sense or so-called axioms. For example, heat flows from a hot body to a relatively cold body once they are in thermal contact—not vice versa!— as if, there is a *preferred direction* for energy transfer. Actually Carnot, as mentioned previously, used this simple axiom in conjunction with a thought experiment for his ideal heat engine to arrive at three important results: a) the fall of heat can generate work, b) continuous work can be generated for a cycle, and c) a reversible cycle is the most efficient one. In other words, the efficiency of heat engines only depends on the reservoirs' temperatures, and for an ideal cycle it has the maximum possible value, but not 100%. Any heat engine's efficiency can be written as $\eta = 1 - \dfrac{Q_2}{Q_1}$, where Q_2 is rejected heat and Q_1 is heat input. For Carnot's ideal engine it can be shown that $\eta = 1 - \dfrac{T_2}{T_1}$, where T_1 and T_2 are hot and cold reservoirs temperatures, respectively [3], [30], [23]. Therefore, we have $\dfrac{Q_1}{T_1} = \dfrac{Q_2}{T_2}$. This result leads to a very important equation in thermodynamics which was discovered by Clausius, or $\sum_i \left(\dfrac{Q_i}{T_i} \right)_{rev} = 0$, considering the sign convention for heat exchanges (i.e., heat input as positive and heat output to a system as negative quantities). In other words, for a reversible ideal cycle the algebraic sum of heat exchanged divided by their corresponding temperatures is equal to zero. Therefore, the quantity $\dfrac{Q}{T}$ is a state variable, that is, it does not depend on involved process paths (recall that Q is a path-dependent variable). Clausius called this function *Entropy*[4] [14], [21], [31]. For irreversible cycles, with at least one irreversible path included in, the sum of heat exchanges scaled by T^{-1} is not equal to zero, or $\sum_i \left(\dfrac{Q_i}{T_i} \right)_{irr} < 0$. To better describe and help in understanding these relations, we consider a cycle which consists of an irreversible path from point 1 to point 2, that is, $1 \rightarrow 2$, and a reversible path from 2 to 1, that is, $2 \rightarrow 1$, as shown in Figure 2.2 by solid curves. Note that the path from 2 to 1 could be an irreversible one as well.

At point 1 or 2, we have $\oint dS = \Delta S = S_1 - S_1 = S_2 - S_2 = 0$. Now we travel along the irreversible path while measuring the quantity $\dfrac{\Delta Q_{irr}}{T}\bigg|_{1 \rightarrow 2}$.

For this derivation, we use symbol T to represent the corresponding temperature. Also we calculate the change in entropy between these points, or $\int_1^2 dS = \Delta S_{1\rightarrow 2} = S_2 - S_1$. If we compare these two values, we find out that

$$S_2 - S_1 > \left.\frac{\Delta Q_{irr}}{T}\right|_{1\rightarrow 2} \tag{2.2}$$

This inequality is the key concept for understanding the 2nd law. It tells us that our measurement unit, $\left.\dfrac{\Delta Q_{irr}}{T}\right|_{1\rightarrow 2}$ is not sufficient for measuring the total change in entropy from point 1 to point 2, due to irreversibility. Let's turn this inequality to an equality, by subtracting a

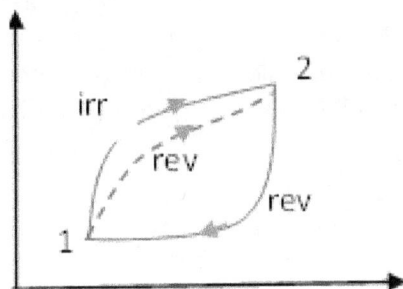

FIGURE 2.2 A cycle with reversible and irreversible paths.

positive quantity, S_{Gen} (sometimes referred to as entropy generation) from $\Delta S_{1\rightarrow 2}$ or $\Delta S_{1\rightarrow 2} = S_2 - S_1 - S_{Gen} = \left.\dfrac{\Delta Q_{irr}}{T}\right|_{1\rightarrow 2}$ which can be written as $\Delta S_{1\rightarrow 2} = \left.\dfrac{\Delta Q_{irr}}{T}\right|_{1\rightarrow 2} + S_{Gen}$. The entropy generated during the heat transfer along the irreversible path increases with the level of irreversibility. In other

---

[4] "*We might call S the transformational content of the body, just as we termed the magnitude U its thermal and ergonal content. But as I hold it better to borrow terms for important magnitudes from the ancient languages, so that they may be adopted unchanged in all modern languages, of the body, from the Greek word [πρoπη trope], transformation. I have intentionally formed the word entropy so as to be as similar as possible to the word energy; for the two magnitudes to their physical meanings, that a certain similarity in designation appears to be desirable.*"

words, the value of S_{Gen} is a measure of deviation of an irreversible process from an ideal reversible one (imagine that we have another reversible path from point 1 to point 2 as well, shown in Figure 2.2 with a dashed curve). Now we return from point 2 to the original point 1 through the reversible path $2 \to 1$. Similarly, we measure $\left. \dfrac{\Delta Q_{rev}}{T} \right|_{2 \to 1}$ along this path and compare

it with the entropy change, or $\int_{2}^{1} dS = \Delta S_{2 \to 1} = S_1 - S_2$. We find out that they are equal, or

$$S_1 - S_2 = \left. \frac{\Delta Q_{rev}}{T} \right|_{2 \to 1} \tag{2.3}$$

Therefore, the measurement unit of $\left. \dfrac{\Delta Q_{rev}}{T} \right|_{2 \to 1}$ is sufficient for measuring

the change in entropy from point 2 to point 1, since the path is a reversible one. We can also conclude that along a reversible path $\left. S_{Gen} \right|_{rev} = 0$. Now we combine Equation 2.2 and Equation 2.3, after multiplying the latter by -1, to get

$$\left. \begin{array}{l} S_2 - S_1 > \left. \dfrac{\Delta Q_{irr}}{T} \right|_{1 \to 2} \\[4mm] S_2 - S_1 = - \left. \dfrac{\Delta Q_{rev}}{T} \right|_{2 \to 1} \end{array} \right\} \overset{yields}{\to} \quad \left. \frac{\Delta Q_{irr}}{T} \right|_{1 \to 2} + \left. \frac{\Delta Q_{rev}}{T} \right|_{2 \to 1} < 0 \tag{2.4}$$

Now we consider the third path (shown by a dashed curve in Figure 2.2), being a reversible one joining points 1 and 2. Using a similar argument as mentioned previously, we can write $S_2 - S_1 = \left. \dfrac{\Delta Q_{rev}}{T} \right|$. Note that the value of $\Delta S_{1 \to 2} = S_2 - S_1$ remains the same, since entropy is a state function, and hence it does not matter which path we select to move from point 1 to point 2. Now we combine this relation with the one previously obtained for reversible path $2 \to 1$, that is, Equation 2.3, to arrive at

$$\left. \begin{array}{l} S_2 - S_1 = \left. \dfrac{\Delta Q_{rev}}{T} \right|_{1 \to 2} \\[4mm] S_1 - S_2 = \left. \dfrac{\Delta Q_{rev}}{T} \right|_{2 \to 1} \end{array} \right\} \overset{yields}{\to} \quad \left. \frac{\Delta Q_{rev}}{T} \right|_{1 \to 2} + \left. \frac{\Delta Q_{rev}}{T} \right|_{2 \to 1} = 0 \tag{2.5}$$

By combining Equation 2.4 with Equation 2.5, we arrive at the celebrated Claudius inequality, which is a mathematical statement of the 2nd law, $\begin{cases} \dfrac{\Delta Q_{rev}}{T}\bigg|_{1\to 2} + \dfrac{\Delta Q_{rev}}{T}\bigg|_{2\to 1} = 0 \\[2ex] \dfrac{\Delta Q_{irr}}{T}\bigg|_{1\to 2} + \dfrac{\Delta Q_{rev}}{T}\bigg|_{2\to 1} < 0 \end{cases}$. These relations are usually written in a more

compact form using closed-path-integral notation (i.e. $\oint = \int_1^2 + \int_2^1$) instead of summation, as

$$\oint \frac{\delta Q}{T} \leq 0 \tag{2.6}$$

In Equation 2.6 the equal sign is for reversible cycles and the unequal sign is for a cycle with at least one irreversible path. Further, using $\oint dS = \Delta S = S_1 - S_1 = S_2 - S_2 = 0$, we can write $\oint \frac{\delta Q}{T} \leq \oint dS$ which requires us to have

$$dS \geq \frac{\delta Q}{T} \tag{2.7}$$

Also, using the Clausius inequality we can write

$$\oint \frac{\delta Q}{T} + S_{Gen} = 0 \tag{2.8}$$

where $S_{Gen} = 0$ for fully reversible cycles and $S_{Gen} > 0$ for cycles with at least one irreversible path.

Further, we expand the analysis to provide a reasonably clearer physical meaning for entropy [32]. We already mentioned entropy generation for irreversible processes and entropy increase concepts. For a reversible cycle we can write the 1st law as $dU = TdS - \delta W_{rev}$, assuming the engineering sign convention for work. Similarly, for an irreversible cycle we can write $dU = \delta Q_{irr} - \delta W_{irr}$. Now combining these two relations, we get $TdS - \delta W_{rev} = \delta Q_{irr} - \delta W_{irr}$, or

$$dS = \frac{\delta Q_{irr}}{T} + \frac{1}{T}\left(\delta W_{rev} - \delta W_{irr}\right) \tag{2.9}$$

but using Carnot's principles (i.e., the efficiency of a reversible cycle is the maximum possible one) we have $\delta W_{rev} > \delta W_{irr}$. Therefore, $dS > \frac{\delta Q_{irr}}{T}$.

We can interpret the quantity $(\delta W_{rev} - \delta W_{irr})$ as lost work, δW_{lost} due to irreversibility. Note that the previously defined entropy generation can be written as

$$S_{Gen} = \frac{1}{T}\left(\delta W_{rev} - \delta W_{irr}\right) = \frac{\delta W_{lost}}{T} \tag{2.10}$$

Equation 2.9 shows that entropy change depends on heat transfer and lost work. This has implications in the design of thermal systems, like heat exchangers. One may assume that by minimization of S_{Gen}, or similarly lost work, we can optimize the performance of heat exchangers, but it turns out that this is not always the case [33].

In conclusion and for practical applications, we can define entropy as a function of a quantity of heat which shows the possibility of conversion of that heat into work. The increase in entropy is small when heat is added at a high temperature and is greater when heat addition is made at a lower temperature. Thus for maximum entropy there is a minimum availability for conversion into work, and for minimum entropy there is a maximum availability for conversion into work.

The calculation of entropy is needed in the engineering design of a thermal system. Through the following examples we will calculate entropy for an ideal gas, and more examples will be given in further chapters.

2.3.1. Entropy for an Ideal Gas

For one mole of an ideal gas, we have $pV = RT$ and $dU = C_v dT$. After substitution for $dU = TdS - PdV$ and rearranging the terms, we get $dS = C_v \frac{dT}{T} + R \frac{dV}{V}$. Integrating between two state points we get

$$\Delta S = S_2 - S_1 = \int_1^2 C_v \frac{dT}{T} + R \int_1^2 \frac{dV}{V} \tag{2.11}$$

For constant heat capacity we have $\Delta S = S_2 - S_1 = C_v ln\frac{T_2}{T_1} + Rln\frac{V_2}{V_1}$. However, in general C_v is a function of temperature, and the value of $\int_1^2 C_v \frac{dT}{T}$ can be calculated once the functional relation for C_v is given.

Entropy change in terms of temperature and pressure can be obtained by using $C_v = C_p - R$ or $\Delta S = S_2 - S_1 = \int_1^2 C_p \frac{dT}{T} + R \int_1^2 \left(\frac{dV}{V} - \frac{dT}{T}\right)$. But $d(pV) = RdT$ or $pdV + Vdp = RdT$, and after dividing L.H.S. by pV

and R.H.S. by RT (recall that $pV = RT$), we get $\dfrac{pdV + Vdp}{pV} = \dfrac{RdT}{RT}$ or $\dfrac{dV}{V} - \dfrac{dT}{T} = -\dfrac{dp}{p}$. Substituting into the previous relation gives

$$\Delta S = S_2 - S_1 = \int_1^2 C_p \frac{dT}{T} - R \int_1^2 \frac{dp}{p} \tag{2.12}$$

For constant C_p we have $\Delta S = S_2 - S_1 = C_p ln \dfrac{T_2}{T_1} - Rln \dfrac{p_2}{p_1}$.

For cases when heat capacity is not assumed as a constant and is a function of temperature, we can write $\int_{T_1}^{T_2} C_p \dfrac{dT}{T} = \int_0^{T_2} C_p \dfrac{dT}{T} - \int_0^{T_1} C_p \dfrac{dT}{T} = S_2^0 - S_1^0$. The values of $S^0(T)$ for commonly used gases are tabulated (e.g., see Appendix 1 of reference [19]). Finally, using Equation 2.12, we have

$$\Delta S = S(T_2, p_2) - S(T_1, p_1) = S^0(T_2) - S^0(T_1) - Rln \frac{p_2}{p_1} \tag{2.13}$$

Example: Entropy Change for CO_2

Calculate the entropy change for 1 kmol of CO_2 as an ideal gas when it is heated from 100°C and 1 atm to 250°C and 3 atm.

■ Solution:

From the property table we read for CO_2; S^0 (523K) = 236.834 kJ/kmol.K and S^0 (373K) = 222.367 kJ/kmol.K. Using Equation 2.13, we get

$$\Delta S = 236.834 - 222.367 - 8.314 ln \frac{3}{1} = 5.333 \text{ kJ/kmol.K.}$$

2.3.2. Entropy Change for an Open System

In an open system mass and associated energy exchanges happen between the system and the surroundings, in addition to possible heat and work exchanges. The entropy changes satisfy the following inequality relation

$$dS \geq \frac{\delta Q}{T} + \sum_i \left(s_i dm_i - s_0 dm_0 \right)$$

where dm is the mass exchange.

2.4. Microscopic Boltzmann Entropy

Although the macroscopic definition of entropy in relation to the 2nd law does not necessarily require the atomic nature of matter, it is this latter viewpoint that helps understanding entropy more clearly and more naturally. When atoms in a given volume of a gas at a certain pressure wander around, they can be at different configurations (i.e., space and momentum/velocity), whereas the macroscopic volume, pressure, and temperature of the gas remain constant for a given set of boundary conditions. The main question is then what are the chances or probability of any possible configuration to happen. The number of configurations is a huge number, since we have a very large number of atoms (order of Avogadro's number), but the most probable ones are more likely to happen rather than the less probable ones to achieve equilibrium. Boltzmann [16] used statistical mechanics and the concept of ensemble average to arrive at his celebrated formula that says that entropy is actually a function of the number of probable microscopic configurations (see Equation 1.2) or $S = k_B \ln W$.

To help in understanding this concept, let's assume a hot body and a cold body. The atoms in the hot body have a higher kinetic energy represented by the temperature of the body at the macroscopic level. Similarly, the atoms of the colder body have a lower kinetic energy. If we put these two bodies in contact, the probability of all atoms having a lower average temperature distributed in the domain of the two bodies is much higher than the probability of having the colder body getting relatively colder and the hot body getting hotter through exchanging the same amount of heat. Therefore, we conclude that heat is transferred from hot body to cold body, not vice versa (at least the probability of the latter happening is practically zero and theoretically extremely small!). As another example, let's imagine a container with just a limited number of identical particles, say three (without loss of generality). We put these particles in one half of the container, say the left side, and the other half, say the right side, is empty, with a physical barrier separating the compartments. We measure the number of probable configurations based on particles being in the left side of the container or in the right side of it. Now we remove the barrier. There are eight possible configurations (see Chapter 3) for the distribution of particles between the two sides of the container. Therefore, entropy is $S = k_B \ln 8$. Note that $8 = 2^3$, with 2 being the number of volumes/compartments in the container and 3 the number of particles. In general we can show by deduction that for n moles of gas we have $N = nN_A$, and hence

$$S = nN_A k_B \ln 2 \tag{2.14}$$

where N_A is Avogadro's number.

In the next chapter we expand on these results along with a summary of statistical thermodynamics and a relatively more complete treatment of the topic as a background and introduction of statistical thermodynamics.

2.5. Summary of Relations for Common Thermodynamic Processes

- There can be no machine which would continuously supply mechanical work without some form of energy disappearing simultaneously. Such a fictitious machine is called a *perpetual motion machine of the first kind*, or in brief, is thus impossible.

- The energy of an isolated system is always constant.

- In case of:

(*i*) Reversible constant volume process (V = constant)

$$\Delta U = mC_V \left(T_2 - T_1\right); W = 0; Q = mC_V \left(T_2 - T_1\right)$$

(*ii*) Reversible constant pressure process (p = constant)

$$\Delta U = mC_V \left(T_2 - T_1\right); W = p\left(V_2 - V_1\right); Q = mC_p \left(T_2 - T_1\right)$$

(*iii*) Reversible temperature or isothermal process (pV = constant)

$$\Delta U = 0; W = p_1 V_1 \ln r; Q = W$$

where r is the expansion or compression ratio.

(iv) Reversible adiabatic process (pV^γ = constant)

$$\pm\Delta U = \mp W = \frac{mR(T_1 - T_2)}{\gamma - 1}; Q = 0; \frac{T_2}{T_1} = \left(\frac{V_1}{V_2}\right)^{\gamma-1} = \left(\frac{p_2}{p_1}\right)^{\frac{\gamma-1}{\gamma}}$$

(v) Polytropic reversible process (pV^n = constant)

$$\Delta U = mC_V(T_2 - T_1); W = \frac{mR(T_2 - T_1)}{n - 1}; Q = \Delta U + W$$

$$\text{and } \frac{T_2}{T_1} = \left(\frac{V_1}{V_2}\right)^{n-1} = \left(\frac{p_2}{p_1}\right)^{\frac{n-1}{n}} Q = \left(\frac{\gamma - n}{\gamma - 1}\right)W$$

• The steady flow equation can be expressed as follows:

$$u_1 + \frac{C_1^2}{2} + Z_1 g + p_1 v_1 + Q = u_2 + \frac{C_2^2}{2} + Z_2 g + p_2 v_2 + W$$

$$h_1 + \frac{C_1^2}{2} + Q = h_2 + \frac{C_2^2}{2} + W, \text{ neglecting } Z_1 \text{ and } Z_2$$

where, Q = Heat supplied per kg of fluid; W = Word done by 1 kg of fluid;
C = Velocity of fluid; Z = Height above datum;
p = Pressure of the fluid; u = Internal energy per kg of fluid;
pv = Energy required per kg of fluid.

This equation is applicable to any medium in any steady flow.

• In unsteady-flow processes, the rates at which mass and energy enter the control volume may not be the same as the rate of flow of mass and energy moving out of the control volume. The filling of a tank is an example of an unsteady-flow process.

2.6. Exercises

1. Explain path and state functions with examples.

2. Describe the definitions for thermodynamic state point, process, and cycle.

3. Describe the 1st law and the related sign convetions for work and heat.

4. Explain why we cannot define heat and work at a state point.

5. Describe the 2nd law statements as defined by Carnot, Kelvin/Planck, and Clausius.

6. Describe the relations of reversible and irreversible cycles/processes with the quantity $\Sigma \dfrac{Q}{T}$

7. What does entropy generation mean in a process/cycle? Explain entropy generation using Equation 2.8.

8. Explain the relation between lost work and entropy generation using Equation 2.10.

9. Calculate the entropy change for 1 kmol of the following gases. Assume the ideal gas law for each gas when changing from (110°C, 1 atm) to (150°C, 2.5 atm).

    a) H_2O (vapor)   b) CO_2 c) N_2 d) Ar

CHAPTER 3

STATISTICAL THERMODYNAMICS FOR ENGINEERS: A SUMMARY

In This Chapter

- Overview
- Distribution and Configuration Weight
- Boltzmann Factor and Partition Function
- Macroscopic Thermodynamic Property Relations
- Physical Meaning of Thermodynamic Partition and Probability Functions
- Gibbs Statistical Thermodynamics
- Exercises

3.1. Overview

In this chapter we give a brief introduction to statistical thermodynamics concepts to provide a common background for readers and to help the understanding of thermodynamics concepts and quantities in relation to their microscopic and macroscopic viewpoints. Readers who are interested to learn more can refer to the reference section and a selection of sources given at the end of this chapter. We refer to atoms and molecules as particles synonymously in this book, specifically in this chapter.

Statistical thermodynamics, or in general statistical mechanics, considers the atomic/molecular structures of matter for the derivation of governing equations or laws and material *average* behavior. One can say that this approach links "molecules" to "engines" or "particles" to "continuum." In statistical thermodynamics a probabilistic approach is taken to calculate the average properties, at equilibrium, for a large number of particles in a system. For example, pressure is the average force exerted on a unit area of the wall of a container by striking particles, with a possible variation in the number of strikes and their amplitude fluctuations. In other words, if we can measure pressure with very high accuracy, we can capture its fluctuations about an average value. For a given set of conditions, at an equilibrium state, the average value of the pressure can be treated as the pressure of the system while neglecting its fluctuations, which are relatively very small (typically about $10^{-12} = (degrees\ of\ freedom)^{-0.5}$ of the average value, [6]).

The general idea of studying and analyzing the dynamics of particles using classical mechanics (e.g., Newtonian), and consequently their macroscopic continuum quantities, is appealing. This approach should include the detail of the mechanics of particles in motion (i.e., position and momentum) and their interactions with one another and calculate the macroscopic value of quantities like pressure, temperature, and entropy. Although this approach seems appealing, it faces two major roadblocks: a) the volume of calculations and data required to trace the trajectories of particles and their dynamics is overwhelming (for the current computers and those in the foreseeable future) due to the huge number of particles involved (e.g., in a mole of a gas it is of the order of 10^{23}), and b) laws of classical mechanics are insensitive to the direction of time (Maxwell's demon [34], [35]). In other words, if we reverse the directions of the applied "forces" on the particles, then they should move in the reverse direction as well—a phenomenon that we don't observe in practice. As an example, consider a glass of water falling from the table and shattering on the floor. Now suppose that if we reverse all the forces applied on the particles, imagine playing a video capture of the scene backward, then the shattered particles should gather and rearrange themselves on the table where they were originally located, as a glass of water. The chance of this happening by *itself* is practically zero! Therefore, there seems to be an inherent preference in the way that phenomena happen and energy transfers from one form to another. We need to have new "laws" which are more inclusive than classical mechanics laws of motion, in particular considering a statistical average behavior of particles to solve practical problems.

3.2. Distribution and Configuration Weight

In statistical mechanics we need to employ probability theory when applied to a large number of particles in a system or assembly. To summarize it here, we consider a simple system and expand our discussion based on the results obtained to more real practical systems which have a very large number of particles (order of Avogadro's number ~10^{23}). Let's assume that we have five glass marbles which are identical and labeled from 1 to 5 (i.e., distinguishable), and we are asked to distribute these marbles into three empty cups labeled A, B, and C, as shown in Figure 3.1.

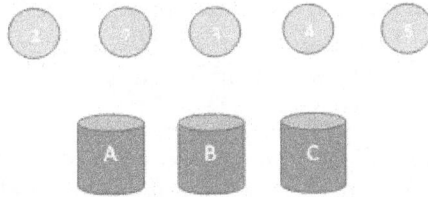

FIGURE 3.1 A sketch of marbles and cups for considering possible distributions of marbles.

There are many possible configurations; for example, we can put all the marbles in cup A for having the configuration (5,0,0) or we can put 2 in cup A, 3 in cup B, and 0 in cup C to have the configuration (2,3,0). For every possible configuration we can have several distributions of marbles. Table 3.1 shows possible distributions for configuration (2,3,0), for example.

Cup A ($n_A = 2$)	Cup B ($n_B = 3$)	Cup C ($n_C = 0$)
Distribution of marbles, distinguished by their labels		
1,2	3,4,5	null
1,3	2,4,5	null
1,4	2,3,5	null
1,5	2,3,4	null
2,3	1,4,5	null
2,4	1,3,5	null
2,5	1,3,4	null
3,4	1,2,5	null
3,5	1,2,4	null
4,5	1,2,3	null

TABLE 3.1. Distribution of Five Particles into Three Containers with Configuration (2,3,0)

The first column in Table 3.1 gives the possible selections of marbles which could be placed in cup A, assuming that only two marbles are allowed in cup A. Note that the marbles are identical in size, properties, and so on, but they are distinguishable because they are numbered from 1 to 5. Each row of Table 3.1 shows possible selections of marbles for having the configuration (2,3,0). The total number of rows or possible distributions for a given configuration is called the *weight of configuration*, W. In other words, W is the number of *microstates* or *distributions* corresponding to a given *macrostate* or *configuration*. For this example W = 10 (i.e., the corresponding number of rows in Table 3.1).

To expand from this example to a more general case of physical systems, with a large number of particles we use combinatorial formula [7]

$$W = \frac{N!}{\prod_i^C n_i!} = \frac{N!}{n_1! n_2! n_3! \cdots n_c!} \tag{3.1}$$

where N is the total number of particles, C is the total number of containers or energy levels, and n_i is the number of particles in container i. Note that N and C can have different values and they are not necessarily equal. For generality, we replaced the letter-index symbol for the number of particles in each container to a number-index symbol, for example $n_1 \equiv n_A$, $n_2 \equiv n_B$, $n_3 \equiv n_C$, and so forth. Therefore, for our example given in Table 3.1, we have $W = \frac{5!}{2! \times 3! \times 0!} = \frac{5 \times 4 \times 3 \times 2 \times 1}{2 \times 1 \times 3 \times 2 \times 1 \times 1} = 10$. Similarly, for configuration (1,3,1) we can calculate $W = \frac{5!}{1! \times 3! \times 1!} = 20$, and for configuration (2,2,1) we get $W = \frac{5!}{2! \times 2! \times 1!} = 30$. Other possible configurations are (5,0,0) with $W = 1$, and (4,1,0) with $W = 5$. These configurations and their corresponding weights are given in Table 3.2. Note that different configurations can have the same weight.

Configurations	No. of configurations (Total = 21)	Weight (W)	Remark
(5,0,0), (0,5,0), (0,0,5)	3	1	Minimum weight
(4,1,0), (4,0,1), (1,4,0), (1,0,4) (0,4,1), (0,1,4)	6	5	
(3,2,0), (3,0,2), (2,3,0), (2,0,3) (0,3,2), (0,2,3)	6	10	
(3,1,1), (1,3,1), (1,1,3)	3	20	
(2,2,1), (2,1,2), (1,2,2)	3	30	Maximum weight

TABLE 3.2 Possible Configurations and Corresponding Weights for the Example Considered

The maximum number of configurations for a given number of particles and containers is given as

$$W_t = \frac{(N+C-1)!}{(C-1)!N!} \tag{3.2}$$

For this example, we can list the total number of configurations (i.e., $W_t = \frac{(5+3-1)!}{(3-1)!5!} = 21$, or the number of configurations as listed in the first column of Table 3.2) and calculate the maximum weight which corresponds to any of the configurations.

For this example we had two constraints: having the total number of marbles and containers as constants. In real systems we have similar constraints, simply by replacing the containers with discrete energy levels and the constraint of energy conservation to have a unique (i.e., equilibrium) configuration corresponding to maximum W. Calculating the maximum value of the configuration weight is very important from the statistical mechanics point of view. This can be interpreted as the degrees-of-freedom for the particles to be distributed and hence result in a final specific configuration. The higher the numerical value of the configuration weight the more likely that we get the corresponding configuration as the final result of particle distribution. For our example the distributions corresponding to W = 30 have the highest probability to happen.

3.3. Boltzmann Factor and Partition Function

Manual calculation of configuration weight is not practically feasible for real systems with a very large number of particles. Hence, we need to develop and use a more systematic approach based on the mathematics of probability and a system's physical properties. Nevertheless, there are a few useful points that we can learn from the simple example mentioned in the previous section to help develop more relevant calculations.

It is obvious from the example that the calculation of the configuration weight W requires desired or allowable configuration n_i, (see Equation 3.1). For example, for the given configuration (2,3,0) we can calculate its weight to be equal to 10. However, for real systems we don't know the desired configuration of the particles in advance. In addition we also need to apply the constraint of having the total number of particles as a constant, or

$$\sum_i n_i = N \tag{3.3}$$

and for a real system (e.g., a gas) we are required to satisfy the energy conservation of the system, according to the 1st law. The total energy of a large collection of particles is defined (note that index i is a counter for energy levels, here)

$$\sum_i n_i \, \epsilon_i = E \qquad (3.4)$$

where ϵ_i is the energy of energy of each particle at level i and E is the total energy. Therefore, for example, we have n_0 particles with energy ϵ_0, n_1 particles with energy ϵ_1, and so on. The probability is given as $P_i = \frac{n_i}{N}$, by definition. Now we substitute for n_i in Equation 3.3 and Equation 3.4 to get

$$\sum_i P_i = 1 \text{ and } \sum_i P_i \, \epsilon_i = \frac{E}{N} = \langle E \rangle \qquad (3.5)$$

where <E> is the average energy per particle.

We continue the discussion with an important question: that is, at what values of n_i do we have the corresponding maximum value of W? (or 30 in our example, see the last row in Table 3.2) This is a mathematical question of finding the maximum value of the function $W(n_i)$, for given N (see Equation 3.1). It is useful to mention here that according to Maxwell-Boltzmann statistics, the maximum W corresponds to the equilibrium state distribution [36]. In order to calculate the maximum value of W, we equivalently consider its logarithm instead, $\ln W$ and let $\frac{\partial \ln W}{\partial n_i}\Big|_N = 0$. We use Sterling's approximation for the factorials, or $N! \cong N^N e^{-N}$ (more accurate as N increases) to get, using Equation 3.1, $\ln W = \ln \frac{N!}{\prod_i n_i!} = \ln \left(N^N e^{-N} \right)$, $-\ln \left(n_1{}^{n_1} e^{-n_1} \times n_2{}^{n_2} e^{-n_2} \times ... \right) = N \ln N - N - \left(n_1 \ln n_1 - n_1 + n_2 \ln n_2 - n_2 + \cdots \right)$, and after substituting for $(n_1 + n_2 + ...) = N$ (see Equation 3.3), we get

$$\ln W = N \ln N - \sum_i n_i \ln n_i \qquad (3.6)$$

Now we differentiate Equation 3.6 w.r.t. n_i (recall that summation is over entire available energy levels, i) $\frac{\partial \ln W}{\partial n_i}\Big|_N = \underbrace{\frac{\partial}{\partial n_i} (N \ln N)}_{=0} - \frac{\partial}{\partial n_i} \left(\sum_i n_i \ln n_i \right) =$

$-\left(\ln n_i + 1 \right)$. Since we want to calculate the maximum W, we let $\frac{\partial \ln W}{\partial n_i}$ be equal to zero, to have

$$\ln n_i + 1 = 0 \qquad (3.7)$$

Equation 3.7 gives the value of n_i that maximizes ln W (or equivalently W). However, maximum W should be subjected to two constraints for real systems: conservation of particles ($\Sigma_i n_i - N = 0$) and conservation of energy, according to the 1st law ($\Sigma_i n_i \epsilon_i - E = 0$); see Equation 3.3 and Equation 3.4, respectively.

Using the Lagrangian multipliers method [37], we form a linear combination of Equation 3.3, Equation 3.4, and Equation 3.7 (after differentiating Equation 3.3 and Equation 3.4 w.r.t. n_i, since Equation 3.7 is the result of differentiating w.r.t. n_i as well) to have

$$\left(\ln n_i + 1\right) + \alpha \frac{\partial}{\partial n_i}\left(\sum_i n_i - N\right) + \beta \frac{\partial}{\partial n_i}\left(\sum_i n_i \in_i -E\right) = 0 \qquad (3.8)$$

where α and β are parameters to be determined. After performing the differentiation operations, we get ln $n_i + 1 + \alpha + \beta\epsilon_i = 0$. Some authors [17] neglect "1" in this relation, since ln n_i is relatively large. From this relation we can calculate n_i as

$$n_i = e^{-(1+\alpha)}e^{-\beta_i \in_i} \qquad (3.9)$$

Equation 3.9 gives the value of $n_i = n_i^*$ that maximizes W subjected to the two constraints as mentioned above, and is a very important result in statistical mechanics. We still need to calculate the parameters α and β.

3.3.1. Statistical Entropy: Calculation of Lagrange Multipliers

In order to calculate α we use the constraint $\Sigma_i \, n_i = N$, and after substituting for n_i into Equation 3.9, we have $e^{-(1+\alpha)} \Sigma_i e^{-\beta\epsilon_i} = N$. Therefore, $e^{-(1+\alpha)} = \frac{N}{\Sigma_i e^{-\beta\epsilon_i}}$. Back substituting into Equation 3.9 for n_i, we get n_i^*:

$$n_i^* = \frac{Ne^{-\beta\epsilon_i}}{\sum_i e^{-\beta\epsilon_i}} \qquad (3.10)$$

The sum $\mathcal{z} = \Sigma_i\left(e^{-\beta\epsilon_i}\right)$ is the *partition function* which gives the distribution of particles energies over various available energy levels. In other words, referring to the quantity $e^{-\beta\epsilon_i}$ as the *Boltzmann factor*, the partition function is the sum of the Boltzmann factor. This $\mathcal{z}$ function is rich with information about the system and is a very important quantity in statistical mechanics. We will discuss the physical meaning of this function and other parameters in further sections.

In order to calculate α we make use of the relation $e^{-(1+\alpha)} = \frac{N}{\Sigma_i e^{-\beta\epsilon_i}} = \frac{N}{Z}$ to get

$$\alpha = \ln Z - \ln N - 1 \tag{3.11}$$

Equation 3.11 clearly shows that α is a linear function of Z on a semilogarithmic plot. Also, form Equation 3.10 for n_i^* and recalling $P_i = \frac{n_i}{N}$, we have

$$P_i^* = \frac{e^{-\beta\epsilon_i}}{\sum_i e^{-\beta\epsilon_i}} = \frac{e^{-\beta\epsilon_i}}{Z} \tag{3.12}$$

or the probability, that is the Boltzmann distribution, that corresponds to maximum W.

The calculation of β is more involved. β is related to the general level of particle energy. The relation for the partition function, or $Z = \Sigma_i e^{-\beta\epsilon_i}$, shows that Z is a function of β and ϵ_i, or $Z = Z(\beta, \epsilon_i)$. Physically, this means that the partition function depends on volume through ϵ_i and temperature through β. This leads us to the state function entropy, $S = S(T, V)$. Using this hint we can use the microscopic or statistical entropy to relate to its macroscopic or Clausius entropy. The macroscopic entropy is that of the statistical one multiplied by the Boltzmann constant (see Equation 1.2). For derivation, we first calculate the microscopic entropy relation and then use it to get our macroscopic one.

Using Equation 3.6, after substituting for $n_i = NP_i$, we get $\ln W = N \ln N - N\ln N \underbrace{\Sigma_i P_i}_{=1} - N\Sigma_i P_i \ln P_i$, or finally, we define

$$S_{mic} = -N\sum_i P_i \ln P_i \tag{3.13}$$

Multiplying Equation 3.13 by k_B (Boltzmann constant, 1.381×10^{-2} J/K) and substituting for P_i from Equation 3.12, we get the macroscopic entropy $S =$

$$k_B S_{mic} = -k_B N\sum_i P_i \ln \frac{e^{-\beta\epsilon_i}}{z} = -k_B N\left\{\sum_i P_i(-\beta\epsilon_i - \ln Z)\right\} = k_B N\beta \underbrace{\sum_i P_i \epsilon_i}_{=\frac{E}{N}} + \underbrace{k_B N \ln Z}_{using \Sigma_i P_i=1}$$

After simplifying we have

$$S = k_B\beta E + k_B N \ln Z \tag{3.14}$$

The function $\psi = \ln Z$ is the *Massieu function*, which shows up in relations for thermodynamic properties (see Table 3.3). A quick dimensional analysis of Equation 3.14 shows that β has the unit of inverse of energy or it should be inversely proportional to absolute temperature. To get the exact full relation for β, we make use of the equation for the 1st law in its differential form, or $dE = TdS - pdV$. Note, here we use symbol E for internal energy or total energy, synonymously. Therefore, at a constant volume and a fixed number of particles we have $\frac{\partial S}{\partial E} = \frac{1}{T}$. Using Equation 3.14, we have

$\frac{\partial S}{\partial E} = \frac{\partial}{\partial E}(k_B \beta E) + \frac{\partial}{\partial E}(k_B N \ln \mathcal{Z})$. It can be shown that $\frac{\partial}{\partial E}(k_B N \ln \mathcal{Z}) = 0$,

[17]. Finally, by equating the two relations for entropy we get $\frac{\partial S}{\partial E} = \frac{1}{T} = k_B \beta$, or

$$\beta = \frac{1}{k_B T} \tag{3.15}$$

Therefore, from Equation 3.9 we get $n_i = \frac{N}{\mathcal{Z}} e^{-\epsilon_i / k_B T} = N e^{-(\epsilon_i / k_B T + \psi)}$.

This completes the calculation of α and β and entropy along the way. An alternative relation for S can be derived by substituting for β, using Equation 3.15, into the equation for entropy, that is, Equation 3.14, or

$$S = \frac{E}{T} + k_B N \ln \mathcal{Z} \tag{3.16}$$

Readers may note that using the statistical approach, the relation for macroscopic entropy and statistical microscopic entropy was established and found, as given by Equation 3.14, or equivalently by Equation 3.16.

3.4. Macroscopic Thermodynamic Property Relations

In this section we use the Massieu function, $\psi = \ln \mathcal{Z}$, in order to derive average macroscopic thermodynamic properties like energy, pressure, temperature, Helmholtz energy, and entropy. The methodology for the derivations is given in detail by Tribus [30] based on the works of Jaynes [38], [39]. The objective in all of the derivations is to write these macroscopic properties in terms of ψ.

- Average Energy is given by $\langle E \rangle = \Sigma_i P_i \epsilon_i$ (see Equation 3.5). But by

  Equation 3.10, after using $\mathcal{Z} = e^\psi$, we have $P_i = \frac{e^{-\beta \epsilon_i}}{\mathcal{Z}} = e^{-(\psi + \beta \epsilon_i)}$.

  Substituting for P_i, we get $\langle E \rangle = \sum_i P_i \epsilon_i = \sum_i e^{-(\psi + \beta \epsilon_i)} \epsilon_i = \epsilon_1 e^{-(\psi + \beta \epsilon_1)} +$

  $\epsilon_2 e^{-(\psi + \beta \epsilon_2)} + \cdots$. This relation is actually equal to $-\frac{\partial \psi}{\partial \beta}$, since

$$-\frac{\partial \psi}{\partial \beta} = -\frac{\partial}{\partial \beta}(\ln \mathcal{Z}) = -\frac{\partial}{\partial \beta}\ln\left[\sum_i(e^{-\beta\epsilon_i})\right] = -\frac{\partial}{\partial \beta}\left[\ln\left(e^{-\beta\epsilon_1} + e^{-\beta\epsilon_2} + \cdots\right)\right] =$$

$$\underbrace{\frac{\left(\epsilon_1 e^{-\beta\epsilon_1} + \epsilon_2 e^{-\beta\epsilon_{2+\cdots}}\right)}{\left(e^{-\beta\epsilon_1} + e^{-\beta\epsilon_{2+\cdots}}\right)}}_{=\mathcal{Z}} = \frac{\sum_i e^{-(\beta\epsilon_i)}\epsilon_i}{e^\psi} = \sum_i \epsilon_i\, e^{-(\psi+\beta\epsilon_i)}. \text{ Therefore, we}$$

conclude the following relation for the average macroscopic energy

$$\langle E \rangle = -\frac{\partial \psi}{\partial \beta} = -\frac{\partial}{\partial \beta}(\ln \mathcal{Z}) \tag{3.17}$$

Note that $\frac{\partial \langle E \rangle}{\partial \beta} = -\frac{\partial^2 \psi}{\partial \beta^2}$, which shows that $\frac{\partial \langle E \rangle}{\partial \beta}$ is always negative, since

an increase in β (or decrease in temperature) means a decrease in energy.

- Average Entropy (i.e., entropy for a closed system) is $\langle S \rangle = \frac{\langle E \rangle}{T} + k_B \underbrace{\ln \mathcal{Z}}_{\psi}$

  (see Equation 3.16). Substituting for $\langle E \rangle$ from Equation 3.17 we get
  $\langle S \rangle = -\frac{1}{T}\frac{\partial \psi}{\partial \beta} + k_B\psi$. After substituting for $1/T = \beta k_B$ (see Equation 3.15)
  we get

$$\langle S \rangle = -\beta k_B \frac{\partial \psi}{\partial \beta} + k_B\psi \tag{3.18}$$

- Pressure is defined as the statistical average of the forces (i.e., rate of momenta) exerted by the particles on a unit surface. In a system of particles pressure variation changes the energy levels, that is, ϵ_i, but not the population distribution of particles, that is, P_i, among available energy levels. On the contrary, temperature variations, or more precisely β, reshuffle the particle distributions among the energy levels. In other words, and using an analogy, if the system is considered as a flexible rubber staircase with a given particle distribution on each step of the staircase, the pressure would just change the heights of the stairs (equivalent to stretching the rubber staircase) and the temperature juggles, up or down, the particles over the unstretched staircase steps. The related mathematical relation is obtained by writing the total differential of average energy (i.e., Equation 3.5), or $d\langle E \rangle = \sum_i \epsilon_i\, dP_i + \sum_i P_i d\,\epsilon_i$, where $\sum_i \epsilon_i\, dP_i = \delta Q$ is the heat exchange and $\sum_i P_i d\,\epsilon_i = \delta W$ is the work exchange.

It can be shown (see [30], Chapter 6) that the $\delta W = -\langle P \rangle dV = \sum_i P_i d\,\epsilon_i$ results in the following relation for average pressure $p = \langle P \rangle$ in relation to volume, V:

$$p = \frac{1}{\beta}\frac{\partial \psi}{\partial V} \tag{3.19}$$

• The average enthalpy of a system can be obtained using the relation $\langle H \rangle = \langle E \rangle + pV$. After substituting for energy and pressure, using Equation 3.17 and Equation 3.19, we get

$$H = -\frac{\partial \psi}{\partial \beta} + \frac{V}{\beta}\frac{\partial \psi}{\partial V} \tag{3.20}$$

• In Chapter 4 we will define two additional energy functions: Helmholtz and Gibbs functions (see Equation 4.4 and Equation 4.8). Here we derive their macroscopic average relations. For the Helmholtz function, we have $\langle A \rangle = \langle E \rangle - T\langle S \rangle$. Using Equation 3.15, Equation 3.17, and Equation 3.18 we get $\langle A \rangle = -\frac{\partial \psi}{\partial \beta} - T\left(-\beta k_B \frac{\partial \psi}{\partial \beta} + k_B \psi\right) = -\frac{\psi}{\beta}$. For the Gibbs energy function, we have $\langle G \rangle = \langle A \rangle + pV$. Using Equation 3.19 we get $\langle G \rangle = -\frac{\psi}{\beta} + \frac{V}{\beta}\frac{\partial \psi}{\partial V}$. Finally, we have

$$\begin{cases} \langle A \rangle = -\dfrac{\psi}{\beta} \\[2mm] \langle G \rangle = -\dfrac{\psi}{\beta} + \dfrac{V}{\beta}\dfrac{\partial \psi}{\partial V} \end{cases} \tag{3.21}$$

We list the key relations obtained so far in this section in Table 3.3 for reference (note, angle brackets are dropped for simplicity).

$n_i = \dfrac{Ne^{-\beta \epsilon_i}}{\sum_i e^{-\beta \epsilon_i}} = Ne^{-\psi - \beta \epsilon_i}$	$z = \sum_i e^{-\beta \epsilon_i} = e^{\psi}$	$P_i = \dfrac{e^{-\beta \epsilon_i}}{\sum_i e^{-\beta \epsilon_i}} = e^{-\psi - \beta \epsilon_i}$
$A = -\dfrac{\psi}{\beta}$	$S = k_B \psi - \beta k_B \dfrac{\partial \psi}{\partial \beta}$	$E = -\dfrac{\partial \psi}{\partial \beta} = -\dfrac{N}{z}\dfrac{\partial z}{\partial \beta}$
$p = \dfrac{1}{\beta}\dfrac{\partial \psi}{\partial V}$	$\beta = \dfrac{1}{k_B T}$	$\alpha = \psi - \ln N - 1$

TABLE 3.3. Summary of Main Relations for Average Statistical Thermodynamic Properties

The practical application of relations given in Table 3.3 becomes evident when the function $\psi = \psi(\beta, V)$ is available. By manipulating the relations given in Table 3.3, one can find additional relations among macroscopic properties[30]. These relations actually recall classical thermodynamics laws, for example the 1st and 2nd laws and principles, like those of Carnot and Clausius. Experimental measurements could be used for finding the macroscopic average properties.

3.5. Physical Meaning of Thermodynamic Partition and Probability Functions

For a system of N particles (e.g., an ideal gas) with total energy E we are not interested in determining the amount of energy associated with each particle, but we would like to know how the energy is shared among the particles or, equivalently, how particles occupy available energy levels while satisfying total energy conservation. Considering the glass marbles example, we would like to know the possible distributions of five particles among three containers that give us the highest configuration weight, correspondingly. To answer this question we should recall from quantum mechanics that energy levels are quantized [7]. In other words, the particles in our system, under given constraints, can exist in a finite number of allowed microscopic states. This number can be very big, but it is finite, and every particle can occupy one of the available energy levels ϵ_i. By definition the lowest energy level is $\epsilon_0 = 0$, therefore we have $\epsilon_i = 0$, ϵ_1, $\epsilon_2 \dots$. The higher the temperature the more energy levels are available to particles for occupation. At a given time the particles will be in a particular configuration with, for example, n_0 particles occupying energy level ϵ_0, and n_1 particles at energy level ϵ_1, and so forth. The distribution n_i can change, or particles can jump up or down among the energy levels w.r.t. time while their total number and their total energy is conserved. The maximum configuration weight corresponds to the most probable distribution, which is the equilibrium state of the system.

To clarify further, let's assume a system of $N = 3$ particles with total energy $N = 3\Delta$ and four energy levels available to the particles which are equally spaced at Δ, or $\epsilon_i = 0$, Δ, 2Δ, 3Δ. The partition function is the sum of the Boltzmann factor, hence $\mathcal{Z} = \sum_i e^{-\beta \epsilon_i} = 1 + e^{-\beta\Delta} + e^{-2\beta\Delta} + e^{-3\beta\Delta}$.

$\beta\Delta$	3	1	0.7	0.3
$\mathcal{Z}$	1.052389	1.553002	1.865639	2.6962
$\psi = \ln \mathcal{Z}$	0.051063	0.4401898	0.6236036	0.9918434

TABLE 3.4. Values of Partion and Massieu Functions for a Four-Level Energy System

The dimensionless quantity $\beta\Delta = \frac{\Delta}{k_B T}$ can take different values based on values of temperature of the system. Table 3.4 shows the values of $\mathcal{Z}$ for several arbitrarily chosen values of $\beta\Delta$. Note that for decreasing values of β, temperature increases.

The probability of a given particle occupying energy level ϵ_i at equilibrium distribution (corresponding to maximum configuration weight) is $P_i^* = \frac{e^{-\beta\epsilon_i}}{\mathcal{Z}}$ (see Equation 3.12). For example, for $\beta\Delta = 1$ and the corresponding $\mathcal{Z} \cong 1.553$ the Boltzmann probability distribution is $P_0^* = \frac{1}{1.553} = 0.644$, $P_1^* = \frac{e^{-1}}{1.553} = 0.237$, $P_2^* = \frac{e^{-2}}{1.553} = 0.087$, and $P_3^* = \frac{e^{-3}}{1.553} = 0.032$. These results satisfy the constraint given by Equation 3.5, $\sum_i P_i^* = 0.644 + 0.237 + 0.087 + 0.032 = 1$, as expected.

Now, for generality, we extend this example to a system with possible infinite energy levels. Therefore, the partition function reads $\mathcal{Z} = \sum_i e^{-\beta\epsilon_i} = 1 + e^{-\beta\Delta} + e^{-2\beta\Delta} + e^{-3\beta\Delta} + \cdots$. This is a geometric series and therefore we have $\mathcal{Z} = \frac{1}{1-e^{-\beta\Delta}}$. Table 3.5 shows the corresponding values of $\mathcal{Z}$ for given values of $\beta\Delta$ as well as the distribution of probabilities P_i^* for the first twenty energy levels. The higher the temperature (i.e., lower β) the more levels of energy are populated by the particles, as expected. It is useful to graph the particle probability distributions at various energy levels, that is, energy level populations. For comparison we make, using an Excel™ spreadsheet, a graph for the first twenty energy levels, as shown in Figure 3.2. In this figure the graphs show particle probability distributions in order of increasing temperature, from the top-left figure to the right-bottom one.

$\beta\Delta$		3	1	0.7	0.3
Z		1.052396	1.581977	1.986434	3.858296
P0	0	0.950213	0.632121	0.503415	0.259182
P1	1	0.047308	0.232544	0.249988	0.192007
P2	2	0.002355	0.085548	0.124141	0.142242
P3	3	0.000117	0.031471	0.061646	0.105375
P4	4	5.84E-06	0.011578	0.030613	0.078064
P5	5	2.91E-07	0.004259	0.015202	0.057831
P6	6	1.45E-08	0.001567	0.007549	0.042842
P7	7	7.21E-10	0.000576	0.003749	0.031738
P8	8	3.59E-11	0.000212	0.001862	0.023512
P9	9	1.79E-12	7.8E-05	0.000924	0.017418
P10	10	8.89E-14	2.87E-05	0.000459	0.012904
P11	11	4.43E-15	1.06E-05	0.000228	0.009559
P12	12	2.2E-16	3.88E-06	0.000113	0.007082
P13	13	1.1E-17	1.43E-06	5.62E-05	0.005246
P14	14	5.46E-19	5.26E-07	2.79E-05	0.003887
P15	15	2.72E-20	1.93E-07	1.39E-05	0.002879
P16	16	1.35E-21	7.11E-08	6.88E-06	0.002133
P17	17	6.74E-23	2.62E-08	3.42E-06	0.00158
P18	18	3.36E-24	9.63E-09	1.7E-06	0.001171
P19	19	1.67E-25	3.54E-09	8.43E-07	0.000867
P20	20	8.32E-27	1.3E-09	4.19E-07	0.000642

TABLE 3.5. Values of Z for Given Values of $\beta\Delta$ and Distribution of Probabilities P_i^* for the First Twenty Energy Levels

Now we calculate the total energy of this system. Using the relation for energy (see Table 3.3) $E = -\frac{N}{Z}\frac{\partial Z}{\partial \beta}$ and substituting for $\frac{\partial Z}{\partial \beta} = \frac{\partial\left(1-e^{-\beta\Delta}\right)^{-1}}{\partial \beta} - \Delta e^{-\beta\Delta}Z^2$, we get $E = NZ\Delta e^{-\beta\Delta}$. Therefore, the values of dimensionless average energy can be calculated using $E' = \frac{E}{N\Delta} = Ze^{-\beta\Delta} = e^{\psi-\beta\Delta}$, as shown in Table 3.6.

$\beta\Delta$	3	1	0.7	0.3
Z	1.052395696	1.581976707	1.986433864	3.858295914
$\psi = \ln Z$	0.051063	0.4401898	0.6236036	0.9918434
E'	0.052395696	0.581976707	0.986433864	2.858295914

TABLE 3.6. Average Dimensionless Energy for an Assembly of Particles.

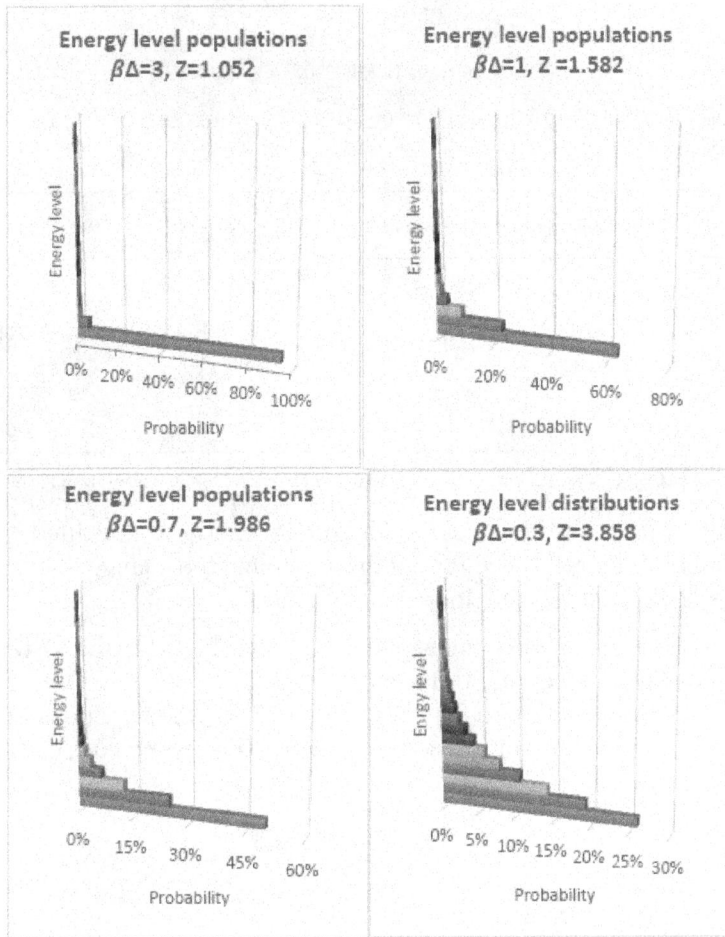

FIGURE 3.2. Energy populations at different values of temperature (using $\beta\Delta$) and corresponding partition function $\mathcal{Z}$.

Now we calculate the entropy of this system. Using Equation 3.14, $S = k_B N \ln \mathcal{Z} + \beta k_B E$, we divide it by N to get its average value and nondimensionalize it by dividing by $\left(\frac{\Delta}{T}\right)$, or $S' = \frac{ST}{N\Delta} = \frac{k_B T}{\Delta} \ln \mathcal{Z} + \frac{\beta k_B T}{\Delta} \frac{E}{N} = \frac{1}{\beta\Delta} \ln \mathcal{Z} + \frac{E}{N\Delta} = \frac{\psi}{\beta\Delta} + \frac{E}{N\Delta}$. Using the values for $E' = \frac{E}{N\Delta}$, from Table 3.6, we can calculate the values of S', as shown in Table 3.7.

Now we calculate the Helmholtz energy of the system. Using $A = -k_B T N \ln \mathcal{Z}$ we divide it by $N\Delta$ to get its average dimensionless value, or $A' = \frac{A}{N\Delta} = -\frac{k_B T N}{N\Delta} \ln \mathcal{Z} = -\frac{1}{\beta\Delta} \ln \mathcal{Z} = -\frac{\psi}{\beta\Delta}$.

Table 3.8 shows these values.

$\beta\Delta$	3	1	0.7	0.3
$\mathcal{Z}$	1.052395696	1.581976707	1.986433864	3.858295914
$\psi = \ln \mathcal{Z}$	0.051063	0.4401898	0.6236036	0.9918434
S'	0.069419	1.040652	1.966921	7.359048

TABLE 3.7. Average Dimensionless Entropy S' for an Assembly of Particles

$\beta\Delta$	3	1	0.7	0.3
$\mathcal{Z}$	1.052395696	1.581976707	1.986433864	3.858295914
$\psi = \ln \mathcal{Z}$	0.051063	0.4401898	0.6236036	0.9918434
A'	−0.01702	−0.45868	−0.98049	−4.50075

TABLE 3.8. Average Dimensionless Helmholtz Energy A', for an Assembly of Particles

Note that negative values for Helmholtz energy are justified with consideration of the reference for energy, or simply the change in energy with reference to a datum (see, for example, [40]).

All of the thermodynamics quantities calculated for the system given previously are at its equilibrium state.

Figure 3.3 and Figure 3.4 show, respectively, the variations of dimensionless energy E' and dimensionless entropy S' versus values of $\beta\Delta$.

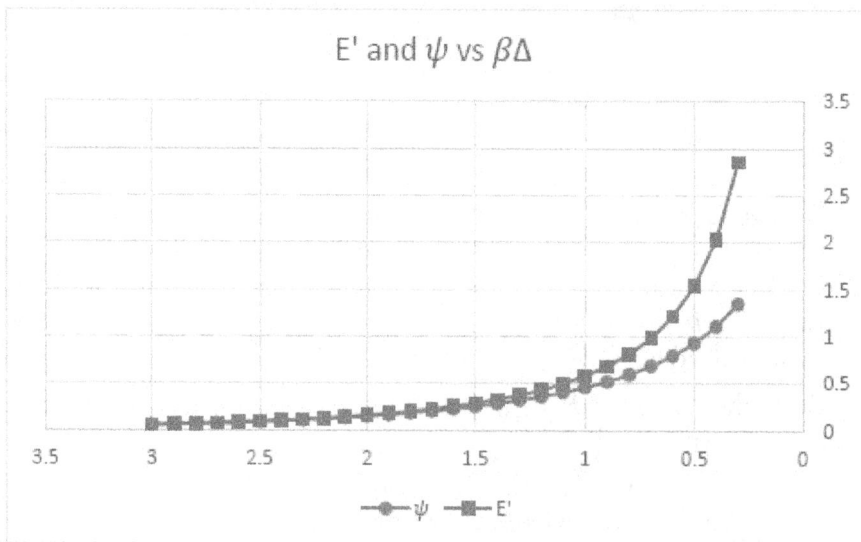

FIGURE 3.3. Variations of energy and Massieu functions.

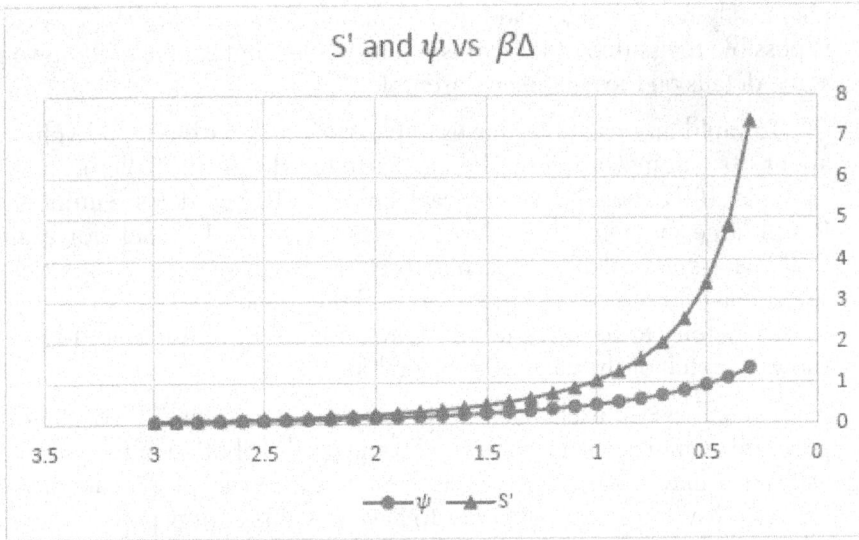

FIGURE 3.4. Variations of entropy and Massieu functions.

3.6. Gibbs Statistical Thermodynamics

All the relations given in the previous sections (i.e., Boltzmann's statistical model) are based on the assumption that the particles are independent and are in a closed, isolated system, with the constraints of constant number and energy at equilibrium state: for example, an ideal gas in an isolated container without heat/work exchanges with the surroundings. Although these assumptions served us to develop and discuss the foundation of Boltzmann's statistical thermodynamics, they are severely restrictive and mainly applicable to limited systems like ideal diluted gases. Other systems like dense gases, liquids, and solids with interactive particles are some examples of more complex types of systems. The extension from Boltzmann's statistical model to more inclusive realistic systems is made possible by a hypothesis proposed and developed by Gibbs [8] in 1902; hence it is called the Gibbs statistical model. In brief, Gibbs proposed the hypothesis of *ergodic ensemble average* definition and application. To understand the merits of Gibbs's ensemble hypothesis, we can relate it to Boltzmann's model when each "particle" is analogously replaced by a "subsystem" of the system under study. The system is then assumed to be identically replicated to form the ensemble. The subsystem in an ensemble may be interacting, although weakly, energy (so-called canonical ensemble), having constant temperature (so-called micro-canonical ensemble), or it may be open and chemically reactive (so-called grand canonical ensemble, [4], [7], [41]). It is possible to assume other forms of ensembles, but in this section we give some details on the canonical ensemble.

To build an ensemble, for example, let's start with a liquid volume of about 10^{25} atoms/particles. We can virtually divide the volume into, for example, 10^{10} cubes. Therefore, each cube has 10^{15} particles, a number that is still large enough to enable us to use a statistical model (for example Boltzmann's) for calculating its macroscopic thermodynamic properties. We select one of the cubes as our system and the remainder $(1 - N_B)$ as replicas of this system, to have a total of N_B systems for our ensemble. Figure 3.5 shows a sketch of the canonical ensemble.

The cubes can exchange energy, weakly, so as to have a constant temperature overall in the ensemble. The energy distribution of the particles in each cube may be different, but each cube is macroscopically identical with the rest of systems. This allows us to have possible energy population of the system somewhere in the assembly.

Having our hypothetical assembly of replicas of our system and two postulates (definitions to follow), we can use our Boltzmann's model (i.e., interpret system 1 and its replicas as constituting "particles" of the ensemble which consists of N_B "particles") to calculate thermodynamic properties of real systems like solids, liquids, and dense gases.

The two fundamental postulates are [7]:

1. Ergodic hypothesis: The time average of a system thermodynamic variable is equal to its ensemble space (i.e., phase space) average, which is the average over instantaneous values of the variable in each member of the ensemble as $N_B \to \infty$.

2. Principle of equal a priori probability: For an isolated thermodynamic system, the members of the ensemble are distributed with equal probability over the possible system energy states (i.e., quantum energies) defined by specification of the type of ensemble, for example canonical or (N, V, E); see Figure 3.5.

The first postulate gives us the mathematical legitimacy of calculating a simpler space-average of a thermodynamic variable for an ensemble instead of a more relatively complicated time average of its members, i.e., the system in question (see [42]). The second postulate ensures that we have enough members in our ensemble with equal probability of having

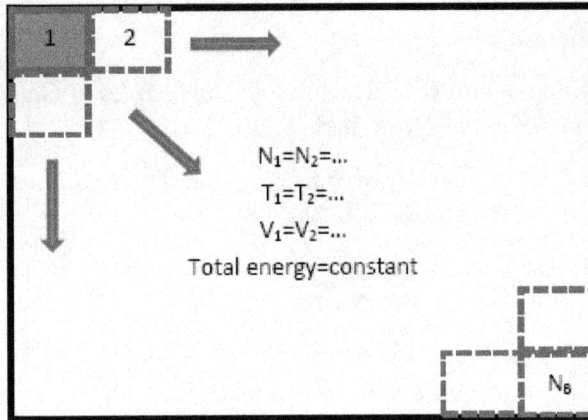

FIGURE 3.5. A canonical ensemble sketch.

energy populations. In other words, for any possible microstates of the system we have a member somewhere in the ensemble with the same probability. As an analogy for the second postulate, assume that we have an infinite number of tuning forks, each tuned to a frequency. Now music is played which is a combination of signals with many frequencies. At each instant of time some of the forks will vibrate, with equal probability, due to being resonated by their corresponding applied natural frequency (i.e., it does not matter which frequency is played by the signals; all have the same probability of resonating the tuning forks). In this analogy, the tuning forks are the quantum energy levels or microstates and the music signals are like particles populating these energy levels.

It can be shown (see [43]) that the relationships for ensemble partition function $\mathbb{Z}$ and that of Boltzmann are $\mathbb{Z} = z^{N_B}$ for distinguishable particles and $\mathbb{Z} = z^{N_B}/N_B!$ for indistinguishable particles.

To make the calculations of complex systems more feasible and relatively simpler, Tribus [30] explains asimpler approach based on the work by Jaynes [38], [39]. Readers interested in further study on Gibbs statistical thermodynamics are referred to the cited references.

The following references are collected here for readers who are interested in reading in more depth on the topic of statistical thermodynamics.

- R. C. Tolman, *The Principle of Statistical Mechanics* (London:Oxford University Press, 1938).

- L.D. Landau and E.M. Lifshitz, *Statistical Physics* (New York: Pergamon Press, 1958).

- J.H. Keenan and G.N. Hatsopoulos, *Principles of General Thermodynamics* (New York: Wiley, 1965).

- F. Reif, *Fundamentals of Statistical and Thermal Physics* (New York: McGraw-Hill, 1965).

- M. Tribus, *Thermostatics and Thermodynamics* (Princeton, NJ: D. Van Nostrand, 1961).

- E.A. Guggenheim, *Thermodynamics: An Advanced Treatment for Chemists and Physicists* (Elsevier, 1967).

- T. M. Reed and K.E. Gubbins, *Applied Statistical Mechanics* (New York: MacGraw-Hill, 1973, reprinted by Butterworth-Heinemann, 1991).

- Stanley I. Sandler, *An Introduction to Applied Statistical Thermodynamics* (New York: Wiley, 2011).

- Andrew Cooksky, *Physical Chemistry Thermodynamics, Statistical Mechanics, & Kinetics* (Upper Saddle River: Pearson, 2014).

3.7. Exercises

1. Describe the configuration weight of a given distribution (see Equation 3.1).

2. With reference to Table 3.2, calculate the weight of given distributions.

3. With reference to Table 3.1, if we increase the number of particles to $N = 7$ calculate: a) the weights for distributions (3,3,1), (2,4,1), and (2,2,3), and b) the maximum weight and associated distributions (see Equation 3.2).

4. Redo the example given in the text and make a spreadsheet for calculating and making graphs similar to Table 3.5 and Figure 3.2 for values of $\beta\Delta = 5, 2, 1, 0.5, 0.2, 0.1$.

4

MATHEMATICS OF THERMODYNAMIC FUNCTIONS AND PROPERTY RELATIONS

In This Chapter

- Overview
- Natural and Conjugate Variables
- Legendre Transformation
- Enthalpy, Helmholtz, and Gibbs Energy Functions
- Thermodynamic Equilibrium
- Maxwell Relations
- Alternative Relations for Entropy, Internal Energy, and Enthalpy
- Measuring Thermodynamic Properties and Experiments
- A Complete List of 336 Derivatives for 168 Thermodynamic Functions
- Exercises

4.1 Overview

In this chapter we define the mainly used thermodynamic functions and derive their related relations and applications. In order to derive these functions and their inter - relations we need to have some mathematical tools available. The mathematical fundamentals of these tools are discussed and used in relation to some specific thermodynamic functions. It is, however, assumed that the readers are familiar with calculus, at the level of a college/university introductory course, specifically topics related to total and partial differentiation of multivariable functions.

4.2. Natural and Conjugate Variables

Let's start with a motivational example. The differential form of the 1st law is usually given as $dU = \delta Q + \delta W$, where U is internal energy, Q is heat, and W is work. For a reversible system, for example a piston – cylinder containing an ideal gas, we can write the 1st law as

$$dU = TdS - pdV \qquad (4.1)$$

where T is absolute temperature, S is entropy, p is pressure, and V is volume of the gas. This relation suggests that internal energy can be written as a function of S and V, or $U = U(S,V)$. Therefore, we can write its total derivative (since U is a thermodynamic state function) as

$$dU = \frac{\partial U}{\partial S}\bigg|_V dS + \frac{\partial U}{\partial V}\bigg|_S dV \qquad (4.2)$$

where, for example, $\dfrac{\partial U}{\partial S}\bigg|_V$ means that differentiation w.r.t. S is taken assuming V as a constant. By comparing Equation 4.1 and Equation 4.2, we can write the following relations

$$T = \frac{\partial U}{\partial S}\bigg|_V, \; p = -\frac{\partial U}{\partial V}\bigg|_S \qquad (4.3)$$

These relations are very useful for defining and calculating temperature and pressure of a gas, provided we have the functional form of internal energy or $U = U(S, V)$ at hand, which can, in principle, be obtained through laboratory measurements. Variables S and V are called *natural variables* for internal energy, since we can define other thermodynamics properties (i.e., pressure and temperature, see Equation 4.3) using the analysis given above with function U containing information about the system. Also, we define *Conjugate variables* (T, p) and pairs (T, S) and (p, V) based on Equation 4.1. Notice that the product of members of a conjugate pair (i.e., TS and pV) has the dimension of energy, and each pair is composed of an extensive (S or V) and an intensive (T or p) variable.

In order to find the functional form of the internal energy, we need to measure entropy, which is a relatively more difficult quantity to be measured as compared to temperature, pressure, or volume. Therefore, one can ask this question: "Is it possible to find another thermodynamic state function which can be as representative as internal energy for the system and also is a function of those thermodynamic properties that are more easily measurable?" The short answer to this question is "Yes." However, in order to find these desired system – representative functions in a systematic way we need a proper mathematical tool that can be used to guide the derivation. Fortunately, this tool exists and is referred to as the *Legendre transformation*. In the next section we discuss the Legendre transformation and its application to thermodynamic functions and variables.

4.3. Legendre Transformation

Legendre transformation is a mathematical procedural technique that could be used to represent the information provided by an originally given original function with construction of a new function [44], [45]. To demonstrate the procedure, and without losing generality, let's consider a function of two variables, $f = f(x,y)$. The total derivative of f is $df = \dfrac{\partial f}{\partial x} dx + \dfrac{\partial f}{\partial y} dy$.

Let's define the notations $\dfrac{\partial f}{\partial x} = u(x,y)$ and $\dfrac{\partial f}{\partial y} = w(x,y)$. Now we define conjugate pairs (u, x) and (w, y) with the objective of finding a new function, instead of f, that depends on u or w as one of its independent variables. Usually, the original variables, for example, (x,y), are called natural variables and their conjugate pair, for example, (u,w), are called conjugate variables. The steps for calculating the desired new function are as follows:

1. Choose a conjugate pair. For this example we choose the pair (u, x).

2. Calculate the total derivative of the product of the selected pair. For this example, we get $d(ux) = udx + xdu$.

3. Subtract the total derivative of the conjugate product from that of the main function. For this example, we have $df - d(ux) = [udx + wdy] - [udx + xdu] = -xdu + wdy$. We can also write $df - d(ux) = d(f - ux)$.

4. Define the new function by subtracting the product of the conjugate pairs from the original function. For this example we get the new function as $h = f - ux$. Therefore, $dh = -xdu + wdy$. This suggests that the new function h is a function of u and y or $h = h(u, y)$, as aimed for.

By following the previously mentioned steps, we obtained a new function $h = h(u,y) = h\left(\dfrac{\partial f}{\partial x}, y\right)$, using Legendre transformation, that can be used instead of the original function $f = f(x,y)$. Similarly, we could select the conjugate pair (w,y) and use the previously mentioned steps to find another function, $g = g(x,w) = g\left(x, \dfrac{\partial f}{\partial y}\right)$. We can further use any of the two functions h or g, and again use the Legendre transformation to find a third function $l = l(u,w) = l\left(\dfrac{\partial f}{\partial x}, \dfrac{\partial f}{\partial y}\right)$.

In general, having a function of N independent variables we can get $(2^N - 1)$ new functions by using Legendre transformation. For our example $N = 2$, $(2^2 - 1) = 3$ new functions were derived (i.e., h, g, l) in addition to the original function f.

Note that the new functions (in our example, h, g, l) do not provide new information; rather the original functional information is represented by new functions.

4.4. Enthalpy, Helmholtz, and Gibbs Energy Functions

In this section, we use Legendre transformation to explore new thermodynamic state functions. We start with the 1st law relation, $dU = TdS - pdV$, which in its functional form reads $U = U(S,V)$. We would like to replace the variable S with its conjugate T and find a new function instead of U. Following Legendre transformation procedure, as mentioned in the previous section, we have $d(ST) = TdS + SdT$, which leads to $d(U - TS) = -pdV - SdT$. We then get our new function $A = A(V,T)$ defined as

$$A = U - TS \qquad (4.4)$$

Function A is called the Helmholtz energy function. We can also use its total derivative $dA = -pdV - SdT$ and the relation $dA = \left.\dfrac{\partial A}{\partial V}\right|_T dV + \left.\dfrac{\partial A}{\partial T}\right|_V dT$ to get by comparison the following relations

$$p = -\left.\dfrac{\partial A}{\partial V}\right|_T \quad \text{and} \quad S = -\left.\dfrac{\partial A}{\partial T}\right|_V \qquad (4.5)$$

The Helmholtz function has several useful properties. It measures the capacity of a system to perform useful work for a closed thermodynamic system at a constant temperature. Also, we can find entropy using the Helmholtz partial derivative w.r.t. temperature (see Equation 4.5).

Now we consider the internal energy function again, but we use the conjugate pair $(-p, V)$. By applying the Legendre transformation steps we have $d(-pV) = -pdV - Vdp$. After subtracting this relation from dU we get $d(U + pV) = TdS + Vdp$, from which we get another new function $H = H(S, p)$ defined as

$$H = U + pV \tag{4.6}$$

Function H is the Enthalpy. We can also write its total derivative relation as $dH = \dfrac{\partial H}{\partial S}\bigg|_p dS + \dfrac{\partial H}{\partial p}\bigg|_s dp$. By comparing this relation with $dH = TdS + Vdp$, we get

$$T = \frac{\partial H}{\partial S}\bigg|_p \quad \text{and} \quad V = \frac{\partial H}{\partial p}\bigg|_s \tag{4.7}$$

Another function of interest is Gibbs energy. Using the enthalpy function and the conjugate pair (T, S), we have $d(H - TS) = TdS + Vdp - TdS - SdT = Vdp - SdT$. The new function is the Gibbs function, or

$$G = H - TS = U + pV - TS = A + pV \tag{4,8}$$

Therefore, we can write $dG = Vdp - SdT$, which is a function of pressure and temperature, or $G = G(p, T)$. Using the total derivative of G, or $dG = \dfrac{\partial G}{\partial p}dp + \dfrac{\partial G}{\partial T}dT$, and comparing it with the relation given above, we get another pair of relations, as

$$V = \frac{\partial G}{\partial p}\bigg|_T \quad \text{and} \quad S = -\frac{\partial G}{\partial T}\bigg|_p \tag{4.9}$$

The Gibbs energy function is useful for constant pressure and/or temperature process and has many applications in thermodynamic processes with chemical reactions involved.

Expanding on the relations given above for closed systems where thermodynamic properties are functions of two variables, we can also consider open systems where the number of moles also changes due to mass exchange. For systems with material exchange and production, like chemical processes, another term should be added to the internal energy related to the chemical potentials of the reactive components. In this case the internal energy reads

$$dU = TdS - pdV + \mu dN \tag{4.10}$$

Where μ is chemical potential and N is the particle number for a single species. The additional energy term related to chemical potential will contribute to the energy of the system. With the additional chemical potential term, the Helmholtz function reads

$$dA = -SdT - pdV + \mu dN \tag{4.11}$$

And the Gibbs function reads

$$dG = -SdT + Vdp + \mu dN \tag{4.12}$$

By comparison, we get

$$\mu = \left(\frac{\partial U}{\partial N}\right)_{S,V}, \ \mu = \left(\frac{\partial A}{\partial N}\right)_{T,V}, \ \mu = \left(\frac{\partial G}{\partial N}\right)_{T,p} \tag{4.13}$$

To summarize the relations mentioned above and their inter-relations, we show them collectively in Figure 4.1.

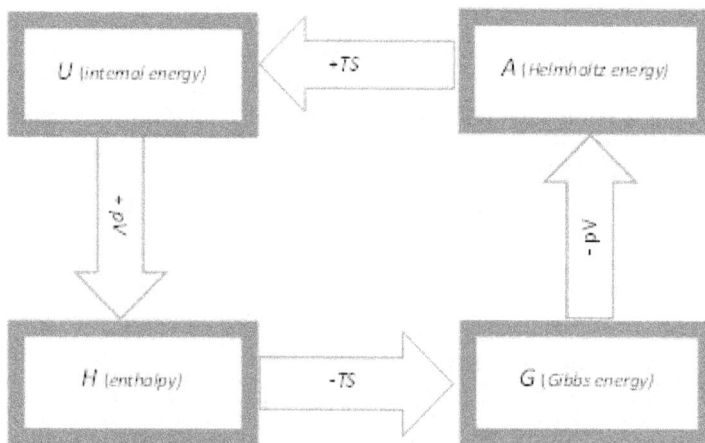

FIGURE 4.1 Internal energy, enthalpy, Helmholtz, and Gibbs thermodynamic functions and their inter - relations.

4.5. Thermodynamic Equilibrium

In the previous section, we derived the relations for enthalpy, Helmholtz, and Gibbs thermodynamic energy functions. In this section we use these functions, along with the 1st and 2nd laws including entropy function, to show that their extremum (maximum or minimum) values provide the necessary conditions for equilibrium, required for a spontaneous change through a process.

We have the 1st law relation, $dU = \delta Q - pdV$, and the 2nd law relation, $dS \geq \dfrac{\delta Q}{T}$. Combining these two relations by eliminating δQ, we get

$$dU + pdV - TdS \leq 0 \qquad (4.14)$$

Equation 4.14 is very useful in applications, since it expresses the 1st and 2nd laws of thermodynamics combined. Any process/phenomenon should satisfy this relation for spontaneous change to happen, and it could be used even to predict the possibility of a process to happen or not, for example the direction of a chemical reaction.

To start with a series of examples with specific conditions, we consider a process at constant pressure and entropy, or $dp = dS = 0$. By applying these conditions to Equation 4.14, we get

$dU + pdV = dU + \underbrace{d(pV)}_{Vdp=0} = d(U + pV) \leq 0$. But $H = U + pV$, and hence $dH \leq 0$ for a process at constant pressure and entropy. In other words, a process reaches its thermodynamic equilibrium when enthalpy is at its minimum value, given that constant pressure and entropy process conditions apply.

Similarly, we consider a process at constant volume and temperature, or $dV = dT = 0$.

By applying these conditions to Equation 4.14 we get $dU - TdS = dU - \underbrace{d(TS)}_{SdT=0} = d(U - TS) \leq 0$. But $A = U - TS$ is the Helmholtz energy, hence $dA \leq 0$ for a process at constant volume and temperature. In other words, a process reaches its thermodynamic equilibrium when Helmholtz energy is minimized and when constant volume and temperature conditions apply.

Further we consider a process at constant pressure and temperature, or $dp = dT = 0$. By applying these conditions to Equation 4.14 we get $dU + \underbrace{d(pV)}_{Vdp=0} - \underbrace{d(TS)}_{SdT=0} = d(U + pV - TS) \le 0$. But $G = U + pV - TS$ is Gibbs energy, hence $dG \le 0$ for a process at constant pressure and temperature. In other words, a process reaches its thermodynamic equilibrium when Gibbs energy is minimized and when constant pressure and temperature conditions apply. As last, let's consider an isolated system (the whole universe is considered to be an isolated system!) which has constant energy and volume, without any heat and work exchanges to happen with its surroundings. Therefore, for an isolated system we have: $dU = dV = \delta Q = \delta W = 0$. After applying these conditions to Equation 4.14 we get $-TdS \le 0$, or $TdS \ge 0$, but absolute temperature cannot be negative, hence $dS \ge 0$. In other words, for an isolated system (like the universe) the entropy tends to maximize itself and at equilibrium it is at its stationary maximum value. Combining this condition with that of minimum energy (see the Clausius statement in Chapter 2), we can conclude that the universe reaches its minimum energy while conserved and maximum entropy values at equilibrium, a condition sometimes referred to as thermal/heat death of the universe, after Lord Kelvin [46].

Table 4.1 shows the condition of equilibrium using the thermodynamic functions discussed previously.

TABLE 4.1 Equilibrium Conditions and Corresponding Thermodynamic Functions

Conditions		Thermodynamic function used for equilibrium state
$dp = dS = 0$	$dH \le 0$	Enthalpy is minimum
$dV = dT = 0$	$dA \le 0$	Helmholtz energy is minimum
$dp = dT = 0$	$dG \le 0$	Gibbs energy is minimum
$dU = dV = 0$	$dS \ge 0$	Entropy is maximum

4.6. Maxwell Relations

In the previous section we derived the differential forms of the state functions U, H, A, G and their corresponding natural and conjugate variables. We summarize these relations, for ease of reference, in Table 4.2.

TABLE 4.2   Thermodynamic Energy Functions, Their Natural and Conjugate Variables, and Inter – Relations

Thermodynamic energy function	Natural variables	Conjugate variables	Differential form	Inter – relations
U	S,V	$T,-p$	$dU = TdS - pdV$	$T = \left.\dfrac{\partial U}{\partial S}\right\|_V , \ -p = \left.\dfrac{\partial U}{\partial V}\right\|_S$
H	S,p	T,V	$dH = TdS + Vdp$	$T = \left.\dfrac{\partial H}{\partial S}\right\|_p , \ V = \left.\dfrac{\partial H}{\partial p}\right\|_S$
A	T,V	$-S,-p$	$dA = -SdT - pdV$	$-S = \left.\dfrac{\partial A}{\partial T}\right\|_V , \ -p = \left.\dfrac{\partial A}{\partial V}\right\|_T$
G	T,p	$-S,V$	$dG = -SdT + Vdp$	$-S = \left.\dfrac{\partial G}{\partial T}\right\|_p , \ V = \left.\dfrac{\partial G}{\partial p}\right\|_T$

We observe that all thermodynamic state functions listed in Table 4.2 are a function of four independent variables (S, V, T, p) in total, either as their corresponding natural or conjugate variables. For example, for the differential form of internal energy $U(S,V)$ we have its conjugate variables $(T,-p)$ appearing as the coefficients of differentials of its natural variables (S,V). This "symmetry" appears in the relations for the remaining functions as well. In addition, and from a physical point of view, experimental measurements of variables T, p, V are more easily accessible than S. This motivates seeking relations that can eliminate the need for measurement of S. The solution comes from Maxwell relations, which make use of this "symmetry" and the total differential properties of the thermodynamic function to derive relations among the natural and conjugate variables themselves.

To derive Maxwell relations, we start with the internal energy U, where its inter – relations (see Table 4.2) are $T = \left.\dfrac{\partial U}{\partial S}\right|_V , \ -p = \left.\dfrac{\partial U}{\partial V}\right|_S$. Assuming that function U is a smooth function (i.e., its higher order differentials are available), we differentiate T w.r.t. V and p w.r.t. S, or $\left.\dfrac{\partial T}{\partial V}\right|_S$ and $-\left.\dfrac{\partial p}{\partial S}\right|_V$. But, using the inter – relations for U, we find that these two relations are equivalent, since $\left.\dfrac{\partial}{\partial V}\left(\left.\dfrac{\partial U}{\partial S}\right|_V\right)\right|_S = \left.\dfrac{\partial}{\partial S}\left(\left.\dfrac{\partial U}{\partial V}\right|_S\right)\right|_V$, and since the differentiation order could be interchanged for smooth functions. Finally, we get $\left.\dfrac{\partial T}{\partial V}\right|_S = -\left.\dfrac{\partial p}{\partial S}\right|_V$.

By repeating similar operations for inter – relations corresponding to functions H, A, and G, we get $\left.\frac{\partial T}{\partial p}\right|_S = \left.\frac{\partial V}{\partial S}\right|_p$, $\left.\frac{\partial S}{\partial V}\right|_T = \left.\frac{\partial p}{\partial T}\right|_V$, and $-\left.\frac{\partial S}{\partial p}\right|_T = \left.\frac{\partial V}{\partial T}\right|_p$.

These relations are called *Maxwell relations* and are applicable to thermodynamic state functions at equilibrium. Note that using a reciprocal relation for partial derivatives (i.e., $\left(\frac{\partial y}{\partial x}\right)_z = 1 / \left(\frac{\partial x}{\partial y}\right)_z$), we can write the Maxwell relations given previously in their equivalent reciprocal forms as well. For example, $\left.\frac{\partial p}{\partial T}\right|_S = \left.\frac{\partial S}{\partial V}\right|_p$, and so on.

We present a procedure to build/recall Maxwell relations using the definition of a "*determinant*" of a matrix with having the differential relations for a desired thermodynamic function for which we seek the corresponding Maxwell relations. The procedure is as follows: We build a matrix consisting of the function and its natural and conjugate variables and let the determinant of such matrix be equal to zero. For example, for Helmholtz functions, we form $\begin{vmatrix} "V" & "T" \\ \dfrac{\partial}{\partial T} & \dfrac{\partial}{\partial V} \\ -S & -p \end{vmatrix} = \left.\frac{\partial}{\partial T}(-p)\right|_V - \left.\frac{\partial}{\partial V}(-S)\right|_T = 0$. This gives the relation $\left.\frac{\partial p}{\partial T}\right|_V = \left.\frac{\partial S}{\partial V}\right|_T$, as expected. The procedure is shown in Figure 4.2.

$$\begin{vmatrix} natural2 & natural1 \\ \dfrac{\partial}{\partial(natural1)} & \dfrac{\partial}{\partial(natural2)} \\ conjugate1 & conjugate2 \end{vmatrix} = 0$$

FIGURE 4.2 Sketch for building a Maxwell relation using a "determinant" operation.

As an another example, we use the suggested procedure for Gibbs energy, having *natural1* = T, *natural2* = p, *conjugate1* = $-S$, and *conjugate2* = V; therefore we have $\begin{vmatrix} "p" & "T" \\ \dfrac{\partial}{\partial T} & \dfrac{\partial}{\partial p} \\ -S & V \end{vmatrix} = \left.\frac{\partial}{\partial T}(V)\right|_p - \left.\frac{\partial}{\partial p}(-S)\right|_T = 0$, or $\left.\frac{\partial V}{\partial T}\right|_p = -\left.\frac{\partial S}{\partial p}\right|_T$. For reference, we list all the Maxwell relations in Table 4.3.

Thermodynamic function	Natural variables	Conjugate variables	Construction 'Curl' structure	Maxwell relations
U	S,V	$T,-p$	$\begin{vmatrix} "V" & "S" \\ \dfrac{\partial}{\partial S} & \dfrac{\partial}{\partial V} \\ T & -p \end{vmatrix} = 0$	$-\dfrac{\partial p}{\partial S}\bigg|_V = \dfrac{\partial T}{\partial V}\bigg|_S$
H	S,p	T,V	$\begin{vmatrix} "p" & "S" \\ \dfrac{\partial}{\partial S} & \dfrac{\partial}{\partial p} \\ T & V \end{vmatrix} = 0$	$\dfrac{\partial V}{\partial S}\bigg|_p = \dfrac{\partial T}{\partial p}\bigg|_S$
A	T,V	$-S,-p$	$\begin{vmatrix} "V" & "T" \\ \dfrac{\partial}{\partial T} & \dfrac{\partial}{\partial V} \\ -S & -p \end{vmatrix} = 0$	$\dfrac{\partial p}{\partial T}\bigg|_V = \dfrac{\partial S}{\partial V}\bigg|_T$
G	T,p	$-S,V$	$\begin{vmatrix} "p" & "T" \\ \dfrac{\partial}{\partial T} & \dfrac{\partial}{\partial p} \\ -S & V \end{vmatrix} = 0$	$\dfrac{\partial V}{\partial T}\bigg|_p = -\dfrac{\partial S}{\partial p}\bigg|_T$

TABLE 4.3   Maxwell Relations and Their Corresponding "Curl" Construction Forms

4.7. Alternative Relations for Entropy, Internal Energy, and Enthalpy

Calculation and measurement of thermodynamic state functions S,U, and H is desirable and required in many practical applications. However, it is not easy to measure them by experiments in the laboratory and usually relations in terms of other more – easily measurable thermodynamic properties (e.g., T, V, p) are used. In this section we derive alternative relations for entropy, internal energy, and enthalpy to facilitate the calculations and measurement of these functions.

Let's define entropy as a function of temperature and volume, or $S = S(T,V)$. Therefore, we can write $dS = \dfrac{\partial S}{\partial T}\bigg|_V dT + \dfrac{\partial S}{\partial V}\bigg|_T dV$, and after multiplying by T we get $TdS = T\dfrac{\partial S}{\partial T}\bigg|_V dT + T\dfrac{\partial S}{\partial V}\bigg|_T dV$. In addition, for a reversible process, we have $dU = TdS - pdV$, and for constant volume processes (i.e., $dV = 0$) the relation reduces to $dU = C_V dT = TdS$, where C_V

is the specific heat at a constant volume. Therefore, we have $C_V = T \dfrac{\partial S}{\partial T}\bigg|_V$.

Note, we are using specific quantities for energy and entropy. Using the Maxwell relation $\dfrac{\partial S}{\partial V}\bigg|_T = \dfrac{\partial p}{\partial T}\bigg|_V$ we get the 1st *equation for entropy*, as

$$TdS = C_V dT + T \frac{\partial p}{\partial T}\bigg|_V dV \qquad (4.15)$$

Note that Equation 4.15 is useful for calculation and measurement of entropy, since C_V, V, and T are relatively easier to measure experimentally.

Similarly, we define entropy as a function of temperature and pressure, or $S = S(T, p)$. Therefore, we can write $dS = \dfrac{\partial S}{\partial T}\bigg|_p dT + \dfrac{\partial S}{\partial p}\bigg|_T dp$ and after

multiplying by T we get $TdS = T\dfrac{\partial S}{\partial T}\bigg|_p dT + T\dfrac{\partial S}{\partial p}\bigg|_T dp$. In addition, we have

$dH = dU + d(pV)$, and for constant pressure processes (i.e., $dp = 0$) the relation for a reversible process reduces to $dH = TdS - pdV + pdV + V\underbrace{dp}_{=0} = C_p dT$,

where C_p is a specific heat at a constant pressure. Therefore, we have $C_p = T\dfrac{\partial S}{\partial T}\bigg|_p$. Using the Maxwell relation $\dfrac{\partial S}{\partial p}\bigg|_T = -\dfrac{\partial V}{\partial T}\bigg|_p$, we get the 2nd *equation for entropy*, as

$$TdS = C_p dT - T\frac{\partial V}{\partial T}\bigg|_p dp \qquad (4.16)$$

Note that Equation 4.16 is useful for calculation and measurement of entropy, since p and T are relatively easier to measure experimentally.

Now we derive the alternative equation for the internal energy function. Let's define U as a function of temperature and volume, or $U = U(T, V)$. Therefore, we can write $dU = \dfrac{\partial U}{\partial T}\bigg|_V dT + \dfrac{\partial U}{\partial V}\bigg|_T dV$, but $\dfrac{\partial U}{\partial T}\bigg|_V = C_V$, and after substitution we get

$$dU = C_V dT + \frac{\partial U}{\partial V}\bigg|_T dV \qquad (4.17)$$

Observing Equation 4.17, we conclude that a relation for the partial derivative term $\dfrac{\partial U}{\partial V}\bigg|_T$ is required. To derive this relation we use the function

$U = U(S,V)$ with its total differential $dU = \dfrac{\partial U}{\partial S}\bigg|_V \, dS + \dfrac{\partial U}{\partial V}\bigg|_S \, dV$. After equating

this relation with Equation 4.17, we get $C_V dT + \dfrac{\partial U}{\partial V}\bigg|_T \, dV = \dfrac{\partial U}{\partial S}\bigg|_V \, dS + \dfrac{\partial U}{\partial V}\bigg|_S \, dV$.

In this relation, we substitute for $dS = \dfrac{\partial S}{\partial T}\bigg|_V \, dT + \dfrac{\partial S}{\partial V}\bigg|_T \, dV$ (assum-

ing the entropy function $S = S(V, T)$) to get $C_V dT + \dfrac{\partial U}{\partial V}\bigg|_T \, dV = \dfrac{\partial U}{\partial S}\bigg|_V$

$\left[\dfrac{\partial S}{\partial T}\bigg|_V \, dT + \dfrac{\partial S}{\partial V}\bigg|_T \, dV \right] + \dfrac{\partial U}{\partial V}\bigg|_S \, dV$. After expanding this relation, we get

$C_V dT + \dfrac{\partial U}{\partial V}\bigg|_T \, dV = \underbrace{\dfrac{\partial U}{\partial S}\bigg|_V \dfrac{\partial S}{\partial T}\bigg|_V}_{\frac{\partial U}{\partial T}\big|_V = C_V} \, dT + \dfrac{\partial U}{\partial S}\bigg|_V \dfrac{\partial S}{\partial V}\bigg|_T \, dV + \dfrac{\partial U}{\partial V}\bigg|_S \, dV$. Finally, we

get $\dfrac{\partial U}{\partial V}\bigg|_T = \dfrac{\partial U}{\partial S}\bigg|_V \dfrac{\partial S}{\partial V}\bigg|_T + \dfrac{\partial U}{\partial V}\bigg|_S$. But $\dfrac{\partial U}{\partial S}\bigg|_V = T$ and $\dfrac{\partial U}{\partial V}\bigg|_S = -p$ (see Table 4.2)

and $\dfrac{\partial S}{\partial V}\bigg|_T = \dfrac{\partial p}{\partial T}\bigg|_V$ using a Maxwell relation (see Table 4.3). Therefore, we get

$\dfrac{\partial U}{\partial V}\bigg|_T = T\dfrac{\partial p}{\partial T}\bigg|_V - p$. Now we plug this relation into Equation 4.17 to get

$$dU = C_V dT + \left[T\dfrac{\partial p}{\partial T}\bigg|_V - p \right] dV \qquad (4.18)$$

Equation 4.18 is the *alternative relation for internal energy*, expressed as a function of more easily experimentally measurable variables T, p, and V.

Similarly, we derive an alternative relation for enthalpy function. Let's define H as a function of temperature and volume, or $H = H(T, p)$.

Therefore, we can write $dH = \dfrac{\partial H}{\partial T}\bigg|_p \, dT + \dfrac{\partial U}{\partial p}\bigg|_T \, dp$, but $\dfrac{\partial H}{\partial T}\bigg|_p = C_p$, and after substitution we get

$$dH = C_p dT + \dfrac{\partial H}{\partial p}\bigg|_T \, dp \qquad (4.19)$$

Observing Equation 4.19 we conclude that a relation for the partial derivative term $\dfrac{\partial H}{\partial p}\Big|_T$ is required. To derive this relation we use the function $H = H(S, p)$ with its total differential $dH = \dfrac{\partial H}{\partial S}\Big|_p dS + \dfrac{\partial U}{\partial p}\Big|_S dp$. After equating the two relations we get $C_p dT + \dfrac{\partial H}{\partial p}\Big|_T dp = \dfrac{\partial H}{\partial S}\Big|_p dS + \dfrac{\partial H}{\partial p}\Big|_S dp$.

In this relation, we substitute for $dS = \dfrac{\partial S}{\partial p}\Big|_T dp + \dfrac{\partial S}{\partial T}\Big|_p dT$ (assuming the entropy function $S = S(p, T)$) to get

$$C_p dT + \frac{\partial H}{\partial p}\Big|_T dp = \frac{\partial H}{\partial S}\Big|_p \left[\frac{\partial S}{\partial T}\Big|_p dT + \frac{\partial S}{\partial p}\Big|_T dp \right] + \frac{\partial H}{\partial p}\Big|_S dp.$$ After expanding this relation we get $C_p dT + \dfrac{\partial H}{\partial p}\Big|_T dp = \underbrace{\dfrac{\partial H}{\partial S}\Big|_p \dfrac{\partial S}{\partial T}\Big|_p}_{\frac{\partial H}{\partial T}\big|_p = C_p} dT + \dfrac{\partial U}{\partial S}\Big|_p \dfrac{\partial S}{\partial p}\Big|_T dp + \dfrac{\partial H}{\partial p}\Big|_S dp.$

Finally, we get $\dfrac{\partial H}{\partial p}\Big|_T = \dfrac{\partial H}{\partial S}\Big|_p \dfrac{\partial S}{\partial p}\Big|_T + \dfrac{\partial H}{\partial p}\Big|_S$. But $\dfrac{\partial H}{\partial S}\Big|_p = T$ and $\dfrac{\partial H}{\partial p}\Big|_S = V$, since $dH = TdS + Vdp$, and $\dfrac{\partial S}{\partial p}\Big|_T = -\dfrac{\partial V}{\partial T}\Big|_p$ using a Maxwell relation (see Table 4.3). Therefore, we get $\dfrac{\partial H}{\partial p}\Big|_T = V - T\dfrac{\partial V}{\partial T}\Big|_p$. Now we plug this relation into Equation 4.19 to get

$$dH = C_p dT + \left[V - T\frac{\partial V}{\partial T}\Big|_p \right] dp \tag{4.20}$$

Equation 4.20 is the *alternative relation for enthalpy*, expressed as a function of experimentally measurable variables T, p, and V.

All the alternative relations derived above apply to real gases, and of course to ideal gases. Note that for an ideal gas, the terms in the bracket on the R.H.S. of Equation 4.18 and Equation 4.20 are equal to zero, and we recover the familiar relations for ideal gas internal energy and enthalpy.

4.7.1. Cyclic Mathematical Relations for Partial Derivatives

Another useful mathematical relation in thermodynamics can be obtained using the cyclic relationship that exists among partial derivatives of a function, also referred to as *Euler's chain rule*. We derive this relation, in its generic mathematical form, in this section and apply it to the thermodynamics relations in further sections.

Let's assume a function, $f = f(x, y, z)$, with permissible first partial derivatives. This function can be written as the equation of surfaces in, for example, a Cartesian coordinates system (x, y, z) as $x = x(y, z)$, $y = y(x, z)$, and $z = z(x, y)$. For each function we can write the total derivative as: $dx = \dfrac{\partial x}{\partial y}\bigg|_z dy + \dfrac{\partial x}{\partial z}\bigg|_y dz$, similarly $dy = \dfrac{\partial y}{\partial x}\bigg|_z dx + \dfrac{\partial y}{\partial z}\bigg|_x dz$, and

$dz = \dfrac{\partial z}{\partial x}\bigg|_y dx + \dfrac{\partial z}{\partial y}\bigg|_x dy$. Now we plug in for dz in the relation for dy to get after some manipulations: $dy = \left[\dfrac{\partial y}{\partial x}\bigg|_z + \dfrac{\partial y}{\partial z}\bigg|_x \dfrac{\partial z}{\partial x}\bigg|_y\right]dx + \dfrac{\partial y}{\partial z}\bigg|_x \dfrac{\partial z}{\partial y}\bigg|_x dy$. But

using the chain rule the last term simplifies to $\underbrace{\dfrac{\partial y}{\partial z}\bigg|_x \dfrac{\partial z}{\partial y}\bigg|_x}_{=\frac{\partial y}{\partial y}} dy = dy$. Therefore,

we have $dy = \left[\dfrac{\partial y}{\partial x}\bigg|_z + \dfrac{\partial y}{\partial z}\bigg|_x \dfrac{\partial z}{\partial x}\bigg|_y\right]dx + dy$, or $\dfrac{\partial y}{\partial z}\bigg|_x \dfrac{\partial z}{\partial x}\bigg|_y = -\dfrac{\partial y}{\partial x}\bigg|_z$. After multi-

plying both sides by $\dfrac{\partial x}{\partial y}\bigg|_z$ we get

$$\left(\dfrac{\partial x}{\partial y}\bigg|_z\right)\left(\dfrac{\partial y}{\partial z}\bigg|_x\right)\left(\dfrac{\partial z}{\partial x}\bigg|_y\right) = -1 \tag{4.21}$$

In order to make it easier to recall this relation, readers may use Figure 4.3.

With a similar analysis, writing the derivatives in the opposite direction of what is shown in Figure 4.3, we can write

$$\left(\dfrac{\partial x}{\partial z}\bigg|_y\right)\left(\dfrac{\partial z}{\partial y}\bigg|_x\right)\left(\dfrac{\partial y}{\partial x}\bigg|_z\right) = -1 \tag{4.22}$$

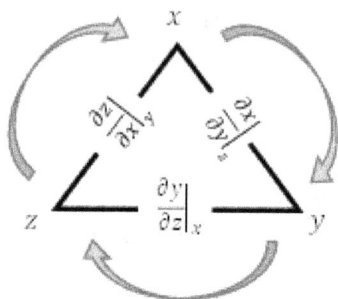

FIGURE 4.3 Sketch for Euler's chain rule.

In terms of thermodynamics variables, for example, letting $p \equiv x$, $V \equiv y$, and $T \equiv z$, we can use Equation 4.21 or Equation 4.22 to write the relations among the partial derivatives for pressure, volume, and temperature as:

$$\left(\frac{\partial p}{\partial V}\bigg|_T\right)\left(\frac{\partial V}{\partial T}\bigg|_p\right)\left(\frac{\partial T}{\partial p}\bigg|_V\right) = -1 \tag{4.23}$$

and

$$\left(\frac{\partial p}{\partial T}\bigg|_V\right)\left(\frac{\partial T}{\partial V}\bigg|_p\right)\left(\frac{\partial V}{\partial p}\bigg|_T\right) = -1 \tag{4.24}$$

4.8. Measuring Thermodynamic Properties and Experiments

In the previous sections we derived several relations for thermodynamic functions. These functions are applicable for calculations and measurement of thermodynamic quantities based on other available measured data. However, some thermodynamic quantities are easier to measure than others in practice. The three quantities that are usually preferred in experimental measurement are pressure, volume, and temperature. Having these quantities, we can calculate different thermodynamic quantities. In this section we discuss the procedure and some related experiments, including

- Joule – Thomson coefficient
- Coefficient of expansion (isobaric compressibility)
- Coefficient of compressibility (isothermal compressibility)
- Specific heats

4.8.1. Joule – Thomson Coefficient and Experiment

Adiabatic irreversible expansion of a gas experiment is named after Joule, who used it for his own famous experiment (mechanical – energy equivalent of heat, [47]). In this experiment, Joule showed that (within the accuracy of his experimental setting) internal energy of an ideal gas is only a function of its temperature. However, for real gases this is not true (see Equation 4.18). Expanding on these results, another important experiment is that of change of temperature of a gas versus its pressure when enthalpy is kept constant. This experiment and its important outcome relates to the definition of the *Joule – Thomson coefficient* as

$$\mu_{JT} = \frac{\partial T}{\partial p}\bigg|_H \tag{4.25}$$

This quantity is the slope of a constant – enthalpy curve (i.e., enthalpic) at a given point representing the variation of temperature versus pressure, as schematically shown in Figure 4.4. The constant enthalpy process is usually obtained by so – called *throttling*, using a porous medium or a throttling valve.

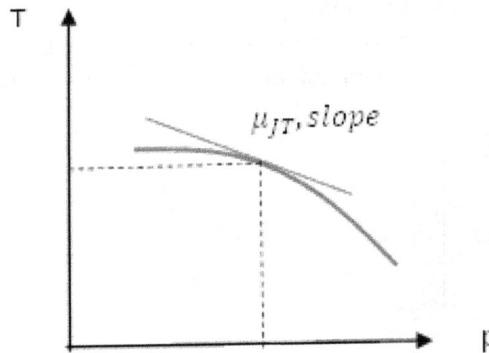

FIGURE 4.4 Sketch for Joule - Thomson coefficient definition.

We have $dH = TdS + Vdp$ after plugging in for TdS using Equation 4.16, and when we let $dH = 0$, after some manipulation we get $\left(C_p dT\right)_H = \left[\left(T\frac{\partial V}{\partial T}\bigg|_p - V\right)dp\right]_H$. Therefore, using the definition given by Equation 4.25 we have

$$\mu_{JT} = \frac{1}{C_p}\left(T\frac{\partial V}{\partial T}\bigg|_p - V\right) \tag{4.26}$$

Equation 4.26 gives the Joule – Thomson coefficient in terms of measurable variables V,T,p. For an ideal gas we can show that $\mu_{JT} = 0$ (since the expression in the bracket is zero) which means that when an ideal gas is throttled its temperature remains constant.

In order to measure the value of μ_{JT}, an experiment was performed by Joule and Thomson ([23], [17]). In this experiment the variation of temperature versus pressure for a constant enthalpy process is measured. The Joule – Thomson experiment apparatus consists of an insulated tube (i.e., $\delta Q = 0$) and a porous medium, as sketched in Figure 4.5. By applying the 1st law, we have $U_2 - U_1 = p_1 V_1 - p_2 V_2$ or $U_2 + p_2 V_2 = U_1 + p_1 V_1$, that is, constant enthalpy.

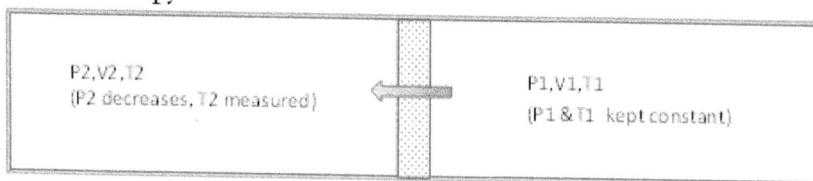

FIGURE 4.5 Joule - Thomson experiment sketch.

For given values of p_1 and T_1 at the upstream of the porous plug, we measure T_2 at the downstream for a series of decreasing pressure values, that is, p_2. The result is schematically shown in Figure 4.6.

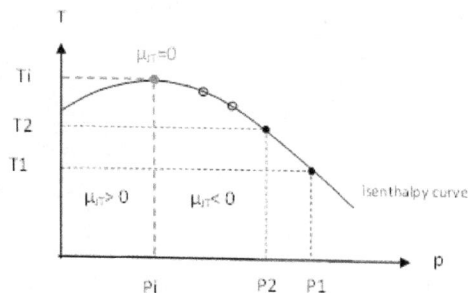

FIGURE 4.6 Joule - Thomson coefficient; a sketch of their experiment results.

Note that the dimension of μ_{JT} is $\left[\dfrac{K}{Pa}\right]$, for example in *SI* units. It is therefore the change of temperature per unit of pressure. From Figure 4.6 we observe that as the pressure in the downstream is reduced, the gas moves under a higher pressure difference, since upstream pressure is kept at a fixed value to the downstream side. The temperature can be calculated for each pressure point using $\mu_{JT} = \dfrac{\partial T}{\partial p} \cong \dfrac{T_2 - T_1}{p_2 - p_1}$, for a small change in pressure. Therefore, we have

$$T_2 = T_1 - \mu_{JT}\left(p_1 - p_2\right)$$

Since $p_1 - p_2 > 0$, for $\mu_{JT} < 0$ we get $T_2 > T_1$, that is, gas is heated, and for $\mu_{JT} > 0$ we get $T_2 < T_1$, that is, gas is cooled. These results indicate that when a real gas expands through a constant enthalpy process, its temperature changes and depends on the sign of μ_{JT}. Therefore, as the flow rate increases due to increase in pressure difference, the gas heats up until μ_{JT} changes sign from $\mu_{JT} < 0$ to $\mu_{JT} > 0$, and then the gas cools down. The throttling cooling technique has industrial applications for cooling and liquefying gases.

As shown in Figure 4.6, an important point of the curve is where $\mu_{JT} = 0$. Since μ_{JT} changes sign, therefore there should be a point (possibly within the range of Δp) for which the slope is zero. At this point, the so – called the *inversion point*, the curve reaches its maximum value and the corresponding temperature is called the *inversion temperature*, T_i. A family of isenthalpic curves can be drawn, for the same gas, for several values of pressure, and the locus of resulting inversion temperatures can form another curve by connecting the corresponding inversion points. This curve is called an *inversion curve*. The region inside the inversion curve is the cooling region, and the outside of it is the heating region. Figure 4.7 shows a sketch of family of isenthalpic curves and corresponding inversion curve for typical materials.

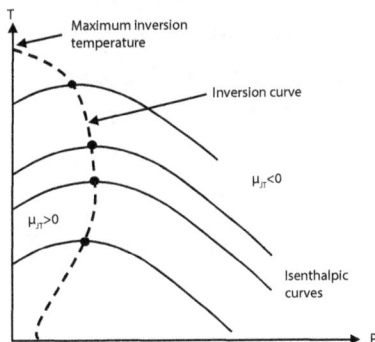

FIGURE 4.7 Isenthapic and inversion curves for nitrogen

The maximum inversion temperature T_{imax} is the point where the inversion curve coincides with the vertical axis or $p = 0$ line. Mathematically, we have $T_{imax} = \dfrac{\partial T}{\partial p}\Big|_{at\ p=0} = 0$. If a gas is at a temperature higher than its T_{imax}, then it cannot be cooled by throttling. Therefore, when $T_{Gas} > T_{imax}$, cooling is not possible by throttling. However, the gas temperature may be lowered by other methods and then be cooled through throttling. Table 4.4 shows the values of T_{imax} for some gases.

Gas at room condition (293K, atm.)	T_{imax}	Cooling by throttling
H2	205K	Not possible
Air	603K	Possible
O2	761K	Possible
N2	621K	Possible
CO	652K	Possible

TABLE 4.4 Maximum Inversion Temperature for Some Common Gases

In summary, we can list the following important conclusions, among others, for the Joule-Thomson experiment:

• The isenthalpic curve is a nonlinear function of pressure, or $T = T(p)$.

• The Joule-Thomson coefficient, $\mu_{JT} = \dfrac{\partial T}{\partial p}$, is a function of pressure as well.

• A real gas behaves similar to a pseudo-ideal gas at its inversion point, for a given pressure, since $\mu_{JT} = 0$.

In practice, the Hampson-Linde, Claude, and Joule-Thomson cycles could be used for liquefying a gas [48], [49].

Example: Calculate a Joule-Thomson coefficient, inversion temperature, and its maximum value for a real gas using the van der Waals model.

Solution: The van der Waals equation reads $\left(p + \dfrac{a}{V^2}\right)(V - b) = RT$, where V is molar volume, and a and b are constants. We use Equation 4.26, $\mu_{JT} = \dfrac{1}{c_p}\left(T\dfrac{\partial V}{\partial T}\Big|_p - V\right)$. We need to calculate $\dfrac{\partial V}{\partial T}\Big|_p$ using van der Waals equation and chain rule; we have $\dfrac{\partial}{\partial V}\left[pV - pb + \dfrac{a}{V} - \dfrac{ab}{V^2}\right]\dfrac{\partial V}{\partial T} = \dfrac{\partial}{\partial T}(RT)$,

which results in $\left.\dfrac{\partial V}{\partial T}\right|_p = R / \left(p - \dfrac{a}{V^2} + \dfrac{2ab}{V^3}\right)$. Back substituting into the

equation for μ, after some manipulation, we get

$$\mu_{JT} = \dfrac{\left(\dfrac{2a}{RT}\right)\left(1 - \dfrac{b}{V}\right)^2 - b}{c_p\left[1 - \dfrac{2}{VRT}\left(1 - \dfrac{b}{V}\right)^2\right]}$$

For inversion temperature we should have $\mu_{JT} = 0$, or $\left(\dfrac{2a}{RT_i}\right)\left(1 - \dfrac{b}{V}\right)^2$
$- b = 0$. After solving for T_i we get

$$T_i = \left(\dfrac{2a}{Rb}\right)\left(1 - \dfrac{b}{V}\right)^2$$

Also we can calculate T_{imax} using the relation for T_i when $\dfrac{b}{V} = 0$ (since $p \to 0$ when $V \to \infty$). Therefore

$$T_{imax} = \dfrac{2a}{Rb}$$

For helium we have $a = 3.46 \times 10^{-3}$ Pa.m^6/mol^2 and $b = 23.71 \times 10^{-6}$ m^3/mol,
$R = 8.314$ J/K.mol. Therefore $T_{imax} = \dfrac{2a}{Rb} = \dfrac{2 \times 3.46 \times 10^{-3}}{8.314 \times 23.71 \times 10^{-6}} = 35.1$.
experimental value [50] is 45 K (-228 ^{0}C).

4.8.2. Isobaric and Isothermal Compressibility

Matter expands and contracts due to changes in temperature and pressure. Change in volume under constant temperature or pressure is a thermodynamic property of interest with practical applications. In this section we derive the relations for three relevant properties: the thermal expansion coefficient, the coefficient of compressibility, and their ratio.

Assume volume as a function of temperature and pressure, or $V = V(T, p)$,
and then we can write $dV = \left.\dfrac{\partial V}{\partial T}\right|_p dT + \left.\dfrac{\partial V}{\partial p}\right|_T dp$. The partial derivative
$\left.\dfrac{\partial V}{\partial T}\right|_p$ is the slope, at a given point, of an isobaric curve in the $V - T$ plane.

Similarly, the partial derivative $\dfrac{\partial V}{\partial p}\Big|_T$ is the slope of an isothermal curve on a $V - p$ plane. Both of these quantities (i.e., $\dfrac{\partial V}{\partial T}\Big|_p$ and $\dfrac{\partial V}{\partial p}\Big|_T$) are intensive type properties.

For practical applications, we are interested to calculate how much the volume of a gas, for example, changes with temperature when pressure is kept constant. Identifying this behavior by *coefficient of thermal expansion* α, the mathematical definition of α is given as

$$\alpha = \frac{1}{V}\frac{\partial V}{\partial T}\Big|_p \tag{4.25}$$

For example, for an ideal gas $(pV = nRT)$ we have $\alpha = \dfrac{1}{V}\left(\dfrac{nR}{p}\right) = \dfrac{nR}{nRT} = \dfrac{1}{T}$. This result shows that the higher the temperature for an ideal gas, the less it is expanded or the less responsive it is through its volume to a change in temperature. The linear coefficient of expansion, normally used for homogeneous solids, is approximately equal to $\alpha/3$.

Equivalently, we are interested to calculate how much volume of a gas, for example, changes with pressure when temperature is kept constant. Identifying this behavior by coefficient of compressibility κ, or isothermal compressibility, the mathematical definition of κ is given as

$$\kappa = -\frac{1}{V}\frac{\partial V}{\partial p}\Big|_T \tag{4.28}$$

For example, for an ideal gas $(pV = nRT)$ we have $\kappa = -\dfrac{1}{V}\left(-\dfrac{nRT}{p^2}\right) = \dfrac{nRT}{Vp^2} = \dfrac{1}{p}$. This result shows that the higher the pressure for an ideal gas, the less it is compressible or responsive through its volume to a change in pressure. The bulk modulus of elasticity, normally used for homogeneous solids, is the inverse of the compressibility coefficient (i.e., $= -V\dfrac{\partial p}{\partial V}\Big|_T$) and Young's modulus of elasticity is related to κ (i.e., $\dfrac{3(1-2v)}{\kappa}$, where v is Poisson's ratio).

Another useful relation can be obtained by back substituting Equation 4.27 and Equation 4.28 into $dV = \left.\dfrac{\partial V}{\partial T}\right|_p dT + \left.\dfrac{\partial V}{\partial p}\right|_T dp$ or $dV = \alpha V dT - \kappa V dp$. After dividing both sides by V and integrating between state points 1 and 2, we get

$$\ln\left(\frac{V_2}{V_1}\right) = \alpha\left(T_2 - T_1\right) - \kappa\left(p_2 - p_1\right) \tag{4.29}$$

Another relation can be derived by applying Euler's chain rule on the relations for α and κ, that is, Equation 4.27 and Equation 4.28.

Recall Equation 4.21, derived for $z = z(x, y)$. Equivalently, we have $V = V(p, T)$; note that p and T are independent variables. By substituting for $x \equiv p$, $y \equiv T$, and $z \equiv V$, we have $\left(\left.\dfrac{\partial p}{\partial T}\right|_V\right)\left(\left.\dfrac{\partial T}{\partial V}\right|_p\right)\left(\left.\dfrac{\partial V}{\partial p}\right|_T\right) = -1$.

But $\left.\dfrac{\partial V}{\partial p}\right|_T = -\kappa V$ and $\left.\dfrac{\partial V}{\partial T}\right|_p = \alpha V$, and after substitution we get $\left.\dfrac{\partial p}{\partial T}\right|_V (\alpha V)^{-1}(-\kappa V) = -1$. Therefore, we get

$$\left.\frac{\partial p}{\partial T}\right|_V = \frac{\alpha}{\kappa} \tag{4.30}$$

Equation 4.30 is useful for calculating how much pressure changes with temperature for a constant volume (or isochoric) process.

Table 4.5 Shows the relations obtained for α, κ, $\left.\dfrac{\partial p}{\partial T}\right|_V$.

Process	Plane – coordinates	Quantity
Isobaric	$V - T$	$\alpha = \dfrac{1}{V}\left.\dfrac{\partial V}{\partial T}\right|_p$
Isothermal	$V - p$	$\kappa = -\dfrac{1}{V}\left.\dfrac{\partial V}{\partial p}\right|_T$
Isochoric	$p - T$	$\left.\dfrac{\partial p}{\partial T}\right|_V = \dfrac{\alpha}{\kappa}$

TABLE 4.5 Relations for Isobaric and Isothermal Compressibility Coefficients and Their Rati

Example:

For a van der Waals gas, calculate the isobaric and isothermal compressibility coefficients α and κ and their ratio.

■ **Solution:**

From the previous example we have $\left.\dfrac{\partial v}{\partial T}\right|_p = R \Big/ \left(p - \dfrac{a}{V^2} + \dfrac{2ab}{V^3} \right)$.

Therefore, $\alpha = \dfrac{R}{V\left(p - \dfrac{a}{V^2} + \dfrac{2ab}{V^3} \right)} = \dfrac{R}{\left(pV - \dfrac{a}{V} + \dfrac{2ab}{V^2} \right)}$. For calculating

κ we need to calculate $\left.\dfrac{\partial V}{\partial p}\right|_T$ first. This can be done by differentiating van der Waals's equation w.r.t. p while keeping T constant,

or $\dfrac{\partial}{\partial p}\left[\left(p + \dfrac{a}{V^2} \right)(V - b) - RT \right]_T = 0$. Working out the differentiation gives $\left(1 - \dfrac{2a}{V^3}\dfrac{\partial V}{\partial p} \right)(V - b) + \left(p + \dfrac{a}{V^2} \right)\dfrac{\partial V}{\partial p} = 0$. Finally, we get

$\left.\dfrac{\partial V}{\partial p}\right|_T = \dfrac{(V - b)}{\left(\dfrac{a}{V^2} - \dfrac{2ab}{V^3} - p \right)}$. Therefore, $\kappa = -\dfrac{1}{V}\left.\dfrac{\partial V}{\partial p}\right|_T = \dfrac{(V - b)}{\left(pV - \dfrac{a}{V} + \dfrac{2ab}{V^2} \right)}$. Using

these results, we have $\dfrac{\alpha}{\kappa} = \dfrac{R}{V - b}$.

4.8.3. Specific Heats

Specific heats are thermodynamic properties of matter. The amount of heat required to measure heating up substances is useful information in practice. The heating could be done under a constant volume or pressure process. In this section we derive the relevant relations, using internal energy and enthalpy equations.

Having internal energy as a function of temperature and volume, or $U = U(T, V)$, we can write its total differential as $dU = \left.\dfrac{\partial U}{\partial T}\right|_V dT + \left.\dfrac{\partial U}{\partial V}\right|_T dV$. The

partial derivative $\left.\dfrac{\partial U}{\partial T}\right|_V$, or change in internal energy due to a temperature

at a constant volume process, relates to a *specific heat at constant volume* C_V,

$$C_V = \left.\frac{\partial U}{\partial T}\right|_V$$

For an ideal gas, since internal energy is only a function of temperature and $\left.\frac{\partial U}{\partial V}\right|_T = 0$, we can write $dU = C_V dT$. Note that the mass of the substance could be absorbed in the U or C_V, for units homogeneity.

Similarly, we define enthalpy as a function of temperature and pressure, or $H = H(T, p)$. We write its total differential as $dH = \left.\frac{\partial H}{\partial T}\right|_p dT + \left.\frac{\partial H}{\partial V}\right|_T dp$. The partial derivative $\left.\frac{\partial H}{\partial T}\right|_p$, or change in enthaply due to temperature at a constant pressure process, relates to a *specific heat at constant pressure C_p*,

$$C_p = \left.\frac{\partial H}{\partial T}\right|_p$$

For an ideal gas, since enthalpy is only a function of temperature and $\left.\frac{\partial H}{\partial p}\right|_T = 0$, we can write $dH = CpdT$. Both C_V and C_p are material properties. Note that the mass of the substance could be absorbed in the H or C_p. It is useful to seek a relation for C_V and C_p in terms of α and κ, as defined in the previous section. To do this, we use the 1st law, or $\delta Q = dU + pdV$. But substituting for $dU = \left.\frac{\partial U}{\partial T}\right|_V dT + \left.\frac{\partial U}{\partial V}\right|_T dV$, we get

$\delta Q = \left.\frac{\partial U}{\partial T}\right|_V dT + \left(\left.\frac{\partial U}{\partial V}\right|_T + p\right)dV = C_V dT + \left(\left.\frac{\partial U}{\partial V}\right|_T + p\right)dV$. We substitute for

$\delta Q = C_p dT$ and combine the last two relations, and after rearranging terms, we get $C_p - C_V = \left(\left.\frac{\partial U}{\partial V}\right|_T + p\right)\left.\frac{\partial V}{\partial T}\right|_p$. It can be shown that $\left.\frac{\partial U}{\partial V}\right|_T + p = T\left.\frac{\partial p}{\partial T}\right|_V$.

Therefore, we have $C_p - C_V = T\left.\frac{\partial p}{\partial T}\right|_V \left.\frac{\partial V}{\partial T}\right|_p$. After substituting for the partial derivatives using Equation 4.27 and Equation 4.30, we get $C_p - C_V = T\left(\frac{\alpha}{\kappa}\right)(\alpha V)$, or

$$C_p - C_V = \frac{\alpha^2 VT}{\kappa} \qquad (4.31)$$

For an ideal gas, this relation reduces to $C_p - C_V = \frac{T^{-2}VT}{p^{-1}} = R$, as expected.

4.8.4. Thermodynamic Relations Derivation Using the Jacobian Method

As shown so far, there are a large number of derivatives involved and needed for thermodynamics relations. For many practical problems, it would be useful to know the relevant derivatives for calculations and the measurement of thermodynamic properties. However, a systematic methodology would be required to find these derivatives. In this section we discuss the Jacobian method as a tool to derive any desired or required thermodynamic derivative.

The Jacobian matrix is a mathematical arrangement of a partial derivative of a multivariable function. Specifically, the determinant of the matrix containing the partial derivatives is defined and referred to as the *Jacobian*. Geometrically, the Jacobian is the relative change of a geometrical entity (e.g., line, area, volume) with reference to a transformation of coordinates. For incompressible materials the Jacobian is equal to unity and for all materials and transformations cannot be negative [51],[52]. The Jacobian method enables us to derive thermodynamic relations in a systematic way, as well as 1st and 2nd order partial derivatives involving thermodynamic variables. In further sections we will develop key relations for Jacobian algebra with its application to thermodynamic functions and derivatives.

To help the discussion we present a simple example to demonstrate the application and meaning of the Jacobian and its application for calculating partial derivatives related to thermodynamic functions.

4.8.5. The Jacobian of a 2D Coordinate System Rotation

Consider a 2D Cartesian coordinate system (y_1, y_2). This system rotates about the positive y_3-axis (using R.H. rule) through an angle θ. The new coordinates system is designated by (x_1, x_2), as shown in Figure 4.9.

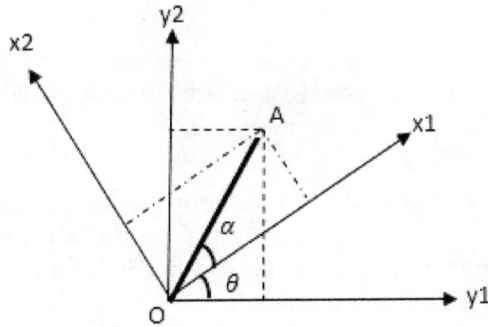

FIGURE 4.9 Transformation through the rotation of 2D Cartesian coordinates.

Using the components of an arbitrary line $\left|\overline{OA}\right| = r$, making an angle α with x_1 axis, we can write $y_1 = r\cos(\theta + \alpha)$, $y_2 = r\sin(\theta + \alpha)$, $x_1 = r\cos(\alpha)$, and $x_2 = r\sin(\alpha)$. After expanding the trigonometric functions and some manipulations, we get

$$\begin{Bmatrix} x_1 \\ x_2 \end{Bmatrix} = \begin{bmatrix} \cos\theta & \sin\theta \\ -\sin\theta & \cos\theta \end{bmatrix} \begin{Bmatrix} y_1 \\ y_2 \end{Bmatrix}.$$ The Jacobian matrix of transformation is

defined as $\begin{bmatrix} \cos\theta & \sin\theta \\ -\sin\theta & \cos\theta \end{bmatrix}$ and its determinant as the *Jacobian* of the trans-

formation, or $\mathcal{J} = \begin{vmatrix} \cos\theta & \sin\theta \\ -\sin\theta & \cos\theta \end{vmatrix} = 1$. In general, we can write the length

of a differential line in the rotated coordinates as $dx_i = \dfrac{\partial x_i}{\partial y_j} dy_j = F_{ij} dy_j$

or in matrix notation for 2D coordinates (i.e., i,j = 1,2), without los-

ing generality, as $\begin{Bmatrix} dx_1 \\ dx_2 \end{Bmatrix} = \begin{bmatrix} \dfrac{\partial x_1}{\partial y_1} & \dfrac{\partial x_1}{\partial y_2} \\ \dfrac{\partial x_2}{\partial y_1} & \dfrac{\partial x_2}{\partial y_2} \end{bmatrix} \begin{Bmatrix} dy_1 \\ dy_2 \end{Bmatrix}$. Therefore the Jacobian is

$\mathcal{J} = \begin{vmatrix} \dfrac{\partial x_1}{\partial y_1} & \dfrac{\partial x_1}{\partial y_2} \\ \dfrac{\partial x_2}{\partial y_1} & \dfrac{\partial x_2}{\partial y_2} \end{vmatrix} = \dfrac{\partial x_1}{\partial y_1} \cdot \dfrac{\partial x_2}{\partial y_2} - \dfrac{\partial x_1}{\partial y_2} \cdot \dfrac{\partial x_2}{\partial y_1}$. In continuum mechanics the matrix

$$\begin{bmatrix} \dfrac{\partial x_1}{\partial y_1} & \dfrac{\partial x_1}{\partial y_2} \\ \dfrac{\partial x_2}{\partial y_1} & \dfrac{\partial x_2}{\partial y_2} \end{bmatrix}$$ is referred to as a deformation gradient; hence the Jacobian is

the determinant of the deformation gradient matrix. In index notation, we can write $\mathcal{J} = \mathcal{E}_{ijk} F_{i1} F_{i2} F_{i3}$, where $\mathcal{E}_{ijk}$ is the permutation symbol (i.e., for an even number of permutations its value is 1, for an odd number of permutations it is -1, otherwise it is zero).

Wolfram Alpha® [53] provides 2D and 3D Jacobian calculators. Interested readers may use these tools for more exercises and calculation of Jacobians.

4.8.6. Key Jacobian Properties Used in Thermodynamics

First, we define a symbol for expressing the Jacobian, in order to facilitate its algebraic manipulation and prevent needing to write down the whole determinant. The mostly used symbol for Jacobian is $\mathcal{J} = \dfrac{\partial\left(x_1, x_2, \ ..., \ x_n\right)}{\partial\left(y_1, y_2, \ ..., \ y_n\right)}$,

where $x_i = x_i\left(y_1, y_2, ..., y_n\right)$ with i = 1,2, ... , n. For thermodynamics relations we use functions with two variables (for single species-phase substances). Therefore, we modify the symbol for the purpose of ease of writing when manipulating thermodynamic relations using the Jacobian method. Therefore, in this book, for a function of two variables, we use the following symbol and definition for the Jacobian:

$$J\left(x,y\right) = \frac{\partial\left(x,y\right)}{\partial\left(a,b\right)} = \frac{\left[x,y\right]}{\left[a,b\right]} = \left(\frac{\partial x}{\partial a}\right)_b \cdot \left(\frac{\partial y}{\partial b}\right)_a - \left(\frac{\partial x}{\partial b}\right)_a \cdot \left(\frac{\partial y}{\partial a}\right)_b \qquad (4.32)$$

where x and y are functions of a and b. Using this definition, we can write down the following list of properties which are mostly useful for thermodynamic relations and derivations. For a more complete list, interested readers are referred to the classic paper of Shaw [54].

1) $\dfrac{\left[x,y\right]}{\left[a,b\right]} \cdot \dfrac{\left[a,b\right]}{\left[c,d\right]} = \dfrac{\left[x,y\right]}{\left[c,d\right]}$, which can be shown using the chain rule of differentiation.

2) $\left[x, y\right] = -\left[y, x\right] = \left[y, -x\right] = \left[-y, x\right]$

3) $[x, x] = 0$

4) $\dfrac{[x,z]}{[y,z]} = \left(\dfrac{\partial x}{\partial y}\right)_z$

5) If we have a function of three variables, $f(x, y, z) = 0$, like an equation of state in thermodynamics, then we can write $z = z(x, y)$, with its total differential as $dz = \left(\dfrac{\partial z}{\partial x}\right)_y dx + \left(\dfrac{\partial z}{\partial y}\right)_x dy$. We define $M = \left(\dfrac{\partial z}{\partial x}\right)_y = \dfrac{[z,y]}{[x,y]}$ and

$N = \left(\dfrac{\partial z}{\partial y}\right)_x = \dfrac{[z,x]}{[y,x]}$, using property (4). Therefore, we have

$$dz = \underbrace{\frac{[z,y]}{[x,y]}}_{M} dx + \underbrace{\frac{[z,x]}{[y,x]}}_{N} dy \qquad (4.33)$$

Similarly, for functions $x = x(y, z)$ and $= y(x, z)$, we can write, by inspection, $dx = \dfrac{[x,z]}{[y,z]} dy + \dfrac{[x,y]}{[z,y]} dz$ and $dy = \dfrac{[y,z]}{[x,z]} dx + \dfrac{[y,x]}{[z,x]} dz$.

Equation 4.33 can be written, after multiplying both sides by $[x,y]$ and rearranging terms, as

$$[x,y]dz + [y,z]dx + [z,x]dy = 0 \qquad (4.34)$$

Similarly, we can write relations like $[y,z]dx + [z,x]dy + [x,y]dz = 0$ and $[z,x]dy + [y,z]dx + [x,y]dz = 0$.

6) If $c = func(a,b)$, using Equation 4.34 we can write $[x,y] [z,c] + [y,z]$ $[x,c] + [z,x] [y,c] = 0$. Then after dividing by $[x,y]$ and using the definitions for M and N (see Equation 4.33), we have

$$[z,c] = M[x,c] + N[y,c] \qquad (4.35)$$

Equation 4.33 and Equation 4.35 are key Jacobian properties used for thermodynamics relations and derivations.

4.8.7. Thermodynamic Energy Functions and Maxwell Relations in Jacobian Form

In the previous section (see Table 4.2) we derived the relations for four thermodynamic energy/potential functions (i.e., internal energy, enthalpy, Helmholtz, and Gibbs). These functions are usually expressed in terms of four thermodynamic variables (i.e., pressure, volume, temperature, and entropy). Therefore, in total we have eight quantities that can have functional inter-relations. Using properties of the Jacobian (see Equation 4.35) we can rewrite these energy functions as follows (note, c is a dummy variable).

$$\begin{cases} dU = TdS - pdV \xrightarrow{\;Yields\;} [U,c] = T[S,c] - p[V,c] \\ dH = TdS + Vdp \xrightarrow{\;Yields\;} [H,c] = T[S,c] + V[p,c] \\ dA = -SdT - pdV \xrightarrow{\;Yields\;} [A,c] = -S[T,c] - p[V,c] \\ dG = -SdT + Vdp \xrightarrow{\;Yields\;} [G,c] = -S[T,c] + V[p,c] \end{cases} \quad (4.36)$$

Also all four Maxwell relations (see Table 4.3) can be written in terms of the Jacobian as a single relation. For example, we can write, using property (4), $\left.\dfrac{\partial T}{\partial V}\right|_S = \dfrac{[T,S]}{[V,S]}$ and $-\left.\dfrac{\partial p}{\partial S}\right|_V = -\dfrac{[p,V]}{[S,V]}$, and hence (see Table 4.3) $\dfrac{[T,S]}{[V,S]} = -\dfrac{[p,V]}{[S,V]} = \dfrac{[p,V]}{[V,S]}$, which yields $[T,S] = [p,V]$.

$$\begin{cases} \left.\dfrac{\partial T}{\partial V}\right|_S = -\left.\dfrac{\partial p}{\partial S}\right|_v \\ \left.\dfrac{\partial V}{\partial S}\right|_p = \left.\dfrac{\partial T}{\partial p}\right|_S \\ \left.\dfrac{\partial p}{\partial T}\right|_V = \left.\dfrac{\partial S}{\partial V}\right|_T \\ \left.\dfrac{\partial V}{\partial T}\right|_p = -\left.\dfrac{\partial S}{\partial p}\right|_T \end{cases} \xrightarrow{\;reduce\ to\;} [T,S] = [p,V] \quad (4.37)$$

This reduction of four apparently different Maxwell relations into one has a physical meaning. Considering the Gibbs relation for internal energy (see Equation 4.1), or $TdS = pdV + dU$, and integrating it around a closed path, that is, a cycle, we get $\oint TdS = \oint pdV$, since $\oint dU = 0$ (recall U is a state function). This means that the area enclosed by a closed path in a $T - S$ plane is equal to that of a $p - V$ plane (or the balance of work and heat), and hence $[T, S] = [p, V]$ is recovered.

In the previous section, we discussed other thermodynamic properties such as the isobaric thermal expansion coefficient (see Equation 4.27) and the isothermal compressibility coefficient (see Equation 4.28). In terms of the Jacobians, these relations can be written as

$$\begin{cases} \alpha = \dfrac{1}{V}\dfrac{\partial V}{\partial T}\bigg|_p & \xrightarrow{\textit{Yields}} \alpha = \dfrac{1}{V}\dfrac{\left[V,p\right]}{\left[T,p\right]} \\[4mm] \kappa = -\dfrac{1}{V}\dfrac{\partial V}{\partial p}\bigg|_T & \xrightarrow{\textit{Yields}} \kappa = \dfrac{1}{V}\dfrac{\left[V,T\right]}{\left[T,p\right]} \end{cases} \tag{4.38}$$

For further reference, specific heats also can be written as

$$\begin{cases} C_p = T\left(\dfrac{\partial S}{\partial T}\right)_p = T\dfrac{\left[S,p\right]}{\left[T,p\right]} \\[4mm] C_V = T\left(\dfrac{\partial S}{\partial T}\right)_V = T\dfrac{\left[S,V\right]}{\left[T,V\right]} \end{cases} \tag{4.39}$$

4.8.8. Jacobian Application to Thermodynamic Functions

As mentioned, in thermodynamics we have eight variables that are normally used for describing a closed system (there are ten or more if chemical potential and number of particles are involved for open systems, see [30]). The four energy functions are internal energy, enthalpy, Helmholtz energy, and Gibbs energy; the four thermodynamic properties are pressure, volume, temperature, and entropy. By choosing any three variables from these eight variables (e.g., symbolically x,y,z), we can form a function (e.g., $x = x(y,z)$) and derive the corresponding partial derivatives, that is, $\left(\dfrac{\partial x}{\partial y}\right)_z$ and $\left(\dfrac{\partial x}{\partial z}\right)_y$, in the form of the Jacobian. Therefore, we can build $C(8,3) = \dfrac{8!}{3! \times (8-3)!} = 56$ sets of triple-variable combinations out of eight variables. For example, consider an enthalpy-entropy-pressure

(H,S,p) set with the corresponding equation for its total derivative, or $dH = \left(\dfrac{\partial H}{\partial S}\right)_p dS + \left(\dfrac{\partial H}{\partial p}\right)_S dp$, which requires us to calculate expressions for the two partial derivatives involved. Therefore, in total we would be required to calculate $56 \times 2 = 112$ independent partial derivatives. One should note that the two possible permutations for each selected triple-variable set do not contain independent partial derivatives. For example, permuting (H,S,p) gives (p,H,S) and (S,p,H) as apparently new sets of triple-variable selections. However, it is possible to derive the corresponding partial derivatives for the permuted sets (i.e., $\left(\dfrac{\partial p}{\partial H}\right)_S$ and $\left(\dfrac{\partial p}{\partial S}\right)_H$ corresponding to $dp = \left(\dfrac{\partial p}{\partial H}\right)_S dH + \left(\dfrac{\partial p}{\partial S}\right)_H dS$ and similarly, $\left(\dfrac{\partial S}{\partial p}\right)_H$ and $\left(\dfrac{\partial S}{\partial H}\right)_p$ corresponding to $dS = \left(\dfrac{\partial S}{\partial p}\right)_H dp + \left(\dfrac{\partial S}{\partial H}\right)_p dH$)) from those of the original ones (i.e., $\left(\dfrac{\partial H}{\partial S}\right)_p$ and $\left(\dfrac{\partial H}{\partial p}\right)_S$). This realization could potentially reduce the number of calculations from $3 \times 112 = 336$ to 112 for the required partial derivatives.

Systematic calculation of these derivatives (i.e., all 336 of them) is possible using the Jacobian method. Original work on this topic goes back to Rankine [55] and more recent work by Bryan [56], Bridgman [57], Tribus [30], and Shaw [54], among others, who provided comprehensive lists in the form of tables. Modern computers and symbolic-based computational software packages have provided an effective tool for the calculations of thermodynamic derivatives using a Jacobian-method-based algorithm. For example, Perk [58] has provided the Maple™ code and Mikhailov has provided the Mathematica™ code and an app [59]. These tools are valuable for fast and accurate computations of thermodynamic derivatives. The Jacobian method for calculating thermodynamic derivatives is described in the next section.

4.8.9. Jacobian Methodology

1) Select any three variables out of the eight-variable set (U,H,A,G,P,V,T,S).

2) Write down the total derivative of one of the variables from the selected triple set as a function of the other two. The choice of the functional form is arbitrary, but usually reference can be made to those given in Equation 4.36, but not exclusively. Permutation of the selected triple set is not required in principle.

3) Express the required partial derivatives involved in terms of the Jacobian, using Jacobian properties.

4) If the Jacobian contains thermodynamic energy functions, then they should be replaced by their corresponding relations (see Equation 4.36).

5) If the Jacobian contains an entropy function, then it should be replaced by making use of the Maxwell relation in its Jacobian form (see Equation 4.37) and/or using specific heats (see Equation 4.39).

6) Any remaining Jacobian should be replaced by other measurable thermodynamic properties, like p, V, T, α, κ (see Equation 4.38) to get the resulting expression free of any Jacobian terms and expressed in terms of measurable thermodynamic quantities.

A complete survey of the remaining variables (besides the already selected ones) following the above-mentioned methodology should provide 336 partial derivatives, of which 112 are independent. In other words, each triple set variable requires two independent derivatives. In the remainder of this section we present several worked-out examples to demonstrate the application of the Jacobian method as described above.

4.8.10. Derivatives for Enthalpy, H=H(T,p)

Given enthalpy as a function of temperature and pressure, or $H = H(T,P)$, calculate the total differential dH and its corresponding partial derivatives, $\left(\dfrac{\partial H}{\partial T}\right)_p$ and $\left(\dfrac{\partial H}{\partial p}\right)_T$. Also show that the derivatives for the permuted functions can be calculated from these derivatives, without making use of the systematic Jacobian method.

■ Solution:

Following the steps mentioned above, we have

Steps 1–2: In this example, our triple variable set is (H,T,P) with the given functional form for $dH = dH(T,P)$. Therefore, we can write

$$dH = \left(\frac{\partial H}{\partial T}\right)_p dT + \left(\frac{\partial H}{\partial p}\right)_T dp.$$

Step 3: Writing the derivatives in Jacobian form using Equation 4.33, we get

$$dH = \frac{[H,p]}{[T,p]}dT + \frac{[H,T]}{[p,T]}dp \qquad (4.40)$$

Steps 4–5: For replacing Jacobian terms containing H, we use Equation 4.36 by writing $[H,p] = T[S,p] + \underbrace{V[p,p]}_{=0} = T[S,p]$. Note here that we select the dummy variable to be p, or let $c = p$ in Equation 4.36, related to the equation for enthalpy (i.e., $dH = TdS + Vdp$). Also for $[H,T]$, we can write $[H,T] = T[S,T] + V[p,T]$. Substituting for the Jacobian terms into the relation for dH, given in Equation 4.40, we get $dH = \frac{T[S,p]}{[T,p]}dT + \left\{ \frac{T[S,T]}{[p,T]} + V \right\}dp$. Now we eliminate Jacobian terms containing entropy S, using the Maxwell relation (i.e., $[S,T] = -[p,V]$ (see Equation 4.37), and $C_p = T\frac{[S,p]}{[T,p]}$ (see Equation 4.39). Therefore, we get $dH = C_p dT + \left\{ -T\frac{[p,V]}{[p,T]} + V \right\}dp$, but $\frac{[p,V]}{[p,T]} = \frac{[V,p]}{[T,p]} = \left(\frac{\partial V}{\partial T} \right)_p = \alpha V$, using Equation 4.38.

Step 6: Finally, we get

$$dH = C_p dT + (V - T\alpha V)dp \qquad (4.41)$$

Readers, please note in this example that by using the systematic Jacobian method, we derived an equation for changes in enthalpy equivalent to Equation 4.20. We can conclude, by comparison, that

$$\left(\frac{\partial H}{\partial T} \right)_p = C_p \text{ and } \left(\frac{\partial H}{\partial p} \right)_T = (V - T\alpha V) \qquad (4.42)$$

Readers are encouraged to compare these results with those provided by running Mikhailov's app to find out that they are equivalent.

Now we investigate the dependency of other partial derivatives that are required for permutations of the given functional form. These functions are $p = p(H,T)$ and $T = T(p,H)$, related to sets (p,H,T) and (T,p,h), respectively. First, we consider the total derivative

$dp = \left(\dfrac{\partial p}{\partial H}\right)_T dH + \left(\dfrac{\partial p}{\partial T}\right)_H dT = \dfrac{[p,T]}{[H,T]} dH + \dfrac{[p,H]}{[T,H]} dT$. From here we can follow the Jacobian systematic methodology to find the partial derivatives. However, since we have already done that for the functional form $H = H(T,P)$, we can benefit from it to calculate $\left(\dfrac{\partial p}{\partial H}\right)_T$ and $\left(\dfrac{\partial p}{\partial T}\right)_H$. We have already calculated $\left(\dfrac{\partial H}{\partial p}\right)_T = (V - T\alpha V)$, hence simply $\left(\dfrac{\partial p}{\partial H}\right)_T = \dfrac{1}{(V - T\alpha V)}$, using the reciprocal rule for partial derivatives. For $\left(\dfrac{\partial p}{\partial T}\right)_H$ we use the obtained results for ratio $\left(\dfrac{\partial H}{\partial T}\right)_p \Big/ \left(\dfrac{\partial H}{\partial p}\right)_T = \dfrac{C_p}{V - T\alpha V}$. But in Jacobian form we have $\dfrac{\left(\dfrac{\partial H}{\partial T}\right)_p}{\left(\dfrac{\partial H}{\partial p}\right)_T} = \dfrac{[H,p]}{[T,p]} \cdot \dfrac{[p,T]}{[H,T]} = \dfrac{-[p,H]}{-[T,H]} \cdot \underbrace{\dfrac{[p,T]}{[T,p]}}_{=-1} = -\left(\dfrac{\partial p}{\partial T}\right)_H$.

Hence $\left(\dfrac{\partial p}{\partial T}\right)_H = \dfrac{C_p}{T\alpha V - V}$ Similarly, for the functional form $T = T(p,H)$, we can use the partial derivatives calculated so far in order to calculate $\left(\dfrac{\partial T}{\partial p}\right)_H = \dfrac{V(\alpha T - 1)}{C_p}$ and $\left(\dfrac{\partial T}{\partial H}\right)_p = \dfrac{1}{C_p}$.

This example demonstrates the Jacobian methodology application for calculating thermodynamic partial derivatives, as well as their interrelations for a selected triple variable set.

4.8.11. Derivatives for Entropy, $S = S(G,p)$

Given entropy as a function of Gibbs energy and pressure, or $S = S(G,p)$, calculate the total differential dS and its corresponding partial derivatives, $\left(\dfrac{\partial S}{\partial G}\right)_p$ and $\left(\dfrac{\partial S}{\partial p}\right)_G$.

■ **Solution:**

following the Jacobian method and steps, we have

Steps 1–2: In this example, our selected triple variable set (S,G,P) is with the given functional form for $S = S(G,P)$. Therefore, we can write

$$dS = \left(\frac{\partial S}{\partial G}\right)_p dG + \left(\frac{\partial S}{\partial p}\right)_G dp.$$

Step 3: writing the derivatives in Jacobian form, using Equation 4.33, we get

$$dS = \frac{[S,p]}{[G,p]} dG + \frac{[S,G]}{[p,G]} dp \qquad (4.43)$$

Steps 4–5: For replacing Jacobian terms containing G, we use Equation 4.36 by writing $[G,p] = -S[T,p] + \underbrace{V[p,p]}_{=0} = -S[T, p]$. Similarly, $[G,S] = -S[T,S] + V[p,S]$. Note that we select the dummy variable to be p and S, respectively, in Equation 4.36, related to the equation for enthalpy (i.e., $dG = -SdT + Vdp$). Substituting for the Jacobian terms into the relation for dS given in Equation 4.43, we get $dS = -\frac{[S,p]}{S[T,p]} dG + \left\{ \frac{S[T,S]}{[p,G]} - V\frac{[p,S]}{S[p,G]} \right\} dp$. Now we eliminate the Jacobian terms containing entropy S using the Maxwell relation (i.e., $[T,S] = [p,V]$ (see Equation 4.37), and $C_p = T\frac{[S,p]}{[T,p]}$ (see Equation 4.39). Therefore, we get $dS = -\frac{C_p}{TS} dG + \left\{ \frac{S[T,S]}{S[T,p]} - V\frac{[p,S]}{S[T,p]} \right\} dp$, but $\frac{S[T,S]}{S[T,p]} = -\frac{[V,p]}{[T,p]} = -\alpha V$ using Equation 4.38, and $V\frac{[p,S]}{S[T,p]} = -\frac{V}{S}\frac{C_p}{T}$.

Therefore, we get

$$dS = -\frac{C_p}{TS} dG + \left(\frac{VC_p}{TS} - \alpha V\right) dp \qquad (4.44)$$

By comparison, we can write

$$\left(\frac{\partial S}{\partial G}\right)_p = -\frac{C_p}{TS} \quad and \quad \left(\frac{\partial S}{\partial p}\right)_G = \left(\frac{VC_p}{TS} - \alpha V\right) \qquad (4.45)$$

Example: Derive an expression for the Joule-Thomson coefficient

In this example we demonstrate the application of the Jacobian method for calculating the Joule-Thomson coefficient $\mu_{J-T} = \left(\dfrac{\partial T}{\partial p}\right)_H$.

■ **Solution:**

Writing the partial derivative in terms of the Jacobian, we get $\left(\dfrac{\partial T}{\partial p}\right)_H = \dfrac{[T,H]}{[p,H]} = \dfrac{[H,T]}{[H,p]}$. But we have $[H,p] = T[S,p]$ and $[H,T] = T[S,T] + V[p,T]$. Substituting, we get $\mu_{J-T} = \dfrac{T[S,T] + V[p,T]}{T[S,p]}$.

Eliminating terms involving S, we get $\mu_{J-T} = \dfrac{T[V,P] + V[p,T]}{C_p[T,p]}$. Using Equation 4.38, we have $T[V,p] = \alpha V$, and finally can write the Joule-Thomson coefficient as

$$\mu_{J-T} = \frac{V}{C_p}(\alpha T - 1) \qquad (4.46)$$

4.8.12 Derivatives for Enthalpy, H = H(G,p)

Given enthalpy as a function of Gibbs energy and pressure, or $H = H(G,P)$, calculate the total differential dH and its corresponding partial derivatives, $\left(\dfrac{\partial H}{\partial G}\right)_p$ and $\left(\dfrac{\partial H}{\partial p}\right)_G$.

■ **Solution:**

Following the steps mentioned above, we have

Steps 1–2: In this example, our triple variable set is (H,G,P) with the given functional form for $H = H(G,P)$. Therefore, we can write

$$dH = \left(\frac{\partial H}{\partial G}\right)_p dG + \left(\frac{\partial H}{\partial p}\right)_G dp.$$

Step 3: Writing the derivatives in Jacobian form using Equation 4.33, we get

$$dH = \frac{[H,p]}{[G,p]} dG + \frac{[H,G]}{[p,G]} dp$$

Steps 4–5: For replacing Jacobian terms containing H, we use Equation 4.36 by writing $[H,p] = T[S,p] + V\underbrace{[p,p]}_{=0}$. Note here that we select the dummy variable to be p, or let $c = p$ in Equation 4.36, related to the equation for enthalpy (i.e., $dH = TdS + Vdp$). Also for $[G,p]$, we can write $[G,p] = -S[T,p] + V[p,p] = -S[T,p]$, and similarly $[G,H] = -S[T,H] + V[p,H]$. Substituting for the Jacobian terms into the relation for dH, we get $dH = \dfrac{T[S,p]}{-S[T,p]} dG + \left\{ \dfrac{[T,H]}{[T,p]} - \dfrac{V}{S} \dfrac{[p,H]}{[T,p]} \right\} dp$. But, using $C_p = T \dfrac{[S,p]}{[T,p]}$ and manipulating the terms in the curly bracket, we can write $dH = -\dfrac{C_p}{S} dG + \left\{ \dfrac{[H,T]}{[p,T]} - \dfrac{V}{S} \underbrace{\dfrac{T[S,p]}{[T,p]}}_{=C_p} \right\} dp$. Now we eliminate the Jacobian term involving enthalpy, using $[H,T] = T[S,T] + V[p,T]$, or

$$dH = -\frac{C_p}{S} dG + \left\{ \frac{T[S,T]}{[p,T]} + V + \frac{V}{S} C_p \right\} dp. \text{ But } \frac{[S,T]}{[p,T]} = \frac{[S,T]}{[V,p]} \cdot \frac{[V,p]}{[p,T]} \text{ using}$$

the chain rule, and $[S,T] = [V,p]$ using Maxwell relation, and $\dfrac{[V,p]}{[p,T]} = -\alpha V$ (see Equation 4.38).

Step 6: Hence, after substitution we arrive at

$$dH = -\frac{C_p}{S} dG + V \left(1 - \alpha T + \frac{C_p}{S} \right) dp \tag{4.47}$$

Which yields, by comparison:

$$\left(\frac{\partial H}{\partial G} \right)_p = -\frac{C_p}{S} \quad and \quad \left(\frac{\partial H}{\partial p} \right)_G = V \left(1 - \alpha T + \frac{C_p}{S} \right) \tag{4.48}$$

4.9. A Complete List of 336 Derivatives for 168 Thermodynamic Functions

It is tedious but useful to calculate all 336 partial derivatives associated with 168 thermodynamic functions. These calculations are similar to the derivations given in the previous sections in terms of the Jacobian methodology application. Table 4.6 shows $56 \times 3 = 168$ triple-variable sets and functions with their corresponding $56 \times (2 + 4) = 336$ derivatives.

Note that if we add two more variables, that is, chemical potential μ and particle number N (at least for one-component systems), to the list of our eight variables considered above, then we have a total of ten variables. Therefore, we should select four-variable sets to form a thermodynamic function (see, for example, Equation 4.10-Equation 4.13). This gives $C(10,4) = \dfrac{10!}{4! \times (10-4)!} = 210$ functions. Since each function requires three derivatives, then we need to calculate $210 \times 3 = 630$ derivatives. If we include four possible permutations for each selected four-variable set, then we have $210 \times 3 \times 4 = 2520$ associated derivatives.

TABLE 4.6. A Complete List of Expressions for 336 Derivatives Related to 168 Thermodynamic Functions, Derivatives Derived Using Jacobian Method (α, κ, C_v, C_p Are Given in Table 4.5 and by Equation 4.39).

Thermodynamic Variables			Function	Derivatives	Permuted Functions	Derivatives
U	H	A	$U = U(H, A)$	$(\partial U/\partial H)_A =$ $(\kappa P C_V + \kappa P S - \alpha$ $S T)/(\kappa P C_V + \alpha P$ $V - \alpha S T + S)$	$A = A(U, H)$	$(\partial A/\partial U)_H = (\kappa P C_V$ $+ \alpha P V - \alpha S T + S)/$ $(-Cp + \kappa P Cv + \alpha$ $P V)$
						$(\partial A/\partial H)_U = (\kappa P C v +$ $\kappa P S - \alpha \, ST)/(Cp - \kappa$ $P C_V + \alpha P V)$
				$(\partial U/\partial A)_H =$ $(-C_p + \kappa P C_V + \alpha$ $P V)/(\kappa P C_V + \alpha P$ $V - \alpha S T + S)$	$H = H(A, U)$	$(\partial H/\partial A)_U = (Cp - P \kappa$ $C_V + \alpha P V))/(\kappa P C v +$ $\kappa P S - \alpha ST)$
						$(\partial H/\partial U)_A = (\kappa P C v$ $+ \alpha P V - \alpha S T + S)/$ $(\kappa P C v + \kappa P S - \alpha$ $S T)$

U H G $U = U(H, G)$ $\quad (\partial U/\partial H)_G = (Cp + \kappa PS - \alpha PV - \alpha ST)/(Cp - \alpha ST + S)$ $\quad G = G(U, H)$ $\quad (\partial G/\partial U)_H = (Cp - \alpha ST + S)/(-Cp + \kappa P Cv + \alpha PV)$

$\quad (\partial G/\partial H)_U = (Cp + \kappa PS - \alpha PV - \alpha ST)/(Cp - \kappa P Cv - \alpha PV)$

$\quad (\partial U/\partial G)_H = (-Cp + \kappa P Cv + \alpha PV)/(Cp - \alpha ST + S)$ $\quad H = H(G, U)$ $\quad (\partial H/\partial G)_U = (Cp - \kappa P Cv - \alpha PV)/(Cp + \kappa PS - \alpha PV - \alpha ST)$

$\quad (\partial H/\partial U)_G = (Cp - \alpha ST + S)/(Cp + \kappa PS - \alpha PV - \alpha ST)$

U H P $U = U(H, P)$ $\quad (\partial U/\partial H)_P = 1 - (\alpha PV)/Cp$ $\quad P = P(U, H)$ $\quad (\partial P/\partial U)_H = Cp/(-VCp + \kappa PV Cv + \alpha PV^2)$

$\quad (\partial P/\partial H)_U = (\alpha PV - Cp)/(-VCp + \kappa PV Cv + \alpha PV^2)$

$\quad (\partial U/\partial P)_H = (-VCp + \kappa PV Cv + \alpha PV^2)/Cp$ $\quad H = H(P, U)$ $\quad (\partial H/\partial P)_U = (-VCp + \kappa PV Cv + \alpha PV^2)/(\alpha PV - Cp)$

$\quad (\partial H/\partial U)_P = Cp/(Cp - \alpha PV)$

U H V $U = U(H, V)$ $\quad (\partial U/\partial H)_V = (\kappa Cv)/(\kappa Cv + \alpha V)$ $\quad V = V(U, H)$ $\quad (\partial V/\partial U)_H = (\kappa Cv + \alpha V)/(Cp - \kappa P Cv - \alpha PV)$

$\quad (\partial V/\partial H)_U = (\kappa Cv)/(-Cp + \kappa P Cv + \alpha PV)$

$\quad (\partial U/\partial V)_H = (Cp - \kappa P Cv - \alpha PV)/(\kappa Cv + \alpha V)$ $\quad H = H(V, U)$ $\quad (\partial H/\partial V)_U = (-Cp + \kappa P Cv + \alpha PV)/(\kappa Cv)$

$\quad (\partial H/\partial U)_V = (\alpha V)/(\kappa Cv) + 1$

U H T $U = U(H, T)$ $\quad (\partial U/\partial H)_T = (\alpha T - \kappa P)/(\alpha T - 1)$ $\quad T = T(U, H)$ $\quad (\partial T/\partial U)_H = (\alpha T - 1)/(-Cp + \kappa P Cv + \alpha PV)$

$\quad (\partial T/\partial H)_U = (\kappa P - \alpha T)/(-Cp + \kappa P Cv + \alpha PV)$

| U | H | S | $U = U(H, S)$ | $(\partial U/\partial T)_H = (-Cp + \kappa P Cv + \alpha P V)/(\alpha T - 1)$ | $H = H(T, U)$ | $(\partial H/\partial T)_U = (-Cp + \kappa P Cv + \alpha P V)/(\kappa P - \alpha T)$ |

$$(\partial U/\partial T)_H = (-Cp + \kappa P Cv + \alpha P V)/(\alpha T - 1)$$

$H = H(T, U)$

$$(\partial H/\partial T)_U = (-Cp + \kappa P Cv + \alpha P V)/(\kappa P - \alpha T)$$

$$(\partial H/\partial U)_T = (\alpha T - 1)/(\alpha T - \kappa P)$$

U   H   S $U = U(H, S)$

$$(\partial U/\partial H)_S = (\kappa P Cv)/Cp$$

$S = S(U, H)$

$$(\partial S/\partial U)_H = Cp/(T (cp - P (\kappa Cv + \alpha V)))$$

$$(\partial S/\partial H)_U = (\kappa P Cv)/(-T Cp + \kappa P T Cv + \alpha P T V)$$

$$(\partial U/\partial S)_H = (T (Cp - P (\kappa Cv + \alpha V)))/Cp$$

$H = H(S, U)$

$$(\partial H/\partial S)_U = (-T Cp + \kappa P T Cv + \alpha P T V)/(\kappa P Cv)$$

$$(\partial H/\partial U)_S = Cp/(\kappa P Cv)$$

U   A   G $U = U(A, G)$

$$(\partial U/\partial A)_G = (-Cp - \kappa P S + \alpha P V + \alpha S T)/(-\kappa P S + \alpha P V + S)$$

$G = G(U, A)$

$$(\partial G/\partial U)_A = (-\kappa P S + \alpha P V + S)/(\kappa P Cv + \kappa P S - \alpha S T)$$

$$(\partial G/\partial A)_U = (Cp + \kappa P S - \alpha P V - \alpha S T)/(\kappa P Cv + \kappa P S - \alpha S T)$$

$$(\partial U/\partial G)_A = (\kappa P C_v + \kappa P S - \alpha S T)/(-\kappa P S + \alpha P V + S)$$

$A = A(G, U)$

$$(\partial A/\partial G)_U = (\kappa P Cv + \kappa P S - \alpha S T)/(Cp + \kappa P S - \alpha P V - \alpha S T)$$

$$(\partial A/\partial U)_G = (-\kappa P S + \alpha P V + S)/(-Cp - \kappa P S + \alpha P V + \alpha S T)$$

U   A   P $U = U(A, P)$

$$(\partial U/\partial A)_P = (\alpha P V - Cp)/(\alpha P V + S)$$

$P = P(U, A)$

$$(\partial P/\partial U)_A = (\alpha P V + S)/(\kappa P V Cv + \kappa P S V - \alpha S T V)$$

$$(\partial P/\partial A)_U = (\alpha P V - Cp)/(-\kappa P V Cv - \kappa P S V + \alpha S T V)$$

$$(\partial U/\partial P)_A = (\kappa P V Cv + \kappa P S V - \alpha S T V)/(\alpha P V + S)$$

$A = A(P, U)$

$$(\partial A/\partial P)_U = (-\kappa P V Cv - \kappa P S V + \alpha S T V)/(\alpha P V - Cp)$$

$$(\partial A/\partial U)_P = (\alpha P V + S)/(\alpha P V - Cp)$$

$U \quad A \quad V \quad U = U(A, V)$

$(\partial U/\partial A)_V = -Cv/S$

$V = V(U, A)$

$(\partial V/\partial U)_A = (\kappa S)/(-\kappa P Cv - \kappa P S + \alpha S T)$

$(\partial V/\partial A)_U = (\kappa Cv)/(-\kappa P Cv - \kappa P S + \alpha S T)$

$(\partial U/\partial V)_A = -((P Cv)/S) - P + (\alpha T)/\kappa$

$A = A(V, U)$

$(\partial A/\partial V)_U = (-\kappa P Cv - \kappa P S + \alpha S T)/(\kappa Cv)$

$(\partial A/\partial U)_V = -S/Cv$

$U \quad A \quad T \quad U = U(A, T)$

$(\partial U/\partial A)_T = 1 - (\alpha T)/(\kappa P)$

$T = T(U, A)$

$(\partial T/\partial U)_A = (\kappa P)/(\kappa P Cv + \kappa P S - \alpha S T)$

$(\partial T/\partial A)_U = (\alpha T - \kappa P)/(\kappa P Cv + \kappa P S - \alpha S T)$

$(\partial U/\partial T)_A = Cv - (\alpha S T)/(\kappa P) + S$

$A = A(T, U)$

$(\partial A/\partial T)_U = (\kappa P Cv + \kappa P S - \alpha S T)/(\alpha T - \kappa P)$

$(\partial A/\partial U)_T = (\kappa P)/(\kappa P - \alpha T)$

$U \quad A \quad S \quad U = U(A, S)$

$(\partial U/\partial A)_S = (\kappa P Cv)/(\kappa P Cv - \alpha S T)$

$S = S(U, A)$

$(\partial S/\partial U)_A = (\alpha S T - \kappa P Cv)/(-\kappa P T Cv - \kappa P S T + \alpha S T^2)$

$(\partial S/\partial A)_U = (\kappa P Cv)/(-\kappa P T Cv - \kappa P S T + \alpha S T^2)$

$(\partial U/\partial S)_A = (-\kappa P T Cv - \kappa P S T + \alpha S T^2)/(\alpha S T - \kappa P Cv)$

$A = A(S, U)$

$(\partial A/\partial S)_U = (-\kappa P T Cv - \kappa P S T + \alpha S T^2)/(\kappa P Cv)$

$(\partial A/\partial U)_S = 1 - (\alpha S T)/(\kappa P Cv)$

$U \quad G \quad P \quad U = U(G, P)$

$(\partial U/\partial G)_P = (\alpha P V - Cp)/S$

$P = P(U, G)$

$(\partial P/\partial U)_G = S/(V (Cp + \kappa P S - \alpha P V - \alpha S T))$

$(\partial P/\partial G)_U = (\alpha P V - Cp)/(V (-Cp - \kappa P S + \alpha P V + \alpha S T))$

$(\partial U/\partial P)_G = (V (Cp + \kappa P S - \alpha P V - \alpha S T))/S$

$G = G(P, U)$

$(\partial G/\partial P)_U = (V (-Cp - \kappa P S + \alpha P V + \alpha S T))/(\alpha P V - Cp)$

$(\partial G/\partial U)_P = S/(\alpha P V - Cp)$

U	G	V	$U = U(G, V)$	$(\partial U/\partial G)_V = (\kappa Cv)/(\alpha V - \kappa S)$	$V = V(U, G)$	$(\partial V/\partial U)_G = (\alpha V - \kappa S)/ (Cp + \kappa P S - \alpha P V - \alpha S T)$
						$(\partial V/\partial G)_U = (\kappa Cv)/ (-Cp - \kappa P S + \alpha P V + \alpha S T)$
				$(\partial U/\partial V)_G = (Cp + \kappa P S - \alpha P V - \alpha S T)/(\alpha V - \kappa S)$	$G = G(V, U)$	$(\partial G/\partial V)_U = (-Cp - \kappa P S + \alpha P V + \alpha S T)/(\kappa Cv)$
						$(\partial G/\partial U)_V = (\alpha V - \kappa S)/(\kappa Cv)$
U	G	T	$U = U(G, T)$	$(\partial U/\partial G)_T = \kappa P - \alpha T$	$T = T(U, G)$	$(\partial T/\partial U)_G = 1/(Cp + \kappa P S - \alpha P V - \alpha S T)$
						$(\partial T/\partial G)_U = (\alpha T - \kappa P)/(Cp + \kappa P S - \alpha P V - \alpha S T)$
				$(\partial U/\partial T)_G = Cp + \kappa P S - \alpha P V - \alpha S T$	$G = G(T, U)$	$(\partial G/\partial T)_U = (Cp + \kappa P S - \alpha P V - \alpha S T)/(\alpha T - \kappa P)$
						$(\partial G/\partial U)_T = 1/(\kappa P - \alpha T)$
U	G	S	$U = U(G, S)$	$(\partial U/\partial G)_S = (\kappa P Cv)/(Cp - \alpha S T)$	$S = S(U, G)$	$(\partial S/\partial U)_G = (\alpha S T - Cp)/ (T(-Cp - \kappa P S + \alpha P V + \alpha S T))$
						$((\partial S/\partial G)_U = (\kappa P Cv)/ (T(-Cp - \kappa P S + \alpha P V + \alpha S T))$
				$(\partial U/\partial S)_G = (T(-Cp - \kappa P S + \alpha P V + \alpha S T))/(\alpha S T - Cp)$	$G = G(S, U)$	$(\partial G/\partial S)_U = (T(-Cp - \kappa P S + \alpha P V + \alpha S T))/(\kappa P Cv)$
						$(\partial G/\partial U)_S = (Cp - \alpha S T)/(\kappa P Cv)$
U	P	V	$U = U(P, V)$	$(\partial U/\partial P)_V = (\kappa Cv)/\alpha$	$V = V(U, P)$	$(\partial V/\partial U)_P = (\alpha V)/ (Cp - \alpha PV)$
						$(\partial V/\partial P)_U = (\kappa V Cv)/ (\alpha P V - Cp)$
				$(\partial U/\partial V)_P = Cp/(\alpha V) - P$	$P = P(V, U)$	$(\partial P/\partial V)_U = (\alpha P V - Cp)/(\kappa V Cv)$
						$(\partial P/\partial U)_V = \alpha/(\kappa Cv)$

U	P	T	$U = U(P, T)$	$(\partial U/\partial P)_T = \kappa P V - \alpha T V$	$T = T(U, P)$	$(\partial T/\partial U)_P = 1/(Cp - \alpha P V)$
						$(\partial T/\partial P)_U = (\kappa P V - \alpha T V)/(\alpha P V - Cp)$
				$(\partial U/\partial T)_P = Cp - \alpha P V$	$P = P(T, U)$	$(\partial P/\partial T)_U = (\alpha P V - Cp)/(\kappa P V - \alpha T V)$
						$(\partial P/\partial U)_T = 1/(\kappa P V - \alpha T V)$
U	P	S	$U = U(P, S)$	$(\partial U/\partial P)_S = (\kappa P V Cv)/Cp$	$S = S(U, P)$	$(\partial S/\partial U)_P = Cp/(TCp - \alpha P T V)$
						$(\partial S/\partial P)_U = (\kappa P V Cv)/(\alpha P T V - T Cp)$
				$(\partial U/\partial S)_P = T - (\alpha P T V)/Cp$	$P = P(S, U)$	$(\partial P/\partial S)_U = (\alpha P T V - T Cp)/(\kappa P V Cv)$
						$(\partial P/\partial U)_S = Cp/(\kappa P V Cv)$
U	V	T	$U = U(V, T)$	$(\partial U/\partial V)_T = (\alpha T)/\kappa - P$	$T = T(U, V)$	$(\partial T/\partial U)_V = 1/Cv$
						$(\partial T/\partial V)_U = (\kappa P - \alpha T)/(\kappa Cv)$
				$(\partial U/\partial T)_V = Cv$	$V = V(T, U)$	$(\partial V/\partial T)_U = (\kappa Cv)/(\kappa P - \alpha T)$
						$(\partial V/\partial U)_T = \kappa/(\alpha T - \kappa P)$
U	V	S	$U = U(V, S)$	$(\partial U/\partial V)_S = -P$	$S = S(U, V)$	$(\partial S/\partial U)_V = 1/T$
						$(\partial S/\partial V)_U = P/T$
				$(\partial U/\partial S)_V = T$	$V = V(S, U)$	$(\partial V/\partial S)_U = T/P$
						$(\partial V/\partial U)_S = -1/P$
U	T	S	$U = U(T, S)$	$(\partial U/\partial T)_S = (\kappa P Cv)/(\alpha T)$	$S = S(U, T)$	$(\partial S/\partial U)_T = \alpha/(\alpha T - \kappa P)$
						$(\partial S/\partial T)_U = -(\kappa P Cv)/(\alpha T^2 - \kappa P T)$
				$(\partial U/\partial S)_T = T - (\kappa P)/\alpha$	$T = T(S, U)$	$(\partial T/\partial S)_U = -(\alpha T^2 - \kappa P T)/(\kappa P Cv)$
						$(\partial T/\partial U)_S = (\alpha T)/(\kappa P Cv)$

H	A	G	$H = H(A, G)$	$(\partial H/\partial A)_G = (Cp - \alpha S T + S)/(\kappa P S - \alpha P V - S)$	$G = G(H, A)$	$(\partial G/\partial H)_A = (-\kappa P S + \alpha P V + S)/(\kappa P Cv + \alpha P V - \alpha S T + S)$
						$(\partial G/\partial A)_H = (Cp - \alpha S T + S)/(\kappa P Cv + \alpha P V - \alpha S T + S)$
				$(\partial H/\partial G)_A = (\kappa P Cv + \alpha P V - \alpha S T + S)/(-\kappa P S + \alpha P V + S)$	$A = A(G, H)$	$(\partial A/\partial G)_H = (\kappa P Cv + \alpha P V - \alpha S T + S)/(Cp - \alpha S T + S)$
						$(\partial A/\partial H)_G = (\kappa P S - \alpha P V - S)/(Cp - \alpha S T + S)$
H	A	P	$H = H(A, P)$	$(\partial H/\partial A)_P = -Cp/(\alpha P V + S)$	$P = P(H, A)$	$(\partial P/\partial H)_A = (\alpha P V + S)/(V(\kappa P Cv + \alpha P V - \alpha S T + S))$
						$(\partial P/\partial A)_H = Cp/(V(\kappa P Cv + \alpha P V - \alpha S T + S))$
				$(\partial H/\partial P)_A = (V(\kappa P Cv + \alpha P V - \alpha S T + S))/(\alpha P V + S)$	$A = A(P, H)$	$(\partial A/\partial P)_H = (V(\kappa P Cv + \alpha P V - \alpha S T + S))/Cp$
						$(\partial A/\partial H)_P = -(\alpha P V + S)/Cp$
H	A	V	$H = H(A, V)$	$(\partial H/\partial A)_V = -(\kappa Cv + \alpha V)/(\kappa S)$	$V = V(H, A)$	$(\partial V/\partial H)_A = -(\kappa S)/(\kappa P Cv + \alpha P V - \alpha S T + S)$
						$(\partial V/\partial A)_H = -(\kappa Cv + \alpha V)/(\kappa P Cv + \alpha P V - \alpha S T + S)$
				$(\partial H/\partial V)_A = -(\kappa P Cv + \alpha P V - \alpha S T + S)/(\kappa S)$	$A = A(V, H)$	$(\partial A/\partial V)_H = -(\kappa P Cv + \alpha P V - \alpha S T + S)/(\kappa Cv + \alpha V)$
						$(\partial A/\partial H)_V = -(\kappa S)/(\kappa Cv + \alpha V)$

H	A	T	$H = H(A, T)$	$(\partial H/\partial A)_T = (1 - \alpha\, T)/(\kappa\, P)$	$T = T(H, A)$	$(\partial T/\partial H)_A = (\kappa\, P)/(\kappa\, P\, Cv + \alpha\, P\, V - \alpha\, S\, T + S)$
						$(\partial T/\partial A)_H = (\alpha\, T - 1)/(\kappa\, P\, Cv + \alpha\, P\, V - \alpha\, S\, T + S)$
				$(\partial H/\partial T)_A = (\kappa\, P\, Cv + \alpha\, P\, V - \alpha\, S\, T + S)/(\kappa\, P)$	$A = A(T, H)$	$(\partial A/\partial T)_H = (\kappa\, P\, Cv + \alpha\, P\, V - \alpha\, S\, T + S)/(\alpha\, T - 1)$
						$(\partial A/\partial H)_T = (\kappa\, P)/(1 - \alpha\, T)$
H	A	S	$H = H(A, S)$	$(\partial H/\partial A)_S = Cp/(\kappa\, P\, Cv - \alpha\, S\, T)$	$S = S(H, A)$	$(\partial S/\partial H)_A = -(\alpha\, S\, T - \kappa\, P\, Cv)/(T\,(\kappa\, P\, Cv + \alpha\, P\, V - \alpha\, S\, T + S))$
						$(\partial S/\partial A)_H = Cp/(T\,(-\kappa\, P\, Cv - \alpha\, P\, V + S\,(\alpha\, T - 1)))$
				$(\partial H/\partial S)_A = -(T\,(\kappa\, P\, Cv + \alpha\, P\, V - \alpha\, S\, T + S))/(\alpha\, S\, T - \kappa\, P\, Cv)$	$A = A(S, H)$	$(\partial A/\partial S)_H = (T\,(-\kappa\, P\, Cv - \alpha\, P\, V + S\,(\alpha\, T - 1)))/Cp$
						$(\partial A/\partial H)_S = (\kappa\, P\, Cv - \alpha\, S\, T)/Cp$
H	G	P	$H = H(G, P)$	$(\partial H/\partial G)_P = -(Cp/S)$	$P = P(H, G)$	$(\partial P/\partial H)_G = S/(V\,(Cp - \alpha\, TVS + VS))$
						$(\partial P/\partial G)_H = Cp/(V\,(Cp - \alpha\, S\, T + S))$
				$(\partial H/\partial P)_G = (V\, Cp)/S - \alpha\, T\, V + V$	$G = G(P, H)$	$(\partial G/\partial P)_H = (V\,(Cp - \alpha\, S\, T + S))/Cp$
						$(\partial G/\partial H)_P = -(S/Cp)$
H	G	V	$H = H(G, V)$	$(\partial H/\partial G)_V = (\kappa\, Cv + \alpha\, V)/(\alpha\, V - \kappa\, S)$	$V = V(H, G)$	$(\partial V/\partial H)_G = (\alpha\, V - \kappa\, S)/(Cp - \alpha\, S\, T + S)$
						$(\partial V/\partial G)_H = -(\kappa\, Cv + \alpha\, V)/(Cp + \alpha\, S\, T - S)$
				$(\partial H/\partial V)_G = (Cp - \alpha\, S\, T + S)/(\alpha\, V - \kappa\, S)$	$G = G(V, H)$	$(\partial G/\partial V)_H = -(Cp + \alpha\, S\, T - S)/(\kappa\, Cv + \alpha\, V)$
						$(\partial G/\partial H)_V = (\alpha\, V - \kappa\, S)/(\kappa\, Cv + \alpha\, V)$
H	G	T	$H = H(G, T)$	$(\partial H/\partial G)_T = 1 - \alpha\, T$	$T = T(H, G)$	$(\partial T/\partial H)_G = 1/(Cp - \alpha\, S\, T + S)$
						$(\partial T/\partial G)_H = (\alpha\, T - 1)/(Cp - \alpha\, S\, T + S)$

H	G	S	$H = H(G, S)$	$(\partial H/\partial T)_G = Cp - \alpha S T + S$	$G = G(T, H)$	$(\partial G/\partial T)_H = (Cp - \alpha S T + S)/(\alpha T - 1)$
						$(\partial G/\partial H)_T = 1/(1 - \alpha T)$
				$(\partial H/\partial G)_S = -Cp/(\alpha S T - Cp)$	$S = S(H, G)$	$(\partial S/\partial H)_G = (\alpha S T - Cp)/(T (S (\alpha T - 1) - Cp))$
						$(\partial S/\partial G)_H = Cp/(T (S (\alpha T - 1) - Cp))$
				$(\partial H/\partial S)_G = (T (S (\alpha T - 1) - Cp))/(\alpha S T - Cp)$	$G = G(S, H)$	$(\partial G/\partial S)_H = (T (S (\alpha T - 1) - Cp))/Cp$
						$(\partial G/\partial H)_S = -\alpha S T/Cp + 1$
H	P	V	$H = H(P, V)$	$(\partial H/\partial P)_V = (\kappa Cv + \alpha V)/\alpha$	$V = V(H, P)$	$(\partial V/\partial H)_P = (\alpha V)/Cp$
						$(\partial V/\partial P)_H = -(\kappa V Cv + \alpha V^2)/Cp$
				$(\partial H/\partial V)_P = Cp/(\alpha V)$	$P = P(V, H)$	$(\partial P/\partial V)_H = -Cp/(\kappa V Cv + \alpha V^2)$
						$(\partial P/\partial H)_V = \alpha / (\kappa Cv + \alpha V)$
H	P	T	$H = H(P, T)$	$(\partial H/\partial P)_T = V - \alpha T V$	$T = T(H, P)$	$(\partial T/\partial H)_P = 1/Cp$
						$(\partial T/\partial P)_H = (\alpha T V - V)/Cp$
				$(\partial H/\partial T)_P = Cp$	$P = P(T, H)$	$(\partial P/\partial T)_H = Cp/(\alpha T V - V)$
						$(\partial P/\partial H)_T = 1/(V - \alpha T V)$
H	P	S	$H = H(P, S)$	$(\partial H/\partial P)_S = V$	$S = S(H, P)$	$(\partial S/\partial H)_P = 1/T$
						$(\partial S/\partial P)_H = -(V/T)$
				$(\partial H/\partial S)_P = T$	$P = P(S, H)$	$(\partial P/\partial S)_H = -(T/V)$
						$(\partial P/\partial H)_S = 1/V$
H	V	T	$H = H(V, T)$	$(\partial H/\partial V)_T = (\alpha T - 1)/\kappa$	$T = T(H, V)$	$(\partial T/\partial H)_V = \kappa/(\kappa Cv + \alpha V)$
						$(\partial T/\partial V)_H = (1 - \alpha T)/(\kappa Cv + \alpha V)$
				$(\partial H/\partial T)_V = Cv + (\alpha V)/\kappa$	$V = V(T, H)$	$(\partial V/\partial T)_H = (\kappa Cv + \alpha V)/(1 - \alpha T)$
						$(\partial V/\partial H)_T = \kappa/(\alpha T - 1)$

H	V	S	$H = H(V, S)$	$(\partial H/\partial V)_S = -Cp/(\kappa Cv)$	$S = S(H, V)$	$(\partial S/\partial H)_V = (\kappa Cv)/(\alpha T V + \kappa T Cv)$
						$(\partial S/\partial V)_H = Cp/(T(\kappa Cv + \alpha V))$
				$(\partial H/\partial S)_V = (\alpha T V)/(\kappa Cv) + T$	$V = V(S, H)$	$(\partial V/\partial S)_H = (T(\kappa Cv + \alpha V))/Cp$
						$(\partial V/\partial H)_S = -(\kappa Cv)/Cp$
H	T	S	$H = H(T, S)$	$(\partial H/\partial T)_S = Cp/(\alpha T)$	$S = S(H, T)$	$(\partial S/\partial H)_T = \alpha/(\alpha T - 1)$
						$(\partial S/\partial T)_H = Cp/(T - \alpha T^2)$
				$(\partial H/\partial S)_T = T - 1/\alpha$	$T = T(S, H)$	$(\partial T/\partial S)_H = (T - \alpha T^2)/Cp$
						$(\partial T/\partial H)_S = (\alpha T)/Cp$
A	G	P	$A = A(G, P)$	$(\partial A/\partial G)_P = (\alpha P V)/S + 1$	$P = P(A, G)$	$(\partial P/\partial A)_G = S/(V(\kappa P S - \alpha PV - S))$
						$(\partial P/\partial G)_A = (\alpha P V + S)/(V(-\kappa P S + \alpha P V + S))$
				$(\partial A/\partial P)_G = (V(\kappa P S - \alpha P V - S))/S$	$G = G(P, A)$	$(\partial G/\partial P)_A = (V(-\kappa P S + \alpha P V + S))/(\alpha P V + S)$
						$(\partial G/\partial A)_P = S/(\alpha P V + S)$
A	G	V	$A = A(G, V)$	$(\partial A/\partial G)_V = (\kappa S)/(\kappa S - \alpha V)$	$V = V(A, G)$	$(\partial V/\partial A)_G = (\kappa S - \alpha V)/(-\kappa P S + \alpha P V + S)$
						$(\partial V/\partial G)_A = (\kappa S)/(\kappa P S - \alpha P V - S)$
				$(\partial A/\partial V)_G = (-\kappa P S + \alpha P V + S)/(\kappa S - v V)$	$G = G(V, A)$	$(\partial G/\partial V)_A = (\kappa P S - \alpha P V - S)/(\kappa S)$
						$(\partial G/\partial A)_V = 1 - (\alpha V)/(\kappa S)$
A	G	T	$A = A(G, T)$	$(\partial A/\partial G)_T = \kappa P$	$T = T(A, G)$	$(\partial T/\partial A)_G = 1/(S \kappa P - S - \alpha P V)$
						$(\partial T/\partial G)_A = (\kappa P)/(-\kappa P S + \alpha P V + S)$

A	G	S	$A = A(G, S)$	$(\partial A/\partial T)_G = S\,(\kappa P - 1) - \alpha\,P\,V$	$G = G(T, A)$	$(\partial G/\partial T)_A = (-\kappa P S + \alpha\,P\,V + S)/(\kappa P)$
						$(\partial G/\partial A)_T = 1/(\kappa P)$
				$(\partial A/\partial G)_S = (\alpha\,S\,T - \kappa P\,Cv)/(\alpha\,S\,T - Cp)$	$S = S(A, G)$	$(\partial S/\partial A)_G = (\alpha\,S\,T - Cp)/(T\,(-\kappa PS + \alpha\,P\,V + S))$
						$(\partial S/\partial G)_A = (\alpha\,S\,T - \kappa P\,Cv)/(T\,(\kappa P S - \alpha\,P\,V - S))$
				$(\partial A/\partial S)_G = (T\,(-\kappa PS + \alpha\,P\,V + S))/(\alpha\,S\,T - Cp)$	$G = G(S, A)$	$(\partial G/\partial S)_A = (T\,(\kappa P S - \alpha\,P\,V - S))/(\alpha\,S\,T - \kappa P\,Cv)$
						$(\partial G/\partial A)_S = (\alpha\,S\,T - Cp)/(\alpha\,S\,T - \kappa P\,Cv)$
A	P	V	$A = A(P, V)$	$(\partial A/\partial P)_V = -(\kappa\,S)/\alpha$	$V = V(A, P)$	$(\partial V/\partial A)_P = -(\alpha\,V)/(\alpha\,P\,V + S)$
						$(\partial V/\partial P)_A = -(\kappa\,S\,V)/(\alpha\,P\,V + S)$
				$(\partial A/\partial V)_P = -(\alpha\,P\,V + S)/(\alpha\,V)$	$P = P(V, A)$	$(\partial P/\partial V)_A = -(\alpha\,P\,V + S)/(\kappa\,S\,V)$
						$(\partial P/\partial A)_V = -\alpha/(\kappa\,S)$
A	P	T	$A = A(P, T)$	$(\partial A/\partial P)_T = \kappa\,P\,V$	$T = T(A, P)$	$(\partial T/\partial A)_P = -1/(\alpha\,P\,V + S)$
						$(\partial T/\partial P)_A = (\kappa\,P\,V)/(\alpha\,P\,V + S)$
				$(\partial A/\partial T)_P = -(\alpha\,P\,V + S)$	$P = P(T, A)$	$(\partial P/\partial T)_A = (\alpha\,P\,V + S)/(\kappa\,P\,V)$
						$(\partial P/\partial A)_T = 1/(\kappa\,P\,V)$
A	P	S	$A = A(P, S)$	$(\partial A/\partial P)_S = (\kappa\,P\,V\,Cv - \alpha\,S\,T\,V)/Cp$	$S = S(A, P)$	$(\partial S/\partial A)_P = -Cp/(T\,(\alpha\,P\,V + S))$
						$(\partial S/\partial P)_A = ((\kappa\,P\,V\,Cv)/T - \alpha\,S\,V)/(\alpha\,P\,V + S)$
				$(\partial A/\partial S)_P = -(T\,(\alpha\,P\,V + S))/Cp$	$P = P(S, A)$	$(\partial P/\partial S)_A = (\alpha\,P\,V + S)/((\kappa\,P\,V\,Cv)/T - \alpha\,S\,V)$
						$(\partial P/\partial A)_S = Cp/(\kappa\,P\,V\,Cv - \alpha\,S\,T\,V)$

A	V	T	$A = A(V, T)$	$(\partial A/\partial V)_T = -P$	$T = T(A, V)$	$(\partial T/\partial A)_V = -1/S$	
						$(\partial T/\partial V)_A = -(P/S)$	
				$(\partial A/\partial T)_V = -S$	$V = V(T, A)$	$(\partial V/\partial T)_A = -(S/P)$	
						$(\partial V/\partial A)_T = -1/P$	
A	V	S	$A = A(V, S)$	$(\partial A/\partial V)_S = (\alpha\, S\, T)/(\kappa\, Cv) - P$	$S = S(A, V)$	$(\partial S/\partial A)_V = -Cv/(S\, T)$	
						$(\partial S/\partial V)_A = (\alpha\, S\, T - \kappa\, P\, Cv)/(\kappa\, S\, T)$	
				$(\partial A/\partial S)_V = -(S\, T)/Cv$	$V = V(S, A)$	$(\partial V/\partial S)_A = (\kappa\, S\, T)/(\alpha\, S\, T - \kappa\, P\, Cv)$	
						$(\partial V/\partial A)_S = (\kappa\, Cv)/(\alpha\, S\, T - P\, Cv)$	
A	T	S	$A = A(T, S)$	$(\partial A/\partial T)_S = (\kappa\, P\, Cv)/(\alpha\, T) - S$	$S = S(A, T)$	$(\partial S/\partial A)_T = -\alpha/(\kappa\, P)$	
						$(\partial S/\partial T)_A = (\kappa\, P\, Cv - \alpha\, S\, T)/(\kappa\, P\, T)$	
				$(\partial A/\partial S)_T = -(\kappa\, P)/\alpha$	$T = T(S, A)$	$(\partial T/\partial S)_A = (\kappa\, P\, T)/(\kappa\, P\, Cv - \alpha\, S\, T)$	
						$(\partial T/\partial A)_S = (\alpha\, T)/(\kappa\, P\, Cv - \alpha\, T\, S)$	
G	P	V	$G = G(P, V)$	$(\partial G/\partial P)_V = V - (\kappa\, S)/\alpha$	$V = V(G, P)$	$(\partial V/\partial G)_P = -(\alpha\, V)/S$	
						$(\partial V/\partial P)_G = (\alpha\, V^2 - \kappa\, S\, V)/S$	
				$(\partial G/\partial V)_P = -S/(\alpha\, V)$	$P = P(V, G)$	$(\partial P/\partial V)_G = S/(\alpha\, V^2 - \kappa\, S\, V)$	
						$(\partial P/\partial G)_V = \alpha/(\alpha V - \kappa\, S)$	
G	P	T	$G = G(P, T)$	$(\partial G/\partial P)_T = V$	$T = T(G, P)$	$(\partial T/\partial G)_P = -1/S$	
						$(\partial T/\partial P)_G = V/S$	
				$(\partial G/\partial T)_P = -S$	$P = P(T, G)$	$(\partial P/\partial T)_G = S/V$	
						$(\partial P/\partial G)_T = 1/V$	
G	P	S	$G = G(P, S)$	$(\partial G/\partial P)_S = V - (\alpha\, S\, T\, V)/Cp$	$S = S(G, P)$	$(\partial S/\partial G)_P = -Cp/(S\, T)$	
						$(\partial S/\partial P)_G = (V\, Cp - \alpha\, S\, T\, V)/(S\, T)$	
				$(\partial G/\partial S)_P = -(S\, T)/Cp$	$P = P(S, G)$	$(\partial P/\partial S)_G = (S\, T)/(V\, Cp - \alpha\, S\, T\, V)$	
						$(\partial P/\partial G)_S = Cp/(VCp - \alpha\, S\, T\, V)$	

			Function	Partial derivatives	Inverse	Inverse partial derivatives
G	V	T	$G = G(V, T)$	$(\partial G/\partial V)_T = -(1/\kappa)$	$T = T(G, V)$	$(\partial T/\partial G)_V = \kappa/(\alpha V - \kappa S)$
						$(\partial T/\partial V)_G = 1/(\alpha V - \kappa S)$
				$(\partial G/\partial T)_V = (\alpha V)/\kappa - S$	$V = V(T, G)$	$(\partial V/\partial T)_G = \alpha V - \kappa S$
						$(\partial V/\partial G)_T = -\kappa$
G	V	S	$G = G(V, S)$	$(\partial G/\partial V)_S = (\alpha S\, T - Cp)/(\kappa\, Cv)$	$S = S(G, V)$	$(\partial S/\partial G)_V = (\kappa\, Cv)/(T\,(\alpha V - \kappa S))$
						$(\partial S/\partial V)_G = (\alpha S\, T - Cp)/(T\,(\kappa S - \alpha V))$
				$(\partial G/\partial S)_V = (T\,(\alpha V - \kappa S))/(\kappa\, Cv)$	$V = V(S, G)$	$(\partial V/\partial S)_G = (T\,(\kappa S - \alpha V))/(\alpha S\, T - Cp)$
						$(\partial V/\partial G)_S = (\kappa\, Cv)/(\alpha S\, T - Cp)$
G	T	S	$G = G(T, S)$	$(\partial G/\partial T)_S = Cp/(\alpha T) - S$	$S = S(G, T)$	$(\partial S/\partial G)_T = -\alpha$
						$(\partial S/\partial T)_G = -(\alpha S\, T - Cp)/T$
				$(\partial G/\partial S)_T = -(1/\alpha)$	$T = T(S, G)$	$(\partial T/\partial S)_G = -T/(\alpha S\, T - Cp)$
						$(\partial T/\partial G)_S = (\alpha T)/(Cp - \alpha T S)$
P	V	T	$P = P(V, T)$	$(\partial P/\partial V)_T = -1/(\kappa V)$	$T = T(P, V)$	$(\partial T/\partial P)_V = \kappa/\alpha$
						$(\partial T/\partial V)_P = 1/(\alpha V)$
				$(\partial P/\partial T)_V = \alpha/\kappa$	$V = V(T, P)$	$(\partial V/\partial T)_P = \alpha V$
						$(\partial V/\partial P)_T = -\kappa V$
P	V	S	$P = P(V, S)$	$(\partial P/\partial V)_S = -Cp/(\kappa V\, Cv)$	$S = S(P, V)$	$(\partial S/\partial P)_V = (\kappa\, Cv)/(\alpha T)$
						$(\partial S/\partial V)_P = Cp/(\alpha T V)$
				$(\partial P/\partial S)_V = (\alpha T)/(\kappa\, Cv)$	$V = V(S, P)$	$(\partial V/\partial S)_P = (\alpha T V)/Cp$
						$(\partial V/\partial P)_S = -(\kappa V\, Cv)/Cp$
P	T	S	$P = P(T, S)$	$(\partial P/\partial T)_S = Cp/(\alpha T V)$	$S = S(P, T)$	$(\partial S/\partial P)_T = -\alpha V$
						$(\partial S/\partial T)_P = Cp/T$
				$(\partial P/\partial S)_T = -1/(\alpha V)$	$T = T(S, P)$	$(\partial T/\partial S)_P = T/Cp$
						$(\partial T/\partial P)_S = (\alpha T V)/Cp$

V	T	S	$V = V(T, S)$	$(\partial V/\partial T)_S = -(\kappa C\upsilon)/(\alpha T)$	$S = S(V, T)$	$(\partial S/\partial V)_T = \alpha/\kappa$
				$(\partial V/\partial S)_T = \kappa/\alpha$	$T = T(S, V)$	$(\partial S/\partial T)_V = C\upsilon/T$
						$(\partial T/\partial S)_V = T/C\upsilon$
						$(\partial T/\partial V)_S = -(\alpha T)/(\kappa C\upsilon)$

4.10 Exercises

1. For the following functions construct their differential relations and determine: a) natural variables, b) conjugate variables, and c) possible conjugate pairs.

   $z = z(x, y)$, $U = U(S, V)$, $H = H(S, p)$, $A = A(T, V)$, $G = G(T, p)$

2. For the relation $dG = -SdT + Vdp$, a) determine natural and conjugate variables, and b) derive alternate functions using the Legendre transformation.

3. Repeat problem #2 for the relation $du = TdS - pdV + \mu dN$.

4. With reference to Figure 4.1 extend it by adding the additional variable N and sketch the result in a 3D flowchart relating the functions.

5. With reference to Table 4.3, derive all Maxwell relations using the "curl" operational procedure.

6. Using Equation 4.18 and Equation 4.20 show that for an ideal gas they reduce to $dU = C_V \, dT$ and $dH = C_p \, dT$, respectively.

7. Show that having a function given as $f = f(x, y, z)$, the following relations are valid: $\dfrac{\partial x}{\partial y} = \dfrac{1}{\dfrac{\partial y}{\partial x}}$ and $\left(\dfrac{\partial x}{\partial y}\right)_z = -\dfrac{\left(\dfrac{\partial z}{\partial y}\right)_x}{\left(\dfrac{\partial z}{\partial x}\right)_y}$.

8. Derive the Joule–Thomson coefficient for a) an ideal gas and b) a gas with model equation $\left(p + \dfrac{a}{V^2}\right)V = RT$, where a and b are constants.

9. Show that Joule-Thomson coefficient can be written in terms of compressibility Z-factor $\left(Z = \dfrac{pV}{RT} \right)$ as $\mu_{JT} = \dfrac{RT^2}{pC_p}\left(\dfrac{\partial z}{\partial T} \right)_p$. (hint: use Equation 4.26)

10. Derive the thermal expansion coefficient α and compressibility coefficient κ for an ideal gas.

11. Show that for a substance at critical condition the isothermal compressibility is infinite. Using this result discuss the following cases for the density of substance:
    a) Under zero gravitational field, and
    b) Under non-zero gravitational field (e.g., on Earth)

12. Prove the following Jacobian properties: a) $\dfrac{[x,z]}{[y,z]} = \left(\dfrac{\partial x}{\partial y} \right)_z$ and b) $[x, x] = 0$.

13. Given the function $z = z(x, y)$, show that
    $$[z,c] = \left(\dfrac{\partial z}{\partial x} \right)_y [x,c] + \left(\dfrac{\partial z}{\partial y} \right)_x [y,c]$$ using Jacobian properties.

14. Show that all Maxwell relations reduce to $[T, S] = [p, V]$, using Jacobian properties.

15. Prove the following relations, using the Jacobian method:
    a) $\left(\dfrac{\partial T}{\partial S} \right)_U = \left(\dfrac{T}{C_V} \right)\left(1 - \dfrac{\alpha T}{kp} \right)$
    b) $\left(\dfrac{\partial A}{\partial S} \right)_T = -\left(\dfrac{kp}{\alpha} \right)$
    c) $\left(\dfrac{\partial G}{\partial V} \right)_T = -\dfrac{1}{\kappa}$
    d) $\left(\dfrac{\partial G}{\partial S} \right)_V = \dfrac{T(\alpha V - \kappa S)}{\kappa C_V}$

16. Show that $\left(\dfrac{\partial U}{\partial V}\right)_T + p = T\left(\dfrac{\partial p}{\partial T}\right)_V$, using the Jacobian method.

17. Show that the thermal expansion α, and the compressibility κ coefficients are related through the following relation: $\dfrac{\partial \alpha}{\partial p} = -\dfrac{\partial \kappa}{\partial T}$.

CHAPTER 5

MATERIAL BEHAVIOR AND MODELS

In This Chapter

- Overview
- Phase Diagram
- Ideal Gas Law
- Real Gas Laws
- Departure Functions and Factors
- Application of Departure Functions
- The Clapeyron Equation
- Equations of State: Material Behavior Models
- Exercises

5.1. Overview

Microstructure and molecule/atom composition of substances define their behavior and response to applied load and boundary conditions. Generally speaking, when molecules form a stable structure and each one remains in the close proximity of others during deformation, we have a solid material (e.g., iron). When molecules move more freely away from one another and slide through the medium with strong attractive forces among themselves, we have a liquid (e.g., water). Finally, when molecules are relatively freer to occupy the whole space available to them and inter-active forces become relatively less strong, then we have a gas (e.g., air). These are most common forms of materials, so-called *phases*, which we encounter with our everyday experience and use in practice. However, there are other phases, somehow combined, like super fluids and plasma that we do not include in this book. This wide range of materials makes it very challenging to have a unified mathematical model or equation of state which can predict their behavior under a wide range of conditions. This challenge is tackled by mainly three approaches: a) experimental measurements and creating databases for different materials (e.g., see NIST, ASM, MatWeb), b) semi-empirical techniques and curve fitting, and c) theoretical approaches based on quantum statistical methods. In engineering and practical applications we use model equations, valid for a range of thermodynamic variable variations, as well as material property databases, for example *steam tables* or *property tables*, and also semi-empirical models. The equation-based approach is sometimes referred to as *constitutive/state equations*. The ultimate reference, though, is the experimental data for verification and validation of all modeling results.

Although thermodynamics laws apply to all material phases (i.e., solid, liquid, gas, etc.), in practice we mostly focus on compressible materials, like gases and liquids, when discussing thermodynamics-related topics. In general, properties are divided into two groups; intensive thermodynamic properties (i.e., properties that are independent of the amount of substance) and extensive properties (i.e., properties that depend on the amount of substance, like volume, entropy, energy, etc.). The intensive properties like temperature, pressure, and specific volume are usually considered as independent variables as thermodynamic coordinates. For pure substances we need only two of these independent variables (according to Gibbs phase rule, to be discussed further), and the third one can be determined as a dependent variable or as a function of the selected independent variables. The number of required independent variables or degrees of freedom can be determined using the Gibbs phase rule [3]. For C number of components with φ number of phases, the Gibbs phase rule gives

$$f = (C - \varphi) + 2 \tag{5.1}$$

where f is the degrees of freedom or number of independent variables required to define a thermodynamic system state. For example, for a pure substance, like water in a liquid phase, we have $C = 1$ and $\varphi = 1$, which gives the number of independent variables needed as $f = 2$, using Equation 5.1. Therefore, two thermodynamic properties, like pressure and temperature, are sufficient to determine water liquid properties at any state point. Note that the function of independent variables is the equation of state, for example an ideal gas. If we have water in a state with two phases in equilibrium, like a liquid-vapor mixture, then we get $C = 1$ and $\varphi = 2$, which gives $f = 1$ degree of freedom, using Equation 5.1. Therefore, one property is sufficient, like pressure or temperature, to determine the water liquid-vapor mixture. Another important result of Gibbs's rule is the maximum possible phases with the constraint of $f \geq 1$. This requirement gives $\varphi_{max} = C + 2$. Therefore, for a single component substance we get $\varphi_{max} = 3$. This uniquely defines the three-phase equilibrium at triple point for single species material (see Figure 5.1).

5.2. Phase Diagram

Considering a system of three coordinates (p, T, V) we can map material properties of a matter, as shown typically in Figure 5.1. This figure is called a phase space diagram.

Usually 2D projection of the surface phase diagram is used for simplicity and application. For example, on $p - T$ or $p - V$ coordinates, see Figure 5.2. The phase diagram is extensively discussed in introductory thermodynamics textbooks (see [19], [23]). Here in this chapter, we recall the main features and extend the discussion to the more advanced topics of state equations or models. Several models are suggested with practical applications in engineering and design. We start with ideal gases and continue with real gases.

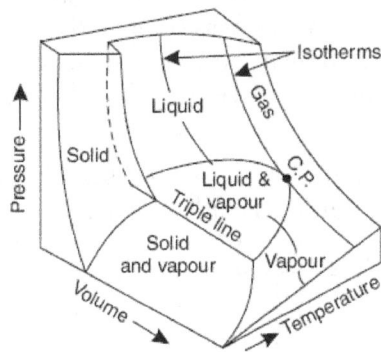

FIGURE 5.1. 3D phase space diagram.

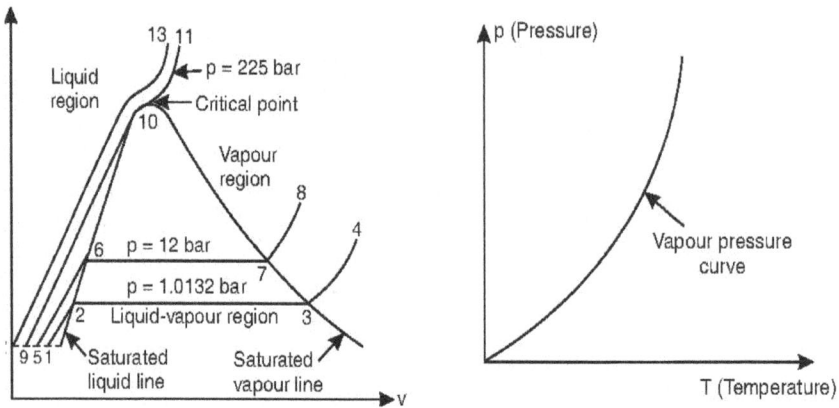

FIGURE 5.2. Projections of property space onto p-V and p-T planes.

5.3. Ideal Gas Law

An ideal gas is defined as a collection of hard spherical-shaped gas molecules which are in motion but only collide elastically with one another and boundaries; that is, other intermolecular forces are absent. Also the distance between molecules, on average, is much larger than the size of the molecules, such that the volume occupied by the molecules themselves is negligible compared to the volume of the gas container. The mathematical relation among pressure, temperature, and volume of an ideal gas is referred to as the *ideal gas law* [13]. Please see Figure 5.1 and Figure 5.3.

This model can be used in practice for many gases, with good accuracy, at low pressure and high temperature as compared to those at critical point. There are several ways to derive the model equation based on ideal gas assumptions. In this section we make use of Boyle's and Charles's laws to derive the ideal gas model.

We assume that the volume of the ideal gas is a function of its pressure and temperature, or $V = V(p, T)$. Therefore, we have

$$dV = \frac{\partial V}{\partial p}\bigg|_T dp + \frac{\partial V}{\partial T}\bigg|_p dT \tag{5.2}$$

Boyle's law [60], states that at constant temperature, for a given amount of gas, the volume and pressure are inversely proportional, or $V = \dfrac{k_1}{p}$.

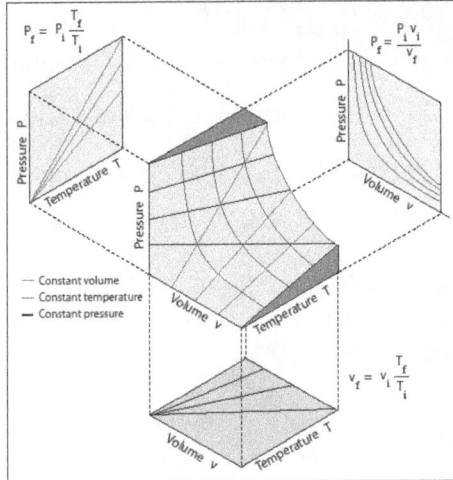

FIGURE 5.3. Phase diagram based on ideal gas law assumption. (from HyperPhysics, (c) Rod Nave, Georgia State University, with permission)

Therefore, we have $\dfrac{\partial V}{\partial p}\bigg|_T = \dfrac{-k_1}{p^2}$, where k_1 is a constant. In addition, Charles's law [61] states that at constant pressure, for a given amount of gas, the volume and temperature are linearly proportional, or $V = k_2 T$. Therefore, we have $\dfrac{\partial V}{\partial T}\bigg|_p = k_2$, where k_2 is a constant. After substituting for the partial derivatives into Equation 5.2, we get $dV = -\dfrac{k_1}{p^2}dp + k_2 dT = -\dfrac{Vp}{p^2}dp + \dfrac{V}{T}dT$.

After dividing both sides by V and integrating, we have $\int \dfrac{dV}{V} = -\int \dfrac{dp}{p} + \int \dfrac{dT}{T}$.

Performing the integrals, we get $\ln V + \ln p = \ln T + \ln k_3$, where k_3 is the constant of integration. Finly, we get $pV = k_3 T$. Through experiment [17], the value of k_3 can be obtained and is equal to nR, where n is number of moles and R is the universal gas constant (i.e., $8.314\,J\,/\,K.mol$ in SI units). Therefore, the ideal gas model reads

$$PV = nRT \qquad (5.3)$$

Using Kinetic theory [7], [62] we can show that for monoatomic ideal gases, $PV = \frac{3}{2}k_B T$, where k_B is the Boltzmann constant ($1.3806488 \times 10^{23}\,J\,/\,K$). Also we can write Equation 5.3 as $PV = mR_g T$, where m is mass and R_g is the gas constant, or $R_g = \dfrac{R}{M_W}$, where M_W is the atomic mass of the gas. Although the ideal gas model has many applications in practice, under high pressure and low temperature gases behave non-ideally and the ideal gas model gives less accurate results.

5.4. Real Gas Laws

Real gases don't behave close to that of an ideal gas, except at low pressure and high temperature. However, an ideal gas could be a kind of reference that we can use to compare against the real gas behavior. In this section we discuss several approaches to modeling real gases, along with some model equations or Equations of State (EOS).

Note that we refer to a gas as being "real" or "ideal" based on its behavior under given thermodynamic conditions.

5.4.1. Z-Factor and Generalized Compressibility Chart

Molecules in a real gas interact non-elastically and exert forces on one another, either attractive or repulsive. When the distances among molecules are small (order of molecule size), the repulsive forces dominate, and when molecules are farther apart, the attractive forces dominate. In addition, when the volume of the gas decreases due to high pressure, the volume occupied by the molecules themselves becomes comparable to the volume of the gas and should be considered as well. Therefore, we cannot use the ideal gas model for modeling real gases. As mentioned, one way to model real gas behavior is to compare them in reference to ideal gas behavior. For an ideal gas we have (see Equation 5.3), per mole, $\dfrac{pv}{RT} = Z = 1$, where v is molar volume. We define Z as the *compressibility factor* or Z-factor. For real gases Z deviates from unity. In other words

$$\frac{v}{RT/p} = \frac{molar\ volume\ of\ real\ gas}{molar\ volume\ if\ the\ gas\ behave\ ideally} = Z$$

When $Z = 1$, the gas behaves like an ideal gas. When $Z > 1$ the gas does not behave like an ideal gas; rather, the inter-molecular forces are such that the molecules occupy a larger volume than that of an equivalent ideal gas, or the pressure is greater than that of an ideal gas. When $Z < 1$ the gas does not behave like an ideal gas, rather the inter-molecular forces are such that the molecules occupy a smaller volume than that of an equivalent ideal gas, or the pressure is less than that of an ideal gas. If we have the data for the Z-factor, then we can use the following equation (Equation 5.4) to model the gas behavior.

$$pV = nZRT \qquad (5.4)$$

The Z-factor is determined experimentally or semi-empirically using an EOS. In order to organize and group these experimental data for practical applications, the *Principle of Corresponding States* (PCS) is usually used, proposed by van der Waals [63]. According to this principle fluids behave similarly when their pressure and temperature are scaled with respect to their

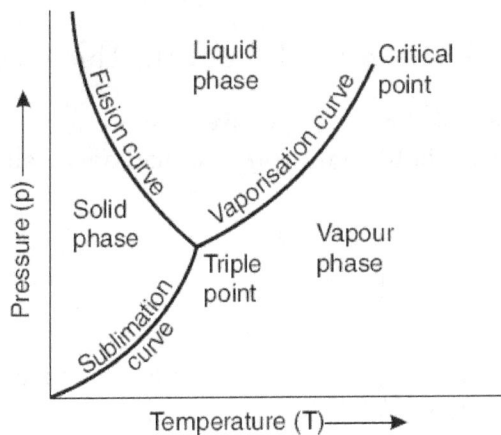

FIGURE 5.4. Sketch of phase diagram on a p-T plane with triple and critical points.

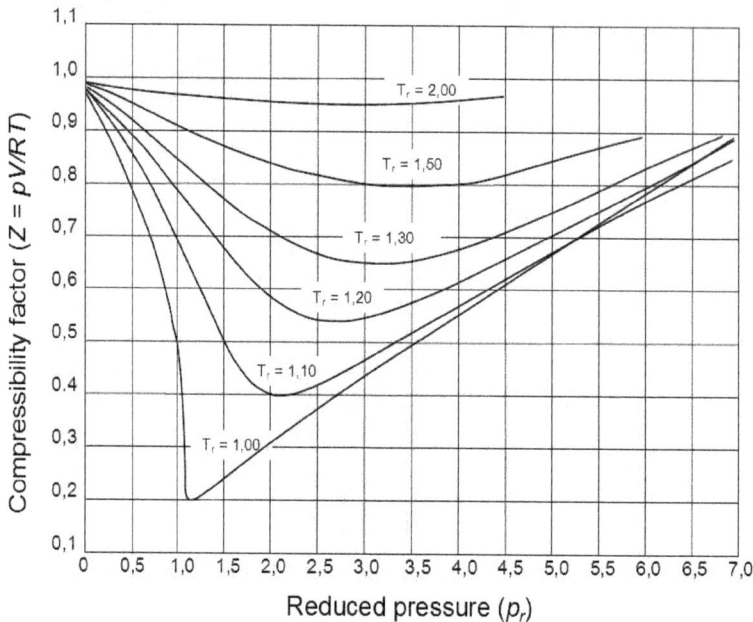

FIGURE 5.5. Generalized diagram of compressibility factor (Z=pv/RT), Source: File:Diagramma generalizzato fattore di compressibilità.jpg, Daniele Pugliesi.

corresponding critical values or they have the same compressibility factor.

The scaled quantities are called *reduced pressure*, $p_r = \dfrac{p}{p_c}$, and *reduced temperature*, $T_r = \dfrac{T}{T_c}$, where p_c and T_c are critical pressure and temperature values of the substance, respectively (see Figure 5.1 and Figure 5.4). In other words, the PCS is expressed mathematically as

$$Z = Z\left(p_r, T_r\right) \tag{5.5}$$

Therefore, when the values of p_r and T_r are the same for given gases, we should get the same value of Z for those gases.

Critical values are unique for each substance; for example, for water $p_c = 217.7$ atm, and $T_c = 373.946°C$.

Fluids at their reduced pressure and temperature are at the so-called corresponding states. Therefore, PCS can be stated as: *"substances behave alike at the same reduced states and substances at the same reduced states are at corresponding states*" [63]. Experimental data is in agreement with PCS [19], [64]. We will show in a later section (see Equation 5.47, for example), that equations of states are the same for any gas when written in terms of the corresponding reduced variables, that is, reduced pressure, temperature, and volume.

Now, using the PCS or Equation 5.5, we can make a graph/chart of Z versus p_r for different values of T_r. This chart is called the *Generalized Compressibility Chart*, as shown in Figure 5.5.

In general there are two types of generalized compressibility charts available: a) Nelson-Obert charts, which are based on average experimental data and are applicable to fluids with asymmetric molecules, for example, water, and b) Lee-Kesler charts, which are based on ideal fluid assumption with spherical symmetric molecules, for example, O_2 and N_2.

We can conclude the following important points by studying a generalized compressibility chart:

- At very low pressure ($p_r \ll 1$), gases behave as an ideal gas at any temperature.

- At high temperatures ($T_r \gtrsim 2$), gases behave similar or close to an ideal gas.

- At a critical point state the deviation of a gas from ideal gas behavior is greatest. Some authors proposed considering $Z = Z(p_r, T_r, Z_c)$ to achieve greater accuracy, where Z_c is the critical compressibility factor [65].

- As the pressure increases, the Z factor first decreases to a minimum and then increases.

- For temperatures of 1.5 times the critical temperature, the minimum Z factor is approximately 0.77, and for temperatures of twice the critical temperature, the minimum Z factor is about 0.94.

- At high pressures, the Z factor increases above 1, where the gas is no longer compressible. In these conditions, the specific volume of the gas is becoming so small, and the distance between molecules is much smaller, so that the density is more strongly affected by the volume occupied by the individual molecules. Hence, the Z factor continues to increase above unity as the pressure increases.

The generalized compressibility charts also provide a family of curves for pseudo-reduced specific volume $\hat{v}_r = \dfrac{V}{RT_c / p_c}$. This is useful when the gas pressure is unknown, as prior. The value of RT_c/p_c is used as the scale for V, instead of v_c, and hence the designation of pseudo-reduced specific volume applies.

Example:

Air at 200K and 132 bar (= 13200 kPa) is given. Calculate its specific molar volume: a) assuming air as an ideal gas, and b) using the generalized compressibility chart, that is, Z-chart.

■ **Solution:**

a) For an ideal gas we have $pV = RT$, hence $V = \dfrac{(0.287)(200)}{13200} =$ 0.0043485 m³/kg.

b) Using critical pressure for air at 37.69 bar and its critical temperature at 132.5K, we have $p_r = \dfrac{132}{37.69} = 3.5$ and $T_r = \dfrac{200}{132.5} = 1.51$.

From the Z-chart we get $Z = 0.8$. Hence, we calculate

$$V = Z\left(\frac{RT}{p}\right) = (0.8)(0.0043485) = 0.0034788 \text{ m}^3/\text{kg. This result}$$

shows that the specific volume is 80% of that of the ideal gas model. This is consistent with the assumption that ideal gas gives a higher value for a specific volume, since the inter-molecular forces are absent in that model.

Example:

Water vapor at 600 °F and 0.51431 ft³/lbm is given. Determine its pressure: a) using the steam table, b) the ideal gas model, and c) a Z-chart. Use $R = 0.5956$ Psia.ft³/lbm.R, $p_c = 3200$ Psia, $T_c = 1164.8$ R.

■ **Solution:**

a) From the steam table (see Appendix 2 of [19].) we read the pressure for superheated water vapor given as $p = 1000$ Psia. Since steam table data is experimental data, it is relatively the most accurate data.

b) Using the ideal gas model, we have $p = \dfrac{RT}{V} = \dfrac{(0.5956)(459+600)}{0.51431} =$
1228 Psia, an error of 22.8% as compared to results from using the steam table.

c) For this example we cannot calculate the reduced pressure, since we don't have it. But we can calculate and use
$$\hat{v}_r = \frac{0.51431}{(0.5956)(1164.8)/3200} = 2.372 \text{ and } T_r = \frac{1060}{1164.8} = 0.91.$$
Using the Z-chart (see Figure 5.5.) we get $Z = 0.82$ and $p_r = 0.33$.

Finally, we have $p = 0.33 \times 3200 = 1056$ Psia, an error of 5.6% as compared to the results from using the steam table.

5.5. Departure Functions and Factors

In order to calculate state variables like internal energy, enthalpy, entropy, Helmholtz, and Gibbs functions for real gases, we calculate the deviation of a real gas behavior from that of an ideal gas at the same thermodynamic conditions, for example, pressure, temperature, or volume. The deviations are represented by so-called *departure functions* (also referred to as residue functions). Using a departure function, for example for enthalpy, we can define a related Z-factor and use related available charts for calculating the enthalpy of a real gas. In this section we give details of derivation for several departure functions, define their corresponding Z-factors, and demonstrate their implementations through some worked-out numerical examples.

5.5.1. Departure Function and Factor for Internal Energy

We start with the calculations related to internal energy. As previously mentioned, we know that the internal energy of an ideal gas is a function of temperature but that of a real gas is a function of temperature and pressure (or volume). We designate internal energy for the ideal gas as U^* and that of the real gas as U; or $U^* = U^*(T)$ and $U = U(T, p)$, or similarly $U = U(T, V)$. Therefore, the departure function U' (given at the same pressure and temperature) is defined as

$$U'(T, p) = U(T, p) - U^*(T) \tag{5.6}$$

In Chapter 4, we derived an equation (see Equation 4.18.) for change in the internal energy of a gas, as cited here again for convenience:

$dU = C_V dT + \left[T \dfrac{\partial p}{\partial T}\bigg|_V - p \right] dV$. This relation applies to real gas as well as to

an ideal gas. It includes relatively straightforward calculations to show that the sum of the terms in the bracket is equal to zero for an ideal gas, hence $dU^* = C_V dT$. Therefore, the change for departure function for internal

energy reads $dU' = dU - dU^* = \left[T \dfrac{\partial p}{\partial T}\bigg|_V - p \right] dV$. We normalize this relation

by dividing both sides by RT and integrate to get the departure factor $Z_{U'}$ as

$$Z_{U'} = \frac{U'}{RT} = \int_{V^*=\infty}^{V} \left[\frac{1}{R} \frac{\partial p}{\partial T}\bigg|_V - \frac{p}{RT} \right] dV \tag{5.7}$$

Note that the lower limit of the integral corresponds to a very low-density condition (i.e., very high specific volume) representing an ideal gas condition. It is desirable and more convenient to change the variable from V, taken here as specific volume, to density ρ, and also write the integrand in terms of Z, using the real gas equation $pV = ZRT$. Therefore, using $V = \rho^{-1}$, $dV = -\rho^{-2}d\rho$ we can write Equation 5.7 in terms of density as the independent variable, as

$$Z_{U'} = \frac{U'}{RT} = \int_0^\rho \left[\frac{p}{\rho RT} - \frac{1}{\rho R} \frac{\partial p}{\partial T}\bigg|_\rho \right] \frac{d\rho}{\rho} \tag{5.8}$$

For the integrand we need to calculate the term $\dfrac{\partial p}{\partial T}\bigg|_\rho$. But we have

$p = Z\rho RT$, which gives the pressure as a function of Z, T, and ρ. Therefore,

we can write $dp = \dfrac{\partial p}{\partial Z}\bigg|_{T,\rho} dZ + \dfrac{\partial p}{\partial T}\bigg|_{Z,\rho} dT + \dfrac{\partial p}{\partial \rho}\bigg|_{T,Z} d\rho$. However, for con-

stant ρ the last term vanishes (i.e., $\dfrac{\partial p}{\partial \rho}\bigg|_{T,Z} d\rho = 0$) and hence we get, after dividing both sides by dT and using $\left.\dfrac{\partial p}{\partial Z}\right|_{T,\rho} = \rho RT$ and $\left.\dfrac{\partial p}{\partial T}\right|_{Z,\rho} = Z\rho R$,

$\left.\dfrac{\partial p}{\partial T}\right|_{\rho} = \rho RT \left.\dfrac{\partial Z}{\partial T}\right|_{\rho} + \rho RZ$. Now we substitute this relation into the integrand of Equation 5.8 to get $\left[\dfrac{p}{\rho RT} - \dfrac{1}{\rho R}\left.\dfrac{\partial p}{\partial T}\right|_{\rho}\right] = \left[\dfrac{Z\rho RT}{\rho RT} - \dfrac{1}{\rho R}\right.$

$\left.\left(\rho RT \left.\dfrac{\partial Z}{\partial T}\right|_{\rho} + \rho RZ\right)\right] = \left(Z - T\left.\dfrac{\partial Z}{\partial T}\right|_{\rho} - Z\right)$. Finally, we have the relation for

the dimensionless departure function or *Departure factor* $Z_{U'}$, as

$$Z_{U'} = \dfrac{U'}{RT} = -\int_0^{\rho} T \left.\dfrac{\partial Z}{\partial T}\right|_{\rho} \dfrac{d\rho}{\rho} \tag{5.9}$$

The value of $Z_{U'}$ for given thermodynamic conditions can be calculated either with having an equation of state for the gas or by using the departure function/factor charts [19], [30].

Using Equation 5.6 and Equation 5.9 (i.e., $U' = RTZ_{U'}$) for a process 1–2 connecting two thermodynamic state points (T_1, p_1) and (T_2, p_2) respectively, we can write (taking the temperature scale to be the critical temperature)

$$(U_2 - U_1) = (U_2^* - U_1^*) + RT_c \left(Z_{U_2} - Z_{U_1}\right) \tag{5.10}$$

It is sometimes useful to have the relation for the departure function for internal energy in terms of pressure (see Table 5.1).

5.5.2. Departure Function and Factor for Enthalpy

As previously mentioned, we know that enthalpy of an ideal gas is a function of temperature but that of a real gas is a function of temperature and pressure or volume. We designate enthalpy for the ideal gas as H^* and that of the real gas as H; or $H^* = H^*(T)$ and $H = H(T,p)$, or similarly $H = H(T,V)$. Therefore, the departure function H' (given at the same pressure and temperature) is defined as

$$H'(T,p) = H(T,p) - H^*(T) \tag{5.11}$$

In Chapter 4, we derived an equation (see Equation 4.20) for change in enthalpy of a gas, as cited here again for convenience, $dH = C_p dT + \left[V - T \dfrac{\partial V}{\partial T}\Big|_p \right] dp$. This relation applies to real gas as well as to an ideal gas. It is a straightforward calculation to show that the sum of terms in the bracket is zero for an ideal gas, hence $dH^* = C_p dT$. Therefore, the change for the departure function for enthalpy reads $dH' = dH - dH^* = \left[V - T \dfrac{\partial V}{\partial T}\Big|_p \right] dp$. We normalize this relation by dividing both sides with RT and integrate to get the departure factor $Z_{H'}$ as

$$Z_{H'} = \frac{H'}{RT} = \int_0^p \left[\frac{V}{RT} - \frac{1}{R}\frac{\partial V}{\partial T}\Big|_p \right] dp \tag{5.12}$$

Note that the lower limit of the integral corresponds to a vanishingly low-pressure condition compatible with an ideal gas assumption. In the integrand we need to calculate the term $\dfrac{\partial V}{\partial T}\Big|_p$. But we have $V = ZRT/p$, which gives the specific volume as a function of Z, T, and p. Therefore, we can write $dV = \dfrac{\partial V}{\partial Z}\Big|_{T,p} dZ + \dfrac{\partial V}{\partial T}\Big|_{Z,p} dT + \dfrac{\partial V}{\partial p}\Big|_{T,Z} dp$. However, for constant p the last term vanishes (i.e., $\dfrac{\partial V}{\partial p}\Big|_{T,Z} dp = 0$) and after dividing both sides by dT and using $\dfrac{\partial V}{\partial Z}\Big|_{T,p} = \dfrac{RT}{p}$ and $\dfrac{\partial V}{\partial T}\Big|_{Z,p} = \dfrac{RZ}{p}$, we have $\dfrac{\partial V}{\partial T}\Big|_p = \dfrac{RT}{p}\dfrac{\partial Z}{\partial T}\Big|_p + \dfrac{RZ}{p}$. Now we substitute this relation into the integrand of Equation 5.12 to get $\left[\dfrac{V}{RT} - \dfrac{1}{R}\dfrac{\partial V}{\partial T}\Big|_p \right] = \left[\dfrac{Z}{p} - \dfrac{1}{R}\left(\dfrac{RT}{p}\dfrac{\partial Z}{\partial T}\Big|_p + \dfrac{RZ}{p} \right) \right] = \dfrac{1}{p}\left(Z - T\dfrac{\partial Z}{\partial T}\Big|_p - Z \right)$. Finally, we have the relation for the dimensionless departure function and *Departure factor $Z_{H'}$*, as

$$Z_{H'} = \frac{H'}{RT} = -\int_0^p T \frac{\partial Z}{\partial T}\Big|_p \frac{dp}{p} \tag{5.13}$$

The value of $Z_{H'}$ for a given thermodynamic state condition can be calculated either with having an equation of state for the gas or by using the so-called departurve function/factor charts. In order to use a departure function/factor chart for real gases, it is desirable to write this integral in terms of reduced pressure and temperature. To this end, after substituting for $p = p_r p_c$ and $T = T_r T_c$, we get $\dfrac{H'}{RT_r T_c} = -\displaystyle\int_0^{p_r} T_r T_c \left. \dfrac{\partial Z}{\partial (T_r T_c)}\right|_{p_r} \dfrac{dp_r}{p_r}$, or

$$Z_{H'} = \frac{H'}{RT_c} = -\int_0^{p_r} T_r^2 \left.\frac{\partial Z}{\partial T_r}\right|_{p_r} \frac{dp_r}{p_r} \tag{5.14}$$

Using Equation 5.11 and Equation 5.14 (i.e., $H' = RT_c Z_{H'}$), for a process $1-2$ connecting two state points (T_1, p_1) and (T_2, p_2) respectively, we can write

$$\left(H_2 - H_1\right) = \left(H_2^* - H_1^*\right) + RT_c\left(Z_{H_2} - Z_{H_1}\right) \tag{5.15}$$

Using the relation $H = U + pV = U + ZRT$ and Equation 5.15, we can write an alternate relation for internal energy change for real gases, as

$$\left(U_2 - U_1\right) = \left(U_2^* - U_1^*\right) + RT_2\left(1 - Z_2\right) - RT_1\left(1 - Z_1\right) + RT_c\left(Z_{H_2} - Z_{H_1}\right) \tag{5.16}$$

Note that $(pV)' = (pV) - (pV)^* = ZRT - RT = RT(Z-1)$.

5.5.3. Departure Function and Factor for Entropy

As previously mentioned, we know that entropy of both ideal and real gases is a function of temperature and pressure (or volume). We designate entropy for the ideal gas as S^* and that of the real gas as S; or $S^* = S^*(T,p)$ and $S = S(T,p)$, or similarly $S = S(T,V)$. Therefore, the departure function S' (at the same pressure and temperature) is defined as

$$S'(T,p) = S(T,p) - S^*(T,p) \tag{5.17}$$

We can write $dS' = \left.\dfrac{\partial S'}{\partial T}\right|_p dT + \left.\dfrac{\partial S'}{\partial p}\right|_T dp$. Therefore, we can write the departure function at a constant temperature as $S' = \displaystyle\int_0^p \left.\dfrac{\partial S'}{\partial p}\right|_T dp$, since at a constant temperature the first term on the R.H.S. vanishes. We normalize this relation by dividing both sides by R to have

$$\frac{S'}{R} = \int_0^p \frac{1}{R} \frac{\partial S'}{\partial p}\bigg|_T dp \qquad (5.18)$$

Note that the lower limit of the integral corresponds to a very low-pressure condition consistent with an ideal gas assumption. In the integrand we need to calculate the term $\dfrac{\partial S'}{\partial p}\bigg|_T$. But using the Maxwell relation,

$\dfrac{\partial S'}{\partial p}\bigg|_T = -\dfrac{\partial V'}{\partial T}\bigg|_p$, where $V' = V - V^* = \dfrac{ZRT}{p} - \dfrac{RT}{p} = \dfrac{RT}{p}(Z-1)$. Therefore,

$\dfrac{\partial V'}{\partial T}\bigg|_p = \dfrac{\partial}{\partial T}\left[\dfrac{RT}{p}(Z-1)\right]_p = \dfrac{R}{p}\left(T\dfrac{\partial Z}{\partial T}\bigg|_p + Z - 1\right)$. After substitution into the

integrand of Equation 5.18, we get $\dfrac{1}{R}\dfrac{\partial S'}{\partial p}\bigg|_T = -\dfrac{1}{R}\dfrac{\partial V'}{\partial T}\bigg|_p = -\dfrac{1}{p}\left(T\dfrac{\partial Z}{\partial T}\bigg|_p + Z - 1\right)$.

Finally, we have the relation for the dimensionless entropy departure function or departure factor $Z_{S'}$, after substituting into Equation 5.18, as

$$Z_{S'} = \frac{S'}{R} = -\int_0^p T \frac{\partial Z}{\partial T}\bigg|_p \frac{dp}{p} - \int_0^p (Z-1)\frac{dp}{p} \qquad (5.19)$$

The value of $Z_{S'}$ for a given condition can be calculated by either having an equation of state for the gas or by using the so-called departure function/factor charts. In order to use departure functions/factors charts for real gases, it is desirable to write Equation 5.19 in terms of reduced pressure and temperature. After substituting for $p = p_r p_c$ and $T = T_r T_c$, we get

$$Z_{S'} = \frac{S'}{R} = -\int_0^{p_r} T_r \frac{\partial Z}{\partial T_r}\bigg|_{p_r} \frac{dp_r}{p_r} - \int_0^{p_r} (Z-1)\frac{dp_r}{p_r} \qquad (5.20)$$

Using Equation 5.17 and Equation 5.20 (i.e., $S' = RZ_{S'}$) for a process $1-2$ connecting two state points (T_1, p_1) and $(T_2\ p_2)$ respectively, we can write

$$\left(S_2 - S_1\right) = \left(S_2^* - S_1^*\right) + R\left(Z_{S_2} - Z_{S_1}\right) \tag{5.21}$$

5.5.4. Departure Function and Factor for Gibbs Energy

Gibbs free energy function, G is very useful for thermodynamic calculations, specifically at constant pressure. For derivation of a Gibbs departure function we make use of its relation in terms of enthalpy and entropy, or $G = H - TS$. Therefore, we can write $G' = G - G^* = (H - TS) - \left(H^* - TS^*\right) = H' - TS'$. We normalize G' by dividing both sides with RT to get

$$\frac{G'}{RT} = \frac{H'}{RT} - \frac{S'}{R} \tag{5.22}$$

After substituting Equation 5.13 and Equation 5.19 into Equation 5.22, we have

$$\frac{G'}{RT} = \int_0^p (Z - 1)\frac{dp}{p} \tag{5.23}$$

In order to use the departure function/factor chart for real gases, it is desirable to write Equation 5.23 in terms of reduced pressure and temperature. After substituting for $p = p_r p_c$ and $T = T_r T_c$, we get

$$\frac{G'}{RT_c} = \int_0^{p_r} T_r (Z - 1)\frac{dp_r}{p_r} \tag{5.24}$$

5.5.5. Departure Function and Factor for Helmholtz Energy

The Helmholtz free energy function, A is very useful for thermodynamic calculations, specifically at a constant volume. For derivation of a Helmholtz departure function, we make use of its relation in terms of Gibbs energy, or $A = G - pV$. Therefore, we can write $A' = A - A^* = (G - pV) - \left(G^* - (pV)^*\right) = G' - (pV)'$. We normalize A' by dividing both sides with RT to get

$$\frac{A'}{RT} = \frac{G'}{RT} - \frac{(pV)'}{RT} \tag{5.25}$$

But $\dfrac{(pV)^{'}}{RT} = \dfrac{pV - (pV)^{*}}{RT} = \dfrac{ZRT - RT}{RT} = Z - 1.$ After substituting this relation and Equation 5.23 into Equation 5.25 we get

$$\frac{A'}{RT} = -(Z-1) + \int_0^P (Z-1)\frac{dp}{p} \tag{5.26}$$

In order to use a departure function/factor chart for real gases, it is desirable to write this integral in terms of reduced pressure and temperature. After substituting for $p = p_r p_c$ and $T = T_r T_c$, we get

$$\frac{A'}{RT_c} = -T_r(Z-1) + \int_0^{p_r} T_r(Z-1)\frac{dp_r}{p_r} \tag{5.27}$$

So far, we have derived the relations for departure functions for entropy and four potential thermodynamic functions. Table 5.1 summarizes these departure functions, both in terms of pressure and density, and corresponding departure Z-factors.

5.6. Application of Departure Functions

The integrals listed in Table 5.1 need to be worked out in order to calculate the related departure function or corresponding factor. As mentioned, this can be done when the equation of the state is assumed or given. An alternative method is by using available departure Z-charts (see e.g., [19]) or so-called generalized departure charts. One can read the value of a departure factor for a given reduced pressure and temperature (or a pseudo-specific volume), and then calculate the corresponding state function for a real gas. In this section we provide a few examples to demonstrate the relevant calculation steps.

5.6.1. Departure Functions for a Real Gas

Let's assume that for a real gas the equation of state is given as $Z = 1 + Cp_r / T_r$, for $pV = ZRT$, where C is a dimensionless constant. Calculate the departure functions U', H', S', A', and G'.

Thermodynamic Function	Departure factor	Pressure-form	Density-form	Critical pressure-temperature form
Internal energy	$Z_{U'} = \dfrac{U - U^*}{RT}$	$\displaystyle\int_0^p T \left.\dfrac{\partial Z}{\partial T}\right|_p \dfrac{dp}{p} - (Z-1)$	$\displaystyle -\int_0^\rho T \left.\dfrac{\partial Z}{\partial T}\right|_\rho \dfrac{d\rho}{\rho}$	$\dfrac{U'}{RT_c} = -T_r^2 \displaystyle\int_0^{p_r} \left.\dfrac{\partial Z}{\partial T_r}\right|_{p_r} \dfrac{dp_r}{p_r} - (Z-1)$
Enthalpy	$Z_{H'} = \dfrac{H - H^*}{RT}$	$\displaystyle -\int_0^p T \left.\dfrac{\partial Z}{\partial T}\right|_p \dfrac{dp}{p}$	$\displaystyle -\int_0^\rho T \left.\dfrac{\partial Z}{\partial T}\right|_\rho \dfrac{d\rho}{\rho} + (Z-1)$	$\dfrac{H'}{RT_c} = -T_r^2 \displaystyle\int_0^{p_r} \left.\dfrac{\partial Z}{\partial T_r}\right|_{p_r} \dfrac{dp_r}{p_r}$
Entropy	$Z_{S'} = \dfrac{S - S^*}{R}$	$\displaystyle -\int_0^p \left(T \left.\dfrac{\partial Z}{\partial T}\right|_p + Z - 1 \right) \dfrac{dp}{p}$	$\displaystyle -\int_0^\rho \left(T \left.\dfrac{\partial Z}{\partial T}\right|_\rho \right) \dfrac{d\rho}{\rho} - \int_0^\rho (Z-1) \dfrac{d\rho}{\rho} + \ln Z$	$\dfrac{S'}{R} = -T_r \displaystyle\int_0^{p_r} \left.\dfrac{\partial Z}{\partial T_r}\right|_{p_r} \dfrac{dp_r}{p_r} - \int_0^{p_r} (Z-1) \dfrac{dp_r}{p_r}$
Helmholtz energy	$Z_{A'} = \dfrac{A - A^*}{RT}$	$\displaystyle\int_0^p (Z-1) \dfrac{dp}{p} - (Z-1)$	$\displaystyle\int_0^\rho (Z-1) \dfrac{d\rho}{\rho} - \ln Z$	$\dfrac{A'}{RT_c} = -T_r(Z-1) + \displaystyle\int_0^{p_r} (Z-1) \dfrac{dp_r}{p_r}$
Gibbs energy	$Z_{G'} = \dfrac{G - G^*}{RT}$	$\displaystyle\int_0^p (Z-1) \dfrac{dp}{p}$	$\displaystyle\int_0^\rho (Z-1) \dfrac{d\rho}{\rho} + (Z-1) - \ln Z$	$\dfrac{G'}{RT_c} = T_r \displaystyle\int_0^{p_r} (Z-1) \dfrac{dp_r}{p_r}$

TABLE 5.1. Departure Functions for Thermodynamic State Functions

■ **Solution:**

For U' (see Table 5.1) we need the term $\left.\dfrac{\partial Z}{\partial T_r}\right|_{p_r} = -Cp_r / T_r^2$.

Substituting into $\dfrac{U'}{RT_c} = -\int_r^{p_r} T_r^2 \left.\dfrac{\partial Z}{\partial T_r}\right|_{p_r} \dfrac{dp_r}{p_r} - (Z-1) = Cp_r - Cp_r / T_r$. Hence,

$U' = RT_c Cp_r \left(1 - 1/T_r\right)$. Note that for an ideal gas or $Z = 1$, we should have $C = 0$, which gives $U' = 0$, as expected.

For an enthalpy departure function, we have $\dfrac{H'}{RT_c} = -\int_0^{p_r} T_r^2 \left.\dfrac{\partial Z}{\partial T_r}\right|_{p_r} \dfrac{dp_r}{p_r} = Cp_r$. Hence, $H' = RT_c Cp_r$.

For an entropy departure function, we have

$$\dfrac{S'}{R} = -\int_0^{p_r} T_r \left.\dfrac{\partial Z}{\partial T_r}\right|_{p_r} \dfrac{dp_r}{p_r} - \int_0^{p_r} (Z-1)\dfrac{dp_r}{p_r} = Cp_r / T_r - Cp_r / T_r = 0.$$

For a Helmholtz departure function, we have

$$\dfrac{A'}{RT_c} = -T_r (Z-1) + \int_0^{p_r} T_r (Z-1)\dfrac{dp_r}{p_r} = -Cp_r + Cp_r = 0.$$

For a Gibbs departure function, we have $\dfrac{G'}{RT_c} = \int_0^{p_r} T_r (Z-1)\dfrac{dp_r}{p_r} = Cp_r$. Hence, $G' = RT_c Cp_r$.

For this example, we can calculate the numerical values of all departure functions for given values of C, p_r, and T_r related to a gas.

5.6.2. Departure Functions for a van der Waals Gas

Calculate the following energy functions for a vdW gas:

(a) Enthalpy departure function, (b) Internal energy departure function, (c) Entropy departure function, (d) Helmholtz energy departure function, and (e) Gibbs energy departure function for a van der Waals fluid, with EOS given as $p = RT (V - b)^{-1} - aV^{-2}$. Note, V is the specific volume. Summarize the results in a table.

■ Solution:

(a) From relations given for the enthalpy departure factor in Table 5.1, we have three choices. For this example, we use the relation in terms of density, which makes it easier to calculate the partial derivative involved. Hence, we have $Z_{H'} = \dfrac{H - H^*}{RT} = -\int_0^\rho T \dfrac{\partial Z}{\partial T}\Big|_\rho \dfrac{d\rho}{\rho} + (Z - 1)$.

We need the relation for Z for the vdW fluid. But $Z = pV / RT = p(\rho RT)^{-1}$, where we substitute for specific volume in terms of the density, or $V = \rho^{-1}$. Therefore, for a vdW fluid with $p = RT(\rho^{-1} - b)^{-1} - a\rho^2$, we get

$$Z = \left[RT(\rho^{-1} - b)^{-1} - a\rho^2 \right](\rho RT)^{-1} = 1/(1 - b\rho) - a\rho / RT.$$

Now, by differentiation w.r.t. T while keeping ρ constant, we get $\dfrac{\partial Z}{\partial T}\Big|_\rho = \dfrac{a\rho}{RT^2}$. After substitution into the relation for $Z_{H'}$

(see Table 5.1) we get $Z_{H'} = -\int_0^\rho T \dfrac{\partial Z}{\partial T}\Big|_\rho \dfrac{d\rho}{\rho} + (Z - 1) = -\dfrac{a}{RT}$

$\int_0^\rho d\rho + \dfrac{1}{1 - b\rho} - \dfrac{a\rho}{RT} - 1 = -2\dfrac{a\rho}{RT} + \dfrac{b\rho}{1 - b\rho}$. Writing this result in

terms of V, we have $Z_{H'} = -2\dfrac{a}{RTV} + \dfrac{b}{V - b}$. From these results, we can conclude that $Z_{H'} = 0$ for an ideal gas when $a = b = 0$, as expected.

(b) We use the relation for internal energy and enthalpy, or $U = H - pV$. Therefore, we have $U' = H' - (pV)'$, by simply subtracting $U^* = H^* - (pV)^*$. But

$$(pV)' = pV - (pV)^* = ZRT - RT = RT(Z - 1). \quad \dfrac{(pV)'}{RT} = Z - 1.$$

Now, using the result obtained in part (a) for the enthalpy departure function, we get $\dfrac{U'}{RT} = \dfrac{U - U^*}{RT} = -\int_0^\rho T \dfrac{\partial Z}{\partial T}\Big|_{\tilde{n}} \dfrac{d\rho}{\rho} = -\dfrac{a\rho}{RT}.$

(c) For the entropy departure function, we use the relation (see Table 5.1) $\dfrac{S'}{R} = -\int_0^\rho T\dfrac{\partial Z}{\partial T}\bigg|_{\rho\tilde{n}}\dfrac{d\rho}{\rho} - \int_0^\rho (Z-1)\dfrac{d\rho}{\rho} + \ln Z$. We have already calculated the first integral for a vdW fluid in part (a). The second integral reads, after substituting for $Z-1$,

$$-\int_0^\rho (Z-1)\dfrac{d\rho}{\rho} = -\int_0^\rho \dfrac{d\rho}{\rho(1-b\rho)} + \int_0^\rho \dfrac{a}{RT}d\rho + \int_0^\rho \dfrac{d\rho}{\rho}.$$ After combin-

ing the first and last integrals and performing the integration,

we get $-\int_0^\rho (Z-1)\dfrac{d\rho}{\rho} = \int_0^\rho \dfrac{-b}{1-b\rho}d\rho + \dfrac{a\rho}{RT} = \ln(1-b\rho) + \dfrac{a\rho}{RT}.$

Collecting all terms, we get $\dfrac{S'}{R} = \ln(1-b\rho) + \ln\left(\dfrac{1}{1-b\rho} - \dfrac{a\rho}{RT}\right).$

(d) We use the relation for internal energy and the Helmholtz function, or $A = U - TS$. Therefore, we have $A' = U' - TS$, by simply subtracting $A^* = U^* - TS^*$. But we have the results for U' and S'. Therefore, $\dfrac{A'}{RT} = \dfrac{U'}{RT} - \dfrac{S'}{R} = -\dfrac{a\rho}{RT} - \ln(1-b\rho) - \ln\left(\dfrac{1}{1-b\rho} - \dfrac{a\rho}{RT}\right).$

(e) We use the relation for Gibbs energy and the Helmholtz function, or $G = A + pV$. Therefore, we have $A' = U' - TS'$, by simply subtracting $A^* = U^* - TS^*$. But we have the results for A' and (pV).

Therefore, $\dfrac{G'}{RT} = \dfrac{A'}{RT} + \dfrac{(pV)'}{RT} = -2\dfrac{a\rho}{RT} - \ln(1-b\rho) - \ln\left(\dfrac{1}{1-b\rho} - \dfrac{a\rho}{RT}\right) + \dfrac{1}{1-b\rho} - 1.$

We collect all the results obtained in Table 5.2.

The following example demonstrates the application of the Z-factor for departure functions.

TABLE 5.2. Departure Factors for Several Thermodynamic Functions for the van der Waals Gas

Departure functions for a vdW fluid $(p = RT(V-b)^{-1} - aV^{-2})$, $V = \rho^{-1}$ is the specific volume, ρ density, T absolute temperature, R universal gas constant, a and b are the fluid's constants.

Departure function, normalized	Density form	Volume form
$\dfrac{H'}{RT}$	$-2\dfrac{a\rho}{RT} + \dfrac{b\rho}{1-b\rho}$	$-2\dfrac{a}{RTV} + \dfrac{b}{V-b}$
$\dfrac{U'}{RT}$	$-\dfrac{a\rho}{RT}$	$-\dfrac{a}{RTV}$
$\dfrac{S'}{R}$	$\ln(1-b\rho) + \ln\left(\dfrac{1}{1-b\rho} - \dfrac{a\rho}{RT}\right)$	$\ln(1-b/V) + \ln\left(\dfrac{V}{V-b} - \dfrac{a}{RTV}\right)$
$\dfrac{A'}{RT}$	$-\dfrac{a\rho}{RT} - \ln(1-b\rho) - \ln\left(\dfrac{1}{1-b\rho} - \dfrac{a\rho}{RT}\right)$	$-\dfrac{a}{RTV} - \ln(1-b/V) - \ln\left(\dfrac{V}{V-b} - \dfrac{a}{RTV}\right)$
$\dfrac{G'}{RT}$	$-2\dfrac{a\rho}{RT} - \ln(1-b\rho) - \ln\left(\dfrac{1}{1-b\rho} - \dfrac{a\rho}{RT}\right) +$ $\dfrac{1}{1-b\rho} - 1$	$-2\dfrac{a}{RTV} - \ln(1-b/V) - \ln\left(\dfrac{V}{V-b} - \dfrac{a}{RTV}\right)$ $+\dfrac{V}{V-b} - 1$

5.6.3. Isothermal Compression of a Real Gas

Propane (C_3H_8) is compressed isothermally from $T_1 = T_2 = 94^0C$ and $p_1 = 14$ atm to $p_2 = 55$ atm. Calculate the work done on the gas w_{in} and heat rejected q_{out} during the process. $R = 188.5$ J/kg.K.

Solution:

The critical pressure and temperature of propane are $p_{cr} = 42\,atm$ (or 4.26 MPa) absolute, and $T_{cr} = 96.85$ °C (or 370 K). Since the gas condition is close to its critical values, we cannot assume ideal gas behavior and should consider it as a real gas. We use the Z-factor for the gas model and departure factor for its energy functions.

At state 1: $T_{r1} = \dfrac{273+94}{370} = 0.991$ and $P_{r1} = \dfrac{14}{42} = 0.33$. From tables (see e.g., Cengel's Figures A-15 and 29[19]) we read $Z_1 = 0.88$ and the enthalpy departure factor $Z_{h1} = 0.37$. Similarly,

At state 2: $T_{r2} = \dfrac{273+94}{370} = 0.991$ and $P_{r2} = \dfrac{55}{42} = 1.31$. From departure tables (e.g., [19]) we read $Z_2 = 0.22$ and the enthalpy departure factor $Z_{h2} = 4.2$. Note that the enthalpy departure factor is deviation from that if the gas behaves like an ideal gas (see Table A-29 of Cengel [19]), or $Z_h = \dfrac{h^*-h}{RT_{cr}}$.

We calculate the change in enthalpy as $h_2 - h_1 = \underbrace{\left(h_2^* - h_1^*\right)}_{=0,\ isothermal} +$

$\left(Z_{h1} - Z_{h2}\right)RT_{cr} = (0.37-4.2)(188.5 \times 370) = -267.123$ kJ/kg. To calculate the change in internal energy, we use the relation $u = h - pv = h - ZRT$. Therefore $u_2 - u_1 = \left(h_2 - h_1\right) + \left(Z_1 - Z_2\right)RT_1 = -267.123 + (0.88-0.22)$ $(273+94) = -221.465$ kJ/kg. Applying the 1st law to the gas control volume gives $u_2 - u_1 = -q_{out} + w_{in}$. We can calculate the work using $w_{in} = -\int_1^2 pdv = -RT_1 \int_1^2 Z\dfrac{dv}{v}$. But the variation of Z is between two constant values of $Z_1 = 0.88$ and $Z_2 = 0.22$. Therefore, as an approximation, we use its average value $\dfrac{0.88+0.22}{2} = 0.55$ for performing the integration. Hence $w_{in} = -\left((0.1885)(273+94)\right) \times 0.55 \ln\dfrac{v_2}{v_1} = -38.05 \ln\left(\dfrac{Z_2 p_1}{Z_1 p_2}\right) =$

$-38.05 \ln\left(\dfrac{0.22 \times 14}{0.88 \times 55}\right) = 104.811$ kJ/kg. Substituting into the energy balance equation, we get $-221.465 = -q_{out} + 104.811$, $q_{out} = 326.276$ kJ/kg.

5.7. The Clapeyron Equation

Clapeyron published a paper in 1834 [13]. This paper is significant in the history of thermodynamics and its development. In this paper he drew the attention of the scientific community of the time to the past work of Carnot [12], which was published about ten years earlier and was almost totally ignored. This development led to the mathematical formulation of the 2nd law by Clausius, using the work of Carnot. In the same paper, Clapeyron also derived an equation for the temperature dependence of the vapor pressure of a liquid, or the so-called Clapeyron equation. This equation gives a relation for thermodynamics quantities during a phase change—specifically enthalpy change associated with a phase change. In this section we discuss the derivation and physical meaning and present a few numerical examples demonstrating applications of the Clapeyron equation (see Equation 5.29).

As schematically shown in Figure 5.6, the saturation curves (or coexisting-phase lines) are not straight lines in the p-T plane, hence the slope of a saturation curve changes. The sublimation (solid-gas boundary) and vaporization (liquid-gas) curves have a positive slope, but the melting (solid-liquid) curve could have a negative (for a less dense solid than corresponding liquid materials, like water) or positive slope (for a more dense solid than corresponding liquid materials). The value of the slope changes with pressure and temperature (referred to as saturation pressure and temperature) and the Clapeyron equation (see Equation 5.28) can be used to calculate the change in enthalpy associated with phase change.

There are several possible ways to derive the Clapeyron equation. Here, we use entropy and Maxwell relations and consider the vaporization curve, for example. At the saturation line, for a given temperature, we can read the corresponding saturation pressure from the phase diagram. Let's assume entropy to be a function of volume and temperature, or $S = S(V,T)$, with its total derivative $dS = \left.\frac{\partial S}{\partial V}\right|_T dV + \underbrace{\left.\frac{\partial S}{\partial T}\right|_V dT}_{=0} = \left.\frac{\partial S}{\partial V}\right|_T dV$. The last term vanishes, since temperature is constant during a phase change, hence $dT = 0$. But from Maxwell relations (see Table 4.3) we have $\left.\frac{\partial S}{\partial V}\right|_T = \left.\frac{\partial p}{\partial T}\right|_V$.

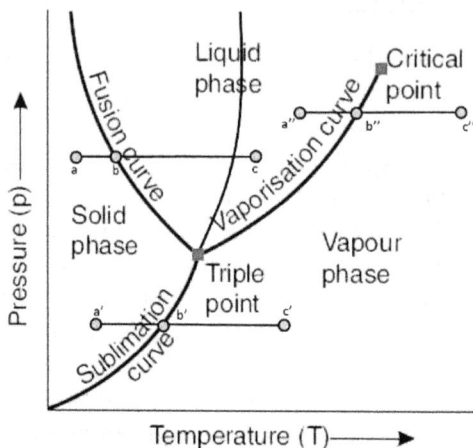

FIGURE 5.6 Phase diagram on p-T plane for typical materials, schematic.

Therefore, after substitution we have $dS = \dfrac{\partial p}{\partial T}\bigg|_V dV = \dfrac{dp}{dT}dV$. Note that we can write a total differentiation instead of a partial one, since during the phase change pressure is only a function of temperature (see phase diagram). Integrating for a phase change, for example along the line a″-c″ (as shown in Figure 5.6) we have $\displaystyle\int_1^2 dS = \dfrac{dp}{dT}\int_1^2 dV$, since $\dfrac{dp}{dT}$ is a constant at the point of interest where the phase change is occurring (i.e., the slope at point b″, shown in Figure 5.6). Working out the integration gives, after rearranging terms

$$\left(\frac{dp}{dT}\right)_{sat} = \frac{S_2 - S_1}{V_2 - V_1} \tag{5.28}$$

Note that points 1 and 2 correspond to phases of the substance, for this example vapor and liquid, that is, $S_2 = S_g$, $S_1 = S_f$, $V_2 = V_g$, $V_1 = V_f$.

Now using the 1st and 2nd laws, we have

$$dH = dU + d(pV) = T_{sat}dS - pdV + pdV + \underbrace{V\,dp}_{=0} = T_{sat}dS .\qquad \text{Integrating}$$

this relation gives $S_2 - S_1 = \dfrac{H_2 - H_1}{T_{sat}}$. After substituting for $(S_2 - S_1)$ into Equation 5.28, we get

$$\left(\frac{dp}{dT}\right)_{sat} = \frac{H_2 - H_1}{T_{sat}(V_2 - V_1)} \tag{5.29}$$

Equation 5.29 is the Clapeyron equation. If we divide the R.H.S. fraction by mass of substance involved in the phase change, then we get the Clapeyron equation in terms of specific enthapies and volumes, or

$$\left(\frac{dp}{dT}\right)_{sat} = \frac{h_2 - h_1}{T_{sat}(v_2 - v_1)} \tag{5.30}$$

The quantity $\Delta h = h_2 - h_1$ is the latent heat of transformation for the phase change. For example, for evaporation of liquid water, we usually use the notation $\Delta h = h_2 - h_1 = h_g - h_f = h_{fg}$, with data available from steam tables. Similarly $\Delta v = v_2 - v_1 = v_g - v_f = v_{fg}$. The Clapeyron equation applies to phase change from solid to vapor (sublimation) and solid to liquid (melting) equally as well.

An approximation of the Clapeyron equation (also referred to as the Clausius-Clapeyron equation, see Equation 5.31) can be obtained, using ideal gas law. We can assume, with good approximation, that when a liquid, for example water, changes phase to its vapor, then a specific volume in the gas phase is much larger than in its liquid phase, or $v_g \gg v_f$, hence $v_2 - v_1 = v_g - v_f \cong v_g$. Now assuming the ideal gas law for the vapor, we have $v_g = \dfrac{RT}{p}$. Note that we are using a specific volume here. Substituting for $v_2 - v_1 = \dfrac{RT}{p}$ into the Clapeyron equation, that is, Equation 5.30, we have $\dfrac{dp}{dT} = \dfrac{p(h_2 - h_1)}{RT^2}$. Assuming negligible change in enthalpies, that is, $h_2 - h_1 \cong$ constant, or using the averaged value of the enthalpies for each phase, after rearranging terms and integrating we have $\int_1^2 \dfrac{dp}{p} = \dfrac{\Delta h}{R}\int_1^2 \dfrac{dT}{T^2}$. Working out the integration we get

$$\ln\left(\frac{p_2}{p_1}\right) = \frac{\Delta h}{R}\left(\frac{1}{T_1} - \frac{1}{T_2}\right)$$ (5.31)

Equation 5.31 shows that $\ln p$ is a linear function of $\frac{1}{T}$. This relationship is validated by experimental results (see [67]).

In order to demonstrate the Clapeyron equation's application, we present a few numerical examples.

Example: Calculate the pressure at which water boils at 200 °C.

■ **Solution:**

The phase change is a liquid-vapor transition. We assume the boiling temperature of water at 1 atm to be $100\,^{0}\text{C}$ and $\Delta h = 44.03\,kJ\,/\,mol$. Substituting data into Equation 5.31, we get $\ln\left(\frac{p_2}{1\,atm}\right) = \frac{44.03 \times 10^3}{8.314}\left(\frac{1}{373.15} - \frac{1}{473.15}\right)$, or $p_2 = 20.1$ atm. Therefore, twenty times the pressure of the original pressure is required to double the temperature.

Example: Vapor pressure change for Mercury

A container has 4 g of mercury at a room temperature of 25 °C. Calculate the temperature at which the mercury vapor pressure doubles. The enthalpy of vaporization for mercury is 59.11 kJ/mol.

■ **Solution:**

we have $p_1 = 1\,unit$, $p_2 = 2\,unit$, $T_1 = 298\,K$. Using Equation 5.31, we have $ln\left(\frac{2}{1}\right) = \frac{59.11 \times 10^3}{8.314}\left(\frac{1}{298} - \frac{1}{T_2}\right)$, or $T_2 = 306.92\,K = 33.8\,^{\circ}\text{C}$. Note that the vapor pressure at room temperature is not needed, since we are only interested in the ratio of the pressures, not their absolute values.

5.8. Equations of State: Material Behavior Models

In order to model material behavior, we need to have reliable and validated equations which relate their thermodynamic properties to one another. These models are called *Equations of State* (EOS), which are functional relationships among the measurable macroscopic thermodynamic variables. For example, for three variables p, V, T we get $f(p, V, T) = 0$. Note that function f as a EOS reduces the number of independent variables by one and imposes a constraint on the gas behavior as a model. Therefore, we can arbitrarily define two of the three variables p, V, T as independent variables and the third one is defined by the EOS. There exists a large list of proposed EOSs with a different range of applicability for materials [68]. As mentioned in the previous section, experimental data (e.g., property tables, steam tables; see www.nist.gov, http://steamtablesonline.com) are the ultimate references available for material thermodynamic properties. An alternative resource is the compressibility factor chart through the application of the Z-factor, a semi-empirical tool based on corresponding state principle. However, validated EOSs are very useful and could be used as well for modeling and calculation when consideration of their limitations and range of applicability is addressed adequately. In general, an EOS must satisfy the following properties to become representative: a) it should recover the ideal gas law as pressure approaches zero, and b) the critical isotherm resulting from the EOS must have zero slope and an inflection point at the critical point. Additional properties, like constant pressure prediction during the phase change for isotherms, adds to the validity and applicability of the EOS.

In this section we discuss some of the proposed real gas models, not necessarily in a chronological historical order of their development, along with related numerical examples for their applications.

The ideal gas model, as discussed previously, has a wide range of applications for gases at low pressure or high temperature. When considering the fundamental assumption of the ideal gas model (i.e., molecules are hard spheres with negligible volume/size which interact *only* through elastic collisions), we deduce that any realistic behavior modeling should include the molecular interactions (i.e., attractive and repulsive forces), the size of the molecules and their collective volume, and also the deviation of the molecules' shape from being perfect spheres.

5.8.1. The Virial Equation Model

One way to extend the ideal gas model for real gases is to consider a series expansion which gives pressure in terms of volume (or equivalently specific volume). The mathematical foundation for the validity of this expansion is based on the application of the Massieu and Lennard-Jones potential functions (see [30],[69]). This approach is called the *Virial*

Equation of State (proposed by H. K. Onnes [70],[71]). This series can be written, for example, as

$$p = \frac{RT}{V} + \frac{f_1(T)}{V^2} + \frac{f_2(T)}{V^3} + \frac{f_3(T)}{V^4} + \cdots \qquad (5.32)$$

where $f_i(T)$ are functions of temperature and represent the material specific molecular effects including their interaction forces [69],[72]. It is clear that considering only the first term in the series given by Equation 5.32 will recover the ideal gas model, as expected. The larger the number of terms involved in the series, the more accurate the model is, especially when it is close to its critical condition. In other words, more terms in the series are needed to represent the gas material behavior as the pressure increases and temperature decreases. The coefficient functions $f_i(T)$ are usually determined experimentally or by using theoretical approaches like statistical mechanics [30],[73]. However, in some thermodynamic conditions second and third coefficients (i.e., $f_1(T)$ representing two-particle interactions and $f_2(T)$ representing three-particle interactions) are normally used for modeling. Higher order coefficients are more challenging to be measured or determined. For example, Table 5.3 shows the data related to second and third coefficients for several commonly used gases [74]:

Pitzer [75] proposed a generalized formulation for calculating the 2nd coefficient. Consider two terms of the virial expansion in Equation 5.32 and write the virial equation as $\frac{pV}{RT} = 1 + \frac{f_1(T)/RT}{V} = 1 + \frac{Bp}{RT}$, where $B(T) = f_1(T)/RT$ defined for a modified second coefficient as a function of temperature. Using critical temperature (i.e., $T = T_c T_r$) and pressure (i.e., $p = p_c p_r$), we can write $Z = 1 + \left(\frac{Bp_c}{RT_c}\right)\left(\frac{p_r}{T_r}\right)$. According to Pitzer, $B_c = \frac{Bp_c}{RT_c}$ is computed according to Equation 5.33

$$B_c = B_0 + \omega B_1 \qquad (5.33)$$

where $B_0 = 0.083 - 0.422 T_r^{-1.6}$, $B_1 = 0.139 - 0.172 T_r^{-4.2}$, and $\omega = -1 - \log_{10}\left(p_r^{sat}\right)\Big|_{T_r=0.7}$, where ω is a measure of gas molecules polarity, and $\left(p_r^{sat}\right)\Big|_{T_r=0.7}$ is the reduced saturation vapor pressure calculated at a reduced temperature of 0.7.

Gas	$f_1(T = 298.15K), [10^{-6}\, m^3/mol]$	$f_2(T = 298.15K), [10^{-12}\, m^6/mol^2]$
H2	14.1	350
He	11.8	121
N2	−4.5	1100
O2	−16.1	1200
Ar	−15.8	1160
CO	−8.6	1550

TABLE 5.3 Second and Third Virial Coefficients at 298K

Virial EOSs are usually categorized based on the number of terms used in the corresponding series, as shown in Equation 5.32. For example, 2nd order models use two terms and 3rd order models use three terms (a so-called *cubic* EOS). We will continue by discussing a number of the cubic models.

Considering Equation 5.32 we can write the expansion of the Z function in terms of volume and pressure as well, or

$$
\begin{cases}
Z = \dfrac{pV}{RT} = 1 + \dfrac{f_1'}{V} + \dfrac{f_2'}{V^2} + \dfrac{f_3'}{V^3} + \cdots \\[2mm]
Z = \dfrac{pV}{RT} = 1 + g_1' p + g_2' p^2 + g_3' p^3 + \cdots
\end{cases}
\tag{5.34}
$$

where g_i' are functions of temperature as well. It can be shown [23] that

$$
\begin{cases}
g_1' = \dfrac{f_1'}{RT} \\[3mm]
g_2' = \dfrac{f_2' - f_1'^2}{(RT)^2} \\[3mm]
g_3' = \dfrac{f_3' - 3f_1' f_2' + 2f_1'^3}{(RT)^3} \\[3mm]
\vdots
\end{cases}
\tag{5.35}
$$

This can be done by calculating p from the expansion in terms of $\left(\dfrac{1}{V}\right)$ from Equation 5.34 or $p = \dfrac{RT}{V} + \dfrac{RTf_1'}{V^2} + \dfrac{RTf_2'}{V^3} + \cdots$, into the expansion in terms of p, or $Z = 1 + g_1' \left[\dfrac{RT}{V} + \dfrac{RTf_1'}{V^2} + \dfrac{RTf_2'}{V^3} + \cdots \right] + g_2' \left[\dfrac{RT}{V} + \dfrac{RTf_1'}{V^2} + \dfrac{RTf_2'}{V^3} + \cdots \right]^2 + g_3' \left[\dfrac{RT}{V} + \dfrac{RTf_1'}{V^2} + \dfrac{RTf_2'}{V^3} + \cdots \right]^3$. This relation can be compared with the first relation given in Equation 5.34 for balancing the coefficients of powers of $\left(\dfrac{1}{V}\right)$. Therefore, balancing coefficients of terms involving $\left(\dfrac{1}{V}\right)$ gives the first relation in Equation 5.35 balancing coefficients of terms involving $\left(\dfrac{1}{V^2}\right)$ gives the second relation, and balancing coefficients of terms involving $\left(\dfrac{1}{V^3}\right)$ gives the third relation, and so on. The values of g_i' are temperature dependent and gas specific. For example, for methane at 400 bar we have:

$$g_1'\left(0°C\right) = -2.349 \times 10^{-3}\ 1/\text{bar},\ g_1'\left(100°C\right) = -0.677 \times 10^{-3},$$
$$g_1'\left(200°C\right) = -0.106 \times 10^{-3}$$

$$g_2'\left(0^0C\right) = -0.877 \times 10^{-6}\ 1/\text{bar}^2,\ g_2'\left(100^0C\right) = 1.447 \times 10^{-6},$$
$$g_2'\left(200^0C\right) = 0.967 \times 10^{-6}$$

$$g_3'\left(0^0C\right) = 29 \times 10^{-9}\ 1/\text{bar}^3,\ g_3'\left(100^0C\right) = 4.1 \times 10^{-9},\ g_3'\left(200^0C\right) = 0.99 \times 10^{-9}.$$

Several cubic EOSs with higher accuracy have been developed with the aim of improving the modeling results for higher pressures. Some of these models are listed below, which are described in further sections along with their merits. We also compare the results of these models through a numerical example, presented at the end of this chapter, and their predictions for the Joule-Thomson inversion curve.

- Van der Waals model
- Dieterici model
- Redlich-Kwong (RK) model

- Berthelot model

- Peng-Robinson model

- Beattie-Bridgeman model

- Benedict-Webb-Rubin model

5.8.2. The van der Waals Model

Johannes van der Waals (referred to as vdW) proposed a model for real gas behavior in his PhD dissertation [76], before virial EOS formulation was discovered. His approach for derivation of his equation was based on physics and thermodynamic principles for real gas behavior [17]. Briefly explained here, he argued that the volume of the molecules occupied total volume should be subtracted from the gas volume as well as the pressure reduction due to intermolecular attractive forces. The former is a correction to the ideal gas model when the pressure increases and hence gas-specific volume decreases and repulsive forces (or Coulomb forces) are significant. The latter may be more clear when noticing that free molecules hit a surface, for example, more often both in terms of number of them and the frequency as compared to when molecules are more constrained by their intermolecular attractive forces. Hence, and considering ideal gas equations as $pV = nRT$ for n number of moles, we have $p = nRT / (V - bn) - a(n / V)^2$; the second term on the R.H.S. is due to pressure reduction, which is proportional to the square of the gas density. The vdW equation is written as

$$\left(p + a\frac{n^2}{V^2} \right)(V - bn) = nRT \tag{5.36}$$

For $n = 1$, we have the vdW equation in terms of the molar volume $v = V / n$, as

$$\left(p + \frac{a}{v^2} \right)(v - b) = RT$$

In this case the SI units of constants a and b are $Pa \cdot m^6 /kmole^2$ and $m^3 /kmole$, respectively. Note that if volume is expressed as a specific volume (e.g., $\frac{m^3}{kg}$) in Equation 5.36, then the right-hand side should be divided by molecular weight/atomic mass of the gas considered. Therefore, the SI units of constants a and b are $Pa \cdot m^6 / kg^2$ and m^3/kg, respectively.

Expanding and rewriting the vdW equation in terms of the powers of v, we get a cubic equation, as given by Equation 5.37, and hence it is categorized as a cubic model.

$$v^3 - \left(b + \frac{RT}{P}\right)v^2 + \left(\frac{a}{p}\right)v - \frac{ab}{p} = 0 \qquad (5.37)$$

A detailed discussion on the validity of three possible real roots of Equation 5.37 in comparison to experimental data can be found in the reference section [30]. In brief, Equation 5.37 gives three possible solutions for volume at a given pressure, when temperature is kept constant or an isotherm. In other words, in a $p - v$ coordinates system we can draw isotherms for a substance. It turns out that at temperatures below critical temperature (or the region of the phase diagram in which liquid and vapor coexist), the vdW model does not predict the experimental data accurately.

The vdW constants a and b are gas specific; for example, for CO2 $a = 3.7 \times 10^5 \, Pa \cdot m^6 / kmole^2$ and $b = 4.3 \times 10^{-2} \, m^3 / kmole$, and for water $a = 5.5 \times 10^5 \, Pa \cdot m^6 / kmole^2$ and $b = 3.1 \times 10^{-2} \, m^3 / kmole$ [77]. In order to derive a relation for their values for a gas we can consider modeling phase-change behavior and calculate the vdW prediction for the critical compressibility factor (or the inverse of the *critical coefficient*), defined at a critical point for a gas as

$$Z_c = \frac{p_c v_c}{RT_c} \qquad (5.38)$$

Considering the isotherm that passes through the critical point, or the T_c-isotherm curve on the $p - v$ plane, we know that the slope and the curvature of the isotherm are both zero at the critical point (i.e., $p = p_c, T = T_c, V = V_c$), or $\left.\frac{\partial p}{\partial V}\right|_{T_c} = \left.\frac{\partial^2 p}{\partial V^2}\right|_{T_c} = 0$. Therefore, using the vdW equation we have

$$\begin{cases} \left.\dfrac{\partial p}{\partial V}\right|_{T=T_c} = \dfrac{-RT_c}{\left(V_c - b\right)^2} + \dfrac{2a}{V_c^3} = 0 \\[4mm] \left.\dfrac{\partial^2 p}{\partial V^2}\right|_{T=T_c} = \dfrac{2RT_c}{\left(V_c - b\right)^3} - \dfrac{6a}{V_c^4} = 0 \end{cases} \tag{5.39}$$

Solving Equation 5.39 for V_c and T_c gives:

$$V_c = 3b \tag{5.40}$$

$$T_c = \frac{8a}{27Rb} \tag{5.41}$$

By substituting for V_c and T_c into Equation 5.36 for $n = 1$, we can calculate the critical pressure as

$$p_c = \frac{a}{27b^2} \tag{5.42}$$

Using relations given by Equation.5.40-Equation 5.42 and $Z_c = \dfrac{p_c V_c}{RT_c}$, we have

$$Z_c\big|_{vdW} = \frac{3}{8} = 0.375 \tag{5.43}$$

The calculated value of $Z_c\big|_{vdW}$ can be checked against experimental results, as shown in Table 5.4.

We mentioned that vdW constants a and b (see Equation 5.36) are gas specific. Using Equation 5.41-Equation 5.42, and solving for a and b, we can write the general relations for a and b in terms of critical pressure and temperature as

Gas	Z_c	p_c(atm)	T_c(K)	V_c(cm³/mol)
He	0.305	2.26	5.2	57.8
Ar	0.292	48	150.7	75.3
N₂	0.292	33.5	126.2	90.1
CO₂	0.274	72.9	304.2	94
CH₄	0.286	45.4	190.6	98.6

TABLE 5.4. Experimental Values of Critical Coefficient for Several Gases, as Compared to the Theoretical Value of 0.375

$$a = \frac{27}{64} \frac{(RT_c)^2}{p_c}, \qquad b = \frac{1}{8} \frac{RT_c}{p_c} \qquad (5.44)$$

Equation 5.44 can be used for calculating vdW constants a and b for a specified gas, having its critical pressure and temperature.

Boyle Temperature

As previously mentioned, the coefficients of virial expansion are functions of temperature. For cubic equations we have $Z = 1 + \frac{Bp}{RT}$ with $B = B(T)$. For example, for N₂ at 273K and 600K, $B = -10.5$ cm³/mol and $B = 21.7$ cm³/mol, respectively. Therefore, there might be a temperature at which B vanishes and hence gas behaves ideally, since $Z = 1$. This temperature is called the *Boyle temperature*, T_B. In other words, the Boyle temperature is the temperature at which a real gas behaves like an ideal gas. Mathematically speaking, if we draw the curve representing gas behavior on the Z-p coordinates system, the slope of such a curve is zero at the origin (or $p = 0$). This requirement can be written as $\lim\limits_{p \to 0} \left(\dfrac{\partial Z}{\partial p} \right)_T = 0$. For a vdW gas, letting $B = b - \dfrac{a}{RT} = 0$ gives $T = T_B = \dfrac{a}{Rb}$. Using Equation 5.41, we get

$$T_B\big|_{\text{vdW}} = \frac{27}{8} T_c \qquad (5.45)$$

For example, for CO₂ the Boyle temperature is $T_B = 714.8$K and $T_B\big|_{\text{vdW}} = \dfrac{27}{8} \times 304.2 = 1026.7$K, whereas for Helium the Boyle temperature is $T_B = 22.6\text{K}$ and $T_B|_{vdW} = \dfrac{27}{8} \times 5.2 = 17.6\text{K}$. The error in these results shows the limitation of the vdW model.

We can also elaborate on the vdW model limitations by writing the vdW equation in terms of dimensionless reduced pressure $p_r = \dfrac{p}{p_c}$, reduced volume $V_r = \dfrac{V}{V_c}$, and reduced temperature $T_r = \dfrac{T}{T_c}$. By substituting these variables into the vdW equation (Equation 5.36), for $n=1$) and using Equation 5.40-Equation 5.42 for critical volume, temperature, and pressure, we get

$$\left(p_r + \frac{3}{V_r^2} \right)\left(V_r - \frac{1}{3} \right) = \frac{8}{3}T_r \tag{5.46}$$

Equation 5.46 shows the limitation of vdW model, that is, for $V_r = \dfrac{1}{3}$ then $p_r \to \infty$. In other words, in order to compress a gas to one-third of its critical volume, an infinite pressure is needed, which is unrealistic! Equation 5.46 is also a useful form of the vdW model, since it does not contain any constant coefficients and can be applied to all gases when the gas critical properties are given.

The vdW equation model can be written in terms of compressibility factor, Z and hence serves the proof for the principle of corresponding states:

$$\left(Z + \frac{27p_r}{64ZT_r^2} \right)\left(1 - \frac{p_r}{8ZT_r} \right) = 1 \tag{5.47}$$

Example:

Show that the locus of extremum points (i.e., maximum and minimum) of constant temperature curves in the (p_r, V_r) coordinates system for a vdW gas is given by the equation; $p_r = \dfrac{3}{V_r^2} - \dfrac{2}{V_r^3}$.

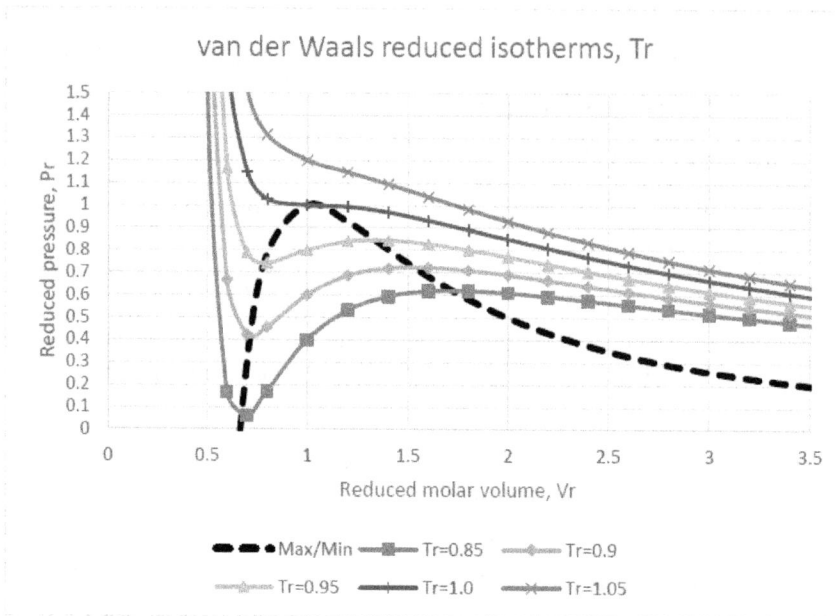

FIGURE 5.7. van der Waals isotherms (solid lines) with locus of their maxima/minima (dash line) on a reduced coordinates system.

■ Solution:

We rearrange the terms in Equation 5.16 to get $p_r = \dfrac{8}{3}T_r\left(V_r - \dfrac{1}{3}\right)^{-1} - \dfrac{3}{V_r^2}$. This is the equation of family of p_r vs V_r curves for given constant values of temperature T_r. The maxima and minima points are where $\left(\dfrac{\partial p_r}{\partial V_r}\right)_{T_r} = 0$. Performing the differentiation we get $\left(\dfrac{\partial p_r}{\partial V_r}\right)_{T_r} = -\dfrac{24T_r}{(3V_r - 1)^2} + \dfrac{6}{V_r^3} = 0$, which yields $T_r = \dfrac{(3V_r - 1)^2}{4V_r^3}$. Back substituting for T_r into the equation for p_r gives, after simplification, $p_r = \dfrac{3}{V_r^2} - \dfrac{2}{V_r^3}$.

Figure 5.7 shows these results.

5.8.3. The Maxwell Construct and Equation

As shown in Figure 5.7, those vdW isotherms below critical point (i.e., $p_r = V_r = T_r = 1$) oscillate in the two-phase region (note that the dash line in Figure 5.7 does not represent the "dome" or the region where liquid and vapor coexist, see Figure 5.8). This oscillatory behavior of vdW isotherms is referred to as a *vdW loop*. This oscillation is unrealistic, compared to experimental data [78], [79] which gives a constant pressure during the phase change along an isotherm at $T < T_c$. As shown in Figure 5.8, moving along isotherm Tr = 0.85, for example, in the direction of decreasing volume the gas behavior follows the path E-C-A in reality as compared to E-D-C-B-A suggested by vdW. The reason for this behavior can be shown, [80] based on the requirement of having minimum Gibbs energy for stability of the gas phase transition with isothermal and isobaric conditions, i.e. $dG = -SdT + Vdp = 0$ when $dT = dp = 0$, (see Table 4.1).

Usually the method of the *Maxwell construct* (or Maxwell's equal-area rule) is used to fix the vdw loop problem. Maxwell suggested this method, [81] based on the invariance of Gibbs energy during condensation as a requirement for stability. The Maxwell construction is simply drawing a horizontal line parallel to V_r axis connecting two points on an isotherm such that the two regions produced by intersection with the van der Waals isotherm have equal areas as shown in Figure 5.8 for two isotherms at $T_r = 0.85$ and $T_r = 0.95$ with their corresponding Maxwell constructs. Mathematically, this can be derived [80], using the minimum Gibbs energy requirement during phase change, or $\int_E^A dG_r = \int_E^D dG_r + \int_D^C dG_r + \int_C^B dG_r + \int_B^A dG_r = 0$, where dimensionless Gibbs energy is $G_r = \dfrac{G}{V_c p_c}$. But $dGr_{|T_r = \text{constant}} = V_r dp_r$, hence

$$\underbrace{\int_E^D V_r dp_r - \int_C^D V_r dp_r}_{\text{area eclosed by EDCE}} = \underbrace{\int_B^C V_r dp_r - \int_B^A V_r dp_r}_{\text{area eclosed by CBAC}}.$$

The two ends of each Maxwell construct line (e.g., points A and E) are at the cross-section of the isotherm $T_r = 0.85$ and the phase dome, that is, the region that separates the pure liquid and pure gas phases. Therefore, point A is at a saturated liquid phase with a reduced liquid molar volume V_{rf}, and point E is at a saturated vapor phase with a reduced gas molar volume V_{rg}. Both points A and E are at the same reduced pressure p_m according to the Maxwell construct requirement. Recall that line AE should be a horizontal line. An app (through the Wolfram Demonstration Project) is available [78]. Users may make use of this tool after installing the Wolfram CDF Player.

It is useful to discuss the mathematics behind the Maxwell construct and derive the related equations, including corresponding pressure and volume for a given isotherm. We start with Equation 5.46, which yields $p_r = \dfrac{8T_r}{3V_r - 1} - \dfrac{3}{V_r^2}$. Pressure and volume at points A and E should satisfy this equation; hence we get

$$p_m = \frac{8T_r}{3V_{rf} - 1} - \frac{3}{V_{rf}^2} \quad \text{and} \quad p_m = \frac{8T_r}{3V_{rg} - 1} - \frac{3}{V_{rg}^2} \tag{5.48}$$

Eliminating p_m gives

$$\frac{8T_r}{3V_{rf} - 1} - \frac{3}{V_{rf}^2} = \frac{8T_r}{3V_{rg} - 1} - \frac{3}{V_{rg}^2} \tag{5.49}$$

Also, the Maxwell construct requires having equal areas enclosed between the line AE and the isotherm curve. In other words, the area under the line AE and the curve ABCDE should be equal. This requirement is equivalent to having the same $p - v$ type work during the phase change process, considering the vdW isotherm and the Maxwell construct. Therefore, we can write $\displaystyle\int_{V_{rf}}^{V_{rg}} p_r dV_r = p_m \left(V_{rg} - V_{rf} \right)$. Substituting for p_r from Equation 5.46, we get $\displaystyle\int_{V_{rf}}^{V_{rg}} \left(\frac{8T_r}{3V_r - 1} - \frac{3}{V_r^2} \right) dV_r = p_m \left(V_{rg} - V_{rf} \right)$. Performing the integration, we get $p_m = \dfrac{1}{V_{rg} - V_{rf}} \left[\dfrac{8T_r}{3} \ln\left(\dfrac{3V_{rg} - 1}{3V_{rf} - 1} \right) + 3\left(\dfrac{1}{V_{rg}} - \dfrac{1}{V_{rf}} \right) \right]$.

Eliminating p_m, using the first relation given with Equation 5.48, we get

$$\frac{8T_r}{3V_{rf} - 1} - \frac{3}{V_{rf}^2} = \frac{1}{V_{rg} - V_{rf}} \left[\frac{8T_r}{3} \ln\left(\frac{3V_{rg} - 1}{3V_{rf} - 1} \right) + 3\left(\frac{1}{V_{rg}} - \frac{1}{V_{rf}} \right) \right] \tag{5.50}$$

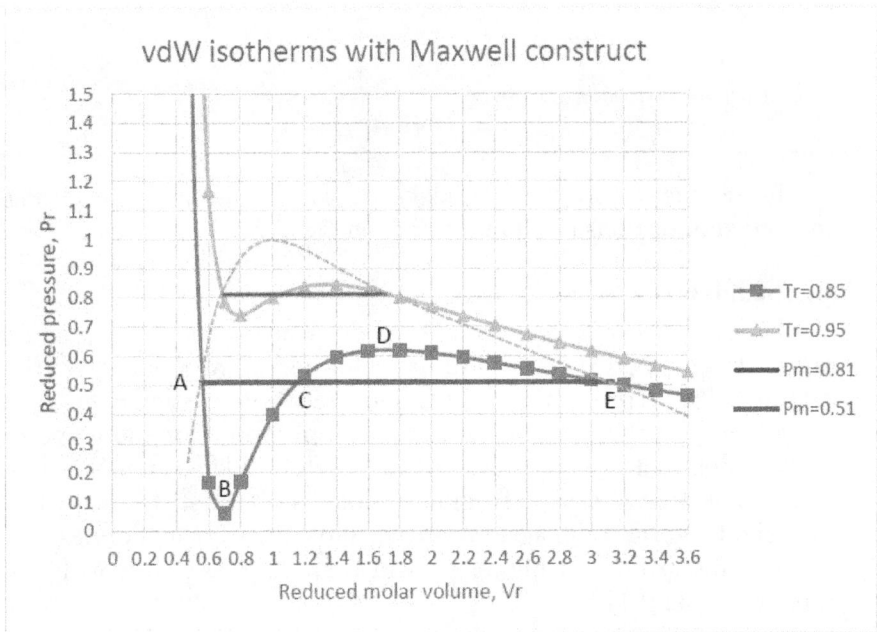

FIGURE 5.8. VdW reduced isotherms (0.85, 0.95) with their Maxwell constructs (reduced pressures 0.51, 0.81).

Equation 5.49 and Equation 5.50 form a two-equation system for two unknowns V_{rf} and V_{rg}, for a given isotherm T_r. A symbolic mathematical software tool can be used to solve this system for explicit algebraic expressions. Performing the manipulation gives the following relations for a given vdW isotherm[82]:

$$
\begin{cases}
V_{rf} = 1 - 2659.84(1 - T_r)^3 + 11440.55(1 - T_r)^4 - 17766.11(1 - T_r)^5 + \\
\quad 14.9451(T_r - 1) + 279.1356(T_r - 1)^2 \\
V_{rg} = 1 - 18.8392(T_r - 1) - 133.412(T_r - 1)^2 - 1111.96(T_r - 1)^3 - \\
\quad 3661.78(T_r - 1)^4 - 5325.71(T_r - 1)^5 \\
p_m = 1 - 0.864592(1 - T_r)^3 - 0.946667(1 - T_r)^4 + 3.999832(T_r - 1) + \\
\quad 4.793989(T_r - 1)^2
\end{cases}
\qquad (5.51)
$$

Abbott [83] uses the Wolfram Mathematica™ software tool and reports numerical values given in the table below for several vdW isotherms.

T_r	V_{rf}	V_{rg}	p_m (Equation 5.48)
0.85	0.55336	3.12764	0.5045
0.9	0.603402	2.34884	0.647
0.95	0.684122	1.72707	0.812

Readers may want to try to make graphs of other isotherms using the method and relations given in this section.

5.8.4. The Dieterici Model

Following the classical work of van der Waals, many models and similar work appeared in the literature. One of the alternatives is that of Dieterici [80]. He treated the volume reduction as similar to the vdW model, and for the pressure reduction he proposed an energy potential type model, that is, molecules should reach to a kinetic energy level in order to be able to escape from intermolecular attractive forces. In general, the Dieterici EOS agrees with the vdW EOS for moderate pressure but deviates from it at higher pressures and agrees better with experimental results [81].

The Dieterici EOS for n number of moles in a volume V reads

$$p(V - nb) = nRTe^{-\frac{na}{RTV}} \tag{5.52}$$

where a and b are constants. The constants a and b are gas specific; for example, for CO2, $a = 3.59\,bar.L^2/mol^2$ and $b = 0.0427\,L/mol$ ([77]).

For $n = 1$, Equation 5.52 reduces to, in terms of the molar volume $v = V/n$,

$$p(v - b) = RTe^{-\frac{a}{RTv}} \tag{5.53}$$

In order to derive a relation for the values of model constants we can consider modeling phase-change behavior and calculate the Dieterici prediction for the critical compression factor (also, as mentioned, referred to as the *critical coefficient*) $Z_c = \dfrac{p_c V_c}{RT_c}$ defined at the critical point for a gas. Considering a T_c-isotherm curve on the $p - V$ plane, we know that the slope and the curvature are both zero at the critical point (i.e., $p = p_c, T = T_c, V = V_c$), or $\dfrac{\partial p}{\partial V}\Big|_{T_c} = \dfrac{\partial^2 p}{\partial V^2}\Big|_{T_c} = 0$. Therefore, using the Dieterici equation we have

$$p_c = \frac{a}{4e^2 b^2}, \quad T_c = \frac{a}{4Rb}, \quad v_c = \frac{V_c}{n} = 2b \qquad (5.54)$$

After substituting for V_c and T_c into $Z_c = \frac{p_c V_c}{R T_c}$, we can calculate the Dieterici critical coefficient as

$$Z_c \Big|_{\text{Dietrici}} = 2e^{-2} = 0.271 \qquad (5.55)$$

The calculated value of $Z_c \big|_{\text{Dietrici}}$ can be checked against experimental results, as shown in Table 5.4, along with those of the vdW model. Recall that $Z_c \big|_{\text{vdW}} = 0.375$ (see Equation 5.43.). The Dieterici model's results show better agreement with those of the experimental ones.

Using the relations given by Equation 5.54 and solving for a and b, we can write the general relations for a and b in terms of critical pressure and temperature as

$$a = \frac{4}{e^2} \frac{R^2 T_c^2}{p_c}, \quad b = \frac{1}{e^2} \frac{R T_c}{p_c} \qquad (5.56)$$

The Dieterici equation in terms of dimensionless reduced variables can be obtained by substituting Equation 5.54 into Equation 5.53, which yields, after some manipulations:

$$p_r (2v_r - 1) = T_r e^{2\left(1 - \frac{1}{T_r v_r}\right)} \qquad (5.57)$$

Example:

Show that the locus of extremum points (i.e., maximum and minimum) of constant temperature curves in the (p_r, v_r) coordinates system for a Dieterici gas is given by the equation $p_r = \frac{1}{v_r^2} e^{\left(\frac{v_r - 1}{v_r - 0.5}\right)}$.

■ **Solution:**

We rearrange the terms in Equation 5.57 to get $p_r = \frac{T_r}{(2v_r - 1)} e^{\left(2 - \frac{2}{T_r v_r}\right)}$.

This is the equation of the family of p_r vs v_r curves for given constant values of temperature T_r. The maxima and minima points are where

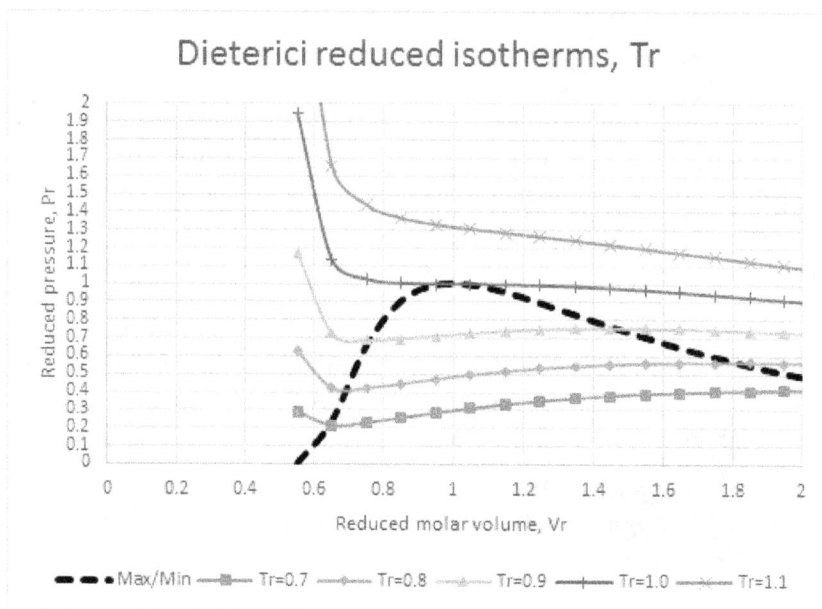

FIGURE 5.9 Dieterici isotherms (solid lines) with locus of their maxima/minima (dash line) on a reduced coordinates system.

$\left(\dfrac{\partial p_r}{\partial V_r}\right)_{T_r} = 0$. Performing the differentiation we get, after simplification,

$\left(\dfrac{1}{T_r v_r^2} - \dfrac{1}{2v_r - 1}\right) = 0$, which yields $T_r = \dfrac{2v_r - 1}{v_r^2}$. Back substituting for T_r into the equation for p_r gives, after simplification, $p_r = \dfrac{1}{v_r^2} e^{\left(\frac{v_r - 1}{v_r - 0.5}\right)}$.

Figure 5.9 shows these results. Readers may note that the Dieterici isotherms in the two-phase region below the critical point are more flat and show less oscillations compared to those of vdW. This is a closer behavior to experimental data and can be interpreted as the quality of this model in terms of prediction of the behavior of real gases.

5.8.5. The Redlich-Kwong Model

Redlich and Kwong published a paper in 1949 [82]. In this paper they proposed a model (referred to as the RK model) for real gases behavior. The RK model is a cubic equation model similar to vdW, but gives more accurate results, specifically for non-polar hydrocarbon vapors, and further modifications for the liquid-vapor equilibrium phase [83]. In this model the volume reduction, due to the molecules themselves, is treated as in the vdW model, and for the pressure reduction Redlich-Kwong proposed a temperature-dependent term for the effects of intermolecular attractive forces on the gas pressure. In general, the RK model agrees with experimental results for reduced pressure of about half of reduced temperature (i.e., $p_r \lesssim 0.5T_r$). The RK model for n number of moles in a volume V reads

$$\left(p + \frac{a\sqrt{n^4/T}}{V(V+nb)} \right)(V - bn) = nRT \tag{5.58}$$

where a and b are constants. For $n = 1$, Equation 5.58 reduces to, in terms of the molar volume $v = V/n$,

$$\left(p + \frac{a}{v(v+b)\sqrt{T}} \right)(v - b) = RT \tag{5.59}$$

The constants a and b are gas specific; for example, for CO2, $a = 63.81^6\, atm.K^{1/2}.L^2/kmol^2$ and $b = 29.7\, L/kmol$, and for water vapor, $a = 140.910^6\, atm.K^{1/2}.L^2/kmol^2$ and $b = 21.1\, L/kmol$ [77].

In order to derive a relation for their values for a gas we can consider modeling phase-change behavior and calculate the RK prediction for the critical compression factor (also, as mentioned, referred to as the *critical coefficient*) $Z_c = \dfrac{p_c V_c}{RT_c}$ defined at the critical point for a gas. Considering a T_c-isotherm curve on the $p-V$ plane, we know that the slope and the curvature are both zero at the critical point (i.e., $p = p_c, T = T_c, V = V_c$) or $\left. \dfrac{\partial p}{\partial V} \right|_{T_c} = \left. \dfrac{\partial^2 p}{\partial V^2} \right|_{T_c} = 0$. Therefore, using the RK equation we have [84]:

$$p_c = \frac{3 - 5\left(\sqrt[3]{4}-1\right)}{\sqrt[3]{4}-2+\sqrt[3]{16}} \left(\frac{a^2 R}{b^5} \right)^{1/3} \ , \ T_c = \left(\sqrt[3]{4}-1\right)^2 \left(\frac{a}{bR} \right)^{2/3} \ , \ v_c = \frac{V_c}{n} = \frac{b}{\sqrt[3]{2}-1} \tag{5.60}$$

After substituting for V_c and T_c into $Z_c = \dfrac{p_c V_c}{RT_c}$, we can calculate the RK critical coefficient as

$$Z_c\big|_{\text{RK}} = \frac{1}{3}$$

The calculated value of $Z_c\big|_{\text{RK}}$ can be checked against experimental results, as shown in Table 5.4, along with those of the vdW model. Recall that $Z_c\big|_{\text{vdW}} = 0.375$ (see Equation 5.43).

Using the relations given by Equation 5.60 and solving for a and b, we can write the general relations for a and b in terms of critical pressure and temperature as

$$a = \frac{1}{9\left(\sqrt[3]{2}-1\right)}\frac{R^2 T_c^{2.5}}{p_c} \quad , \quad b = \frac{\left(\sqrt[3]{2}-1\right)}{3}\frac{RT_c}{p_c} \tag{5.61}$$

The RK equation in terms of dimensionless reduced variables can be obtained by substituting Equation 5.60 into Equation 5.59, which yields

$$p_r = \frac{\beta T_r}{\alpha\left(\gamma v_r - 1\right)} - \frac{\left(\beta T_r\right)^{-0.5}}{\alpha\gamma v_r\left(\gamma v_r + 1\right)} \tag{5.62}$$

where $\alpha = \dfrac{3-5\left(\sqrt[3]{4}-1\right)}{\sqrt[3]{4}-2+\sqrt[3]{16}} \cong 0.0299$, $\beta = \left(\sqrt[3]{4}-1\right)^2 \cong 0.345$, and

$\gamma = \dfrac{1}{\sqrt[3]{2}-1} \cong 3.8473$.

Following similar treatment to that for vdW, we can derive the Maxwell construct relations for the RK model. This operation results in the following two-equation system for the v_{rf} and v_{rg}:

$$\begin{cases} \dfrac{\beta T_r}{\alpha\left(\gamma v_{rf}-1\right)} - \dfrac{\left(\beta T_r\right)^{-0.5}}{\alpha\gamma v_{rf}\left(\gamma v_{rf}+1\right)} = \dfrac{\beta T_r}{\alpha\left(\gamma v_{rg}-1\right)} - \dfrac{\left(\beta T_r\right)^{-0.5}}{\alpha\gamma v_{rg}\left(\gamma v_{rg}+1\right)} \\[4mm] \dfrac{\beta T_r}{\alpha\left(\gamma v_{rf}-1\right)} - \dfrac{\left(\beta T_r\right)^{-0.5}}{\alpha\gamma v_{rf}\left(\gamma v_{rf}+1\right)}\left(v_{rg}-v_{rf}\right) = \dfrac{\beta T_r}{\alpha\gamma}\ln\left(\dfrac{\gamma v_{rg}-1}{\gamma v_{rf}-1}\right) - \dfrac{\left(\beta T_r\right)^{-0.5}}{\alpha\gamma}\left[\ln\left(\dfrac{v_{rg}}{v_{rf}}\right) - \ln\left(\dfrac{\gamma v_{rg}+1}{\gamma v_{rf}+1}\right)\right] \end{cases} \tag{5.63}$$

After solving Equation 5.63 for v_{rf} and v_{rg}, we can calculate the Maxwell construct pressure for the RK model as

$$p_m = \frac{\beta T_r}{\alpha\left(\gamma v_{rf} - 1\right)} - \frac{\left(\beta T_r\right)^{-0.5}}{\alpha\gamma v_{rf}\left(\gamma v_{rf} + 1\right)}.$$

Example:

Derive the locus of extremum points (i.e., maximum and minimum) of constant temperature curves (i.e., isotherms) in the (p_r, v_r) coordinates system for a Redlich-Kwong gas.

■ Solution:

Equation 5.62 is the equation of the family of p_r vs v_r curves for given constant values of temperature T_r. The maxima and minima points are where $\left(\dfrac{\partial p_r}{\partial V_r}\right)_{T_r} = 0$. Performing the differentiation we get, after simplification, $T_r = \dfrac{\gamma v_r - 1}{\beta\gamma v_r\left(\gamma v_r + 1\right)}\sqrt[3]{\Gamma}$, where $\Gamma = \dfrac{\left(2\gamma v_r + 1\right)^2\left(\gamma v_r - 1\right)}{\gamma v_r\left(\gamma v_r + 1\right)}$. Back

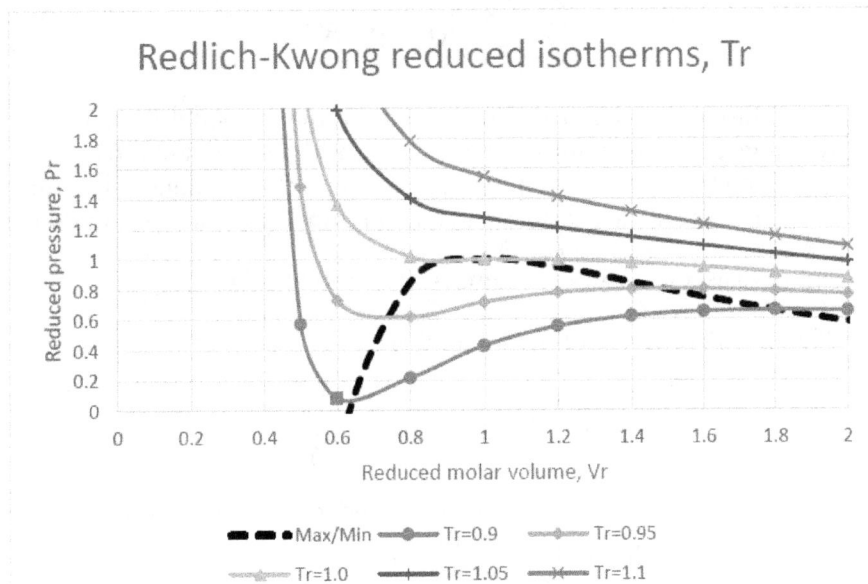

FIGURE 5.10 Redlich-Kwong isotherms (solid lines) with locus of their maxima/minima (dash line) on a reduced coordinates system.

substituting for T_r into the equation for p_r given by Equation 5.62 after simplification, we get $p_r = \dfrac{\sqrt[3]{\Gamma}}{\alpha \gamma v_r \left(\gamma v_r + 1\right)} - \dfrac{\sqrt[6]{1/\Gamma}}{\alpha \left[\sqrt{\gamma v_r \left(\gamma^2 v_r^2 - 1\right)}\right]}.$

α, β, and γ are constants given through definitions under Equation 5.62. Figure 5.10 shows these results.

5.8.6. The Berthelot Model

Berthelot published a paper [85],[86], in which he proposed an EOS. His equation is very similar to that of vdW with an improvement on the treatment of pressure related constant a being a function of temperature, or $a = a(T)$, similar to the RK model. The Berthelot equation model reads as

$$\left(p + \frac{an^2}{TV^2}\right)(V - nb) = nRT \tag{5.64}$$

where $a = a(T)$ and b are model constants. For $n = 1$, Equation 5.64 reduces to, in terms of the molar volume $v = V / n$,

$$\left(p + \frac{a}{Tv^2}\right)(v - b) = RT \tag{5.65}$$

A further modified version of the Berthelot model is also proposed which gives better results in comparison with experimental data. In fact, in the modified model the pressure is scaled with a dimensionless quantity using the gas critical pressure and temperature values (i.e., p_c and T_c) for the desired gas, or $p^{\bullet} = \dfrac{RT}{v}$, where $p^{\bullet} = p / \mathcal{P}$ and $\mathcal{P} = \left[1 + \dfrac{9p_r}{128 T_r}\left(1 - \dfrac{6}{T_r^2}\right)\right].$
The modified Berthelot model reads as

$$p = \frac{RT}{v}\left[1 + \frac{9p_r}{128 T_r}\left(1 - \frac{6}{T_r^2}\right)\right] \tag{5.66}$$

where $p_r = p / p_c$ and $T_r = T / T_c$ are reduced pressure and temperature.

We focus on the original Berthelot model, or Equation 5.65. Considering an isotherm curve on the $p - V$ plane, we know that the slope and the curvature are both zero at the critical point (i.e., $p = p_c, T = T_c, V = V_c$), or $\left.\frac{\partial p}{\partial V}\right|_{T_c} = \left.\frac{\partial^2 p}{\partial V^2}\right|_{T_c} = 0$. Therefore, using the Berthelot equation we have

$$p_c = \frac{1}{b}\sqrt{\frac{aR}{216b}} \quad, \quad T_c = \sqrt{\frac{8a}{27bR}} \quad, \quad v_c = \frac{V_c}{n} = 3b \qquad (5.67)$$

After substituting for V_c and T_c into $Z_c = \frac{p_c V_c}{R T_c}$, we can calculate the Berthelot critical coefficient as

$$Z_c\big|_{\text{Berthelot}} = 0.375 \qquad (5.68)$$

The calculated value of $Z_c\big|_{\text{Berthelot}}$ is equal to that of vdW (see Equation 5.43).

Using the relations given by Equation 5.67 and solving for a and b, we can write the general relations for a and b in terms of critical pressure and temperature as

$$a = \frac{27R^2 T_c^3}{64p_c} \quad, \quad b = \frac{RT_c}{8p_c} \qquad (5.69)$$

The Berthelot equation in terms of dimensionless reduced variables can be obtained by substituting Equation 5.67 into Equation 5.65, which yields, after manipulation:

$$\left(p_r + \frac{3}{T_r v_r^2}\right)(3v_r - 1) = 8T_r \qquad (5.70)$$

Example:

Show that the locus of extremum points (i.e., maximum and minimum) of constant temperature curves in the (p_r, v_r) coordinates system for a Berthelot gas is given by the equation $p_r = \frac{1}{\sqrt{v_r}}\left(\frac{4}{v_r} - \frac{6}{3v_r - 1}\right)$.

■ **Solution:**

We rearrange the terms in Equation 5.70 to get $p_r = \frac{8T_r}{3v_r - 1} - \frac{3}{T_r v_r^2}$.

This is the equation of the family of p_r vs v_r curves for given constant values

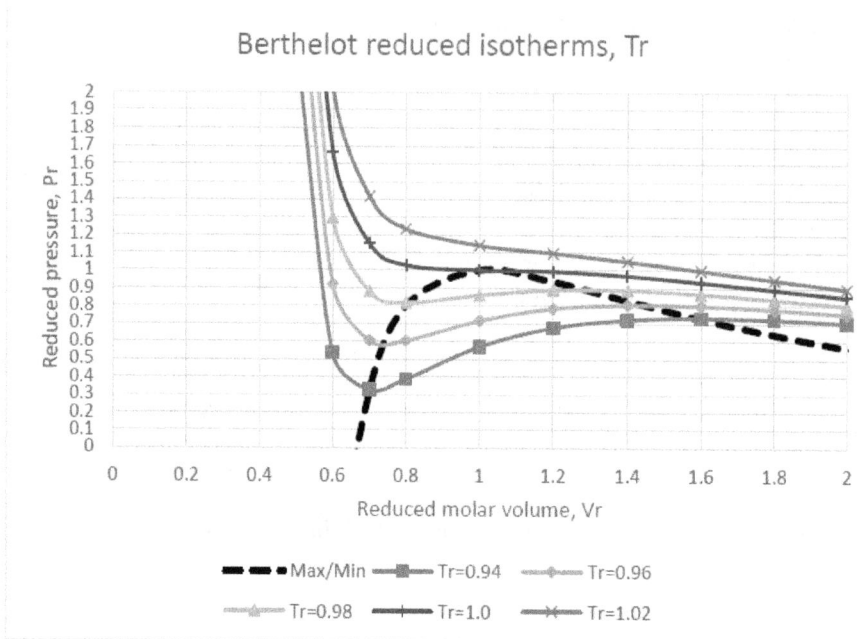

FIGURE 5.11 Berthelot isotherms (solid lines) with locus of their maxima/minima (dash line) on a reduced coordinates system.

of temperature T_r. The maxima and minima points are where $\left(\dfrac{\partial p_r}{\partial V_r}\right)_{T_r} = 0$.

Performing the differentiation we get, after simplification, $T_r = \dfrac{3v_r - 1}{2v_r\sqrt{v_r}}$.

Back substituting for T_r into the equation for p_r gives, after simplification,

$$p_r = \frac{1}{\sqrt{v_r}}\left(\frac{4}{v_r} - \frac{6}{3v_r - 1}\right).$$ Figure 5.11 shows these results.

5.8.7. The Peng-Robinson Model

The Peng-Robinson model [87] is an extension to the vdW model with modifications to the pressure changes due to intermolecular forces. This model considers pressure correction, w.r.t. ideal gas, to the acentric factor of the gas molecules, ω. For each substance the acentric factor is the deviation of the molecules from being a perfect sphere. For example, $\omega|_{Ar} = 0$, $\omega|_{CO2} = 0.228$, and $\omega|_{O2} = 0.022$. The acentric factor is proportional to the logarithm of reduced saturation pressure at the reduced temperature of 0.7, or

$$\omega = -\log_{10}\left(p_r^{sat}\big|_{at\ T_r=0.7}\right) - 1$$

The model is expressed by following equation and parameters

$$\begin{cases}
\left(p + \dfrac{\alpha a}{V^2 + 2bV - b^2}\right)(V - b) = RT \\[2mm]
a = 0.45724\,\dfrac{R^2 T_c^2}{p_c} \\[2mm]
b = 0.0778\,\dfrac{RT_c}{p_c} \\[2mm]
\alpha = \left[1 + \kappa\left(1 - \sqrt{T/T_c}\,\right)\right]^2 \\[2mm]
\kappa = 0.37464 + 1.54226\omega - 0.26992\omega^2
\end{cases} \tag{5.71}$$

The following models use more than two empirical gas specific constants for fitting a gas behavior curve with the experimental data. They usually give more accurate results, but at the cost of more constants needed.

5.8.8. The Beattie-Bridgeman Model

Beattie and Bridgeman published their paper in 1928 [88]. This model has five empirical constants.

TABLE 5.5. Beattie-Bridgeman Model Constants for Several Gases

Gas	A_0 $[kPa \cdot m^6/kmol^2]$	a $[m^3/kmol]$	B_0 $[m^3/kmol]$	b $[m^3/kmol]$	c $[K^3 \cdot m^3/kmol]$
Air	131.8441	0.01931	0.04611	-0.001101	43400
Argon, Ar	30.7802	0.02328	0.03931	0.0	59900
Carbon dioxide, CO2	507.2836	0.07132	0.10476	0.07235	660000
Helium, He	2.1886	0.05984	0.01400	0.0	40
Hydrogen, H2	20.0117	-0.00506	0.02096	-0.04359	504
Nitrogen, N2	136.2315	0.02617	0.05046	-0.00691	42000
Oxygen, O2	151.0857	0.02562	004624	0.004208	48000

The model is expressed by the following equation and parameters [19]:

$$pv^2 = RT\left(1 - \frac{c}{vT^3}\right)\left(v + B_0 - \frac{bB_0}{v}\right) - A_0\left(1 - \frac{a}{v}\right) \tag{5.72}$$

where A_0, B_0, a, b, and c are gas specific constants. The five constants of the Beattie-Bridgeman model are available in several references, for example [89], [19], as given in Table 5.5.

5.8.9. The Benedict-Webb-Rubin Model

Benedict, Webb, and Rubin published their paper in 1940 [90]. This model is an extension of the Beattie-Bridgemen model with eight empirical constants.

The model is expressed by the following equation and parameters, as shown in Table 5.6 [19]:

$$p = \frac{RT}{v} + \frac{1}{v^2}\left(B_0RT - A_0 - \frac{C_0}{T^2}\right) + \frac{1}{v^3}\left(bRT - a + \frac{c}{T^2}e^{-\gamma/v^2}\right) + \frac{1}{v^5}\frac{c\gamma}{T^2}e^{-\gamma/v^2} + \frac{1}{v^6}a\alpha \tag{5.73}$$

5.8.10. The Joule-Thomson Inversion Curve: A Measure for Model Equations

It is recommended, by researchers to use Joule-Thomson inversion curve (see Equation 4.26 and Figure 4.6) obtained based on direct measurement or empirically correlated using experimental data as a performance measure of model equations. This method of comparison is used extensively in literature [95], [96], [97]. As an example, in this section we use vdW model and calculate its corresponding Joule-Thomson inversion curve for the pupose of comparison against a sug)gested least-squared curve fit equation based on 89 experimental data points, reported by Gunn et. al., [98]. Their empirical equation can be considered as a generalized inversion curve, mostly applicable to simple fluids/gases, in terms of reduced variable as

$$p_r = 0.091167T_r^5 - 1.6721T_r^4 + 11.826T_r^3 - 41.567T_r^2 + 71.598T_r - 36.275 \tag{5.74}$$

In order to calculate vdW invesion curve we use Equation 4.26 and rewrite it using Euler's chain rule (see Equation 4.23) as follow;

$$\mu_{JT} = \frac{1}{C_p}\left[T\left(\frac{\partial V}{\partial T}\right)_p - V\right] = 0, \text{ or } T\left(\frac{\partial V}{\partial T}\right)_p - V = 0. \text{ But } \left(\frac{\partial V}{\partial T}\right)_p = -\left(\frac{\partial p}{\partial T}\right)_V\left(\frac{\partial V}{\partial p}\right)_T, \text{ Therefore}$$

$$T\left(\frac{\partial V}{\partial T}\right)_p - V = T\left[\left(\frac{\partial p}{\partial T}\right)_V\left(\frac{\partial V}{\partial p}\right)_T\right] + V = 0. \text{ Now multiply by } \left(\frac{\partial p}{\partial V}\right)_T \text{ to get}$$

Gas	a $[kPa \cdot m^9/ kmol^3]$	A_o $[kPa \cdot m^6/ kmol^2]$	b $[m^6/kmol^2]$	B_o $[m^3/kmol]$	$c \times 10^{-4}$ $[kPa \cdot m^9 \cdot K^2/ kmol]$	$C_o \times 10^{-5}$ $[kPa \cdot m^6 \cdot K^2/ kmol^2]$	$\alpha \times 10^5$ $[m^9/kmol^3]$	γ $[m^6/kmol^2]$
n-Butane, C4H10	190.68	1021.6	0.039998	0.12436	3205	1006	110.1	0.0340
Carbon dioxide, CO2	13.86	277.30	0.007210	0.04991	151.1	140.4	8.470	0.0054
Carbon monoxide, CO	3.71	135.87	0.002632	0.05454	10.54	8.673	13.50	0.0060
Methane, CH4	5.00	187.91	0.003380	0.04260	25.78	22.86	12.44	0.0060
Nitrogen, N2	2.54	106.73	0.002328	0.04074	7.379	8.164	12.72	0.0053

TABLE 5.6 Benedict-Webb-Rubin Model Constants for Several Gases

$$T\left(\frac{\partial \mathrm{p}}{\partial \mathrm{T}}\right)_{\mathrm{V}} + V\left(\frac{\partial \mathrm{p}}{\partial \mathrm{V}}\right)_{\mathrm{T}} = 0 \qquad (5.75)$$

Equation 5.75 is an alternative relation for Joule-Thomson inversion curve which could be applied to equations of state. We can write Equation 5.75 in terms of reduced variables, after sunstituting for $T = T_r T_c$ and $p = p_r p_c$:

$$T_r\left(\frac{\partial \mathrm{p}_r}{\partial \mathrm{T}_r}\right)_{\mathrm{Vr}} + Vr\left(\frac{\partial \mathrm{p}_r}{\partial \mathrm{V}_r}\right)_{\mathrm{Tr}} = 0 \qquad (5.76)$$

For vdW model we use its reduced form given by Equation 5.46, or rewritten as

$$p_r = \frac{8T_r}{3V_r - 1} - \frac{3}{V_r^2} \qquad (5.77)$$

Now using Equation 5.76 in conjunction with Equation 5.77, we have

$$T_r\left(\frac{8}{3V_r - 1}\right) + V_r\left(\frac{6}{V_r^3} - \frac{24T_r}{(3V_r - 1)^2}\right) = 0 \text{ or}$$

$$\frac{8T_r}{3V_r - 1} - \frac{24V_r T_r}{(3V_r - 1)^2} + \frac{6}{V_r^2} = 0 \qquad (5.78)$$

Eliminating V_r between Equation 5.77 and Equation 5.78 (using a symbolic equations solver, for example PTC MathCad™) gives the vdW inversion curve in terms of reduced pressure and temperature as

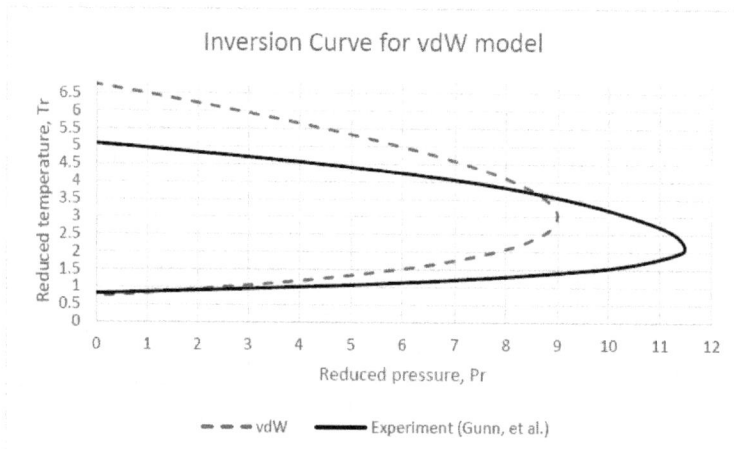

FIGURE 5.12 Joule-Thomson inversion curve for vdW model in comparison with experimental data.

$$p_r = 24\sqrt{3T_r} - 12T_r - 27 \tag{5.79}$$

Figure 5.12 shows the grapgh for vdw inversion curve based on Equation 5.79, in comparison with experimental data, given by Equation 5.74. Based on this comparison, we can conclude that vdW model performance is more realistic for low pressure and temperature region but is poor near the peak (occurs at $T_r \approx 3$ vs 2 for Gunn's et. al., [98]) and higher temperatures region.

Similar analyses for other model equations (e.g. Berthelot, Dieterici, Redlich-Kwong, Beatie-Brigeman, Virial equation, Martine-Hou) are reported, [95] , [97]. The overall research conclusion favours the Redlich-Kwong performance, as a relatively accurate two-constant cubic equation model for practical applications. Figure 5.13 shows the inversion curves for vdW, Dieterici, and Redlich-Kwong models.

The RK model inversion curve are given as

$$\begin{cases} T_r = 5.33857\left(\dfrac{v_r - 0.25992}{v_r + 0.25992}\right)\sqrt[3]{\left(\dfrac{v_r - 0.25992}{v_r + 0.25992}\right)}\left(v_r + 0.155952\right)^2 \\[4mm] p_r = \dfrac{2.9991T_r}{v_r - 0.25992} - \dfrac{3.84687}{v_r\left(v_r + 0.25992\right)\sqrt{T_r}} \end{cases} \tag{5.80}$$

The system of equations (i.e. Equation 5.80) can be solved parametrically, considering v_r as the parameter. We used Excel™ to solve and graph Equation 5.80 using initial value of 0.001 for v_r, with increments of 0.05 for lower values and 0.5 for higher values, up to maximum value of 2000. Equation for p_r is simply a rewrite of Equation 5.62, after plugging in the constants (i.e. α, β, and γ) and some algebraic manipulations. Equation for T_r can be obtained by solving Equation 5.76 for T_r after substituting for

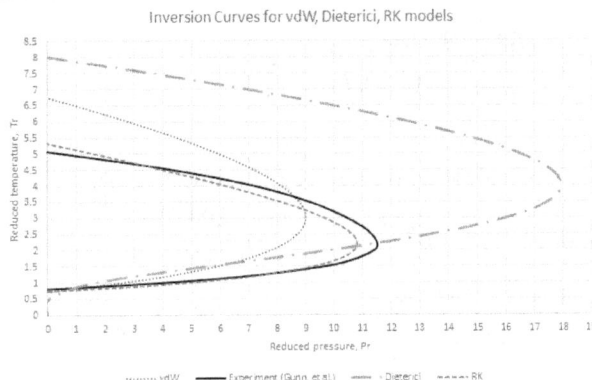

FIGURE 5.13 Joule-Thomson inversion curves for vdW, Dieterici, and Redlich-Kwong models in comparison with experimental data.

$$\left(\frac{\partial p_r}{\partial T_r}\right)_{v_r} = \frac{3}{v_r - 0.25992} + \frac{1.92366}{v_r\left(v_r + 0.25992\right)T_r\sqrt{T_r}} \quad and \quad \left(\frac{\partial p_r}{\partial v_r}\right)_{T_r} =$$

$$-\frac{3T_r}{\left(v_r - 0.25992\right)^2} - \frac{3.84732\left(2v_r + 0.25992\right)}{\sqrt{T_r}}\left(\frac{1}{v_r\left(v_r + 0.26992\right)}\right)^2$$

which in turn are derived using Equation 5.62.

Similar analysis would result the inversion curve for Dieterici inversion curve, as

$$p_r = \left(8 - T_r\right)e^{\left(2.5 - 4/T_r\right)} \tag{5.81}$$

Example:

Molar Volume for CO2 Gas Using Several Real Gas Models

In order to compare the predictions of several models, we calculate the molar volume of a gas at a condition above its critical point. We expect at least one real solution from each model considered. The results will be checked against the experimental result and the corresponding error.

Calculate the predictions of the following models for the molar volume of a Carbon Dioxide (CO2) gas at 400K and 10132.5 kPa. Compare the result of each model with the experimental value ([91]) of 0.268L/mol. Models are: a) ideal gas, b) van der Waals, c) Dieterici, d) Redlich-Kwong (RK), e) Berthelot, f) Beattie-Bridgeman, and g) Benedict-Webb-Rubin (BWR).

■ Solution:

The critical temperature and pressure for COS are: $T_c = 304.25K$ and $p_c = 72.9\,atm$ (7.39 MPa), R=8.31447 J/mol.K.

1. Ideal gas: $v = \dfrac{RT}{p} = \dfrac{8.31447 \times 400}{10132500} = 3.2823 \times 10^{-4}\,\dfrac{m^3}{mol} = 0.32823\,\dfrac{L}{mol}$.

2. vdW gas: we can read the model constants from a database, but here we use the critical point data for calculating the vdW constants. Using

Equation 5.44 we get $a = \dfrac{27\left(8.31447 \times 304.25\right)^2}{64 \times 7.39 \times 10^6} = 0.3653\dfrac{Pa \cdot m^6}{mol^2}$

and $b = \dfrac{8.31447 \times 304.25}{8 \times 7.39 \times 10^6} = 4.279 \times 10^{-5}\,\dfrac{m^3}{mol}$. Plugging into the vdW model equation, we get $\left(10132500 + \dfrac{0.3653}{v^2} \right)\left(v - 4.279 \times 10^{-5} \right) -$

$8.31447 \times 400 = 0$. This is a cubic equation in terms of v. The solution can be obtained analytically or using a calculator. We expect one real solution, since the temperature of the gas is higher than its critical one. We used WolframAlpha™ to find the solution, as

$$v = 2.524 \times 10^{-4} \frac{m^3}{mol} = 0.2524 \frac{L}{mol}.$$

3. Dieterici gas: Using Equation 5.56, we get $a = \dfrac{4\left(8.31447 \times 304.25 \right)^2}{e^2 \times 7.39 \times 10^6} =$

$0.46877 \dfrac{Pa \cdot m^6}{mol^2}$ and $b = \dfrac{8.31447 \times 304.25}{e^2 \times 7.39 \times 10^6} = 4.6327 \times 10^{-5} \dfrac{m^3}{mol}.$

Plugging into the Dieterici model equation, we get

$$v = 2.1855 \times 10^{-4} \frac{m^3}{mol} = 0.21855 \frac{L}{mol}.$$

4. Redlich-Kwong gas: Using Equation 5.61, we get

$a = \dfrac{\left(8.31447 \right)^2 \left(304.25 \right)^{2.5}}{9\left(\sqrt[3]{2} - 1 \right) 7.39 \times 10^6} = 6.4568 \dfrac{Pa \cdot m^6 \cdot K^{0.5}}{mol^2}$ and

$b = \dfrac{\left(\sqrt[3]{2} - 1 \right) 8.31447 \times 304.25}{3 \times 7.39 \times 10^6} = 2.9658 \times 10^{-5} \dfrac{m^3}{mol}.$ Plugging into the RK

model equation, we get $v = 2.606 \times 10^{-4} \dfrac{m^3}{mol} = 0.2606 \dfrac{L}{mol}.$

5. Berthelot gas: Using Equation 5.69, we get $a = \dfrac{27\left(8.31447 \right)^2 \left(304.25 \right)^3}{647.39 \times 10^6} =$

$111.148 \dfrac{Pa \cdot m^6 \cdot K}{mol^2}$ and $b = \dfrac{8.31447 \times 304.25}{8 \times 7.39 \times 10^6} = 4.2786 \times 10^{-5} \dfrac{m^3}{mol}.$

Plugging into the Berthelot model equation, we get

$$v = 2.9053 \times 10^{-4} \frac{m^3}{mol} = 0.29053 \frac{L}{mol}.$$

6. Beattie-Bridgeman gas: The five constants of the Beattie-Bridgeman model are available in several references [89]. Here we use the database provided by Cengel ([19]), their Table 3-4).

Gas	A_0	a	B_0	b	c
	$[kPa \cdot m^6/kmol^2]$	$[m^3/kmol]$	$[m^3/kmol]$	$[m^3/kmol]$	$[K^3 \cdot m^3/kmol]$
CO2	$[kPa \cdot m^6/kmol^2]$	0.07132	0.10476	0.07235	6.6×10^5

The units of the thermodynamic inputs should be pressure in kPa and temperature in K, which result in volume in m^3/kmol. After plugging into the BWR model equation and solving for v, we get

$$v = 2.673 \times 10^{-4} \frac{m^3}{mol} = 0.2673 \frac{L}{mol}.$$

7. Benedict-Webb-Rubin gas: The eight constants of the BWR model are available in several references. Originally published in [92]. Here we use the database provided by Cengel ([19], their Table 3-4).

Gas	A	A_0	B	B_0	c	C_0	α	γ
CO2	13.86	277.3	0.00721	0.04991	1.511×10^6	1.404×10^7	8.47×10^{-5}	0.00539

The units of the thermodynamic inputs should be pressure in kPa and temperature in K, which result in volume in m^3/kmol. After plugging into the BWR model equation and solving for v, we get

$$v = 2.82952 \times 10^{-4} \frac{m^3}{mol} = 0.282952 \frac{L}{mol}.$$

We summarize these results in Table 5.7 with corresponding errors for each mode. Each error is calculated against the experimental value of v = 0.268 L/mol.

TABLE 5.7 Results for Molar Volume of Carbon Dioxide Using Several Real Gas Models

Model/Experiment	Molar volume (L/mol)	Error (%)	No. of model constants
Ideal gas	0.32823	22.5	0
vdW	0.2524	5.8	2
Dieterici	0.21855	18.5	2
Redlich-Kwong	0.2606	2.8	2
Berthelot	0.29053	8.4	2
Beattie-Bridgeman	0.2673	0.3	5
Benedict-Webb-Rubin	0.283	5.6	8
Experimental	0.268	0.0	N/A

Exercises

1. Using the Gibbs phase rule determine the number of independent variables (i.e., d.o.f) to define the following thermodynamic systems: a) liquid water, b) mixture of liquid-vapor water, c) water at its triple point, and d) melting ice.

2. Calculate the (a) Enthalpy departure function, (b) Internal energy departure function, (c) Entropy departure function (d) Helmholtz energy departure function, and (e) Gibbs energy departure function for a fluid, with EOS given as $p = RT(V)^{-1} - aV^{-2}$. Note, V is the specific volume. Summarize the results in a table.

3. Calculate how much pressure is needed to melt ice at −7°C, using the Clapeyron equation.

4. Show that the 2nd virial coefficient for the vdW equation reads $B = b - a/RT$.

5. Derive expressions for enthalpy and entropy for a gas which is going through an isothermal process with EOS: $p(V - b) = RT$.

6. For a vdW gas, draw an sketch in p-V plane to show; an isotherm below critical temperature, the lucas of maxima/minima, the saturation dom, Maxwell contruct, and ciritical point. Discuss the significance vdW loop and Maxwell construct.

7. For a Dieterici gas, calculate the relations for critical pressure, temperature, and volume. Also calculate the Boyle temperature in terms of critical temperature.

8. Derive the relations for critical pressure, temperature, and volume (i.e., p_c, T_c, and V_c) for the Berthelot model, in terms of model constants. Similarly, calculate the model constants in terms of critical properties.

9. Derive the relations for critical pressure, temperature, and volume (i.e., p_c, T_c, and V_c) for the Redlich-Kwong model in terms of model constants. Similarly, calculate the model constants in terms of critical properties.

10. Derive the relations for critical pressure, temperature, and volume (i.e., p_c, T_c, and V_c) for the Dieterici model in terms of model constants. Similarly, calculate the model constants in terms of critical properties.

11. Reproduce the charts given in Figure 5.7 to Figure 5.11 for the van der Waals, Berthelot, Redlich-Kwong, and Dieterici models.

12. Investigate the physical significance of the points at the cross-section of a Maxwell construct and focus curve for maxima/minima of a vdW isotherm (hint: see Figure 5.8.).

13. investigate the physical significance of the points at the cross-section of the vdW isotherms and focus curve for their maxima/minima (hint: see Figure 5.7, Figure 5.8).

14. Determine the units of the Beattie-Bridgeman model constant in the SI system.

15. Determine the units of the Benedict-Webb-Rubin model constant in the SI system.

16. Calculate the nitrogen molar volume [in L/mol] at a temperature of 140K and pressure of 4 MPa and compare your results against the experimental data. Consider the following models: a) ideal gas, b) van der Waals, c) Dieterici, d) Redlich-Kwong (RK), e) Berthelot, f) Beattie-Bridgeman, and g) Benedict-Webb-Rubin (BWR).

CHAPTER 6

GAS MIXTURES: NON-REACTIVE AND REACTIVE

In This Chapter

- Overview
- Mixture Relations
- Mixture Models and Mixing Rules
- Thermodynamic Properties of Ideal Gas Mixtures
- Mixture of Ideal Gases at Different Temperatures
- Reactive Mixtures: Combustion
- Chemical Potential
- Gibbs Energy and Entropy Changes of an Ideal Gas Mixture
- Chemical Reactions Equilibrium
- Chemical Equation for Combustion
- 1stLaw Application to Combustion
- 2nd Law Application to Combustion
- Exercises

6.1. Overview

In the previous chapters, we discussed thermodynamic properties of single pure substances. However, in practice many substances, like gases or liquids, are not composed of just one pure substance, but rather, are in the form of mixtures. Examples are: air (a mixture of several gases), water vapor and liquid, and industrial liquid/gas mixtures. In this chapter we discuss related definitions, major types, and calculations of thermo-dynamic properties of mixtures with a focus on gas mixtures. We first analyze non-reactive mixtures, and then consider combustion mechanics for reactive mixtures.

6.2. Mixture Relations

To categorize mixtures, we divide them into two major types: *non-reactive* and *reactive*. A non-reactive mixture is composed of two or more pure substances, which could be in different phases, at equilibrium state. We can define thermodynamic properties, like pressure, volume, tem-perature, and internal energy, for a mixture. In a non-reactive mixture, the substances can have mechanical and thermal interactions with one another but are chemically neutral. Reactive mixtures can have all the properties of non-reactive mixtures in addition to chemical interactions; as a result some new materials may form. For example, in a combustion reaction of methane, new substances like carbon dioxide and water are created among the products.

For each mixture type, we may have the constituents at different phases. For example, water solid/ice, liquid, and vapor can exist at its triple point (0.01°C, 0.006 atm); air at atmospheric pressure and a wide range of tem-peratures contains several gases (mainly oxygen and nitrogen). For a pure substance, we can have phases of the substance at equilibrium, for example, boiling water or melting ice. In addition, we define *a homogeneous* mixture as having a single phase of all components, whereas a *heterogeneou* mixture has components in more than one phase. A *pure substance* is one that has a homogeneous and invariable chemical composition, even though there is a change of phase.

For our discussion, for ease and clarity of presentation without los-ing generality, we consider binary mixtures and extend the derivation to multi-component mixtures, considering related definitions and governing equations.

Let's consider a mixture made of two components A and B, where the number of identical particles (could be atoms or molecules) for each component is N_A and N_B and the mass of each particle is m_A and m_B, respectively. Figure 5.13 shows a sketch of a binary mixture.

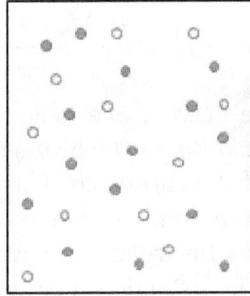

FIGURE 6.1 A sketch for a binary mixture of gases.

Then the *total mass* of the mixture is the sum of the mass of components A and B, $M = \underbrace{m_A N_A}_{M_A} + \underbrace{m_B N_B}_{M_B}$. In practice, the number of molecules of a substance is very large; therefore, we scale them using Avogadro's number,

$N_G \cong 6.022 \times 10^{23}$, and define the *mole number* as $n_A = \dfrac{N_A}{N_G}$ and $n_B = \dfrac{N_B}{N_G}$ for the corresponding components. The total mole number is then defined as $n = n_A + n_B$. The atomic mass (also referred to as molecular weight) is the mass of N_G number of particles, or $M_{WA} = m_A N_G$ and $M_{WB} = m_B N_G$. Substituting for N_G we get $M_{WA} = m_A \dfrac{N_A}{n_A} = \dfrac{M_A}{n_A}$, and similarly, $M_{WB} = m_B \dfrac{N_B}{n_B} = \dfrac{M_B}{n_B}$. Now we define the *mass fractions* as the ratio of each component's mass over the mixture mass, as $y_A = \dfrac{M_A}{M}$ and $y_B = \dfrac{M_B}{M}$.

Also *mole fractions* are defined as $x_A = \dfrac{n_A}{n}$ and $x_B = \dfrac{n_B}{n}$. Obviously, the relations $y_A + y_B = 1$ and $x_A + x_B = 1$ hold. The *molar mass* of the mixture is defined as: $\hat{M} = \dfrac{M}{n}$, which can be rewritten in terms of mole fractions of the components as: $\hat{M} = \dfrac{M_A + M_B}{n} = \dfrac{n_A M_{WA} + n_B M_{WB}}{n} = x_A M_{WA} + x_B M_{WB}$.

Molar mass, as defined, is the mass of one mole of the substance. For fluids, concentration is defined as *molality*, or the inverse of molar mass, as the number of moles per unit mass of the substance. This definition is different from *molarity*, which is defined as the number of moles per unit volume of the substance. Using the relations for a binary mixture, we can define similar ones for multi-component mixtures. The extended relations are listed below, as given by Equation 6.1. We use index i representing each component, or $i = A, B, \cdots$.

$$\begin{cases} M = \sum_i m_i N_i = \sum_i M_i \\[2mm] n_i = \dfrac{N_i}{N_G} \\[2mm] M_{Wi} = m_i N_G \\[2mm] M_i = n_i M_{Wi} \\[2mm] y_i = \dfrac{M_i}{M} \quad , \quad x_i = \dfrac{n_i}{n} \\[2mm] \hat{M} = \sum_i x_i M_{Wi} \end{cases} \tag{6.1}$$

Example: Mole and Mass Fractions of Air

Calculate the mole fractions of air components assuming air is a mixture of 21% oxygen and 79% nitrogen, by volume. Also calculate air molal mass. Assume ideal gas behavior for the air mixture.

■ Solution

One mole of O_2 is $32\,g$ and one mole of N_2 is $28\,g$. Then one mole of air has $32 \times 0.21 = 6.72g$ oxygen and $28 \times 0.79 = 22.12g$ nitrogen. Then the molal mass of air is $22.12 + 6.72 = 28.84\,g$.

The mass fraction of O_2 is $y_{O_2} = \dfrac{32 \times 0.21}{28.84} = 0.233$ and for N_2 is

$y_{N_2} = \dfrac{28 \times 0.79}{28.84} = 0.767$. The ratio is sometimes used as $\dfrac{0.767}{0.233} = 3.29$, or for each gram of oxygen there are 3.29 grams of nitrogen in 4.29 grams of air. The mole fraction for oxygen $n_{O_2} = 0.21$ and for nitrogen $y_{N_2} = 0.79$. The ratio is sometimes used as $\dfrac{0.79}{0.21} = 3.76$, or for each mole of oxygen there are 3.76 moles of nitrogen in 4.76 moles of air. Note that the measured molar mass (or molecular weight) of air is 28.97 g/gmole, which is due to extra traces of other gases (Argon, CO_2, etc.) in the air mixture.

6.3. Mixture Models and Mixing Rules

With modeling mixtures, for example gases, we are faced with two requirements: a) choosing a suitable model or equation of state for each component of the mixture, which could be the same model for all, and b) defining a mixing rule in order to couple the selected models in order to have a unified model for the mixture. To deal with the first requirement, one could use any one of the models discussed in Chapter 5–for example, the ideal gas model, non-ideal gas models, or generalized compressibility factor method. In this section we will consider the ideal gas model first and then discuss other models for real gas mixtures. However, for the latter requirement a mixing rule is needed to know how the selected models for the mixture components are coupled with one another to represent the mixture behavior as a whole. A rule could be a mathematical equation relating the thermodynamic properties of the components (e.g., volumes, pressures) with consideration of the physics of the mixture.

We start the discussion with the assumption of an ideal gas model for a binary mixture, without losing generality, and introduce two very common mixing rules: so-called *Dalton's law* and *Amagat's law*. The mixture rule assumptions are not based on thermodynamic arguments but should consider the physics of intermolecular interactions, either among those molecules of the same species or those of a different species. It does not make sense to discuss or prove that a mixing rule is "correct" or "wrong"; rather the results could be validated against experimental data or possibly theoretical data, perhaps using a statistical thermodynamic approach. Several references are available for further readings on this topic; for example, see [43], [93], [94], [95].

6.3.1. Dalton's Mixing Rule

Considering the ideal gas model for components A and B forming a mixture in a container with volume V at temperature T, we can write $p_A = \dfrac{n_A R T_A}{V_A}$ and $p_B = \dfrac{n_B R T_B}{V_B}$, where R is a universal gas constant. In other words, we assume that each gas behaves as an ideal gas even with the presence of the other gas. Similarly, we assume that "any gas is a vacuum to any other gas mixed with it" [96]. According to Dalton's rule each gas fills the whole volume of the container maintaining the thermal equilibrium, hence they have the same temperature. Therefore, we have the Dalton's mixing rule as

$$V_A = V_B = V \text{ and } T_A = T_B = T \qquad (6.2)$$

Substituting Equation 6.2 into the ideal gas equations for the components, we get $p_A = \dfrac{n_A R T}{V}$ and $p_B = \dfrac{n_B R T}{V}$. Now assuming the ideal gas model for the whole mixture as well, we have $pV = nRT$. Therefore, after substituting for $n = n_A + n_B$ we get $pV = (n_A + n_B)RT$. Rearranging the terms we have $p = \dfrac{n_A R T}{V} + \dfrac{n_B R T}{V} = p_A + p_B$. This relation or additive-pressure law is the result of applying Dalton's mixing rule to a mixture of ideal gases and states that the pressure of a mixture is the sum of the pressures of its components, when each component fills the whole volume available at a thermal equilibrium condition. For a multi-component mixture, we can write Dalton's law as given by Equation 6.3

$$p = \sum_{i=A,B,\cdots} p_i \qquad (6.3)$$

Note that Dalton's law is exact for ideal gases, but it is approximate for real gases (as will be shown in a further section). That is, when the Dalton's mixing rule is applied to a real gas, we don't get, exactly, the additive-pressure relation as given by Equation 6.3. This emphasizes the difference between a mixing rule and selecting a model for the mixture's (and its components') behavior. In other words, the same mixing rule can be selected for a mixture along with considering different models for the mixture, but the results of the mixture model would depend on the model selected for the components.

Partial Pressure

Contribution of each component to the total pressure of a mixture is defined as the partial pressures p_A and p_B. Using Dalton's mixing rule, we have $\dfrac{p_A}{p_B} = \dfrac{n_A RT / V}{n_B RT / V} = \dfrac{n_A}{n_B}$. Therefore, $p_A / p = \dfrac{p_A}{p_A + p_B} = \dfrac{p_A}{p_A + \dfrac{n_B p_A}{n_A}}$

$= \dfrac{n_A}{n_A + n_B} = \dfrac{n_A}{n} = x_A$. Similarly, we can show that $p_B / p = x_B$. Extending the derivation to multi-component gas mixtures, we have

$$p_i = x_i p \tag{6.4}$$

Note that in Equation 6.3 and Equation 6.4, p is the mixture pressure.

6.3.2. Amagat's Mixing Rule

According to Amagat's mixing rule each gas in the mixture has the same pressure as that of the mixture and also the same temperature. Therefore, for our binary mixture, composed of components A and B, we have Amagat's rule as

$$p_A = p_B = p \quad \text{and} \quad T_A = T_B = T \tag{6.5}$$

Therefore, the ideal gas models for the components can be written as $V_A = \dfrac{n_A RT}{p}$ and $V_B = \dfrac{n_B RT}{p}$. Substituting into the ideal gas equation for the whole mixture $pV = (n_A + n_B) RT$, we get $V = \dfrac{n_A RT}{p} + \dfrac{n_B RT}{p} = V_A + V_B$.

This is Amagat's law, or the additive volume rule, which states that the volume of a mixture is the sum of the volumes of its components, when each component is at the same pressure and temperature of the mixture. For a multi-component mixture, we can write Amagat's law as given by Equation 6.6, as

$$V = \sum_{i=A,B,\cdots} V_i \tag{6.6}$$

Note that Amagat's law is exact for ideal gases, but it is approximate for real gases (as will be shown in a further section).

Partial Volume

Contribution of each component to the total volume of the mixture is defined as the partial volume, or V_A and V_B. Writing the volume ratios as V_A/V and V_B/V and using Amagat's mixing rule, we have $\dfrac{V_A}{V_B} =$

$\dfrac{n_A RT/p}{n_B RT/p} = \dfrac{n_A}{n_B}$. Therefore, $V_A/V = \dfrac{V_A}{V_A + V_B} = \dfrac{V_A}{V_A + \dfrac{n_B V_A}{n_A}} = \dfrac{n_A}{n_A + n_B} =$

$\dfrac{n_A}{n} = x_A$. Similarly, it can be shown that $V_B/V = x_B$. Extending the derivation to multi-component gas mixtures, we have

$$V_i = x_i V \tag{6.7}$$

Figure 6.2 shows Dalton's and Amagat's rules, schematically:

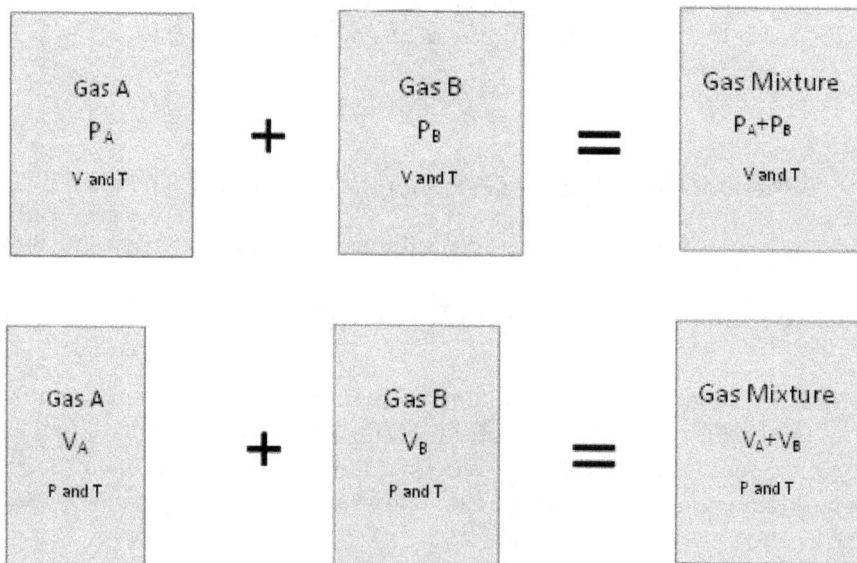

FIGURE 6.2 Dalton's law of additive pressure (top) and Amagat's law of additive volume for a gas mixture.

Example: Dalton's and Amagat's Rules Applied to Air Mixture

Using Dalton's and Amagat's rules, calculate partial pressure and partial volumes for air at 1 atm and 25°C. Assume ideal gas and mole fractions of 0.21 for oxygen and 0.79 for nitrogen and $R = 0.08206 \dfrac{L.atm}{mole.K}$.

■ **Solution**

We have $x_{O_2} = 0.21$ and $x_{N_2} = 0.79$. Using Equation 6.4, we have $p_{O_2} = 0.21 \times 1 = 0.21\,atm$, and $p_{N_2} = 0.79 \times 1 = 0.79\,atm$, or the partial pressure and mole fractions have the same numerical values. The volume of oxygen and nitrogen is equal to the volume of the mixture, according to Dalton's law. Using Equation 6.7 for Amagat's mixing rule, we have $V_i = x_i V$,

but $V = \dfrac{1 \times 0.08206 \times (273.15 + 25)}{1} = 24.47\,L$. This is the volume of one mole of air. Therefore, $V_{O_2} = 0.21 \times 24.47 = 5.14\,L$ and $V_{N_2} = 0.79 \times$. $24.47 = 19.33\,L$ Alternatively, we could use the definition of partial volume, or Amagat's rule (the volume of the component at the mixture pressure and temperature) to get

$$V_{O_2} = \frac{n_{O_2} RT}{p} = \frac{0.21 \times 0.08206 \times 298.15}{1} = 5.14\,L, \text{ and } V_{N_2} = \frac{n_{N_2} RT}{p} :$$

$$= \frac{0.79 \times 0.08206 \times 298.15}{1} = 19.33\,L.$$

6.3.3. Real Gas Mixtures

For real gas mixtures non-ideal gas models could be used. Any of the real gas models/approaches as discussed in Chapter 5 can be used for the mixture and its components. For example, the van der Waals or the Z-factor/compressibility approach could be used. However, in addition a mixture rule is required to calculate the mixture model coefficients in terms of those of its constituent gases. As mentioned in the previous sections Dalton's and Amagat's rules are examples of mixing rules. Another example is Kay's rule [97], a subset of the Z-factor method, which we will explain and discuss in the next section.

6.3.4. *Z*-factor and Kay's Mixing Rules

For a real gas we can assume the compressibility factor approach, or the $PV = ZnRT$ model. Applying this model to a mixture and its components gives $P_i V_i = Z_i n_i RT_i$. Assuming a mixing rule, in principle, we can relate Z to Z_i and other mixture components' properties, like n_i, $(p, V, T)_i$. One can assume Dalton's mixing rule results for an ideal gas (see Equation 6.3) and apply it in conjunction with a real gas model to get $\dfrac{ZnRT}{V} = \sum_i \dfrac{Z_i n_i RT}{V}$, which yields

$$Z = \sum_i x_i Z_i \qquad (6.8)$$

Similarly, applying Amagat's mixing rule (see Equation 6.6) will give the same relation as given by Equation 6.8, since $\dfrac{ZnRT}{p} = \sum_i \dfrac{Z_i n_i RT}{p}$. The application of Dalton's law gives less accurate results as compared to that of Amagat's, specifically when the mixture pressure is high. The reason is that Amagat's law involves the mixture pressure whereas Dalton's law does not consider it [19]. One of the shortcomings of the Z-factor approach for a mixture is its lack of consideration for dissimilar molecule interactions of the gas components. Kay's rule tries to address this issue by modifying this approach. In Kay's model [97], the mixture is considered as a whole to be a substance by itself with its critical pressure P_{cm} and critical temperature T_{cm}. These variables are then related to those of the mixture constituents, p_{ci} and T_{ci}, as given by Equation 6.9.

$$P_{cm} = \sum_i x_i p_{ci} \text{ and } T_{cm} = \sum_i x_i T_{ci} \qquad (6.9)$$

Having P_{cm} and T_{cm}, we can use a generalized compressibility chart to calculate the Z-factor for the mixture.

6.3.5. Compressibility Z-factor with Dalton's and Amagat's Mixing Rules

In the previous section, we applied Dalton's and Amagat's results based on the ideal gas approximation to the real gas mixture. However, the more accurate application of these two rules should include the fundamental assumptions associated with each one (see Equation 6.2 and Equation 6.5).

We consider a binary mixture with real gases A and B. Therefore, we have $p_A V_A = Z_A n_A R T_A$ and $p_B V_B = Z_B n_B R T_B$ for the components. Assuming Dalton's mixing rule (i.e., $V_A = V_B = V$ and $T_A = T_B = T$) we get, after substituting $\dfrac{p_A}{Z_A} = \dfrac{n_A R T}{V}$ and $\dfrac{p_B}{Z_B} = \dfrac{n_B R T}{V}$ into $pV = ZnRT = Z(n_A + n_B) RT$ for the mixture

$$p = Z\left(\frac{p_A}{Z_A} + \frac{p_B}{Z_B}\right) \tag{6.10}$$

Note that for $Z = Z_A = Z_B = 1$, Equation 6.10 reduces to Equation 6.3, that is, Dalton's law for ideal gases.

Similarly, if we assume Amagat's rule (i.e., $p_A = p_B = p$ and $T_A = T_B = T$), we get, after substitution for the components $\dfrac{V_A}{Z_A} = \dfrac{n_A R T}{p}$ and $\dfrac{V_B}{Z_B} = \dfrac{n_B R T}{p}$ into the mixture model $pV = ZnRT = Z(n_A + n_B) RT$, the mixture volume

$$V = Z\left(\frac{V_A}{Z_A} + \frac{V_B}{Z_B}\right) \tag{6.11}$$

Note that for $Z = Z_A = Z_B = 1$, Equation 6.11 reduces to Equation 6.6, that is, Amagat's law for ideal gases.

Equation 6.10 and Equation 6.11 clearly show that Dalton's (i.e. $p = \sum p_i$,) and Amagat's (i.e., $V = \sum V_i$) laws are approximate relations for real gases.

We can extend the derivation for mixtures with more than two components, and by a similar analysis, we can write the general relations equivalent to Equation 6.10 and Equation 6.11, as given by Equation 6.12

$$p = Z \sum_{i=A,B,\cdots} \frac{p_i}{Z_i} \text{ and } V = Z \sum_{i=A,B,\cdots} \frac{V_i}{Z_i} \tag{6.12}$$

Now we demonstrate the applications of these mixing rules and models with a series of worked out examples.

6.3.6. A vdW Mixture with Algebraic and Kay's Mixing Rules

In a container, there are 3 kmol of N_2 (i.e., component 1) and 5 kmol of CO_2 (i.e., component 2) gases. The temperature and pressure of the mixture are 300K and 15MPa, respectively. Calculate the volume of the mixture assuming the van der Waals model and the following mixing rules:

mixing rule 1: $\begin{cases} a = x_1 a_1 + x_2 a_2 \\ b = x_1 b_1 + x_2 b_2 \end{cases}$ and

mixing rule 2: $\begin{cases} a = (x_1)^2 a_1 + (x_2)^2 a_2 + 2 x_1 x_2 \sqrt{a_1 a_2} \\ b = x_1 b_1 + x_2 b_2 \end{cases}$. Compare your

answers against Kay's mixing rule. x_1 and x_2 are mole fractions of component 1 and 2, respectively $R = 8.314 \, J / mol.K$. and a and b are vdW constants.

■ **Solution**

We have $n_1 = 3$ and $n_2 = 5$, then $n = 3 + 5 = 8$ kmol, the total moles of the mixture. Therefore, the mole fractions are $x_1 = \frac{3}{8} = 0.375$ and $x_2 = \frac{5}{8} = 0.625$. The coefficients of the vdW equation are given in

terms of critical values, as $a = \frac{27}{64} \frac{R^2 T_c^2}{p_c}$ and $b = \frac{1}{8} \frac{RT_c}{p_c}$. From gas tables

[19]we get $T_{c1} = 126.2 \, K$, $p_{c1} = 3.39 \, MPa$, $T_{c2} = 304.2 \, K$, and $p_{c2} = 7.39 \, MPa$.

Therefore, $a_1 = \frac{27}{64} \frac{(8.314 \times 126.2)^2}{3.39 \times 10^3} = 137.001 \, kPa.m^6/kmol^2$ and

$b_1 = \frac{1}{8} \frac{8.314 \times 126.2}{3.39 \times 10^3} = 3.869 \times 10^{-2} \, m^3/Kmol$. Similarly,

$a_2 = \frac{27}{64} \frac{(8.314 \times 304.2)^2}{7.39 \times 10^3} = 365.155 \, kPa.m^6/kmol^2$ and

$$b_2 = \frac{1}{8}\frac{8.314 \times 304.2}{7.39 \times 10^3} = 4.278 \times 10^{-2} \text{ m}^3/\text{Kmol}.$$

Using mixing rule 1, we get $\begin{cases} a = 0.375 \times 137.001 + 0.625 \times 365.155 = 279.597 \\ b = 0.375 \times 3.869 \times 10^{-2} + 0.625 \times 4.278 \times 10^{-2} = 0.0412 \end{cases}$.

Using a and b, we can write the vdW equation for the mixture using

$$\left(p + \frac{an^2}{V^2}\right)(V - bn) = nRT \text{, to get } \left(15 \times 10^3 + 279.597 \frac{8^2}{V^2}\right)(V - 8 \times 0.0412) =$$

$8 \times 8.314 \times 300$. Solving for V and selecting the acceptable result, we get

$$V = 0.7753 \text{ m}^3.$$

Using mixing rule 2, we get

$$\begin{cases} a = 0.375^2 \times 137.001 + 0.625^2 \times 365.155 + 2 \times 0.375 \times 0.625\sqrt{137.001 \times 365.155} = 266.748 \\ b = 0.0412. \end{cases}$$

Using a and b, we can write the vdW equation for the mixture as

$$\left(15 \times 10^3 + 266.748 \frac{8^2}{V^2}\right)(V - 8 \times 0.0412) = 8 \times 8.314 \times 300 \text{ . Solving for}$$

V, we get $V = 0.835 \text{ m}^3$.

For the Kay's rule, we use Equation 6.9, $p_{cm} = 0.375 \times 3.39 + 0.625 \times$

$7.39 = 5.89 \text{ MPa}$ and $T_{cm} = 0.375 \times 126.2 + 0.625 \times 304.2 = 237.45 \text{ K}$.

These are the critical pressure and temperature of the mixture. The reduced

pressure and temperature are: $p_R = \frac{15}{5.89} = 2.55$ and $T_R = \frac{300}{237.45} = 1.263$;

using a generalized compressibility chart ([19]Table A-15), we get $Z = 0.6$,

so therefore the mixture volume is $V = \frac{ZnRT}{p} = \frac{0.6 \times 8 \times 8.314 \times 300}{15 \times 10^3} = 0.798 \text{ m}^3$.

This example shows that Kay's mixing rule result for the mixture volume is approximately equal to the average of those values that were calculated using algebraic rules 1 and 2. In addition, these results show that although the same model (i.e., vdW) was considered and used for the gas mixture, we got different results when different mixing rules were employed.

6.3.7. An Ideal Gas Mixture with Amagat's Mixing Rule

Repeat the previous example using ideal gas law and compressibility factor with Amagat's rules, to calculate the mixture volume.

■ **Solution**

Assuming ideal gas law, we have $V = \dfrac{8 \times 8.314 \times 300}{15000} = 1.33$ m³.

Assuming Amagat's law and compressibility factor approach, we need to calculate the Z-factor for each component. For component 1, we have

$p_r = \dfrac{15}{7.39} = 2.03$ and $T_r = \dfrac{300}{126.2} = 2.38$. Therefore, from the compressibility chart [19] we read $Z_1 = 1.02$. Similarly for component 2, we

have $p_r = \dfrac{15}{7.39} = 2.03$ and $T_r = \dfrac{300}{304.2} = 0.99$. Therefore, from the compressibility chart we read $Z_2 = 0.3$. Now, using Equation 6.8 we have for

the mixture $Z = 0.375 \times 1.02 + 0.625 \times 0.3 = 0.57$. Therefore, $V = \dfrac{ZnRT}{p}$

$$= \dfrac{0.57 \times 8 \times 8.314 \times 300}{15000} = 0.758 \text{ m}^3.$$ From these results, we see that the

ideal gas law overestimates the mixture volume, since it ignores the intermolecular forces among the molecules.

6.3.8. van der Waals Mixtures with Dalton's Mixing Rule

In this section we derive the relation for a mixture of vdW gases in terms of its components. We consider a binary mixture, without losing the generality, and then expand the formulation for a multi-component mixture. We assume that components A and B are also vdW gases. Using the vdW

model (see Equation 5.36) we have for a binary gas mixture $\left(p + a\dfrac{n^2}{V^2} \right)$

$(V - bn) = nRT$ and its two components $\left(p_A + a_A \dfrac{n_A^2}{V_A^2} \right)(V_A - b_A n_A) = n_A RT_A$

and $\left(p_B + a_B \dfrac{n_B^2}{V_B^2} \right)(V_B - b_B n_B) = n_B RT_B$. Now we apply Dalton's rule (i.e. $V_A = V_B = V$, and $T_A = T_B = T$), and hence we can write the equations for each component as

$$\begin{cases} \left(p_A + a_A \dfrac{n_A^2}{V^2} \right)\left(V - b_A n_A \right) = n_A RT \\[3mm] \left(p_B + a_B \dfrac{n_B^2}{V^2} \right)\left(V - b_B n_B \right) = n_B RT \end{cases} \tag{6.13}$$

After substituting $n = n_A + n_B$ in the equation for the mixture, also assumed to be an vdW gas, we have

$$\left(p + a \dfrac{\left(n_A + n_B \right)^2}{V^2} \right)\left(V - bn_A - bn_B \right) = n_A RT + n_B RT \tag{6.14}$$

After substituting Equation 6.13 into Equation 6.14 and rearranging the terms, we get

$$pV^3 - \left(pbn_A + pbn_B \right)V^2 + a\left(n_A + n_B \right)^2 V - ab\left(n_A + n_B \right)^3 = \left(p_A + p_B \right)V^3$$

$$- \left(p_A b_A n_A + p_A b_B n_B \right)V^2 + \left(a_A n_A^2 + a_B n_B^2 \right)V - a_A b_A n_A^3 - a_B b_B n_B^3$$

Equating the coefficients of the terms involving the powers of volume V, we get the following relations:

Coefficients of $V^3 \xrightarrow{\text{yields}} p = p_A + p_B$, i.e., Dalton's law for an ideal gas.

Coefficients of $V^2 \xrightarrow{\text{yields}} pbn_A + pbn_B = p_A b_A n_A + p_B b_B n_B$

Coefficients of $V \xrightarrow{\text{yields}} a\left(\underbrace{n_A + n_B}_{n} \right)^2 = a_A n_A^2 + a_B n_B^2$ or

$$a = a_A x_A^2 + a_B x_B^2 \tag{6.15}$$

$$\text{Coefficients of } V^0 \xrightarrow{yields} ab \underbrace{\left(n_A + n_B \right)}_{n}^{3} = a_A b_A n_A^3 + a_B b_B n_B^3 \quad \text{or}$$

$$b = \frac{a_A b_A x_A^3 + a_B b_B x_B^3}{a} \tag{6.16}$$

Finally, we have the relations for a and b of the mixture in terms of those components as given by Equation 6.15 and Equation 6.16. Therefore, having the critical pressure and temperature of each component, we can calculate the mixture properties. These relations can be expanded for a multi-component mixture, as given in Equation 6.17.

$$a = \sum_{i=A,B,\cdots} a_i x_i^2 \quad \text{and} \quad b = \frac{1}{a} \sum_{i=A,B,\cdots} a_i b_i x_i^3 \tag{6.17}$$

The following example demonstrates application of the relations obtained in this section.

Example: vdW Mixture with Dalton's Mixing Rule

Use data given in the previous example, calculate the mixture volume assuming the vdW model and Dalton's mixing rule.

■ **Solution**

From the previous example solution, we have $a_1 = 137.001$ kPa.m^6/kmol^2, $b_1 = 3.869 \times 10^{-2}$ m^3/Kmol, $a_2 = 365.155$ kPa.m^6/kmol^2, and $b_2 = 4.278 \times 10^{-2}$ m^3/Kmol. The mole fractions are $x_1 = 0.375$ and $x_2 = 0.625$ Now, using Equation 6.17 we get $a = 137.001 \times 0.375^2 + 365.155 \times 0.625^2 = 161.904$ kPa.m^6/kmol^2 and

$$b = \frac{137.001 \times 3.869 \times 10^{-2} \times 0.375^3 + 365.155 \times 4.278 \times 10^{-2} \times 0.625^3}{161.904} = 0.0253 \text{ m}^3/\text{Kmol}$$

Now having the values of a and b for the mixture, we can write the vdW equation as $\left(15000 + 161.904 \times \frac{8^2}{V^2} \right)(V - 0.0253 \times 8) = 8 \times 8.314 \times 300$.

Solving for volume gives $V = 0.968$ m^3.

6.4. Thermodynamic Properties of Ideal Gas Mixtures

Gibbs extended Dalton's law in order to calculate other thermodynamic properties of ideal gas mixtures [98]. The Gibbs-Dalton law states that total thermodynamic properties like internal energy, enthalpy, entropy, and so on, of a mixture of ideal gases is equal to the sum of those of its constituents when each ideal gas occupies the volume of the mixture at the same temperature of the mixture. According to the Gibbs-Dalton law the internal energy of the mixture is the sum of the internal energies of the components. For example, we consider the internal energy of an ideal gas mixture, or

$U = \dfrac{3}{2}nRT = \dfrac{3}{2}pV$. By using Dalton's law $p = p_A + p_B$, then we have

$U = \dfrac{3}{2}p_A V + \dfrac{3}{2}p_B V = U_A + U_B$. For multi-component mixtures we have

$U = \sum_i U_i$. Similarly, for enthalpy H and entropy of S the mixture, we can write similar relations as listed in Equation 6.18.

$$\begin{cases} U = \sum_{i=A,B,\cdots} U_i \\ H = \sum_{i=A,B,\cdots} H_i \\ S = \sum_{i=A,B,\cdots} S_i \end{cases} \qquad (6.18)$$

As mentioned, sometimes we use the properties per unit mole of substance, hence we define $\hat{u}_i = \dfrac{U_i}{n_i}$, $\hat{h}_i = \dfrac{H_i}{n_i}$, and $\hat{s}_i = \dfrac{S_i}{n_i}$ as the molar internal energy, enthalpy, and entropy of the mixture components, respectively. Substituting into Equation 6.18, we get

$$
\begin{cases}
U = \displaystyle\sum_{i=A,B,\cdots} n_i \hat{u}_i \\[2ex]
H = \displaystyle\sum_{i=A,B,\cdots} n_i \hat{h}_i \\[2ex]
S = \displaystyle\sum_{i=A,B,\cdots} n_i \hat{s}_i
\end{cases}
\tag{6.19}
$$

Dividing both sides of Equation 6.19 by the total mole number of mixture n, we get

$$
\begin{cases}
\hat{U} = \displaystyle\sum_i x_i \hat{u}_i \\[2ex]
\hat{H} = \displaystyle\sum_i x_i \hat{h}_i \\[2ex]
\hat{S} = \displaystyle\sum_i x_i \hat{s}_i
\end{cases}
\tag{6.20}
$$

where, $\hat{U} = \dfrac{U}{n}$, $\hat{H} = \dfrac{H}{n}$ $\hat{S} = \dfrac{S}{n}$, and are the molar internal energy, enthalpy, and entropy of the mixture, respectively.

We know that for ideal gases, internal energy and enthalpy are functions of temperature only. Therefore, using the relations $\hat{C}_v = \dfrac{\partial \hat{U}}{\partial T}$, $\hat{C}_p = \dfrac{\partial \hat{H}}{\partial T}$, and Equation 6.20, we can write the molar specific heats of the ideal gas mixture in terms of those of its components, as given by Equation 6.21

$$
\begin{cases}
\hat{C}_v = \displaystyle\sum_i x_i \hat{c}_{vi} \\[2ex]
\hat{C}_p = \displaystyle\sum_i x_i \hat{c}_{pi}
\end{cases}
\tag{6.21}
$$

Recall that each component is assumed to be at the same volume and temperature of the mixture (i.e., Dalton's mixing rule) for all the relations given by Equation 6.18-Equation 6.21.

6.5. Mixture of Ideal Gases at Different Temperatures

In the previous sections, we discussed Dalton's and Amagat's laws, which have their common assumption that in a mixture, the gas constituents are at the same temperature of the mixture. In this section, we remove this assumption and derive equations for a mixture when its constituents are at different temperatures before mixing happen, and calculate the mixture temperature. In this analysis, we follow the path given by Hawkins [99].

Consider two containers with V_A volume V_B and containing gases at temperatures T_A and T_B, respectively. The containers are then connected with a pipe which has negligible volume compared to the vessels, and the whole system is thermally insulated. We let the gases mix and the system reaches its equilibrium state, without any work exchange happening between the system and the environment. We are interested to find the mixture temperature in terms of its component thermodynamic properties.

FIGURE 6.3 Sketch of the gas containers, before and after mixing.

For the mixture we can write $V = V_A + V_B$ and $M = M_A + M_B$, where M_A and M_B are the gas components' mass. By applying the 1st law to the system, we have the specific internal energy of the gases before mixing, u_A and u_B, that after mixing u are equal to each other (since there is no work or heat exchanges), or $Mu = M_A u_A + M_B u_B$. For ideal gases we have $u = c_V T$, $u_A = c_{VA} T_A$, and $u_B = c_{VB} T_B$, where each c_V is a specific heat at a constant volume.

Hence, $Mc_V T = M_A c_{VA} T_A + M_B c_{VB} T_B$. Therefore, solving for T we get

$$T = \frac{M_A c_{VA} T_A + M_B c_{VB} T_B}{M c_V} \tag{6.22}$$

However, we have $M c_V = M_A c_{VA} + M_B c_{VB}$. After substituting into Equation 6.22, we have Equation 6.23

$$T = \frac{M_A c_{VA} T_A + M_B c_{VB} T_B}{M_A c_{VA} + M_B c_{VB}} \tag{6.23}$$

Equation 6.23 can further be manipulated by the elimination of

$M_A T_A = \dfrac{p_A V_A}{R_A}$ and $M_B T_B = \dfrac{p_B V_B}{R_B}$, using the ideal gas law. p_A, p_B

and R_A, R_B are pressures and gas contents of the components. Therefore, we get

$$T = \frac{c_{VA} \dfrac{p_A V_A}{R_A} + c_{VB} \dfrac{p_B V_B}{R_B}}{c_{VA} \dfrac{p_A V_A}{T_A R_A} + c_{VB} \dfrac{p_B V_B}{T_B R_B}} \tag{6.24}$$

For multi-component mixtures, we can expand Equation 6.24 to

$$T = \frac{\sum_i \left(c_{Vi} \dfrac{p_i V_i}{R_i} \right)}{\sum_i \left(c_{Vi} \dfrac{p_i V_i}{T_i R_i} \right)} \tag{6.25}$$

For similar-type gases, like monatomic or diatomic, we can simplify Equation 6.24 by using universal gas constant $R = M_{WA} R_A = M_{WB} R_B$ to eliminate R_A and R_B in Equation 6.24. Also, we use the relation for molar heat, or $M_{WA} c_{VA} = M_{WB} c_{VB}$, where M_{WA} and M_{WB} are the molecular weight (atomic mass) of the gas components, respectively. After substituting these relations into Equation 6.24 and working out the algebra, we get

$$T = \frac{p_A V_A + p_B V_B}{p_A V_A / T_A + p_B V_B / T_B} \qquad (6.26)$$

Relations for specific conditions can be obtained from Equation 6.26. For example, for equal-pressure gases we have $p_A = p_B$, so therefore we get $T = \dfrac{V_A + V_B}{V_A / T_A + V_B / T_B}$. Similarly, for equal-volume gases we have $V_A = V_B$, so therefore we get $T = \dfrac{p_A + p_B}{p_A / T_A + p_B / T_B}$.

We now can expand on the results obtained, that is, Equation 6.26, and write the general relation for a multi-component mixture, or

$$T = \frac{\sum_i (p_i V_i)}{\sum_i (p_i V_i / T_i)} \qquad (6.27)$$

Example: Mixture of Ar and N$_2$ at Different Temperatures

A mixture is made of 3 m³ of nitrogen at temperature of 40 °C and pressure of 550 kPa and 2 m³ of argon at temperature of 10 °C and pressure of 690 kPa. Calculate the temperature of the mixture.

■ **Solution**

Using Equation 6.26, we get $T = \dfrac{(550)(3) + (690)(2)}{\dfrac{(550)(3)}{273 + 40} + \dfrac{(690)(2)}{273 + 10}} = 298.58K = 25.5^{\circ}C$.

Note that the conversion of units is not required, since the conversion factors would appear in each term.

It is interesting to find the solution using Equation 6.24, with specific heats of nitrogen and argon being 0.743 kJ/kg.K and 0.312 kJ/kg.K, respectively.

$$T = \frac{(0.743)\dfrac{(550)(3)}{(296.8)} + (0.312)\dfrac{(690)(2)}{(208)}}{(0.743)\dfrac{(550)(3)}{(273 + 40)(296.8)} + (0.312)\dfrac{(690)(2)}{(273 + 10)(208)}} = 302.3K = 29.2^{\circ}C.$$

The discrepancy in the results is due to the approximation that exists in Equation 6.26 for dissimilar-type molecules, that is, the difference in their molar heats.

6.6. Reactive Mixtures: Combustion

In previous sections we discussed gas mixtures and mixing rules for ideal and real gases. In a non-reactive mixture, each component contributes to the mixture's thermodynamic properties (e.g., pressure, volume, enthalpy, and entropy), but there is no chemical interaction occurring among the molecules of different components. Yet in a reactive mixture the amount of substances may change as a result of possible chemical reactions happening. In addition, in a non-reactive mixture the amount of substances may change due to mass transfer, for example in open systems. In any case, the calculation of thermodynamic properties resulting from the changes in the amount of materials is needed. In this section, we discuss and derive relevant relations with a focus on reactive mixtures. For practical applications, we are interested in calculating the mechanics and chemistry of chemical reactions related to combustion phenomena. In further sections, we will give definitions and concepts related to chemical reactions, chemical potentials, and combustions, including some worked out numerical examples.

6.7. Chemical Potential

Let's consider a system of N identical particles (i.e., not a mixture) with internal energy U, entropy S, and volume V. The internal energy as a function of its natural variables S, V, and N can be written as

$$dU = TdS - pdV + \underbrace{\left.\frac{\partial U}{\partial N}\right|_{S,V}}_{new\ term} dN \tag{6.28}$$

The last term (i.e., $\dfrac{\partial U}{\partial N} dN$) is due to changes in the number of particles (e.g., a concentration variation in non-reactive systems or chemical reactions in reactive systems). This term is proportional to changes in internal energy per infinitesimal change in the number of particles, when entropy and volume are kept constant (i.e., $\left.\dfrac{\partial U}{\partial N}\right|_{S,V}$). By comparing the last term to the remaining

terms in Equation 6.28, we may interpret the quantity $\left.\dfrac{\partial U}{\partial N}\right|_{S,V}$ as a kind of

"chemical pressure/force" driving the dN or the chemical reaction ([100], [101]) the same way that T is driving the changes in entropy dS and p is

driving the changes in volume dV. It is customary to represent $\left.\dfrac{\partial U}{\partial N}\right|_{S,V}$ by

symbol μ and define it as *chemical potential* (we will explain the reason in a further section, [102]). Therefore, we have

$$dU = TdS - pdV + \mu dN \qquad (6.29)$$

As mentioned in Chapter 4, we use the relations for enthalpy ($H = U + pV$), Helmholtz ($F = U - TS$), and Gibbs functions ($G = U - TS + pV$) (see Figure 4.1).Repeated here for convenience and derivation purposes([103]), we write the total derivative of these quantities as

$$\begin{cases} dH = dU + pdV + Vdp \\ dF = dU - TdS - SdT \\ dG = dU - TdS - SdT + pdV + Vdp \end{cases} \qquad (6.30)$$

After substituting Equation 6.29 for dU into Equation 6.30, we get

$$\begin{cases} dH = TdS + Vdp + \mu dN \\ dF = -SdT - pdV + \mu dN \\ dG = -SdT + Vdp + \mu dN \end{cases} \qquad (6.31)$$

But the total derivatives of these functions, that is, $H = H(S, p, N)$, $F = F(T, V, N)$, and $G = G(T, p, N)$ can be written as

$$
\left\{
\begin{aligned}
dH &= \left.\frac{\partial H}{\partial S}\right|_{p,N} ds + \left.\frac{\partial H}{\partial p}\right|_{S,N} dp + \left.\frac{\partial H}{\partial N}\right|_{S,p} dN \\
dF &= \left.\frac{\partial F}{\partial T}\right|_{V,N} dT + \left.\frac{\partial F}{\partial V}\right|_{T,N} dV + \left.\frac{\partial F}{\partial N}\right|_{T,V} dN \\
dG &= \left.\frac{\partial G}{\partial T}\right|_{p,N} dT + \left.\frac{\partial G}{\partial p}\right|_{T,N} dp + \left.\frac{\partial G}{\partial N}\right|_{T,p} dN
\end{aligned}
\right. \qquad (6.32)
$$

Now, by comparing the R.H.S. of Equation 6.31 and Equation 6.32 term by term along with Equation 6.29, we conclude that chemical potential can be written as

$$
\mu = \left.\frac{\partial U}{\partial N}\right|_{S,V} = \left.\frac{\partial H}{\partial N}\right|_{S,p} = \left.\frac{\partial F}{\partial N}\right|_{T,V} = \left.\frac{\partial G}{\partial N}\right|_{T,p} \qquad (6.33)
$$

Equation 6.33 gives the relations for chemical potential using any of the thermodynamic functions $U, H, F,$ or G at their specified state conditions. Among these, and for calculations related to chemical reactions of combustion, we are mostly interested in the definition of chemical potential in terms of the Gibbs energy function, since it is defined at a constant pressure and temperature.

Now we focus on the relation $\left.\dfrac{\partial G}{\partial N}\right|_{T,p} dN$. This is actually the total change in the Gibbs energy function when temperature and pressure are kept constant (see Equation 6.32). Since N is a very large number (order of Avogadro's number, N_G) we write the amount of substances measured in the number of moles n. Therefore, using Equation 6.1, we have $\left.\dfrac{\partial G}{\partial (nN_G)}\right|_{T,p}$

$d(nN_G) = \left.\dfrac{\partial G}{\partial n}\right|_{T,p} dn$. Therefore, we can write chemical potential as

$\mu = \left.\dfrac{\partial G}{\partial n}\right|_{T,p}$, which shows that chemical potential is the same as molar

Gibbs energy[104], $\hat{G} = \dfrac{G}{n}$ or

$$\mu = \left.\frac{\partial G}{\partial n}\right|_{T,p} = \hat{G} \tag{6.34}$$

In other words, chemical potential is Gibbs free energy per unit mole of the substance. The concept of chemical potential is no different from that of electric potentials (electrical energy per unit charge), or pressure as the energy per unit volume, or gravitational potential (gravitational energy per unit mass), or similar other potentials defined in physics.

Now we extend the concept of chemical potential to multi-component mixtures. The quantity $\left.\frac{\partial G}{\partial n}\right|_{T,p} dn$ can be written as the sum of each component's contribution, or $\mu_i = \left.\frac{\partial G}{\partial n_i}\right|_{p,T,n_{j \ne i}}$. Therefore, the change in the Gibbs energy function for a multi-component system is written as

$$dG = -SdT + Vdp + \sum_i \mu_i dn_i \tag{6.35}$$

For example, for a binary mixture we have the last term expanded as,

$$\sum_{i=1}^{2} \mu_i dn_i = \mu_1 dn_1 + \mu_2 dn_2 = \left.\frac{\partial G}{\partial n_1}\right|_{p,T,n_2} dn_1 + \left.\frac{\partial G}{\partial n_2}\right|_{p,T,n_1} dn_2 \ .$$

6.7.1. Chemical Potential and Gibbs Energy for an Ideal Gas

As a demonstration example, we calculate the chemical potential of an ideal gas. From Equation 6.34, we conclude that $\mu = \mu(T,p)$. In other words chemical potential can change with pressure and temperature, hence for a given set of values of pressure and temperature μ, only changes when the amount of substance changes. But what is the relation for change in chemical potential when pressure changes? To answer this question, we consider an ideal gas, or $pV = nRT$. Therefore, we can write using Equation 6.34,

$$\left.\frac{\partial \mu}{\partial p}\right|_T = \frac{\partial}{\partial p}\underbrace{\left(\frac{G}{n}\right)}_{\hat{G}} = \frac{1}{n}\frac{\partial G}{\partial p} \ . \text{ But we have } \left.\frac{\partial G}{\partial p}\right|_{T,N} = V \ , \text{ see Equation 6.31}$$

and Equation 6.32. Therefore, $\dfrac{\partial \mu}{\partial p} = \dfrac{V}{n}$. Finally, using the ideal gas equation,

we get $\dfrac{\partial \mu}{\partial p} = \dfrac{RT}{p}$. To calculate μ, we integrate this relation between refer-

ence pressure p_0 (usually set at the standard value of 1 atm) and the pressure

of the gas. Hence, $\mu(T, p)\big|_{p_0}^{p} = RT \ln p\big|_{p_0}^{p}$, or

$$\mu(T, p) = \mu_0 + RT \ln\left(\frac{p}{p_0}\right) \tag{6.36}$$

where $\mu_0 = \mu(T, p_0)$ is a function of temperature and is the chemical potential of the gas at the reference pressure. The values of μ_0 are tabulated for different values of temperature [105]. Equation 6.36 shows that chemical potential is proportional to the logarithm of pressure for ideal gases. Using Equation 6.36 along with Equation 6.34, we can write

$$G(T, p) = G_0 + nRT \ln\left(\frac{p}{p_0}\right) \tag{6.37}$$

where $G_0 = G(T, p_0) = n\mu_0$. Equation 6.37 implies that for $p = 0$,

the value of $(G - G_0) \to -\infty$, a strange result! Actually not, since when pressure approaches zero, then volume would increase and hence entropy would as well. Therefore, Gibbs energy should decrease, since $G = H - TS$, see [41].

6.7.2. Concept of Fugacity

Equation 6.36 is based on an ideal gas assumption. A graph (see Figure 6.4) of this equation on a semi-logarithmic plot shows the linear variation of chemical potential versus the logarithm of pressure. One can ask this question: what quantity could replace pressure (i.e., an intensive property) for real gases while the linear relationship still holds? In other words, we are interested in a thermodynamic intensive quantity that can replace pressure but approaches pressure when gas behaves like an ideal gas, for example, under very low pressure.

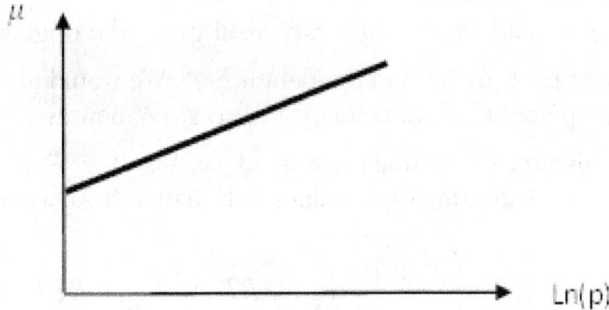

FIGURE 6.4 Chemical potential vs. pressure, a sketch.

Lewis [106] proposed a function f, or *fugacity* as the quantity of interest. Therefore, we have the chemical potential for real gases, analogous to Equation 6.36, as

$$\mu(T,p) = \mu_0 + RT \ln\left(\frac{f}{f_0}\right) \tag{6.38}$$

where $f_0 = f(T, p_0)$. Obviously, fugacity should approach the value of pressure as the real gas behavior tends toward that of ideal gas. Therefore, the condition for the fugacity function is

$$\lim_{p \to 0}\left(\frac{f}{p}\right) \to 1$$

For example, the fugacity of nitrogen (gas) at 273.15K (0° C) and 10132.5kPa (100 atm) is 9831.56 kPa (97.03 atm). This means that the chemical potential of nitrogen gas is related to the fugacity of 97.03 atm (not 100 atm!) as a real gas. In other words, if we consider the nitrogen gas at 97.03 atm and as an ideal gas, then we get the same fugacity and hence the same chemical potential as the real nitrogen gas at 100 atm. For more detail discussion and measurements of fugacity see [17], [62]. It can be shown (see [62]) that the ratio $\frac{f}{p} > 1$ when $\frac{Z-1}{p} > 0$ for all pressures up

to p, otherwise $\frac{f}{p} < 1$ when $\frac{Z-1}{p} < 0$. Z is the compressibility factor, as defined in Chapter 4.

6.7.3. Ideal Gas Mixture

Now we add another arbitrary ideal gas to the original one to make a mixture at pressure p and temperature T. We would like to calculate the chemical potential of our primary gas, say gas A, now as a component of our binary mixture. The partial pressure of gas A is $p_A = x_A p$ is (see Equation 6.4). After substituting these values in Equation 6.36, we have

$$\mu_A = \mu_{0,A} + RT \ln\left(\frac{x_A p}{p_0}\right) = \mu_{0,A} + RT \ln\left(\frac{p}{p_0}\right) + RT \ln x_A \qquad (6.39)$$

But we know that $x_A < 1$ (see Equation 6.1), hence the last term on the R.H.S. of Equation 6.39 is negative (since $\ln x_A < 0$). Therefore, the chemical potential of gas A in the mixture, as given by Equation 6.39, is smaller than that of when gas A is pure (i.e., not mixed with another gas) and kept at the total pressure (and temperature) of the mixture. Note that Equation 6.39 holds for any component in a multi-component mixture of ideal gases, as well for

$$\mu_i = \mu_{0,i} + RT \ln\left(\frac{p}{p_0}\right) + RT \ln x_i \qquad (6.40)$$

where $\mu_{0,i}$ is the chemical potential of component i at pressure p_0 ([23]).

6.8. Gibbs Energy and Entropy Changes of an Ideal Gas Mixture

Changes in Gibbs energy and entropy functions are among the desired quantities in thermodynamics. As we will discuss in the next section, if changes in Gibbs energy is negative for a mixture of gases, then mixing and/or a chemical reaction would be spontaneous. In this section, we calculate these changes for a multi-component mixture. To facilitate the derivation, we transform the reference to the standard pressure (and temperature, see previous section) and define $G' = G - G_0$. Therefore, for $G = G_0$ we have $G' = 0$. Now we can write Equation 6.37 as:

$$G'(T,p)\big|_{pure} = nRT\ln\left(\frac{p}{p_0}\right) \tag{6.41}$$

Recall that this is the Gibbs energy for a pure single component gas. For the case when a gas is a component of a mixture, say component i, the Gibbs energy is then written, using Equation 6.41, as

$$G_i'(T,p)\big|_{mix} = n_i RT\ln\left(\frac{x_i p}{p_0}\right) \tag{6.42}$$

The change for Gibbs energy (i.e., Gibbs energy of the component i in the mixture minus Gibbs free energy of the same component when not in the mixture, as a pure substance) is defined as $\Delta G_i = G_i'(T,p)\big|_{mix} - G'(T,p)\big|_{pure}$. Hence, we have, using Equation 6.41 and Equation 6.42, $\Delta G_i = n_i RT\ln\left(\frac{x_i p}{p_0}\right) - n_i RT\ln\left(\frac{p}{p_0}\right) = n_i RT\ln(x_i)$. The change in Gibbs energy for a multi-component mixture is the sum of those of its components, or $\Delta G_{mix} = \sum_i \Delta G_i = RT\sum_i n_i \ln(x_i)$. Multiplying the R.H.S. by $\frac{n}{n}$ and using $x_i = \frac{n_i}{n}$ we get our final equation for change in Gibbs free energy for an ideal gas mixture as

$$\Delta G_{mix} = nRT\sum_i x_i \ln(x_i) \tag{6.43}$$

Note that for spontaneous change (in a non-reactive or reactive mixture) to happen, Equation 6.43 gives $\Delta G_{mix} < 0$, as expected. Also, it is noteworthy that ΔG_{mix} is proportional to temperature but independent of pressure. Therefore, two (or more) ideal gases at temperature T and pressure p mix spontaneously and the Gibbs energy change is independent of pressure.

Now we calculate the entropy change for the same mixture mentioned above. We use the relation $G = H - TS$, therefore for a mixture the changes are $\Delta G = \Delta H - T\Delta S$. But $\Delta H = 0$ for the ideal gas mixture at constant temperature, so hence we have $\Delta G = -T\Delta S$. Therefore, using Equation 6.43, we can write the change in entropy for an ideal gas mixture as

$$\Delta S_{mix} = nR \sum_i x_i \ln\left(\frac{1}{x_i}\right) \tag{6.44}$$

Note that for spontaneous change (in a non-reactive or reactive mixture) to happen, Equation 6.44 gives $\Delta S_{mix} > 0$, as expected. Other thermodynamic functions of an ideal gas mixture are zero. For example, $\Delta H_{mix} = 0$, $\Delta V_{mix} = 0$, $\Delta U_{mix} = 0$, for given initial T and p as constants.

6.9. Chemical Reactions Equilibrium

Combustion phenomena have many applications in engineering and industry for power production, transportation, heavy equipment engines, and so forth. A reactive mixture of substances is the focus of this section. A reaction, or series of reactions, may occur in a mixture when certain conditions are satisfied. For example, the right temperature, pressure, and amount of reactants are required for the possible direction of the reactions toward equilibrium. When a reaction happens, conservation of mass and energy apply. That is, all atoms in the reactants and those produced by the reaction should be conserved. Both the 1st and 2nd laws apply to a reaction; that is, the 1st law provides us a tool to calculate the energy (e.g., enthalpy, thermal, and work) gained or lost during a reaction, and the 2nd law provides us a means of calculating the reaction entropy and to know, in advance, its possible direction. One may imagine a reaction to be analogous to a "chemical machine" that would crank and use the fuels to produce some products if all required conditions are fulfilled. In the remainder of this section we discuss reaction equilibrium, combustion reaction mass balance, air-fuel ratio, the heat of reaction, and adiabatic flame temperature, among other topics.

In Chapter 4, we discussed the topic of thermodynamic equilibrium. In summary, for convenience of referencing, for a mixture to be in equilibrium, we have its energy decreasing and tending toward a minimum value and its entropy increasing toward a maximum value. However, under different thermodynamic conditions certain thermodynamic functions describe the process more suitably than others. General equilibrium can be expressed in the following equation (see Chapter 4)

$$TdS - dU - pdV \geq 0 \tag{6.45}$$

Now let's examine special cases; for example, if reactions are happening in an insulated closed system with $dU = dV = 0$, then $dS \geq 0$. Similarly, in the case of having a mixture reaction under constant volume and temperature, then Helmholtz energy is the suitable function to consider. This becomes clear by using the relation $A = U - TS$ which gives $dA = dU - TdS - SdT$. After substituting for dU from Equation 6.45, we get

$$dA + SdT + pdV \leq 0 \tag{6.46}$$

and for $dT = dV = 0$, we get $dA \leq 0$.

In processes involving chemical reactions, we usually deal with mixtures under constant temperature and pressure. For these conditions the Gibbs energy function would be used. To see this, we use the relation $G = U - TS + pV$ which gives $dG = dU - TdS - SdT + pdV + Vdp$. After substituting for dU from Equation 6.45, we get

$$dG + SdT - Vdp \leq 0 \tag{6.47}$$

and for or $dT = dp = 0$, we get $dG \leq 0$.

These equations (i.e., Equation 6.45-Equation 6.47) provide us the relations for calculations of thermodynamic properties of reactions in mixtures under the conditions specified.

6.10. Chemical Equation for Combustion

In a chemical reaction, *reactants* or fuels chemically and thermally interact and new substances or *products* are produced. If the reaction also delivers heat, then it is called *exothermic*, otherwise it is called *endothermic*, that is, the reaction absorbs heat from the surroundings. For a combustion reaction, fuels could be in solid phase (e.g., wood, coal), liquid phase (e.g., petroleum), or gas phase (e.g., natural gas). The fuels are oxidized (i.e., atoms lose electrons) by an oxidizer, which is usually oxygen present in the atmospheric air, and equivalently the oxidizer is reduced (i.e., atoms gain electrons). Symbolically, a reaction is written as:

$$A + B = C + D$$

the equal sign indicates the balance of the masses involved in the reaction. When $A + B$ react to produce $C + D$, then we can write $\underbrace{A + B}_{reactants} \rightarrow \underbrace{C + D}_{products}$. Otherwise, we have $\underbrace{A + B}_{products} \leftarrow \underbrace{C + D}_{reactants}$. The direction of the reaction should satisfy the reaction's entropy increase and the Gibbs energy decrease. Combustion calculations are based on Avogadro's law, or "At the same pressure and temperature, equal volumes of all perfect gases contain the same number of molecules"[107]. Therefore, the mass of equal volumes of gases are proportional to the molecular weight (or atomic mass number). Table 6.1 shows typical materials' molecular weight.

H_2	O_2	N_2	C	S	H_2O	CO_2	SO_2	CH_4
2	32	28	12	32	18	44	64	16

TABLE 6.1. Atomic Mass (Rounded) for Common Substances

Using Avogadro's law, if we get 2g of H_2, then we have 1 mole of hydrogen. Also, 2g of H_2 and 12g of carbon contain the same number of molecules, equal to Avogadro's number ($\sim 6 \times 10^{23}$).

For combustion calculations, we use volumetric measure of non-solid reactants and mass-based (gravimetric) measure for solid ones. For example, the equation for the combustion of hydrogen is

$$2H_2 + O_2 = 2H_2O$$

In this reaction, hydrogen is oxidized and oxygen is reduced. In terms of the number of moles (or volumes), we have $2\,moles\,H_2 + 1\,mole\,O_2 = 2\,moles\,H_2O$. Or $\left(2 \times \dfrac{2g}{gmole}\right)H_2 + \left(1 \times \dfrac{32g}{gmole}\right)O_2 = \left(2 \times \dfrac{18g}{gmole}\right)H_2O$. Therefore, the number of moles do not necessarily balance out, but the mass of the substances' atoms/molecules should balance (i.e., 1 kg of hydrogen reacting with 8 kg of oxygen produces 9 kg of water, Since substances have different atomic masses or molecular weights).

For practical applications, pure oxygen is not normally used; rather, air is the oxidizer. If we consider the major components of a natural air mixture, that is, 21% oxygen and 79% nitrogen, by volume measure, then for 1 mole of oxygen we have $\dfrac{79}{21} = 3.76$ moles of nitrogen, or

$O_2 + 3.76 N_2 = 4.76$ moles of air. Using this relation, we can write the combustion of hydrogen when oxidized by air as

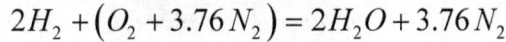

$$2H_2 + \left(O_2 + 3.76 N_2\right) = 2H_2O + 3.76 N_2$$

For combustion, we consider the number of moles of the reactants and products per unit mole of the fuel. Therefore

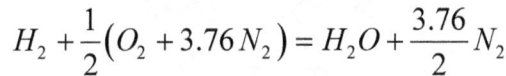

$$H_2 + \frac{1}{2}\left(O_2 + 3.76 N_2\right) = H_2O + \frac{3.76}{2} N_2$$

More complex combustion reactions involve hydrocarbons as fuels. In further sections we present worked out examples of such reactions.

6.10.1. Complete Combustion Reaction and Air-Fuel Ratio

A combustion reaction is complete when all fuel constituents are oxidized; for example, C, H_2, and S_2 oxidize to the products CO_2, H_2O, and SO_2, respectively. However, for having a complete combustion several conditions should be fulfilled, including enough air for burning the fuel. This is usually measured by the ratio of air mass with respect to fuel mass and is called the *Air-Fuel Ratio* (AFR). The minimum amount of air mass required is calculated based on the related combustion reaction equation, and the AFR is then referred to as the *stoichiometric* AFR. However, in practice extra air is supplied to make sure that the combustion reaction is complete. This extra air is referred to as *excess air* and is expressed as a percentage of stoichiometric air. For example, 50% excess air means that 1.5 times stoichiometric air is supplied to the reaction.

Example: AFR Calculations for Common Hydrocarbons

Calculate the AFRs for combustion of Methane (CH_4), Octane (C_8H_{18}), and Butane (C_4H_{10}). Express the results as molar-based and mass-based for each AFR.

■ **Solution**

For methane, we have

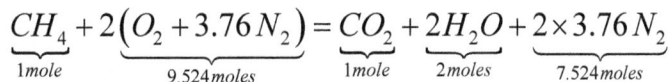

$$\underbrace{CH_4}_{1\,mole} + \underbrace{2\left(O_2 + 3.76 N_2\right)}_{9.524\,moles} = \underbrace{CO_2}_{1\,mole} + \underbrace{2H_2O}_{2\,moles} + \underbrace{2 \times 3.76 N_2}_{7.524\,moles}$$

therefore, the AFR (molar) is $\dfrac{9.524}{1} = 9.524$. For the mass-based AFR we use molecular weights of substances: $M_{w-air} = 28.97\,g\,/\,gmole$ and $M_{w-CH4} = 16\,g\,/\,gmole$. Therefore, the AFR (mass-based) is $9.524 \times \dfrac{28.97}{16} = 17.2$ grams (or kg) of air per 1 gram (or kg) of methane.

■ **Solution**

For Octane, we have

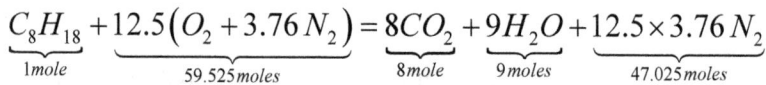

$$\underbrace{C_8 H_{18}}_{1\,mole} + \underbrace{12.5\left(O_2 + 3.76\,N_2\right)}_{59.525\,moles} = \underbrace{8CO_2}_{8\,mole} + \underbrace{9H_2O}_{9\,moles} + \underbrace{12.5 \times 3.76\,N_2}_{47.025\,moles}$$

therefore, the AFR (molar) is $\dfrac{59.525}{1} = 59.525$. For the mass-based AFR we use molecular weights of substances: $M_{w-air} = 28.97\,g\,/\,gmole$ and $M_{w-C8H18} = 114\,g\,/\,gmole$. Therefore, the AFR (mass-based) is $59.525 \times \dfrac{28.97}{114} = 15.12$ grams (or kg) of air per 1 gram (or kg) of octane.

■ **Solution**

For Butane, we have

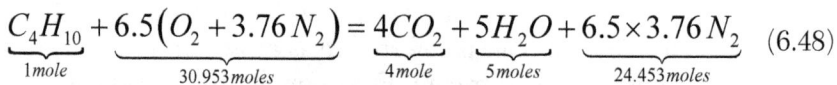

$$\underbrace{C_4 H_{10}}_{1\,mole} + \underbrace{6.5\left(O_2 + 3.76\,N_2\right)}_{30.953\,moles} = \underbrace{4CO_2}_{4\,mole} + \underbrace{5H_2O}_{5\,moles} + \underbrace{6.5 \times 3.76\,N_2}_{24.453\,moles} \quad (6.48)$$

therefore, the AFR (molar) is $\dfrac{30.953}{1} = 30.953$. For the mass-based AFR we use molecular weights of substances: $M_{w-air} = 28.97\,g\,/\,gmole$ and $M_{w-C4H10} = 58\,g\,/\,gmole$. Therefore, the AFR (mass-based) is $30.953 \times \dfrac{28.97}{58} = 15.454$ grams (or kg) of air per 1 gram (or kg) of butane.

6.10.2. Heat of Combustion

For practical applications, we are interested in the amount of energy, mostly in the form of heat, generated by a combustion reaction: for example, in a boiler furnace, external combustion engine, or thermal engine, we would like to know how much heat is available to/from the equipment.

So far, we have analyzed a chemical reaction from its equilibrium point of view. Now we extend the analysis to calculate the amount of heat generated by using the 1st law and the possibility of the reaction to happen using the 2nd law. In order to express the related topics, we use an example with its relevant extensions, during the discussion, after establishing some definitions.

For a reaction the fuel constituents can be at a certain pressure and temperature. This will determine the state that the fuel is at before the combustion. The products of the combustion may also be in the same thermodynamic state as that of the fuel or at a different state. We discussed previously that determination of the absolute value of energy is not feasible and we usually measure or calculate its changes. Therefore, we need to have a reference for our calculations related to energy. For combustion, enthalpy is the quantity of interest rather than internal energy. One reason could be stated as that when enthalpy is available and set to zero at the reference point, then we can also calculate internal energy, knowing or measuring the pressure and volume change (recall $U = H - pV$).

The standard reference state (STP) is set at 25 °C (77 °F, 298.15K) and 1 atm (14.6 psia, 101325 Pa). By definition, we set the enthalpy of elemental substances, which are stable at their equilibrium at the reference condition, to be zero. For example, enthalpies of liquid mercury, oxygen, and nitrogen are set to zero at 25 °C and 1 atm.

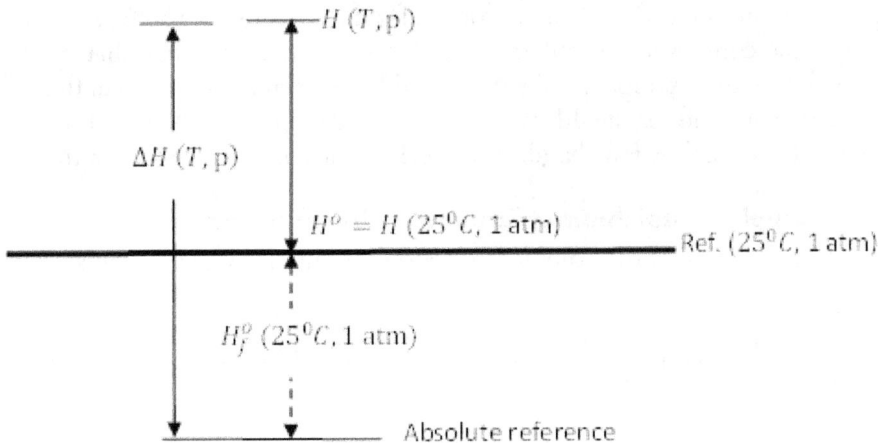

FIGURE 6.5. Definition for enthalpy, energy potentials, calculations.

6.10.3. Enthalpy of Formation

During a combustion reaction several compounds are produced; the fuel may also be a compound of elemental substances, for example hydrocarbons (CH_4, C_2H_{10}, C_8H_{18}, etc.). Enthalpy of formation, H_{form} is defined as the difference between the enthalpy of the compound, H_{comp} and that of its constituents, measured at the standard condition. Mathematically, we can write this definition as(see Figure 6.5)

$$\Delta H = \left(H - H^0 \right) + H_f^0$$

For example, the enthalpy of formation of hydrogen is zero, by definition, and that of CO_2 is $-393.492\,kJ\,/\,gmole$. Usually, the enthalpy of formation at STP and different values of temperatures are tabulated ([19]). Figure 6.5, shows the concept and the definitions explained previously.

6.11. 1st Law Application to Combustion

We use combustion of a common hydrocarbon, like butane, as an example for demonstrating combustion-related calculations for application of the 1st law and adiabatic flame temperature.

We consider fueling a combustion chamber with butane, C_4H_{10} and dry air. The products of the combustion and the reactants can be at any given thermodynamic state, in principle. For the first example, we consider complete combustion with stoichiometric air, for the 2nd example 50% excess air enters the combustion chamber at a different temperature than that of the fuel, for the 3rd example products are at different temperatures than that of the reactants, and for the 4th example we calculate product temperature for an adiabatic combustion chamber, a so-called adiabatic flame temperature.

6.11.1. Example 1: Stoichiometric Combustion of Butane

The equation of the complete reaction for butane (C_4H_{10}) reads

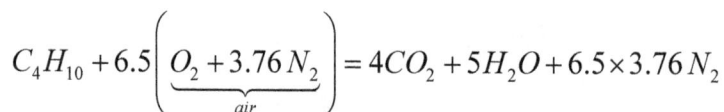

$$C_4H_{10} + 6.5\left(\underbrace{O_2 + 3.76\,N_2}_{air} \right) = 4CO_2 + 5H_2O + 6.5 \times 3.76\,N_2$$

We assume that water produced is in liquid form, and therefore generated heat is considered to be the high-heating value. The sketch of the combustion chamber is shown in Figure 6.6, with H_R as the enthalpy of the reactants and H_P as enthalpy of the products, and Q as heat produced (or $-Q$ as absorbed heat).

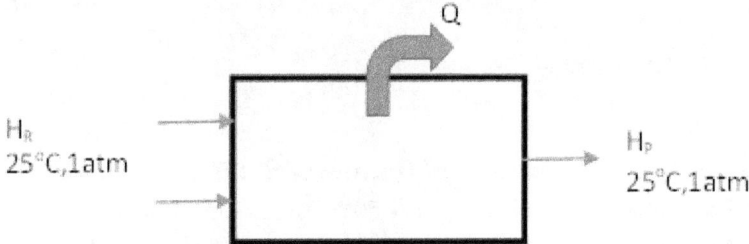

FIGURE 6.6. Sketch of the combustion for butane at STP, Example 1.

By applying the 1st law to the closed system of the chamber, we have $Q = H_R - H_P = \left(1 \times h_{C4H10} + 6.5 \times h_{O2} + 6.5 \times 3.76 h_{N2}\right) - \left(4 \times h_{CO2} + 5 \times h_{H2O} + 24.44 \times h_{N2}\right)$. The molar enthalpy of each substance is $\Delta h_i = \left(h_i - h_i^0\right) + h_f^0$. We make a mini-table (see Table 6.2) containing the data required for the calculations. Since both reactants and products are at standard conditions (i.e., 25 °C, 1atm), we have $\left(h_i - h_i^0\right) = 0$. Also $h_f^0 = 0$ for oxygen and nitrogen, as elemental substances, by definition. The values of h_f^0 are read from the properties table (e.g., see [19]).

Substance	C_4H_{10}	O2	N2	CO2	H2O(l)	H2O(g)
h_f^0 (kJ/kmol)	−126150	0	0	−393520	−285830	−241820

TABLE 6.2. Data for Stoichiometric Combustion of Butane, Example 1

For products, we have $H_P = \left(4 h_f^0\right)_{CO2} + \left(5 h_f^0\right)_{H2O} = -4 \times 393520 - 5 \times 285830 = -3003230$ kJ/kmol of fuel (C_4H_{10}). Similarly for reactants, we have $H_R = \left(h_f^0\right)_{C4H10} + \left(6.5 \underbrace{h_f^0}_{=0}\right)_{O2} + \left(24.44 \underbrace{h_f^0}_{=0}\right)_{N2} = -126150$.

Applying the 1st law, we get $Q = -126150 - (-3003230) = 2877080$ kJ/kmol of C_4H_{10}. This is the amount of heat generated by burning butane using air, which is equivalent to $(2877080 \, kJ \, / \, kmol) \left(\dfrac{1 \, kmol}{58.123 \, kg} \right) =$

$49500 \, kJ \, / \, kg$ of butane.

6.11.2. Example 2: Combustion of Butane with 50% Excess Air

We repeat Example 1, but with 50% excess air as oxidizer. Air enters the combustion chamber at 7 °C. All reactions are at a constant pressure of 1 atm, as shown in Figure 6.7. The equation of the reaction for butane (C_4H_{10}) with 50% excess air reads

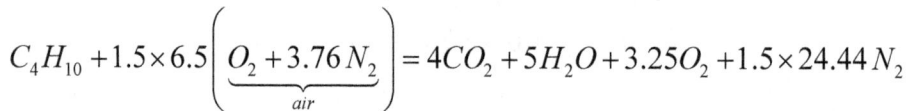

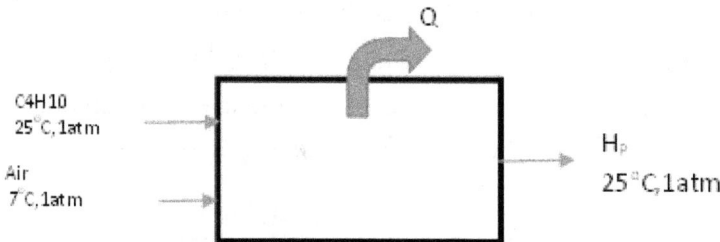

$$C_4H_{10} + 1.5 \times 6.5 \underbrace{\left(O_2 + 3.76 \, N_2 \right)}_{air} = 4CO_2 + 5H_2O + 3.25O_2 + 1.5 \times 24.44 \, N_2$$

FIGURE 6.7. Sketch of the Combustion for Butane, Example 2.

We make a mini-table (see Table 6.3) containing the data required for the calculations; all enthalpies are in kJ/kmol.

For products, we have the same as in the previous example: $H_P = -3003230$ kJ/kmol of fuel (C_4H_{10}).

For reactants, we have $H_R = \left(h_f^0 \right)_{C4H10} + 1.5 \times 6.5 \left(h \big|_{7°C} - h^0 + {}_{f}^{0} \right)$ $h_f^0 \Big)_{O2} + 1.5 \times 6.5 \times 3.76 \left(h \big|_{7°C} - h^0 + h_f^0 \right)_{N2} = -150693.5$ kJ/kmol of C_4H_{10}. Applying the 1st law, we get $Q = -150693.5 - (-3003230) =$

2852536.5 kJ/kmol of C_4H_{10}. This is the amount of heat generated by burning butane using air, which is equivalent to

$$\left(2852536.5\,kJ\,/\,kmol\right)\left(\frac{1\,kmol}{58.123\,kg}\right) = 49078\,kJ\,/\,kg \text{ of butane.}$$

It is clear from these results that excess air cools down the reaction and hence the amount of heat generated by the combustion is lower than that in Example 1.

Substance	$h^0 = h\big|_{25^0 C}$	$h\big|_{7^0 C}$	h_f^0
C_4H_{10}	–	–	–126150
O2	8682	8150	0
N2	8669	8141	0
CO2	–	–	–393520
H20	–	–	–285830

TABLE 6.3. Data for Example 2

6.11.3. Example 3: Combustion of Butane with Products at 1000 K

We repeat Example 2, but consider the products to exit at a temperature of 1000 K. The combustion reaction is exactly similar to that given in Example 2, with the water as vapor in the product mixture. The molar enthalpies of the substances are given in Table 6.4.

Substance	$h^0 = h\big|_{25^0 C}$	$h\big|_{7^0 C}$	$h\big|_{1000 K}$	h_f^0
C_4H_{10}	–	–		–126150
O2	8682	8150	31389	0
N2	8669	8141	30129	0
CO2	9364	–	42769	–393520
H20	9904	–	35882	–241820

TABLE 6.4. Data for Example 3

For products, we have $H_P = 4\left(h\big|_{1000K} - h^0 + h_f^0\right)_{CO2} + 5\left(h\big|_{1000K} - h^0 + h_f^0\right)_{H2O} + 3.25\left(h\big|_{1000K} - h^0 + h_f^0\right)_{O2} + 36.66\left(h\big|_{1000K} - h^0 + h_f^0\right)_{N2} = -1659149$ kJ/kmol of fuel (C_4H_{10}).

For reactants, we have the same as the previous example (i.e. Example 2) $H_R = -150693.5$ kJ/kmol of C_4H_{10}. Applying the 1st law, we get $Q = -150693.5 - (-1659149) = 1508456$ kJ/kmol of C_4H_{10}. This is the amount of heat generated by burning butane using air at the conditions given, which is equivalent to $(1508456\,kJ/kmol)\left(\dfrac{1\,kmol}{58.123\,kg}\right) = 25953\,kJ/kg$ of butane.

6.11.4. Example 4: Adiabatic Flame Temperature of Butane

We extend our butane combustion example by changing the burner condition by adding thermal insulation to it. Therefore, the generated heat will not be lost to the environment nor gained from the surroundings, as sketched in Figure 6.8. This will affect the products' temperature, or so-called *adiabatic flame temperature*, T_{ad}.

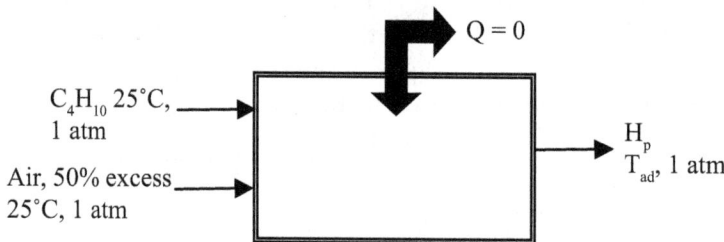

FIGURE 6.8. Sketch of combustion of butane inside an insulated burner at STP, Example 4.

The combustion reaction is similar to that given in Example 2.

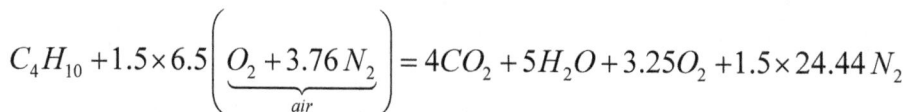

$$C_4H_{10} + 1.5 \times 6.5\left(\underbrace{O_2 + 3.76\,N_2}_{air}\right) = 4CO_2 + 5H_2O + 3.25O_2 + 1.5 \times 24.44\,N_2$$

Using the molar enthalpies from the mini-table in Example 2, we can write for the reactants, $H_R = -126150$ kJ/kmol of fuel ($C_4 H_{10}$).

For products, we have $H_P = 4\left(h\big|_{CO2} - 9364 - 393520\right) + 5\left(h\big|_{H2O} - \right.$
$9904 - 241820\left.\right) + 3.25\left(h\big|_{O2} - 8682 + 0\right) + 36.66\left(h\big|_{N2} - 8689 + 0\right)$. Note that since the temperature at the outlet is not given, we cannot read the enthalpies $h\big|_{CO2}$ and $h\big|_{H2O}$ explicitly from properties tables. After applying the 1st law and rearranging the terms, we get

$$4h\big|_{CO2} + 5h\big|_{H2O} + 3.25h\big|_{O2} + 36.66h\big|_{N2} = 3090028 \text{ kJ / kmol of } C_4 H_{10}$$

This equation should be satisfied knowing the enthalpies of the products at the exit temperature, T_{ad}. Assuming an ideal gas (justified with having the products at high temperature), we can calculate the enthalpies using specific heats (either as constants or polynomial functions of temperature), or their tabulated values (e.g., [19]). The exact result is when tabulated enthalpies are used with a trial-error approach. Following this approach, we use the average value of the molar enthalpy, that is, $\dfrac{3090028}{(4 + 5 + 3.25 + 36.66)} = 63178$

kJ/kmol. This corresponds to the enthalpy of nitrogen at about 1940 K. We use nitrogen for our estimate, since we have a relatively large amount of it in the product. Therefore, we use a lower value for our first-guessed temperature, $\tilde{T}_{ad} = 1840$ K. Using data for substances in the products, at 1840 K,

we get $\left(4h\big|_{CO2} + 5h\big|_{H2O} + 3.25h\big|_{O2} + 36.66h\big|_{N2}\right)_{at\ 1840\ K} = 4(91196) +$
$5(74506) + 3.25(61866) + 36.66(59075) = 3104068$ kJ/kmol. This is larger than 3090028 kJ/kmol, therefore we should reduce the temperature, and hence we take our second guess to be $\tilde{T}_{ad} = 1820$ K. This

gives $\left(4h\big|_{CO2} + 5h\big|_{H2O} + 3.25h\big|_{O2} + 36.66h\big|_{N2}\right)_{at\ 1820\ K} = 4(90000) +$
$5(73507) + 3.25(61118) + 36.66(58363) = 3065756$ kJ/kmol. This is smaller than 3090028 kJ/kmol, therefore we accept the value of $T_{ad} = 1835$ K. Readers may want to continue the trial-error calculations to arrive at a more exact temperature.

In this example, we used 50% excess air, which has a cooling effect on the products' temperature. If stoichiometric air is used, the maximum adiabatic temperature can be calculated using a similar approach, which is 2243 K.

In real-world practical applications, the combustion chamber may not be fully insulated neither the chamber material is ideally thermally conductive. Therefore T_{ad} could be useful for material selection, since it is the maximum possible temperature as a result of combustion. The actual flame temperature would be a value between ambient and T_{ad}. Several other factors, however, will affect the combustion reaction and consequently the product temperature: for example, non-standard reactants' condition, excess air, incomplete combustion reaction, and turbulence. Table 6.5 shows the adiabatic flame temperature for some common gaseous fuels. This table shows that hydrocarbons have a narrow range of adiabatic flame temperature.

Fuel	Oxidizer	
	O_2	Air
H_2 (Hydrogen)	3079	2384
CH_4 (Methane)	3054	2227
C_3H_6 (Propene)	3095	2268
C_8H_{18} (Octane)	3108	2277
C_4H_{10} (Butane)	–	2246

TABLE 6.5. Adiabatic Flame Temperature (K) for Some Fuels

6.11.5. Example 5: Adiabatic Flame Temperature of Butane with 50% Excess Moist Air

Extending on Example 4, and since air usually contains water vapor, in this example we consider the effect of moisture on the butane combustion with air entering the combustion chamber with relative humidity, $\varphi = 80\%$.

All reactions are at a constant pressure of 1 atm, as shown in Figure 6.9. The equation of the reaction for butane (C_4H_{10}) with 50% excess dry air reads

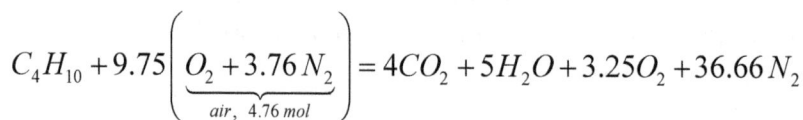

$$C_4H_{10} + 9.75 \underbrace{\left(O_2 + 3.76 N_2 \right)}_{air, \ 4.76 \, mol} = 4CO_2 + 5H_2O + 3.25O_2 + 36.66 N_2$$

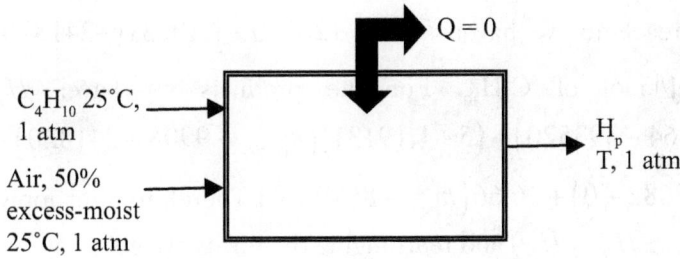

FIGURE 6.9. Sketch of combustion of butane inside insulated burner at STP, Example 5.

To include the water vapor contents of the air, we calculate the number of moles of vapor entering the combustion with the air. Assuming the ideal gas law, the ratio of the water vapor, n_v over the total number, n_{air} moles of moist air (i.e., dry air plus water vapor content) is proportional to the partial pressure, p_v over total pressure, or $n_v = n_{air}\left(\dfrac{p_v}{p_{atm}}\right)$. But partial pressure of the vapor is proportional to water saturation pressure at 25 °C, or $p_v = \varphi\, p_{sat}|_{25°C} = (0.8)(3.1698\,kPa) = 2.53584\,kPa$. The total number of moles of dry air is $9.75 \times 4.76 = 46.41\,kmol$. Therefore, $n_{air} = 46.41 + n_v$. After substitution, we get $n_v = (46.41 + n_v)\left(\dfrac{2.53584}{101.325}\right)$, or $n_v = 1.19131$ kmol. This is the number of moles of water vapor entering the combustion as the moisture content of the air and should be added to the reactants, and equivalently to the products in the reaction equation. Hence, the combustion reactions for moist air reads

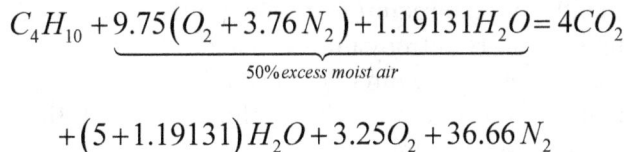

$$C_4H_{10} + \underbrace{9.75(O_2 + 3.76\,N_2) + 1.19131 H_2O}_{50\%\,excess\,moist\,air} = 4CO_2$$

$$+ (5 + 1.19131) H_2O + 3.25 O_2 + 36.66\,N_2$$

For the reactants we have: $H_R = -126150 + 1.19131(-241820) =$ -414233 kJ/kmol of C_4H_{10}. For the products we have: $H_P =$ $4\left(h\big|_{CO2} - 9364 - 393520\right) + (5 + 1.19131)\left(h\big|_{H2O} - 9904 - 241820\right) +$ $3.25\left(h\big|_{O2} - 8682 + 0\right) + 36.66\left(h\big|_{N2} - 8689 + 0\right)$. Therefore, after applying the 1st law (i.e., $H_R = H_P$) and rearranging the terms, we get

$$4h\big|_{CO2} + 6.19131 h\big|_{H2O} + 3.25 h\big|_{O2} + 36.66 h\big|_{N2} = 3101826 \text{ kJ} / \text{kmol of } C_4H_{10}$$

Similar to the methodology mentioned in Example 4, we use the trial-error method to calculate the adiabatic flame temperature. We use $\tilde{T}_{ad} = 1820$ K as our first guess. Therefore, we get $4h\big|_{CO2} + 6.19131 h\big|_{H2O} + 3.25 h\big|_{O2} + 36.66 h\big|_{N2} = 4(90000) +$ $6.19131(73507) + 3.25(61118) + 36.66(58363) = 3153326 > 3101826$.

Thus, we lower the temperature and use $\tilde{T}_{ad} = 1760$, which gives the value of $3035026 < 3101826$ kJ/kmol. The average value, or $T_{ad} = \dfrac{1820 + 1760}{2} = 1790\,K$, gives the result of 3094130. Therefore, we accept $\tilde{T}_{ad} = 1795$ K as our final answer. In comparison to the result obtained in Example 4, this example shows that the moisture content of the air acts as a coolant and decreases the adiabatic flame temperature.

6.12. 2nd Law Application to Combustion

As mentioned, the requirement for spontaneous change for a reaction is the increase in its entropy until it reaches the equilibrium state. For a reaction to happen in the direction desired or assumed, the entropy of the products should be larger than that of the reactants. If the entropies of the products and reactants are equal, then the reaction is reversible and can happen in both directions(see following sketch). Similarly, we can calculate the total chemical potentials associated with both sides (i.e., reactants and products) and identify the reaction direction from higher to lower values of chemical potentials[101].

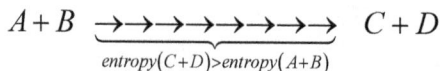

$$A + B \underbrace{\rightarrow\rightarrow\rightarrow\rightarrow\rightarrow\rightarrow\rightarrow}_{entropy(C+D)>entropy(A+B)} C + D$$

$$A + B \quad \longleftarrow\longleftarrow\longleftarrow\longleftarrow\longleftarrow\longleftarrow \quad C + D$$
$$\underbrace{\qquad\qquad\qquad\qquad\qquad}_{entropy(A+B)>entropy(C+D)}$$

$$A + B \underbrace{\rightleftharpoons\rightleftharpoons\rightleftharpoons\rightleftharpoons\rightleftharpoons\rightleftharpoons\rightleftharpoons\rightleftharpoons\rightleftharpoons\rightleftharpoons}_{entropy(C+D)=entropy(A+B)} C + D$$

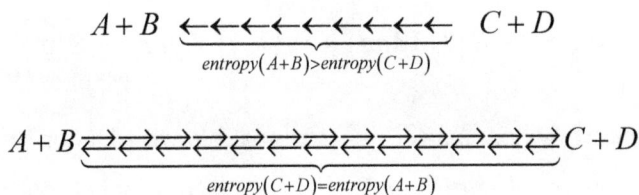

In the previous examples, we assumed the reaction would be possible and that products are created from associated reactants. We used the 1st law for energy/enthalpy calculations for the reactions considered. In this section we calculate the entropies involved in chemical reactions related to each substance. There are two aspects when dealing with entropy calculation related to combustion, as compared to that of enthalpy calculations. First, we can calculate a substance's absolute entropy value, according to the 3rd law, and second, the pressure under which the reaction happens and partial pressure when we have mixture of gases need to be included in the calculations, since entropy is a function of temperature and pressure for all gases, including ideal gases. These requirements make entropy calculations of a combustion more complex than those of enthalpy calculations.

As mentioned in Chapter 2, for an ideal gas the change in entropy (per unit mole) between two state points, like (T_1, p_1) and (T_2, p_2), is given by

$$\Delta S = S(T_2, p_2) - S(T_1, p_1) = S_0(T_2) - S_0(T_1) - R \ln \frac{p_2}{p_1} \qquad (6.49)$$

where $S_0(T_i) = \int_0^{T_i} C_p \frac{dT}{T}$ is absolute entropy (defined using the 3rd law) and the values for a wide range of temperatures at atmospheric pressure are tabulated (e.g., see [19]). For combustion reactions, we have to use the partial pressure of each component in a gas mixture, for example air as an oxidizer. As presented in the previous examples, usually combustion reactants and products are at atmospheric pressure when entering and/or exiting the combustion chamber, or $p_1 = 1\,atm$. In addition to the assumption of an ideal gas, the partial pressure p_i of each component is a fraction of mixture pressure, p_m or $p_2 = p_i = x_i p_m$ (see Equation 6.3). After substituting for p_1 and p_2 into Equation 6.49 we get

$$S(T_2, p_i) - S(T_1, 1atm) = S_0(T_2) - S_0(T_1) - R \ln \frac{x_i p_m}{1 atm} \qquad (6.50)$$

For cases when $T_1 = T_2 = T$, compatible with Dalton's assumption for mixtures, we get (note $S_0(T) - S_0(T) = 0$ and cancels out):

$$S(T, p_i) = \underbrace{S(T, 1atm)}_{S_0(T)} - R \ln \frac{x_i p_m}{1 atm} \qquad (6.51)$$

Equation 6.51 is useful for calculating the entropy of component i of a mixture at its corresponding partial pressure. To demonstrate its application, a couple of worked out examples are presented.

6.12.1. Example 6: Entropy of Air at Standard Condition

Air at standard condition (i.e., $T = 25^0 C$ and $p_m = 1 atm$) is given. Assuming that the mixture contains 3 kmol of oxygen and 3×3.76 kmol of nitrogen, calculate its entropy.

■ **Solution**

Air mixture $= 3(O_2 + 3.76 N_2)$ or 3×4.76 kmol of air. The mole fractions are $x_{O2} = \dfrac{3}{3 \times 4.76} = 0.21$ and $x_{N2} = \dfrac{3 \times 3.76}{3 \times 4.76} = 0.79$.

From the property table we read: $S_0(25^0 C)\big|_{O2} = 205.04$ kJ/kmol.K and $S_0(25^0 C)\big|_{N2} = 191.61$ kJ/kmol.K. Substituting into Equation 6.51, we get $S\big|_{O2} = 205.04 - 8.314 \ln \dfrac{(0.21 \times 1)}{1} = 218.02$ kJ/kmol.K,

and $S\big|_{N2} = 191.61 - 8.314 \ln \dfrac{(0.79 \times 1)}{1} = 193.57$ kJ/kmol.K. Entropy of the air is then the sum of total entropies, or $S(25^0 C, 1atm)\big|_{Air}$

$$= (3 \times 218.02) + (11.28 \times 193.57) = 2837.53 \text{ kJ/K, or the specific entropy}$$

of $\dfrac{2837.53}{28.97 \times 3 \times 4.76} = 6.86 \text{ kJ/kg.K.}$

6.12.2. Example 7: Entropy of a Multi-Component Gas Mixture

Combustion of a hydrocarbon produces a mixture of $CO2(10.02\%)$, $O2(5.62\%)$, $CO(0.88\%)$, and $N2(83.48\%)$ by volume concentration. Calculate the entropy of the mixture at 25 °C and 1 atm, assuming the ideal gas law applies.

■ **Solution**

With the ideal gas assumption, the mole fractions are equal to the volume concentration. Hence, $x_{CO2} = 10.02 \times 10^{-2}$, $x_{O2} = 5.62 \times 10^{-2}$, $x_{CO} = 0.88 \times 10^{-2}$, and $x_{N2} = 83.48 \times 10^{-2}$. Reading absolute entropies for each component or $S_0 \left(25^0 C, 1 atm \right)$ from the property table and using Equation 6.51, we get the entropy of each component in the mixture (in *kJ/ kmol.K*) as:

$$S\big|_{CO2} = 213.8 - 8.314 \ln \left(0.1002 \right) = 232.93$$

$$S\big|_{O2} = 205.04 - 8.314 \ln \left(0.0562 \right) = 228.97$$

$$S\big|_{CO} = 197.65 - 8.314 \ln \left(0.0088 \right) = 237$$

$$S\big|_{N2} = 153.3 - 8.314 \ln \left(0.8348 \right) = 154.8$$

We consider 100 kmol of the mixture, and then its entropy is $S \left(25^0 C, 1 atm \right) \big|_{Air} = (10 \times 232.93 + 5.62 \times 228.97 + 0.88 \times 237 + 83.48 \times 154.8) = 16752 \text{ kJ/K.}$

Now we have all the required tools ready in order to apply the 2nd law to a combustion reaction. Considering a combustion chamber, the entropy of the products, S_P should be larger than that of the reactants, S_R according to the 2nd law (see Chapter 2) for irreversible reactions, or

$S_P - S_R > 0$. The amount of entropy generated S_{Gen}, due to irreversibility, and the entropy due to heat generated $\sum \dfrac{Q}{T}$ as the result of combustion, compensate for the total entropy increase of the combustion reactions. Therefore, we can write

$$S_P - S_R - \left(S_{Gen} + \sum \frac{Q}{T} \right) = 0 \qquad (6.52)$$

For reversible reactions $S_{Gen} = 0$ and for adiabatic combustion $Q = 0$. Note that in Equation 6.52, Q and T should be considered at the same boundary between the system and the environment with the sign convention $Q_{out} < 0$ and $Q_{in} > 0$. Knowing the S_{Gen} for a reaction, we can calculate the amount of lost work, or TS_{Gen} (see Chapter 2), where T is the temperature at the interface of the system and environment. This temperature is the ambient temperature for combustion. In order to demonstrate the application of the relations mentioned above, we consider a couple of worked out examples.

6.12.3. Example 8: Entropy Generation during an Adiabatic Combustion

We extend the calculations presented in Example 4 in order to calculate the entropy generated during the combustion. We calculated the adiabatic flame temperature of butane, with 50% excess air, to be 1835 K. For convenience we rewrite the combustion reaction,

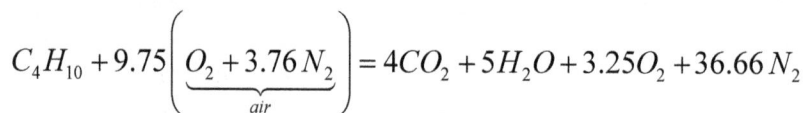

$$C_4 H_{10} + 9.75 \left(\underbrace{O_2 + 3.76 N_2}_{air} \right) = 4 CO_2 + 5 H_2 O + 3.25 O_2 + 36.66 N_2$$

We need the mole fractions of the reactant and product mixtures, along with absolute entropies of each substance. To facilitate and organize our calculations, we build a mini-table (see Table 6.6) of required data using Equation 6.51 and Equation 6.52. Recall that reactants are at 25 °C, but products are at 1835 K, and all reactions are at atmospheric pressure.

Using data given in Table 6.6 and using Equation 6.52 (with $Q = 0$) we get $S_{Gen} = 12843.34 - 9532.1 = 3311.24$ kJ/K per kmol of C_4H_{10}. The amount of lost work is then $3311.24 \times 298 = 986.75$ MJ per kmol of C_4H_{10}. This is the amount of work that could be obtained by the reaction process, but it is lost due to the irreversibility of the combustion reaction.

It would be useful to compare the entropy generation for the same reaction, but when the combustion chamber is not insulated. For the next example, we calculate the enthalpy and the entropy generation.

Substance		n_i (kmol)	$x_i = \dfrac{n_i}{\sum n_i}$	$S_i^o \, (298K)$ (kJ/kmol.K)	$R \ln x_i$ (kJ/kmol.k)	$S_i = S_i^o - R \ln x_i$ (kJ/kmol.k)	$H_i = n_i S_i$ (kJ/k)
Fuel	C_4H_{10}	1	1	310.12	0	310.12	310.12
Air	O_2	9.75	0.21	205.04	$8.314 \ln 0.21 = -12.98$	218.02	2125.7
	N_2	36.66	0.79	191.61	$8.314 \ln 0.79 = -1.96$	193.57	7096.28
				$S_i^o (1835K)$			$\sum_{react.} = 9532.1$
Products	CO_2	4	0.082	304.03	$8.314 \ln 0.082 = -20.8$	324.83	1299.32
	H_2O	5	0.102	260.22	$8.314 \ln 0.102 = -18.98$	279.2	1396
	O_2	3.25	0.066	265.42	$8.314 \ln 0.066 = -22.6$	288.02	936.1
	N_2	36.66	0.75	248.88	$8.314 \ln 0.75 = -2.4$	251.28	9211.92
		$\sum_{prod.} = 48.91$					$\sum_{prod} = 12843.34$

TABLE 6.6. Data for Example 8

6.12.4. Example 9: Entropy Generation of Butane Combustion with 50% Excess

We repeat the problem statement of Example 8, but with no insulation, and products exit the combustion chamber to the ambient standard conditions. Hence heat is generated as the result of combustion, as shown in Figure 6.10. The combustion reaction is similar to that given in Example 9, repeated here for convenience.

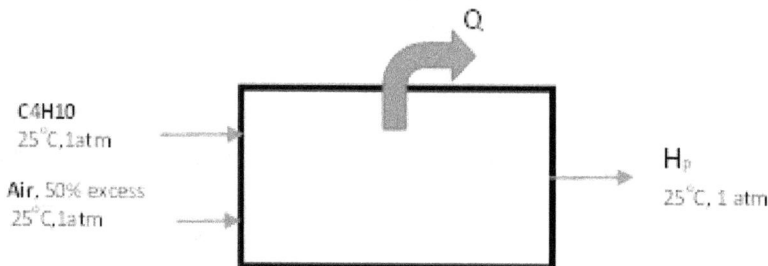

FIGURE 6.10. Sketch of combustion of butane inside burner with 50% excess air, Example 9.

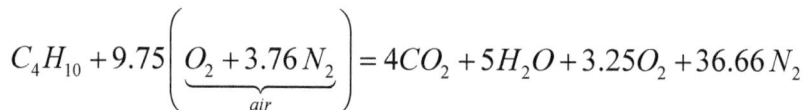

$$C_4H_{10} + 9.75\left(\underbrace{O_2 + 3.76\,N_2}_{air} \right) = 4CO_2 + 5H_2O + 3.25O_2 + 36.66\,N_2$$

The product mixture contains water. For calculating partial pressures, we are required to determine the phase composition of water, since part of the water will condense at 25 °C. In the product mixture we have a total of 48.91 moles of gases including 5 moles of water. A fraction of these 5 moles is vapor and the rest is liquid water, due to condensation which does not contribute to the gaseous part of the product mixture. In order to calculate how many moles of water vapor n_v exit in the mixture, we can write, assuming the product mixture as ideal gases, $n_v = \left((48.91 - 5) + n_v \right)\left(\underbrace{3.1698}_{p_{sat}\,@\,25^0 C} / 101.325 \right)$,

or $n_v = 1.418$ kmol. Therefore, we have $(5 - 1.418) = 3.582$ kmol of water in liquid phase existing in the combustion product. To calculate the amount of heat, we apply the 1st law using the property table (see Table 6.2),

or $\quad Q_{out} = H_R - H_P = 1(-126150) - 4(-393520) - \underbrace{3.582(-285830)}_{H2O(l)}$

$-\underbrace{1.418(-241820)}_{H2O(g)} = 2814674 \quad$ kJ/kmol of C_4H_{10}.

Now we calculate the entropy generated during the combustion. However, the entropy of the reactants is exactly the same as that of Example 8, or $S_R = 9532.1$ kJ/K. For calculating the entropy related to the products, we organize our calculations in a mini-table (see Table 6.7).

Substance	n_i (kmol)	$x_i = \dfrac{n_i}{\sum n_i}$	$S_i^o(298K)$ (kJ/kmol.K)	$R\ln x_i$ (kJ/kmol.k)	$S_i = S_i^o - R\ln x_i$ (kJ/kmol.k)	$H_i = n_i S_i$ (kJ/k)
CO2	4	0.082	213.8	8.314 ln 0.082 = −20.8	233.8	935.2
H2O(g)	1.418	0.029	188.83	8.314 ln 0.029 = −29.44	218.27	781.84
H2O(l)	3.582	0.073	69.92	8.314 ln 0.073 = −21.76	91.68	328.4
O2	3.25	0.066	205.04	8.314 ln 0.066 = −22.6	227.64	739.83
N2	36.66	0.75	191.61	8.314 ln 0.75 = −2.4	194.01	7112.41
	$\sum = 48.91$					$\sum = 9897.68$

(Left margin label: Products)

TABLE 6.7. Data and Calculations for Example 9

Therefore, using Equation 6.52 we get $S_{Gen} = S_P - S_R - \left(-\dfrac{Q_{out}}{298}\right) =$

$9897.68 - 9532.1 + \dfrac{2814674}{298} = 9810.8$ kJ/K per kmol of C_4H_{10}. The

amount of lost work is then $9810.8 \times 298 = 2923.62$ MJ per kmol of C_4H_{10}. This is the amount of work that could be obtained by the reaction process, but it is lost due to irreversibility of the combustion reaction. Comparing these results with those obtained in Example 8, we see that adiabatic combustion is less irreversible than that of a non-insulated one. In other words, more entropy is generated in the non-insulated combustion reaction or more work is lost as compared to the adiabatic combustion.

6.13. Exercises

1. Repeat Example 1 for propane gas (C_3H_6).

2. Repeat Example 2 for propane gas (C_3H_6).

3. Repeat Example 3 for propane gas (C_3H_6).

4. Repeat Example 4 for propane gas (C_3H_6).

5. Repeat Example 5 for propane gas (C_3H_6).

6. Repeat Example 6 for propane gas (C_3H_6).

7. Repeat Example 7 for propane gas (C_3H_6).

8. Repeat Example 8 for propane gas (C_3H_6).

9. Repeat Example 9 for propane gas (C_3H_6).

CHAPTER 7

PROPERTIES OF STEAM

In This Chapter

- Overview
- Definition of a Pure Substance
- Phase Change of a Pure Substance
- Formation of Steam
- Thermodynamic Properties of Steam and Steam Tables
- Determination of the Dryness Fraction of Steam
- Exercises

7.1. Overview

As mentioned in a previous section, pure substances can go through phase changes and coexist in different phases. One of the mostly used pure substances in thermodynamic systems is water. In this chapter we discuss in more detail the definition of a pure substance with emphasis on properties of water in liquid and vapor phases, along with its applications in thermal systems and worked out examples.

7.2. Definition of a Pure Substance

A pure substance is a system which is (i) *homogeneous in composition*, (ii) *homogeneous in chemical aggregation*, and (iii) *invariable in chemical aggregation*. By definition, homogeneous in composition

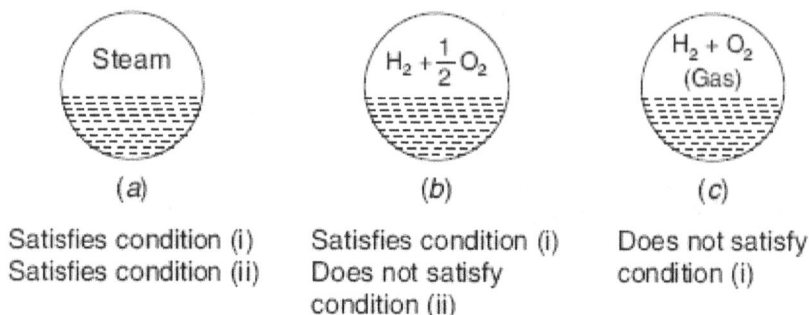

FIGURE 7.1. Illustrating the definition of a pure substance.

means that the composition of each part of the system is the same as the composition of every other part. Composition means the relative proportions of the chemical elements into which the sample can be analyzed, but it does not matter how these elements are combined. For example, in Figure 7.1 system (a), comprising steam and water, is homogeneous in composition, since chemical analysis would reveal that hydrogen and oxygen atoms are present in the ratio 2 : 1 whether the sample is taken from the steam or from the water. The same is true of system (b) containing uncombined hydrogen and oxygen gas in the atomic ratio 2 : 1 in the upper part, and water in the lower part. System (c) however, is not homogeneous in composition, for the hydrogen and oxygen are present in the ratio 1 : 1 in the upper part, but in the ratio 2 : 1 (as water) in the lower part. Homogeneous in chemical aggregation means that the chemical elements must be combined chemically in the same way in all parts of the system. Consideration of Figure 7.1 again shows that system (a) satisfies this condition also, for steam and water consist of identical molecules. System (b) on the other hand is not homogeneous in chemical aggregation, since in the upper part of the system, the hydrogen and oxygen are not combined chemically (individual atoms of hydrogen and oxygen are not uniquely associated), whereas in the lower part of the system the hydrogen and oxygen are combined to form water. Note however that a uniform mixture of steam, hydrogen gas, and oxygen gas would be regarded as homogeneous in both composition and chemical aggregation whatever the relative proportions of the components.

Invariable in chemical aggregation means that the state of chemical combination of the system does not change with time (condition (ii) referred to variation with position). Thus, a mixture of hydrogen and oxygen, which changed into steam during the time that the system was under consideration, would not be a pure substance.

7.3. Phase Change of a Pure Substance

A pure substance may change phase under a given pressure and temperature. Identifying the phase of water is very important in application in order to know the material behavior. Let us consider 1 kg (or similarly 1 mol) of liquid water at a temperature of 20°C in a cylinder fitted with a piston, which exerts on the water a constant pressure of one atmosphere (1.0132 bar) as shown in Figure 7.2 (i).

As the water is heated slowly its temperature rises until the temperature of the liquid water becomes 100°C. During the process of heating, the volume slightly increases as indicated by the line 1–2 on the temperature-specific volume diagram (see Figure 7.3). The piston starts moving upward.

If the heating of the liquid, after it attains a temperature of 100°C, is continued it undergoes a change in phase. A portion of the liquid water changes into vapor as shown in Figure 7.2(ii). This state is described by the line 2–3 in Figure 7.3. The amount of heat required to convert the liquid water completely into vapor under this

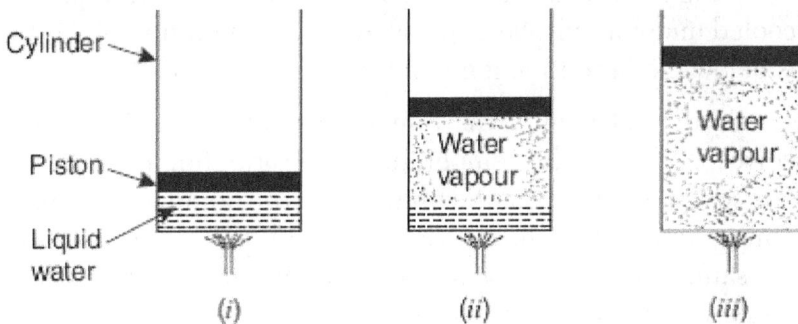

FIGURE 7.2. Phase change of water at a constant pressure from liquid to vapor phase.

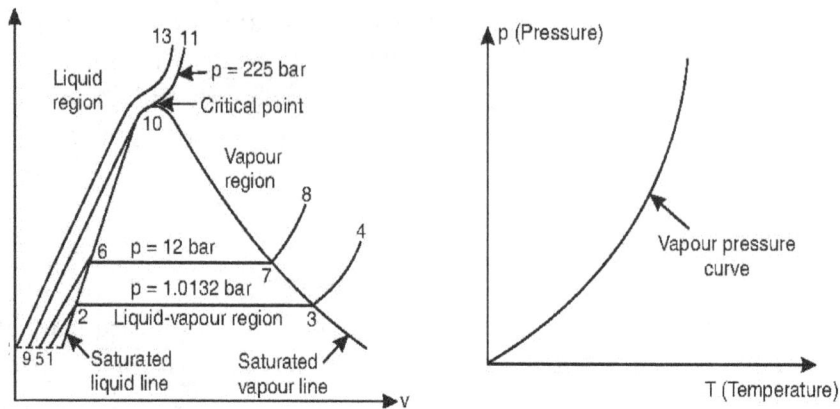

FIGURE 7.3. Phase change on p-v plane (left) and vapor pressure curve for water (right).

condition is called the heat of vaporization. The temperature at which vaporization takes place at a given pressure is called the saturation temperature and the given pressure is called the saturation pressure.

During the process represented by the line 2–3 (Figure 7.3) the volume increases rapidly and the piston moves upward Figure 7.2(iii).

For a pure substance, a definite relationship exists between the saturation pressure and saturation temperature as shown in Figure 7.3, the curve so obtained is called the vapor pressure curve.

It may be noted that if the temperature of the liquid water on cooling becomes lower than the saturation temperature for the given pressure, the liquid water is called a sub-cooled liquid. The point 1 (in Figure 7.3) illustrates this situation when the liquid water is cooled under atmospheric pressure to a temperature of 20 °C, which is below the saturation temperature (100 °C).

Further, at point 1 the temperature of the liquid is 20 °C, and corresponding to this temperature, the saturation pressure is 0.0234 bar, which is lower than the pressure on the liquid water, which is one atmosphere. Thus, the pressure on the liquid water is greater than the saturation pressure at a given temperature. In this condition, the liquid water is known as the compressed liquid.

The term compressed liquid or sub-cooled liquid is used to distinguish it from saturated liquid. All points in the liquid region indicate the states of the compressed liquid.

When all the liquid has been evaporated completely and further heat is added, the temperature of the vapor increases. The curve 3-4 in Figure 7.3 describes the process. When the temperature increases above the saturation temperature (in this case 100 °C), the vapor is known as the superheated vapor and the temperature at this state is called the superheated temperature. There is rapid increase in volume and the piston moves upwards (Figure 7.2 [iii]). The difference between the superheated temperature and the saturation temperature at the given pressure is called the degree of superheat.

If the above-mentioned heating process is repeated at different pressures, a number of curves similar to 1-2-3-4 are obtained. Thus, if the heating of the liquid water in the piston cylinder arrangement takes place under a constant pressure of 12 bar with an initial temperature of 20 °C until the liquid water is converted into superheated steam, then curve 5-6-7-8 will represent the process. In the above heating process, it may be noted that as the pressure increases, the length of constant temperature vaporization gets reduced.

From the heating process at a constant pressure of 225 bar represented by the curve 9-10-11 in Figure 7.3, it can be seen that there is no constant temperature vaporization line. The specific volume of the saturated liquid and of the saturated vapor is the same, that is, $v_f = v_g$. Such a state of the substance is called the critical state. The parameters like temperature, pressure, volume, and so on, at such a state are called critical parameters.

The curve 12-13 (Figure 7.3) represents a constant pressure heating process, when the pressure is greater than the critical pressure. At this state the liquid water is directly converted into superheated steam. As there is no definite point at which the liquid water changes into superheated steam, it is generally called liquid water when the temperature is less than the critical temperature and superheated steam when the temperature is above the critical temperature.

7.3.1. Phase Diagram on a $p - T$ Plane for a Pure Substance

If the vapor pressure of a solid is measured at various temperatures until the triple point is reached and then that of the liquid is measured until the critical point is reached, the result when plotted on a $p - T$ diagram appears as in Figure 7.4.

If the substance at the triple point (see next section) is compressed until there is no vapor left and the pressure on the resulting mixture of liquid and solid is increased, the temperature will have to be changed for equilibrium to exist between the solid and the liquid.

Measurements of these pressures and temperatures give rise to a third curve on the $p - T$ diagram, starting at the triple point and continuing indefinitely.

The points representing the coexistence of (i) solid and vapor lie on the "sublimation curve," (ii) liquid and vapor lie on the "vaporization curve," and (iii) liquid and solid lie on the "fusion curve." In the particular case of water, the sublimation curve is called the frost line, the vaporization curve is called the steam line, and the fusion curve is called the ice line.

The slopes of sublimation and the vaporization curves for all substances are positive. The slope of the fusion curve, however, may be positive or negative. The fusion curves of most substances have a positive slope, but water is one of the important exceptions with negative slope.

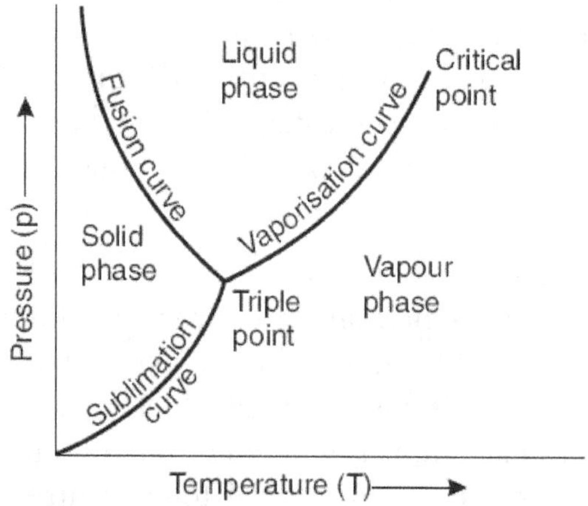

FIGURE 7.4. p-T diagram for a substance such as water.

7.3.2. Triple Point

The triple point is merely the point of intersection of sublimation and vaporization curves. It must be understood that only on a p-T diagram is the triple point represented by a point. On a p-V diagram it is a line, and on a U-V diagram it is a triangle. The pressure and temperature at which all three phases of a pure substance coexist may be measured with the apparatus that is used to measure vapor pressure. Triple-point data for some interesting substances are given in Table 7.1.

TABLE 7.1. Triple-Point Data for Some Gases

Substance	Temp., K	Pressure, mm Hg
Hydrogen (normal)	13.96	54.1
Deuterium (normal)	18.63	128
Neon	24.57	324
Nitrogen	63.18	94
Oxygen	54.36	1.14
Ammonia	195.40	45.57
Carbon dioxide	216.55	3.880
Sulphur dioxide	197.68	1.256
Water	273.16	4.58

7.3.3. 3D p-V-T Surface

We repeat the definitions given in the previous section on a $p - V - T$ surface diagram. A detailed study of the heating process reveals that the temperature of the solid rises and then during the change of phase from solid to liquid (or solid to vapor) the temperature remains constant. This phenomenon is common to all phase changes. Since the temperature is constant, pressure and temperature are not independent properties and cannot be used to specify state during a change of phase.

The combined picture of change of pressure, specific volume, and temperature may be shown on a three-dimensional state model. Figure 7.5 illustrates the equilibrium states for a pure substance which expands on fusion. Water is an example of a substance that exhibits this phenomenon.

All the equilibrium states lie on the surface of the model. States represented by the space above or below the surface are not possible.

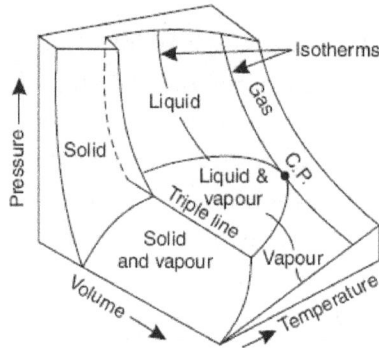

FIGURE 7.5. A pressure-volume-temperature (p-V-T) surface.

It may be seen that the triple point appears as a line in this representation. The point C.P. represents the critical point and no liquid phase exists at temperatures above the isotherms through this point. The term evaporation is meaningless in this situation.

At the critical point the temperature and pressure are called the critical temperature and the critical pressure, respectively, and when the temperature of a substance is above the critical value, it is called a gas. It is not possible to cause a phase change in a gas unless the temperature is lowered to a value less than the critical temperature. Oxygen and nitrogen are examples of gases that have critical temperatures below normal atmospheric temperature.

7.3.4. Phase Change Terminology and Definitions

With reference to Figure 7.6, we define the following terminologies as shown in Table 7.2.

Suffices:	Solid (*i*)	Liquid (*f*)	Vapor (*g*)
Phase change	**Name**	**Process**	**Process suffix**
Solid-liquid	Fusion	Freezing, melting	*if*
Solid-vapor	Sublimation	Frosting, defrosting	*ig*
Liquid-vapor	Evaporation	Evaporating, Condensing	*fg*

Triple point—The only state at which the solid, liquid, and vapor phases coexist in equilibrium.

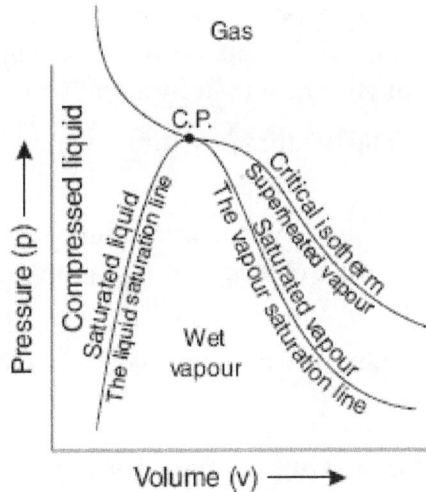

FIGURE 7.6. Phase change terminologies.

Critical point (C.P.)—The limit of distinction between a liquid and vapor.

Critical pressure—The pressure at the critical point.

Critical temperature—The temperature at the critical point.

Gas—A vapor whose temperature is greater than the critical temperature.

Saturation temperature—The phase change temperature corresponding to the saturation pressure. Sometimes called the boiling temperature.

Saturation pressure—The phase change pressure.

Compressed liquid—Liquid whose temperature is lower than the saturation temperature. Sometimes called a sub-cooled liquid.

Saturated liquid—Liquid at the saturation temperature corresponding to the saturation pressure. That is liquid about to commence evaporating, represented by the point f on a diagram.

Saturated vapor—A term including wet and dry vapor.

Dry (saturated) vapor—Vapor which has just completed evaporation. The pressure and temperature of the vapor are the saturation values. Dry vapor is represented by a point g on a diagram.

Wet vapor—The mixture of saturated liquid and dry vapor during the phase change.

Superheated vapor—Vapor with a temperature greater than the saturation temperature corresponding to the pressure of the vapor.

Degree of superheat—The term used for the numerical amount by which the temperature of a superheated vapor exceeds the saturation temperature.

7.3.5. Property Diagrams in Common Use

Besides the $p - V$ diagram which is useful because pressure and volume are easily visualized, and temperature-entropy, the $T - s$ chart which is used in general thermodynamic work, there are other charts which are of practical use for particular applications. The specific enthalpy-specific entropy $h - s$ chart is used for steam plant work and the pressure-specific enthalpy $p - h$ chart is used in refrigeration work. Sketches of these charts are shown in Figure 7.7. These charts are drawn for H_2O (water and steam) and represent the correct shape of the curves for this substance.

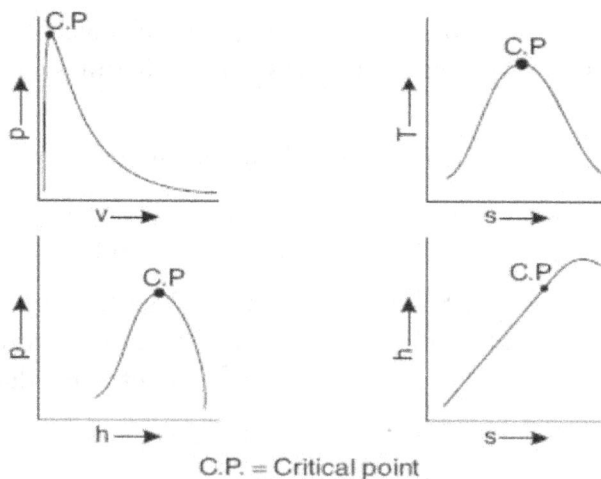

C.P. = Critical point

FIGURE 7.7. Sketches for different phase diagrams for H_2O.

7.4. Formation of Steam

In this section we discuss the process of formation of steam. Consider a cylinder fitted with a piston which can move freely upward and downward in it. For the sake of simplicity, let there be 1 kg of water at 0 °C with volume v_f m^3 under the piston (Figure 7.8[i]). Further let the piston be loaded with load W to ensure heating at a constant pressure. Now if the heat is imparted to water, a rise in temperature will be noticed and this rise will continue till the boiling point is reached. The temperature at which water starts boiling depends upon the pressure and, as such, for each pressure (under which water is heated) there is a different boiling point. This boiling temperature is known as the temperature of formation of steam or saturation temperature.

It may be noted during heating up to boiling point that there will be a slight increase in the volume of water, due to which the piston moves up and hence work is obtained as shown in Figure 7.8(ii). This work, however, is so small that it can be neglected.

Now, if the supply of heat to water is continued, it will be noticed that the rise of temperature after the boiling point is reached is nil but the piston starts moving upward, which indicates that there is increase in volume which is only possible if steam formation occurs. The heat being supplied does not show any rise of temperature but changes water into a vapor state (steam) and is known as latent heat or hidden heat. So long as the steam is in contact with water, it is called wet steam (Figure 7.8[iii]), and if heating of steam is further progressed (as shown in [Figure 7.8(iv)]) such that all the water particles associated with steam are evaporated, the steam so obtained is called dry and saturated steam. If v_g m^3 is the volume of 1 kg of dry and saturated steam, then work done on the piston will be $p(v_g - v_f)$. where p is the constant pressure (due to weight 'W' on the piston).

Again, if the supply of heat to the dry and saturated steam is continued at constant pressure there will be an increase in temperature and volume of steam. The steam so obtained is called superheated

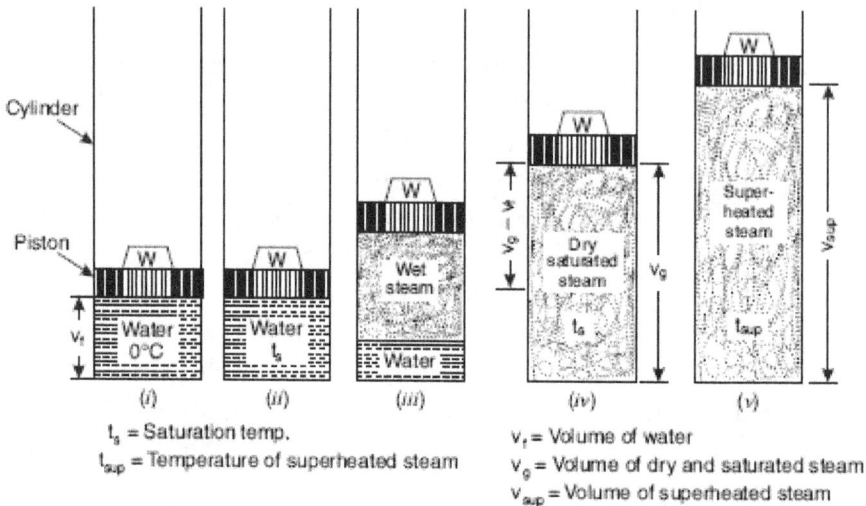

t_s = Saturation temp.

t_{sup} = Temperature of superheated steam

v_f = Volume of water

v_g = Volume of dry and saturated steam

v_{sup} = Volume of superheated steam

FIGURE 7.8. Formation of steam processes (i) to (v).

steam, and it behaves like a perfect gas. This phase of steam formation is illustrated in Figure 7.8 (*v*). Figure 7.9 shows the graphical representation of the formation of steam.

7.4.1. Important Quantities Relating to Steam

Sensible heat of water (h_f). It is defined as the quantity of heat absorbed by 1 kg of water when it is heated from 0 °C (freezing point) to the boiling point. It is also called the total heat (or enthalpy) of water or liquid heat. It is reckoned from 0 °C where sensible heat is taken as zero. If 1 kg of water is heated from 0 °C to 100 °C, the sensible heat added to it will be 4.18 × 100 = 418 kJ but if water is at, say, 20 °C initially, then the sensible heat added will be 4.18 × (100 – 20) = 334.4 kJ. This type of heat is denoted by the letter h_f, and its value can be directly read from the steam tables.

Note that the value of the specific heat of water may be taken as 4.18 kJ/kg K at low pressures, but at high pressures it is different from this value.

Latent heat or hidden heat (h_{fg}). It is the amount of heat required to convert water at a given temperature and pressure into steam at

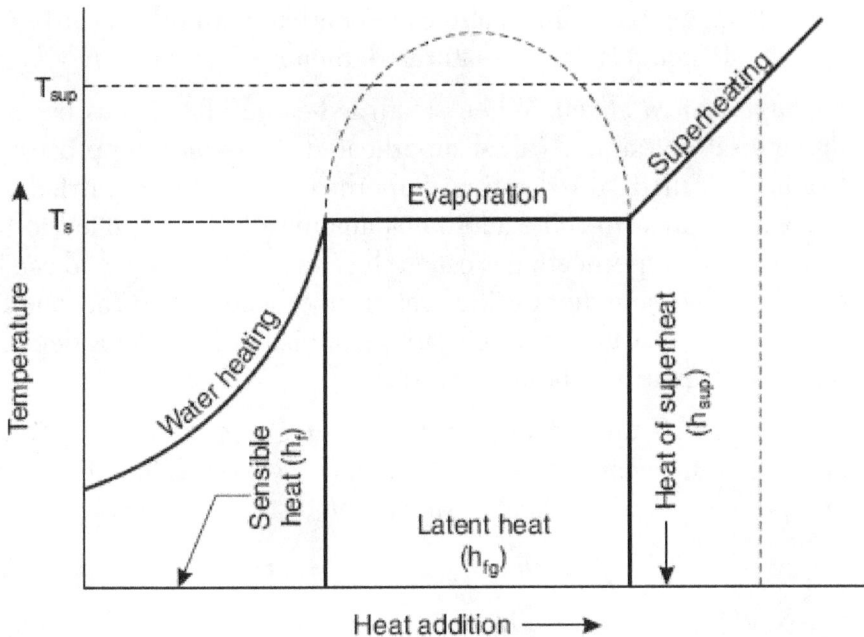

FIGURE 7.9. Graphical represen tation of formation of steam.

the same temperature and pressure. It is expressed by the symbol hfg and its value is available from steam tables. The value of latent heat is not constant and varies according to pressure variation.

Dryness fraction (x) or steam quality. The term dryness fraction is related with to steam. It is defined as the ratio of the mass of actual dry steam to the mass of steam containing it. It is usually expressed by the symbol "x" or "q." If m_s is the mass of dry steam contained in the steam considered and if m_w is the weight of water vapor in the steam considered, then $x = \dfrac{m_s}{m_s + m_w}$. Thus, if in 1 kg of wet steam we have 0.9 kg of dry steam and 0.1 kg water particles then $x = 0.9$. Note: No steam can be completely dry and saturated, so long as it is in contact with the water from which it is being formed.

Total heat or enthalpy of wet steam (h). It is defined as the quantity of heat required to convert 1 kg of water at 0 °C into wet steam at constant pressure. It is the sum of the total heat of water and the latent heat, and this sum is also called enthalpy. In other words, $h = h_f + xh_{fg}$. If steam is dry and saturated, then $x = 1$ and $h_g = h_f + h_{fg}$.

Superheated steam. When steam is heated after it has become dry and saturated, it is called superheated steam and the process of heating is called superheating. Superheating is always carried out at constant pressure. The additional amount of heat supplied to the steam during superheating is called the *Heat of superheat* and can be calculated by using the specific heat of superheated steam at constant pressure, c_{ps} the value of which varies from 2.0 to 2.1 kJ/kg.K depending upon pressure and temperature.

If T_{sup} and T_s are the temperatures of superheated steam in Kelvin and wet or dry steam, respectively, then $(T_{sup} - T_s)$ is called the *degree of superheat*. The total heat of superheated steam is given by

$$h_{sup} = h_f + h_{fg} + c_{ps}\left(T_{sup} - T_s\right) \tag{7.1}$$

Superheated steam behaves like a gas and therefore it follows the gas laws. The value of n for this type of steam is 1.3 and the law for the adiabatic expansion is $pv^{1.3} = constant$.

The advantages obtained by using superheated steam are as follows:

1. By superheating steam, its heat content and hence its capacity to do work is increased without having to increase its pressure.

2. Superheating is done in a superheater which obtains its heat from waste furnace gases which would have otherwise passed uselessly up the chimney.

3. The high temperature of superheated steam results in an increase in thermal efficiency.

4. Since the superheated steam is at a temperature above that corresponding to its pressure, it can be considerably cooled during expansion in an engine before its temperature falls below that at which it will condense and thereby become wet. Hence, heat losses due to condensation of steam on cylinder walls, and so on, are avoided to a great extent.

Volume of wet and dry steam: If the steam has a dryness fraction of x, then 1 kg of this steam will contain x kg of dry steam and $(1 - x)$ kg of water. If v_f is the volume of 1 kg of water and v_g is the volume of 1 kg of perfect dry steam (also known as specific volume), then the volume of 1 kg of wet steam is $xv_g + (1 - x)v_f$.

Note that the volume of v_f at low pressures is very small and is generally neglected. Thus, in general, the volume of 1 kg of wet steam is given by xv_g and density $\dfrac{1}{xv_g}$.

Volume of superheated steam: As superheated steam behaves like a perfect gas, its volume can be found out in the same way as the gases. Then $\dfrac{pv_g}{T_s} = \dfrac{pv_{sup}}{T_{sup}}$ or $v_{sup} = \dfrac{v_g T_{sup}}{T_s}$.

7.5. Thermodynamic Properties of Steam and Steam Tables

In engineering problems, for any fluid which is used as working fluid, the six basic thermodynamic properties required are: p (pressure), T (temperature), v (volume), u (internal energy), h (enthalpy), and s (entropy). These properties must be known at different pressures for analyzing the thermodynamic cycles used for work producing devices. The values of these properties are determined theoretically or experimentally and are tabulated in the form of tables which are known as "Steam Tables." The properties of wet steam are then computed from such tabulated data. Tabulated values are also available for superheated steam. It may be noted that steam has only one saturation temperature at each pressure.

Following are the thermodynamic properties of steam which are tabulated in the form of a table:

p = Absolute pressure (bar or kPa);

t_s = Saturation temperature (°C);

h_f = Enthalpy of saturated liquid (kJ / kg);

h_{fg} = Enthalpy or latent heat of vapourisation (kJ / kg);

h_g = Enthalpy of saturated vapour (steam) (kJ / kg);

s_f = Entropy of saturated liquid $(\text{kJ} / \text{kg K})$;

s_{fg} = Entropy of vapourisation $(\text{kJ} / \text{kg K})$;

s_g = Entropy of saturated vapour (steam) $(\text{kJ} / \text{kg K})$;

v_f = Specific volume of saturated liquid (m^3 / kg);

v_g = Specific volume of saturated vapour (steam) (m^3 / kg);

$h_{fg} = h_g - h_f$; Change of enthalpy during evaporation ;

$s_{fg} = s_g - s_f$; Change of entropy during evaporation ;

$v_{fg} = v_g - v_f$; Change of volume during evaporation.

The above-mentioned properties at different pressures are tabulated in the form of tables, as a sample given in the below table:

The internal energy of steam $u = h - pv$ is also tabulated in some steam tables.

STEAM TABLES

Absolute pressure bar, p	Temperature °C	Specific enthalpy kJ/kg			Specific entropy kJ/kg K sf			Specific volume m³/kg	
	t_s	h_f	h_{fg}	h_g	s_f	s_{fg}	s_g	v_f	v_g
1.0	99.6	417.5	2257.9	2675.4	1.3027	6.0571	7.3598	0.001043	1.6934
50.0	263.9	1154.9	1639.7	2794.2	2.9206	3.0529	5.9735	0.001286	0.00394
100.0	311.1	1408.0	1319.7	2727.7	3.3605	2.2593	5.6198	0.001452	0.01811

7.5.1. External Work during Evaporation

When water is evaporated to form saturated steam, its volume increases from v_f to v_g at a constant pressure, and thus external work is done by steam due to an increase in volume. The energy for doing the work is obtained during the absorption of latent heat. This work is called the external work of evaporation and is given by $p(v_g - v_f)$.

As at low pressure v_f is very small and hence neglected, the work of evaporation is $p \cdot v_f$.

In case of wet steam with dryness fraction x, the work of evaporation will be $p \cdot xv_f$.

7.5.2. Internal Latent Heat

The latent heat consists of true latent heat and the work of evaporation. This true latent heat is called the internal latent heat and may also be given as $h_{fg} - \dfrac{pv_g}{J}$, where $J = 1$ in SI units.

7.5.3. Internal Energy of Steam

It is defined as the actual energy stored in the steam. As per previous articles, the total heat of steam is the sum of sensible heat, internal latent heat, and the external work of evaporation. Work of evaporation is not stored in the steam as it is utilized in doing external work. Hence the internal energy of steam could be found by subtracting the work of evaporation from the total heat. In other words, $h = \dfrac{pv_g}{J} + u$, where $u = h - \dfrac{pv_g}{J}$, or the internal energy of 1 kg of steam at pressure p.

In case of wet steam with a dryness fraction, we have $u = h - \dfrac{pxv_g}{J}$ and if steam is superheated to a volume of v_{sup} per kg, then $h_{sup} = h_f + h_{fg} + c_{ps}\left(T_{sup} - T_s\right)$ and $u = h_{sup} - \dfrac{pv_{sup}}{J}$.

7.5.4. Entropy of Water

Consider 1 kg of water being heated from temperature T_1 to T_2 at a constant pressure. The change in entropy will be given by $ds = \dfrac{dQ}{T} = c_{pw}\dfrac{dT}{T}$. Integrating both sides, we get $s_2 - s_1 = c_{pw}\ln\dfrac{T_2}{T_1}$.

If 0 °C is taken as datum, then the entropy of water per kg at any temperature T above this datum will be $s_f = c_{pw}\ln\dfrac{T}{273}$. Readers may refer to Chapter 4 for a more detailed discussion on this topic.

7.5.5. Entropy of Evaporation

When water is evaporated to steam completely the heat absorbed is the latent heat and this heat goes into water without showing any rise of temperature. Then $Q = h_{fg}$ and $s_{evap} = \dfrac{h_{fg}}{T_s}$.

However, in case of wet steam with dryness fraction x, the evaporation will be partial and the heat absorbed will be xh_{fg} per kg of steam. The change of entropy will be $\dfrac{xh_{fg}}{T_s}$.

7.5.6. Entropy of Wet Steam

The total entropy of wet steam is the sum of entropy of water s_f and entropy of evaporation s_{fg}. In other words $s_{wet} = s_f + \dfrac{xh_{fg}}{T_s}$, where s_{wet} is the total entropy of wet steam, s_f is the entropy of water, and $\dfrac{xh_{fg}}{T_s}$ is the entropy of evaporation. If steam is dry and saturated, that is, = 1, then $s_g = s_f + \dfrac{h_{fg}}{T_s}$.

7.5.7. Entropy of Superheated Steam

Let 1 kg of dry saturated steam at T_s (saturation temperature of steam) be heated to Tsup. If a specific heat at a constant pressure is c_p, then the change of entropy during superheating at a constant pressure p is $c_p \ln \dfrac{T_{sup}}{T_s}$. Total entropy of superheated steam above the freezing point of water can be written as s_{sup} = Entropy of dry saturated steam + change of entropy during superheating, or

$$s_{sup} = s_f + \frac{h_{fg}}{T_s} + c_{ps} \log_e \left(\frac{T_{sup}}{T_s} \right) = s_g + c_{ps} \log_e \left(\frac{T_{sup}}{T_s} \right)$$

$$(7.2)$$

7.5.8. Enthalpy-Entropy Chart—Mollier Diagram

In 1904, Mollier [115], [116] conceived the idea of plotting enthalpy against entropy. His diagram is more widely used than any other entropy diagram, since the work done on vapor cycles can be scaled from this diagram directly as a length, whereas on a $T-s$ diagram it is represented by an area.

A sketch of the h-s chart is shown in Figure 7.10.

Lines of constant pressure are indicated by p_1, p_2 and so on, lines of constant temperature by T_1, T_2, and so on.

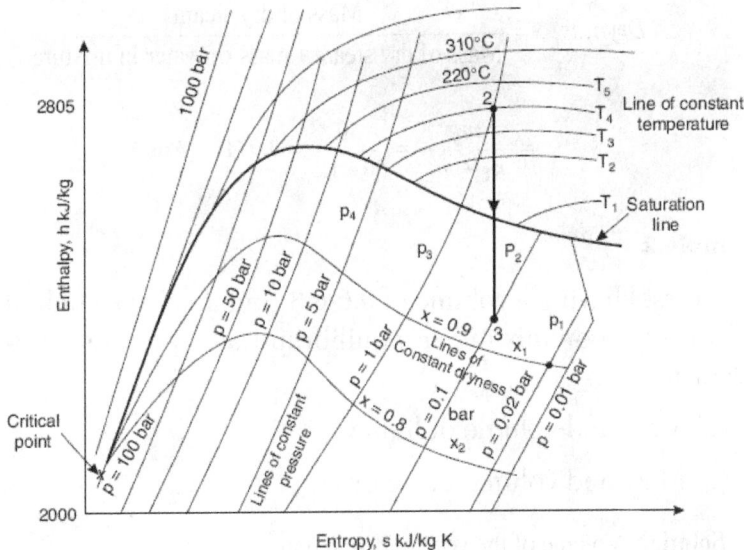

FIGURE 7.10. Enthalpy-entropy (h-s) chart.

Any two independent properties which appear on the chart are sufficient to define the state (e.g., p_1 and x_1 define state 1 and h can be read off the vertical axis).

In the superheat region, pressure and temperature can define the state (e.g., p_3 and T_4 define the state 2, and h_2 can be read off). A line of constant entropy between two state points 2 and 3 defines the properties at all points during an isentropic process between the two states.

In the remaining sections of this chapter, we present several worked out numerical examples related to the topic discussed so far.

Example 1: Steam Dryness

Calculate the dryness fraction (quality) of steam which has 1.5 kg of water in suspension with 50 kg of steam.

■ **Solution.**

Mass of dry steam, $m_S = 50$ kg

Mass of water in suspension, $m_W = 1.5$ kg

$$\textit{Dryness, } x = \frac{\text{Mass of dry steam}}{\text{Mass of dry steam} + \text{mass of water in mixture}}$$

$$= \frac{m_S}{m_S + m_W} = \frac{50}{50 + 1.5} = \textbf{0.971. (Ans.)}$$

Example 2

A vessel having a volume of 0.6 m3 contains 3.0 kg of liquid water and water vapor mixture in equilibrium at a pressure of 0.5 MPa. Calculate:

(1) Mass and volume of liquid;

(2) Mass and volume of vapor.

Solution. Volume of the vessel, $V = 0.6 \text{ m}^3$

Mass of liquid water and water vapour, $m = 3.0$ kg

Pressure, $p = 0.5$ MPa $= 5$ bar

Thus, specific volume, $v = \dfrac{V}{m} = \dfrac{0.6}{3.0} = 0.2 \text{ m}^3/\text{kg}$

At 5 bar: From steam tables,

$$v_{fg} = v_g - v_f = 0.375 - 0.00109 = 0.3739 \text{ m}^3/\text{kg}$$

We know that, $\quad v = v_g - (1-x)\, v_{fg}$, where x = quality of the vapour.

$$0.2 = 0.375 - (1-x) \times 0.3739$$

$$\therefore \qquad (1-x) = \frac{(0.375 - 0.2)}{0.3739} = 0.468$$

$$x = 0.532$$

(1) Mass and volume of liquid,

$$\mathbf{m_{liq.}} = m(1-x) = 3.0 \times 0.468 = \textbf{1.404 kg.} \quad \textbf{(Ans.)}$$

$$\mathbf{V_{liq.}} = m_{liq.}\, v_f = 1.404 \times 0.00109 = \textbf{0.0015 m}^3. \quad \textbf{(Ans.)}$$

(2) Mass and volume of vapor

$$\mathbf{m_{vap.}} = m.x = 3.0 \times 0.532 = \mathbf{1.596\ kg.} \quad (\mathbf{Ans.})$$

$$\mathbf{V_{vap.}} = m_{vap.}\ v_g = 1.596 \times 0.375 = \mathbf{0.5985\ m^3}. \quad (\mathbf{Ans.})$$

Example 3

A vessel having a capacity of 0.05 m³ contains a mixture of saturated water and saturated steam at a temperature of 245 °C. The mass of the liquid present is 10 kg. Calculate the following:

(i) The pressure   (ii) The mass

(iii) The specific volume   (iv) The specific enthalpy

(v) The specific entropy and   (vi) The specific internal energy.

■ **Solution.**

From steam tables, corresponding to 245 °C:

$$p_{sat} = 36.5\ \text{bar},\ v_f = 0.001239\ \text{m}^3/\text{kg},\ v = 0.0546\ \text{m}^3/\text{kg}$$

$$h_f = 1061.4\ \text{kJ/kg},\ h_{fg} = 1740.2\ \text{kJ/kg},\ s_f = 2.7474\ \text{kJ/kg K}$$

$$s_{fg} = 3.3585\ \text{kJ/kg K}.$$

(i) **The pressure = 36.5 bar** (or 3.65 MPa).   (**Ans.**)

(ii)   The mass, m:

$$\text{Volume of liquid,} \quad V_f = m_f v_f$$

$$= 10 \times 0.001239 = 0.01239\ \text{m}^3$$

$$\text{Volume of vapour,} \quad V_g = 0.05 - 0.01239 = 0.03761\ \text{m}^3$$

$$\therefore \text{ Mass of vapour,} \quad m_g = \frac{V_g}{U_g} = \frac{003761}{00546} = 0.688\ \text{kg}$$

$$\therefore \text{ The total mass of mixture,}$$

$$\mathbf{m} = m_f + m_g = 10 + 0.688 = \mathbf{10.688\ kg.} \quad (\mathbf{Ans.})$$

(iii)   The specific volume, v:

Quality of the mixture,

$$x = \frac{m_g}{m_g + m_f} = \frac{0688}{0688 + 10} = 0.064$$

$\therefore$ $$v = v_f + xv_{fg}$$

$$= 0.001239 + 0.064 \times (0.0546 - 0.001239) \qquad (\because \; v_{fg} = v_g - v_f)$$

$$= \textbf{0.004654 m}^3 \textbf{/ kg. \quad (Ans.)}$$

(iv) The specific enthalpy, h:

$$\mathbf{h} = h_f + xh_{fg}$$

$$= 1061.4 + 0.064 \times 1740.2 = \textbf{1172.77 kJ / kg. \quad (Ans.)}$$

(v) The specific entropy, s:

$$\mathbf{s} = s_f + xs_{fg}$$
$$= 2.7474 + 0.064 \times 3.3585 = \textbf{2.9623 kJ / kg K. \quad (Ans.)}$$

(vi)   The specific internal energy, u:

$$\mathbf{u} = h - pv$$

$$= 1172.77 - \frac{36.5 \times 10^5 \times 0.004654}{1000} = \textbf{1155.78 kJ / kg. \quad (Ans.)}$$

Example 4

Determine the amount of heat, which should be supplied to 2 kg of water at 25 °C to convert it into steam at 5 bar and 0.9 dry.

■ **Solution.**

Mass of water to be converted to steam, $m_w = 2$ kg

Temperature of water, $\qquad\qquad\qquad\qquad\qquad tw = 25°C$

Pressure and dryness fraction of steam $\qquad = 5$ bar, 0.9 dry

At 5 bar: From steam tables,

$h_f = 640.1$ kJ/kg ; $h_{fg} = 2107.4$ kJ/kg

Enthalpy of 1 kg of steam (above 0 °C)

$$h = h_f + xh_{fg}$$

$$= 640.1 + 0.9 \times 2107.4 = 2536.76 \text{ kJ/kg}$$

Sensible heat associated with 1 kg of water

$$= m_w \times c_{pw} \times (T_w - 0)$$

$$= 1 \times 4.18 \times (25 - 0) = 104.5 \text{ kJ}$$

Net quantity of heat to be supplied per kg of water

$$= 2536.76 - 104.5 = 2432.26 \text{ kJ}$$

Total amount of heat to be supplied

$$= 2 \times 2432.26 = \textbf{4864.52 kJ.} \quad \textbf{(Ans.)}$$

Example 5

1000 kg of steam at a pressure of 16 bar and 0.9 dry is generated by a boiler per hour. The steam passes through a superheater via boiler stop valve where its temperature is raised to 380 °C. If the temperature of feed water is 30 °C, determine:

(i)   The total heat supplied to feed water per hour to produce wet steam.

(ii)   The total heat absorbed per hour in the superheater.

Take specific heat for superheated steam as 2.2 kJ/kg K.

■ **Solution.**

Mass of steam generated, $m = 1000$ kg/h

Pressure of steam, $p = 16$ bar

Dryness fraction, $x = 0.9$

Temperature of superheated steam,

$$T_{sup} = 380 + 273 = 653 \text{ K}$$

Temperature of feed water = 30 °C

Specific heat of superheated steam, $c_{ps} = 2.2$ kJ/kg K.

At 16 bar. From steam tables,

$t_s = 201.4°C \ (T_s = 201.4 + 273 = 474.4 \text{ K})$;

$h_f = 858.6$ kJ/kg ; $h_{fg} = 1933.2$ kJ/kg

(*i*) *Heat supplied to feed water per hour* to produce wet steam is given by:

$$H = m\left[(h_f + xh_{fg}) - 1 \times 4.18 \times (30 - 0)\right]$$

$$= 1000\left[(858.6 + 0.9 \times 1933.2) - 4.18 \times 30\right]$$

$$= 1000(858.6 + 1739.88 - 125.4)$$

$$= \mathbf{2473.08 \times 10^3 \ kJ.} \quad \textbf{(Ans.)}$$

(*ii*) *Heat absorbed by superheater per hour*

$$= m\left[(1 - x)h_{fg} + c_{ps}(T_{sup} - T_s)\right]$$

$$= 1000[(1 - 0.9) \times 1933.2 + 2.2(653 - 474.4)]$$

$$= 1000(193.32 + 392.92)$$

$$= \mathbf{586.24 \times 103 \ kJ.} \quad \textbf{(Ans.)}$$

Example 6

Using steam tables, determine the mean specific heat for super-heated steam:

(i)   at 0.75 bar, between 100 °C and 150 °C;

(*ii*)   *at 0.5 bar, between 300 °C and 400 °C.*

■ **Solution.**

(*i*) **At 0.75 bar.** From steam tables;

At 100°C, $h_{sup} = 2679.4$ kJ/kg

At 150°C, $h_{sup} = 2778.2$ kJ/kg

∴ $2778.2 = 2679.4 + c_{ps}(150 - 100)$

i. e., $c_{ps} = \dfrac{2778.2 - 2679.4}{50} = \mathbf{1.976.} \quad \textbf{(Ans.)}$

(*ii*)   **At 0.5 bar.** From steam tables;

At 300°C, $h_{sup} = 3075.5$ kJ/kg

At 400°C, $h_{sup} = 3278.9$ kJ/kg

$\therefore$ $3278.9 = 3075.5 + c_{ps} \, (400 - 300)$

i.e., $c_{ps} = \dfrac{3278.9 - 3075.5}{100} = \textbf{2.034. (Ans.)}$

Example 7

A pressure cooker contains 1.5 kg of saturated steam at 5 bar. Find the quantity of heat which must be rejected so as to reduce the quality to 60% dry. Determine the pressure and temperature of the steam at the new state.

Solution. Mass of steam in the cooker = 15 kg

Pressure of steam, $p = 5$ bar

Initial dryness fraction of steam, $x_1 = 1$

Final dryness fraction of steam, $x_2 = 0.6$

Heat to be rejected:

Pressure and temperature of the steam at the new state:

At 5 bar. From steam tables,

$t_s = 51.8°C$; $h_f = 640.1$ kJ/kg ;

$h_{fg} = 2107.4$ kJ/kg ; $v_g = 0.375$ m3/kg

Thus, the volume of pressure cooker

$= 1.5 \times 0.375 = 0.5625$ m^3

Internal energy of steam per kg at initial point 1,

$$u_1 = h_1 - p_1 v_1$$

$= (h_f + h_{fg}) - p_1 v_{g_1}$ $(\because v_1 = v_{g1})$

$$= (640.1 + 2107.4) - 5 \times 10^5 \times 0.375 \times 10^{-3}$$

$$= 2747.5 - 187.5 = 2560 \, \text{kJ/kg}$$

Also, $\qquad V_1 = V_2 \, (V_2 = \text{volume at final condition})$

i.e., $\qquad 0.5625 = 1.5[(1 - x_2) \, v_{f2} + x_2 v_{g2}]$

$$= 1.5 \, x_2 v_{g2} \qquad (\because \, v_{f2} \text{ is negligible})$$

$$= 1.5 \times 0.6 \times v_{g2}$$

$\therefore \qquad\qquad\qquad v_{g2} = \dfrac{0.5625}{1.5 \times 0.6} = 0.625 \, \text{m}^3/\text{kg}.$

From steam tables corresponding to 0.625 m^3/kg,

$p_2 \approx \mathbf{2.9 \, bar}, \; t_s = \mathbf{132.4°C}, \; h_f = 556.5 \, \text{kJ/kg}, \; h_{fg} = 2166.6 \, \text{kJ/kg}$

Internal energy of steam per kg, at final point 2,

$u_2 = h_2 - p_2 v_2$

$$= (h_{f2} + x_2 h_{fg_2}) - p_2 x v_{g2} \qquad\qquad (\because \, v_2 = x v_{g2})$$

$$= (556.5 + 0.6 \times 2166.6) - 2.9 \times 10^5 \times 0.6 \times 0.625 \times 10^{-3}$$

$$= 1856.46 - 108.75 = 1747.71 \, \text{kJ/kg}.$$

Heat transferred at constant volume per kg

$$= u_2 - u_1 = 1747.71 - 2560 = -812.29 \, \text{kJ/kg}$$

Thus, total heat transferred

$$= -812.29 \times 1.5 = \mathbf{-1218.43 \, kJ}. \quad \textbf{(Ans.)}$$

Negative sign indicates that heat has been rejected.

Example 8

A spherical vessel of 0.9 m3 capacity contains steam at 8 bar and 0.9 dryness fraction. Steam is blown off until the pressure drops to

4 bar. The valve is then closed and the steam is allowed to cool until the pressure falls to 3 bar. Assuming that the enthalpy of steam in the vessel remains constant during blowing off periods, determine:

(i)   The mass of steam blown off;

(ii)   The dryness fraction of steam in the vessel after cooling;

(iii)   The heat lost by steam per kg during cooling.

Solution. Capacity of the spherical vessel, $V = 0.9 \text{ m}^3$

Pressure of the steam,	$p_1 = 8$ bar
Dryness fraction of steam,	$x_1 = 0.9$
Pressure of steam after blow off,	$p_2 = 4$ bar.
Final pressure of steam,	$p_3 = 3$ bar.

(*i*)   The mass of steam blown off:

The mass of steam in the vessel

$$m_1 = \frac{V}{x_1 v_{g_1}} = \frac{0.9}{0.9 \times 0.24} = 4.167 \text{ kg} \qquad (\because \text{At 8 bar} : v_g = 0.24 \text{ m}^3/\text{kg})$$

The enthalpy of steam before blowing off (per kg) is

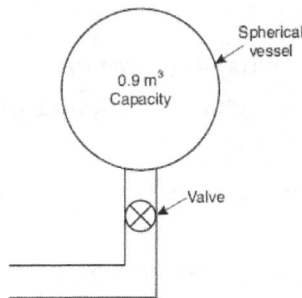

Spherical vessel

0.9 m³ Capacity

Valve

$= h_{f_1} + x_1 h_{fg1} = 720.9 + 0.9 \times 2046.5 \text{ ...... at pressure } 8 \text{ bar}$

$= 2562.75 \text{ kJ/kg}$

Enthalpy before blowing off = Enthalpy after blowing off

$\therefore \qquad\qquad 2562.75 = (h_{f_2} + x_2 h_{fg_2})$ at pressure 4 bar

$= 604.7 + x_2 \times 2133$ ...... at pressure 4 *bar*

$\therefore \qquad\qquad x_2 = \dfrac{2562.75 - 604.7}{2133} = 0.918$

Now the mass of steam in the vessel after blowing off,

$m_2 = \dfrac{0.9}{0.918 \times 0.462} = 2.122$ kg $\qquad [v_{g2} = 0.462 \text{ m}^3/\text{kg}.......\text{at 4 bar}]$

Mass of steam blown off, $m = m_1 - m_2 = 4.167 - 2.122$

$= 2.045$ **kg.** **(Ans.)**

(*ii*)**Dryness fraction of steam in the vessel after cooling,** x_3 :

As it is constant volume cooling

$\therefore \qquad\qquad x_2 v_{g2}$ (at 4 bar)$= x_3 v_{g3}$ (at 3 bar)

$0.918 \times 0.462 = x_3 \times 0.606$

$\therefore \qquad\qquad x_3 = \dfrac{0.918 \times 0.462}{0,606} = 0.699.$ **(Ans.)**

(*iii*) Heat lost during cooling:

Heat lost during cooling $= m(u_3 - u_2)$, where u_2 and u_3 are the internal energies of steam before starting cooling or after blowing and at the end of the cooling.

$\therefore \qquad\qquad u_2 = h_2 - p_2 x_2 v_{g_2} = (h_{f_2} + x_2 h_{fg2}) - p_2 x_2 v_{g2}$

$= (604.7 + 0.918 \times 2133) - 4 \times 10^5 \times 0.918 \times 0.462 \times 10^{-3}$

$= 2562.79 - 169.65 = 2393.14$ kJ/kg

$u_3 = h_3 - p_3 x_3 v_{g3} = (h_{f_3} + x_3 h_{fg3}) - p_3 x_3 v_{g3}$

$= (561.4 + 0.669 \times 2163.2) - 3 \times 10^5 \times 0.699 \times 0.606 \times 10^{-3}$

$= 2073.47 - 127.07 = 1946.4$ kJ/kg

$\therefore$ Heat transferred during cooling

$= 2.045 \, (1946.4 - 2393.14) = -913.6$ kJ.

    i.e.,   **Heat lost during cooling = 913.6 kJ.**   **(Ans.)**

Example 9

If a certain amount of steam is produced at a pressure of 8 bar and dryness fraction 0.8. Calculate:

(i) External work done during evaporation.

(ii) Internal latent heat of steam.

■ **Solution.**

Pressure of steam, $p = 8$ bar

Dryness fraction, $x = 0.8$

At 8 bar. From steam tables,

$v_g = 0.240 \, \text{m}^3/\text{kg}, \; h_{fg} = 2046.5 \, \text{kJ/kg}$

(i) External work done during evaporation

$= pxv_g = 8 \times 10^5 \times 0.8 \times 0.24 \; \text{Nm}$

$= \dfrac{8 \times 10^5 \times 0.8 \times 0.24}{10^3} = \textbf{153.6 kJ.}$ **(Ans.)**

(ii) **Internal latent heat** $= xh_{fg} -$ external work done

$= 0.8 \times 2046.5 - 153.6$

$= \textbf{1483.6 kJ. (Ans.)}$

Example 10

Steam at 120 bar has a specific volume of 0.01721 m³/kg; find the temperature, enthalpy, and the internal energy.

Solution. Pressure of steam, $p = 120$ bar

Specific volume, $v = 0.01721$ m^3 / kg

(*i*) **Temperature:**

First it is necessary to decide whether the steam is wet, dry saturated, or superheated.

At 120 bar, $v_g = 0.0143$ m^3 / kg, which is less than the actual specific volume of 0.01721 m^3/kg, and hence the steam is **superheated.**

From the superheat tables at 120 bar, the specific volume is 0.01721 m^3/kg at a temperature of **350°C.** (Ans.)

(*ii*) **Enthalpy:**

From the steam tables the specific enthalpy at 120 bar, 350°C,

$$h = 2847.7 \text{ kJ / kg.} \quad \textbf{(Ans.)}$$

(*iii*) **Internal energy:**

To find internal energy, using the equation,

$$u = h - pv$$

$$= 2847.7 - \frac{120 \times 10^5 \times 0.01721}{10^3}$$

$$= 2641.18 \text{ kJ / kg. } \left(\textbf{Ans.}\right)$$

Example 11

Calculate the internal energy per kg of superheated steam at a pressure of 10 bar and a temperature of 300 °C. Also find the change of internal energy if this steam is expanded to 1.4 bar and dryness fraction 0.8.

■ **Solution.**

At 10 bar, 300°C. From steam tables for superheated steam:

$$h_{sup} = 3051.2 \text{ kJ / kg } \left(T_{sup} = 300 + 273 = 573 \text{ K}\right)$$

and corresponding to 10 bar (from tables of dry saturated steam)

$$T_S = 179.9 + 273 = 452.9 \text{ K} ; v_g = 0.194 \text{ m}^3 / \text{kg}$$

To find v_{sup}, using the relation,

$$\frac{v_g}{T_S} = \frac{v_{sup}}{T_{sup}}$$

$$\therefore \qquad v_{sup} = \frac{v_g \times T_{sup}}{T_S} = \frac{0.194 \times 573}{452.9} = 0.25 \text{ m}^3 /\text{kg}$$

Internal energy of superheated steam at 10 bar,

$$u_1 = h_{sup} - pv_{sup}$$

$$= 3051.2 - 10 \times 10^5 \times 0.245 \times 10^{-3}$$

$$= \textbf{2806.2 kJ / kg.} \quad \textbf{(Ans.)}$$

At 1.4 bar. From steam tables:

$$h_f = 458.4 \text{ kJ / kg}, \ h_{fg} = 2231.9 \text{ kJ / kg} ; \ v_g = 1.236 \text{ m}^3 / \text{kg}$$

Enthalpy of wet steam (after expansion)

$$h = h_f + x h_{fg}$$

$$= 458.4 + 0.8 \times 2231.9 = 2243.92 \text{ kJ.}$$

Internal energy of this steam,

$$u_2 = h - pxv_g$$

$$= 2243.92 - 1.4 \times 10^5 \times 0.8 \times 1.236 \times 10^{-3}$$

$$= 2105.49 \text{ kJ}$$

Hence change of internal energy per kg

$$u_2 - u_1 = 2105.49 - 2806.2$$

$$= \textbf{-700.7 kJ.} \ \big(\textbf{Ans.}\big)$$

Negative sign indicates decrease in internal energy.

Example 12

Two boilers, one with a superheater and other without a super-heater, are delivering equal quantities of steam into a common main. The pressure in the boilers and the main is 20 bar. The temperature of steam from a boiler with a superheater is 350 °C and the temperature of the steam in the main is 250 °C.

Determine the quality of steam supplied by the other boiler.
Take $c_{ps} = 2.25\ kJ/kg$.

■ **Solution.**
Boiler B$_1$. 20 bar, 350 °C:

Enthalpy, $\quad h_1 = h_{g1} + c_{ps}\ (T_{sup} - T_s)$

$= 2797.2\ + 2.25(350 - 212.4)$

$= 3106.8\ kJ/kg \qquad\qquad ...(i)$

Boiler B$_2$. 20 bar (temperature not known):

$h_2 = h_{f_2} + x_2 h_{fg_2}$

$= (908.6 + x_2 \times 1888.6)\ kJ/kg \qquad ...(ii)$

Main. 20 bar, 250 °C.

Total heat of 2 kg of steam in the steam main

$$= 2\Big[h_g + c_{ps}\ (T_{sup} - T_s)\Big]$$

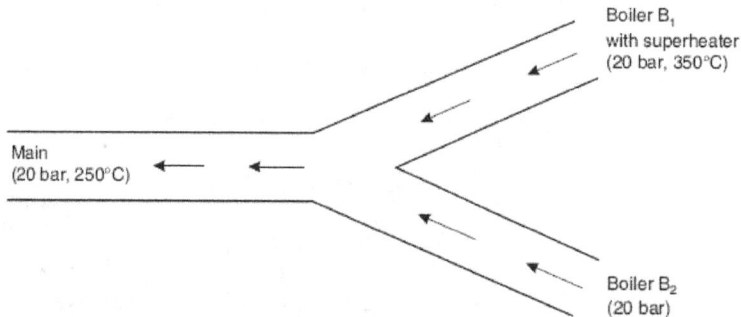

Boiler B$_1$
with superheater
(20 bar, 350°C)

Main
(20 bar, 250°C)

Boiler B$_2$
(20 bar)

$= 2\Big[2797.2 + 2.25\ (250 - 212.4)\Big] = 5763.6\ kJ \qquad ...(iii)$

Equating (*i*) and (*ii*) with (*iii*), we get

$$3106.8 + 908.6 + x_2 \times 1888.6 = 5763.6$$

$$4015.4 + 1888.6x_2 = 5763.6$$

$$\therefore \qquad x_2 = \frac{5763.6 - 4015.4}{1888.6} = 0.925$$

Hence, **quality of steam supplied by the other boiler = 0.925. (Ans.)**

Example 13

Determine the entropy of 1 kg of wet steam at a pressure of 6 bar and 0.8 dry, reckoned from the freezing point (0 °C).

Solution. Mass of wet steam, $m = 1$ kg

Pressure of steam, $\qquad\qquad p = 6$ bar

Dryness fraction, $\qquad\qquad x = 0.8$

At 6 bar. From steam tables,

$$t_s = 158.8°C, \ h_{fg} = 2085 \ kJ/kg$$

Entropy of wet steam is given by

$$S_{wet} = C_{pw} \log_e \frac{T_s}{273} + \frac{xh_{fg}}{T_s} \quad \text{(Where } C_{pw} = \text{Specific heat of water)}$$

$$= 4.18 \log_e \left(\frac{158.8 + 273}{273} \right) + \frac{0.8 \times 2085}{(158.8 + 273)}$$

$$= 1.9165 + 3.8700 = 5.7865 kJ/kgK$$

Hence, **entropy of wet steam = 5.7865 kJ / kg K. (Ans.)**

Example 14

Steam enters an engine at a pressure of 10 bar absolute and 400 °C. It is exhausted at 0.2 bar. The steam at exhaust is 0.9 dry. Find:

(i) Drop in enthalpy;

(ii) Change in entropy.

Solution. Initial pressure of steam, $\quad p_1 = 10$ bar

Initial temperature of steam, $\qquad t_{sup} = 400°C$

Final pressure of steam, $\qquad p_2 = 0.2$ bar

Final condition of steam, $\qquad x_2 = 0.9$

At 10 bar, 400 °C. From steam tables,

$$h_{sup} = 3263.9 \text{ kJ / kg} ; \; s_{sup} = 7.465 \text{ kJ / kg K}$$

i.e., $\qquad h_1 = h_{sup} = 3263.9 \text{ kJ / kg and } s_1 = s_{sup} = 7.465 \text{ kJ / kg K}$

At 0.2 bar. From steam tables,

$$h_f = 251.5 \text{ kJ / kg} ; \; h_{fg} = 2358.4 \text{ kJ / kg} ;$$

$$s_f = 0.8321 \text{ kJ / kg K} ; s_g = 7.9094 \text{ kJ / kg K}$$

Also, $\qquad h_2 = h_{f_2} + x_2 h_{fg_2} = 251.5 + 0.9 \times 2358.4$

$$= 2374 \text{ kJ / kg.}$$

Also, $\qquad s_2 = s_{f_2} + x_2 s_{fg_2}$

$$= s_{f_2} + x_2 \left(s_{g_2} - s_{f_2} \right)$$

$$= 0.8321 + 0.9 (7.9094 - 0.8321)$$

$$= 7.2016 \text{ kJ / kg K}$$

Hence, (i) **Drop in enthalpy,**

$$= h_1 - h_2 = 3263.9 - 2374 = \textbf{889.9 kJ / kg. (Ans.)}$$

(ii) Change in entropy

$$= s_1 - s_2 = 7.465 - 7.2016$$

$$= \textbf{0.2634 kJ / kg K (decrease). (Ans.)}$$

Example 15

Find the entropy of 1 kg of superheated steam at a pressure of 12 bar and a temperature of 250 °C. Take the specific heat of superheated steam as 2.1 kJ/kg K.

Solution. Mass of steam, $\qquad m = 1$ kg

Pressure of steam, $\qquad\qquad p = 12$

bar Temperature of steam, $\qquad T_{sup} = 250 + 273 = 523$ K

Specific heat of superheated steam, $c_{ps} = 2.1$ kJ / kg K

At 12 bar. From steam tables,

$$T_s = 188 + 273 = 461 \text{ K}, \; h_{fg} = 1984.3 \text{ kJ / kg}$$

$\therefore$ Entropy of 1 kg of superheated steam,

$$s_{sup} = c_{pw} \log_e \frac{T_s}{273} + \frac{h_{fg}}{T_s} + c_{ps} \log_e \frac{T_{sup}}{T_s}$$

$$= 4.18 \log_e \left(\frac{461}{273}\right) + \frac{1984.3}{461} + 2.1 \times \log_e \left(\frac{523}{461}\right)$$

$$= 2.190 + 4.304 + 0.265$$

$$= \textbf{6.759 kJ / kg. (Ans.)}$$

7.6. Determination of the Dryness Fraction of Steam

The measuring of steam quality or dryness is required for practical applications. The dryness fraction of steam can be measured by using the following calorimeters:

1. Tank or bucket calorimeter

2. Throttling calorimeter

3. Separating and throttling calorimeter.

Description of these measurement methods are given in the following sections along with some numerical examples.

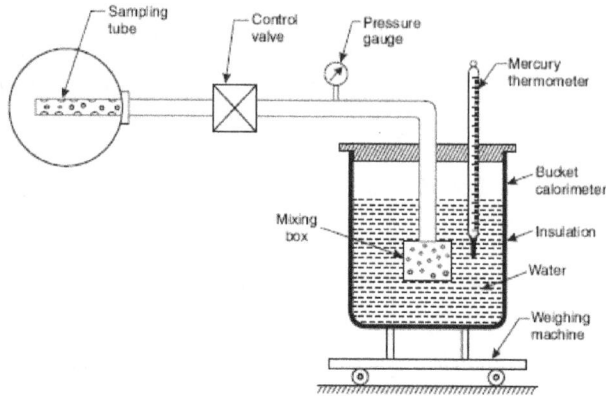

FIGURE 7.11. Tank or bucket calorimeter.

7.6.1. Tank or Bucket Calorimeter

The dryness fraction of steam can be found with the help of a tank calorimeter as follows:

A known mass of steam is passed through a known mass of water and steam is completely condensed. The heat lost by steam is equated to heat gained by the water.

Figure 7.11 shows the arrangement of this calorimeter.

The steam is passed through the sampling tube into the bucket calorimeter containing a known mass of water.

The weights of the calorimeter with water before mixing with steam and after mixing the steam are obtained by weighing.

The temperature of water before and after mixing the steam are measured by a mercury thermometer.

The pressure of steam passed through the sampling tube is measured with the help of a pressure gauge.

Let, p_s = Gauge pressure of steam (bar),

p_a = Atmospheric pressure (bar),

t_s = Daturation temperature of steam known from steam table at pressure $(p_s + p_a)$,

h_{fg} = Latent heat of steam,

x = Dryness fraction of steam,

c_{pw} = Specific heat of water,

c_{pc} = Specific heat of calorimeter,

m_c = Mass of calorimeter, kg,

m_{cw} = Mass of calorimeter and water, kg,

$m_w = (m_{cw} - m_c)$ = Mass of water in calorimeter, kg,

m_{cws} = Mass of calorimeter, water and condensed steam, kg,

$m_s = (m_{csw} - m_{cw})$ = Mass of steam condensed in calorimeter, kg,

t_{cw} = Temperature of water and calorimeter before mixing the steam, °C, and

t_{cws} = Temperature of water and calorimeter after mixing the steam, °C.

Neglecting the losses and assuming that the heat lost by steam is gained by water and the calorimeter, we have

$$\left(m_{cws} - m_{cw}\right) \left[x h_{fg} + c_{pw} \left(t_s - t_{cws}\right)\right]$$

$$= \left(m_{cw} - m_c\right) c_{pw} \left(t_{cws} - t_{cv}\right) + m_c \; c_{pc} \left(t_{cws} - t_{cw}\right)$$

$$\therefore \quad m_s [x h_{fg} + c_{pw} \left(t_s - t_{cws}\right)] = (t_{cws} - t_{cw}) \; [m_{cw} - m_c)(c_{pw} + m_c c_{pc}]$$

or $\qquad m_s [x h_{fg} + c_{pw} (t_s - t_{cws})] = \left(t_{cws} - t_{cw}\right)\left(m_w c_{pw} + m_c c_{pc}\right)$

The $m_c c_{pc}$ is known as the *water equivalent of the calorimeter.*

The value of the dryness fraction, x can be found by solving the above equation.

The value of the dryness fraction found by this method involves some *inaccuracy* since losses due to convection and radiation are *not* taken into account.

The calculated value of the dryness fraction neglecting losses is always less than the actual value of the dryness.

Example 16

Steam at a pressure of 5 bar passes into a tank containing water where it gets condensed. The mass and temperature in the tank before the admission of steam are 50 kg and 20°C, respectively. Calculate the dryness fraction of steam as it enters the tank if 3 kg of steam gets condensed and the resulting temperature of the mixture becomes 40 °C. Take the water equivalent of the tank as 1.5 kg.

■ **Solution.**

Pressure of steam	p = 5 bar
Mass of water in the tank	= 50 kg
Initial temperature of water	= 20 °C
Amount of steam condensed	m_s = 3 kg
Final temperature after condensation of steam	= 40 °C
Water equivalent of tank	= 1.5 kg

Dryness fraction of steam, x:
At 5 bar. From steam tables,

$$h_f = 640.1 \text{ kJ / kg} \; ; \; h_{fg} = 2107.4 \text{ kJ / kg}$$

Total mass of water, m_w = mass of water in the tank + water equivalent of tank

$$= 50 + 1.5 = 51.5 \text{ kg}$$

Also, heat lost by steam = heat gained by water

$$m_s \left[(h_f + x h_{fg}) - 1 \times 4.18 \ (40 - 0) \right] = m_w \left[1 \times 4.18 \ (40 - 20) \right]$$

or
$$3 \left[(640.1 + x \times 2107.4) - 4.18 \times 40 \right] = 51.5 \times 4.18 \times 20$$

or
$$3 (472.9 + 2107.4x) = 4305.4$$

or
$$472.9 + 2107.4x = 1435.13$$

∴
$$x = \frac{1435.13 - 472.9}{2107.4} = 0.456.$$

Hence, the dryness fraction of steam = 0.456. (Ans.)

7.6.2. Throttling Calorimeter

The dryness fraction of wet steam can be determined by using a throttling calorimeter, which is illustrated diagrammatically in Figure 7.12.

The steam to be sampled is taken from the pipe by means of a suitably positioned and dimensioned sampling tube. It passes into an insulated container and is throttled through an orifice to atmospheric pressure. Here the temperature is taken and the steam ideally should have about 5.5 K of superheat.

The throttling process is shown on the h-s diagram in Figure 7.13 by the line 1-2. If steam initially is wet it is throttled through a sufficiently large pressure drop, and then the steam at state 2 will become superheated. State 2 can then be defined by the measured pressure and temperature. The enthalpy, h2 can then be found and hence

$$h_2 = h_1 = \left(h_{f_1} + x_1 h_{fg_1} \right) \text{ at } p_1$$

$$\left[\text{where } h_2 = h_{f2} + h_{fg_2} + c_{ps} \left(T_{sup_2} - T_{s_2} \right) \right]$$

$$\therefore \qquad x_1 = \frac{h_2 - h_{f_1}}{h_{fg_1}}$$

Hence, the dryness fraction is determined and state 1 is defined.

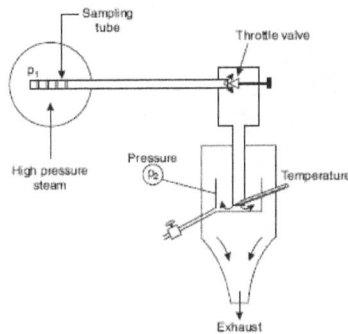

FIGURE 7.12. Throttling calorimeter.

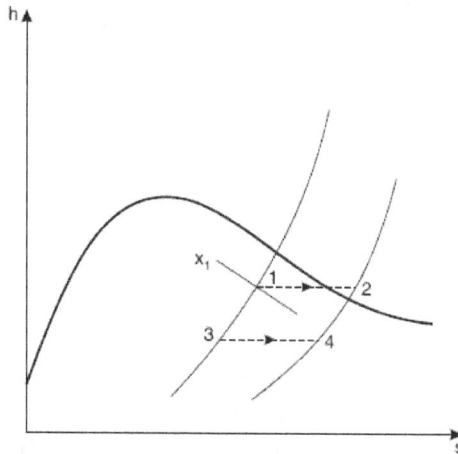

FIGURE 7.13. Throttling process, shown on a h-s plane.

Example 17

A throttling calorimeter is used to measure the dryness fraction of the steam in the steam main which has steam flowing at a pressure of 8 bar. The steam after passing through the calorimeter is at 1 bar pressure and 115 °C.

Calculate the dryness fraction of the steam in the main. Take c_{ps} = 2.1 kJ/kg K.

■ **Solution.**

Condition of steam *before throttling*:

$p_1 = 8$ bar, $x_1 = ?$

Condition of steam *after throttling*:

$p_2 = 1$ bar, $t_2 = t_{sup_2} = 115°C$

As throttling is a *constant enthalpy* process

∴ $$h_1 = h_2$$

i.e., $$h_{f_1} + x_1 h_{g f_1} = h_{f_2} + h_{fg_2} + c_{p_s} \left(T_{sup_2} - T_{s_2} \right)$$

$$[\because T_{sup_2} = 115 + 273 = 388 \text{ K} \quad T_{s_2} = 99.6 + 273 = 372.6 \text{ K} \left(\text{at 1 bar} \right)]$$

$$720.9 + x_1 \times 2046.5 = 417.5 + 2257.9 + 2.1(388 - 372.6)$$

$$720.9 + 2046.5 \, x_1 = 2707.7$$

$$\therefore \qquad x_1 = \frac{2707.7 - 720.9}{2046.5} = 0.97$$

Hence, the dryness fraction of steam in the main = 0.97. (Ans.)

7.6.3. Separating and Throttling Calorimeter

If the steam whose dryness fraction is to be determined is very wet, then throttling to atmospheric pressure may not be sufficient to ensure superheated steam at the exit. In this case it is necessary to dry the steam partially before throttling. This is done by passing the steam sample from the main through a separating calorimeter as shown in Figure 7.14. The steam is made to change direction suddenly, and the water, being denser than the dry steam, is separated out. The quantity of water which is separated out (mw) is measured at the separator; the steam remaining, which now has a higher dryness fraction, is passed through the throttling calorimeter. With the combined separating and throttling calorimeter, it is necessary to condense the steam after throttling and measure the amount of condensate (ms). If only a throttling calorimeter is sufficient, there is no need to measure condensate, as the pressure and temperature measurements at the exit are sufficient.

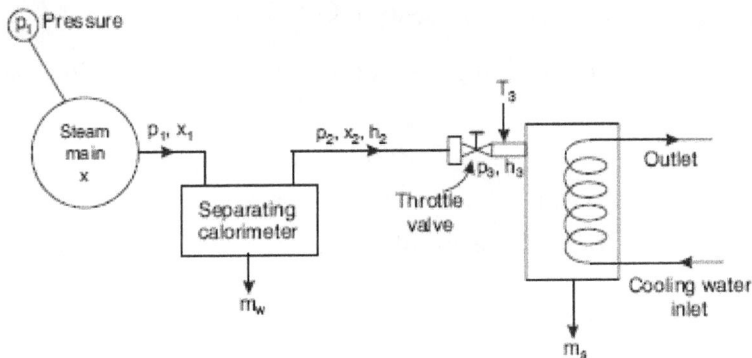

FIGURE 7.14. Separating and throttling calorimeter.

The dryness fraction at 2 is x_2, therefore, the mass of dry steam leaving the separating calorimeter is equal to $x_2 ms$ and this must be the mass of dry vapor in the sample drawn from the main at state 1.

Hence fraction in main, $x_1 = \dfrac{\text{Mass of dry vapour}}{\text{Total mass}} = \dfrac{x_2\, m_s}{m_w + m_s}$.

The dryness fraction, x_2, can be determined as follows:

$$^*h_3 = h_2 = h_{f_2} + x_2 h_{fg_2} \text{ at } p_2 \ [^*h_3 = h_{f_3} + h_{fg_3} + c_{ps}(T_{sup_3} - T_{s_3}) \text{ at pressure } p_3]$$

or,
$$x_2 = \frac{h_3 - h_{f_2}}{h_{fg_2}}$$

The values of hf_2 and hfg_2 are read from steam tables at pressure p_2. The pressure in the separator is small so that p_1 is approximately equal to p_2.

Example 18

The following observations were taken with a separating and a throttling calorimeter arranged in series:

Water separated = *2 kg,* steam discharged from the throttling calorimeter = *20.5 kg,* temperature of steam after throttling= *110°C* , initial pressure = *12 bar abs.,* barometer = *760 mm* of Hg, and final pressure = *5 mm* of Hg.

Estimate the quality of steam supplied.

■ Solution.

Quantity of water separated out, $m_w = 2$ kg

Steam (condensate) discharged from the throttling calorimeter, $m_s = 20.5$ kg

Temperature of steam after throttling, $t_{sup} = 110°C$
Initial pressure of steam, $p_1 = 12$ bar abs.

Final pressure of steam, $p_3 = 760 + 5 = 765$ mm

$$= \frac{765}{1000} \times 1.3366 \qquad\qquad (\because\ 1 \text{ m Hg} = 1.3366 \text{ bar})$$

$\simeq 1$ bar

From steam tables:

At $p_1 = p_2 = 12$ **bar** : $h_f = 798.4$ kJ / kg, $h_{fg} = 1984.3$ kJ / kg

At $p_3 = 1$ **bar** : $t_s = 99.6°C$, $h_f = 417.5$ kJ/kg, $h_{fg} = 2257.9$ kJ/kg

$t_{sup} = 110°C$ (given)

Also, $\qquad\qquad\qquad\qquad h_3 = h_2$

$$\left(h_{f_3} + h_{fg_3}\right) + c_{ps}\left(T_{sup_3} - T_{s_3}\right) = hf_2 + x_2 h_{fg_2}$$

Taking $\qquad\qquad\qquad c_{ps} = 2$ kJ / kg K, we get

$$417.5 + 2257.9 + 2\left[(110 + 273) - (99.6 + 273)\right] = 798.4 + x_2 \times 1984.3$$

$$2696.2 = 798.4 + 1984.3\, x_2$$

$$\therefore \qquad\qquad\qquad\qquad\qquad x_2 = \frac{2696.2 - 798.4}{1984.3} = 0.956$$

Now, the quality of steam supplied,

$$x_1 = \frac{x_2 m_s}{m_w + m_s} = \frac{0.956 \times 20.5}{2 + 20.5}$$

$$= 0.87. \text{(Ans.)}$$

7.7. Exercises

1. What amount of heat would be required to produce 4.4 kg of steam at a pressure of 6 bar and a temperature of 250 °C from water at 30 °C? Take specific heat for superheated steam as 2.2 kJ/kg K.

2. Determine the mass of 0.15 m3 of wet steam at a pressure of 4 bar and dryness fraction 0.8. Also calculate the heat of 1 m^3 of steam.

3. A quantity of steam at 10 bar and 0.85 dryness occupies 0.15 m^3. Determine the heat supplied to raise the temperature of the steam to 300 °C at a constant pressure and percentage of this

heat which appears as external work. Take the specific heat of a superheated steam as 2.2 kJ/kg K.

4. A piston-cylinder contains 3 kg of wet steam at 1.4 bar. The initial volume is 2.25 m³. The steam is heated until its temperature reaches 400 °C. The piston is free to move up or down unless it reaches the stops at the top. When the piston is up against the stops, the cylinder volume is 4.65 m³. Determine the amount of work and heat transfer to or from the steam.

5. Find the internal energy of 1 kg of steam at 20 bar when

   (i) it is superheated, its temperature being 400 °C;

   (ii) it is wet, its dryness being 0.9.

   Assume superheated steam to behave as a perfect gas from the commencement of superheating and thus it obeys Charles's law. Specific heat for steam = 2.3 kJ/kg K.

6. Steam at 140 bar has an enthalpy of 3001.9 kJ/kg; find the temperature, the specific volume, and the internal energy.

7. Find the specific volume, enthalpy, and internal energy of wet steam at 18 bar, dryness fraction 0.85.

8. Find the dryness fraction, specific volume, and internal energy of steam at 7 bar and enthalpy 2550 kJ/kg.

9. Steam at a pressure of 1.1 bar and 0.95 dry is passed into a tank containing 90 kg of water at 25 °C. The mass of the tank is 12.5 kg and the specific heat of the metal is 0.42 kJ/kg K. If the temperature of water rises to 40 °C after the passage of the steam, determine the mass of steam condensed. Neglect radiation and other losses.

10. The following data were obtained in a test on a combined separating and throttling calorimeter:

    Pressure of steam sample $= 15\,bar$, pressure of steam at exit $= 1\,bar$, temperature of steam at the exit $= 150°C$, discharge from separating calorimeter $= 0.5\,kg/min$, discharge from throttling calorimeter $= 10\,kg/min$.

    Determine the dryness fraction of the sample steam.

BASIC STEAM POWER CYCLES

In This Chapter

8.1. Overview

Water as a working fluid is commonly used in thermodynamic systems for power production using power cycles. Through involved processes in a power cycle, water will go through sequences of phases from liquid to vapor under different temperatures and pressures. Therefore, there exist two-phase fluids in parts of the cycle. This creates some challenges in practice; for example, compression of a liquid-vapor mixture is much harder than a single phase. In this chapter we discuss several steam power cycles including the Carnot cycle, the Rankine cycle, and their modified versions with numerical examples.

8.2. Carnot Cycle with Two-Phase Working Fluid

Figure 8.1 shows a Carnot cycle on T-s and p-V diagrams. It consists of (a) two constant pressure processes (4-1) and (2-3) and (b) two frictionless adiabatic processes (1-2) and (3-4). These processes are discussed in the following:

Process (4-1) is a reversible isothermal expansion. One kilogram of boiling water at temperature T_1 is heated to form wet steam of dryness fraction x_1. Thus, heat is absorbed at a constant temperature T_1 and pressure p_1 during this operation.

Process (1-2) is a reversible adiabatic expansion. During this operation steam is expanded isentropically (i.e., entropy is conserved for a reversible adiabatic process) to temperature T_2 and pressure p_2. The point "2" represents the condition of steam after expansion. Generally point "2" is under the dome, where both liquid and vapor are present.

Process (2-3) is a reversible isothermal compression. During this operation heat is rejected at constant pressure p_2 and temperature T_2. As the steam is exhausted it becomes wetter and cooled from 2 to 3.

Process (3-4) is a reversible adiabatic compression. In this operation the wet steam at "3" is compressed isentropically till the steam regains its original state of temperature T_1 and pressure p_1. Thus, the cycle is completed. The same cycle is shown on a T-S diagram in Figure 8.1. With reference to this diagram we can analyze the cycle as follows:

Heat supplied at constant temperature T_1 [operation (4-1)] = area of the rectangle 4-1-b-a = $T_1 (s_1 - s_4)$ or $T_1 (s_2 - s_3)$.

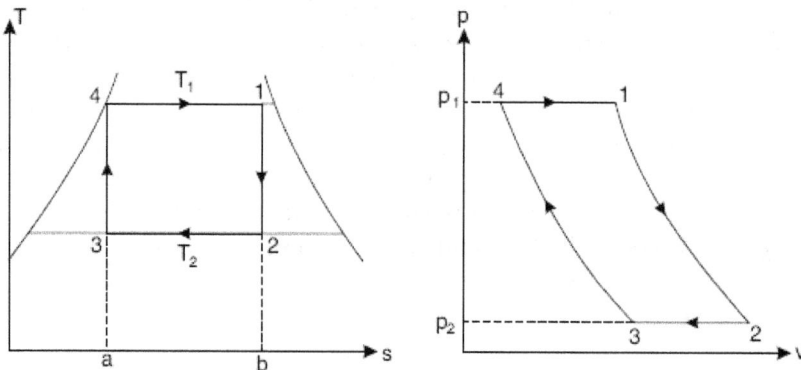

FIGURE 8.1. Carnot cycle on T-s and p-V diagrams.

Heat rejected at a constant temperature T_2 (operation 2-3) = area of the rectangle 2-3-a-b = T_2 (s_2–s_3). Since there is no exchange of heat during isentropic operations (1-2) and (3-4), due to adiabatic and reversibility conditions, then using the 1st law we have: Net work done = Heat supplied – heat rejected Or W_{net} = $T_1(S_2 - S_3) - T_2(S_2 - S_3)$ = $(T_1 - T_2)(S_2 - S_3)$. Efficiency of the cycle is the ratio of net work over heat supplied.

$$\eta_{Carnot} = \frac{(T_1 - T_2)(S_2 - S_3)}{T_1(S_2 - S_3)} = 1 - \frac{T_2}{T_1} \tag{8.1}$$

8.2.1. Limitations of the Carnot Cycle

Though the Carnot cycle is thermodynamically a simple engine and has the highest thermal efficiency for given values of T_1 and T_2, it is extremely difficult to operate in practice because of the following reasons:

- It is difficult to compress a wet vapor isentropically to the saturated state as required by the process 3-4.

- It is difficult to control the quality of the condensate coming out of the condenser so that the state 3 is exactly obtained.

- The efficiency of the Carnot cycle is greatly affected by the temperature T_1 at which heat is transferred to the working fluid. Since the critical temperature for steam is only 374 °C (at critical pressure of 221.2 bar), if the cycle is to be operated in the *wet region*, the maximum possible temperature is severely limited.

- The cycle is still more difficult to operate in practice with superheated steam due to the necessity of supplying the superheat at a constant temperature instead of constant pressure (as is customary).

In a practical cycle, limits of pressure and volume are far more easily realized than limits of temperature so that at present no practical engine operates on the Carnot cycle, although all modern cycles aspire to achieve it.

8.3. Rankine Cycle with Two-Phase Working Fluid

The Rankine cycle is actually a modified version of the Carnot cycle with the purpose of minimizing or removing its practical limitations. In other words, the Ranking cycle is the thermodynamic cycle on which the steam turbine (or engine) works. A sketch of the Rankine cycle's major components is shown in Figure 8.2 with corresponding thermodynamic diagrams as shown in Figure 8.3. It comprises the following processes:

Process 1–2: Reversible adiabatic expansion in the turbine (or steam engine). During this process dry steam is expanded isentropically. Cases for wet steam, process 1'-2', or superheated steam, process 1"-2", are also shown.

Process 2–3: Constant-pressure transfer of heat in the condenser. During this process two-phase fluid enters the condenser and changes to the liquid phase.

Process 3–4: Reversible adiabatic pumping process in the feed pump. During this process liquid water is compressed isentropically.

Process 4–1: Constant-pressure transfer of heat in the boiler. During this process liquid water is expanded.

Figure 8.3 shows the Rankine cycle on p-v, T-s, and h-s diagrams. When the saturated steam enters the turbine, the steam can be wet or superheated also. With reference to this diagram we can analyze the Rankine cycle as follows (considering 1 kg of fluid):

Applying the 1st law to the boiler, turbine, condenser, and pump, we have:

1. **For boiler** (as control volume), we get $Q_2 = h_1 - h_{f4}$.

2. **For turbine** (as control volume), we get $W_T = h_1 - h_2$, where W_T is the turbine work output.

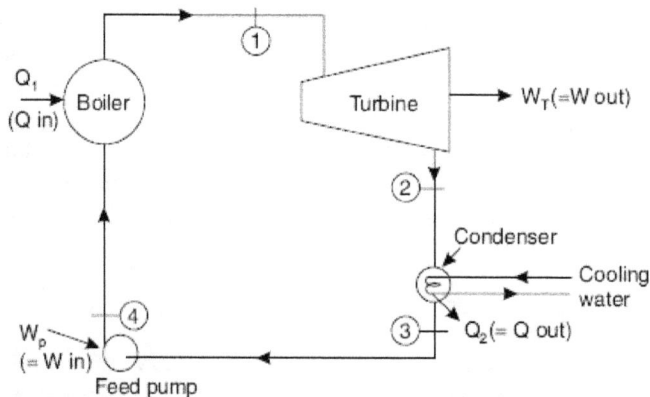

FIGURE 8.2. Rankine cycle processes and its major components.

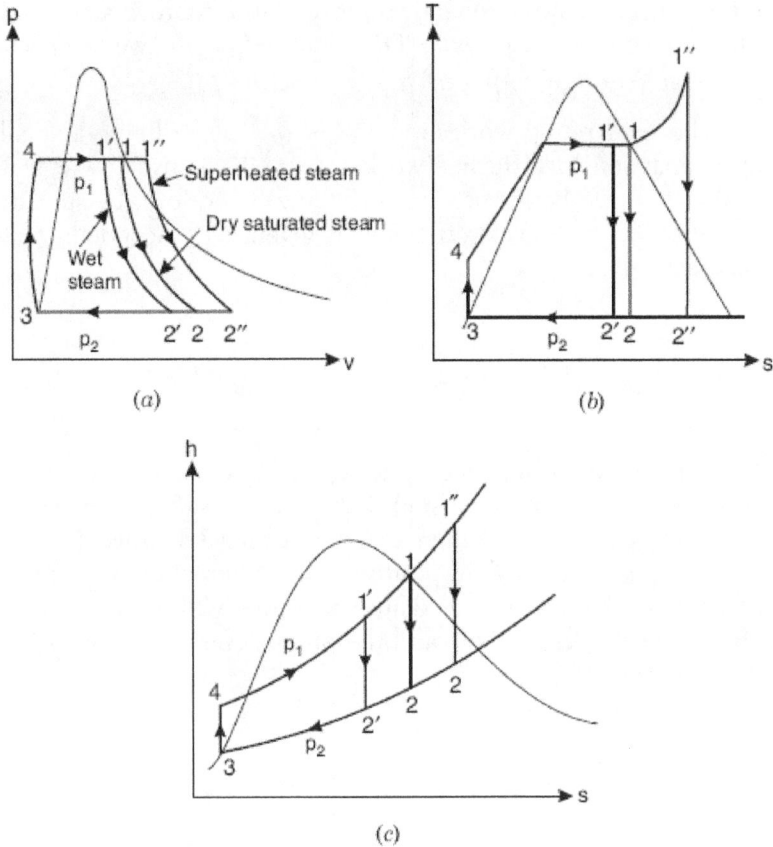

FIGURE 8.3. (a) p-v diagram; (b) T-s diagram; (c) h-s diagram for Rankine cycle.

3. **For condenser**, we get $Q_2 = h_2 - h_{f3}$

4. **For the feed pump**, we get $W_p = h_{f4} - h_{f3}$, where W_p is the pump work input.

Now, the efficiency of the Rankine cycle is given by

$$\eta_{Rankine} = \frac{W_{net}}{Q_1} = \frac{W_T - W_P}{Q_1} = \frac{(h_1 - h_2) - (h_{f_4} - h_{f_3})}{h_1 - h_{f_4}} \tag{8.2}$$

The feed pump handles liquid water, which is incompressible, which means with the increase in pressure its density or specific volume undergoes a little change. Using the general property relation for reversible adiabatic compression (i.e., $dh = du + d(pv) = T\,ds - p\,dv + p\,dv + v\,dp$) we get $dh = vpd$. we get Therefore, $h_{f4} - h_{f3} = v_3(p_1 - p_2)\,h_{f4} - h_{f3} = v_3(p_1 - p_2)$.

The feed pump work input, $W_p = (h_{f4} - h_{f3})$, being a small quantity in comparison with turbine work output, W_T, is usually neglected, especially when the boiler pressures are low. Therefore, by substitution into Equation 8.2, the cycle efficiency is given, with acceptable engineering accuracy, as

$$\eta_{Rankine} \cong \frac{h_1 - h_2}{h_1 - h_{f4}} \tag{8.3}$$

For the actual Rankine cycle, we have to modify the isentropic processes and consider irreversibility due to losses. This can be achieved by considering the entropy increase for 1-2 and 3-4 processes, when points 2 and 4 are shifted toward positive entropy direction to points 2a and 4a, respectively (shown by dashed lines in Figure 8.3b), and defining isentropic efficiencies for the turbine and the pump. The isentropic efficiencies are given as:

$$\eta_{turbine} = \frac{h_1 - h_{2a}}{h_1 - h_2}, \quad \eta_{pump} = \frac{h_4 - h_3}{h_{4a} - h_3} \tag{8.4}$$

8.3.1. Comparison between Rankine and Carnot Cycles

With reference to the previous sections, it is useful to list some the advantages of the Rankine cycle in comparison to the Carnot cycle. The following points are worth noting:

- Between the same temperature limits the Rankine cycle provides a higher specific work output than a Carnot cycle; consequently, the Rankine cycle requires a smaller steam flow rate resulting in a smaller size plant for a given power output. However, the Rankine cycle calls for higher rates of heat transfer in the boiler and condenser.

- Since in the Rankine cycle only part of the heat is supplied isothermally at a constant higher temperature T_1, its efficiency is

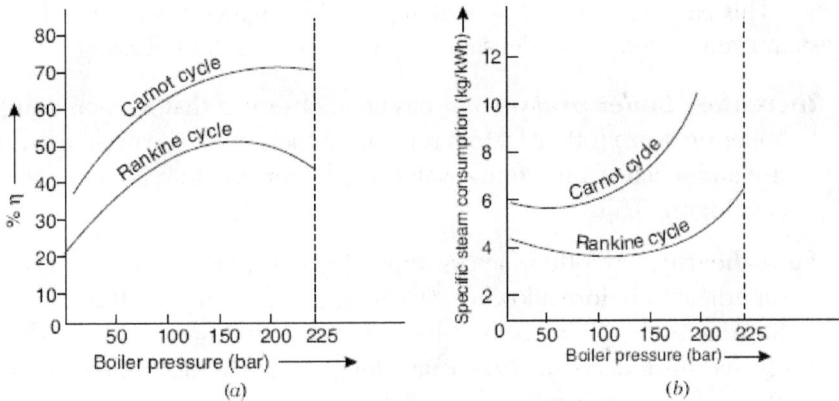

FIGURE 8.4. Efficiency (a) and specific steam consumption (b) against boiler pressure for Carnot and ideal Rankine cycles.

lower than that of the Carnot cycle. The efficiency of the Rankine cycle will approach that of the Carnot cycle more nearly if the superheat temperature rise is reduced.

- The advantage of using a pump to feed liquid to the boiler in the Rankine cycle instead of compressing a wet vapor in the Carnot cycle is obvious in that the latter work for compression is very large compared to the pump.

Figure 8.4 shows the plots between efficiency and specific steam consumption against boiler pressure for Carnot and ideal Rankine cycles.

8.3.2. Effect of Operating Conditions and System Layout on Efficiency

The efficiency of the Rankine cycle, or in general any practical power cycle, can be increased either by a) changing the thermodynamic operation conditions or b) modifying the power system layout and adding auxiliary components. The operation conditions for power cycles using the Rankine cycle can affect its efficiency. The efficiency can be improved by:

- Increasing the average temperature at which heat is supplied.

- Decreasing/reducing the temperature at which heat is rejected.

- Reducing condenser pressure.

This can be achieved by making suitable changes in the conditions of steam generation or condensation, as discussed in the following:

Increasing boiler pressure. It has been observed that by increasing the boiler pressure (other factors remaining the same), the cycle tends to rise and reaches a maximum value at a boiler pressure of about 166 bar (see Figure 8.5[a]).

Superheating. All other factors remaining the same, if the steam is superheated before allowing it to expand in the turbine, then the Rankine cycle efficiency may be increased (see Figure 8.5[b]). The use of superheated steam also ensures longer turbine blade life because of the absence of erosion from high velocity water particles that are suspended in wet vapor.

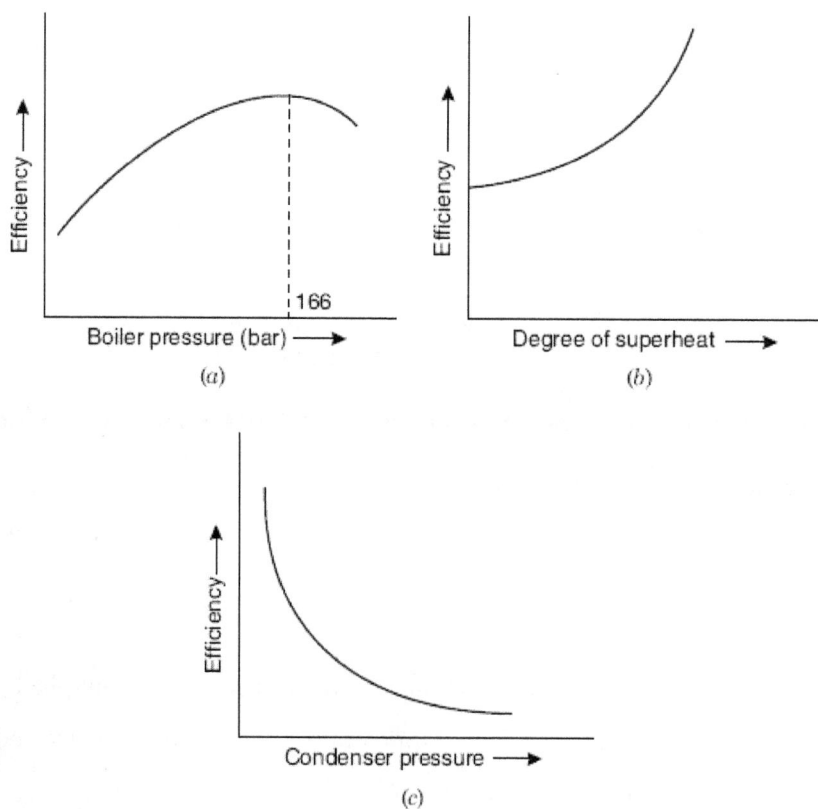

FIGURE 8.5. Effect of operating conditions on the thermal efficiency of the Rankine cycle.

Reducing condenser pressure. The thermal efficiency of the cycle can be amply improved by reducing the condenser pressure (see Figure 8.5[c]), hence reducing the temperature at which heat is rejected, especially in high vacuums. But the increase in efficiency is obtained at the increased cost of the condensation apparatus.

The thermal efficienty of the Rankine cycle is also improved by modifying its system components and layout as listed by the following methods:

- By reheating of steam.
- By regenerative feed heating.
- By water extraction.
- By using binary vapor.

In further sections, we will discuss in detail these methods along with numerical examples.

We use the following numerical examples to demonstrate applications of the Carnot and Rankine cycles.

Example 1

The following data refer to a simple steam power plant:

State No.	Location	Pressure	Quality/Tem.	Velocity
1.	Inlet to turbine	6 MPa (= 60 bar)	380°C	—
2.	Exit from turbine inlet to condenser	10 kPa (= 0.1 bar)	0.9	200 m/s
3.	Exit from condenser and inlet to pump	9 kPa (= 0.09 bar)	Saturated liquid	—
4.	Exit from pump and inlet to boiler	7 MPa (= 70 bar)	—	—
5.	Exit from boiler	6.5 MPa (= 65 bar)	400°C	—
	Rate of steam flow = 10000 kg/h.			

Calculate:

1. *Power output of the turbine.*

2. *Heat transfer per hour in the boiler and condenser separately.*

3. *Mass of cooling water circulated per hour in the condenser. Choose the inlet temperature of cooling water 20 °C and 30 °C at exit from the condenser.*

4. *Diameter of the pipe connecting the turbine with the condenser.*

■ **Solution.** Refer to Figure 8.6.

(a) Power output of the turbine, P:

At 60 bar, 380°C: From steam tables,

$$h_1 = 3043.0 \left(\text{at } 350°\text{C}\right) + \frac{3177.2 - 3043.0}{(400 - 350)} \times 30 \text{ By interpolation}$$

$$= 3123.5 \text{ kJ/kg}$$

At 0.1 bar:

$$h_{f_2} = 191.8 \text{ kJ/kg}, h_{fg_2} = 2392.8 \text{ kJ/kg (from steam tables)}$$

and $\qquad x_2 = 0.9 \,(\text{given})$

∴ $\qquad h_2 = h_{f_2} + x_2 \, h_{fg_2} = 191.8 + 0.9 \times 2392.8 = 2345.3 \text{ kJ/kg}$

Power output of the turbine $= m_s (h_1 - h_2) \text{ kW},$

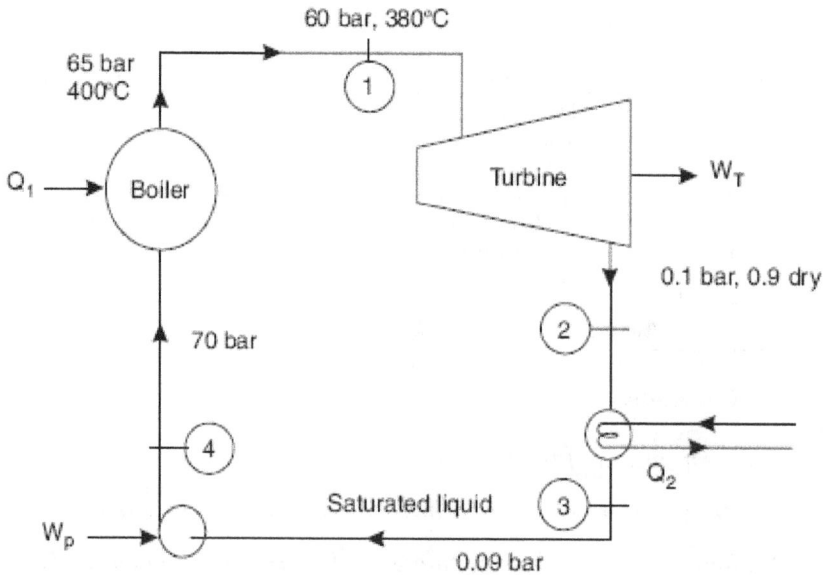

FIGURE 8.6. Cycle for Example 1.

[where m_s = Rate of steam flow in kg/s and h_1, h_2 = Enthalpy of steam in kJ/kg]

$$= \frac{10000}{3600}(3123.5 - 2345.3) = 2162 \text{ kW}$$

Hence power output of the turbine = **2162 kW.**   (**Ans.**)

(b) Heat transfer per hour in the boiler and condenser:

At 70 bar : $h_{f_4} = 1267.4$ kJ/kg

At 65 bar, $400°C$: $h_a = \dfrac{3177.2\,(60 \text{ bar}) + 3158.1\,(70 \text{ bar})}{2} = 3167.6$ kJ/kg

......(By interpolation)

∴ *Heat transfer per hour in the boiler,*

$$Q_1 = 10000\,(h_a - h_{f_4}) \text{ kJ/h}$$

$$= 10000\,(3167.6 - 1267.4) = \mathbf{1.9 \times 10^7\, kJ/h.} \quad (\mathbf{Ans.})$$

At 0.09 bar : $h_{f_3} = 183.3$ kJ/kg

Heat transfer per hour in the condenser,

$$Q_1 = 10000\,(h_2 - h_{f_3})$$

$$= 10000\,(2345.3 - 183.3) = \mathbf{2.16 \times 10^7\, kJ/h.} \quad (\mathbf{Ans.})$$

(c) Mass of cooling water circulated per hour in the condenser, m_w:

Heat lost by steam = Heat gained by the cooling water

$$Q_2 = m_w \times c_{pw}\,(t_2 - t_1)$$

$$2.16 \times 10^7 = m_w \times 4.18\,(30 - 20)$$

∴ $$\mathbf{m}_w = \frac{2.16 \times 10^7}{4.18\,(30 - 20)} = \mathbf{1.116 \times 10^7\, kg/h.} \quad (\mathbf{Ans.})$$

(d) Diameter of the pipe connecting the turbine with the condenser, d:

$$\frac{\pi}{4} d^2 \times C = m_2 x_2 v_{g_2} \qquad\qquad (i)$$

Here, d = Diameter of the pipe (m),

C = Velocity of steam = 200 m/s (given),

m_s = Mass of steam in kg/s,

x_2 = Dryness fraction at '2', and

v_{g_2} = Specific volume at pressure 0.1 bar $(= 14.67 \text{ m}^3/\text{kg})$.

Substituting the various values in eqn. (i), we get

$$\frac{\pi}{4} d^2 \times 200 = \frac{10000}{3600} \times 0.9 \times 14.67$$

$$d = \left(\frac{10000 \times 0.9 \times 14.67 \times 4}{3600 \times \pi \times 200} \right)^{1/2} = \textbf{0.483 m or 483 mm. (Ans.)}$$

Example 2

In a steam power cycle, the steam supply is at 15 bar and dry and saturated. The condenser pressure is 0.4 bar. Calculate the Carnot and Rankine efficiencies of the cycle. Neglect pump work.

■ **Solution.** Steam supply pressure, $\quad p_1 = 15$ bar, $x_1 = 1$

Condenser pressure, $\qquad\qquad\qquad p_2 = 0.4$ bar

Carnot and Rankine efficiencies:

From steam tables:

At 15 bar : $\ t_s = 198.3°C, \qquad h_g = 2789.9$ kJ/kg, $\qquad s_g = 6.4406$ kJ/kg K

At 0.4 bar : $t_s = 75.9°C, \qquad h_f = 317.7$ kJ/kg, $\qquad h_{fg} = 2319.2$ kJ/kg,

$\qquad s_f = 1.0261$ kJ/kg K, $s_{fg} = 6.6448$ kJ/kg K

$\qquad T_1 = 198.3 + 273 = 471.3$ K

$\qquad T_2 = 75.9 + 273 = 348.9$ K

$$\eta_{\text{carnot}} = \frac{T_1 - T_2}{T_1} = \frac{471.3 - 348.9}{471.3}$$

$$= \textbf{0.259 or 25.9\%. (Ans.)}$$

$$\eta_{\text{Rankine}} = \frac{\text{Adiabatic or isentropic heat drop}}{\text{Heat supplied}} = \frac{h_1 - h_2}{h_1 - h_{f_2}}$$

where $\quad h_2 = h_{f_2} + x_2 h_{fg_2} = 317.7 + x_2 \times 2319.2$

Value of x_2:

As the steam expands isentropically,

$$\therefore \qquad\qquad s_1 = s_2$$

$$6.4406 = s_{f_2} + x_2\, s_{fg_2} = 1.0261 + x_2 \times 6.6448$$

$$\therefore \qquad\qquad x_2 = \frac{6.4406 - 1.0261}{6.6448} = 0.815$$

$$\therefore \qquad\qquad h_2 = 317.7 + 0.815 \times 2319.2 = 2207.8 \text{ kJ/kg}$$

$$\text{Hence } \eta_{\text{Rankine}} = \frac{2789.9 - 22.7.8}{2789.9 - 317.7} = 0.2354 \text{ or } 23.54\%. \quad \text{Ans.}$$

Example 3

A steam turbine at 20 bar, 360 °C is expanded to 0.08 bar. It then enters a condenser, where it is condensed to saturated liquid water. The pump feeds back the water into the boiler. Assume ideal processes, find per kg of steam, the net work, and the cycle efficiency.

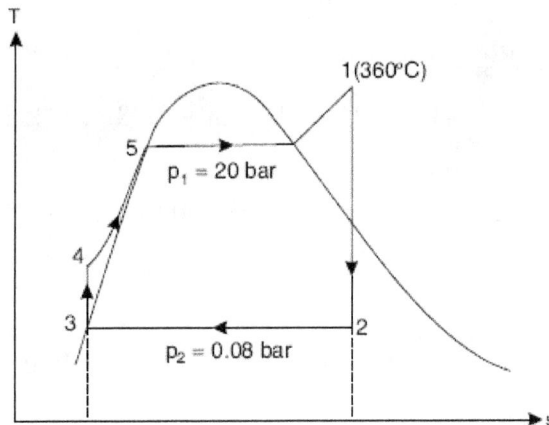

FIGURE 8.7. Steam cycle for Example 3.

■ **Solution:**

Boiler pressure, $\qquad p_1 = 20$ bar (360°C)

Condenser pressure, $\quad p_2 = 0.08$ bar

From steam tables:

At 20 bar (p_1), 360° C : $\qquad\qquad h_1 = 3159.3$ kJ/kg

$\qquad\qquad\qquad\qquad\qquad\qquad s_1 = 6.9917$ kJ/kg K

At 0.08 bar (p_2) : $\qquad\qquad h_3 = h_{f(p_2)} = 173.88$ kJ/kg,

$\qquad\qquad\qquad\qquad\qquad\qquad s_3 = s_{f(p_2)} = 0.5926$ kJ/kg K

$\qquad h_{fg(p_2)} = 2403.1$ kJ/kg, $\qquad s_{g(p_2)} = 8.2287$ kJ/kg K

$\qquad v_{f(p_2)} = 0.001008$ m^3/kg $\qquad \therefore s_{fg(p_2)} = 7.6361$ kJ/kg K

Now $\qquad\qquad\qquad\qquad\qquad s_1 = s_2$

$6.9917 = s_{f(p_2)} + x_2\, s_{fg(p_2)} = 0.5926 + x_2 \times 7.6361$

$\therefore \qquad\qquad x_2 = \dfrac{0.69917 - 0.5926}{7.6361} = 0.838$

$\therefore \qquad\qquad h_2 = h_{f(p_2)} + x_2\, h_{fg(p_2)}$

$\qquad\qquad\qquad = 173.88 + 0.838 \times 2403.1 = 2187.68$ kJ/kg.

Net work, W$_{net}$:

$\qquad W_{net} = W_{turbine} - W_{pump}$

$\qquad W_{pump} = h_{f_4} - h_{f\, p_2} = h_{f_3} = v_{f\, p_2}\ p_1 - p_2$

$\qquad\qquad = 0.00108$ m^3/kg $\times 20 - 0.08 \times 100$ kN/m^2

$\qquad\qquad = 2.008$ kJ/kg

[and $h_{f_4} = 2.008 + h_{f\, p_2} = 2.008 + 173.88 = 175.89$ kJ/kg]

$\qquad W_{turbine} = h_1 - h_2 = 3159.3 - 2187.68 = 971.62$ kJ/kg

$\qquad W_{net} = 971.62 - 2.008 = 969.61$ kJ/kg. Ans.

Cycle efficiency, η_{cycle} :

$$Q_1 = h_1 - h_{f_4} = 3159.3 - 175.89 = 2983.41 \text{ kJ/kg}$$

$$\eta_{cycle} = \frac{W_{net}}{Q_1} = \frac{969.61}{2983.41} = \textbf{0.325 or 32.5\%.} \quad \textbf{Ans.}$$

Example 4

A Rankine cycle operates between pressures of 80 bar and 0.1 bar. The maximum cycle temperature is 600 °C. If the steam turbine and condensate pump efficiencies are 0.9 and 0.8 respectively, calculate the specific work and thermal efficiency. The following is the relevant steam table extract.

p(bar)	t(°C)	Specific volume (m3/kg)		Specific enthalpy (kJ/kg)			Specific entropy (kJ/kg K)		
		v_f	v_g	h_f	h_{fg}	h_g	s_f	s_{fg}	s_g
0.1	45.84	0.0010103	14.68	191.9	2392.3	2584.2	0.6488	7.5006	8.1494
80	295.1	0.001385	0.0235	1317	1440.5	2757.5	3.2073	2.5351	5.7424

80 bar, 600° C	v	0.486 m³ / kg
Superheat	h	3642 kJ / kg
table	s	7.0206 kJ / kg K

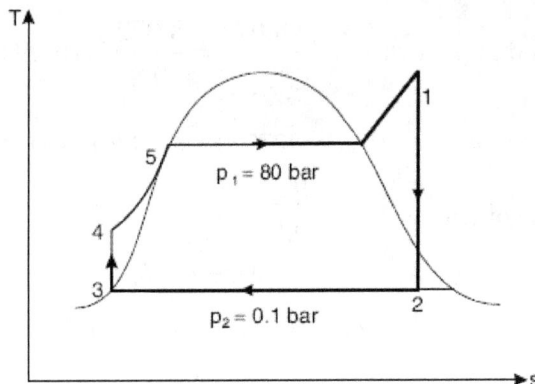

FIGURE 8.8. Rankine cycle for Example 4.

■ **Solution:**

Refer to Figure 8.8.

At 80 bar,   600° C :

$h_1 = 3642$ kJ/kg ;

$s_1 = 7.0206$ kJ/kg K.

Since $s_1 = s_2$

$\therefore 7.0206 - s_{f2} + x_2 \, s_{fg2}$

$\qquad = 0.6488 + x_2 \times 7.5006$

or $\qquad x_2 = \dfrac{7.0206 - 0.6488}{7.5006} = 0.85$

Now, $h_2 = h_{f2} + x_2 \, h_{fg2}$

$\qquad\qquad = 191.9 + 0.85 \times 2392.3$

$\qquad\qquad = 2225.36$ kJ/kg

Actual turbine work

$$= \eta_{\text{turbine}} \times h_1 - h_2$$

$$= 0.9 \, 3642 - 2225.36 = 1275 \text{ kJ/kg}$$

pump work $\qquad = v_{f\,p_2} \; p_1 - p_2$

$$= 0.0010103 \, 80 - 0.1 \times \dfrac{10^5}{10^3} \text{kN/m}^2 = 8.072 \; kJ / kg$$

Actual pump work $\qquad = \dfrac{8.072}{\eta_{\text{pump}}} = \dfrac{8.072}{0.8} = 10.09$ kJ/kg

Specific work $\qquad (W_{\text{net}}) = 1275 - 10.09 = \mathbf{1264.91 \; kJ / kg.}$ **(Ans.)**

Thermal efficiency $\qquad = \dfrac{W_{\text{net}}}{Q_1}$

where, $\qquad\qquad\qquad Q_1 = h_1 - h_{f_4}$

But $\qquad\qquad h_{f_4} = h_{f_3} + \text{pump work} = 191.9 + 10.09 = 202$ kJ/kg

$\therefore$ Thermal efficiency, $\eta_{\text{th}} = \dfrac{1264.91}{3642 - 202} = \mathbf{0.368 \text{ or } 36.8\%.}$ **Ans.**

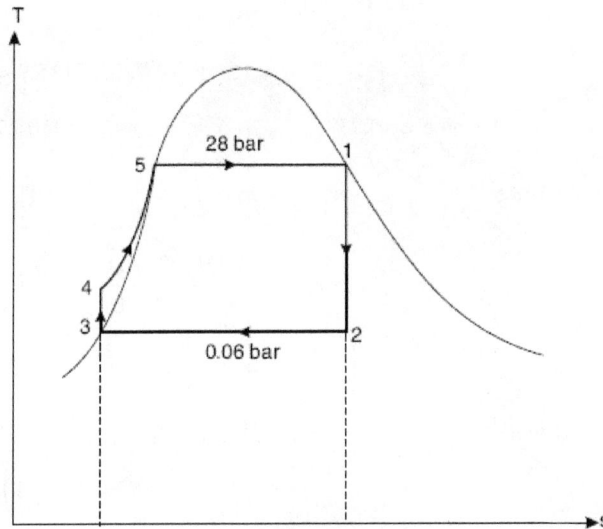

FIGURE 8.9. Rankine cycle for Example 5.

Example 5

A simple Rankine cycle works between pressures 28 bar and 0.06 bar, the initial condition of steam being dry saturated. Calculate the cycle efficiency, work ratio, and specific steam consumption.

■ **Solution:**

From steam tables,

At 28 bar : $\qquad h_1 = 2802$ kJ/kg, $s_1 = 6.2104$ kJ/kg K

At 0.06 bar : $\qquad h_{f_2} = h_{f_3} = 151.5$ kJ/kg, $h_{fg_2} = 2415.9$ kJ/kg,

$$s_{f_2} = 0.521 \text{ kJ/kg K}, \ s_{fg_2} = 7.809 \text{ kJ/kg K}$$

$$v_f = 0.001 \text{ m}^3/\text{kg}$$

Considering turbine *process 1–2*, we have:

$$s_1 = s_2$$

$$6.2104 = s_{f_2} + x_2 \ s_{fg_2} = 0.521 + x_2 \times 7.809$$

∴ $$x_2 = \frac{6.2104 - 0.521}{7.809} = 0.728$$

$$\therefore \qquad\qquad h_2 = h_{f_2} + x_2\, h_{fg_2}$$

$$= 151.5 + 0.728 \times 2415.9 = 1910.27 \text{ kJ/kg}$$

$$\therefore \qquad \text{Turbine work, } W_{\text{turbine}} = h_1 - h_2 = 2802 - 1910.27 = 891.73 \text{ kJ/kg}$$

pump work,
$$W_{\text{pump}} = h_{f_4} - h_{f_3} = v_f \left(p_1 - p_2 \right)$$

$$= \frac{0.001 (28 - 0.06) \times 10^5}{1000} = 2.79 \text{ kJ/kg}$$

$$\left[\because \quad h_{f_4} = h_{f_3} + 2.79 = 151.5 + 2.79 = 154.29 \text{ kJ/kg} \right]$$

$\therefore$ Net work,
$$W_{\text{net}} = W_{\text{turbine}} - W_{\text{pump}}$$

$$= 891.73 - 2.79 = 888.94 \text{ kJ/kg}$$

Cycle efficiency
$$= \frac{W_{\text{net}}}{Q_1} = \frac{888.94}{h_1 - h_{f_4}}$$

$$= \frac{888.94}{2802 - 154.29} = \mathbf{0.3357 \text{ or } 33.57\%.} \quad (\mathbf{Ans.})$$

Work ratio
$$= \frac{W_{\text{net}}}{W_{\text{turbine}}} = \frac{888.94}{891.73} = \mathbf{0.997.} \quad (\mathbf{Ans.})$$

Specific steam consumption
$$= \frac{3600}{W_{\text{net}}} = \frac{3600}{888.94} = \mathbf{4.049 \text{ kg / kWh.}} \quad (\mathbf{Ans.})$$

Example 6

In a Rankine cycle, the steam at the inlet to the turbine is saturated at a pressure of 35 bar and the exhaust pressure is 0.2 bar. Determine:

(*i*) *The pump work,*   (*ii*) *The turbine work,*

(*iii*) *The Rankine efficiency,*   (*iv*) *The condenser heat flow,*

(*v*) *The dryness at the end of expansion.*

  Assume flow rate of 9.5 kg/s.

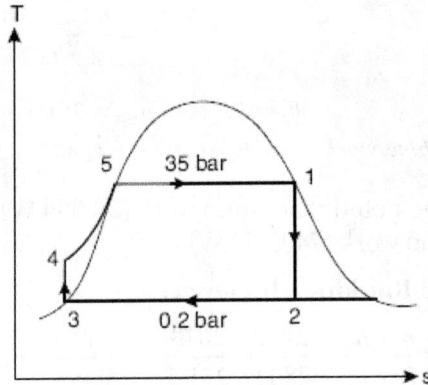

FIGURE 8.10. Rankine cycle for Example 6.

■ **Solution:**

Pressure and condition of steam, at the inlet to the turbine,

$$p_1 = 35 \text{ bar}, x = 1$$

Exhaust pressure, $\quad\quad\quad p_2 = 0.2 \text{ bar}$

Flow rate, $\quad\quad\quad\quad \dot{m} = 9.5 \text{ kg/s}$

From steam tables:

At 35 bar : $\quad\quad\quad\quad h_1 = h_{g_1} = 2802 \text{ kJ/kg}, s_{g_1} = 6.1228 \text{ kJ/kg K}$

At 0.26 bar : $\quad\quad\quad\quad h_f = 251.5 \text{ kJ/kg}, h_{fg} = 2358.4 \text{ kJ/kg},$

$$v_f = 0.001017 \text{ m}^3/\text{kg}, s_f = 0.8321 \text{ kJ/lg K}, s_{fg} = 7.0773 \text{ kJ/kg K}.$$

(*i*) **The pump work:**

Pump work $\quad\quad = (p_4 - p_3)\, v_f = (35 - 0.2) \times 10^5 \times 0.001017 \text{ J or } 3.54 \text{ kJ/kg}$

$$\left[\begin{array}{l} \text{Also } h_{f_4} - h_{f_3} = \text{ pump work } = 3.54 \\ \therefore \quad h_{f_4} = 251.5 + 3.54 = 255.04 \text{ kJ/kg} \end{array} \right]$$

Now power required to drive the pump

$$= 9.5 \times 3.54 \text{ kJ/s or } \textbf{33.63 kW. (Ans.)}$$

(*ii*) **The turbine work:**

$$s_1 = s_2 = s_{f_2} + x_2 \times s_{fg_2}$$

$$6.1228 = 0.8321 + x_2 \times 7.0773$$

$$\therefore \qquad\qquad\qquad x_2 = \frac{6.1228 - 0.8321}{7.0773} = 0.747$$

$$\therefore \qquad\qquad h_2 = h_{f_2} + x_2\, h_{fg_2} = 251.5 + 0.747 \times 2358.4 = 2013 \text{ kJ/kg}$$

$$\therefore \quad \textit{Turbine work} = \dot{m}\,(h_1 - h_2) = 9.5\,(2802 - 2013) = \mathbf{7495.5\,kW.} \quad (\textbf{Ans.})$$

It may be noted that pump work (33.63 kW) is very small as compared to the turbine work (7495.5 kW).

(*iii*) **The Rankine efficiency:**

$$\eta_{\text{ranking}} = \frac{h_1 - h_2}{h_1 - h_2} = \frac{2802 - 2013}{2802 - 251.5} = \frac{789}{2550.5} = \mathbf{0.3093\ or\ 30.93\%.} \quad (\textbf{Ans.})$$

(*iv*) **The condenser heat flow:**

The condenser heat flow $= \dot{m}\,(h_2 - h_{f_3}) = 9.5\,(2013 - 251.5) = \mathbf{16734.25\,kW.}$ (**Ans.**)

(*v*) **The dryness at the end of expansion, x₂:**

The dryness at the end of expansion,

$$x_2 = \mathbf{0.747\ or\ 74.7\%.} \quad (\textbf{Ans.})$$

Example 7

The adiabatic enthalpy drop across the prime mover of the Rankine cycle is 840 kJ/kg. The enthalpy of the steam supplied is 2940 kJ/kg. If the back pressure is 0.1 bar, find the specific steam consumption and thermal efficiency.

■ **Solution:**

Adiabatic enthalpy drop, $h_1 - h_2 = 840$ kJ/kg

Enthalpy of steam supplied, $\qquad\qquad h_1 = 2940$ kJ/kg

Back pressure, $\qquad\qquad\qquad p_2 = 0.1$ bar

From steam tables, corresponding to 0.1 bar : $h_f = 191.8$ kJ/kg

$$\text{Now.} \qquad \eta_{\text{ranking}} = \frac{h_1 - h_2}{h_1 - h_2} = \frac{840}{2940 - 191.8} = \mathbf{0.3056 = 30.56\%.} \quad \textbf{Ans.}$$

Useful work done per kg of steam $= 840$ kJ/kg

$\therefore$ *Specific steam consumption* $= \dfrac{1}{840}$ kg/s $= \dfrac{1}{840} \times 3600 =$ **4.286 kg / kWh.** $\big($**Ans.**$\big)$

Example 8

A 35 kW (I.P.) system engine consumes 284 kg/h at 15 bar and 250 °C. If the condenser pressure is 0.14 bar, determine:

    (i) Final condition of steam;   (ii) Rankine efficiency;

    (iii) Relative efficiency.

  ■ **Solution:**

Power developed by the engine	= 35 kW (I.P.)
Steam consumption	= 284 kg/h
Condenser presure	= 0.14 bar
Steam inlet presure	= 15 bar, 250°C.

From steam tables:

At 15 bar, 250° C : $h = 2923.3$ kJ/kg, $s = 6.709$ kJ/kg K

At 0.14 bar : $h_f = 220$ kJ/kg, $h_{fg} = 2376.6$ kJ/kg,

                          $s_f = 0.737$ kJ/kg K, $s_{fg} = 7.296$ kJ/kg K

(i) **Final condition of steam:**

Since steam expands isentropically,

$\therefore$ $s_1 = s_2 = s_{f_2} + x_2 \, s_{fg_2}$

$6.709 = 0.737 + x_2 \times 7.296$

$\therefore$ $x_2 = \dfrac{6.709 - 0.737}{7.296} =$ **0.818** $\simeq$ **0.82.** $\big($**Ans.**$\big)$

$\therefore$ $h_2 = h_{f_2} + x_2 \, h_{fg_2} = 220 + 0.82 \times 2376.6 = 2168.8$ kJ/kg.

(ii) **Rankine efficiency:**

$\eta_{\text{ranking}} = \dfrac{h_1 - h_2}{h_1 - h_2} = \dfrac{2923.3 - 2168.8}{2923.3 - 220} =$ **0.279 or 27.9%.**   **Ans.**

(*iii*) **Relative efficiency:**

$$\eta_{\text{themal}} = \frac{\text{I.P.}}{\dot{m}\, h_1 - h_2} = \frac{35}{\dfrac{284}{3600}\,2923.3 - 220} = \textbf{0.1641 or } 16.41\%$$

$$\eta_{\text{ranking}} = \frac{\eta_{\text{themal}}}{\eta_{\text{ranking}}} = \frac{0.1641}{0.279}$$

$$= 0.588 \text{ or } 58.8\%. \ (\textbf{Ans.})$$

Example 9

Calculate the fuel oil consumption required in an industrial steam plant to generate 5000 kW at the turbine shaft. The calorific value of the fuel is 40000 kJ/kg and the Rankine cycle efficiency is 50%. Assume appropriate values for isentropic turbine efficiency, boiler heat transfer efficiency, and combustion efficiency.

Solution. Power to be generated at the turbine shaft, $P = 5000$ kW

The calorific value of the fuel, $C = 40000$ kJ/kg

Rankine cycle efficiency, $\eta_{\text{rankine}} = 50\%$

Fuel oil combustion, m_f:

Assume: $\eta_{\text{turbine}} = 90\%$; $\eta_{\text{neat transfer}} = 85\%$; $\eta_{\text{combustion}} = 98\%$

$$\eta_{\text{rankine}} = \frac{\text{Shaft power} / \eta_{\text{turbine}}}{m_f \times C \times \eta_{\text{hear transfer}} \times \eta_{\text{combustion}}}$$

or

$$0.5 = \frac{5000 / 0.9}{m_f \times 40000 \times 0.85 \times 0.98}$$

$$\therefore \quad m_f = \frac{(5000 / 0.9)}{0.5 \times 40000 \times 0.85 \times 0.98} = \textbf{0.3335 kg / s or 1200.6 kg / h.} \ (\textbf{Ans.})$$

8.4. Modified Rankine Cycle

Figure 8.11 and Figure 8.12 show the modified Rankine cycle on *p-V* and *T-s* diagrams (neglecting pump work) respectively. It will be noted that the *p-V* diagram is very narrow at the toe, that is, point 2′, and the work

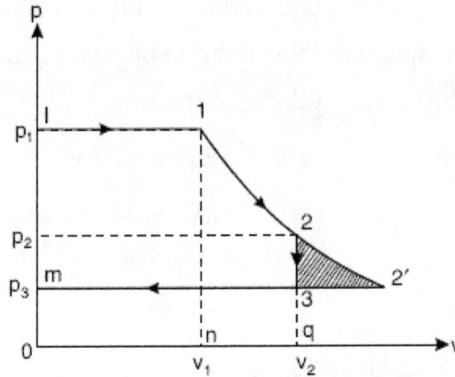

FIGURE 8.11. p-V diagram of modified Rankine cycle.

obtained near to *e* is very small. In fact this work is even too inadequate to overcome friction (due to reciprocating parts). Therefore, the adiabatic is terminated at 2; the pressure drop decreases suddenly while the volume remains constant. This operation is represented by the line 2–3. By doing this the stroke length is reduced; in other words, the cylinder dimensions reduce but at the expense of a small loss of work (area 2-3-2′) which, however, is negligibly small.

The work done during the modified Rankine cycle can be calculated in the following way:

Let p_1, v_1, u_1, and h_1 correspond to initial condition of steam at point 1.

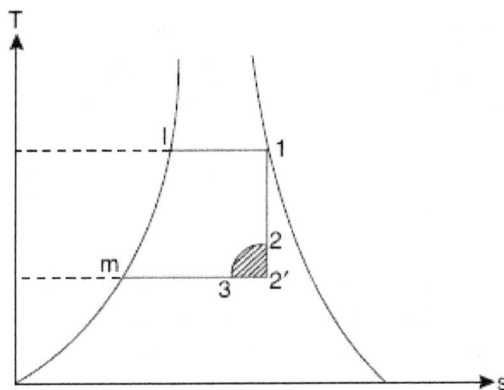

FIGURE 8.12. T-s diagram of modified Rankine cycle.

p_2, v_2, u_2, and h_2 correspond to the condition of steam at point 2.

p_3 and h_3 correspond to the condition of steam at point 3.

Work done during the cycle/kg of steam.

$$= \text{area } l\text{-}1\text{-}2\text{-}3\text{-}m$$

$$= \text{area } `o\text{-}l\text{-}1\text{-}n` + \text{area } `1\text{-}2\text{-}q\text{-}n` - \text{area } `o\text{-}m\text{-}3\text{-}q`$$

$$= p_1 v_1 + (u_1 - u_2) - P_3 v_2$$

Heat supplied

$$= h_1 - h_{f_3}$$

∴ The modified Rankine efficiency

$$= \frac{\text{Work done}}{\text{Heat supplied}}$$

$$= \frac{p_1 v_1 + (u_1 - u_2) - p_3 v_2}{h_1 - h_{f_3}}$$

Alternative method for finding modified Rankine efficiency:

Work done during the cycle/kg of steam

$$= \text{area } `l\text{-}1\text{-}2\text{-}3\text{-}m`$$

$$= \text{area } `l\text{-}1\text{-}2\text{-}s` + \text{area } `s\text{-}2\text{-}3\text{-}m`$$

$$= (h_1 - h_2) + (p_2 - p_3) v_2$$

Heat supplied

$$= h_1 - h_{f_3}$$

Modified Rankine efficiency

$$= \frac{\text{Work done}}{\text{Heat supplied}}$$

$$= \frac{(h_1 - h_2) + (p_2 - p_3) v_2}{h_1 - h_{f_3}}$$

Note that the modified Rankine cycle is used for "reciprocating steam engines" because stroke length and hence cylinder size is reduced with the sacrifice of practically quite a negligible amount of work done.

Example 10

Steam at a pressure of 15 bar and 300 °C is delivered to the throttle of an engine. The steam expands to 2 bar when the release occurs. The steam exhaust takes place at 1.1 bar. A performance test gave the result of the specific steam consumption of 12.8 kg/kWh and a mechanical efficiency of 80 per cent. Determine:

(i) Ideal work or the modified Rankine engine work per kg.

(ii) Efficiency of the modified Rankine engine or ideal thermal efficiency.

(iii) The indicated and brake work per kg.

(iv) The brake thermal efficiency.

(v) The relative efficiency on the basis of indicated work and brake work.

■ **Solution:**

Figure 8.13 shows the p-v and T-s diagrams for a modified Rankine cycle.

From steam tables:

1. *At 15 bar, 300° C :* $h_1 = 3037.6$ kJ/kg, $v_1 = 0.169$ m^3/kg,

    $s_1 = 6.918$ kJ/kg K.

2. *At 2 bar :* $t_{s_2} = 120.2°$C, $h_{f_2} = 504.7$ kJ/kg, $h_{fg_2} = 2201.6$ kJ/kg,

    $s_{f_2} = 1.5301$ kJ/kg k, $s_{fg_2} = 5.5967$ kJ/kg K,

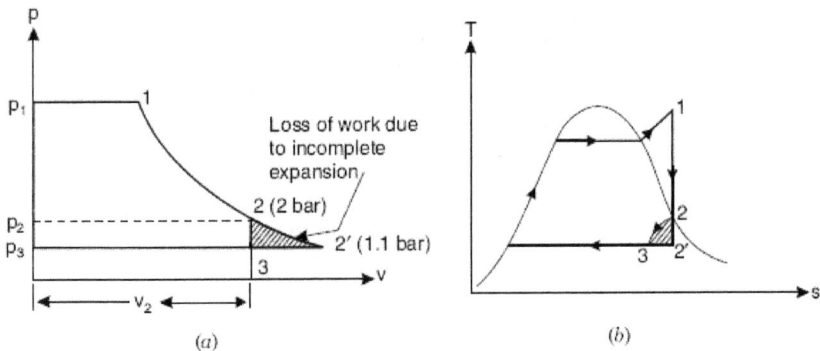

FIGURE 8.13. p-V and T-s diagrams for Example 10.

$$v_{f_2} = 0.00106 \text{ m}^3/\text{kg}, v_{g_2} = 0.885 \text{ m}^3/\text{kg}.$$

3. *At 1.1 bar :* $t_{s_3} = 102.3°\text{C}, h_{f_3} = 428.8 \text{ kJ/kg}, h_{fg_3} = 2250.8 \text{ kJ/kg},$

$$s_{f_3} = 1.333 \text{ kJ/kg K}, s_{fg_3} = 5.9947 \text{ kJ/kg K},$$

$$v_{f_3} = 0.001 \text{ m}^3/\text{kg}, v_{g_3} = 1.549 \text{ m}^3/\text{kg}.$$

During isentropic expansion 1–2, we have

$$s_1 = s_2$$

$$6.918 = s_{f_2} + x_2\, s_{fg_2} = 1.5301 + x_2 \times 5.5967$$

∴ $$x_2 = \frac{6.918 - 1.5301}{5.5967} = 0.96.$$

Then $h_2 = h_{f_2} + x_2\, h_{fg_2} = 504.7 + 0.96 \times 2201.6 = 2618.2 \text{ kJ/kg}$

$$v_2 = x_2\, v_{g_2} + (1 - x_2)\, v_{f_2}$$

$$= 0.96 \times 0.885 + (1 - 0.96) \times 0.00106 = 0.849 \text{ m}^3/\text{kg}.$$

(*i*) **Ideal work:**

Ideal work or modified Rankine engine work/kg,

$$W = (h_1 - h_2) + (p_2 - p_3)v_2$$

$$= (3037.6 - 2618.2) + (2 - 1.1) \times 10^5 \times 0.849/1000$$

$$= 419.4 + 76.41 = \mathbf{495.8\ kJ/kg.}\quad \textbf{(Ans.)}$$

(*ii*) **Rankine engine efficiency:**

$$\eta_{\text{rankine}} = \frac{\text{Work done}}{\text{Heat supplied}} = \frac{495.8}{h_1 - h_{f_3}}$$

$$= \frac{495.8}{3037.6 - 428.8} = \mathbf{0.19\ or\ 19\%.\ Ans.}$$

(*iii*) Indicated and brake work per kg:

Indicated work / kg, $W_{indicated} = \dfrac{I.P.}{\dot{m}}$

$$= \frac{1 \times 3600}{12.8} = \textbf{281.25 kJ / kg. (Ans.)}$$

Brake work / kg, $W_{brake} = \dfrac{B.P.}{\dot{m}} = \dfrac{\eta_{mech.} \times I.P.}{\dot{m}}$

$$= \frac{0.8 \times 1 \times 3600}{12.8} = \textbf{225 kJ/kg. (Ans.)}$$

(*iv*) Brake thermal efficiency:

Brake thermal efficiency $= \dfrac{W_{brake}}{h_1 - h_{f_3}} = \dfrac{225}{3037.6 - 428.8} = \textbf{0.086 or 8.6\%. (Ans.)}$

(*v*) Relative efficiency:

Relative efficiency *on the basis of indicated work*

$$= \frac{\dfrac{W_{indicated}}{h_1 - h_{f_3}}}{\dfrac{W}{h_1 - h_{f_3}}} = \frac{W_{indicated}}{W} = \frac{281.25}{495.8} = \textbf{0.567 or 56.7\%. (Ans.)}$$

Relative efficiency *on the basis of brake work*

$$= \frac{\dfrac{W_{indicated}}{\left(h_1 - h_{f_3}\right)}}{\dfrac{W}{\left(h_1 - h_{f_3}\right)}} = \frac{W_{brake}}{W} = \frac{225}{495.8} = \textbf{0.4538 or 45.38\%. (Ans.)}$$

Example 11

Superheated steam at a pressure of 10 bar and 400 °C is supplied to a steam engine. Adiabatic expansion takes place to the release point at 0.9 bar and it exhausts into a condenser at 0.3 bar. Neglecting clearance, determine for a steam flow rate of 1.5 kg/s:

(i) Quality of steam at the end of expansion and the end of constant volume operation.

(ii) Power developed.

(iii) Specific steam consumption.

(iv) Modified Rankine cycle efficiency.

■ **Solution:**

Figure 8.14 shows the *p-V* and *T-s* diagrams for a modified Rankine cycle.

From steam tables:

1. At 10 bar, 400° C : $h_1 = 3263.9$ kJ/kg, $v_1 = 0.307$ m^3/kg, $s_1 = 7.465$ kJ/kg K

2. At 0.9 bar : $t_{s_2} = 96.7°$C, $h_{g_2} = 2670.9$ kJ/kg, $s_{g_2} = 7.3954$ kJ/kg K,

$v_{g_2} = 1.869$ m^3/kg

3. At 0.3 bar : $h_{f_3} = 289.3$ kJ/kg, $v_{g_3} = 5.229$ m^3/kg

(*i*) Quality of steam at the end of expansion, T$_{sup2:}$

For isentropic expansion 1-2, we have

$$s_1 = s_2$$

$$= s_{g_2} + c_p \, \log_e \frac{T_{sup2}}{T_{s_2}}$$

$$7.465 = 7.3954 + 2.1 \, \log_e \frac{T_{sup2}}{(96.7 + 273)}$$

$$\left(\frac{7.465 - 7.3954}{2.1} \right) = \log_e \frac{T_{sup2}}{369.7} \text{ or } \log_e \frac{T_{sup2}}{369.7} = 0.0033$$

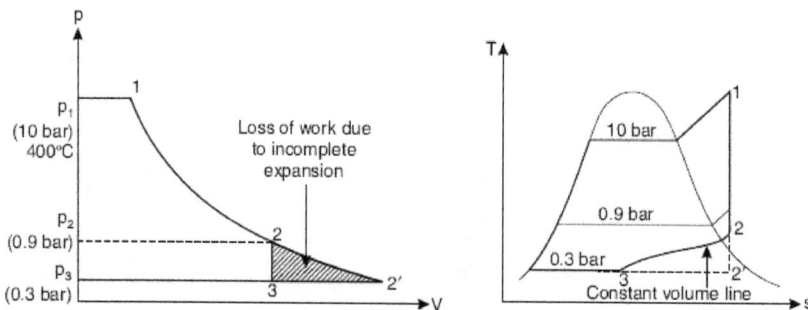

FIGURE 8.14. p-V and T-s diagrams, for Example 11.

$$\frac{T_{sup2}}{369.7} = 1.0337 \text{ or } T_{sup2} = 382 \text{ K}$$

Fig. 14. *p-V* and *T-s* diagrams.

or $\qquad\qquad\qquad t_{sup2} = 382 - 273 = \mathbf{109°C.}$ **(Ans.)**

$\therefore \qquad\qquad\qquad h_2 = h_{g_2} + c_{ps} \left(T_{sup2} - T_{s_2} \right)$

$= 2670.9 + 2.1 \left(382 - 366.5 \right) = 2703.4 \text{ kJ/kg}.$

Quality of steam at the end of constant volume operation, x_3:

For calculating v_2 using the relation

$$\frac{v_{g_2}}{T_{s_2}} = \frac{v_2}{T_{sup2}} \quad \text{(Approximately)}$$

$$\frac{1.869}{369.7} = \frac{v_2}{382}$$

or $\qquad\qquad\qquad v_2 = \frac{1.869 \times 382}{369.7} = 1.931 \text{ m}^3/\text{kg}$

Also $\qquad\qquad\qquad v_2 = v_3 = x_3 \ v_{g_3}$

$$1.931 = x_3 \times 5.229$$

or $\qquad\qquad\qquad \mathbf{x_3} = \frac{1.931}{5.229} = \mathbf{0.37.} \quad \textbf{(Ans.)}$

(*ii*) Power developed, P:

Work done $\qquad\qquad = \left(h_1 - h_2 \right) + \left(p_2 - p_3 \right) v_2$

$$= \left(3263.9 - 2703.4 \right) + \frac{\left(0.75 - 0.3 \right) \times 10^5 \times 1.931}{1000}$$

$$= 560.5 + 86.9 = 647.4 \text{ kJ/kg}$$

$\therefore$ **Power developed** = Steam flow rate × work done (per kg)

$$= 1 \times 647.4 = \mathbf{647.4 \ kW.} \quad \textbf{(Ans.)}$$

(*iii*) **Specific steam consumption, ssc:**

$$ssc = \frac{3600}{Power} = \frac{1 \times 3600}{647.4} = 5.56 \text{ kg / kWh.} \quad (\textbf{Ans.})$$

(*iv*) **Modified Rankine cycle efficiency, η_{mR}:**

$$\eta_{mR} = \frac{(h_1 - h_2) + (p_2 - p_3)v_2}{h_1 - h_{f_3}}$$

$$= \frac{647.4}{3263.9 - 289.3} = 0.217 \text{ or } 21.7\%. \quad (\textbf{Ans.})$$

8.5. Regenerative Rankine Cycle

In the Rankine cycle it is observed that the condensate which is at a fairly low temperature has an irreversible mixing with hot boiler water and this results in a decrease of cycle efficiency. This is usually achieved by bleeding some low-pressure steam from the turbine and mixing it with the feed water in a heat exchanger (or feed-water heater, FWH). This heating method is called regenerative feed heat and the cycle is called a *regenerative cycle*. The resulting heated water then is fed into the boiler. Regenerating improves the cycle efficiency by 4-5%.

The principle of regeneration can be practically utilized by extracting steam from the turbine at several locations and supplying it to the regenerative heaters. The resulting cycle is known as a *regenerative or bleeding cycle*. The heating arrangement comprises: (a) for medium capacity turbines—not more than 3 heaters; (b) for high pressure, high capacity turbines—not more than 5 to 7 heaters; and (c) for turbines of super critical parameters, 8 to 9 heaters. The most advantageous condensate heating temperature is selected depending on the turbine throttle conditions, and this determines the number of heaters to be used. The final condensate heating temperature is kept 50 to 60 °C below the boiler saturated steam temperature so as to prevent evaporation of water in the feed mains following a drop in the boiler drum pressure. The conditions of the steam bled for each heater are so selected that the temperature of saturated steam will be 4 to 10 °C higher than the final condensate temperature.

Figure 8.15 shows a diagrammatic layout of a condensing steam power plant in which a surface condenser is used to condense all the steam that is not extracted for feed water heating. The turbine is double extracting and

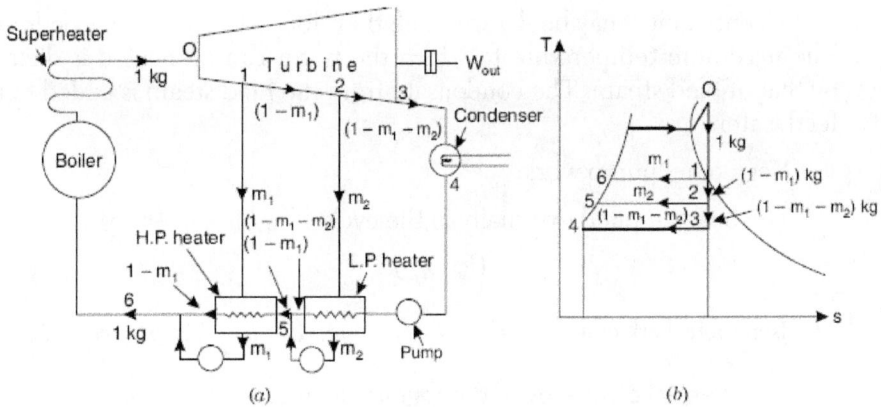

FIGURE 8.15. Regenerative Rankine cycle system and major components.

the boiler is equipped with a super heater. The cycle diagram (T-s) would appear as shown in Figure 8.15(b). This arrangement constitutes a *regenerative cycle*.

Let, m_1 = kg of high pressure (H.P.) steam per kg of steam flow,

m_2 = kg of low pressure (L.P.) steam extracted per kg of steam flow, and

$(1 - m_2 - m_2)$ = kg of steam entering condenser per kg of steam flow.

Energy/Heat balance equation for H.P. heater:

$$m_1 (h_1 - h_{f_6}) = (1 - m_1) (h_{f_6} - h_{f_5})$$

or

$$m_1[(h_1 - h_{f_6}) + (h_{f_6} - h_{f_5})] = (h_{f_6} - h_{f_5})$$

or

$$m_1 = \frac{h_{f_6} - h_{f_5}}{h_1 - h_{f_5}}$$

Energy/Heat balance equation for L.P. heater:

$$m_2 (h_2 - h_{f_5}) = (1 - m_1 - m_2) (h_{f_5} - h_{f_3})$$

or

$$m_2[(h_2 - h_{f_5}) + (h_{f_5} - h_{f_3})] = (1 - m_1) (h_{f_5} - h_{f_3})$$

or

$$m_2 = \frac{(1 - m_1) (h_{f_5} - h_{f_3})}{(h_2 - h_{f_3})}$$

All enthalpies may be determined; therefore, m_1 and m_2 may be found. The maximum temperature to which the water can be heated is dictated by that of bled steam. The condensate from the bled steam is added to the feed water.

Neglecting pump work:

The heat supplied externally in the cycle

$$= \left(h_0 - h_{f_6}\right)$$

Isentropic work done $\quad = m_1\left(h_0 - h_1\right) + m_2\left(h_0 - h_2\right) + \left(1 - m_1 - m_2\right)\left(h_0 - h_3\right)$

The thermal efficiency of the regenerative cycle is

$$\eta_{\text{thermal}} = \frac{\text{Work done}}{\text{Heat supplied}}$$

$$= \frac{m_1\left(h_0 - h_1\right) + m_2\left(h_0 - h_2\right) + \left(1 - m_1 - m_2\right)\left(h_0 - h_3\right)}{\left(h_0 - h_{f_6}\right)}$$

$$\left[\begin{array}{l} \text{The work done by the turbine may also be calculated by summing up the products of the} \\ \text{steam flow and the corresponding heat drop in the turbine stages.} \\ \textit{i.e.,} \qquad \text{Work done} = \left(h_0 - h_1\right) + \left(1 - m_1\right)\left(h_1 - h_2\right) + \left(1 - m_1 - m_2\right)\left(h_2 - h_3\right) \end{array}\right]$$

Advantages of the Regenerative cycle over the Simple Rankine Cycle:

- The heating process in the boiler tends to become reversible.

- The thermal stresses set up in the boiler are minimized. This is due to the fact that temperature ranges in the boiler are reduced.

- The thermal efficiency is improved because the average temperature of heat addition to the cycle is increased.

- Heat rate is reduced.

- The blade height is less due to the reduced amount of steam passed through the low-pressure stages.

- Due to many extractions there is an improvement in the turbine drainage and it reduces erosion due to moisture.

- A small size condenser is required.

Disadvantages of Regenerative cycle over Simple Rankine cycle:

- The plant becomes more complicated.
- Because of addition of heaters greater maintenance is required.
- For given power a large capacity boiler is required.
- The heaters are costly and the gain in thermal efficiency is not much in comparison to the heavier costs.

Note that in the absence of precise information (regarding actual temperature of the feed water entering and leaving the heaters and of the condensate temperatures) the following assumptions should always be made while doing calculations:

- Each heater is ideal and bled steam just condenses.
- The feed water is heated to saturation temperature at the pressure of bled steam.
- Unless otherwise stated the work done by the pumps in the system is considered negligible.
- There is an equal temperature rise in all the heaters (usually 10 °C to 15 °C).

Example 12

A steam turbine is fed with steam having an enthalpy of 3100 kJ/kg. It moves out of the turbine with an enthalpy of 2100 kJ/kg. Feed heating is done at a pressure of 3.2 bar with steam enthalpy of 2500 kJ/kg. The condensate from a condenser with an enthalpy of 125 kJ/kg enters into the feed heater. The quantity of bled steam is 11200 kg/h. Find the power developed by the turbine. Assume that the water leaving the feed heater is saturated liquid at 3.2 bar and the heater is a direct mixing type. Neglect pump work.

■ Solution:

Arrangement of the components is shown in Figure 8.16.

At 3.2 bar, $\qquad h_{f_2} = 570.9$ kJ/kg.

Consider m kg out of 1 kg is taken to the feed heater.

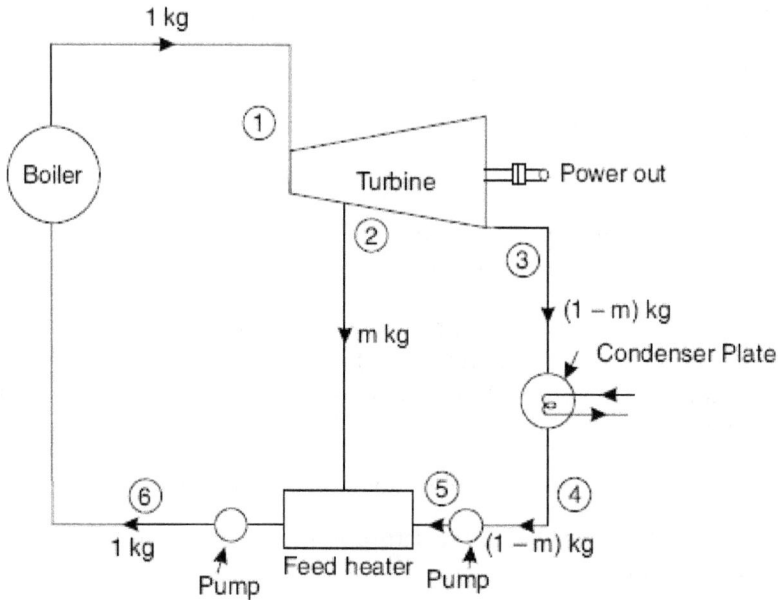

FIGURE 8.16. Steam turbine sketch for Example 12.

Energy balance for the feed heater is written as:

$$mh_2 + (1-m) h_{f_5} = 1 \times h_{f_2}$$
$$m \times 2100 + (1-m) \times 125 = 1 \times 570.9$$
$$2100\, m + 125 - 125\, m = 570.9$$
$$1975\, m = 570.9 - 125$$

∴ $m = 0.226$ kg per kg of steam supplied to the turbine

∴ Steam supplied to the turbine per hour

$$= \frac{11200}{0.226} = 49557.5 \text{ kg/h}$$

Net work developed per kg of steam

$$= (h_1 - h_2) + (1-m)(h_2 - h_3)$$
$$= (3100 - 2500) + (1 - 0.226)(2500 - 2100)$$
$$= 600 + 309.6 = 909.6 \text{ kJ/kg}$$

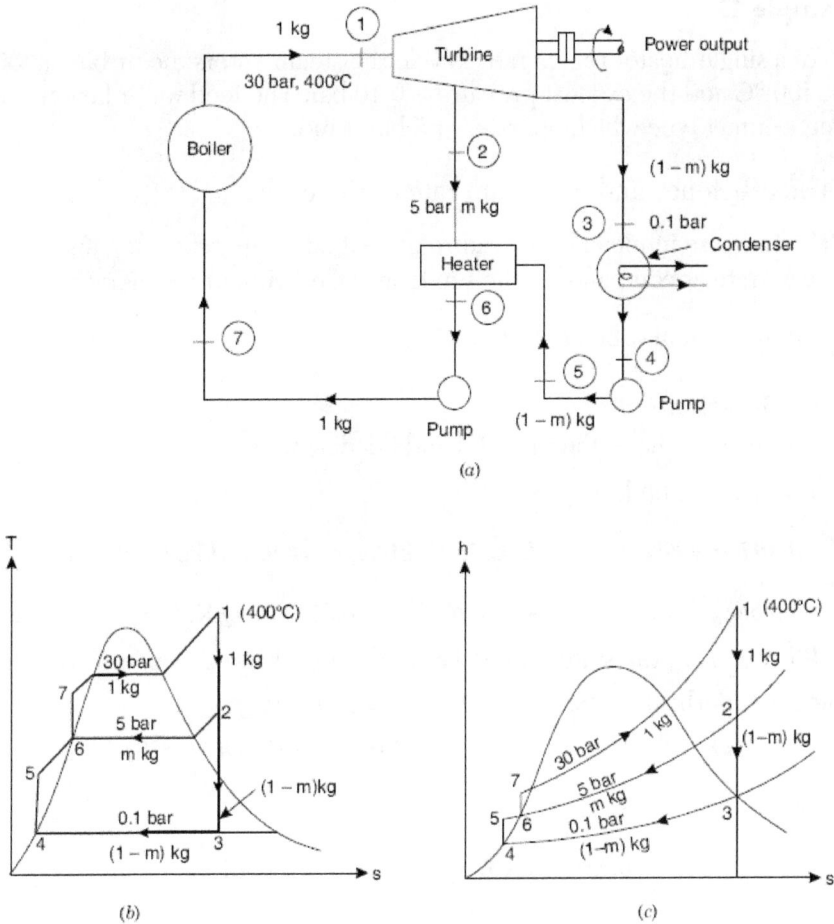

FIGURE 8.17. Regenerative steam cycle for Example 13.

∴ **Power developed by the turbine**

$$= 909.6 \times \frac{49557.5}{3600} \text{ kJ/s}$$

$$= \textbf{12521.5 kW.} \quad \textbf{(Ans.)} \qquad (\because \ 1 \text{ kJ/s} = 1 \text{ kW})$$

Example 13

In a single-heater regenerative cycle the steam enters the turbine at 30 bar, 400 °C and the exhaust pressure is 0.10 bar. The feed water heater is a direct contact type which operates at 5 bar. Find:

1. The efficiency and the steam rate of the cycle.

2. The increase in mean temperature of heat addition, efficiency, and steam rate as compared to the Rankine cycle (without regeneration).

Pump work may be neglected.

■ Solution:

Figure 8.17 shows the flow, T-s and h-s diagrams.

From steam tables:

At 30 bar, 400°C : $h_1 = 3230.9$ kJ/kg, $s_1 = 6.921$ kJ/kg K $= s_2 = s_3$,

At 5 bar : $s_f = 1.8604, s_g = 6.8192$ kJ/kg K, $h_f = 640.1$ kJ/kg

Since $s_2 > s_g$, the state 2 must lie in the superheated region. From the table for superheated steam $t_2 = 172°C, h_2 = 2796$ kJ/kg.

At 0.1 bar : $s_f = 0.649, s_{f_g} = 7.501, h_f = 191.8, h_{f_g} = 2392.8$

Now, $s_2 = s_3$

i.e., $6.921 = s_{f_3} + x_3 \, s_{fg_3} = 0.649 + x_3 \times 7.501$

∴ $x_3 = \dfrac{6.921 - 0.649}{7.501} = 0.836$

∴ $h_3 = h_{f_3} + x_3 \, h_{fg_3} = 191.8 + 0.836 \times 2392.8 = 2192.2$ kJ/kg

Since pump work is neglected

$$h_{f_4} = 191.8 \text{ kJ/kg} = h_{f_5}$$

$$h_{f_6} = 640.1 \text{ kJ/kg (at 5 bar)} = h_{f_7}$$

Energy balance for heater gives

$$m(h_2 - h_{f_6}) = (1-m)(h_{f_6} - h_{f_5})$$

$$m(2796 - 640.1) = (1-m)(640.1 - 191.8) = 448.3(1-m)$$

$$2155.9\,m = 448.3 - 448.3\,m$$

∴ $$m = 0.172\text{ kg}$$

∴ Turbine work, $$W_T = (h_1 - h_2) + (1-m)(h_2 - h_3)$$

$$= (3230.9 - 2796) + (1 - 0.172)(2796 - 2192.2)$$

$$= 434.9 + 499.9 = 934.8\text{ kJ/kg}$$

Heat supplied, $$Q_1 = h_1 - h_{f_6} = 3230.9 - 640.1 = 2590.8\text{ kJ/kg.}$$

(*i*) **Efficiency of cycle, η_{cycle}:**

$$\eta_{\text{cycle}} = \frac{W_T}{Q_1} = \frac{934.8}{2590.8} = 0.3608\text{ or }36.08\%.\quad\textbf{(Ans.)}$$

$$\textbf{Steam rate} = \frac{3600}{934.8} = \textbf{3.85 kg / kWh.}\quad\textbf{(Ans.)}$$

(*ii*) $$T_{m_1} = \frac{h_1 - h_{f_7}}{s_1 - s_7} = \frac{2590.8}{6.921 - 1.8604} = 511.9\text{ K} = 238.9°\text{C.}$$

T_{m_1} (without regeneration)

$$= \frac{h_1 - h_{f_4}}{s_1 - s_4} = \frac{3230.9 - 191.8}{6.921 - 0.649} = \frac{3039.1}{6.272} = 484.5\text{ K} = 211.5°\text{C.}$$

Increase in T_{m_1} due to regeneration

$$= 238.9 - 211.5 = \textbf{27.4°C. (Ans.)}$$

W_T (without regeneration)

$$= h_1 - h_3 = 3230.9 - 2192.2 = 1038.7\text{ kJ/kg}$$

Steam rate without regeneration

$$= \frac{3600}{1038.7} = 3.46\text{ kg/kWh}$$

∴ Increase in steam rate due to regeneration

$$= 3.85 - 3.46 = \textbf{0.39 kg / kWh.}\quad\textbf{(Ans.)}$$

$$\eta_{cycle} \text{ (without regeneration)} = \frac{h_1 - h_3}{h_1 - h_{f_4}} = \frac{1038.7}{3230.9 - 191.8} = 0.3418 \text{ or } 34.18\%. \text{ (Ans.)}$$

Increase in cycle efficiency due to regeneration

$$= 36.08 - 34.18 = 1.9\%. \quad \text{(Ans.)}$$

Example 14

Steam is supplied to a turbine at a pressure of 30 bar and a temperature of 400 °C and is expanded adiabatically to a pressure of 0.04 bar. At a stage of the turbine where the pressure is 3 bar, a connection is made to a surface heater in which the feed water is heated by bled steam to a temperature of 130 °C. The condensed steam from the feed heater is cooled in a drain cooler to 27 °C. The feed water passes through the drain cooler before entering the feed heater. The cooled drain water combines with the condensate in the well of the condenser.

Assuming no heat losses in the steam, calculate the following:

1. Mass of steam used for feed heating per kg of steam entering the turbine;

2. Thermal efficiency of the cycle.

■ **Solution.**

Refer to Figure 8.18.

From steam tables:

At 3 bar : $\quad t_s = 133.5°C$, $h_f = 561.4 \text{ kJ} / \text{kg}$.

At 0.04 bar : $\quad t_s = 29°C, \quad h_f = 121.5 \text{ kJ} / \text{kg}$.

From Mollier chart:

$$h_0 = 3231 \text{ kJ} / \text{kg} \left(\text{at 30 bar, } 400°C\right)$$

$$h_1 = 2700 \text{ kJ} / \text{kg} \left(\text{at 3 bar}\right)$$

$$h_2 = 2085 \text{ kJ} / \text{kg} \left(\text{at 0.04 bar}\right).$$

(*i*) **Mass of steam used, m$_1$:**

Heat lost by the steam = Heat gained by water.

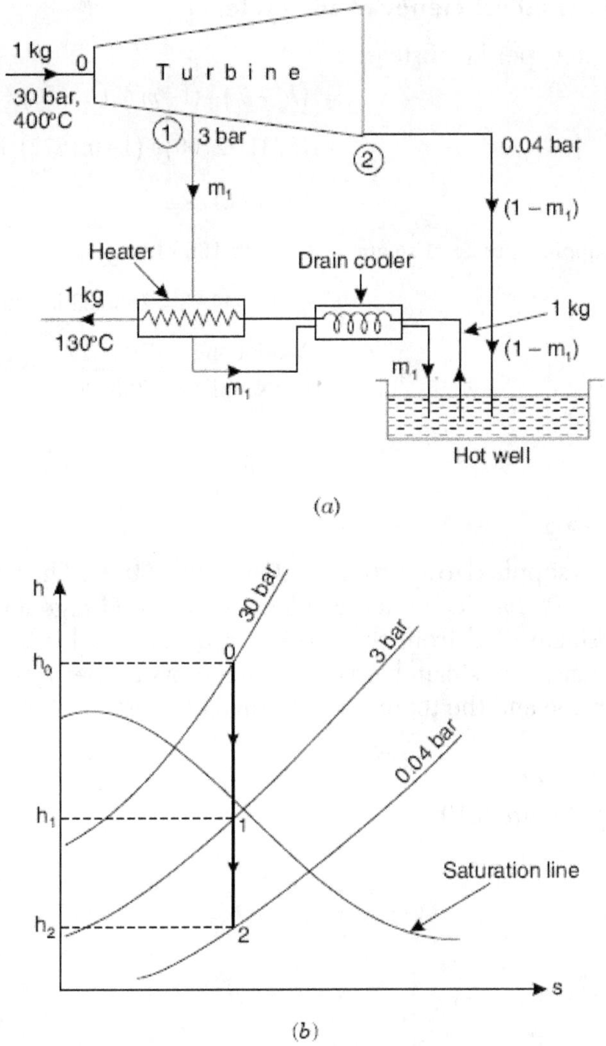

(a)

(b)

FIGURE 8.18. Steam turbine system sketch for Example 14.

Taking the feed-heater and drain-cooler combined, we have:

$$m_1 \left(h_1 - h_{f_2} \right) = 1 \times 4.186 \left(130 - 27 \right)$$

or $\qquad m_1 \left(2700 - 121.5 \right) = 4.186 \left(130 - 27 \right)$

$\therefore \qquad\qquad \mathbf{m_1} = \dfrac{4.186 \left(130 - 27 \right)}{\left(2700 - 121.5 \right)} = \mathbf{0.1672 \ kg.} \quad (\mathbf{Ans.})$

(ii) Thermal efficiency of the cycle:

Work done per kg of steam

$$= 1(h_0 - h_1) + (1 - m_1) \ (h_1 - h_2)$$

$$= 1(3231 - 2700) + (1 - 0.1672) \ (2700 - 2085)$$

$$= 1043.17 \ kJ/kg$$

Heat supplied per kg of steam $= h_0 - 1 \times 4.186 \times 130$

$$= 3231 - 544.18 = 2686.82 \ kJ/kg.$$

$$\eta_{\text{Thermal}} = \frac{\text{Work done}}{\text{Heat supplied}} = \frac{1043.17}{2686.82} = \textbf{0.3882 or 38.82\%.}$$

(Ans.)

Example 15

Steam is supplied to a turbine at 30 bar and 350 °C. The turbine exhaust pressure is 0.08 bar. The main condensate is heated regeneratively in two stages by steam bled from the turbine at 5 bar and 1.0 bar respectively. Calculate masses of steam bled off at each pressure per kg of steam entering the turbine and the theoretical thermal efficiency of the cycle.

■ **Solution:**

Refer to Figure 8.19.

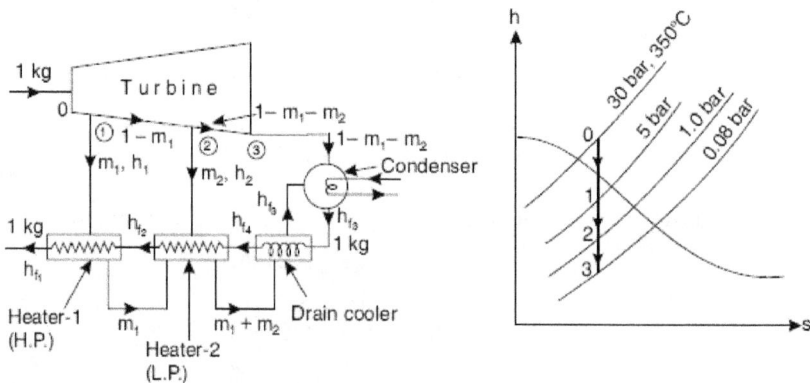

FIGURE 8.19. Steam turbine system sketch for Example 15.

The following assumptions are made:

1. The condensate is heated to the saturation temperature in each heater.

2. The drain water from the H.P. (high pressure) heater passes into the steam space of the L.P. (low pressure) heater without loss of heat.

3. The combined drains from the L.P. heater are cooled in a drain cooler to the condenser temperature.

4. The expansion of the steam in the turbine is adiabatic and frictionless.

> *Enthalpy at 30 bar, 350°C, $h_0 = 3115.3$ kJ / kg.*
>
> After adiabatic expansion (from Mollier chart)

> *Enthalpy at 5 bar,* $h_1 = 2720$ kJ / kg
>
> *Enthalpy at 1.0 bar,* $h_2 = 2450$ kJ / kg
>
> *Enthalpy at 0.08 bar,* $h_3 = 2120$ kJ / kg
>
> From steam tables : $h_{f_1} = 640.1$ kJ / kg $\left(\text{at } 5.0 \text{ bar}\right)$
>
> $h_{f_2} = 417.5$ kJ / kg $\left(\text{at } 1.0 \text{ bar}\right)$
>
> $h_{f_3} = 173.9$ kJ / kg $\left(\text{at } 0.08 \text{ bar}\right)$

At heater No. 1:

$$m_1 h_1 + h_{f_2} = m_1 h_{f_1} + h_{f_1}$$

$$\mathbf{m_1} = \frac{h_{f_1} - h_{f_2}}{h_1 - h_{f_1}} = \frac{640.1 - 417.5}{2720 - 640.1} = \textbf{0.107 kJ / kg of entering steam. (Ans.)}$$

At heater No. 2:

$$m_2 h_2 + m_1 h_{f_1} + h_{f_4} = \left(m_1 + m_2\right) h_{f_2} + h_{f_2} \qquad (i)$$

At drain cooler:

$$\left(m_1 + m_2\right) h_{f_2} + h_{f_3} = h_{f_4} + \left(m_1 + m_2\right) h_{f_3}$$

$$\therefore \qquad\qquad h_{f_4} = \left(m_1 + m_2\right)\left(h_{f_2} - h_{f_3}\right) + h_{f_3} \qquad (ii)$$

Inserting the value of h_{f_4} in eqn. (*i*), we get

$$m_2 h_2 + m_2 h_2 + (m_1 + m_2)(h_{f_2} - h_{f_3}) + h_{f_3} = (m_1 + m_2)h_{f_2} + h_{f_2}$$

$$m_2 h_2 + m_1 h_{f_1} + (m_1 + m_2)h_{f_2} - (m_1 + m_2)h_{f_3} + h_{f_3} = (m_1 + m_2)h_{f_2} + h_{f_2}$$

$$m_2 h_2 + m_1 h_{f_1} - m_1 h_{f_3} - m_2 h_{f_3} + h_{f_3} = h_{f_2}$$

$$m_2 (h_2 - h_{f_3}) = (h_{f_2} - h_{f_3}) - m_1 (h_{f_1} - h_{f_3})$$

$$\mathbf{m_2} = \frac{\left(h_{f_2} - h_{f_3}\right) - m_1\left(h_{f_1} - h_{f_3}\right)}{\left(h_{f_2} - h_{f_3}\right)}$$

$$= \frac{(417.5 - 173.9) - 0.107(640.1 - 173.9)}{(2450 - 173.9)}$$

$$= \frac{193.7}{2276.1} = \mathbf{0.085\ kJ\ /\ kg.\ (Ans.)}$$

Work done $= 1(h_0 - h_1) + (1 - m_1)(h_1 - h_2) + (1 - m_1 - m_2)(h_2 - h_3)$

$$= 1(3115.3 - 2720) + (1 - 0.107)(2720 - 2450)$$
$$+ (1 - 0.107 - 0.085)(2450 - 2120)$$

$$= 395.3 + 241.11 + 266.64 = 903.05\ \text{kJ/kg}$$

Heat supplied / kg
$$= h_0 - h_{f_1}$$
$$= 3115.3 - 640.1 = 2475.2\ \text{kJ/kg}$$

∴ Thermal efficiency of the cycle

$$= \frac{\text{Work done}}{\text{Heat supplied}} = \frac{903.05}{2475.2} = \mathbf{0.3648\ or\ 36.48\%.} \quad \textbf{(Ans.)}$$

Example 16

Steam at a pressure of 20 bar and 250 °C enters a turbine and leaves it finally at a pressure of 0.05 bar. Steam is bled off at pressures of 5.0, 1.5, and 0.3 bar. Assuming (a) that the condensate is heated in each heater up to the saturation temperature of the steam in that heater, (b) that the drain water from each heater is cascaded through a trap into the next heater on the low-pressure side of it, and (c) that the combined drains from the heater operating at 0.3 bar are cooled in a drain cooler to condenser temperature, calculate the following:

1. Mass of bled steam for each heater per kg of steam entering the turbine,

2. Thermal efficiency of the cycle,

3. Thermal efficiency of the Rankine cycle,

4. Theoretical gain due to regenerative feed heating,

5. Steam consumption in kg/kWh with or without regenerative feed heating, and

6. Quantity of steam passing through the last stage nozzle of a 50000 kW turbine with and without regenerative feed heating.

■ **Solution:**
Refer to Figure 8.20(*a*), (*b*).

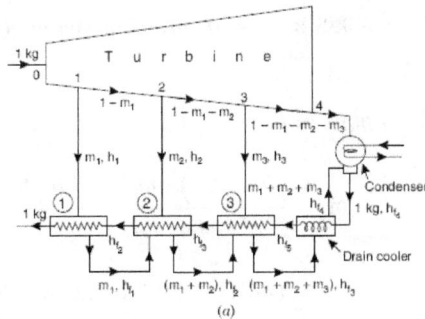

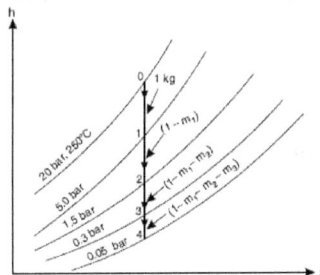

FIGURE 8.20. Steam turbine system sketch for Example 16.

From Mallier Chart : $h_0 = 2905$ kJ/kg, $h_1 = 2600$ kJ/kg, $h_2 = 2430$ kJ/kg

$h_3 = 2210$ kJ/kg, $h_4 = 2000$ kJ/kg

From steam tables:

At 5 bar : $h_{f_1} = 640.1$ kJ/kg

At 1.5 bar : $h_{f_2} = 467.1$ kJ/kg

At 0.3 bar : $h_{f_3} = 289.3$ kJ/kg

At 0.05 bar : $h_{f_4} = 137.8$ kJ/kg.

(i) Mass of bled steam for each heater per kg of steam:

Using heat balance equation:

At heater No. 1:

$$m_1 h_1 + h_{f_2} = m_1 h_{f_1} + h_{f_1}$$

$$\mathbf{m_1} = \frac{h_{f_1} - h_{f_2}}{h_1 - h_{f_1}} = \frac{640.1 - 467.1}{2600 - 647.1}$$

= 0.088 kJ / kg of entering steam. (Ans.)

At heater No. 2:

$$m_2 h_2 + h_{f_3} + m_1 h_{f_1} = h_{f_2} + (m_1 + m_2) h_{f_2}$$

$$\mathbf{m_2} = \frac{\left(h_{f_2} + h_{f_3}\right) - m_1 \left(h_{f_1} - h_{f_2}\right)}{\left(h_2 - h_{f_2}\right)}$$

$$= \frac{(467.1 - 289.3) - 0.088(640.1 - 467.1)}{(2430 - 467.1)} = \frac{162.57}{1962.9}$$

= 0.0828 kJ / kg of entering steam. (Ans.)

At heater No. 3:

$$m_3 h_3 + h_{f_5} + (m_1 + m_2) h_{f_2} = h_{f_3} + (m_1 + m_2 + m_3) h_{f_3} \qquad (i)$$

At drain cooler:

$$(m_1 + m_2 + m_3) h_{f_3} + h_{f_4} = h_{f_5} + (m_1 + m_2 + m_3) h_{f_4}$$

$$\therefore \qquad h_{f_5} = (m_1 + m_2 + m_3)(h_{f_3} - h_{f_4}) + h_{f_4} \qquad (ii)$$

Inserting the value of h_{f_5} in eqn. (i), we get

$$m_3 h_3 + \left(m_1 + m_2 + m_3\right)\left(h_{f_3} - h_{f_4}\right) + h_{f_4} + \left(m_1 + m_2\right)h_{f_2} = h_{f_3} + \left(m_1 + m_2 + m_3\right)h_{f_3}$$

$$\therefore \qquad \mathbf{m_3} = \frac{\left(h_{f_3} - h_{f_4}\right) - \left(m_1 + m_2\right)\left(h_{f_2} - h_{f_4}\right)}{h_3 - h_{f_4}}$$

$$= \frac{\left(289.3 - 137.8\right) - \left(0.088 + 0.0828\right)\ \left(467.1 - 137.8\right)}{\left(2210 - 137.8\right)}$$

$$= \frac{151.5 - 56.24}{2072.2} = \textbf{0.046 kJ / kg of entering steam. (Ans.)}$$

Work done/kg $(neglecting\ pump\ work)$

$$= \left(h_0 - h_1\right) + \left(1 - m_1\right)\ \left(h_1 - h_2\right) + \left(1 - m_1 - m_2\right)\ \left(h_2 - h_3\right) + \left(1 - m_1 - m_2 - m_3\right)\ \left(h_3 - h_4\right)$$

$$= \left(2905 - 2600\right) + \left(1 - 0.088\right)\ \left(2600 - 2430\right) + \left(1 - 0.088 - 0.0828\right)\ \left(2430 - 2210\right)$$

$$\qquad\qquad\qquad\qquad + \left(1 - 0.088 - 0.0828 - 0.046\right)\ \left(2210 - 2000\right)$$

$$= 305 + 155.04 + 182.42 + 164.47 = 806.93 \text{ kJ/kg}$$

Heat supplied/kg $= h_0 - h_{f_1} = 2905 - 640.1 = 2264.9$ kJ/kg.

(ii) Thermal efficiency of the cycle, η_{Thermal}:

$$\eta_{\text{Thermal}} = \frac{\text{Work done}}{\text{Heat supplied}} = \frac{806.93}{2264.9} = \textbf{0.3563 or 35.63\%. (Ans.)}$$

(iii) Thermal efficiency of Rankine cycle, η_{Rankine}:

$$\eta_{\text{Rankine}} = \frac{h_0 - h_4}{h_0 - h_{f_4}} = \frac{2905 - 2000}{2905 - 137.8} = \textbf{0.327 or 32.7\%. (Ans.)}$$

(iv) Theoretical gain due to regenerative feed heating

$$= \frac{35.63 - 32.7}{35.63} = \textbf{0.0822 or 8.22\%. (Ans.)}$$

(v) Steam consumption with regenerative feed heating

$$= \frac{1 \times 3600}{\text{Work done / kg}} = \frac{1 \times 3600}{806.93} = \textbf{4.46 kg / kWh. (Ans.)}$$

Steam consumption without regenerative feed heating

$$= \frac{1 \times 3600}{\text{Work done / kg without regeneration}} = \frac{1 \times 3600}{h_0 - h_4}$$

$$= \frac{1 \times 3600}{2905 - 2000} = \textbf{3.97 kg / kWh. (Ans.)}$$

(*vi*) **Quantity of steam passing through the last stage of a 50000 kW turbine with regenerative feed heating**

$$= 4.46 \left(1 - m_1 - m_2 - m_3\right) \times 50000$$

$$= 4.46 \left(1 - 0.088 - 0.0828 - 0.046\right) \times 50000 = \textbf{174653.6 kg / h.} \quad (\textbf{Ans.})$$

Same without regenerative arrangement

$$= 3.97 \times 50000 = \textbf{198500 kg / h.} \quad (\textbf{Ans.})$$

Example 17

A steam turbine plant developing 120 MW of electrical output is equipped with a reheating and regenerative feed heating arrangement consisting of two feed heaters—one surface type on the H.P. side and the other a direct contact type on L.P. side. The steam conditions before the steam stop valve are 100 bar and 530 °C. A pressure drop of 5 bar takes place due to throttling in valves.

Steam exhausts from the H.P. turbine at 25 bar. A small quantity of steam is bled off at 25 bar for the H.P. surface heater for feed heating and the remaining is reheated in a reheater to 550 °C and the steam enters at 22 bar in the L.P. turbine for further expansion. Another small quantity of steam is bled off at pressure 6 bar for the L.P. heater and the rest of steam expands up to the back pressure of 0.05 bar. The drain from the H.P. heater is led to the L.P. heater and the combined feed from the L.P. heater is pumped to the high-pressure feed heater and finally to the boiler with the help of boiler feed pump.

The component efficiencies are: Turbine efficiency 85%, pump efficiency 90%, generator efficiency 96%, boiler efficiency 90%, and mechanical efficiency 95%. It may be assumed that the feed water is heated up to the saturation temperature at the prevailing pressure in the feed heater. Work out the following:

(i) Sketch the feed heating system and show the process on T-s and h-s diagrams.

(ii) Amounts of steam bled off.

(iii) Overall thermal efficiency of turbo-alternator considering pump work.

(iv) Specific steam consumption in kg/kWh.

■ **Solution:**

(*i*) The schematic arrangement including the feed heating system, and T-s and h-s diagrams of the process are shown in Figure 8.21 and Figure 8.22, respectively.

(*ii*) **Amounts of bled off.** The enthalpies at various state points as read from *h-s* diagram/steam tables, in kJ/kg, are:

$$h_1 = h_2 = 3460$$

$$h_3' = 3050, \text{ and } \therefore \ h_3 = 3460 - 0.85(3460 - 3050) = 3111.5$$

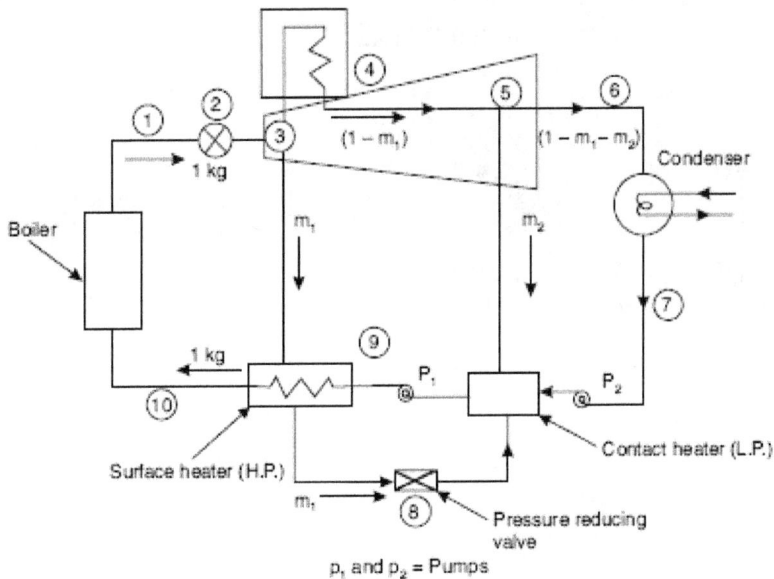

FIGURE 8.21. Steam turbine system sketch for Example 17.

$$h_4 = 3585$$

$$h_5' = 3140, \text{ and } \therefore \ h_5 = 3585 - 0.85(3585 - 3140) = 3207$$

$$h_6' = 2335, \text{ and } \therefore \ h_6 = 3207 - 0.85 (3207 - 2335) = 2466$$

$$h_7 = 137.8 \text{ kJ/kg } (h_f \text{ at } 0.05 \text{ bar})$$

$$h_8 = h_{10} = 962 \text{ kJ/kg } (h_f \text{ at } 25 \text{ bar})$$

and $\quad h_9 = 670.4 \ (h_f \text{ at } 6 \text{ bar}).$

Enthalpy balance for **surface heater:**

$$m_1 h_3 + h_9 = m_1 h_8 + h_{10}, \text{ neglecting pump work}$$

or
$$m_1 = \frac{h_{10} - h_9}{h_3 - h_8} = \frac{962 - 670.4}{3111.5 - 962} = 0.13566 \text{ kg}$$

Enthalpy balance for **contact heater:**

$$m_2 h_5 + (1 - m_1 - m_2) h_7 + m_1 h_8 = h_9, \text{ neglecting pump work}$$

or
$$m_2 \times 3207 + (1 - 0.13566 - m_2) \times 137.8 + 0.13566 \times 962 = 670.4$$

or
$$m_2 = 0.1371 \text{ kg.}$$

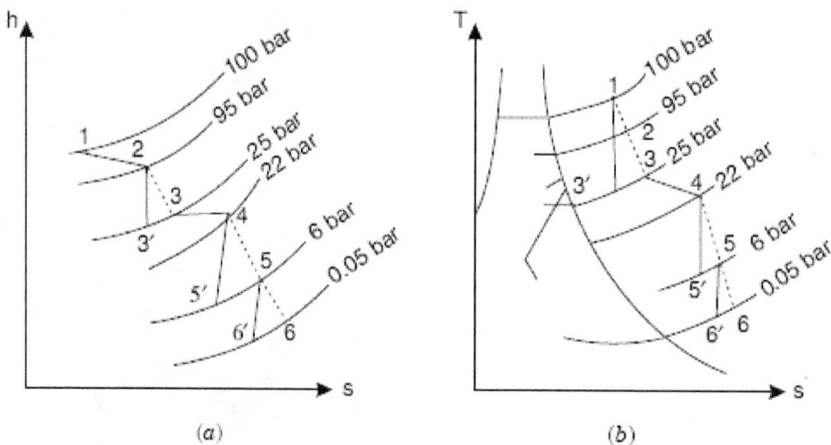

FIGURE 8.22. h-s and T-s diagrams of the process.

Pump Work. Take a specific volume of water as 0.001 m³/kg.

$$(W_{pump})_{L.P.} = (1 - m_1 - m_2)(6 - 0.05) \times 0.001 \times 10^2$$

$$= (1 - 0.13566 - 0.1371) \times 5.95 \times 0.1 = 0.4327 \text{ kJ/kg.}$$

$$(W_{pump})_{H.P.} = 1 \times (100 - 6) \times 0.001 \times 10^2 = 9.4 \text{ kJ/kg}$$

Total pump work (actual) $= \dfrac{0.4327 + 9.4}{0.9} = 10.925$ kJ/kg

Turbine output (indicated) $= (h_2 - h_3) + (1 - m_1)(h_4 - h_5) + (1 - m_1 - m_2)(h_5 - h_6)$

$= (3460 - 3111.5) + (1 - 0.13566)(3585 - 3207) + (1 - 0.13566 - 0.1371)(3207 - 2466)$

$= 1214.105$ kJ/kg

Net electrical output = (Indicated work - pump work) $\times \eta_{mech.} \times \eta_{gen}$.

$$= (1214.105 - 10.925) \times 0.9 \times 0.96 = 1039.55 \text{ kJ/kg}$$

[**Note.** All the above calculations are for 1kg of main (boiler) flow.]

$\therefore$ Main steam flow rate $= \dfrac{120 \times 10^3 \times 3600}{1039.55} = 4.155 \times 10^5$ kJ/h.

Amount of bled off are:

(*a*) Surface (high pressure) heater,

$$= 0.13566 \text{ kg/kg of boiler flow}$$

or $\qquad\qquad\qquad\qquad = 0.13566 \times 4.155 \times 10^5$

i.e., $\qquad\qquad\qquad\qquad = \mathbf{5.6367 \times 10^4 \ kg/h.}$ **(Ans.)**

(*b*) Direct contact (low pressure) heater,

$$= 0.1371 \text{ kg/kg of boiler flow}$$

or $\qquad\qquad\qquad\qquad = 0.1371 \times 4.155 \times 10^5$

i.e., $\qquad\qquad\qquad\qquad = \mathbf{5.697 \times 10^4 \ kg/h.}$ **(Ans.)**

(iii) **Overall thermal efficiency, $\eta_{overall}$:**

Heat input in boiler $= \dfrac{h_1 - h_{10}}{\eta_{boiler}} = \dfrac{3460 - 962}{0.9}$

$= 2775.6$ kJ/kg of boiler flow.

Heat input in reheater $= \dfrac{h_4 - h_3}{\eta_{boiler}} = \dfrac{3585 - 3111.5}{0.9} = 526.1$ kJ/kg of boiler flow

$\therefore$ $\eta_{overall} = \dfrac{1039.55}{2775.6 + 526.1} \times 100 = 31.48\%.$ **(Ans.)**

(iv) **Specific steam consumption :**

Specific steam consumption $= \dfrac{4.155 \times 10^5}{120 \times 10^3} = $ **3.4625 kg / kWh.** **(Ans.)**

8.6. Reheat Rankine Cycle

For attaining greater thermal efficiencies when the initial pressure of steam was raised beyond 42 bar, it was found that the resulting condition of steam after expansion was increasingly wetter and exceeded the safe limit of 12% condensation. Therefore, it became necessary to *reheat* the steam after part of the expansion was over so that the resulting condition after complete expansion fell within the region of permissible wetness. This is achieved by adding more turbines in the system. The steam exiting the first turbine (i.e., high pressure turbine) is fed into the boiler and the low-pressure high-temperature steam is fed into the subsequent turbine, and so on. Reheating is a practical way of achieving efficiency increase without possible material damage of superheating the steam to an unsafe high level.

The reheating or resuperheating of steam is now universally used when high pressure and temperature steam conditions such as 100 to 250 bar and 500°C to 600°C are employed for throttle. For plants of still higher pressures and temperatures, a double reheating may be used.

In practice, reheating *improves* the cycle efficiency by about 5%. A second reheat will give much less of a gain while the initial cost involved would be so high as to prohibit the use of two stage reheating except in cases of very high initial throttle conditions. The cost of reheating equipment consisting of a boiler, piping, and controls may be 5% to 10% more than that of the conventional boilers, and this additional expenditure is justified only

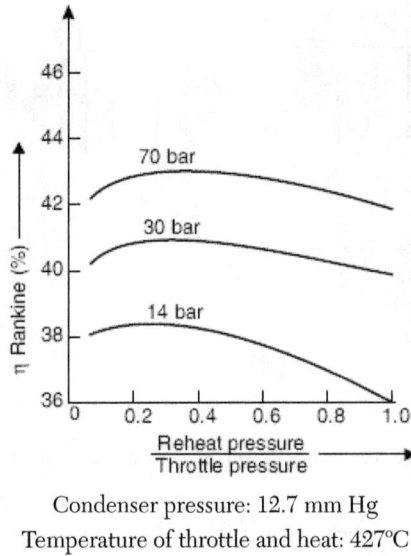

Condenser pressure: 12.7 mm Hg

Temperature of throttle and heat: 427°C

FIGURE 8.23. Regenerative cycle, reheat pressure selection on cycle efficiency.

if gain in thermal efficiency is sufficient to promise a return of this investment. Usually a plant with a base load capacity of 50 MW and initial steam pressure of 42 bar would economically justify the extra cost of reheating.

The improvement in thermal efficiency due to reheating is greatly dependent upon the reheat pressure with respect to the original pressure of steam.

Figure 8.23 shows the reheat pressure selection on cycle efficiency.

Figure 8.24 shows a schematic diagram of a theoretical single-stage reheat cycle. The corresponding representation of an ideal reheating process on a T-s and h-s chart is shown in Figure 8.25 (a and b).

With reference to Figure 8.25(a) we observe that path 5-1 shows the formation of steam in the boiler. The steam at state point 1 (*i.e.*, pressure p_1 and temperature T_1) enters the turbine and expands isentropically to a certain pressure p_2 and temperature T_2. From this state point 2 the whole steam is drawn out of the turbine and is reheated in a reheater to a temperature T_3. (Although there is an optimum pressure at which the steam should be removed for reheating if the highest return is to be obtained, yet, for simplicity, the whole steam is removed from the high-pressure exhaust

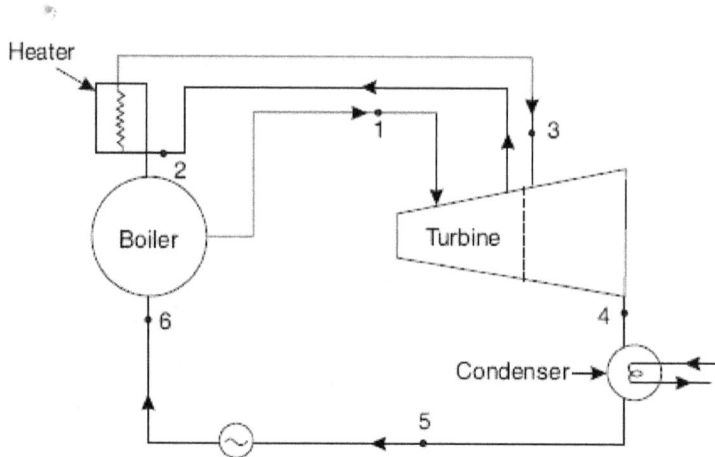

FIGURE 8.24. Schematic diagram of a theoretical single-stage reheat cycle.

where the pressure is about one-fifth of boiler pressure, and after undergoing a 10% pressure drop in circulating through the heater, it is returned to the intermediate pressure or low-pressure turbine). This reheated steam is then readmitted to the turbine where it is expanded to condenser pressure isentropically.

Superheating of steam. The primary object of superheating steam and supplying it to the prime movers is *to avoid too much wetness at the end of expansion.* Use of an inadequate degree of superheat in *steam engines would cause greater condensation* in the engine cylinder, while in case of *turbines* the moisture content of steam would result in *undue blade erosion.* The *maximum wetness in the final condition of steam that may be tolerated without any appreciable harm to the turbine blades is about 12%.* Broadly each *1%* of moisture in steam reduces the efficiency of that part of the turbine in which wet steam passes by 1% to 1.5% and in engines about 2%.

Advantages of superheated steam:

(*i*) Superheating reduces the initial condensation losses in steam engines.

(*ii*) Use of superheated steam results in *improving the plant efficiency by effecting a saving in cost of fuel.* This saving may be on the order of 6% to 7% due to the first 38 °C of superheat, 4% to 5% for the next 38 °C, and so on. This saving is due to the fact that the heat content and consequently the capacity to do work in superheated

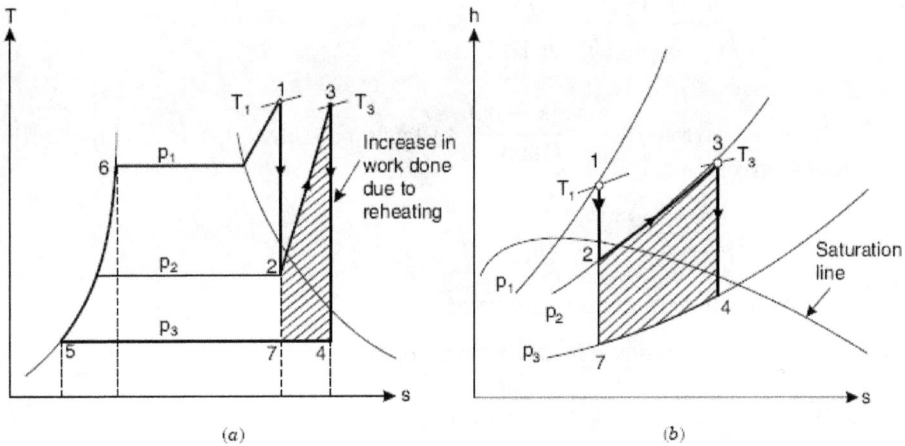

FIGURE 8.25. Ideal reheating process on a T-s and h-s chart.

steam is increased, and the quantity of steam required for a given output of power is reduced. Although additional heat has to be added in the boiler, there is reduction in the work to be done by the feed pump, the condenser pump, and other accessories due to reduction in the quantity of steam used. It is estimated that the *quantity of steam may be reduced by 10% to 15% for the first 38 °C of superheat and somewhat less for the next 38 °C of superheat in the case of condensing turbines.*

(*iii*) When a superheater is used in a boiler, it helps in *reducing the stack temperatures* by extracting heat from the flue gases before these are passed out of chimney.

Thermal efficiency with *"Reheating"* (neglecting pump work):

Heat supplied $= (h_1 - h_{f_4}) + (h_3 - h_2)$

Heat rejected $= h_4 - h_{f_4}$

Work done by the turbine = Heat supplied – heat rejected

$$= (h_1 - h_{f_4}) + (h_3 - h_2) - (h_4 - h_{f_4})$$

$$= (h_1 - h_2) + (h_3 - h_4)$$

Thus, the theoretical thermal efficiency of the reheat cycle is

$$\eta_{\text{thermal}} = \frac{(h_1 - h_2) + (h_3 - h_4)}{(h_1 - h_{f_4}) + (h_3 - h_2)}$$

If pump work, $W_p = \dfrac{v_f(p_1 - p_b)}{1000}$ kJ/kg is considered, the thermal efficiency is given by :

$$\eta_{\text{thermal}} = \frac{[(h_1 - h_4) + (h_3 - h_4)] - W_p}{[(h_1 - h_{f4}) + (h_3 - h_2)] - W_p}$$

W_p is usually small and *neglected.*

Thermal efficiency *without reheating* is

$$\eta_{\text{thermal}} = \frac{h_1 - h_7}{h_1 - h_{f_4}} \qquad (\because h_{f_4} = h_{f_7})$$

1. The reheater may be incorporated in the walls of the main boiler; it may be a separately fired superheater or it may be heated by a coil carrying high-pressure superheated steam, this system being analogous to a steam jacket.*

2. Reheating should be done at *"optimum pressure"* because if the steam is reheated early in its expansion then the additional *quantity of heat supplied will be small* and thus thermal efficiency gain will be small; and if the reheating is done at a fairly low pressure, then, although a large amount of additional heat is supplied, the steam will have a high degree of superheat (as is clear from the Mollier diagram), and thus a large proportion of the heat supplied in the reheating process *will be thrown to waste in the condenser.*

Advantages of Reheating:

1. There is an increased output of the turbine.

2. Erosion and corrosion problems in the steam turbine are eliminated/avoided.

3. There is an improvement in the thermal efficiency of the turbines.

4. The final dryness fraction of steam is improved.

5. There is an increase in the nozzle and blade efficiencies.

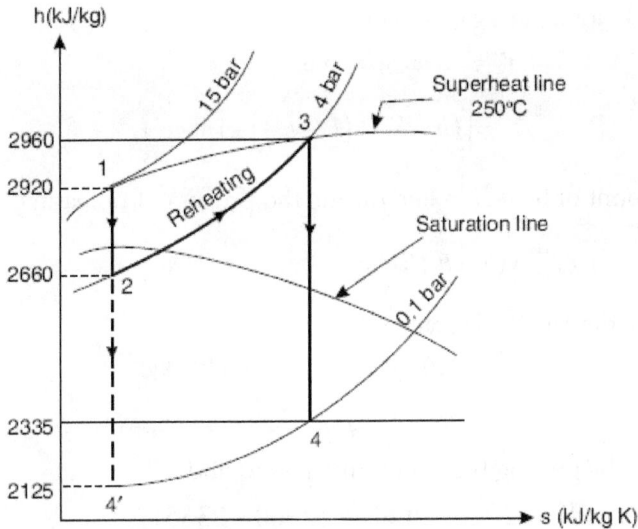

FIGURE 8.26. Cycle h-S diagram for Example 18.

Disadvantages of Reheating:

1. Reheating requires more maintenance.

2. The increase in thermal efficiency is not appreciable in comparison to the expenditure incurred in reheating.

Example 18

Steam at a pressure of 15 bar and 250 °C is expanded through a turbine at first to a pressure of 4 bar. It is then reheated at a constant pressure to the initial temperature of 250 °C and is finally expanded to 0.1 bar. Using a Mollier chart, estimate the work done per kg of steam flowing through the turbine and the amount of heat supplied during the process of reheating. Compare the work output when the expansion is direct from 15 bar to 0.1 bar without any reheat. Assume all expansion processes to be isentropic.

■ **Solution:**

Refer to Figure 8.26.

Pressure, $p_1 = 15$ bar;

$p_2 = 4$ bar ;

$p_4 = 0.1$ bar.

Work done per kg of steam,

$$W = \text{Total heat drop}$$

$$= \left[\left(h_1 - h_2 \right) + \left(h_3 - h_4 \right) \right] \text{kJ / kg} \qquad (i)$$

Amount of heat supplied during the process of reheating,

$$h_{\text{reheat}} = (h_3 - h_2) \text{ kJ/kg} \qquad (ii)$$

From the Mollier diagram or h-s chart,

$$h_1 = 2920 \text{ kJ / kg}, \ h_4 = 2660 \text{ kJ / kg}$$

$$h_3 = 2960 \text{ kJ / kg}, \ h_4 = 2335 \text{ kJ / kg}$$

Now, by putting the values in eqns. (i) and (ii), we get

$$W = (2920 - 2660) + (2960 - 2335)$$

$$= \textbf{885 kJ / kg.} \quad (\textbf{Ans.})$$

Hence *work done per kg of steam* = **885 kJ / kg. (Ans.)**

Amount of heat supplied during reheat,

$$h_{\text{reheat}} = (2960 - 2660) = \textbf{300 kJ / kg.} \quad (\textbf{Ans.})$$

If the expansion would have been continuous without reheating, that is, 1 to 4′, the work output is given by

$$W_1 = h_1 - h_4{}'$$

From the Mollier diagram,

$$h_{4'} = 2125 \text{ kJ / kg}$$

$\therefore$ $\qquad\qquad W_1 = 2920 - 2125 = \textbf{795 kJ / kg.} \quad (\textbf{Ans.})$

Example 19

A steam power plant operates on a theoretical reheat cycle. Steam at boiler at 150 bar, 550 °C expands through the high-pressure turbine. It is reheated at a constant pressure of 40 bar to 550 °C and expands through the low-pressure turbine to a condenser at 0.1 bar. Draw T-s and h-s diagrams. Find:

(i) Quality of steam at turbine exhaust; (ii) Cycle efficiency;

(iii) Steam rate in kg/kWh.

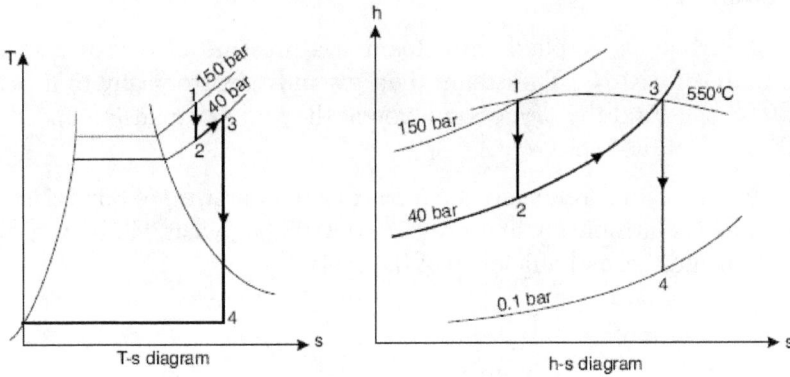

FIGURE 8.27. Theoretical reheat cycle on T-S (a) and h-S (b) diagrams for Example 19.

■ **Solution.** Refer to Figure 8.27. Theoretical reheat cycle on T-S (a) and h-S (b).

From Mollier diagram (*h-s* diagram):

$h_1 = 3450$ kJ / kg ; $h_2 = 3050$ kJ / kg ; $h_3 = 3560$ kJ / kg ; $h_4 = 2300$ kJ / kg

h_{f_4} (from steam tables, at 0.1 bar) $= 191.8$ kJ / kg

(*i*) Quality of steam at turbine exhaust, x_4:

$$x_4 = 0.88 \ (\text{From Mollier diagram})$$

(*ii*) Cycle efficiency, η_{cycle}:

$$\eta_{\text{cycle}} = \frac{(h_1 - h_2) + (h_3 - h_4)}{(h_1 - h_{f_4}) + (h_3 - h_2)}$$

$$= \frac{(3450 - 3050) + (3560 - 2300)}{(3450 - 191.8) + (3560 - 3050)} = \frac{1660}{3768.2} = \textbf{0.4405 or 44.05\%.} \quad \textbf{(Ans.)}$$

(*iii*) Steam rate in kg/kWh:

$$\text{Steam rate} = \frac{3600}{(h_1 - h_2) + (h_3 - h_4)} = \frac{3600}{(3450 - 3050) + (3560 - 2300)}$$

$$= \frac{3600}{1660} = \textbf{2.17 kg / kWh.} \quad \textbf{(Ans.)}$$

Example 20

A turbine is supplied with steam at a pressure of 32 bar and a temperature of 410 °C. The steam then expands isentropically to a pressure of 0.08 bar. Find the dryness fraction at the end of expansion and thermal efficiency of the cycle.

If the steam is reheated at 5.5 bar to a temperature of 395 °C and then expanded isentropically to a pressure of 0.08 bar, what will be the dryness fraction and thermal efficiency of the cycle?

■ **Solution:**

First case. Refer to Figure 8.28.

From the Mollier chart:

$$h_1 = 32650 \text{ kJ / kg}$$

$$h_2 = 2170 \text{ kJ / kg}$$

Heat drop (or work done) $= h_1 - h_2$

$$= 3250 - 2170 = 1080 \text{ kJ / kg}$$

Heat supplied $= h_1 - h_{f_2}$

$$= 3250 - 173.9 \qquad [h_{f_2} = 173.9 \, kJ \, / \, kg \text{ at } 0.08 \text{ bar}]$$

$$= 3076.1 \text{ kJ/kg}$$

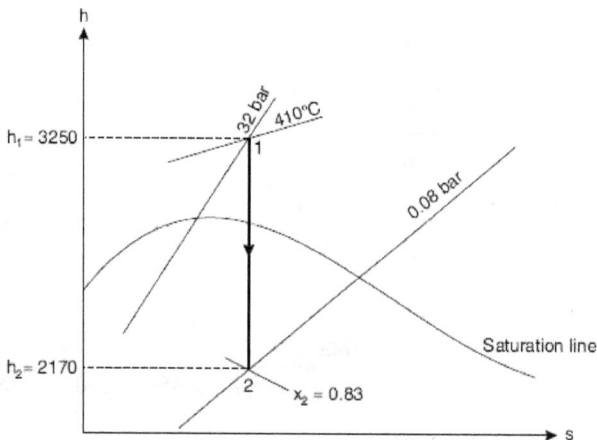

FIGURE 8.28. Cycle for Example 20.

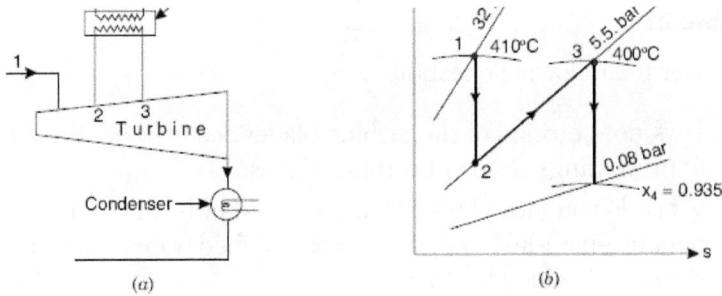

FIGURE 8.29. Schematic of reheat cycle (a), and diagram (b) for Example 20.

Thermal efficiency $= \dfrac{\text{Work done}}{\text{Heat supplied}} = \dfrac{1080}{3076.1} = $ **0.351 or 35.1%. (Ans.)**

Exhaust steam condition, x_2

$$= 0.83 \text{ (From Mollier chart).} \quad \text{(Ans.)}$$

Second case. Refer to Figure 8.29(b).

From the Mollier chart:

$$h_1 = 3250 \text{ kJ / kg ;}$$

$$h_2 = 2807 \text{ kJ / kg ;}$$

$$h_3 = 3263 \text{ kJ / kg ;}$$

$$h_4 = 2426 \text{ kJ / kg.}$$

Work done $= \left(h_1 - h_2\right) + \left(h_3 - h_4\right) = \left(3250 - 2807\right) + \left(3263 - 2426\right) = 1280 \text{ kJ / kg}$

Heat supplied $= \left(h_1 - h_{f_4}\right) + \left(h_3 - h_2\right)$

$$= \left(3250 - 173.9\right) + \left(3263 - 2807\right) = 3532 \text{ kJ / kg}$$

Thermal efficiency $= \dfrac{\text{Work done}}{\text{Heat supplied}} = \dfrac{1280}{3532} = $ **0.362 or 36.2%.** **(Ans.)**

Condition of steam at the exhaust,

$$x_4 = 0.935 \ \left[\textit{From Mollier chart}\right]. \quad \textbf{(Ans.)}$$

Example 21

Answer the following questions:

(a) How does erosion of the turbine blades occur? State the methods of preventing erosion of turbine blades.

(b) What do you mean by TTD of a feed water heater? Draw a temperature-path-line diagram of a closed feed water heater used in a regenerative feed heating cycle.

(c) In a 15 MW steam power plant operating on ideal reheat cycle, steam enters the H.P. turbine at 150 bar and 600 °C. The condenser is maintained at a pressure of 0.1 bar. If the moisture content at the exit of the L.P. turbine is 10.4%, determine:

(i) Reheat pressure; (ii) Thermal efficiency; (iii) Specific steam consumption; and (iv) Rate of pump work in kW. Assume steam to be reheated to the initial temperature.

■ **Solution:**

(a) The erosion of the moving blades is caused by the presence of water particles in (wet) steam in the L.P. stages. The water particles strike the leading surface of the blades. Such impact, if sufficiently heavy, produces severe local stresses in the blade material causing the surface metal to fail and flake off.

The erosion, if any, is more likely to occur in the region where the steam is wettest, that is, in the last one or two stages of the turbine. Moreover, the water droplets are concentrated in the outer parts of the flow annuals where the velocity of impact is highest.

Erosion difficulties due to moisture in the steam may be avoided by reheating (see Figure 8.30). The whole of steam is taken from the turbine at a suitable point 2, and a further supply of heat is given to it along 2-3 after which the steam is readmitted to the turbine and expanded along 3-4 to condenser pressure.

Erosion *may also be reduced by using steam traps in between the stages to separate moisture from the steam.*

(*b*) TTD means *"Terminal temperature difference."* It is the *difference between temperatures of bled steam/condensate and the feed water at the two ends of the feed water heater.*

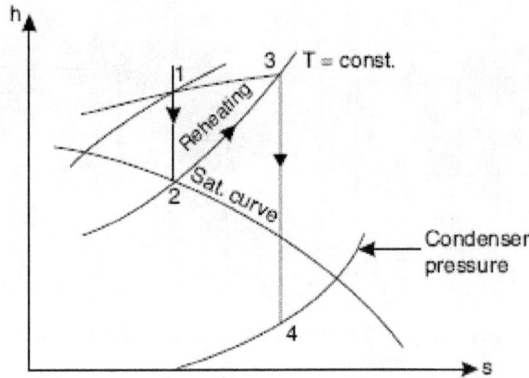

FIGURE 8.30. Reheating details on the h-S diagram for Example 21.

The required temperature-path-line diagram of a closed feed water heater is shown in Figure 8.31.

(c) The cycle is shown on *T-s* and *h-s* diagrams in Figure 8.32. Cycle shown on T-s and h-s diagrams (a) and (b), respectively, for Example 21. respectively. The following values are read from the Mollier diagram:

$$h_1 = 3580 \text{ kJ/kg}, \ h_2 = 3140 \text{ kJ/kg}, \ h_3 = 3675 \text{ kJ/kg}, \ \text{and} \ h_4 = 2335 \text{ kJ/kg}$$

Moisture contents in exit from L.P. turbine $= 10.4\%$

$$x_4 = 1 - 0.104 = 0.896$$

(*i*) **Reheat pressure:** From the Mollier diagram, the **reheat pressure is 40 bar.** (**Ans.**)

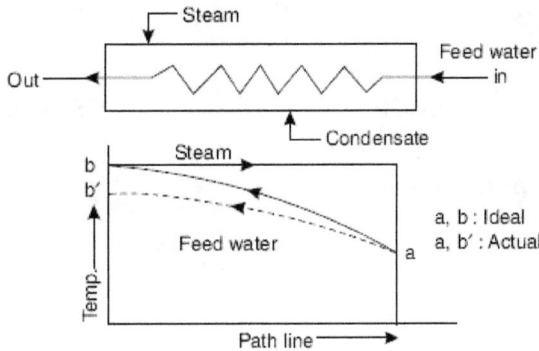

FIGURE 8.31. Temperature-path-line diagram of a closed feed water heater, Example 21.

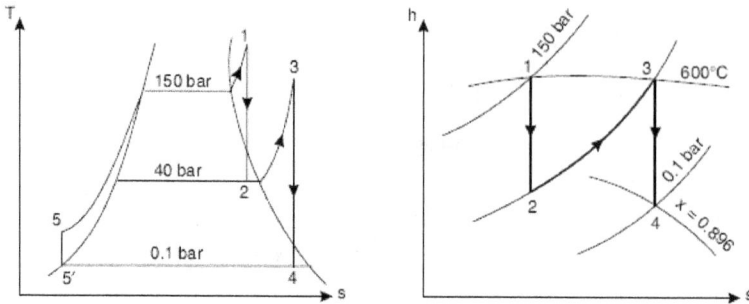

FIGURE 8.32. Cycle shown on T-s and h-s diagrams (a) and (b), respectively, for Example 21.

(ii) Thermal efficiency, η_{th}:

Turbine work $= (h_1 - h_2) + (h_3 - h_4)$

$$= (3580 - 3140) + (3675 - 2335) = 1780 \text{ kJ / kg}.$$

Assuming a specific volume of water $= 10^{-3} \text{ m}^3 / \text{kg}$, the pump work $= 10^{-3}(150 - 0.1) = 0.15 \text{ kJ/kg}$, that is, may be neglected in computing of η_{th}, $h_5 = h_4 = 191.8 \text{ kJ/kg}$, ($h_f$ at 0.1 bar) from steam tables,

$$Q_{\text{input}} = (h_1 - h_5) + (h_3 - h_2)$$

$$= (3580 - 191.8) + (3675 - 3140) = 3923.2 \text{ kJ / kg}$$

$$\%\eta_{th} = \frac{1780}{3923.2} \times 100 = 45.37\%. \quad (\textbf{Ans.})$$

(iii) Specific steam consumption:

Steam consumption $= \dfrac{15 \times 10^3}{1780} = 8.427 \text{ kg / s}$

Specific steam consumption $= \dfrac{8.427 \times 3600}{15 \times 10^3} = \textbf{2.0225 kg / kWh.} \quad (\textbf{Ans.})$

(iv) Rate of pump work:

Rate of pump work $= 8.427 \times 0.15 = \textbf{1.26 kW.} \quad (\textbf{Ans.})$

8.7. Binary Vapor Cycle

The Carnot cycle gives the highest thermal efficiency, which is given by $\left(1-\dfrac{T_1}{T_2}\right)$. To approach this cycle in an actual engine it is necessary that the whole of heat must be supplied at a constant temperature T_1 and rejected at T_2. This can be achieved only by using a vapor in the wet field but not in the superheated. The efficiency depends on temperature T_1 since T_2 is fixed by the natural sink to which heat is rejected. This means that T_1 should be *as large as possible, consistent with the vapor being saturated.*

If we use steam as the working medium, the temperature rise is accompanied by a rise in pressure, and at a critical temperature of 374.15°C the pressure is as high as 225 bar, which will create many difficulties in design, operation, and control. It would be desirable to use some fluid other than steam, which has more desirable thermodynamic properties than water.

An ideal working fluid for a Rankine cycle should have high critical temperature combined with low pressure. Mercury, diphenyl oxide and similar compounds, aluminum bromide, and zinc ammonium chloride are fluids which possess the required properties in varying degrees. Mercury is the only working fluid which has been successfully used in practice. It has high critical temperature (588.4 °C) and correspondingly low critical pressure (21 bar abs.). The mercury alone cannot be used, as its saturation temperature at atmospheric pressure is high (357 °C). Hence binary vapor cycle is generally used to increase the overall efficiency of the plant. Two fluids (mercury and water) are used in cascade in the binary cycle for production of power.

The few more properties required for an ideal binary fluid used in a high temperature limit are as follows:

1. It should have high critical temperature at reasonably low pressure.

2. It should have high heat of vaporization to keep the weight of fluid in the cycle to a minimum.

3. Freezing temperature should be below room temperature.

4. It should have chemical stability through the working cycle.

5. It must be non-corrosive to the metals normally used in power plants.

6. It must have an ability to wet the metal surfaces to promote the heat transfer.

7. The vapor pressure at a desirable condensation temperature should be nearly atmospheric, which will eliminate the requirement of power for the maintenance of a vacuum in the condenser.

8. After expansion through the prime mover the vapor should be nearly saturated so that a desirable heat transfer coefficient can be obtained, which will reduce the size of the condenser required.

9. It must be available in large quantities at reasonable cost.

10. It should not be toxic and, therefore, dangerous to human life.

Although mercury does not have all the required properties, it is more favorable than any other fluid investigated. It is most stable under all operating conditions.

Although mercury does not cause any corrosion to metals, it is extremely dangerous to human life; therefore, elaborate precautions must be taken to prevent the escape of vapor. The major disadvantage associated with mercury is that it does not wet the surface of the metal and forms a serious resistance to heat flow. This difficulty can be considerably reduced by adding magnesium and titanium (2 parts in 100000 parts) in mercury.

Thermal Properties of Mercury

Mercury fulfills practically all the desirable thermodynamic properties stated above.

1. Its freezing point is – 3.3 °C and its boiling point is – 354.4 °C at atmospheric pressure.

2. The pressure required when the temperature of vapor is 540 °C is only 12.5 bar (app.) and, therefore, heavy construction is not required to get high initial temperature.

3. Its liquid saturation curve is very steep, approaching the isentropic of the Carnot cycle.

4. It has no corrosive or erosive effects upon metals commonly used in practice.

5. Its critical temperature is so far removed from any possible upper temperature limit with existing metals as to cause no trouble.

Some *undesirable properties* of mercury are as follows:

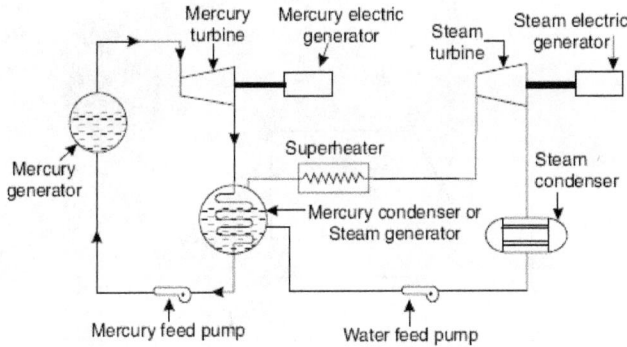

FIGURE 8.33. Line diagram of binary vapor cycle.

1. Since the latent heat of mercury is quite low over a wide range of desirable condensation temperatures, several kg of mercury must be circulated per kg of water evaporated in a binary cycle.

2. The cost is a considerable item as the quantity required is 8 to 10 times the quantity of water circulated in a binary system.

3. Mercury vapor in larger quantities is poisonous, therefore, the system must be perfect and tight.

Figure 8.33 shows the schematic line diagram of a binary vapor cycle using mercury and water as working fluids. The processes are represented on a *T-s* diagram as shown in Figure 8.34.

8.7.1. Analysis of a Binary Vapor Cycle

h_{hg_1} = Heat supplied per kg of Hg (mercury) vapour formed in the mercury boiler.

h_{hg_2} = Heat lost by one kg of Hg vapour in the mercury condenser.

h_s = Heat given per kg of steam generated in the mercury condenser or steam boiler.

W_{hg} = Work done per kg of Hg in the cycle.

W_s = Work done per kg of steam in the steam cycle.

η_s = Thermal efficiency of the steam cycle.

η_{hg} = Thermal efficiency of the Hg cycle.

m = Mass of Hg in the Hg cycle per kg of steam circulated in the steam cycle.

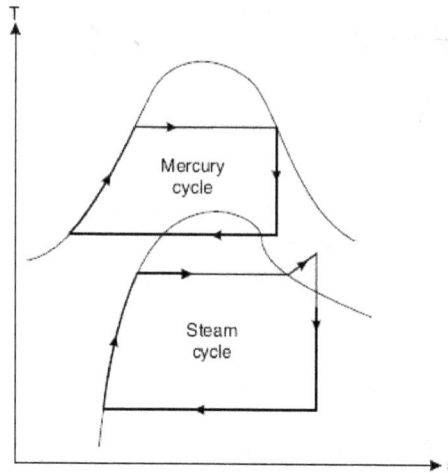

FIGURE 8.34. Binary vapor cycle on a T-s diagram.

The heat losses to the surroundings, in the following analysis, are neglected and steam generated is considered one kg and Hg in the circuit is m kg per kg of water in the steam cycle.

Heat supplied in the Hg boiler

$$h_t = m \times h_{hg_1}$$

Work done in the mercury cycle

$$= m \cdot W_{hg}$$

Work done in the steam cycle

$$= 1 \times W_s$$

Total work done in the binary cycle is given by

$$W_t = m \ W_{hg} + W_s$$

∴ Overall efficiency of the binary cycle is given by

$$\eta = \frac{\text{Work done}}{\text{Heat supplied}} = \frac{W_t}{h_t} = \frac{mW_{hg} + W_s}{mh_{hg_1}} \tag{8.5}$$

Thermal efficiency of the mercury cycle is given by

$$\eta_{hg} = \frac{m W_{hg}}{m h_{hg_1}} \tag{8.6}$$

$$= \frac{W_{hg}}{h_{hg_1}} = \frac{h_{hg_1} - h_{hg_2}}{h_{hg_1}} = 1 - \frac{h_{hg_2}}{h_{hg_1}}$$

$$= \frac{m h_{hg_1} - h_s}{m h_{hg_1}} = 1 - \frac{1}{m} \cdot \frac{h_s}{h_{hg_1}} \tag{8.7}$$

Heat lost by mercury vapor = Heat gained by steam

$$\therefore \qquad\qquad m h_{hg_2} = 1 \times h_s \tag{8.8}$$

Substituting the value of m from eqn. (8.8) into eqn. (8.7), we get

$$\eta_{hg} = 1 - \frac{h_{hg_2}}{h_{hg_1}} \tag{8.9}$$

The thermal efficiency of the steam cycle is given by

$$\eta_s = \frac{W_s}{h_s} = \frac{h_{s_1} - h_{s_2}}{h_{s_1}} = \frac{h_{s_1} - h_{s_2}}{m h_{hg_2}} \tag{8.10}$$

From the eqns. (8.5) and (8.7 – 8.10), we get

$$\eta = \eta_{hg} (1 - \eta_s) + \eta_s \tag{8.11}$$

To solve the problems eqns. (8.6), (8.10), and (8.11) used.

In the design of a binary cycle, another important problem is the limit of exhaust pressure of the mercury (location of optimum exhaust pressure) which will provide maximum work per kg of Hg circulated in the system and high thermal efficiency of the cycle. It is not easy to decide as the number of controlling factors are many.

Example 22

A binary vapor cycle operates on mercury and steam. Standard mercury vapor at 4.5 bar is supplied to the mercury turbine, from which it exhausts at 0.04 bar. The mercury condenser generates saturated steam at 15 bar which is expanded in a steam turbine to 0.04 bar.

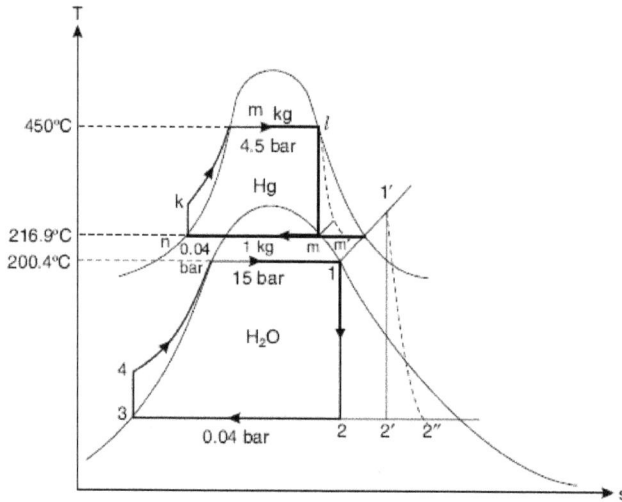

FIGURE 8.35. The binary vapor cycle for Example 22.

(i) Determine the overall efficiency of the cycle.

(ii) If 48000 kg/h of steam flows through the steam turbine, what is the flow through the mercury turbine?

(iii) Assuming that all processes are reversible, what is the useful work done in the binary vapor cycle for the specified steam flow?

(iv) If the steam leaving the mercury condenser is superheated to a temperature of 300 °C in a superheater located in the mercury boiler, and if the internal efficiencies of the mercury and steam turbines are 0.84 and 0.88 respectively, calculate the overall efficiency of the cycle. The properties of standard mercury are given as follows:

p (bar)	T (°C)	h_f(kJ/kg)	h_g(kJ/kg)	s_f(kJ/kg K)	s_g(kJ/kg K)	v_f(m³/kg)	v_g(m³/kg)
4.5	450	62.93	355.98	0.1352	0.5397	79.9×10^{-6}	0.068
0.04	216.9	29.98	329.85	0.0808	0.6925	76.5×10^{-6}	5.178.

■ **Solution:**

The binary vapor cycle is shown in Figure 8.35.

Mercury cycle:

$$h_l = 355.98 \text{ kJ} / \text{kg}$$

$$s_l = 0.5397 = s_m = s_f + x_m \, s_{fg}$$

or
$$0.5397 = 0.0808 + x_m = (0.6925 - 0.0808)$$

∴
$$x_m = \frac{(0.5397 - 0.0808)}{(0.6925 - 0.0808)} = 0.75$$

$$h_m = h_f + x_m\, h_{fg} = 29.98 + 0.75 \times (329.85 - 29.98)$$

$$= 254.88 \text{ kJ / kg}$$

Work obtained from mercury turbine

$$(W_T)_{Hg} = h_1 - h_m = 355.98 - 254.88 = 101.1 \text{ kJ / kg}$$

Pump work in mercury cycle,

$$(W_P)_{Hg} = h_{f_k} - h_{f_n} = 76.5 \times 10^{-6} \times (4.5 - 0.04) \times 100 = 0.0341 \text{ kJ / kg}$$

∴
$$W_{net} = 101.1 - 0.0341 \approx 101.1 \text{ kJ / kg}$$

$$Q_1 = h_1 - h_{f_k} = 355.98 - 29.98 = 326 \text{ kJ / kg} \qquad (\because h_{f_n} \approx h_{f_k})$$

∴
$$\eta_{Hg\ cycle} = \frac{W_{net}}{Q_1} = \frac{101.1}{326} = 0.31 \text{ or } 31\%.$$

Steam cycle:

At 15 bar : $h_1 = 2789.9 \text{ kJ / kg}, \; s_1 = 6.4406 \text{ kJ / kg}$

At 0.04 bar : $h_f = 121.5 \text{ kJ / kg}, \; h_{f_g} = 2432.9 \text{ kJ / kg},$

$s_f = 0.432 \text{ kJ / kg K}, \; s_{fg_2} = 8.052 \text{ kJ / kg K}, \; v_f = 0.0001 \text{ kJ / kg K}$

Now, $s_1 = s_2$

$$6.4406 = s_f + x_2\, s_{fg} = 0.423 + x_2 \times 8.052$$

∴
$$x_2 = \frac{6.4406 - 0.423}{8.052} = 0.747$$

$$h_2 = h_{f_2} + x_2\, h_{fg} = 121.5 + 0.747 \times 2432.9 = 1938.8 \text{ kJ / kg}$$

Work obtained from steam turbine,

$$(W_T)_{steam} = h_1 - h_2 = 2789.9 - 1938.8 = 851.1 \text{ kJ / kg}$$

Pump work in steam cycle,

$$(W_P)_{steam} = h_{f_4} - h_{f_3} = 0.001\,(15 - 0.04) \times 100 = 1.496 \text{ kJ / kg} \approx 1.5 \text{ kJ / kg}$$

or $\qquad h_{f_4} = h_{f_3} + 1.5 = 121.5 + 1.5 = 123 \text{ kJ}/\text{kg}$

$$Q_1 = h_1 - h_{f_4} = 2789.9 - 123 = 2666.9 \text{ kJ}/\text{kg}$$

$$\left(W_{net} \right)_{steam} = 851.1 - 1.5 = 849.6 \text{ kJ}/\text{kg}$$

$\therefore \qquad \eta_{steam\ cycle} = \dfrac{W_{net}}{Q_1} = \dfrac{849.6}{2666.6} = 0.318 \text{ or } 31.8.$

(i) Overall efficiency of the binary cycle:

Overall efficiency of the binary cycle

$$\eta_{Hg\ cycle} + \eta_{steam\ cycle} - \eta_{Hg\ cycle} \times \eta_{steam\ cycle}$$

$$= 0.31 + 0.318 - 0.31 \times 0.318 = 0.5294 \text{ or } 52.94\%$$

Hence overall efficiency of the binary cycle = 52.94%. (Ans)

$\eta_{overall}$ can also be found out as follows :

Energy balance for a mercury condenser-steam boiler gives :

$$m \left(h_m - h_{f_n} \right) = 1 \left(h_1 - h_{f_4} \right)$$

where is the amount of mercury circulating for 1 kg of steam in the bottom cycle

$\therefore \quad m = \dfrac{h_1 - h_{f_4}}{h_m - h_{f_n}} = \dfrac{2666.9}{254.88 - 29.98} = 11.86 \text{ kg}$

$\left(Q_1 \right)_{total} = m \left(h_1 - h_{f_k} \right) = 11.86 \times 326 = 3866.36 \text{ kJ}/\text{kg}$

$\left(W_T \right)_{total} = m \left(h_1 - h_m \right) + \left(h_1 - h_2 \right)$

$\qquad = 11.86 \times 101.1 + 851.1 = 2050.1 \text{ kJ}/\text{kg}$

$\left(W_P \right)_{total}$ may be neglected

$\qquad \eta_{overall} = \dfrac{W_T}{Q_1} = \dfrac{2050.1}{3866.36} = 0.53 \text{ or } 53\%.$

(ii) Flow through mercury turbine:

If 48000 kg/h of steam flows through the steam turbine, the flow rate of mercury,

$$m_{Hg} = 48000 \times 11.86 = \textbf{569280 kg}/\textbf{h. (Ans.)}$$

(iii) Useful work in binary vapor cycle:

Useful work, $\left(W_T\right)_{\text{total}} = 2050.1 \times 48000 = 9840.5 \times 10^4 \text{ kJ / h}$

$$= \frac{9840.5 \times 10^4}{3600} = 27334.7 \text{ kW} = \textbf{27.33 MW. (Ans.)}$$

(iv) Overall efficiency under new conditions:

Considering the efficiencies of turbines, we have:

$$\left(W_T\right)_{\text{Hg}} = h_l - h_m{}' = 0.84 \times 101.1 = 84.92 \text{ kJ / kg}$$

$\therefore \qquad h_{m'} = h_l - 84.92 = 355.98 - 84.92 = 271.06 \text{ kJ / kg}$

$\therefore \qquad m'\,(h_{m'} - \text{h}_{n'}) = (h_1 - h_{f_4})$

or $\qquad\qquad m' = \dfrac{h_1 - h_{f_4}}{h_{m'} - h_{n'}} = \dfrac{2666.9}{271.06 - 29.98} = 11.06 \text{ kg}$

$$\left(Q_1\right)_{\text{total}} = m'\,(h_1 - h_{f_k}) + 1\,(h_1{}' - h_1)$$

$$\Big[\,At\ 15\ bar,\,300°C : h_g = 3037.6 \text{ kJ / kg}, \ s_g = 6.918 \text{ kJ / kg K}\,\Big]$$

$$= 11.06 \times 326 + (3037.6 - 2789.9) = 3853.26 \text{ kJ / kg}$$

$$s_1{}' = 6.918 = s_2{}' = 0.423 + x_2{}' \times 8.052$$

$\therefore \qquad\qquad x_2{}' = \dfrac{6.918 - 0.423}{8.052} = 0.80.$

$$h_2{}' = 121.5 + 0.807 \times 2432.9 = 2084.8 \text{ kJ / kg}$$

$$\left(W_T\right)_{\text{steam}} = h_1{}' - h_2{}' = 0.88\,(3037.6 - 2084.8)$$

$$= 838.46 \text{ kJ / kg}$$

$$\left(W_T\right)_{\text{total}} = 11.06 \times 84.92 + 838.46 = 1777.67 \text{ kJ / kg}$$

Neglecting pump work,

$$\eta_{\text{overall}} = \frac{1777.67}{3853.26} = \textbf{0.4613 or 46.13\%. (Ans.)}$$

8.8 Exercises

1. Rankine cycle efficiency of a good steam power plant may be in the range of

   (a) 15 to 20%   (b) 35 to 45%
   (c) 70 to 80%   (d) 90 to 95%

2. A Rankine cycle operating on a low-pressure limit of p_1 and high-pressure limit of p_2

   (a) has a higher thermal efficiency than the Carnot cycle operating between the same pressure limits
   (b) has a lower thermal efficiency than the Carnot cycle operating between the same pressure limits
   (c) has the same thermal efficiency as the Carnot cycle operating between the same pressure limits
   (d) may be more or less depending upon the magnitudes of p_1 and p_2

3. The Rankine efficiency of a steam power plant

   (a) improves in summer as compared to that in winter
   (b) improves in winter as compared to that in summer
   (c) is unaffected by climatic conditions
   (d) none of the above

4. The Rankine cycle comprises

   (a) two isentropic processes and two constant volume processes
   (b) two isentropic processes and two constant pressure processes
   (c) two isothermal processes and two constant pressure processes
   (d) none of the above

5. In Rankine cycle the work output from the turbine is given by

   (a) change of internal energy between inlet and outlet
   (b) change of enthalpy between inlet and outlet
   (c) change of entropy between inlet and outlet
   (d) change of temperature between inlet and outlet

6. Regenerative heating, that is, bleeding steam to reheat feed water to the boiler

   (a) decreases thermal efficiency of the cycle

   (b) increases thermal efficiency of the cycle

   (c) does not affect thermal efficiency of the cycle

   (d) may increase or decrease thermal efficiency of the cycle depending upon the point of extraction of the steam

7. Regenerative cycle thermal efficiency

   (a) is always greater than simple Rankine thermal efficiency

   (b) is greater than simple Rankine cycle thermal efficiency only when steam is bled at a particular pressure

   (c) is the same as a simple Rankine cycle thermal efficiency

   (d) is always less than a simple Rankine cycle thermal efficiency

8. In a regenerative feed heating cycle, the optimum value of the fraction of steam extracted for feed heating

   (a) decreases with an increase in Rankine cycle efficiency

   (b) increases with an increase in Rankine cycle efficiency

   (c) is unaffected by an increase in Rankine cycle efficiency

   (d) none of the above

9. In a regenerative feed heating cycle, the greatest economy is affected

   (a) when steam is extracted from only one suitable point of a steam turbine

   (b) when steam is extracted from several places in different stages of a steam turbine

   (c) when steam is extracted only from the last stage of a steam turbine

   (d) when steam is extracted only from the first stage of a steam turbine

10. The maximum percentage gain in a regenerative feed heating cycle's thermal efficiency

   (a) increases with the number of feed heaters increasing

   (b) decreases with the number of feed heaters increasing

(c) remains unaffected by the number of feed heaters

(d) none of the above

11. Explain the various operations of a Carnot cycle. Also represent it on *T-s* and *p-V* diagrams.

12. Describe the different operations of a Rankine cycle. Derive also the expression for its efficiency.

13. State the methods of increasing the thermal efficiency of a Rankine cycle.

14. Explain with the help of neat diagram a "Regenerative Cycle." Derive also an expression for its thermal efficiency.

15. State the advantages of a regenerative cycle/simple Rankine cycle.

16. Explain with a neat diagram the working of a Binary vapor cycle.

17. A simple Rankine cycle works between the pressure of 30 bar and 0.04 bar, the initial condition of steam being dry saturated; calculate the cycle efficiency, work ratio, and specific steam consumption. .

18. A steam power plant works between 40 bar and 0.05 bar. If the steam supplied is dry saturated and the cycle of operation is Rankine, find: Cycle efficiency and Specific steam consumption.

19. Compare the Rankine efficiency of a high-pressure plant operating from 80 bar and 400 °C and a low-pressure plant operating from 40 bar 400 °C, if the condenser pressure in both cases is 0.07 bar.

20. A steam power plant working on a Rankine cycle has the range of operation from 40 bar dry saturated to 0.05 bar. Determine: The Cycle efficiency, Work ratio, and Specific fuel consumption.

21. In a Rankine cycle, the steam at inlet to turbine is saturated at a pressure of 30 bar and the exhaust pressure is 0.25 bar. Assume flow rate of 10 kg/s. Determine: [Ans. (i) 30 kW, (ii) 7410 kW, (iii) 29.2%, (iv) 17900 kW, (v) 0.763].

(*i*) The pump work   (*ii*) Turbine work

(*iii*) Rankine efficiency   (*iv*) Condenser heat flow

(*v*) Dryness at the end of expansion

22. In a regenerative cycle the inlet conditions are 40 bar and 400°C. Steam is bled at 10 bar in regenerative heating. The exit pressure is 0.8 bar. Neglecting pump work determine the efficiency of the cycle.

23. A turbine with one bleeding for regenerative heating of feed water is admitted with steam having an enthalpy of 3200 kJ/kg and the exhausted steam having an enthalpy of 2200 kJ/kg. The ideal regenerative feed water heater is fed with 11350 kg/h of bled steam at 3.5 bar (whose enthalpy is 2600 kJ/h). The feed water (condensate from the condenser) with an enthalpy of 134 kJ/kg is pumped to the heater. It leaves the heater dry saturated at 3.5 bar. Determine the power developed by the turbine.

24. A binary-vapor cycle operates on mercury and steam. Saturated mercury vapor at 4.5 bar is supplied to the mercury turbine, from which it exhausts at 0.04 bar. The mercury condenser generates saturated steam at 15 bar which is expanded in a steam turbine to 0.04 bar. Find:

(a) Find the overall efficiency of the cycle.

(b) If 50000 kg/h of steam flows through the steam turbine, what is the flow through the mercury turbine?

(c) Assuming that all processes are reversible, what is the useful work done in the binary vapor cycle for the specified steam flow?

(d) If the steam leaving the mercury condenser is superheated to a temperature of 300 °C in a superheater located in the mercury boiler, and if the internal efficiencies of the mercury and steam turbines are 0.85 and 0.87 respectively, calculate the overall efficiency of the cycle. The properties of saturated mercury are given in the following table:

p (bar)	t (°C)	h_f	h_g	s_f	s_g	v_f	v_g
		(kJ/kg)		(kJ/kg K)		(m^3/kg)	
4.5	450	63.93	355.98	0.1352	0.5397	79.9×10^{-6}	0.068
0.04	216.9	29.98	329.85	0.0808	0.6925	76.5×10^{-3}	5.178

CHAPTER 9

COMPOUND STEAM ENGINES

In This Chapter

* Overview
* Advantage of Compound Steam Engines
* Classification of Compound Steam Engines
* Unified Steam Engines
* Exercises

9.1. Overview

A simple steam engine is one in which each of the engine cylinders receives steam direct from the boiler, and exhausts into the atmosphere or into a condenser. Compound steam engines are those which have two or more cylinders of successively increasing diameters so arranged that exhaust steam from the first and smallest cylinder is passed forward to do work in a second, and sometimes a third or fourth cylinder, before escaping to the condenser.

Compounding is primarily done to secure a large total ratio of expansion without the disadvantages attached to a very early cut-off of steam in a single cylinder. For single-cylinder engine, we can list some of these disadvantages as:

* A large range of temperature, accompanied by excessive condensation and re-evaporation effects unless high superheats are used. The range of temperatures is not merely

the difference between the initial temperature and the exhaust temperature, but includes the fact that the cylinder is heating up during admission and cooling during expansion and exhaust so that with an early cut-off, the ratio of the cooling period to the heating period is large.

- A cut-off earlier than 30% of the stroke, with an ordinary slide valve, is accompanied by an early release and excessive compression.

- With an early cut-off the turning moment on the shaft is more uneven; as a result a greater flywheel effect is required. This effect, however, is reduced if a number of cylinders and cranks are used.

9.2. Advantage of Compound Steam Engines

The compound steam engines claim the following advantages over single-stage steam engines:

- Owing to reduction in temperature range per cylinder, the cylinder condensation is minimized.

- As there is re-evaporation in the early stages of expansion, the steam can be expanded still further in the later stages. Consequently, the loss due to condensation in compound steam engines is not cumulative and is generally restricted to low pressure cylinders.

- The leakage past the valves and piston is minimized since the pressure difference across these components is less.

- Although the total expansion for the whole engine is large, the expansion ratio per cylinder is generally not more than four. This permits the use of a simple type of valve gear.

- The cylinder condensation can be reduced by reheating the steam after expansion in each cylinder.

- In compound steam engines, the turning moment is more uniform (since the pressure difference is less).

- In a single-stage steam engine if the expansion ratio is large, the cylinder has to be made strong enough to withstand the force of high pressure steam. Furthermore, it should have large volume to contain low pressure steam.

- High speeds are possible since mechanical balancing may be made more or less perfect.

- In compound steam engines since the forces are distributed over more components, the forces in the working parts are reduced.

- The engine can be started in any position, and this quality is of great advantage in marine, locomotive, and mining work.

- If the engine undergoes a breakdown, it can be modified to run on a reduced load; this is of much help in marine propulsion.

- For the same power and economy, the cost of the compound engines is less than that of a simple engine.

- Due to lighter reciprocating parts of an engine, the engine vibrations are reduced.

9.3. Classification of Compound Steam Engines

The compound steam engines are usually classified as follows:

1. Tandem compound steam engines

2. Cross compound steam engines

The Cross-compound engines are further grouped in two types: Woolfe type compound steam engines and Receiver type compound steam engines.

9.3.1. Tandem Compound Steam Engines

In this type of engine, the two cylinders have a common piston rod and are fixed in tandem, working on the same crank. These cylinders may be regarded as having cranks at zero degrees to each other (see Figure 9.1).

Steam generated in the boiler is supplied to one side of the piston of the high pressure (H.P.) cylinder (shown by solid arrowheads). On the other side of the high-pressure piston, exhaust takes place simultaneously and this exhaust steam now acts on the piston of the low pressure (L.P.) cylinder (shown by solid arrowheads). The valves should be operated in such a way that there should be continuous admission of steam in the high pressure as well as the low-pressure cylinder simultaneously. After the steam supply is cut off in the high-pressure cylinder, the steam expands, and after the high-pressure exhaust steam is admitted to the low-pressure cylinder up to the cut-off point, it further expands to the condenser pressure if the engine is condensing or to the atmosphere for a non-condensing engine.

Since usually the steam engines are double-acting, the steam flow takes place during the return stroke as shown by thin arrowheads.

As the cycles of the H.P. and L.P. cylinders are in phase (Figure 9.2), the maximum turning moment on the crank-shaft, due to each

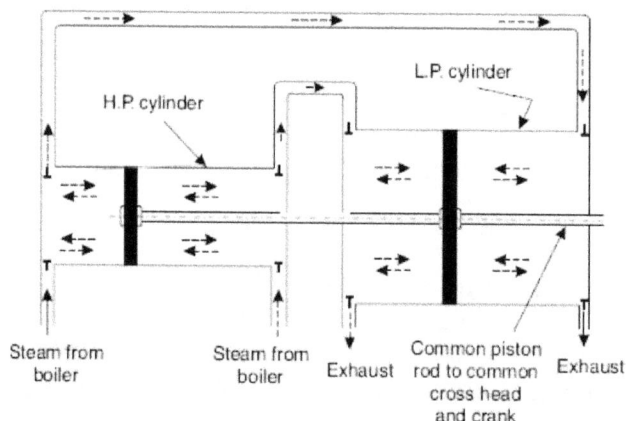

FIGURE 9.1. Tandem compound steam engine.

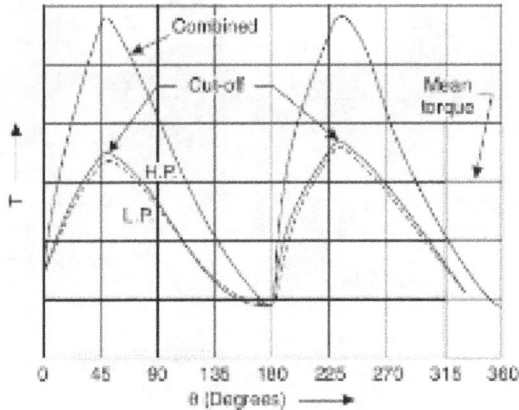

FIGURE 9.2. Torque vs. rotation for a typical Tandem engine.

cylinder, will act at the same instant. This is the disadvantage of this type of compound engine, as a large flywheel is consequently required. Problems arising out of vibrations and balancing exist in this type of engine.

9.3.2. Cross Compound Steam Engines

Woolfe-type compound steam engines are shown schematically in Figure 9.4. In this type of two-cylinder compound engine, the cranks of the cylinders are asynchronized at 180° to each other. The cylinders are arranged side by side, and exhaust steam from the H.P. cylinder passes directly into the L.P. cylinder; the expansion is, therefore, continuous during the stroke. As the cranks are at 180° the two cycles are in phase and cause a large variation in the turning moment on the crankshaft; this is the same disadvantage as in the tandem type of compound steam engines.

Receiver type compound steam engines: A receiver-type compound steam engine is shown in Figure 9.3. In this type, the steam from one cylinder is not directly discharged into the next cylinder but it is discharged into a chamber known as a receiver. So, the receiver is nothing but a reservoir of steam from where the steam is admitted into the L.P. cylinder during its admission stroke. In this arrangement, the crank angles are *less than 180°*. In a two-cylinder compound engine the angle is 90°.

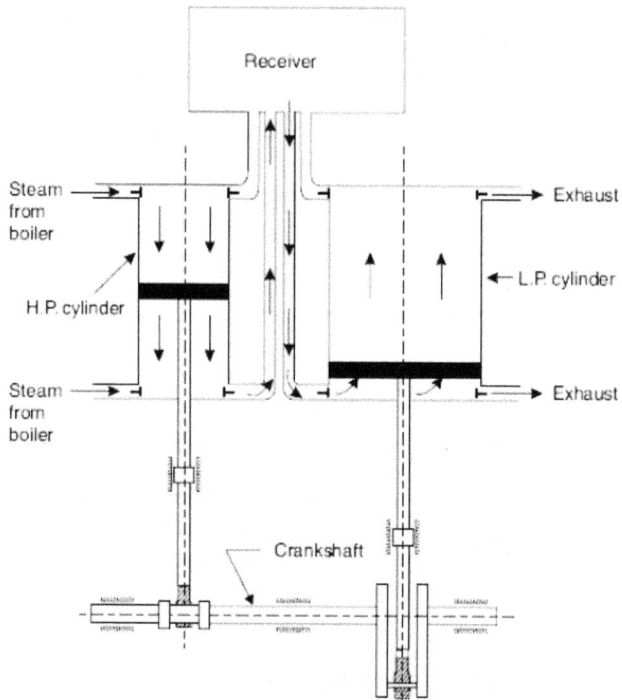

FIGURE 9.3. Receiver-type compound steam engine.

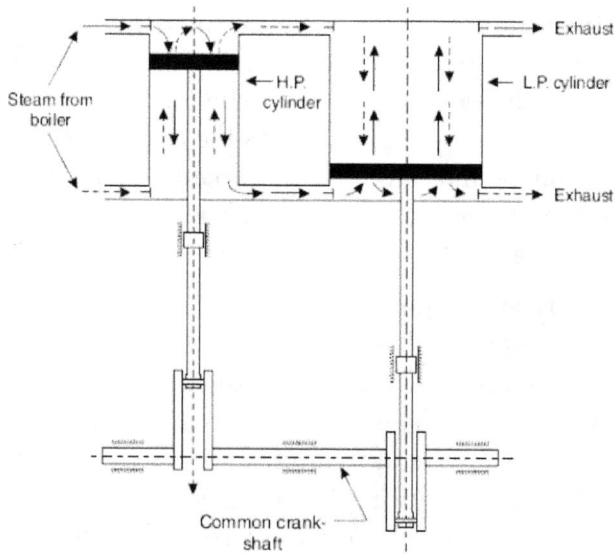

FIGURE 9.4. Woolfe-type compound steam engine.

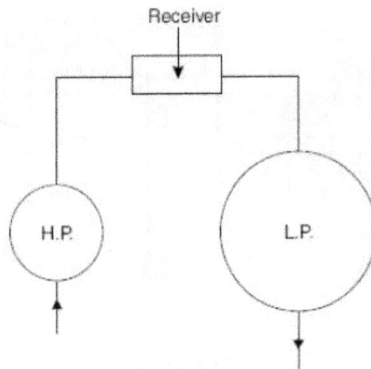

FIGURE 9.5. Double expansion engine.

Figure 9.6 shows a $T - \theta$ curve for a receiver type compound engine. The turning moment is more uniform and, therefore, a lighter flywheel will be required. This type of engine can start in any position. It can also be run at reduced loads, with one cylinder in operation.

There is always an unavoidable pressure drop in the receiver due to condensation of steam, but it can be reduced by steam jacketing the receiver. The receiver should be large enough to keep the pressure in it fairly constant; its volume should be about 1.5 times the H.P. cylinder volume.

According to a number of expansion stages, the compound steam engines may also be classified as follows:

1. Double expansion

2. Triple expansion

1. Double expansion. With reference to Figure 9.5, in a double expansion engine, the expansion of steam takes place in two cylinders. The steam is first expanded in a high-pressure cylinder and then it is discharged into the low-pressure cylinder. Finally, it is exhausted into the condenser.

2. Triple expansion. With reference to Figure 9.7., in this type of engine, the expansion of steam is completed in three cylinders.

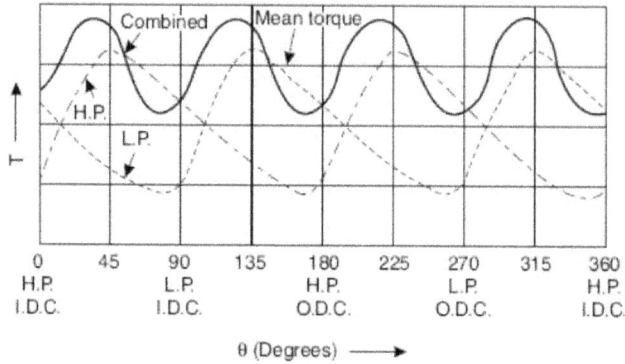

FIGURE 9.6. T - θ curve.

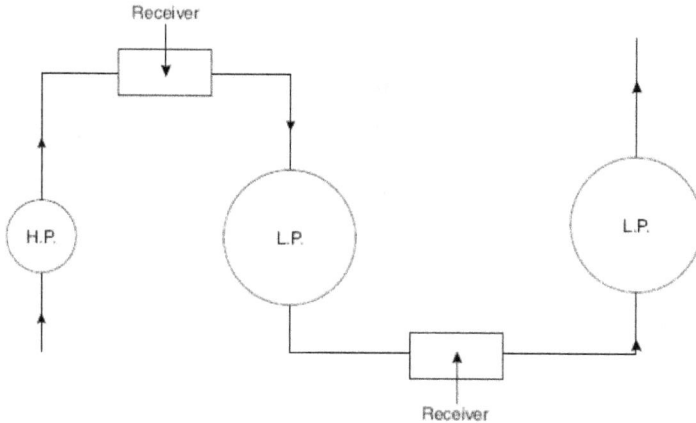

FIGURE 9.7. Triple expansion engine.

The steam from the high-pressure cylinder is exhausted into an intermediate pressure (I.P.) cylinder and then the steam is discharged into the low-pressure cylinder. The steam from the low-pressure cylinder is discharged into the condenser.

9.3.3. Multi-Cylinder Engines

In a multi-cylinder compound steam engine, the number of cylinders to be provided depends upon the total range of pressure through which the steam is to be expanded.

- Simple engines are usually non-condensing with initial pressure ranging from 6 to 7 bar.

- Compound condensing engines with two cylinders have initial pressures ranging from 7 to 10.5 bar.

- Compound condensing engines with three cylinders (i.e., triple expansion engines) have initial pressures ranging from 10.5 to 14 bar.

- The initial pressure in a quadruple expansion engine (i.e., four-cylinder engine) may be from 14 to 21 bar.

- In case of condensing engines, the L.P. (release) pressure may vary from 0.7 to 0.85 bar and for non-condensing engines from 1.4 to 1.75 bar.

- In very large engines, for example, when the diameter of the L.P. cylinder exceeds 2.5 m, it is usual practice to fit two L.P. cylinders, thereby producing a four-cylinder triple expansion engine. One H.P. and two L.P. cylinders are used on high speed engines.

9.3.4. Estimation of Cylinder Dimensions for Compound Engines

The cylinder dimensions, in case of a two-cylinder compound engine, can be determined from the power and the hypothetical indicator diagram (see Figure 9.8). The number of assumptions made will depend on the amount of known data; there may be:

Equal initial loads/forces on the pistons; and

Equal work in the cylinders.

Figure 9.8 represents a hypothetical indicator diagram for a two-cylinder compound engine, and let there be a pressure drop between the H.P. cylinder release and the receiver as shown. This pressure drop causes a loss of work represented by the dark area; in many compound engines this area is a considerable amount and cannot be neglected. Although the pressure drop after release is wasteful, it is partly counterbalanced by the drying effect on the steam which it produces.

Let p_1 = Initial steam pressure (bar),

p_2 = Release pressure in H.P. cylinder (bar),

p_3 = Receiver steam pressure (bar),

p_4 = Steam pressure at release of L.P. cylinder (bar),

p_b = Condenser pressure (bar),

V_1 = Volume at cut-off in H.P. cylinder (m^3),

V_2 = Volume of H.P. cylinder (m^3),

V_3 = Volume at cut-ff of L.P. cylinder (m^3),

V_4 = Volume of L.P. cylinder (m^3),

L = Length of stroke (m),

$D_{\text{H.P.}}$ = Diameter of H.P. cylinder (m),

$D_{\text{L.P.}}$ = Diameter of L.P. cylinder (m),

N = Speed of the engine (r.p.m.), and

$(\text{D.F.})_0$ = Overall diagram factor.

Then, neglecting clearance volume,

H.P. cylinder volume, $V_2 = \dfrac{\pi}{4} D_{\text{H.P.}}{}^2 L$

L.P. cylinder volume, $V_4 = \dfrac{\pi}{4} D_{\text{L.P.}}{}^2 L$

Ratio of expansion in H.P. cylinders, $r_{\text{H.P.}} = \dfrac{V_2}{V_1}$

Ratio of expansion in L.P. cylinder, $r_{\text{L.P.}} = \dfrac{V_4}{V_3}$

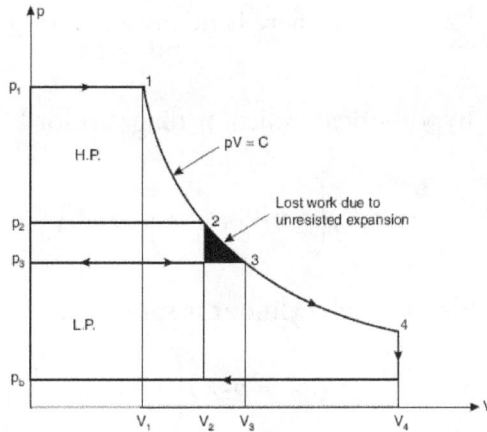

FIGURE 9.8. Hypothetical indicator diagram for a two-cylinder compound engine.

$$\text{Ratio of expansion for whole engine, } R = \frac{V_4}{V_1}$$

As expansion is assumed isothermal,

$$p_1V_1 = p_2V_2 = p_3V_3 = p_4V_4$$

Total work done by engine/stroke

$$= (\text{D.F.})_0 \left[p_1V_1\left(1+\log_e\frac{V_4}{V_1}\right) - p_bV_4 \right]$$

Total power of engine, neglecting dark area

$$= (\text{D.F.})_0 \left[p_1V_1\left(1+\log_e\frac{V_4}{V_1}\right) - p_bV_4 \right] \times \frac{2N \times 10^5}{60 \times 1000} \text{ kW}$$

Area of hypothetical diagram for H.P. cylinder

$$= p_1V_1\left(1+\log_e\frac{V_2}{V_1}\right) - p_3V_2$$

$$= p_1 V_1 \log_e \frac{V_2}{V_1} \qquad \begin{bmatrix} \text{If there is no pressure drop at release} \\ p_1 V_1 = p_3 V_2 \end{bmatrix}$$

Area of the hypothetical indicator diagram for L.P. cylinder

$$= p_3 V_3 \left(1 + \log_e \frac{V_4}{V_3} \right) - p_b V_4.$$

(*i*) If work done in each cylinder is same, then

$$p_1 V_1 \left(1 + \log_e \frac{V_2}{V_1} \right) - p_3 V_2 = p_3 V_3 \left(1 + \log_e \frac{V_4}{V_3} \right) - p_b V_4$$

$$= \frac{1}{2} \times \left[p_1 V_1 \left(1 + \log_e \frac{V_4}{V_1} \right) - p_b V_4 \right]$$

(*ii*) If the initial load/force on the pistons is same, then

$$\left(p_1 - p_3 \right) \frac{\pi}{4} D^2_{\text{H.P}} = \left(p_3 - p_b \right) \frac{\pi}{4} D^2_{\text{L.P}}$$

From the above equations, the required dimensions can be found as per data supplied and assumptions made.

9.3.5. Thermal Efficiency Losses in Compound Engines

Initial condensation becomes increasingly serious as the expansion ratio is increased and as the temperature range across the cylinder becomes greater. The most effective solution adopted in practice is that of compounding, but nevertheless the initial condensation and leakage occur on a reduced scale in each stage of the expansion. The pressure drop in the receiver, which includes the wire-drawing at the exhaust and inlet ports and valves, is a further loss, resulting in an increase of entropy due to un-resisted expansion, and overall available heat is reduced.

9.3.6. The Governing of Compound Engines

There are three methods of governing compound engines. These are:

1. Throttle governing. Reducing the steam supply pressure in the H.P. cylinder;

2. Cut-off governing on the H.P. cylinder. Varying the point of cut-off in the H.P. cylinder; and

3. Cut-off governing on the L.P. cylinder. Varying the point of cut-off in the L.P. cylinder.

For simplicity, consider an engine with a cylinder volume ratio of $5 : 2$, and with cut-off in the H.P. cylinder at a full load of $\frac{1}{2}$. Let the back pressure in the L.P. cylinder be zero.

Figure 9.9 illustrates the previous three methods, the full load diagrams being 1-2-3-4-5-6-1, 1'-2'-3'-4'-5'-6'-1', and 1"-2"-3"- 4"-5"-6"-1", respectively.

Throttle governing. Refer to Figure 9.9 (*a*). Suppose the full load pressure 6-1 is reduced to half by throttling, that is to 6-7, the cut-off in each cylinder remaining the same. The light load diagram will become 7-8-10-5-6-7. The high-pressure work will be reduced from 1-2-3-7-1 to 7-8-9-11-7 and the low-pressure work will be reduced from 7-3-4-5-6-7 to 11-9-10-5-6-11.

High pressure cut-off governing. Refer to Figure 9.9 (*b*). Suppose for the light load the cut-off in high pressure is reduced from $\frac{1}{2}$ to $\frac{1}{4}$, the initial pressure and the cut-off in low pressure remaining constant. The light load diagram will become 1'-8'-10'-5'-6'-1', the high-pressure work becoming 1'-8'-9'-11'-1', which is not very different from 1'-2'-3'-7'-1', while the low-pressure work is reduced from 7'-3'-4'-5'-6'-7' to 11'-9'-10'-5'-6'-11'. Governing by varying the cut-off in the H.P. cylinder will therefore reduce the proportion of work done in the L.P. cylinder at light loads. With condensing engines at very light loads this may cause the average pressure to overcome the back

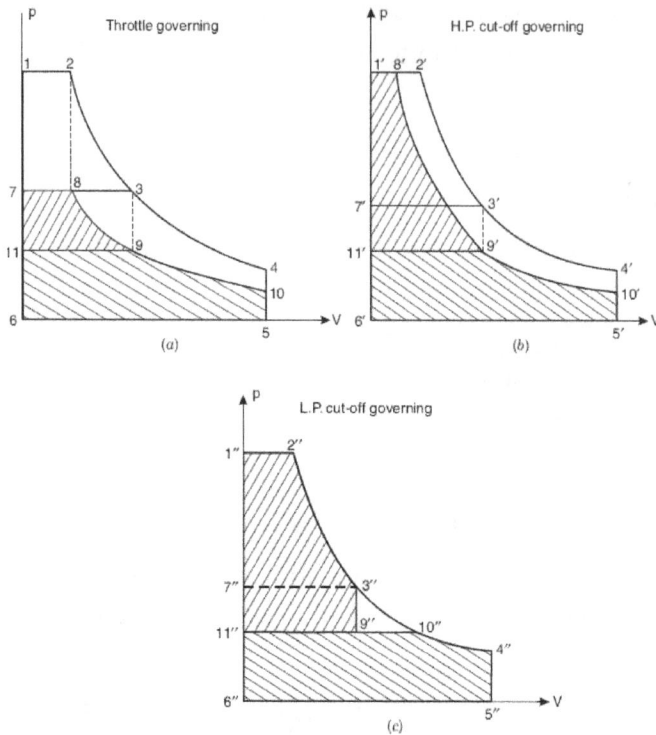

FIGURE 9.9. Governing of compound steam engines.

pressure and frictional resistance, thus reducing the efficiency of the engine.

A comparison of the diagrams Figure 9.9 (a) and Figure 9.9 (b) shows that the high pressure cut-off governing is more economical than throttle governing. The light load diagrams for the L.P. cylinder, namely 11-9-10-5-6-11 and 11'-9'-10'-5'-6'-11', are the same in each case, the release pressure 10-5 or 10'-5' also being the same. Hence the same volume of steam at the same release pressure is exhausted from the engine in each case, whereas the work done with throttle governing is represented by 7-8-10-5-6-7 while the work done with high pressure cut-off governing is represented by the larger area 1'-8'-10'-5'-6'-1'.

Low pressure cut-off governing. Refer to Figure 9.9(c). Suppose the high-pressure cut-off is kept constant at, while the low-pressure cut-off is changed from 2/5 to 3/5. By this change of cut-off, the high-pressure work is increased from 1"-2"-3"-7"-1" to 1"-2"-3"-9"-11"-1" while the low-pressure work is decreased from 7"-3"-4"-5"-6"-7" to 11"-10"-4"-5"-6"-11". Thus, by making the cut-off in the L.P. cylinder earlier, the total work done by the engine is only slightly affected, but the proportion of the total work done in the L.P. cylinder is reduced, while the work done in the H.P. cylinder is increased.

9.4. Unified Steam Engines

The condensation of steam during admission and part of expansion in the reciprocating steam engine is due to the fact that the walls of the cylinder are relatively cool in comparison with the steam. This low temperature is due to the effect of the cylinder having been filled with low pressure, low temperature steam prior to admission. The "Uniflow" engine was designed with a view to reducing the effects of initial condensation of the steam, and this reduction is achieved by virtue of the uni-directional flow of steam in the engine cylinder, so that any particular part of the cylinder wall is maintained at an approximate even temperature.

A uniflow steam engine is illustrated diagrammatically in Figure 9.10. It has a cylinder that is long in comparison with its stroke, and has a piston about 45% as long as the cylinder. Steam enters at each end and flows one way (literally uniflow) toward the middle, where the ring of exhaust ports is uncovered by the piston as it nears the end

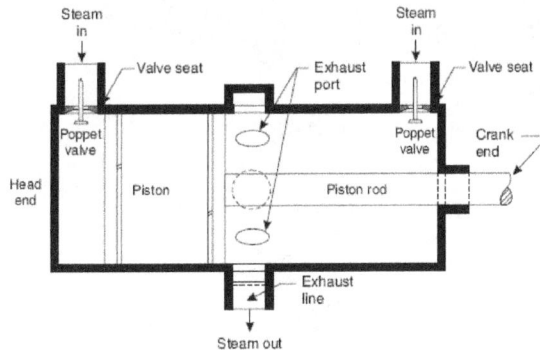

FIGURE 9.10. Uniflow steam engine sketch.

of the stroke. The long piston thus acts as an exhaust valve. On the return stroke, the piston very soon covers the exhaust ports, so that compression (point *L* in Figure 9.11) starts early and can be carried almost to line pressure. By cleverly using adjustable clearance plugs, an engine of this type can be adjusted to compress the cushion steam to the desired point. Other variations on the uniflow type use auxiliary exhaust valves to prohibit excessive compression.

Advantages:

1. Uniflow engines are fairly simple in design and of robust construction.

2. Since condensation losses are very small, the uniflow engine dispenses with the necessity for compounding.

3. Uniflow engines have high efficiency and specific steam consumption is practically constant over its working range of load.

4. Having fewer rubbing parts than the compound engines, it has a higher mechanical efficiency.

Disadvantages:

1. A single cylinder engine of high power is very long, and the large sizes must be horizontal engines.

2. One advantage of the compound engine, the more even turning moment, is lost in the Uniflow Engine, so that it has poor mechanical balance and requires a very large and heavy flywheel.

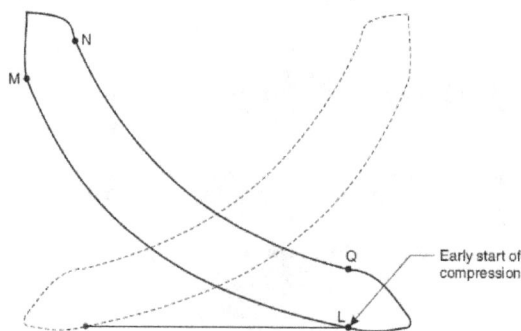

FIGURE 9.11. Indicator diagram for uniflow engine.

These engines must consequently run at low speeds.

3. There is a loss of net output work due to early compression.

4. The cylinder must be very large and robust as it has to withstand high pressure. The volume of the cylinder is nearly equal to the volume of the L.P. cylinder in the compound steam engine.

Example 1

A compound engine is to develop 90 kW at 100 r.p.m. Steam is supplied at 7.5 bar and the condenser pressure is 0.21 bar. Assuming hyperbolic expansion and an expansion ratio of 15, a diagram factor of 0.72 and neglecting clearance and receiver losses, determine the diameters of the cylinders so that they may develop equal powers. Stroke of each piston = L.P. cylinder diameter.

■ **Solution.**

Power to be developed,	I.P. = 90 kW
Engine speed,	$N = 100$ r.p.m.
Admission steam pressure,	$p_1 = 7.5$ bar
Condenser pressure,	$p_b = 0.21$ bar
Expansion ratio,	$R = 15$
Diagram factor,	D.F. = 0.72.

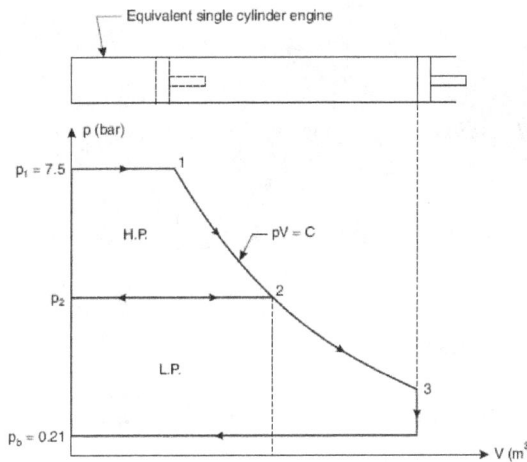

FIGURE 9.12. Compound engine sketch for Example 1.

Cylinder diameters, $D_{L.P.}$, $D_{H.P.}$:

Stroke of each piston   = L.P. cylinder diameter

$p_{m(actual)}$ referred to L.P. cylinder

$$= D.F. \left[\frac{p_1}{R}(1+\log_e R)-p_b \right]$$

$$= 0.72 \left[\frac{7.5}{15}(1+\log_e 15)-0.21 \right] = 1.18 \text{ bar}$$

Indicated power, $I.P. = \dfrac{10\,p_{m(actual)}\,LAN}{3} = \dfrac{10 \times 1.18 \times D_{L.P.} \times \dfrac{\pi}{4} D^2_{L.P.} \times 100}{3}$

$$= 90 = 308.92\, D^3_{L.P.}$$

$\therefore$ $$D_{L.P.} = \left(\frac{90}{308.32} \right)^{1/3} = 0.6629 \text{ m} \simeq 663 \text{ mm}$$

i.e., Diameter of low pressure cylinder = **663 mm.** (**Ans.**)

Work done in H.P. cylinder

$$= p_1 V_1 + p_1 V_1 \log_e \frac{V_2}{V_1} - p_2 V_2$$

But $\qquad\qquad p_1 V_1 = p_2 V_2$

$\therefore$ Work done in *H.P. cylinder* $= p_1 V_1 \log_e r_{H.P.}$ $\qquad \left(\because r_{H.P.} = \dfrac{V_2}{V_1} \right)$

Work done in L.P. cylinder $= p_2 V_2 + p_2 V_2 \log_e \dfrac{V_3}{V_2} \, p_b V_3$

$$= p_2 V_2 + p_2 V_2 \log_e r_{L.P.} - p_b V_3 \qquad \left(\because r_{L.P.} = \frac{V_3}{V_2} \right)$$

Equating work done in H.P. cylinder to that done in L.P. cylinder, we have:

$$p_1 V_1 \log_e r_{H.P.} = p_2 V_2 + p_2 V_2 \log_e r_{L.P.} - p_b V_3$$

$\therefore \qquad p_b V_3 = p_1 V_1 \left(1 + \log_e r_{L.P.} - \log_e r_{H.P.}\right) \qquad \left(\because p_1 V_1 = p_2 V_2\right)$

or $\qquad \dfrac{p_b V_3}{p_1 V_1} - 1 = \log_e r_{L.P.} - \log_e r_{H.P.}$

or $\qquad \log_e \left(\dfrac{r_{L.P.}}{r_{H.P.}}\right) = \dfrac{p_b V_3}{p_1 V_1} - 1$

But $\qquad \dfrac{V_3}{V_1} = R = 15$

$\therefore \qquad \log_e \left(\dfrac{r_{L.P.}}{r_{H.P.}}\right) = \left(\dfrac{0.21}{7.5} \times 15 - 1\right)$

Also $\qquad r_{L.P.} = \dfrac{V_3}{V_2} \text{ and } r_{H.P.} = \dfrac{V_2}{V_1}$

$\qquad \log_e \left(\dfrac{r_{L.P.}}{r_{H.P.}}\right) = \log_e \left(\dfrac{V_3}{V_2}\right) \times \left(\dfrac{V_1}{V_2}\right) = \log_e \left(\dfrac{15 V_1^2}{V_2^2}\right) \quad \left(\because V_3 = 15 V_1\right)$

$\therefore \qquad \log_e \left(\dfrac{15 V_1^2}{V_2^2}\right) = \dfrac{0.21}{7.5} \times 15 - 1$

or $\qquad \log_e \left(\dfrac{V_2^2}{15 V_1^2}\right) = 1 - \dfrac{0.21}{7.5} \times 15 = 0.58$

or $\qquad \dfrac{V_2^2}{15 V_1^2} = 1.786, \quad \therefore \ \dfrac{V_2^2}{V_1^2} = 26.79$

$\therefore$ Ratio of expansion for H.P. cylinder, $r_{H.P.} = \dfrac{V_2}{V_1} = (26.79)^{1/2} = 5.176$

Also $\qquad \dfrac{V_3}{V_2} = \dfrac{V_3}{V_1} \times \dfrac{V_1}{V_2} = \dfrac{15}{(V_2/V_1)} = \dfrac{15}{5.176} \approx 2.9$

Volume of H.P. cylinder $= \dfrac{\pi}{4} \times \left(\dfrac{0.663}{\sqrt{2.9}}\right)^2 \times 0.663 = \dfrac{\pi}{4} D^2_{H.P.} \times 0.663$

$$\therefore \qquad D_{\text{H.P.}} = \left[\frac{(0.663)^3}{2.9 \times 0.663} \right]^{1/2} = 0.3893 \text{ m} = 389 \text{ mm}$$

i.e.,   *Diameter of H.P. cylinder* = **389 mm**. (**Ans.**)

Example 2

On the basis of the following particulars of a double acting compound engine, calculate:

(i) The piston diameters,

(ii) Common stroke, and

(iii) The L.P. cut-off if the initial loads on the piston rods are to be equal.

Indicated power developed	= 309 kW
Speed	= 200 r.p.m.
Steam supply pressure	= 14 bar
Exhaust pressure	= 0.35 bar
Number of expansions	= 9
Ratio of cylinder volumes	= 4.5 : 1
Stroke length	= 0.75 x L.P. piston diameter
Overall diagram factor	= 0.62
Pressure loss in the receiver between the two cylinders	= 0.2 bar
Neglect clearance.	

■ **Solution.**

(*i*) **Piston diameters**:

$$\frac{V_4}{V_1} = R = 9, \quad \frac{V_4}{V_2} = 4.5$$

Theoretical mean effective pressure of the whole engine,

$$p_{m(th.)} = \frac{p_1}{R} \left(1 + \log_e R\right) - p_b$$

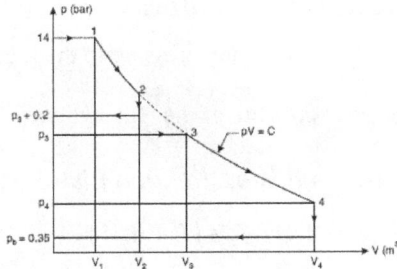

FIGURE 9.13. p-v diagram for double acting compound engine, Example 2.

$$= \frac{14}{9} \left(1 + \log_e 9\right) - 0.35 = 4.623 \text{ bar}$$

Actual mean effective pressure,

$$p_{m(act.)} = \text{Diagram factor} \times p_{m(th.)}$$

$$= 0.62 \times 4.623 = 2.866 \text{ bar}$$

Indicated power, $\qquad \text{I.P.} = \dfrac{10\, p_{m(actual)}\, LAN}{3}$

$$309 = \frac{10 \times 2.866 \times (0.75 \times D_{\text{L.P.}}) \times \dfrac{\pi}{4} \times D^2_{\text{L.P.}} \times 200}{3}$$

$$\left(\therefore \quad L = 0.75\, D_{\text{L.P.}} ....\text{given}\right)$$

$$\therefore \qquad D_{\text{L.P.}} = \left(\frac{309 \times 3 \times 4}{10 \times 2.866 \times 0.75 \times \pi \times 200}\right)^{1/3}$$

$$= 0.65 \text{ m or } 650 \text{ mm}$$

i.e., $\qquad$ *Diameter of L.P. cylinder =* **650 mm.** $\quad$ **(Ans.)**

Diameter of H.P. cylinder $= \dfrac{650}{\sqrt{4.5}} =$ **306 mm.** $\quad \left(\textbf{Ans.}\right)$

(*ii*) Common stroke, L:

$$L = 0.75 \times D_{\text{L.P.}} = 0.75 \times 650 = \textbf{487.5 mm.} \quad \textbf{(Ans.)}$$

(*iii*) **Cut-off in L.P. cylinder:**

Let p_3 = Admission pressure to the L.P. cylinder, and

$p_3 + 0.2$ = Exhaust pressure of the H.P. cylinder.

Since the initial loads on the two piston rods are to be equal,

$$\therefore \qquad \left[14 - \left(p_3 + 0.2\right)\right] V_2 = \left(P_3 - 0.35\right) V_4$$

or $\quad 14 - \left(p_3 + 0.2\right) = \left(P_3 - 0.35\right) \dfrac{V_4}{V_2} = \left(p_3 - 0.35\right) \times 4.5 \qquad \left[\therefore \dfrac{V_4}{V_2} = 4.5\right]$

or $\qquad\qquad 14 - p_3 - 0.2 = 4.5 p_3 - 1.575$ or $5.5 p_3 = 15.375$

$$\therefore \qquad\qquad p_3 = \frac{15.375}{5.5} = 2.795 \text{ bar}$$

As $\qquad\qquad p_1 V_1 = p_4 V_4$

$$\therefore \qquad\qquad p_4 = \frac{p_1 V_1}{V_4} = 14 \times \frac{1}{9} = 1.555 \text{ bar}$$

$\therefore$ *Cut-off in the L.P. cylinder*

$$= \frac{V_3}{V 4} = \frac{p_4}{p_3} = \frac{1.555}{2.795} \qquad\qquad \left(\therefore p_3 V_3 = p_4 V_4\right)$$

$$= \mathbf{0.556.} \quad \textbf{(Ans.)}$$

Example 3

A two-cylinder compound, double-acting steam engine is required to develop 62.5 kW (brake) at 350 r.p.m. when supplied with steam at 20 bar and exhausting to a condenser at 0.2 bar. Cut-off ratio in both cylinders to be 0.4. The stroke length for both cylinders is 250 mm. Estimate the cylinder diameters to develop equal work.

Neglect clearance and assume hyperbolic expansion.

Take, mechanical efficiency = 85% and diagram factor for each cylinder = 0.8.

■ **Solution.**

Power to be developed,   B.P. = 62.5 kW

Engine speed, $N = 350$ r.p.m.

Steam supply pressure, $p_1 = 20$ bar

Condenser/back pressure, $p_2 = 0.2$ bar

Cut-off ratio in both cylinders   = 0.4

Stroke length (for both cylinders), $L = 250$ mm

Mechanical efficiency = 85% $\Big]$ for each cylinder
    Diagram factor = 0.8

Cylinder diameters:

Refer to Figure 9.14

For equal work done in both cylinders,

$$\text{D.F.} \left[p_1 V_1 \left\{ 1 + \log_e \frac{V_2}{V_1} \right\} - p_3 V_2 \right] = \text{D.F.} \left[p_3 V_3 \left\{ 1 + \log_e \left(\frac{V_4}{V_3} \right) \right\} - p_b V_4 \right]$$

Dividing throughout by V2, we get

$$p_1 \frac{V_1}{V_2} \left[1 + \log_e \left(\frac{V_2}{V_1} \right) \right] - p_3 = p_3 \frac{V_3}{V_2} \left[1 + \log_e \left(\frac{V_4}{V_3} \right) \right] - p_b \frac{V_4}{V_2}$$

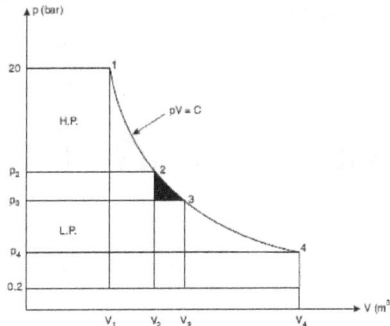

FIGURE 9.14. p-v diagram for two-cylinder compound, double-acting steam engine, Example 3.

Inserting $p_3 V_3 = p_1 V_1$

$$p_1 \cdot \frac{V_1}{V_2} \left[1 + \log_e \left(\frac{V_2}{V_1} \right) \right] - p_3 = p_1 \cdot \frac{V_1}{V_2} \left[1 + \log_e \left(\frac{V_4}{V_3} \right) \right] - p_b \cdot \frac{V_4}{V_2}$$

$$\therefore \qquad p_1 \cdot \frac{V_1}{V_2} \left[\log_e \left(\frac{V_2}{V_1} \right) - \log_e \left(\frac{V_4}{V_3} \right) \right] = p_3 - p_b \cdot \frac{V_4}{V_2}$$

$$\therefore \qquad 20 \times 0.4 \left[\log_e 2.5 - \log_e 2.5 \right] = p_3 - p_b \cdot \frac{V_4}{V_2}$$

or

$$p_3 - p_b \cdot \frac{V_4}{V_2} = 0$$

Considering points of cut-off as on hyperbolic curve

$$p_1 V_1 = p_3 V_3$$

$$\therefore \qquad p_1 \times 0.4 \, V_2 = p_3 \times 0.4 \, V_4 \qquad \left[\frac{V_1}{V_2} = \frac{V_3}{V_4} = 0.4 \ldots\ldots \text{ given} \right]$$

$$\therefore \qquad \frac{V_4}{V_2} = \frac{p_1}{p_3} = \frac{20}{p_3}$$

Inserting the value of V_4 / V_2 in eqn. (i), we get

$$p_3 - p_b \times \frac{20}{p_3} = 0$$

$$\therefore \qquad p_3^2 = 20 \times p_b = 20 \times 0.2 = 4$$

$$\therefore \qquad p_3 = 2 \text{ bar}$$

Work developed/stroke in the H.P. cylinder is

$$\frac{62.5 \times 60 \times 1000}{0.85 \times 350 \times 2} = \text{D.F} \left[p_1 V_1 \left\{ 1 + \log_e \left(\frac{V_2}{V_1} \right) \right\} - p_3 V_2 \right]$$

$$= \text{D.F.} \times V_2 \left[p_1 \cdot \frac{V_1}{V_2} \left\{ 1 + \log_e \left(\frac{V_2}{V_1} \right) \right\} - p_3 \right]$$

$$= 0.8 \times V_2 \left[20 \times 10^5 \times 0.4 \left\{ 1 + \log_e 2.5 \right\} - 2 \times 10^5 \right]$$

$$= 1066426.1 \, V_2$$

$$\therefore \quad V_2 = \frac{62.5 \times 60 \times 1000}{0.85 \times 350 \times 2 \times 1066426.1} = 0.00591 \, \text{m}^3$$

But $V_2 = \pi / 4 D^2_{\text{H.P.}} \times L$

$$0.00591 = \pi/4 \times D^2_{\text{H.P}} \times 0.25$$

$$\therefore \quad D_{\text{H.P.}} = \left(\frac{0.00591 \times 4}{\pi \times 0.25} \right)^{1/2} = 0.173 \, \text{m} = 173 \text{ mm. (Ans.)}$$

As shown earlier, $\dfrac{V_4}{V_2} = \dfrac{20}{p_3}$

But $\dfrac{V_4}{V_2} = \dfrac{\text{Volume of L.P.cylinder}}{\text{Volume of H.P.cylinder}} = \dfrac{\pi / 4 D^2_{\text{L.P.}} \times L}{\pi / 4 D^2_{\text{H.P.}} \times L} = \dfrac{D^2_{\text{L.P.}}}{D^2_{\text{H.P.}}} = \left(\dfrac{D_{\text{L.P.}}}{D_{\text{H.P.}}} \right)^2$

$$\therefore \quad \left(\frac{D_{\text{L.P.}}}{D_{\text{H.P.}}} \right)^2 = \frac{20}{p_3} = \frac{20}{2} = 10$$

$$\therefore \quad \frac{D_{\text{L.P.}}}{D_{\text{H.P.}}} = \sqrt{10} = 3.162$$

i.e., $\quad D_{\text{L.P.}} = 3.162 \times D_{\text{H.P.}} = 3.162 \times 173 = \textbf{547 mm.}$ (**Ans.**)

Example 4

Find the ratio of cylinder diameters for a double-acting condensing steam engine. The steam supplied is at 11.5 bar and the exhaust is at 0.14 bar. Cut-off in each cylinder to be at half stroke. Clearance volume 10% in each case. Total expansion ratio 10, assuming a diagram factor of 75%. Also calculate the indicated power if the steam used per hour was 1090 kg.

■ **Solution.**

Steam supply pressure,	$p_1 = 11.5$ bar
Exhaust pressure,	$p_b = 0.14$ bar
Cut-off in each cylinder	= half stroke
Clearance volume in each case	= 10%
Total expansion ratio	= 10
Diagram factor,	D.F. = 0.75
Steam used	= 1090 kg/h

Refer to Figure 9.15.

(*i*) **Ratio of cylinder diameters:**

If the total ratio of expansion is reckoned on the swept volume, then

$$\frac{V_4'}{V_1'} = 10 \qquad (i)$$

Ratio of cylinder volumes $= \dfrac{V_4'}{V_2'} = \dfrac{V_4'}{2V_1'}$

$$(ii)$$

From (*i*) and (*ii*),

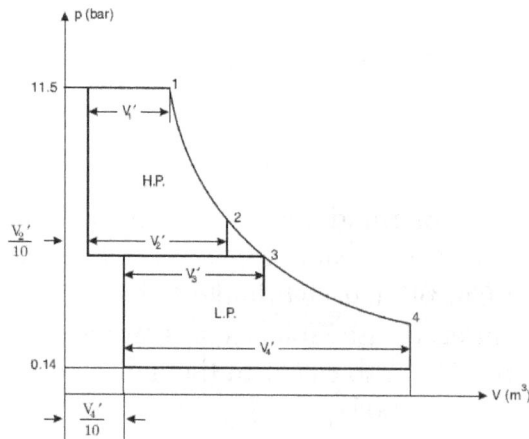

FIGURE 9.15. p-v diagram for double-acting condensing steam engine, Example 4.

Ratio of cylinder volumes $= \dfrac{1}{2} \times 10 = 5$

∴   *Ratio of cylinder diameters* $= \sqrt{5} = $ **2.236.**   **(Ans.)**

(*ii*) Indicated power:

$$p_m \text{ (with clearance)} = p_1 \left(\frac{1}{r} + \left(c + \frac{1}{r} \right) \log_e \left(\frac{c+1}{c+\dfrac{1}{r}} \right) \right) - p_b$$

To determine p_3 the expansion curve is assumed continuous, since one diagram factor is given, so that $P_1 V_1 = P_3 V_3$

∴ $$p_3 = p_1 \cdot \frac{V_1}{V_3} = 11.5 \left[\frac{\left(\dfrac{1}{10} + \dfrac{1}{2} \right) V_2'}{\left(\dfrac{1}{10} + \dfrac{1}{2} \right) V_4'} \right] = \frac{11.5}{5} = 2.3 \text{ bar}$$

i.e., $\qquad\qquad\qquad p_3 = 2.3 \text{ bar}$

$$p_m \text{ of H.P} = 11.5 \left[\frac{1}{2} + \left(\frac{1}{10} + \frac{1}{2} \right) \log_e \left\{ \frac{\dfrac{1}{10}+1}{\dfrac{1}{10}+\dfrac{1}{2}} \right\} \right] - 2.3$$

$= 11.5 \, [0.5 + 0.6 \log_e 1.833] - 2.3 = 7.63 \text{ bar}$

$$p_m \text{ of L.P} = 2.3 \left[\frac{1}{2} + \left(\frac{1}{10} + \frac{1}{2} \right) \log_e \left\{ \frac{\dfrac{1}{10}+1}{\dfrac{1}{10}+\dfrac{1}{2}} \right\} \right] - 0.14$$

$= 1.846 \text{ bar.}$

Since nothing is said about the dryness after cut-off, we must assume the steam is dry and saturated, and if we consider 1 kg of steam is used per stroke, the specific volume at 11.5 bar is 0.17 m³/kg. The mass of steam remaining in the H.P. cylinder clearance at 2.3 bar is

$$\frac{V_2'}{10 \times 0.777} = \frac{V_2'}{7.77}$$

where 0.777 m³/kg is the specific volume at 2.3 bar.

Total volume at cut-off $\quad = \dfrac{V_2'}{10} + 0.5\,V_2' = 0.6\,V_2'$

But $\qquad\qquad 0.6\,V_2' = 0.17\left(1 + \dfrac{V_2'}{7.77}\right) = 0.17 + \dfrac{V_2'}{7.77}$

or $\qquad\qquad 0.6\,V_2' = 0.17 + 0.0218\,V_2'$

∴ $\qquad\qquad V_2' = \dfrac{0.17}{(0.6 - 0.0218)} = 0.294\ \text{m}^3$

and $\qquad\qquad V_4' = 5V_2' = 5 \times 0.294 = 1.47\ \text{m}^3$

Work done / kg of steam $\quad = p_{m(\text{H.P.})}\,V_2' + p_{m(\text{L.P.})} \times V_4'$

$$= 10^5\left(7.63 \times 0.294 + 1.846 \times 1.47\right)\ \text{N-m}$$

$$= 4.957 \times 10^5\ \text{N-m}$$

∴ Ideal indicated power,

$$\text{I.P.} = \frac{4.957 \times 10^5 \times 1090}{3600 \times 1000} = 150\ \text{kW}$$

Actual I.P. allowing for a diagram factor of 0.75

$$= 0.75 \times 150 = \textbf{112.5 kW.}\quad (\textbf{Ans.})$$

Example 5

A triple-expansion engine is required to develop 2940 kW under the following conditions:

Pressure in the steam chest	= 15 bar
Piston speed	= 210 m/min
Exhaust pressure	= 0.15 bar
Cylinder volume ratios	= 1 : 2.4 : 7.2
Total ratio of expansion	= 18
Overall diagram factor	= 0.62.

Assuming equal initial loading on each piston, determine:

(i) Cylinder diameters

(ii) Mean receiver pressures

(iii) Cut-off points in each cylinder

Assume hyperbolic expansion and neglect clearance.

■ **Solution.**

Let $V_2 = 1$ unit.

Since $V_2 : V_4 : V_6 = 1 : 2.4 : 7.2$ (given)

∴ $V_4 = 2.4$ units and $V_6 = 7.2$ units.

Also, total ratio of expansion, $R = V_6 / V_1 = 18$

$$V_1 = \frac{V_6}{18} = \frac{7.2}{18} = 0.4 \text{ units.}$$

(*i*)   Cylinder diameters:

Mean effective pressure referred to L.P. cylinder,

$$p_m = \text{D.F.} \left[\frac{p_1}{R} \left(1 + \log_e R \right) - p_b \right]$$

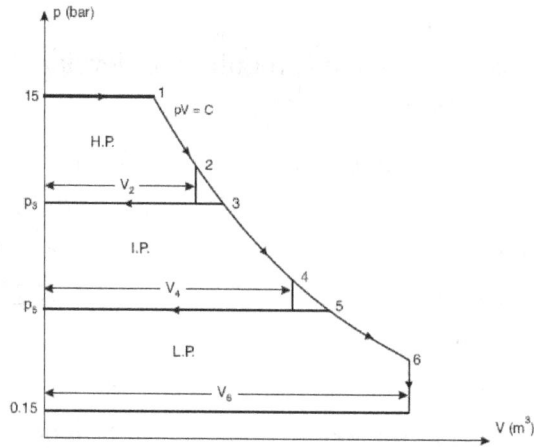

FIGURE 9.16. p-v diagram of triple-expansion engine for Example 5.

$$= 0.62 \left[\frac{15}{18} \left(1 + \log_e 18 \right) - 0.15 \right] = 1.917 \text{ bar}$$

Indicated power, $\quad$ I.P. $= 2940 \quad = \dfrac{10 p_m \, LAN}{3}$

i.e., $\qquad\qquad 2940 = \dfrac{10 \times 1917 \times A \times 105}{3} \qquad \begin{bmatrix} \because \ 2LN = 210 \\ \therefore \ LN = 105 \text{ m / min} \end{bmatrix}$

$\therefore \qquad\qquad A_{\text{L.P.}} = \dfrac{2940 \times 3}{10 \times 1.917 \times 105} = 4.3818 \text{ m}^2$

$$D_{\text{L.P.}} = \left(\frac{4.3818 \times 4}{\pi} \right)^{1/2} = \mathbf{2.362m} \quad \mathbf{or} \quad \mathbf{2362mm.} \quad \mathbf{(Ans)}$$

$$D_{\text{H.P.}} = \frac{D_{\text{L.P.}}}{\sqrt{V_6 / V_2}} = \frac{2362}{\sqrt{7.2 / 1.0}} = \mathbf{880mm.} \quad \mathbf{(Ans.)}$$

$$D_{\text{I.P.}} = \frac{D_{\text{L.P.}}}{\sqrt{V_6 / V_4}} = \frac{2362}{\sqrt{7.2 / 2.4}} = \mathbf{1363.7 \, mm.} \quad \mathbf{(Ans.)}$$

(*ii*) Mean receiver pressures:

For equal initial loads on each piston,

$(p_1 - p_3) \ V_2 = (p_3 - p_5) \ V_4 = (p_5 - p_6) \ V_6$

$$(15 - p_3) \times 1 = (p_3 - p_5) \times 2.4 = (p_5 - 0.15) \times 7.2$$

$$\therefore \qquad 15 - p_3 = 7.2 p_5 - 1.08$$

$$\therefore \qquad p_3 = 16.08 - 7.2 \, p_5 \qquad\qquad\qquad …(i)$$

Also $2.4 \left(p_3 - p_5 \right) = 7.2 \, p_5 - 0.15 \times 7.2$

Inserting the value of p_3 from (i), we get

$$2.4 \left[16.08 - 7.2 \, p_5 - p_5 \right] = 7.2 \, p_5 - 1.08$$

$$38.592 - 19.68 \, p_5 = 7.2 \, p_5 - 1.08$$

$$26.88 \, p_5 = 39.672$$

$$\therefore \qquad\qquad p_5 = \frac{39.672}{26.88} = \textbf{1.476 bar.} \quad \textbf{(Ans.)}$$

Inserting the value of p_5 in (i), we get

$$p_3 = 16.08 - 7.2 \times 1.476 = \textbf{5.45 bar.} \quad (\textbf{Ans.})$$

(iii) Cut-off points in each cylinder:

$$H.P. \; cut \; off = \frac{V_1}{V_2} = \frac{0.4}{1} = \textbf{0.04.} \quad \textbf{(Ans.)}$$

Now $\qquad\qquad\qquad p_1 \, V_1 = p_3 \, V_3$

$$15 \times 0.4 = 5.45 \times V_3$$

$$\therefore \qquad\qquad V_3 = \frac{15 \times 0.4}{5.45} = 1.1 \text{ units.}$$

$$I.P. \; cut \; off = \frac{V_3}{V_4} = \frac{1.1}{2.4} = \textbf{0.458.} \quad \textbf{(Ans.)}$$

Again, $\qquad\qquad p_1 V_1 = p_5 V_5$

$$15 \times 0.4 = 1.476 \times V_5$$

$$\therefore \qquad\qquad V_5 = \frac{15 \times 0.4}{1.476} = 4.065 \text{ units.}$$

$$L.P. \text{ cut off} = \frac{V_5}{V_6} = \frac{4.065}{7.2} = 0.564. \quad \textbf{(Ans.)}$$

Example 6

The following observations are recorded during a trial on a jacketed double-acting compound steam engine supplied with dry and saturated steam:

Steam admission pressure	= 6.2 bar
Diameter of H.P. cylinder	= 250 mm
Diameter of L.P. cylinder	= 375 mm
Stroke	= 600 mm
m.e.p. of H.P. cylinder	= 2.3 bar
m.e.p. of L.P. cylinder	= 2.2 bar
Speed	= 100 r.p.m.
Brake torque	= 4500 Nm
Receiver pressure	= 2.6 bar
Discharge of wet air pump	= 18 kg/min
Discharge from jacket drain	= 2.1 kg/min
Discharge from receiver drain	= 1.25 kg/min
Cooling water from condenser	= 350 kg/min
Temperature rise of cooling water	= 30°C
Temperature of condensate	= 50°C

(i) Draw up a heat balance sheet.

(ii) Find the mechanical and brake thermal efficiencies of the engine.

Neglect mass of air carried by the steam in the condenser.

Solution. Total indicated power of the engine,

$$\text{I.P.} = \frac{10LN}{3} \left[p_{m(\text{H.P.})} \, A_{\text{H.P.}} + p_{m(\text{L.P.})} \, A_{\text{L.P.}} \right]$$

$$= \frac{10 \times 0.6 \times 100}{3} \left[2.3 \times \frac{\pi}{4} \times 0.25^2 + 2.2 \times \frac{\pi}{4} \times 0.375^2 \right]$$

$$= 200 \,(0.1129 + 0.2429) = 71.16 \,\text{kW}$$

Brake power of the engine,

$$\text{B.P.} = \frac{2\pi NT}{60 \times 1000} = \frac{2\pi \times 100 \times 4500}{60 \times 1000} = 47.12\,\text{kW}$$

Heat equivalent of B.P. $= 47.12 \times 60 = 2827.2\ \text{kJ/min}$

Heat in jacket drain $= 2.1 \times 676 = 1419.6\ \text{kJ/min}$

(676 kJ/kg is the enthalpy of water at 6.2 bar)

Heat in receiver drain $= 1.25 \times 540.9 = 676.1\ \text{kJ/min}$

(540.9 kJ/kg is the enthalpy of water at 2.6 bar)

Heat in condensate $= 18 \times 4.18 \times 50 = 3762\ \text{kJ/min}$

Heat in cooling water $= 350 \times 4.18 \times 30 = 43890\ \text{kJ/min}$

Heat supplied to the engine per min.

$$= (\text{Jacket drain} + \text{receiver drain} + \text{air pump discharge}) \times h_1$$

$$= (2.1 + 1.25 + 18) \times 2756.9 = 58859.8\ \text{kJ/min}$$

$$\begin{bmatrix} h_1 = 2756.9\text{kJ/kg} \ (= h_g) \text{ at 6.2 bar from steam tables,} \\ hf_2 = 1 \times 418 \times 50 = 209\ \text{kJ/kg} \end{bmatrix}$$

Heat unaccounted for $= 58859.8 - (2827.2 + 1419.6 + 676.1 + 3762 + 43890)$

$$= 6284.9\ \text{kJ/min.}$$

Heat balance sheet on minute basis

Particulars	Heat (kJ/min)	Percentage (%)
Heat supplied	58859.8	100
(i) Heat equivalent of B.P.	2827.2	4.80
(ii) Heat in cooling water	43890	74.56
(iii) Heat in jacket drain	1419.6	2.41
(iv) Heat in receiver drain	676.1	1.15
(v) Heat in condensate	3762	6.39
(vi) Heat unaccounted for	6284.9	10.67

Example 7

The following data refer to a double-acting compound steam engine:

I.P.	= 365 kW
R.P.M.	= 420
Stroke	= 60 cm
Admission pressure	= 10 bar
Back pressure	= 0.3 bar
Expansion ratio	= 10
Diagram factor	= 0.8

Assuming complete expansion in the H.P. cylinder and equal initial load, expansion follows the law pV = constant, and neglecting clearance, determine:

(i) The admission pressure for the low-pressure cylinder.

(ii) The diameter of each cylinder.

■ **Solution.**

I.P. = 365 kW, N = 420 r.p.m., L = 60 cm = 0.6 m,

$$p_1 = 10 \text{ bar, } p_b = 0.3 \text{ bar, } R = \frac{V_3}{V_1} = 10, \text{ D.F.} = 0.8$$

$(i), (ii): p_2 = ?, D_{\text{L.P.}} = ?, D_{\text{H.P.}} = ?$

Pressure, p_2; $D_{\text{L.P.}}$:

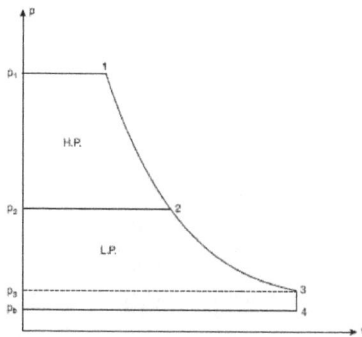

FIGURE 9.17. p-v diagram, double-acting compound steam engine, Example 7.

$$P_{m(\text{actual})} = \text{D.F.} \times \left[\frac{p_1}{r}(1+\log_e R) - p_b \right] = 0.8 \times \left[\frac{10}{10}(1+\log_e 10) - 0.3 \right]$$

$$= 2.402 \text{ bar}$$

$$\text{I.P.} = \frac{10\, p_{m(\text{actual})} \times LAN}{3}$$

$$365 = \frac{10 \times 2.402 \times 0.6 \times \dfrac{\pi}{4} \times D^2_{\text{L.P.}} \times 420}{3}$$

$$\therefore \quad D^2_{\text{L.P.}} = \frac{365 \times 3 \times 4}{10 \times 2.405 \times 0.6 \times \pi \times 420} = 0.2303 \,\text{m}^2$$

$$D_{\text{L.P.}} = 0.4798 \text{ m} \approx 0.48 \text{ m or } 480 \text{ mm}$$

$\therefore$ *Diameter of L.P. cylinder,* $D_{\text{L.P.}}$ = **480 mm.** (**Ans.**)

$$p_1 V_1 = p_3 V_3$$

$$\therefore \qquad\qquad p_3 = \frac{p_1 V_1}{V_3} = 10 \times \frac{1}{10} = 1 \text{ bar} \qquad \left[\therefore \quad \frac{V_1}{V_3} = \frac{1}{10} \right]$$

Also $\qquad\qquad p_2 V_2 = p_3 V_3$

$$p_2 \times \frac{\pi}{4}(D_{\text{H.P.}})^2 \times L = 1 \times \frac{\pi}{4}(D_{\text{L.P.}})^2 \times L$$

or $\qquad\qquad\qquad (D_{\text{H.P.}})^2 = \dfrac{0.48^2}{p_2}$

Since *initial load is the same*, therefore,

$$(p_1 - p_2) \times \frac{\pi}{4} \times (D_{\text{H.P.}})^2 = (p_2 - p_b) \times \frac{\pi}{4} \times (D_{\text{L.P.}})^2$$

$$(10 - p_2) \times \frac{(0.48^2)}{p_2} = (p_2 - 0.3) \times 0.48^2$$

$$(10 - p_2) = p_2\, (p_2 - 0.3) = p_2^{\,2} - 0.3\, p_2$$

or $\qquad p_2^{\,2} + 0.7\,p_2 - 10 = 0$

or $\qquad\qquad\qquad\qquad\qquad p_2 = \dfrac{-0.7 \pm \sqrt{0.7^2 + 4 \times 10}}{2} = 2.83 \text{ bar}$

i.e., $\qquad\qquad\qquad\qquad\qquad p_2 = \textbf{2.83 bar.} \quad (\textbf{Ans.})$

Diameter of H.P. cylinder, $D_{\text{H.P.}}$:

$$D_{\text{H.P.}} = \frac{0.48}{\sqrt{p_2}} = \frac{0.48}{\sqrt{2.83}} = 0.285 \text{ or } 285 \text{ mm}$$

i.e., $\qquad\qquad$ *Diameter of H.P. cylinder,* $D_{\text{H.P.}} = \textbf{285 mm.} \quad (\textbf{Ans.})$

Example 8

A compound steam engine is to develop 260 kW, when taking steam at 8.75 bar and exhausting at 0.15 bar abs. The engine speed is 140 r.p.m. and the piston speed is 150 meters/min, the cylinder cut-off ratio is 0.4 and the cylinder volume ratio is 3.7. Allow a diagram factor of 0.83 for the combined cards, and determine suitable dimensions of the cylinders if the diagram factor for the H.P. cylinders alone is 0.85; determine the separate powers developed in the two cylinders when the L.P. cut-off is arranged to give equal initial loads on the pistons. Assume hyperbolic expansion and neglected clearance effects.

■ Solution.

$p_1 = 8.75$ bar, $p_b = 0.15$ bar, $N = 140$ r.p.m.,

Piston speed $\quad = 150$ m/min, $\dfrac{V_1}{V_2} = 0.4$, $\dfrac{V_4}{V_2} = 3.7$, D.F. = 0.83 ; I.P. = 260 kW

Overall expansion ratio R,

$$\frac{V_4}{V_1} = \frac{V_4}{V_2} \times \frac{V_2}{V_1} = 3.7 \times 2.5 = 9.25$$

If the whole expansion takes place in the L.P. cylinder, then,

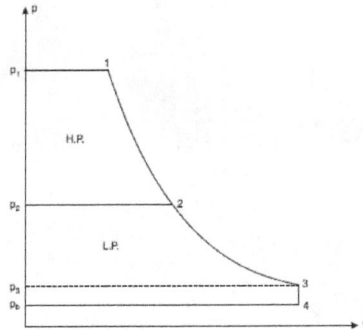

FIGURE 9.18. p-v diagram for compound steam engine, Example 8.

$$\text{I.P.} = \frac{D.F. \times 10 \left[\dfrac{p_1}{R} (1 + \log_e R) - P_b \right] \times V_4 \times N}{6}$$

or

$$260 = \frac{0.83 \times 10 \left[\dfrac{8.75}{9.25} (1 + \log_e 9.25) - 0.15 \right] \times V_4 \times 140}{6}$$

or

$$260 = \frac{0.83 \times 10 \times 2.9 \, V_4 \times 140}{6}$$

$\therefore$ L.P. cylinder volume, $V_4 = 0.4629 \text{ m}^3$

Length of stroke, $L = \dfrac{\text{Piston speed}}{2N} = \dfrac{150}{2 \times 140} = \textbf{0.5357 m. (Ans.)}$

Also, $\dfrac{\pi}{4} D^2_{\text{L.P.}} \times 0.5357 = 0.4629$

$\therefore$ *Diameter of L.P. cylinder,* $D_{\text{L.P.}} \left(\dfrac{0.4629 \times 4}{\pi \times 0.5357} \right)^{1/2} = \textbf{1.049m (Ans.)}$

Volume of H.P. cylinder $= \dfrac{0.4629}{3.7} = \dfrac{\pi}{4} D^2_{\text{H.P.}} \times 0.5357$

$\therefore$ Diameter of H.P. cylinder, $D_{\text{H.P.}} = \textbf{0.545 m. (Ans.)}$

For *equal initial load,*

$$\left(p_1 - p_2\right)\frac{\pi}{4}D^2_{\text{H.P.}} = \left(p_2 - p_b\right)\frac{\pi}{4}D^2_{\text{L.P.}}$$

$$\left(8.75 - p_2\right)\frac{\pi}{4}\times 0.545^2 = \left(p_2 - 0.15\right)\times \frac{\pi}{4}\times\left(1.049^2\right)$$

or $$\left(8.75 - p_2\right) = \left(p_2 - 0.15\right)\times\left(\frac{1.049}{0.545}\right)^2 = \left(p_2 - 0.15\right)\times 3.705$$

or $$8.75 - p_2 = 3.705\,p_2 - 0.556.$$

$$\therefore \qquad p_2 \simeq 1.98\,\text{bar}$$

Power developed in H.P. cylinder

$$= \frac{\text{D.F.}\times 10\left[\dfrac{p_1}{r}\left(1+\log_e r\right)- p_2\right]\times V_2 \times N}{6}$$

$$= \frac{0.85\times 10\left[\dfrac{8.75}{2.5}\left(1+\log_e 2.5\right)-1.98\right]\times \dfrac{\pi}{4}\times\left(0.545\right)^2 \times 0.5357\times 140}{6}$$

$$= 117.16\,\text{kW. (Ans.)}$$

$\therefore$ *Power developed in L.P. cylinder* $= 260 - 117.16 = \mathbf{142.84\,kW.}$ **(Ans.)**

Example 9

A two-cylinder compound steam engine is to develop 92 kW at 110 r.p.m. Steam is supplied at 7.35 bar abs, and the cylinder pressure is 0.21 bar abs. The stroke of each piston is equal to the L.P. cylinder diameter. The total expansion ratio is 15. Allow a diagram factor of 0.7. Assume hyperbolic expansion and neglect clearance and receiver loss. Determine the diameter of the cylinder so that they may develop equal power.

■ **Solution.**

$p_1 = 7.35$ bar, $p_b = 0.21$ bar, $R = \dfrac{V_1}{V_3} = 15$, D.F. $= 0.7$, N $= 110$ r.p.m., I.P. $= 92$

Refer to Figure 17.

$$p_m = \text{D.F.} \left[\frac{p_1}{R} (1 + \log_e R) - p_b \right]$$

$$= 0.7 \left[\frac{7.35}{15} (1 + \log_e 15) - 0.21 \right] = 1.125 \text{ bar}$$

$$\text{I.P.} = \frac{10 \, p_m \, LAN}{3}$$

$$92 = \frac{10 \times 1.125 \times D_{\text{L.P.}} \times \left(\frac{\pi}{4} D^2_{\text{L.P.}} \right) \times 110}{3}$$

∴ *Diameter of L.P. cylinder,* $D_{\text{L.P.}}$ = **0.657 m or 657 mm.** **(Ans.)**

Since power developed in each cylinder is equal, work done in the H.P. cylinder will be equal to the work done in the L.P. cylinder.

$$\therefore \quad p_1 V_1 \left(1 + \log_e \frac{V_2}{V_1} \right) - p_2 V_2 = \frac{1}{2} \left[p_1 V_1 \left(1 + \log_e \frac{V_3}{V_1} \right) - p_b V_3 \right]$$

Dividing both sides by $p_1 V_1$ and considering $p_1 V_1 = p_2 V_2$, we have

$$\text{or} \qquad\qquad \log_e \frac{V_2}{V_1} = \frac{1}{2} \left[1 + \log_e 15 - \frac{0.21}{7.35} \times 5 \right]$$

$$\therefore \qquad\qquad \frac{V_2}{V_1} = 5.154$$

$$\therefore \quad \text{Cylinder volume, } V_2 = 5.154 \times \frac{\left(\frac{\pi}{4} \times 0.657^2 \times 0.657 \right)}{15} = \frac{\pi}{4} D^2_{\text{H.P.}} \times 0.657$$

$$\text{or} \qquad\qquad D_{\text{H.P.}} = 0.385 \text{ m or 385 mm}$$

i.e.,   *Diameter of H.P. cylinder* = **385 mm.** **(Ans.)**

9.5. Exercises

1. The simple steam engine as compared to the compound steam engine for the same output has

(*a*) smaller flywheel

(*b*) large flywheel

(*c*) same size flywheel as output is same

(*d*) same size flywheel as speed is same

2. The leakage past the piston and initial condensation by compounding the steam engine is

(*a*) increased

(*b*) decreased

(*c*) unaffected

(*d*) depends on methods of compounding

3. Woolfe type compound steam engines have

(*a*) two cranks at 180° phase difference

(*b*) two cranks at 90° phase difference

(*c*) only one crank with no phase difference

(*d*) two cranks with no phase difference

4. Tandem compound steam engines have

(*a*) two cranks at 180° phase difference

(*b*) two cranks at 90° phase difference

(*c*) only one crank with no phase difference

(*d*) two cranks with no phase difference

5. Receiver type compound steam engines have

(*a*) two cranks at 180° phase difference

(*b*) two cranks at 90° phase difference

(*c*) only one crank with no phase difference

(*d*) two cranks with no phase difference

6. The compound steam engine that can start in any position is

(*a*) Tandem type

(*b*) Woolfe type

(*c*) Receiver type

(*d*) None of the above

7. The compound steam engine that requires the smallest flywheel for a given output and given speed is

(*a*) Tandem type

(*b*) Woolfe type

(*c*) Receiver type

(*d*) Any one if the speed is the same

8. Governing of compound steam engines is done by

(*a*) throttling steam to high pressure cylinder

(*b*) cut-off variation in high pressure cylinder

(*c*) cut-off variations simultaneously in high pressure and low-pressure cylinders

(*d*) all of the above methods

9. The equation used for expansion in compound steam engines, namely p*V* = *C*, signifies

(*a*) isothermal expansion

(*b*) adiabatic expansion

(*c*) parabolic expansion

(*d*) hyperbolic expansion

10. In a two-cylinder compound steam engine with continuous expansion from initial pressure to back pressure is geometric mean of the initial and back pressure if

    (*a*) work is shared equally by high pressure and low-pressure cylinders

    (*b*) the initial loads on the pistons of the high-pressure cylinder and the low-pressure cylinder are the same

    (*c*) both the cases (a) and (b)

    (*d*) none of the above

11. State the advantages of compound steam engines.

12. How are compound steam engines classified?

13. Explain briefly with neat diagrams the following compound steam engines:

    (*i*)   Woolfe type

    (*ii*)   Receiver type

14. Write a short note on multi-cylinder engines.

15. Discuss the causes of loss of thermal efficiency in compound steam engines.

16. What different methods are used for governing compound steam engines? Discuss their relative advantages and disadvantages.

17. Discuss with a neat sketch the construction and working of a uniflow engine. Also mention its merits and demerits.

18. A double acting compound steam engine is supplied with steam at 14 bar and 0.9 dryness. The condenser working pressure is

0.35 bar. The diagram factor referred to the L.P. cylinder is 0.8. The common stroke is 35 cm. The H.P. cylinder bore is 200 mm, while the L.P. cylinder bore is 300 mm. Expansion in the H.P. cylinder is complete. The engine runs at 300 r.p.m. Neglecting clearance and assuming equal initial piston loads, calculate:

(*i*) The intermediate pressure

(*ii*) The indicated power output

(*iii*) The hourly steam consumption at the operating conditions

19. A two-cylinder double-acting compound steam engine develops 100 kW when running at 240 r.p.m. Dry and saturated steam is supplied to the engine at 16 bar and exhaust from the L.P. cylinder occurs at 0.15 bar. Stroke for both the cylinders is 250 mm. Cut-off occurs at 40% of stroke in the L.P. as well as the H.P. cylinder. The diagram factor for both the cylinders is 0.7. Determine the cylinder diameters for the two cylinders if equal work is done by both the cylinders.

20. A compound steam engine develops 220 kW at a speed of 270 r.p.m., the stroke being 500 mm for both the H.P. and L.P. cylinders. The expansion is hyperbolic. The supply pressure is 12 bar and the exhaust pressure is 0.28 bar. The expansion is complete in the H.P. cylinder and the total expansion ratio is 10. Neglecting clearance and assuming equal power development, determine the cylinder diameters. The diagram factor referred to the L.P. cylinder is 0.75.

21. The diameter and stroke of an L.P. cylinder of a double-acting compound steam engine are each 600 mm. Steam is supplied at a pressure of 11 bar and is exhausted to 0.28 bar. The diagram factor referred to the L.P. cylinder is 0.82. The expansion is complete in the H.P. cylinder. The total expansion ratio in the engine is 8. Assuming equal work developed in each cylinder and engine runs at 240 r.p.m., calculate:

(*i*) The indicated power

(*ii*) The intermediate pressure

(*iii*) The diameter of the H.P. cylinder assuming equal stroke length with that of the L.P. cylinder

22. A two-cylinder compound steam engine receives steam at a pressure of 8 bar and discharges at 0.2 bar. The cylinder volume ratio is 4. Cut-off in the H.P. and L.P. cylinders occurs at 0.4 and 0.5 of the stroke respectively.

   a. Sketch the hypothetical indicator diagram and insert the pressure at cut-off and release in each cylinder, neglecting clearance.

   b. Calculate the mean effective pressure in each cylinder.

   c. Find the ratio of work done in two cylinders.

23. Steam is supplied at a pressure of 10.35 bar to an engine having equal strokes 400 mm for the H.P. and L.P. cylinders and having bore diameters of 330 mm and 630 mm for the H.P. and L.P. cylinders. Cut-off for the H.P. cylinder occurs at one-fourth of stroke. The back pressure is 0.21 bar. The power developed by the engine is 120 kW when the mechanical efficiency is 0.82. If the engine runs at 180 r.p.m., determine the diagram factor. Assume that the engine is double-acting.

24. For a two-cylinder double-acting compound engine, running at 120 r.p.m., developing 120 kW, the data for the H.P. and L.P. cylinders is given as follows:

   H.P. cylinder: Admission pressure = 13 bar (abs); cut-off occurs at 45% of stroke; clearance = 10%; diagram factor = 0.8; law of expansion = hyperbolic.

   L.P. cylinder: Condenser vacuum = 590 mm Hg; barometer = 760 mm Hg; clearance = 12%; cut-off occurs at 55% of stroke; diagram factor = 0.8; law of expansion = hyperbolic.

If the length of stroke for both cylinders is equal to the L.P. cylinder diameter and the total number of expansion is 10, determine:

(*a*) Diameter of H.P. cylinder

(*b*) Diameter of L.P. cylinder

(*c*) Length of stroke

(*d*) Ratio of work done in two cylinders

25. A two-cylinder compound double-acting steam engine develops 92 kW of power at 110 r.p.m. The steam supply is at 7 bar and exhaust is at 0.21 bar. The overall expansion ratio is 15, and the cylinders develop equal power. The expansion may be assumed to be hyperbolic and expansion in the H.P. cylinder is complete. The diagram factor is 0.7 and the common stroke is equal to the diameter of the L.P. cylinder. Neglecting clearance, calculate:

(*a*) Diameter of L.P. cylinder

(*b*) Diameter of H.P. cylinder

(*c*) Length of stroke

26. The bores of a compound steam engine's cylinders are 300 mm and 600 mm while the common stroke is 400 mm. The supply steam is at a pressure of 11 bar and cut-off is at $\frac{1}{3}$ of stroke in the H.P. cylinder. The back pressure in the L.P. cylinder is 0.32 bar. The details of the indicator card are: for the H.P. cylinder, the diagram area is 12.5 cm2 and the spring scale is 2.7 bar/cm; for the L.P. cylinder, the diagram area is 11.4 cm2 and the spring scale is 0.8 bar/cm. The length of both diagrams is 7.5 cm. Determine:

(*a*) The theoretical and actual mean effective pressures referred to the L.P. cylinder.

(*b*) The overall diagram factor and indicated power if the engine runs at 162 r.p.m.

27. A compound steam engine is supplied with steam at 8 bar and exhaust occurs at 0.28 bar. The engine runs at 120 r.p.m. and has a mean speed of 150 meters/min. Cut-off takes place at 40% of stroke in the H.P. cylinder and at 55% of stroke in L.P. cylinder. If the engine cylinder volume ratio is 3.5 and the power developed by the engine is 360 kW, determine:

    (*a*) Cylinder dimensions

    (*b*) L.P. cylinder receiver pressure

    (*c*) Ratio of maximum loads

28. A double-acting compound steam engine has a common stroke of 400 mm. The supply pressure is 17.25 bar while the exhaust pressure is 0.415 bar. The diameter of the L.P. cylinder is 450 mm. The ratio of cylinder volumes is 2.5 : 1. The diagram factor referred to the L.P. cylinder is 0.78. The initial piston loads in the cylinders are equal. The engine runs at 270 r.p.m. The total expansion ratio is 9. Neglecting clearance and assuming hyperbolic expansion, calculate:

    (*a*) The intermediate pressure

    (*b*) The indicated power

    (*c*) The diameter of the H.P. cylinder

29. The following data relate to a two-cylinder double-acting compound steam engine running at 100 r.p.m. *H.P. cylinder:* Diameter = 250 mm, cut-off = 30% of stroke, D.F. = 0.8, clearance = 10% of stroke. *L.P. cylinder:* Diameter = 500 mm, cut-off = 40% of stroke, clearance = 8% of stroke, D.F. = 0.7.

    The admission and exhaust pressures are 6 bar and 0.30 bar, respectively. The common stroke of the engine is 400 mm.

    Determine the power output of the engine.

30. In a triple expansion compound steam engine, the steam is supplied at 12.5 bar. All the cylinders have restricted expansion and the cylinder volumes are in the ratio 1 : 2.5 : 6. If the overall expansion ratio is 12.5, cut-off in the I.P. cylinder occurs at 50%, and cut-off in the L.P. cylinder occurs at 65% of the stroke, determine:

(*a*) The mean effective pressures of three cylinders

(*b*) Loss in power due to restricted expansion in the H.P. and I.P. cylinders

Assume hyperbolic expansion of steam and neglect clearance.

CHAPTER 10

STEAM TURBINES

In This Chapter

- Overview
- Classification of Steam Turbines
- Advantages of the Steam Turbine over Steam Engines
- Description of Common Types of Turbines
- Methods of Reducing Wheel or Rotor Speed
- Difference between Impulse and Reaction Turbines
- Impulse Turbine
- Reaction Turbines
- Types of Power in Steam Turbine Practice
- State Point Locus and Reheat Factor
- Steam Turbine Governing and Control
- Special Forms of Steam Turbines
- Exercises

10.1. Overview

The steam turbine is a prime mover in which the energy of the steam (including thermal or enthalpy and kinetic) is transformed into kinetic energy of the rotor blade, and later in its turn is transformed into the mechanical energy of rotation of the turbine shaft. The turbine shaft, directly or with the help of a reduction gearing, is connected with the driven mechanism.

Depending on the type of the driven mechanism, a steam turbine may be utilized in most diverse fields of industry, for power generation and for transport. Transformation of the input energy of steam into the mechanical energy of rotation of the shaft is brought about by different means.

10.2. Classification of Steam Turbines

There are several ways in which the steam turbines may be classified. The most important and common division is with respect to the action of the steam and the way that it interacts with the turbine blades, as:

1. Impulse: when steam impinges on the turbine blades and is redirected to the stationary nozzles to be furthered to the next stage of rotor blades. In this type of turbine kinetic energy of the steam changes but the enthalpy remains constant.

2. Reaction: when steam flows through airfoil-shaped blades and moves the rotor due to the pressure difference created. Steam flows through the turbine similar to a meandering jet stream, and hence moves the rotors at several stages.

3. Combination of impulse and reaction: as the name indicates, this is a combination of types (1) and (2).

Further, turbines are also classified as follows:

1. According to the number of pressure stages:

- Single-stage turbines with one or more velocity stages are usually of small power capacities; these turbines are mostly used for driving centrifugal compressors, blowers, and other similar machinery.

- Multistage impulse and reaction turbines; they are made in a wide range of power capacities varying from small to large.

2. According to the direction of steam flow:

- Axial turbines in which steam flows in a direction parallel to the axis of the turbine.

- Radial turbines in which steam flows in a direction perpendicular to the axis of the turbine; one or more low-pressure stages in such turbines are made axial.

3. According to the number of cylinders:

- Single cylinder turbines.

- Double cylinder turbines.

- Three cylinder turbines.

- Four cylinder turbines.

Multi-cylinder turbines which have their rotors mounted on one and the same shaft and are coupled to a single generator are known as single shaft turbines; turbines with separate rotor shafts for each cylinder placed parallel to each other are known as multiaxial turbines.

4. According to the method of governing:

- Turbines with throttle governing in which fresh steam enters through one or more (depending on the power developed) simultaneously operated throttle valves.

- Turbines with nozzle governing in which fresh steam enters through two or more consecutively opening regulators.

- Turbines with bypass governing in which steam turbines, besides being fed to the first stage, are also directly fed to one, two, or even three intermediate stages of the turbine.

5. According to heat drop process:

- Condensing turbines with generators; in these turbines steam at a pressure less than atmospheric is directed to a condenser; steam is also extracted from intermediate stages for feed water heating, the number of such extractions usually being from 2–3 to as much 8–9. The latent heat of exhaust steam during the process of condensation is completely lost in these turbines.

- Condensing turbines with one or two intermediate stage extractions at specific pressures for industrial and heating purposes.

- Back-pressure turbines, the exhaust steam from which is utilized for industrial or heating purposes; to this type of turbines can also be added (in a relative sense) turbines with a deteriorated vacuum, the exhaust steam of which may be used for heating and process purposes.

- Topping turbines; these turbines are also of the back-pressure type with the difference that the exhaust steam from these turbines is further utilized in medium and low pressure condensing turbines. These turbines, in general, operate at high initial conditions of steam pressure and temperature, and are mostly used during extension of power station capacities, with a view to obtain better efficiencies.

- Back-pressure turbines with steam extraction from intermediate stages at specific pressure; turbines of this type are meant for supplying the consumer with steam of various pressures and temperature conditions.

- Low pressure turbines in which the exhaust steam from reciprocating steam engines, power hammers, presses, and so on, is utilized for power generation purposes.

- Mixed pressure turbines with two or three pressure stages, with a supply of exhaust steam to its intermediate stages.

6. According to steam conditions at the inlet to the turbine:

- Low pressure turbines, using steam at a pressure of 1.2 to 2 atm.

- Medium pressure turbines, using steam at pressures of up to 40 atm.

- High pressure turbines, utilizing pressures above 40 atm.

- Turbines of very high pressures, utilizing steam at pressures of 170 atm and higher and temperatures of 550°C and higher.

- Turbines of supercritical pressures, using steam at pressures of 225 atm and above.

7. According to their usage in industry:

- Stationary turbines with constant speed of rotation primarily used for driving alternators.

- Stationary steam turbines with variable speed meant for driving turbo-blowers, air circulators, pumps, and so forth.

- Non-stationary turbines with variable speed; turbines of this type are usually employed in steamers, ships, and railway locomotives.

Classifications mentioned above are brief and summarize turbines. Interested readers can refer to KLM Technology Group [110] and the Mechanical Engineers Handbook, [111].

10.3. Advantages of the Steam Turbine over Steam Engines

In the previous chapter, steam engines were discussed. It seems useful to summarize the differences between steam engines and turbines. The following are the principal advantages of steam turbines over steam engines:

1. The thermal efficiency of a steam turbine is much higher than that of a steam engine.

2. The power generation in a steam turbine is at a uniform rate, therefore the necessity of using a flywheel (as in the case of a steam engine) is not felt.

3. Much higher speeds and greater range of speed is possible than in the case of a steam engine.

4. In large thermal stations where we need higher outputs, the steam turbines prove very suitable as these can be made in big sizes.

5. With the absence of reciprocating parts (as in a steam engine), the balancing problem is minimized (balancing is explained in a further section).

6. No internal lubrication is required as there are no rubbing parts in the steam turbine.

7. In a steam turbine there is no loss due to initial condensation of steam.

8. It can utilize a high vacuum very advantageously.

9. Considerable overloads can be carried at the expense of a slight reduction in overall efficiency.

10.4. Description of Common Types of Turbines

The common types of steam turbines are:

- Simple impulse turbine
- Reaction turbine

The main difference between these turbines lies in the way in which the steam is expanded while it moves through them. In the former type steam expands in the nozzles and its pressure does not alter as it moves over the blades, while in the latter type the steam expands continuously as it passes over the blades and thus there is gradual fall in the pressure during expansion.

10.4.1. Simple Impulse Turbines

Figure 10.1 shows a simple impulse turbine diagrammatically. The top portion of the figure exhibits a longitudinal section through the upper half of the turbine, the middle portion shows one set of nozzles which is followed by a ring of moving blades, while lower part of the diagram indicates approximate changes in pressure and velocity during the flow of steam through the turbine. This turbine is called a *"simple"* impulse turbine, since the expansion of the steam takes place in one set of the nozzles.

As the steam flows through the nozzle its pressure falls from steam chest pressure to condenser pressure (or atmospheric pressure if the turbine is non-condensing). Due to this relatively higher ratio of expansion of steam in the nozzles, the steam leaves the nozzle with a very high velocity. Referring to Figure 10.1, it is evident that the velocity of the steam leaving the moving blades is a large portion of the maximum velocity of the steam when leaving the nozzle. The loss of energy due to this higher exit velocity is commonly called the "carryover loss" or "leaving loss."

For animations and graphical descriptions of impulse turbines, the reader can see LearningEngineering.org [112].

The principal example of this turbine is the well-known "De laval turbine," and in this turbine the "exit velocity" or "leaving velocity" or "lost velocity" may amount to 3.3% of the nozzle outlet velocity. Also since all the kinetic energy is to be absorbed by only one ring of the moving blades, the velocity of the wheel is too high (varying from 25000 to 30000 r.p.m.). This wheel or rotor speed, however, can be reduced by different methods, which are discussed in a further section.

10.4.2. Reaction Turbine

In this type of turbine, there is a gradual pressure drop which takes place continuously over the fixed and moving blades. The function of the fixed blades is (the same as the nozzle) that they alter the direction of the steam as well as allow it to expand to a larger velocity. As the steam passes

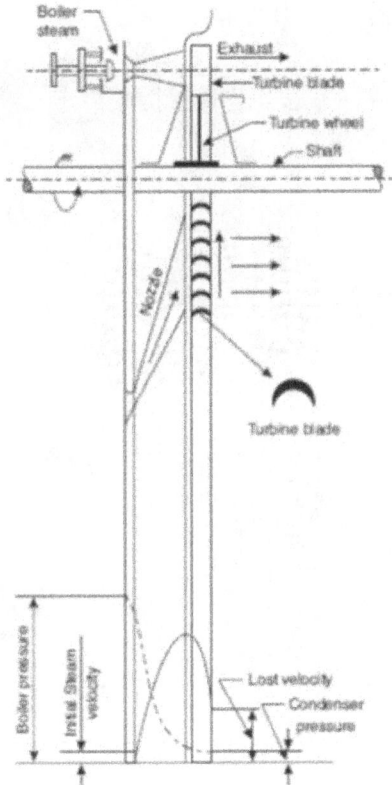

FIGURE 10.1. Sketch for simple impulse turbine.

over the moving blades its kinetic energy (obtained due to fall in pressure) is absorbed by them. Figure 10.2 shows a three-stage reaction turbine. The changes in pressure and velocity are shown as well.

As the volume of steam increases at lower pressures, the diameter of the turbine must increase after each group of blade rings. It may be noted that in this turbine, since the pressure drop per stage is small, the number of stages required is much higher than in an impulse turbine of the same capacity.

10.5. Methods of Reducing Wheel or Rotor Speed

As already discussed under the heading "simple impulse turbine," if the steam is expanded from the boiler pressure to condenser pressure in one stage, the speed of the rotor becomes tremendously high, which brings up

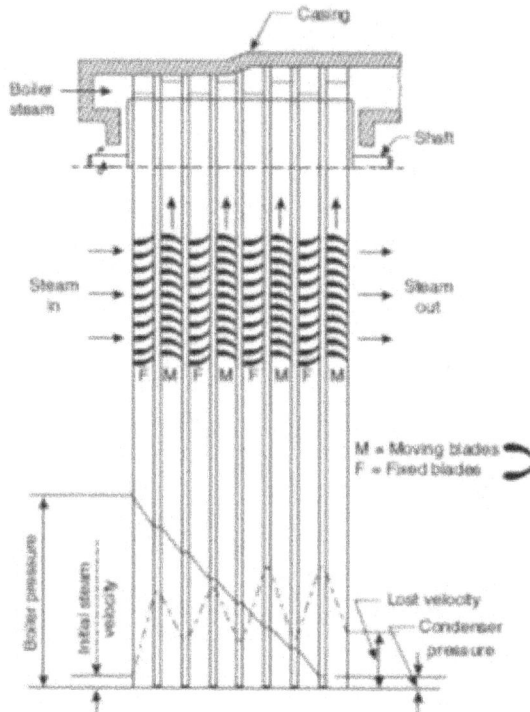

FIGURE 10.2. Reaction turbine (three stage), pressure shown by solid line and velocity by dashed line.

practical complicacies. There are several methods of reducing this speed to a lower value; all these methods utilize a system of multiple rotors in a series, keyed on a common shaft so that the steam pressure or jet velocity is absorbed in stages as the steam flows over the blades. This is known as *compounding*. The different methods of compounding are:

• Velocity compounding.

• Pressure compounding.

• Pressure-velocity compounding.

• Reaction turbine.

10.5.1. Velocity Compounding

Steam is expanded through a stationary nozzle from the boiler or inlet pressure to condenser pressure. So, the pressure in the nozzle drops, and the kinetic energy of the steam increases due to an increase in velocity. A portion of this available energy is absorbed by a row of moving blades. The steam (with a decreased velocity due to moving over the moving blades) then flows through the second row of blades which are fixed. The function of these fixed blades is to redirect the steam flow without altering its velocity to the next row of moving blades where again work is done on them and steam leaves the turbine with a low velocity. Figure 10.3 shows a cutaway section of such a stage and changes in pressure and velocity as the steam passes through the nozzle and fixed and moving blades.

This method has the advantage that the initial cost is low due to a lesser number of stages, yet its efficiency is low.

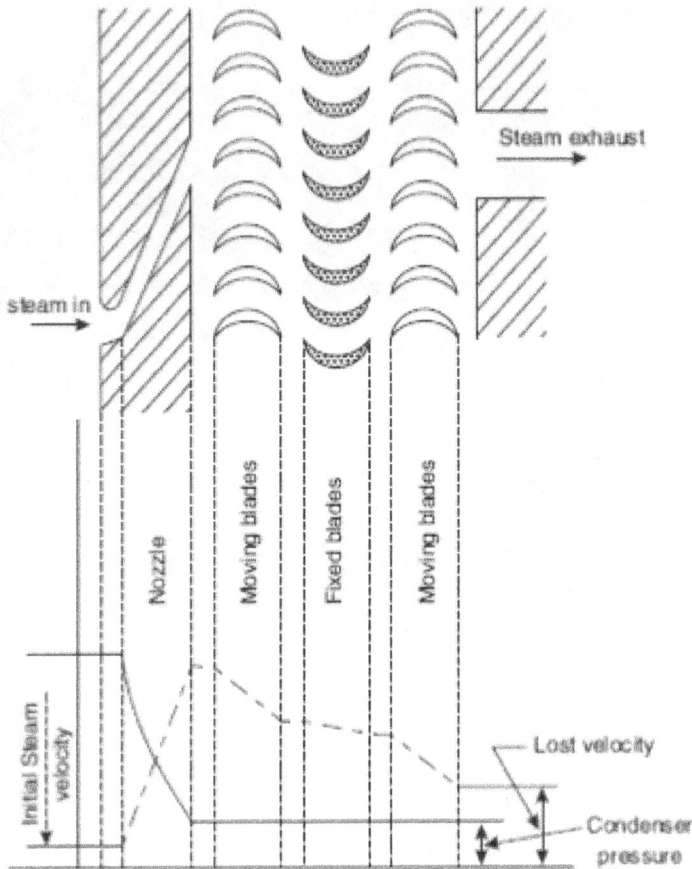

FIGURE 10.3. Velocity compounding, pressure shown by solid line and velocity by dashed line.

10.5.2. Pressure Compounding

Figure 10.4 shows rings of fixed nozzles incorporated between the rings of moving blades. The steam at boiler pressure enters the first set of nozzles and expands partially. The kinetic energy of the steam thus obtained is absorbed by the moving blades (stage 1). The steam then expands partially in the second set of nozzles where its pressure again falls and the velocity increases; the kinetic energy so obtained is absorbed by the second ring of moving blades (stage 2). This is repeated in stage 3 and steam finally leaves the turbine at a low velocity and pressure. The number of stages (or pressure reductions) depends on the number of rows of nozzles through which the steam must pass.

This method of compounding is used in the *Rateau and Zoelly* turbine. This is most efficient turbine since the speed ratio remains constant, but it is expensive owing to a large number of stages [113], [114].

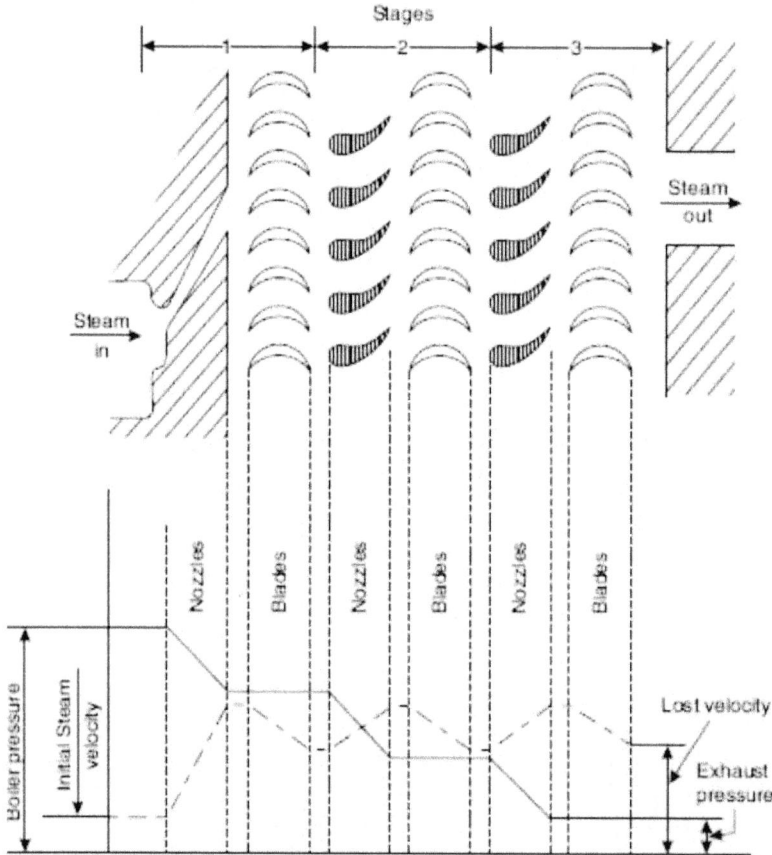

FIGURE 10.4. Pressure compounding, pressure shown by solid line and velocity by dashed line.

10.5.3. Pressure-Velocity Compounding

This method of compounding is the combination of two previously discussed methods. The total drop in steam pressure is divided into stages and the velocity obtained in each stage is also compounded. The rings of nozzles are fixed at the beginning of each stage and pressure remains constant during each stage.

This method of compounding is used in the *Curtis and Moore* turbine [114], [115].

10.6. Difference between Impulse and Reaction Turbines

Table 10.1 gives a list of particulars for Impulse and Reaction turbines.

10.7. Impulse Turbine

10.7.1. Velocity Diagram for a Moving Blade

Figure 10.5 shows the velocity diagram of a *single-stage impulse turbine*.

C_{bl} = Linear velocity of moving blade (m/s)

C_1 = Absolute velocity of steam entering moving blade (m/s)

Particulars	Impulse turbine	Reaction turbine
Pressure drop	Only in nozzles and not in moving blades.	In fixed blades (nozzles) as well as in moving blades.
Area of blade channels	Constant.	Varying (converging type).
Blades	Profile type.	Aerofoil type.
Admission of steam	Not all round or complete.	All round or complete.
Nozzles/fixed blades	Diaphragm contains the nozzle.	Fixed blades similar to moving blades attached to the casing serve as nozzles and guide the steam.
Power	Not much power can be developed.	Much power can be developed.
Space	Requires less space for same power.	Requires more space for same power.
Efficiency	Low.	High.
Suitability	Suitable for small power requirements.	Suitable for medium and higher power requirements.
Blade manufacture	Not difficult.	Difficult.

TABLE 10.1. A Comparison between Impulse and Reaction Turbines

C_0 = Absolute velocity of steam leaving moving blade (m/s)

C_{w_1} = Velocity of whirl at the entrance of moving blade.

= tangential component of C_1.

C_{w_0} = Velocity of whirl at exit of the moving blade.

= tangential component of C_0.

C_{f_1} = Velocity of flow at entrance of moving blade.

= axial component of C_1.

C_{f_0} = Velocity of flow at exit of the moving blade.

= axial component of C_0.

C_{r_1} = Relative velocity of steam to moving blade at entrance.

C_{r_0} = Relative velocity of steam to moving blade at exit.

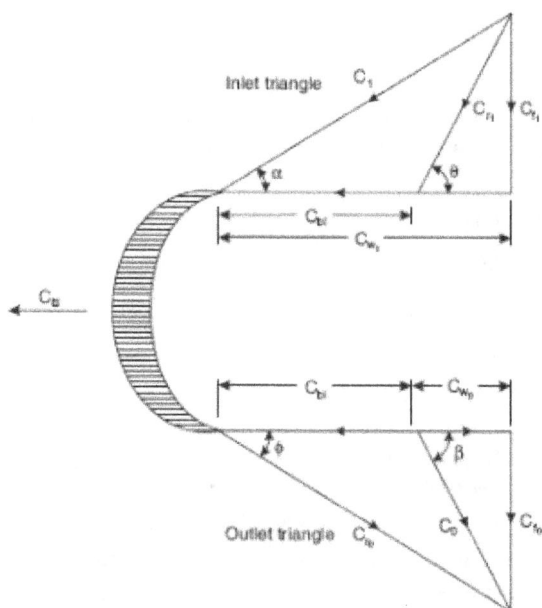

FIGURE 10.5. Velocity diagram for moving blade.

α = Angle with the tangent of the wheel at which the steam with velocity C_1 enters. This is also called *nozzle engle*.

β = Angle which the discharging steam makes with the tangent of the wheel at the exit of moving blade.

θ = Entrance angle of moving blade.

ϕ = Exit angle of moving blade.

The steam jet issuing from the nozzle at a velocity of C_1 impinges on the blade at an angle α. The tangential component of this jet (C_{w_1}) *performs work on the blade*, the axial component (C_{f_1}), however, *does no work but causes the steam to flow through the turbine.* As the blades move with a tangential velocity of C_{bl}, the entering steam jet has a relative velocity C_{r_1} (with respect to the blade) which makes an angle θ with the wheel tangent. The steam then glides over the blade without any shock and discharges at a relative velocity of C_0 at an angle ϕ with the tangent of the blades. The relative velocity at the inlet (C_{r_1}) is the same as the relative velocity at the outlet (C_{r_0}) if *there is no frictional loss at the blade*. The absolute velocity (C_0) of leaving steam makes an angle β to the tangent at the wheel.

To have convenience in solving the problems on turbines it is a common practice to combine the two vector velocity diagrams on a common base which represents the blade velocity (C_{bl}) as shown in Figure 10.6. This diagram has been obtained by superimposing the inlet velocity diagram on the outlet diagram in order for the blade velocity lines C_{bl} to coincide.

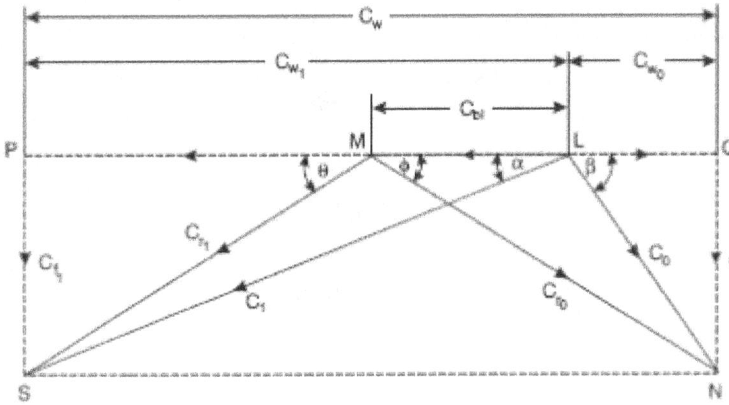

FIGURE 10.6. Combined two vector velocity diagrams on a common base to show blade velocity.

Work Done on the Blade

The work done on the blade may be found out from the change of momentum of the steam jet during its flow over the blade. As discussed previously, it is only the velocity of the whirl which performs work on the blade since it acts in its (the blade's) direction of motion.

From Newton's second law of motion,

Force (tangential) on the wheel = Mass of steam × acceleration

$$= \text{Mass of steam/sec.} \times \text{change of velocity}$$

$$= \dot{m}_s (C_{w_1} - C_{w_0})$$

The value of C_{w_0} is actually negative as the steam is discharged in the *opposite direction* from the blade motion, and therefore due consideration should be given to the fact that *values of* C_{w_1} *and* C_{w_0} *are to be added while doing the solution of the problem* (i.e., when $\beta < 90°$).

Work done on blades / sec. = Force × distance travelled/sec.

$$= \dot{m}_s (C_{w_1} + C_{w_0}) \times C_{bl}$$

Power per wheel
$$= \dot{m}_s (C_{w_1} + C_{w_0}) + C_{bl}$$

$$= \frac{\dot{m}_s C_w C_{bl}}{1000} \text{ kW}$$

$$\left(\because C_w = C_{w_1} + C_{w_0} \right)$$

$$\text{Blade or diagram efficiency} = \frac{\text{Work done the blade}}{\text{Energy supplied to the blade}}$$

$$= \frac{\dot{m}_s \left(C_{w_1} + C_{w_0} \right).C_{bl}}{\dfrac{\dot{m} C_1^2}{2}}$$

$$= \frac{2 C_{bl} \left(C_{w_1} + C_{w_0} \right)}{C_1^2}$$

If h_1 and h_2 are the total heats before and after expansion through the nozzles, then $(h_1 - h_2)$ is the heat drop through a stage of fixed blades rings and moving blades rings.

$$\text{Stage efficiency } \eta_{stage} = \frac{\text{Work done on blade per kg of steam}}{\text{Total energy of supplied per kg of steam}}$$

$$= \frac{C_{b1}\left(C_{w_1} + C_{w_0}\right)}{\left(h_1 - h_2\right)}$$

Now, nozzle efficiency $= \dfrac{C_1^2}{2\left(h_1 - h_2\right)}$

Also $\quad \eta_{stage}$ = Blade efficiency × nozzle efficiency

$$= \frac{2C_{bl}\left(C_{w_1} + C_{w_0}\right)}{C_1^2} \times \frac{C_1^2}{2\left(h_1 - h_2\right)} = \frac{C_{bl}\left(C_{w_1} + C_{w_0}\right)}{\left(h_1 - h_2\right)}$$

The axial thrust on the wheel is due to *difference* between the velocities of flow at entrance and outlet.

Axial force on the wheel $\quad$ = Mass of steam × axial acceleration

$$= \dot{m}_s(C_{f_1} - C_{f_0})$$

The axial force on the wheel must be balanced or must be taken by a thrust bearing.

Energy converted to heat by blade friction

$\qquad$ = loss of kinetic energy during flow over blades

$$= \dot{m}_s\left(C_{r_1}^{\;2} - C_{r_0}^{\;2}\right)$$

10.7.2. Blade Velocity Coefficient

In an impulse turbine, if friction is neglected the relative velocity will remain unaltered as it passes over the blades. In practice the flow of steam over the blades is resisted by friction. The effect of the friction is to reduce the relative velocity of steam as it passes over the blades. In general there is a loss of 10 to 15% in the relative velocity. Owing to friction in the blades, C_{r_0} is less than C_{r_1} and we may write

$$C_{r_0} = K \cdot C_{r_1}$$

where K is termed a blade velocity coefficient.

Expression for Optimum Value of the Ratio of Blade Speed to Steam Speed (for Maximum Efficiency) for a Single-Stage Impulse Turbine

Refer to Figure 10.6.

$$C_w = PQ = MP + MQ = C_{r_1} \cos\theta + C_{r_0} \cos\phi$$

$$C_{r_1} \cos\theta \left[1 + \frac{C_{r_0} \cos\phi}{C_{r_1} \cos\theta}\right]$$

$$C_{r_1} \cos\theta (1 + K.Z) \text{ Where } Z = \frac{\cos\phi}{\cos\theta} \qquad (i)$$

Generally, the angles θ and φ are nearly equal for an impulse turbine and hence it can safely be assumed that Z is a constant.

But, $\qquad\qquad C_{r_1} \cos\theta = MP = LP = LM = C_1 \cos\alpha - C_{bl}$

From eqn. (i), $\qquad\qquad C_w = (C_1 \cos\alpha - C_{bl})(1 + K.Z)$

We know that, Blade efficiency, $\eta_{bl} = \dfrac{2C_{bl} \cdot C_w}{C_1^2} \qquad (ii)$

$$\eta_{bl} = \frac{2C_{bl} \cdot (C_1 \cos\alpha - C_{bl})(1 + KZ)}{C_1^2}$$

$$= 2\left(\rho\cos\alpha - \rho^2\right)(1 + KZ)$$

$$= 2\rho(\cos\alpha - \rho)(1 + KZ) \qquad (iii)$$

where $\rho = \dfrac{C_{bl}}{C_1}$ is the ratio of *blade speed to steam speed* and is commonly called as "**Blade speed ratio**".

For a particular impulse turbine α, K and Z may be assumed to be constant, and from equation (iii) it can be seen clearly that η_{bl} depends on the value of ρ only. Hence differentiating (iii),

$$\frac{d\eta_{bl}}{d\rho} = 2\left(\cos\alpha - 2\rho\right)(1 + KZ)$$

For a maximum or minimum value of η_{bl} this should be zero

$$\cos\alpha - 2\rho = 0, \qquad \therefore \qquad \rho = \frac{\cos\alpha}{2}$$

Now, $\dfrac{d^2\eta_{bl}}{d\rho^2} = 2(-2)(1 + KZ) = -4(1 + KZ)$

which is a negative quantity and thus the value so obtained is the maximum.

Optimum value of ratio of blade speed to steam speed is

$$\rho_{opt} = \frac{\cos \alpha}{2}$$

Substituting this value of ρ in eqn. (*iii*), we get

$$\left(\eta_{bl}\right)_{max} = 2 \times \frac{\cos \alpha}{2}\left(\cos \alpha - \frac{\cos \alpha}{2}\right)(1 + K.Z)$$

$$= \frac{\cos^2 \alpha}{2}(1 + KZ)$$

It is sufficiently accurate to assume symmetrical blades ($\theta = \phi$) and no friction in fluid passage for the purpose of analysis.

$$\therefore \qquad\qquad Z = 1 \text{ and } K = 1$$

$$\therefore \qquad\qquad \left(\eta_{bl}\right)_{max} = \cos^2 \alpha$$

The work done per kg of steam is given by

$$W = \left(C_{w_1} + C_{w_0}\right) C_{bl}$$

Substituting the value of $C_{w_1} + C_{w_0}$ $(= C_w)$

$$W = \left(C_1 \cos \alpha - C_{bl}\right)(1 + KZ) C_{bl} = 2C_{bl}\left(C_1 \cos \alpha - C_{bl}\right) \text{ when } K = 1 \text{ and } Z = 1$$

The maximum value of W can be obtained by substituting the value of $\cos \alpha$

$$\cos \alpha = 2\rho = 2\frac{C_{bl}}{C_1}$$

$$\therefore \qquad\qquad W_{max} = 2C_{bl}\left(2C_{bl} - C_{bl}\right) = 2C_{bl}^2$$

The blade velocity should be *approximately* half of absolute velocity of the steam jet coming out from the nozzle (fixed blade) for the maximum work developed per kg of steam or for maximum efficiency. For the other values of blade speed the absolute velocity at the outlet from the blade will increase; consequently, more energy will be carried away by the steam, and efficiency will decrease.

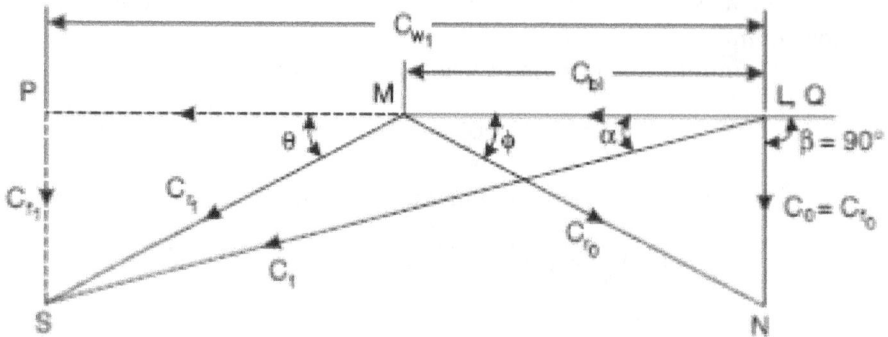

FIGURE 10.7. Equiangular blade velocity vector.

For equiangular blades with no friction losses, the optimum value of $\dfrac{C_{bl}}{C_1}$ corresponds to the case when the outlet absolute velocity is *axial* as shown in Figure 10.7.

Since the discharge is axial $\qquad \beta = 90°, \qquad \therefore C_0 = C_{fo}$ and $C_{wo} = 0.$

The variation of η_{bl} or work developed per kg of steam with $\dfrac{C_{bl}}{C_1}$ is shown in Figure 10.8. This figure shows that:

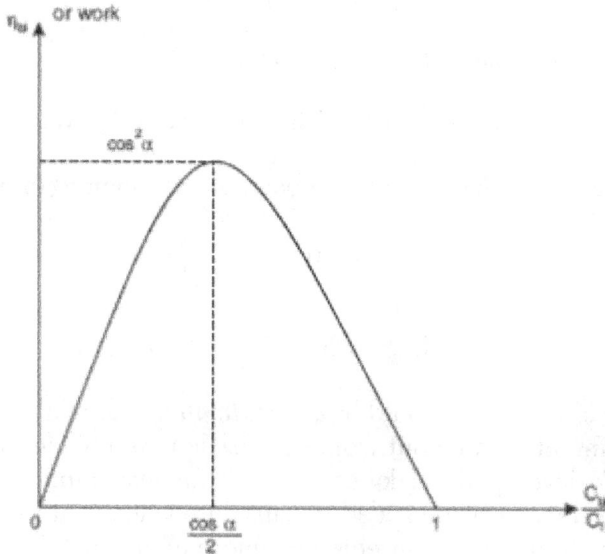

FIGURE 10.8. Work developed per kg of steam with Cbl/C1.

(*i*) When $\dfrac{C_{bl}}{C_1}$ = 0, the work done becomes zero as the distance travelled by the blade (C_{bl}) is zero, even though the torque on the blade is maximum.

(*ii*) The maximum efficiency is $\cos^2 \alpha$ and the maximum work done per kg of steam is $2C^2_{bl}$ when $\dfrac{C_{bl}}{C_1} = \cos \alpha / 2.$

(*iii*) When $\dfrac{C_{bl}}{C_1} = 1$, the work done is zero as the torque acting on the blade becomes zero even though the distance traveled by the blade is maximum.

When the high-pressure steam is expanded from the boiler pressure to the condenser pressure in a single stage of nozzles, the absolute velocity of the steam becomes maximum and blade velocity also becomes tremendously high. In such a case, a velocity compounded stage is used to give a lower blade speed ratio and better utilization of the kinetic energy of the steam. The arrangement of velocity compounding has already been dealt with.

Figure 10.9 shows the velocity diagrams for the first and second row of moving blades of a velocity compounded unit. The speed and angles are such that the final absolute velocity of the steam leaving the second row is

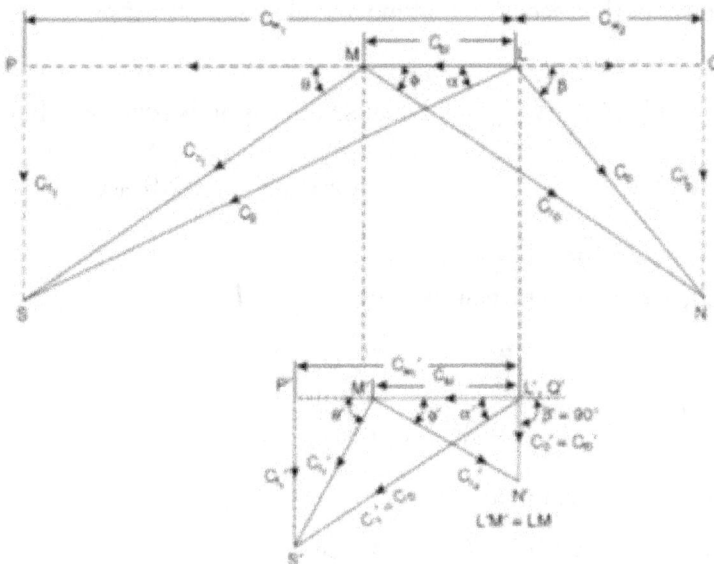

FIGURE 10.9. Velocity diagrams for the first and second row of moving blades of a velocity compounded unit.

axial. With this arrangement, *the K.E. carried by the steam is minimum;* therefore, the efficiency becomes maximum.

The velocity of blades (C_{bl}) is same for both the rows since they are mounted on the same shaft.

Consider the first row of moving blades:

Work done per kg of steam, $W_1 = C_{bl}\left(C_{w_1} + C_{w_2}\right)$

$$= C_{bl}\left[C_{r_1}\cos\theta + C_{r_0}\cos\phi\right]$$

If there is no friction loss and symmetrical blading is used, then

$C_{r_1} = C_{r_0}$ and $\theta = \phi$

∴ $W_1 = C_{bl} \times 2C_{r_1}\cos\theta = 2C_{bl}(C_1\cos\alpha - C_{bl})$

The magnitude of the absolute velocity of steam leaving the first row and entering into the second row of moving blades is the same and only its direction is changed.

∴ $C_1' = C_0$

Consider the second row of moving blades:

Work done per kg,

$W_2 = C_{bl}\,.\,C_{w_1}'$ as $C_{w_2}' = 0$ because discharge is *axial* and $\beta' = 90°$

Alternately, $W'_2 = C_{bl}\,.\left[C'_{r_1}\cos\theta' + C'_{r_0}\phi'\right]$

For symmetrical blades $\theta' = \phi'$

and, if there is no friction loss, then $C'_{r_1} = C'_{r_0}$

∴ $W_2 = 2C_{bl}\,C'_{r_1}\cos\,\theta'$

$$= 2C_{bl}(C_1'\cos\alpha' - C_{bl})$$

Now α′ may be equal to β.

∴ $C_1'\cos\alpha' = C_0\cos\beta = \cos\phi - C_{bl}$

$$= C_{r_1} \cos\theta - C_{bl} = \left(C_1 \cos\alpha - C_{bl}\right) - {}_{bl}$$

$$= C_1 \cos\alpha - 2C_{bl}$$

Substituting the value of $C'_1 \cos\alpha'$ in eqn. (13), we get

$$W_2 = 2C_{bl}\left[\left(C_1 \cos\alpha - 2C_{bl}\right) - C_{bl}\right]$$

$$= 2C_{bl}\left(C_1 \cos\alpha - 3C_{bl}\right)$$

Total work done per kg of steam passing through both stages is given by

$$W_t = W_1 + W_2$$

$$= 2C_{bl}\left[C_1 \cos\alpha - C_{bl}\right] + 2C_{bl}\left[C_1 \cos\alpha - 3C_{bl}\right]$$

$$= 2C_{bl}\left(2C_1 \cos\alpha - 4C_{bl}\right)$$

$$= 4C_{bl}\left(C_1 \cos\alpha - 2C_{bl}\right)$$

The blade efficiency for two-stage impulse turbine is given by

$$\eta_{bl} = \frac{W_t}{\dfrac{C_1^2}{2}} = 4C_{bl}\left[C_1 \cos\alpha_1 - 2C_{bl}\right] \times \frac{2}{C_1^2}$$

$$= \frac{8C_{bl}}{C_1^2}\left(C_1 \cos\alpha - 2C_{bl}\right) = 8\frac{C_{bl}}{C_1}\left(\cos\alpha - 2.\frac{C_{bl}}{C_1}\right)$$

$$= 8\rho\left(\cos\alpha - 2\rho\right)$$

The blade efficiency for a two-stage turbine will be maximum when

$$\frac{d\eta_{bl}}{d\rho} = 0$$

$\therefore$ $\qquad \dfrac{d}{d\rho}\left[8\rho\cos\alpha - 16\rho^2\right] = 0$

$\therefore$ $\qquad 8\cos\alpha - 32\rho = 0$

From which, $\rho = \dfrac{\cos\alpha}{4}$

Substituting this value in equation (16), we get

$$\left(\eta_{bl}\right)_{max} = 8.\frac{\cos\alpha}{4}\left[\cos\alpha - 2.\frac{\cos\alpha}{4}\right] = \cos^2\alpha$$

The maximum work done per kg of steam is obtained by substituting the value of

$$\rho = \frac{C_{bl}}{C_1} = \frac{\cos\alpha}{4}$$

or $\qquad C_1 = \dfrac{4C_{bl}}{\cos\alpha}$ in the eqn. (15).

$\therefore$
$$\left(W_b\right)_{max} = 4C_{bl}\left(\frac{4C_{bl}}{\cos\alpha}.\cos\alpha - 2C_{bl}\right)$$

$$= 8C_{bl}{}^2$$

The present analysis is done for *two stages* only. A similar procedure is adopted for analyzing the problem with three or four stages.

In general, the optimum blade speed ratio for maximum blade efficiency or maximum work done is given by

$$\rho = \frac{\cos\alpha}{2.n}$$

and work in the last row $= \dfrac{1}{2^n}$ of total work

where n is the number of moving/rotating blade rows in a series.

As the number of rows increases, the utility of the last row decreases. In practice, *more than* two rows are hardly preferred.

10.7.3. Advantages of Velocity-Compounded Impulse Turbines

1. Owing to a relatively large heat drop, a velocity-compounded impulse turbine requires a comparatively small number of stages.

2. Due to number of stages being small, its cost is less.

3. As the number of moving blades' rows in a wheel increases, the maximum stage efficiency and optimum value of ρ decreases.

4. Since the steam temperature is sufficiently low in a two- or three-row wheel, a cast-iron cylinder may be used. This will cause savings in material cost.

10.7.4. Disadvantages of a velocity-compounded impulse turbine:

1. It has high steam consumption and low efficiency (see Figure 10.10).

2. In a single row wheel, the steam temperature is high so the cast-iron cylinder cannot be used due to the phenomenon of growth; a cast steel cylinder is used, which is costlier than cast-iron.

Example 1

A stage of a steam turbine is supplied with steam at a pressure of 50 bar and 350 °C, and exhausts at a pressure of 5 bar. The isentropic efficiency of the stage is 0.82 and the steam consumption is 2270 kg/min. Determine the power output of the stage.

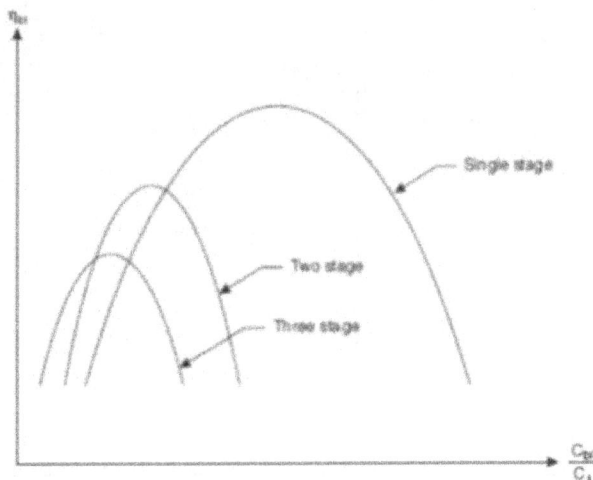

FIGURE 10.10. Sketch 1 for Example 1.

■ **Solution:**

Steam supply pressure, $\qquad p_1 = 50 \text{ bar}, 350°C$

Exhaust pressure $\qquad p_2 = 5 \text{ bar}$

Isentropic efficiency of the stage, $\qquad \eta_{\text{stage}} = 0.82$

Steam consumption, $\qquad m_s = 2270 \text{ kg/min}$

Power output of the stage, P:

Refer to Figure 10.11.

From *Mollier chart:*

$h_1 = 3130.7 \text{ kJ/kg of steam}$

$h_2 = 2640 \text{ kJ/kg of steam}$

Isentropic heat drop $\qquad = h_1 - h_2 = 3130.7 - 2640 = 490.7 \text{ kJ/kg}$

Actual heat drop $\qquad = h_1 - h_2'$

But, $\qquad \eta_{\text{isen. (stage)}} = \dfrac{h_1 - h_2'}{h_1 - h_2}$

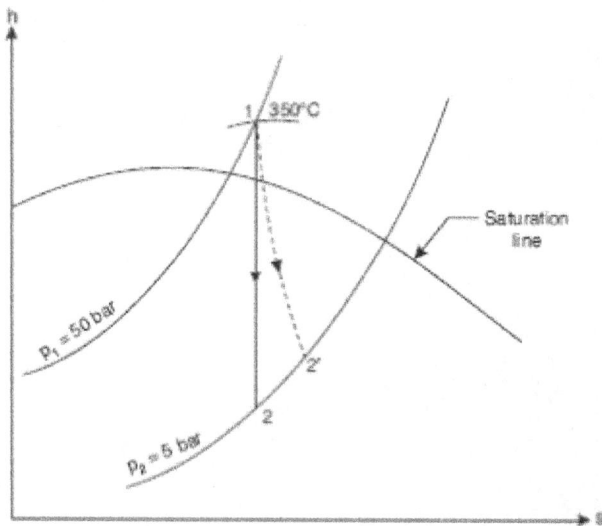

FIGURE 10.11. h-S diagram for Example 1.

or $\qquad\qquad 0.82 = \dfrac{h_1 - h_2'}{490.7} =$ or $h_1 - h_2' = 0.82 \times 490.7 = 402.4$ kJ/kg

∴   Power developed $\qquad\qquad = m_s \left(h_1 - h_2' \right)$

$$= \dfrac{2270}{60} \times 402.4 \text{ kW} = \textbf{15224 kW. (Ans.)}$$

Example 2

In a De Laval turbine steam issues from the nozzle with a velocity of 1200 m/s. The nozzle angle is 20°, the mean blade velocity is 400 m/s, and the inlet and outlet angles of blades are equal. The mass of steam flowing through the turbine per hour is 1000 kg. Calculate:

(i) Blade angles

(ii) Relative velocity of steam entering the blades

(iii) Tangential force on the blades

(iv) Power developed

(v) Blade efficiency

Take blade velocity coefficient as 0.8.

■ **Solution:**

Absolute velocity of steam entering the blade, $C_1 = 1200$ m/s

Nozzle blade, $\qquad\qquad\qquad\qquad \alpha = 20°$

Mean blade velocity, $\qquad\qquad C_{bl} = 400 \,\text{m/s}$

Inlet blade angle, $\qquad\qquad\qquad \theta$=Outlet blade angle, ϕ

Blade velocity co-efficient, $\qquad\qquad K = 0.8$

Mass of steam flowing through the turbine, $m_s = 1000$ kg/h

Refer to Figure 10.12. The procedure of drawing the inlet and outlet triangles (LMS and LMN, respectively) is as follows:

- Select a suitable scale and draw line LM to represent C_{bl}(= 400 m/s).

- At point L make an angle of 20° (α) and cut length LS to represent velocity C_1(= 1200 m/s). Join MS. Produce M to meet the perpendicular drawn from S at P. Thus, an *inlet triangle* is completed.

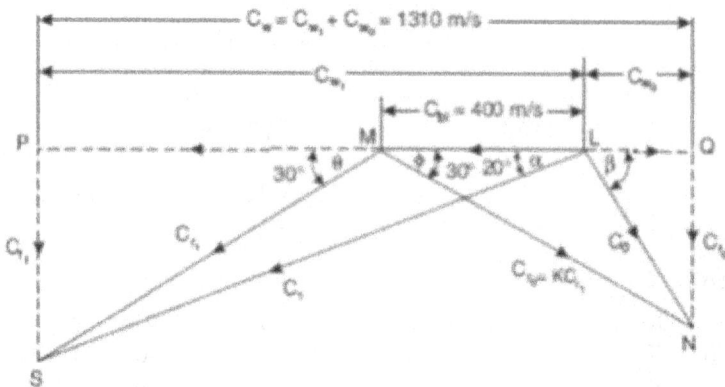

FIGURE 10.12. Sketch for Example 2.

By measurement : $\qquad \theta = 30°, C_{r_1} = MS = 830\,\text{m/s}$

$$\theta = \phi = 30° \qquad \text{(given)}$$

Now, $\qquad C_{r_2} = KC_{r_1} = 0.8 \times 830 = 664\,\text{m/s}$

- At point M make an angle of 30° (ϕ) and cut the length MN to represent . Join LN. Produce L to meet the perpendicular drawn from N at Q. Thus, an *outlet triangle* is completed.

(*i*) **Blade angles** θ, ϕ:

As the blades are symmetrical (given)

$\therefore \qquad \theta = \phi = 30°.$ (Ans.)

(*ii*) **Relative velocity of steam entering the blades,** C_{r_1} :

$C_{r_1} = MS = \mathbf{830\ m/s}.$ **(Ans.)**

(*iii*) **Tangential force on the blades:**

Tangential force $\qquad = \dot{m}_s \left(C_{w_1} + C_{w_0} \right) = \dfrac{1000}{60 \times 60}(1310) = \mathbf{363.8\ N.}$ **(Ans.)**

(*iv*) **Power developed, P:**

$$P = \dot{m} \left(C_{w_1} + C_{w_0} \right) C_{bl} = \frac{1000}{60 \times 60} \times \frac{1310 \times 400}{1000}\ \text{kW} = \mathbf{145.5\ kW.}\ \mathbf{(Ans.)}$$

(v) **Blade efficiency,** η_{bl} :

$$\eta_{bl} = \frac{2C_{bl}\left(C_{w_1} + C_{w_0}\right)}{C_1^2} = \frac{2 \times 400 \times 1310}{1200^2} = \textbf{72.8\%. (Ans.)}$$

Example 3

The velocity of steam exiting the nozzle of the impulse stage of a turbine is 400 m/s. The blades operate close to the maximum blading efficiency. The nozzle angle is 20°. Considering equiangular blades and neglecting blade friction, calculate for a steam flow of 0.6. kg/s the diagram power and the diagram efficiency.

■ **Solution:**

Given: $C_1 = 400\,\text{m/s}$, $= \alpha = 20°$, $\theta = \phi$; $\dot{m}_s = 0.6\,\text{kg/s}$.

For maximum blade efficiency, $\qquad \rho = \dfrac{C_{bl}}{C_1} = \dfrac{\cos\alpha}{2}$

$$\therefore \qquad \frac{C_{bl}}{400} = \frac{\cos 20°}{2} \quad \text{or} \quad C_{bl} = 187.9\,\text{m/s}$$

$$C_{w_1} = C_1 \cos\alpha = 400\cos 20° = 375.9\,\text{m/s}$$

$$C_{f_1} = C_1 \sin\alpha = 400\sin 20° = 136.8\,\text{m/s}$$

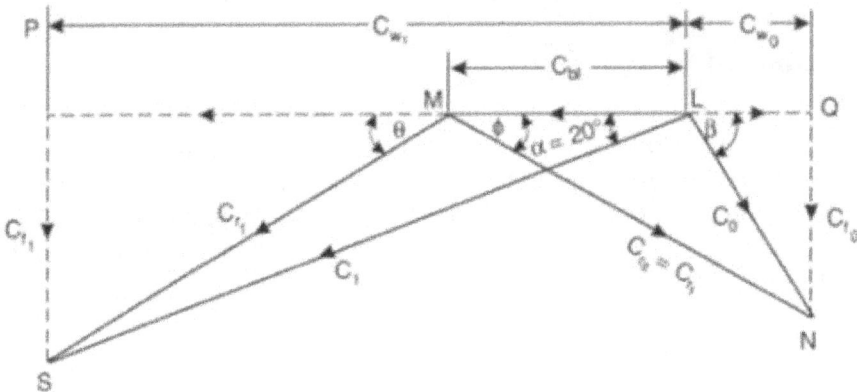

FIGURE 10.13. Sketch for Example 3.

$$\tan\theta = \frac{C_{f_1}}{C_{w_1}-C_{bl}} = \frac{136.8}{375.9-187.9} = 0.727$$

$$\therefore \qquad \theta = \tan^{-1}(0.727) = 36°$$

Now, $C_{r_1}\sin\theta = C_{f_1}$ or $C_{r_1} = \dfrac{C_{f_1}}{\sin\theta}$, $\therefore C_{r_1} = \dfrac{136.8}{\sin 36°} = 232.7\text{m/s}$

Now neglecting friction, $\qquad C_{r_0} = C_{r_1} = 232.7\text{m/s}$

Since the blades are *equiangular*, therefore,
$$\theta = \phi = 36°$$

$$C_{w_0} = C_{r_0}\cos 36° - C_{bl}$$

$$= 237.7\cos 36° - 187.9 = 0.36\text{ m/s}$$

$\therefore \qquad C_w = C_{w_1} + C_{w_0} = 375.9 + 0.36 = 376.26\text{ m/s}$

Diagram power, $\qquad P = \dot{m}_s(C_{w_1}+C_{w_0})\times C_{bl}\times 10^{-3}\text{ kW}$

$$= 0.6\times376.26\times187.9\times10^{-3} = \textbf{42.4 kW. (Ans.)}$$

Blade or diagram efficiency, $\eta_{bl} = \dfrac{2C_{bl}(C_{w_1}+C_{w_0})}{C_1^2}$

$$= \frac{2\times187.9\times376.26}{(400)^2} = 0.884\text{ or }\textbf{88.4\% (Ans.)}$$

Example 4

A single-stage steam turbine is supplied with steam at 5 bar, 200°C at the rate of 50 kg/min. It expands into a condenser at a pressure of 0.2 bar. The blade speed is 400 m/s. The nozzles are inclined at an angle of 20° to the plane of the wheel and the outlet blade angle is 30°. Neglecting friction losses, determine the power developed, blade efficiency, and stage efficiency.

■ **Solution:**

Given: p_1 = 5 bar, 200°C; p_2 = 0.2 bar, m_s = 50 kg/min, C_{bl} = 400 m/s; α = 20°, φ = 30°; $C_{r_1} = C_{r_0}$ (because friction losses are neglected)

Refer to Figure 10.14.

From steam tables :

At 5 bar, 200°C : $h_1 = 2855.4$ kJ/kg ; $s_1 = 7.0592$ kJ / kgK

At 0.2bar : $h_{f_2} = 251.5$ kJ/kg, $h_{fg_2} = 2358.4$ kJ/kg

$s_{f_2} = 0.832$ kJ/kg K ; $s_{fg} = 7.0773$ kJ/kg K

Since the steam expansion takes place *isentropically*,

$\therefore$ $s_1 = s_2$

or $7.0592 = 0.8321 + x_2 \times 7.0773$

or $x_2 = \dfrac{7.0592 - 0.8321}{7.0773} = 0.88$

$\therefore$ Enthalpy of steam at 0.2 bar,

$$h_2 = h_{f_2} + x_2 \, h_{fg_2}$$

$$= 251.5 + 0.88 \times 2358.4 = 2326.9 \,\text{kJ/kg}$$

Enthalpy drop $= h_1 - h_2 = 2855.4 - 2326.9 = 528.5 \text{kJ/kg}$

Velocity of steam entering the blades,

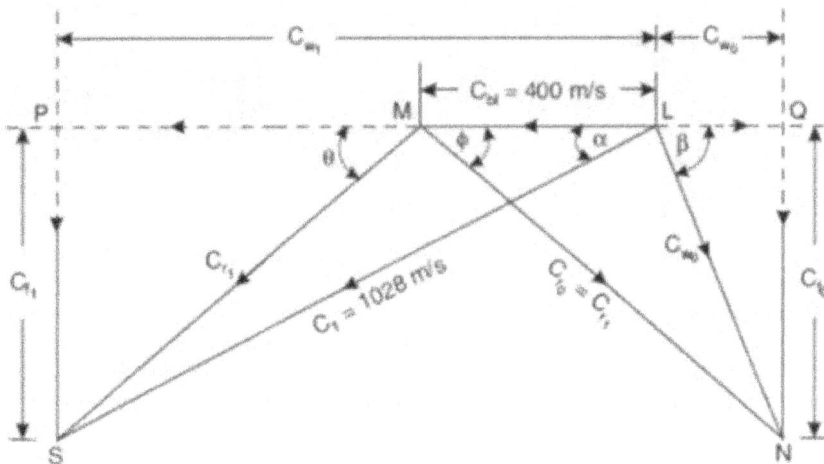

FIGURE 10.14. Sketch for Example 4.

$$C_1 = 44.7\sqrt{h_1 - h_2} = 44.7\sqrt{528.5} \simeq 1028 \text{ m/s}$$

The velocity diagram is shown in Figure 10.16.

Now, $\qquad C_{w_1} = 1028 \cos 20° = 966 \text{ m/s}$

$$C_{f_1} = 1028 \sin 20° = 351.6 \text{ m/s}$$

$$\tan\theta = \frac{C_{f_1}}{C_1 \cos 20° - C_{bl}} = \frac{351.6}{1028 \cos 20° - 400} = 0.6212$$

∴ $\qquad \theta = \tan^{-1}(0.6212) = 31.85°$

Now, $\qquad C_{r_1} \sin 31.85° = C_{f_1} = 351.6$

∴ $\qquad C_{r_1} = \frac{351.6}{\sin 31.85°} = 666 \text{ m/s}$

∴ $\qquad C_{w_0} = C_{r_0} \cos 30° - C_{bl} \qquad (\because C_{r_0} = C_{r_1})$

$$= 666 \cos 30° - 400 \simeq 177 \text{ m/s}$$

Power developed, $\qquad P = \dot{m}s(C_{w_1} + C_{w_0}) C_{bl}$

$$= \frac{50}{60}(966 + 177) \times 400 \times 10^{-3} \text{ kW} = \textbf{381 kW.} \quad (\textbf{Ans.})$$

Blade efficiency, $\qquad \eta_{bl} = \frac{2C_{bl}(C_{w_1} + C_{w_0})}{C_1^2}$

$$= \frac{2 \times 400 \times (966 + 177)}{(1028)^2} = \textbf{0.865 or 86.5\%. (Ans.)}$$

Since there are no losses, therefore,

Stage efficiency $\qquad\qquad\qquad\qquad$ = blade efficiency = **86.5%. (Ans.)**

Example 5

The following data relate to a single-stage impulse turbine:

Steam velocity = 600 m/s $\qquad$ *Blade speed = 250 m/s*

Nozzle angle = 20° $\qquad\qquad$ *Blade outlet angle = 25°*

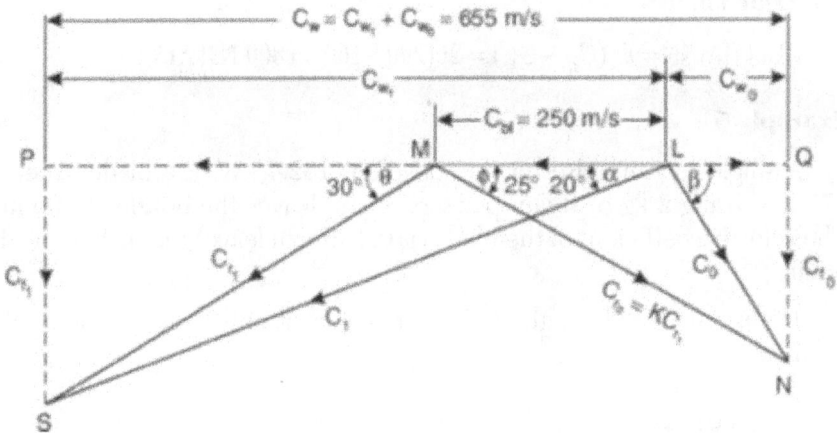

FIGURE 10.15. Sketch for Example 5.

Neglecting the effect of friction, calculate the work developed by the turbine for the steam flow rate of 20 kg/s. Also calculate the axial thrust on the bearings.

■ **Solution:**

Absolute velocity of steam entering the blades, C_1 = 600 m/s

Blade speed, C_{bl} = 250 m / s ; Nozzle angle, $\alpha = 20°$

Blade outlet angle, $\phi = 25°$; Steam flow rate, $\dot{m}_s = 20$ kg / s

Refer to Figure 10.15.

• Triangle *LMS* is drawn with the previous data.

• Then angle *LMN*, *that is,* $\varphi = 25°$, is drawn such that *NM* = *MS* (because the effect of friction is to be neglected, *i.e.,* *K* = 1).

• Join *LN* by vector C_0 which represents the velocity of steam at the outlet from the wheel.

This completes both inlet and outlet triangles.

By measurement:

$C_w = C_{w_1} + C_{w_0} = 655$ m/s ; $C_{f_1} = 200$ m/s ; $C_{f_0} = 160$ m/s.

Work developed, W:

$$W = \dot{m}_s (C_{w_1} + C_{w_0}) C_{bl} = 20 \times 655 \times 250$$

$$= 3275000 \, \text{Nm} / \text{s.} \;\; (\text{Ans.})$$

Axial Thrust:

Axial thrust $= \dot{m}_s (C_{f_1} - C_{f_0}) = 20(200 - 160) = \textbf{800 N. (Ans.)}$

Example 6

A single-row impulse turbine develops 132.4 kW at a blade speed of 175 m/s, using 2 kg of steam per sec. Steam leaves the nozzle at 400 m/s. The velocity coefficient of the blades is 0.9. Steam leaves the turbine blades axially.

Determine nozzle angle and blade angles at entry and exit, assuming no shock.

■ Solution:

Power developed, $P = 132.4$ kW

Blade speed, $\qquad\qquad C_{bl} = 175$ m/s

Steam used, $\qquad\qquad \dot{m}_s = 2$ kg/s

Velocity of steam leaving the nozzle, $C_1 = 400$ m/s

Blade velocity co-efficient, $\qquad K = 0.9$

Power developed, $\qquad\qquad P = \dot{m}_s \dfrac{(C_{w_1} + C_{w_0}) \times C_{bl}}{1000}$ kW

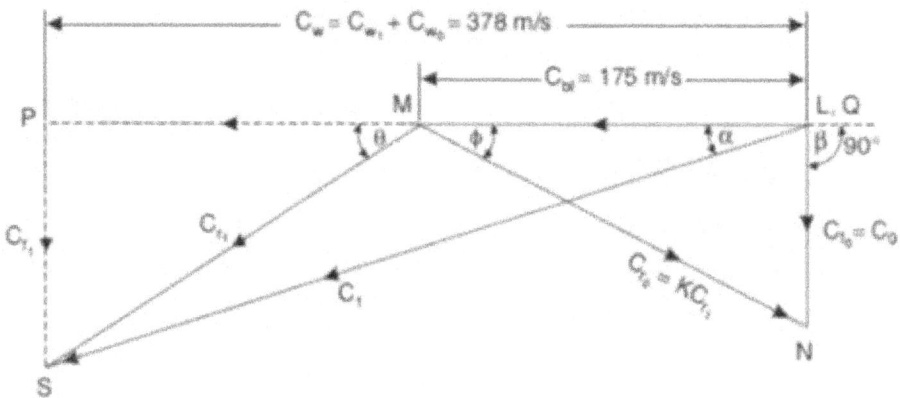

FIGURE 10.16. Sketch for Example 6.

$$132.4 = \frac{2\left(C_{w_1} + C_{w_0}\right) \times 175}{1000} \text{ or } \left(C_{w_1} + C_{w_0}\right) = \frac{132.4 \times 1000}{2 \times 175} = 378 \text{ m/s}$$

$C_{w_0} = 0$, since the *discharge is axial*.

Construct the velocity diagram as shown in Figure 10.16.
In this diagram $\dfrac{C_{r_0}}{C_{r_1}} = 0.9$, $\beta = 90°$, since the *discharge is axial;* and

$$C_{w_1} + C_{w_0} = PL = PQ = 378 \text{ m/s}.$$

From the diagram (by measurement):

Nozzle angle, $\qquad\qquad\qquad \alpha = \mathbf{21°}$. (**Ans.**)

Blade inlet angle, $\qquad\qquad \theta = \mathbf{36°}$. (**Ans.**)

Blade outlet angle, $\qquad\qquad \phi = \mathbf{32°}$. (**Ans.**)

Example 7

A simple impulse turbine has a mean blade speed of 200 m/s. The nozzles are inclined at 20° to the plane of rotation of the blades. The steam velocity from nozzles is 600 m/s. The turbine uses 3500 kg/h of steam. The absolute velocity at the exit is along the axits of the turbine. Determine:

(i) The inlet and exit angles of the blades

(ii) The power output of the turbine

(iii) The diagram efficiency

(iv) The axial thrust (per kg steam per second)

Assume inlet and outlet angles to be equal.

■ **Solution:**

Given: $C_{bl} = 200 \text{ m/s}$; $\alpha = 20°$; $C_1 = 600 \text{ m/s}$; $m_s = 3500 \text{ kg/h}$; $\beta = 90°$; $\theta = \phi$.

(i) Inlet and exit angles of the blades, θ, **P**:

Refer to Figure 10.17.

$$C_{f_1} = C_1 \sin 20° = 600 \sin 20° = 205.2 \text{ m/s}$$

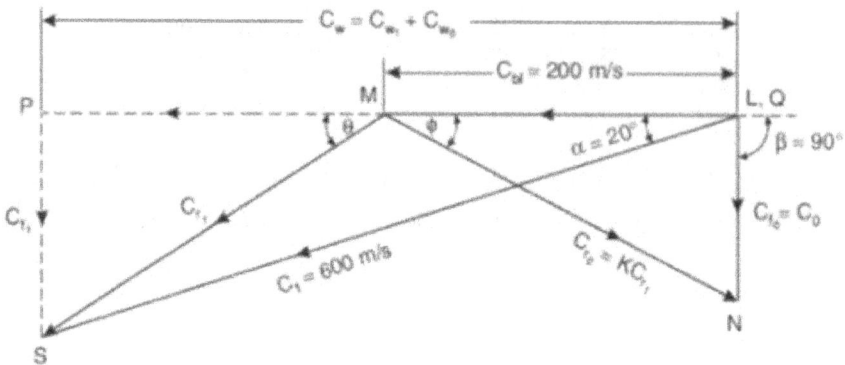

FIGURE 10.17. Sketch for Example 7.

$$\tan \theta = \frac{SP}{PM} = \frac{C_{f_1}}{C_1 \cos 20° - C_{bl}} = \frac{205.2}{600 \cos 20° - 200} = 0.564$$

∴ $$\theta = \tan^{-1}(0.564) = \textbf{29.4°. (Ans.)}$$

Also $$\theta = \phi$$ ...(Given)

∴ $$\phi = \textbf{29.4°. (Ans.)}$$

(ii) The power output of the turbine, P:

$$P = \dot{m}_s (C_{w_1} + C_{w_0}) C_{bl}$$

$$= \frac{3500}{3600}(600 \cos 20° + 0) \times 200 \times 10^{-3} \text{ kW} = \textbf{109.6 kW. (Ans.)}$$

$$(C_{w_0} = 0, \text{ since the discharge is axial})$$

(iii) The blade or diagram efficiency, η_{bl}:

$$\eta_{bl} = \frac{2C_{bl}\left(C_{w_1} + C_{w_0}\right)}{C_1^2}$$

$$= \frac{2 \times 200(600 \cos 20° \times 0)}{600^2} = \textbf{0.626 or 62.6\%. (Ans.)}$$

(iv) The axial thrust (per kg steam per second):

The axial thrust per kg per second

$$= 1 \times (C_{f_1} - C_{f_0}) N$$

where $C_{f_1} = C_1 \sin 20° = 600 \sin 20° = 205.2$ m/s.

Now, $\dfrac{C_{f_0}}{C_{bl}} = \tan 29.4°$

∴ $C_{f_0} = 200 \times \tan 29.4° = 112.69$ m/s

Substituting the values, we get

Axial thrust (per kg steam per second)

$$= 1 \times (205.2 - 112.69) = \textbf{92.51 N.} \quad (\textbf{Ans.})$$

Example 8

Steam with absolute velocity of 300 m/s is supplied through a nozzle to a single-stage impulse turbine. The nozzle angle is 25°. The mean diameter of the blade rotor is 1 meter and it has a speed of 2000 r.p.m. Find suitable blade angles for zero axial thrust. If the blade velocity coefficient is 0.9 and the steam flow rate is 10 kg/s, calculate the power developed.

■ **Solution:**

Absolute velocity of steam entering the blade, $C_1 = 300$ m/s

Nozzle angle, $\alpha = 25°$

Mean diameter of the rotor blade, $D = 1$ m

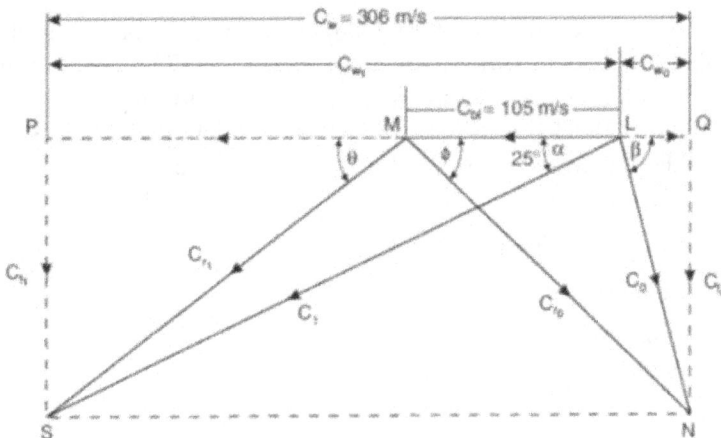

FIGURE 10.18. Sketch for Example 8.

Speed of the rotor,	$N = 2000$ r.p.m.	
Blade velocity co-efficient,	$K = 0.9$	
Steam flow rate,	$\dot{m}_s = 10$ kg/s	

Blade angles:

Blade speed,
$$C_{bl} = \frac{\pi DN}{60} = \frac{\pi \times 1 \times 2000}{60} = 105 \text{ m/s.}$$

— With the above data (*i.e.*, $C_1 = 300$ m/s, $C_{bl} = 105$ m/s and $\alpha = 25°$) draw triangle LMS (Fig. 19). From S draw perpendicular SP on LM produced. Measure C_{r_1}.

— From S draw a line *parallel to LP* ($\because C_{f_1} = C_{f_0}$) and from point M draw an arc equal to C_{r_0} ($= 0.9 C_{r_1}$) to get the point of intersection N. Complete the triangle LMN. From N draw perpendicular NQ on PL produced to get C_{f_0}.

Measure θ and φ (the blade angles) from the velocity diagram.

$\theta = \mathbf{37°}$ and $\phi = \mathbf{42°}$. (**Ans.**)

Power developed, P:

$$P = \frac{\dot{m}_s \left(C_{w_1} + C_{w_0} \right) \times C_{bl}}{1000} = \frac{10 \times 306 \times 105}{1000} = \mathbf{321.3 \text{ kW.}} \text{ (Ans.)}$$

Example 9

In an impulse turbine (with a single row wheel) the mean diameter of the blades is 1.05 m and the speed is 3000 r.p.m. The nozzle angle is 18°, the ratio of blade speed to steam speed is 0.42 and the ratio of the relative velocity at outlet from the blades to that at inlet is 0.84. The outlet angle of the blade is to be made 3° less than the inlet angle. The steam flow is 10 kg/s. Draw the velocity diagram for the blades and derive the following:

(*i*) *Tangential thrust on the blades*

(*ii*) *Axial thrust on the blades*

(*iii*) *Resultant thrust on the blades*

(*iv*) *Power developed in the blades*

(*v*) *Blading efficiency*

■ Solution:

Mean diameter of the blades, $D = 1.05$ m

Speed of the turbine, $\qquad N = 3000$ r.p.m.

Nozzle angle, $\qquad \alpha = 18°$

Ratio of blade speed to steam speed, $\quad \rho = 0.42$

Ratio, $\qquad \dfrac{C_{r_0}}{C_{r_1}} = 0.84$

Outlet blade angle, $\qquad \phi = \theta - 3°$

Steam flow rate $\qquad \dot{m}_s = 10$ kg/s

Blade speed, $\qquad C_{bl} = \dfrac{\pi DN}{60} = \dfrac{\pi \times 1.05 \times 3000}{60} = 164.5$ m/s

But $\qquad \rho = \dfrac{C_{bl}}{C_1} = 0.42$ (given)

$\therefore \qquad C_1 = \dfrac{C_{bl}}{0.42} = \dfrac{164.5}{0.42} = 392$ m/s

With the data, $\qquad C_1 = 392$ m/s ;

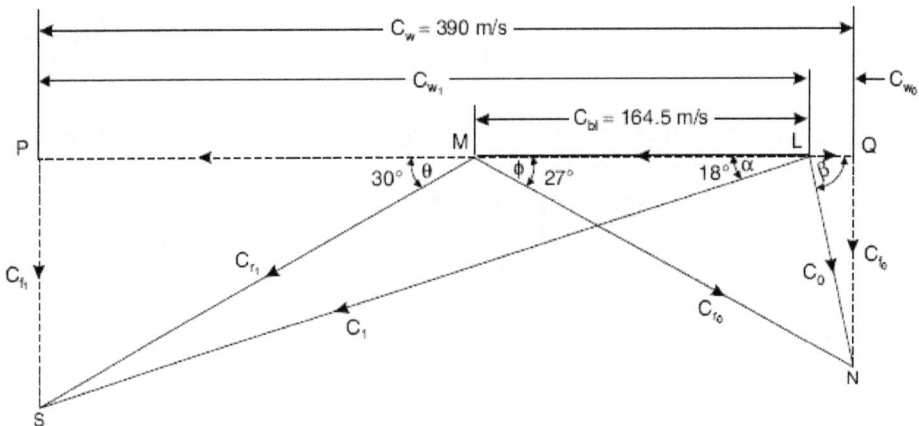

FIGURE 10.19. Sketch for Example 9.

$$\alpha = 18°, \text{complete } \Delta LMS$$

$$\theta = 30° \text{ (on measurement)}$$

$$\therefore \qquad \phi = 30° - 3 = 27°.$$

Now complete the ΔLMN by taking $\phi = 27°$ and $C_{r_0} = 0.84 \, C_{r_1}$.

Finally complete the whole diagram as shown in Figure 10.19.

(i) Tangential thrust on the blades:

Tangential thrust $= \dot{m}_s (C_{w_1} + C_{w_0}) = 10 \times 390 = \mathbf{3900 \, N. \, (Ans.)}$

(ii) Axial thrust:

Axial thrust $= \dot{m}_s (C_{f_1} - C_{f_0}) = 10(120 - 95) = \mathbf{250N. \, (Ans.)}$

(iii) Resultant thrust:

Resultant thrust $= \sqrt{(3900)^2 + (250)^2} = \mathbf{3908 \, N. \, (Ans.)}$

(iv) Power developed, P:

$$P = \frac{\dot{m}_s \left(C_{w_1} + C_{w_0} \right) \times C_{bl}}{1000} = \frac{10 \times 390 \times 164.5}{1000} = \mathbf{641.55 \, kW. \, (Ans.)}$$
0

(v) Blading efficiency, η_{bl}:

$$\eta_{bl} = \frac{2C_{bl} \left(C_{w_1} + C_{w_0} \right)}{C_1^2} = \frac{2 \times 164.5 \times 390}{392^2} = \mathbf{83.5\%. \, (Ans.)}$$

Example 10

In a stage of an impulse reaction turbine provided with a single row wheel, the mean diameter of the blades is 1 m. It runs at 3000 r.p.m. The steam issues from the nozzle at a velocity of 350 m/s and the nozzle angle is 20°. The rotor blades are equiangular. The blade friction factor is 0.86. Determine the power developed if the axial thrust on the end bearing of a rotor is 118 N.

■ Solution:

Mean diameter of the blades, $\qquad D = 1 \, m$

Speed of the turbine, $\qquad\qquad\qquad N = 3000 \, r.p.m.$

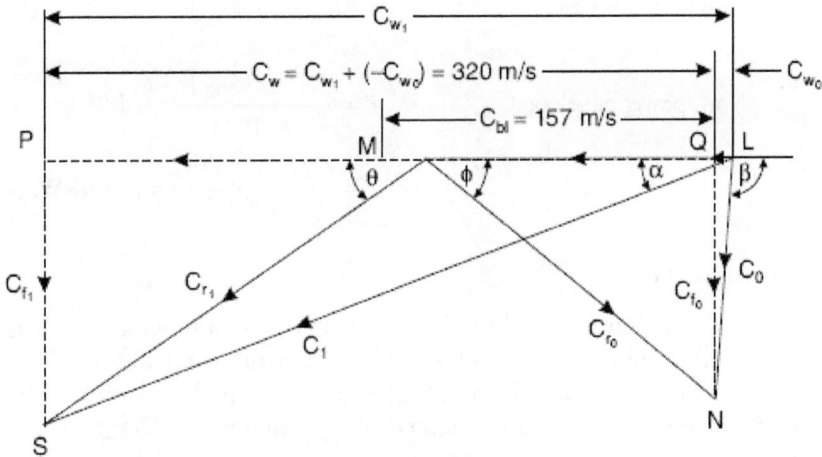

FIGURE 10.20. Sketch for Example 10.

Velocity of steam issuing from the nozzle, $C_1 = 350$ m/s

Nozzle angle, $\qquad\qquad\qquad\qquad \alpha = 20°$

Blade angles, $\qquad\qquad\qquad\qquad \theta = \phi$

Blade friction factor, $\qquad\qquad\qquad K = 0.86$

Axial thrust $\qquad\qquad\qquad\qquad = 118\,N$

Power developed, P:

Blade, velocity, $\qquad\qquad\qquad C_{bl} = \dfrac{\pi DN}{60} = \dfrac{\pi \times 1 \times 3000}{60} = 157$ m/s

— With the data, $C_{bl} = 157$ m / s, $C_1 = 350$ m / s, $\alpha = 20°$, draw the ΔLMS (Fig. 21).

By measurement, $\qquad\qquad\qquad \theta = 35°$

Since the blades are equiangular, $\quad \theta = \phi = 35°$

— Now with $\qquad\qquad\qquad \phi = 35°$ and $C_{r_0} = 0.86\,C_{r_1}$, complete the ΔLMN.

On measurement ; $\qquad\qquad\qquad C_{f_1} = 120$ m/s, $C_{f_0} = 102.5$ m/s

Also, axial thrust $\qquad \dot{m}_s (C_{f_1} - C_{f_0}) = 118$

$\therefore \qquad\qquad\qquad\qquad\qquad \dot{m}_s = \dfrac{118}{C_{f_1} - C_{f_0}} = \dfrac{118}{(120 - 102.5)} = 6.74$ kg/s

Further in this case, $\quad C_w = C_{w_1} + C_{w_0} = C_{w_1} + * \left(-C_{w_0} \right) = 320 \text{ m/s} \quad (*\beta > 90°)$

Now, power developed, $\qquad P = \dfrac{\dot{m}_s \left(C_{w_1} + C_{w_0} \right) \times C_{bl}}{1000} \text{ kW}$

$$= \dfrac{6.74 \times 320 \times 157}{1000} = \textbf{338.6 kW. (Ans.)}$$

Example 11

A simple impulse turbine has one ring of moving blades running at 150 m/s. The absolute velocity of steam at exit from the stage is 85 m/s at an angle of 80° from the tangential direction. The blade velocity coefficient is 0.82 and the rate of steam flowing through the stage is 2.5 kg/s. If the blades are equiangular, determine:

(i) Blade angles;

(ii) Nozzle angle;

(iii) Absolute velocity of steam issuing from the nozzle;

(iv) Axial thrust.

■ Solution:

Blade velocity, $\quad C_{bl} = 150 \text{ m/s}$

Absolute velocity of steam at exit from the stage, $C_0 = 85$ m/s

Angle, $\qquad\qquad\qquad \beta = 80°$

Blade velocity co-efficient, $K = \dfrac{C_{r_0}}{C_{r_0}} = 0.82$

Rate of steam flowing through the stage, $\dot{m}_s = 2.5$ kg/s

Blades are equiangular, i.e., $\theta = \phi$.

— With the above given data the velocity triangle for *exit* can be drawn to a suitable scale. From that, value $\varphi = \theta$ can be obtained. Also, the value of C_{r_0} can be obtained, which helps to get the value of C_{r_1} with the help of given value of "*K*." With these values being known the inlet velocity triangle of the velocity diagram can be completed to get the value of C_1, the absolute velocity of steam issuing from the nozzle, and the value of axial thrust can

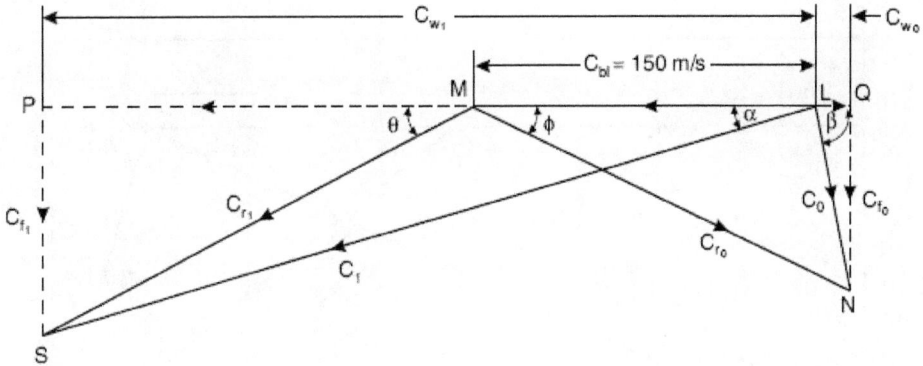

FIGURE 10.21. Sketch for Example 11.

also be calculated. Figure 10.21 gives the velocity diagram of the turbine stage to a suitable scale.

— From the outlet velocity ΔLMN

By measurement, $\quad\quad\quad\quad\quad\quad\quad\quad C_{r_0} = 186$ m/s

$\therefore \quad\quad\quad\quad\quad\quad\quad\quad\quad\quad C_{r_i} = \dfrac{C_{r_0}}{K} = \dfrac{186}{0.82} = 226.8$ m/s

(*i*) **Blades angles θ, φ:**

By measurement $\quad\quad\quad\quad\quad\quad$ θ = φ = blade angles = **27°.** (**Ans.**)

(*ii*) **Nozzle angle, α:**

By measurement ; nozzle angle, $\quad$ α = **16°.** (**Ans.**)

(*iii*) Absolute velocity, C_1:

Absolute velocity of steam issuing from the nozzle,

$C_1 =$ **366 m / s (by measurement). (Ans.)**

(*iv*) Axial thrust:

Also, $\quad\quad\quad\quad\quad\quad\quad\quad\quad \left.\begin{array}{l} C_{f_0} = 84 \text{ m/s} \\ C_{f_1} = 102 \text{m/s} \end{array}\right]$ By measurement.

$\therefore \quad$ Axial thrust $\quad\quad\quad\quad = \dot{m}_s \left(C_{f_1} - C_{f_0} \right)$

$\quad\quad\quad\quad\quad\quad\quad\quad\quad\quad = 2.5 (102 - 84) = $ **45 N. (Ans.)**

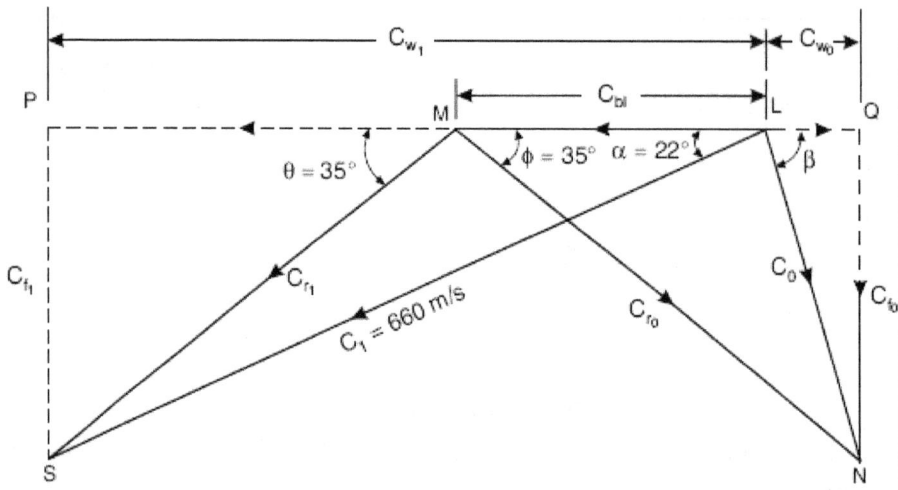

FIGURE 10.22. Sketch for Example 12.

Example 12

One stage of an impulse turbine consists of a converging nozzle ring and one ring of moving blades. The nozzles are inclined at 22° to the blades whose tip angles are both 35°. If the velocity of the steam at exit from the nozzle is 660 m/s, find the blade speed so that the steam passes on without shock. Find the diagram efficiency neglecting losses if the blades are run at this speed.

■ **Solution:**

Given: $\alpha = 22°; \theta = \phi = 35°; C_1 = 660$ m/s.

In case of an impulse turbine, *maximum blade efficiency*,

$$(\eta_{bl})_{max} = \frac{\cos^2 \alpha}{2}(1 + KZ)$$

where K (= blade velocity co-efficient) = 1, (∵ Losses are neglected)

$$Z = \frac{\cos \phi}{\cos \theta} = 1$$ (∵ Blades are *equiangular*)

$$\therefore \quad \eta_{bl} = \frac{\cos^2 \alpha}{2}(1+1) = \cos^2 \alpha = (\cos 22°)^2 = 0.86 \text{ or } 86\%. \text{ (Ans.)}$$

Also,
$$\rho_{opt.} = \frac{\cos \alpha}{2}$$

or
$$\frac{C_{bl}}{C_1} = \frac{\cos 22°}{2} = 0.4636$$

$$\therefore \quad C_{bl} = C_1 \times 0.4636 = 660 \times 0.4636 \simeq 306 \text{ m/s.} \quad \text{(Ans.)}$$

Example 13

In a single-stage impulse turbine the mean diameter of the blade ring is 1 meter and the rotational speed is 3000 r.p.m. The steam is issued from the nozzle at 300 m/s and the nozzle angle is 20°. The blades are equiangular. If the friction loss in the blade channel is 19% of the kinetic energy corresponding to the relative velocity at the inlet to the blades, what is the power developed in the blading when the axial thrust on the blades is 98 N?

■ Solution:

Mean diameter of the bladering D, = 1 m

Speed of the turbine, $\quad\quad\quad\quad N = 3000$ r.p.m.

Absolute velocity of steam issuing from the nozzle, $\quad C_1 = 300$ m/s

Nozzleangle, $\quad\quad\quad\quad\quad\quad\quad\quad \alpha = 20°$

Blade angles are equiangular, $\quad\quad\quad \theta = \phi$

Friction loss in the blade channel $\quad = 19\%$

i.e., $\quad\quad\quad\quad\quad C_{r_0} = (1 - 0.19)\,C_{r_1} = 0.81\,C_{r_1}$

Axial thrust on the blades $\quad\quad\quad\quad = 98$ N

Power developed, P:

Blade speed, $\quad\quad\quad C_{bl} = \frac{\pi D N}{60} = \frac{\pi \times 1 \times 3000}{60} = 157.1$ m/s

Also $\quad\quad\quad\quad\quad\quad\quad \theta = \phi \text{ (given)}$

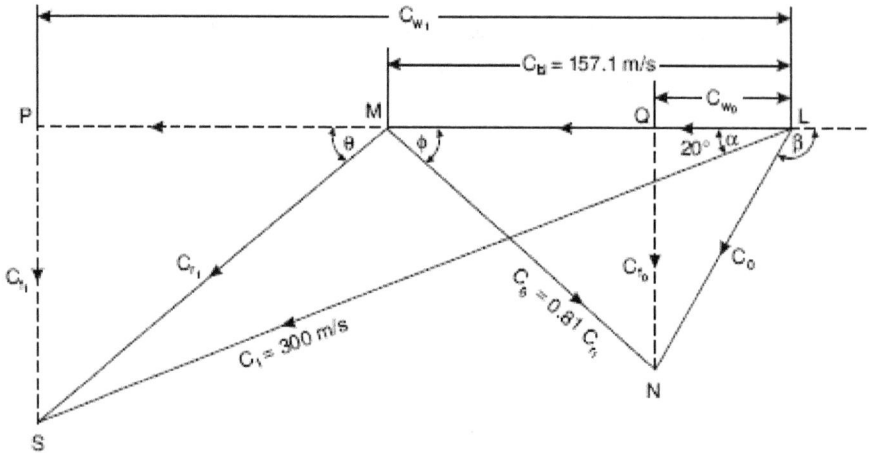

FIGURE 10.23. Sketch for Example 13.

Now, the velocity diagram is drawn to a suitable scale as shown in Figure 10.23.

By measurement (from diagram);

$C_{w_1} = 283.5$ m/s ; $C_{w_0} = 54$ m/s

$C_{f_1} = 100.5$ m/s

$C_{f_0} = 81$ m/s.

Axial thrust $\qquad\qquad\qquad = \dot{m}_s \left(C_{f_1} - C_{f_0} \right)$

$98 = \dot{m}_s \left(100.5 - 81 \right)$ or $\dot{m}_s = \dfrac{98}{100.5 - 81} = 5.025$ kg

Power developed, $p = \dfrac{\dot{m}_s [C_{w_1} + {}^*(-C_{w_0})]}{1000} C_{bl}$ $\qquad$ $({}^*\beta > 90°)$

$= \dfrac{5.025 \left(283.5 - 54 \right) \times 157.1}{1000} = \mathbf{181.2\ kW.\ (Ans.)}$

Example 14

Show that the maximum possible efficiency of a De Laval steam turbine is 88.3% when the nozzle angle is 20°. Deduce the formula used.

■ **Solution:**

Maximum possible efficiency, $\eta_{max} = 88.3\%$

Nozzle angle, $\alpha = 20°$

Maximum possible efficiency of a De Laval turbine (impulse turbine)
$\eta = \cos^2 \alpha$ where α is the nozzle angle.

$$\therefore \quad \eta_{max} = \cos^2 20° = (0.9396)^2 = 0.883 = \textbf{88.3\%} \quad \textbf{(Proved).}$$

Example 15

In a simple impulse turbine, the nozzles are inclined at 20° to the direction of motion of the moving blades. The steam leaves the nozzle at 375 m/s. The blade velocity is 165 m/s. Calculate suitable inlet and outlet angles for the blades in order that the axial thrust is zero. The relative velocity of steam as it flows over the blades is reduced by 15% by friction. Also, determine the power developed for a flow rate of 10 kg/s.

■ **Solution:**

Nozzles, $\quad\quad\quad\quad\quad\quad\quad\quad \alpha = 20°$

Velocity of steam issuing from the nozzles, $C_1 = 375 \text{ m/s}$

Blade speed $\quad\quad\quad\quad\quad\quad\quad C_{bl} = 165 \text{ m/s}$

Axial thrust $\quad\quad\quad\quad\quad\quad = \text{zero } i.e., C_{f_1} = C_{f_0}$

$\dfrac{C_{r_0}}{C_{r_1}} = (1 - 0.15) = 0.85$, *i.e.*, 15% loss due to friction, steam flow rate, $\dot{m}_s = 10 \text{ kg/s}$.

Inlet and outlet angles:

With the above given data, draw a velocity diagram to a suitable scale as shown in Figure 10.24.

By measurement (from velocity diagram),

$$\left.\begin{array}{l} \theta = 35° \\ \phi = 42° \\ \beta = 100° \end{array}\right\} . \text{(Ans.)}$$

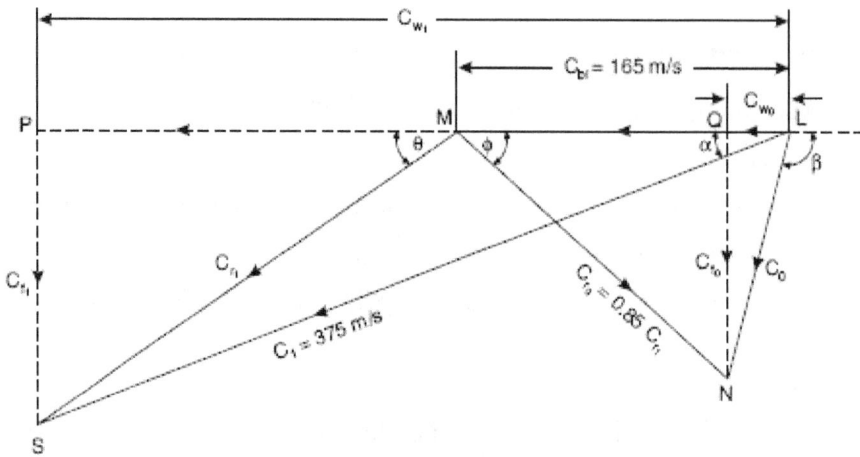

FIGURE 10.24. Sketch for Example 15.

Power developed, P:

Also, $\qquad C_{w_1} = 354$ m/s ; $C_{w_0} = 24$ m/s $\qquad\qquad$ (By measurement)

∴ Power developed $\qquad P = \dfrac{\dot{m}_s \left[C_{w_1} + C_{w_0} \right] \times C_{bl}}{1000}$

$$= \frac{\dot{m}_s \left[C_{w_1} + (-C_{w_0}) \right] \times C_{bl}}{1000} = \frac{10 \left[354 + (-24) \right] \times 165}{1000} = \textbf{544.5 kW. \ (Ans.)}$$

Example 16

In a single-stage impulse turbine, the nozzle angle is 20° and the blade angles are equal. The velocity coefficient for the blade is 0.85. Find the maximum blade efficiency possible. If the actual blade efficiency is 92% of the maximum blade efficiency, find the possible ratio of blade speed to steam speed.

■ **Solution:**

Nozzle angle, $\qquad \alpha = 20°$

Blade angles are equal *i.e.*, $\qquad \theta = \phi$

Blade velocity co-efficient, $\qquad K = 0.85 \left(= \dfrac{C_{r_0}}{C_{r_1}} \right)$

Actual blade efficiency = 92% of maximum blade efficiency

Ratio of blade speed to steam speed, $\rho = \dfrac{C_{bl}}{C_1}$

Maximum blade efficiency is given by:

$$(\eta_{bl})_{max} = \frac{\cos^2 \alpha}{2}(1+KZ)$$

$$= \frac{\cos^2 \alpha}{2}(1+K) \text{ as } Z = \frac{\cos \phi}{\cos \theta} = 1$$

$$(\eta_{bl})_{max} = \frac{\cos^2 20°}{2}(1+0.85) = 0.816 \text{ or } 81.6\%$$

The actual efficiency of the turbine

$$= 0.92 \times 0.816 = 0.75$$

The blade efficiency of a single-stage impulse turbine is given by the relation,

$$\eta_{bl} = 2(1+K)(\rho \times \cos \alpha - \rho^2)$$
$$0.75 = 2(1+0.85)(\rho \times \cos 20° - \rho^2)$$
$$0.75 = 2 \times 1.85(0.94\rho - \rho^2)$$
$$0.203 = 0.94 \rho - \rho^2$$
$$\rho^2 - 0.94 \rho + 0.203 = 0$$

$$\therefore \quad \rho = \frac{0.94 \pm \sqrt{(0.94)^2 - 4 \times 0.203}}{2} = \frac{0.94 \pm 0.267}{2} \text{ or } \rho = 0.603 \text{ or } 0.336$$

Hence possible ratio, $\rho = $ **0.603 or 0.336. (Ans.)**

Example 17

The following data refer to a single-stage impulse turbine:

Isentropic nozzle heat drop =251kJ/kg; nozzle efficiency =90%; nozzle angle =20°; ratio of blade speed to whirl component of steam speed =0.5; blade velocity coefficient =0.9; the velocity of steam entering the nozzle =20 m/s.

Determine: (i) The blade angles at inlet and outlet if the steam enters into the blades without shock and leaves the blades in an axial direction.

(ii) Blade efficiency.

(iii) Power developed and axial thrust if the steam flow is 8 kg/s.

■ **Solution:**

Isentropic heat drop = 251 kJ / kg

Nozzle efficiency, η_{nozzle} = 90%

Nozzle angle, $\alpha = 20°$

Ratio of blade speed to whirl component of steam speed

= 0.5 Blade velocity co-efficient, $K = 0.9$

Velocity of steam entering the nozzle = 20 m / s

(i) Blade angles:

Nozzle efficiency is given by : $\eta_{nozzle} = \dfrac{\text{Useful heat drop}}{\text{Isentropic heat drop}}$

or $0.9 = \dfrac{\text{Useful heat drop}}{251}$

∴ Useful heat drop = $0.9 \times 251 = 225.9$ kJ/kg

Applying the energy equation to the nozzle, we get

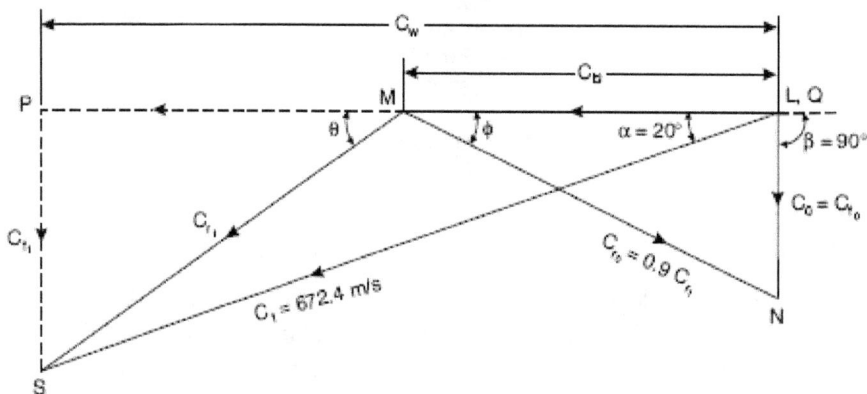

FIGURE 10.25. Sketch for Example 17.

$$\frac{C_1^2 - 20^2}{2} = 225.9 \times 1000$$

$\therefore$ $\qquad\qquad\qquad\qquad C_1^2 = 2 \times 225.9 \times 1000 + 400 = 452200$

i.e., $\qquad\qquad\qquad\qquad\qquad C_1 = 672.4$ m/s

Other data given : $\qquad\qquad \alpha = 20°, \dfrac{C_{bl}}{C_w} = 0.5, \ K = \dfrac{C_{r_0}}{C_{r_1}} = 0.9$

and $\qquad\qquad\qquad\qquad C_{w_0} = 0$, *as the steam leaves the blades axially.*

$\therefore$ $\qquad\qquad\qquad\qquad\qquad C_0 = C_{f_0}$

With the above data construct the velocity triangles as follows:

- Select a suitable scale, say 1 cm = 50 m / s.

- Draw a horizontal line through a point L and angle $\alpha = 20°$. Mark the point along LS as

$$LS = C_1 = 672.4 \text{ m} = \frac{672.4}{50} = 13.5 \text{ cm.}$$

- Draw a line through S which is perpendicular to the horizontal line through L and it cuts at the point P. Measure the distance $LP = 12.7$cm.

$\therefore$ $\qquad\qquad\qquad\qquad C_{w_1} = C_w = 12.7$ cm.

and $\qquad\qquad\qquad\qquad C_{bl} = 0.5 \ C_w = 0.5 \times 12.7 = 6.35$ cm.

- Mark the point M as $LM = C_{bl} = 6.35$ cm

- Join point MS and complete the inlet velocity triangle LMS.

- Measure $MS \ (C_{r_1}) = 7.7$ cm. $\qquad\qquad\qquad C_{r_0} = 0.9 \times 7.7 = 6.93$ cm.

- Draw a perpendicular line through point L to the line LM. From M cut an arc of radius 6.93 cm to cut the vertical line through L and mark the point N and join MN which completes the outlet triangle LMN.

Now find out velocities converting lengths into velocities:

$C_1 = 672.4 \ \text{m/s}$

$C_w = 12.7 \times 50 = 635 \ \text{m/s}$

$C_{bl} = \ 0.5 \ C_w = 0.5 \times 635 = 317.5 \ \text{m/s}$

$C_{r_1} = 7.7 \ \text{cm} = 7.7 \times 50 = 385 \ \text{m/s}$

$C_{r_0} = 0.9 \ C_{r_1} = 0.9 \times 385 = 346.5 \ \text{m/s}$

$C_{f_1} = 4.45 \ \text{cm} = 4.45 \times 50 = 222.5 \ \text{m/s}$

$C_{f_0} = 2.6 \ \text{cm} = 2.6 \times 50 = 130 \ \text{m/s}.$

Blade angles measured from the diagram:

$$\theta = 35°, \phi = 22°. \ \ (\text{Ans.})$$

(ii) Blade efficiency, η_{bl} :

$$\eta_{bl} = \frac{2C_{bl}C_w}{C_1^2} = \frac{2 \times 317.5 \times 635}{(672.4)^2} = \textbf{0.89 or 89\%. \ (Ans.)}$$

(iii) Power developed, P and axial thrust:

$$P = \frac{\dot{m}_s \left(C_{w_1} + C_{w_0}\right) \times C_{bl}}{1000} = \frac{8 \times (635 + 0) \times 317.5}{1000} = \textbf{1612.9 kW.} \ \left(\textbf{Ans.}\right)$$

Axial thrust $= \dot{m}_s \left(C_{f_1} - C_{f_0}\right) = 8(222.5 - 130) = \textbf{740 N.} \ \left(\textbf{Ans.}\right)$

Example 18

In a single-stage steam turbine saturated steam at 10 bar abs. is supplied through a convergent-divergent steam nozzle. The nozzle angle is 20° and the mean blade speed is 400 m/s. The steam pressure leaving the nozzle is 1 bar abs. Find:

(i) The best blade angles if blades are equiangular.

(ii) The maximum power developed by the turbine if the number of nozzles used are 5 and area at the throat of each nozzle is 0.6 cm2.

Assume nozzle efficiency 88% and blade friction coefficient of 0.87.

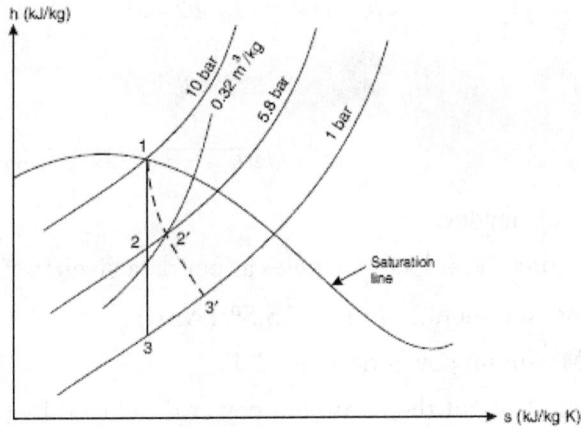

FIGURE 10.26. Sketch for Example 18, h-S diagram.

■ **Solution:**

Supply steam pressure (to nozzles) $= 10$ bar abs.

Nozzle angle,	$\alpha = 20°$
Mean blade speed,	$C_{bl} = 400$ m / s

Steam pressure leaving the nozzle $= 1$ bar abs.

Number of nozzles used	$= 5$
Area of throat at each nozzle	$= 0.6$ cm^2

Nozzle efficiency, $\eta_{nozzle} = 88\%$

Blade friction co-efficient, $K = \dfrac{C_{r_0}}{C_{r_1}} = 0.87.$

The velocity of steam at the outlet of the nozzle is found representing the expansion through the nozzle on the *h-s* chart as shown in Figure 10.26.

From *h-s* chart,

$h_1 - h_3 \approx 402$ kJ/kg

$\eta_{nozzle} = \dfrac{h_1 - h_3{'}}{h_1 - h_3} = 0.88$

$$\therefore \qquad h_1 - h_3' = 0.88 \times 402 = 353.76 \text{ kJ/kg}$$

Also

$$\frac{C_3'^2}{2} = h_1 - h_3'$$

or

$$C_3' = \sqrt{2(h_1 - h_3')} = \sqrt{2 \times 353.76 \times 1000} = 841.14 \text{ m/s}$$

(*i*) Blade angles:

Construct the velocity triangles as per data given in Figure 10.27.

By measurement, $\theta\,(= \phi) = \textbf{35.5°. (Ans.)}$

(*ii*) Maximum power developed, P:

For finding out the maximum power developed by the turbine, let us first find out the maximum mass of steam passing through the nozzle.

The required condition for the maximum mass flow through the nozzle is given by

$$\frac{P_2}{P_1} = \left(\frac{2}{n+1}\right)^{\frac{n}{n-1}}$$

where, p_1 = Pressure of steam at inlet of the nozzle,

p_2 = Pressure of steam at the throat of the nozzle, and

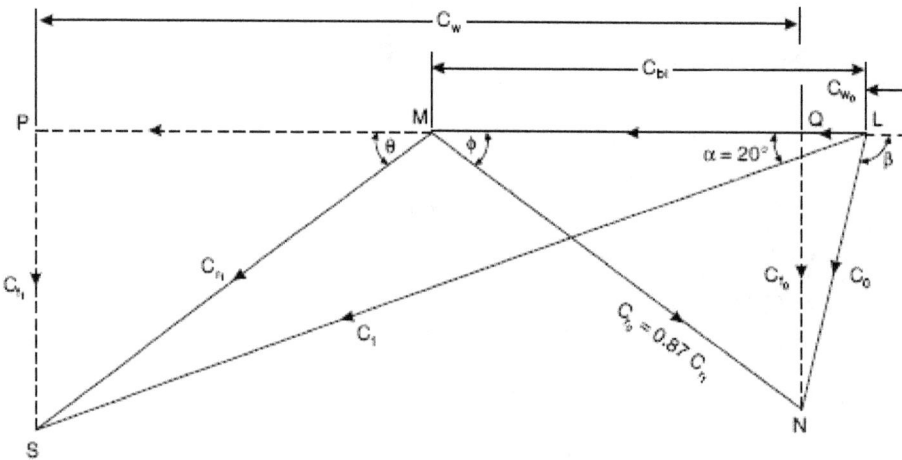

FIGURE 10.27. Sketch for Example 18.

n (index of expansion) = 1.135 as steam is saturated.

$$\therefore \qquad \frac{p_2}{p_1} = \left(\frac{2}{1.135+1}\right)^{\frac{1.135}{0.135}} = 0.58$$

$$\therefore \qquad p_2 = 10 \times 0.58 = 5.8 \text{ bar}$$

From h-s chart (Figure 27)

$$h_1 - h_2 \simeq 105 \text{ kJ/kg}$$

and $\qquad\qquad\qquad h_1 - h_2' = 0.88 \times 105 = 92.5 \text{ kJ/kg}$

$v_2' \bullet$ (specific volume at point 2') $\simeq 0.32 \text{ m}^3/\text{kg}$

The maximum velocity of steam at the throat of the nozzle is given by

$$C = \sqrt{2(h_1 - h_2')} = \sqrt{2 \times 92.4 \times 1000} = 429.88 \text{ m/s}$$

Using the continuity equation at the throat of the nozzle, we can write

$m . v_2' = A \times C$ where A is the area of the nozzle.

$$\therefore \qquad\qquad m \times 0.32 = 0.6 \times 10^{-4} \times 429.88$$

$$\therefore \qquad\qquad m = \frac{0.6 \times 10^{-4} \times 429.88}{0.32} = 0.0806 \text{ kg/s}.$$

Total mass of steam passing through 5 nozzles per second is given by

$m_t = 0.0806 \times 5 = 0.403 \text{ kg/s}$

From velocity diagram, $C_w = 750 \text{ m/s}$ (by measurement)

$$\therefore \quad Power \ developed = \frac{0.403 \times 750 \times 400}{1000} = \mathbf{120.9 \, kW.} \quad \textbf{(Ans.)}$$

Example 19

The first stage of an impulse turbine is compounded for velocity and has two rows of moving blades and one ring of fixed blades. The nozzle angle is 15° and the leaving angles of blades are first-moving 30°, fixed 20°, and second-moving 30°. The velocity of steam leaving the nozzle is 540 m/s. The friction loss in each blade row is 10% of the relative velocity. Steam leaves the second row of moving blades axially.

Find : (i) Blade velocity ;   (ii) Blade efficiency ;

(iii) Specific steam consumption.

■ **Solution:**

Refer to Figure 10.28.

Nozzle angle, $\alpha = 15°\,;\,\alpha' = 20°$

$\beta' = 90°$ [since the steam leaves the blades *axially*]

$\phi = \phi' = 30°$

Velocity of steam leaving the nozzle, C_1 = 540 m/s and $\dfrac{C_{r_0}}{C_{r_1}} = 0.9$

$\dfrac{C_1'}{C_0} = 0.9$ and $\dfrac{C_{r_0}'}{C_{r_1}'} = 0.9$.

Second row of moving blades:

The velocity triangles should be drawn *starting from the second row of moving blades.*

The procedure is as follows: Refer to Figure 10.28.

Draw *LM* to any convenient scale (say 3 cm) as C_{bl} is *not* known.

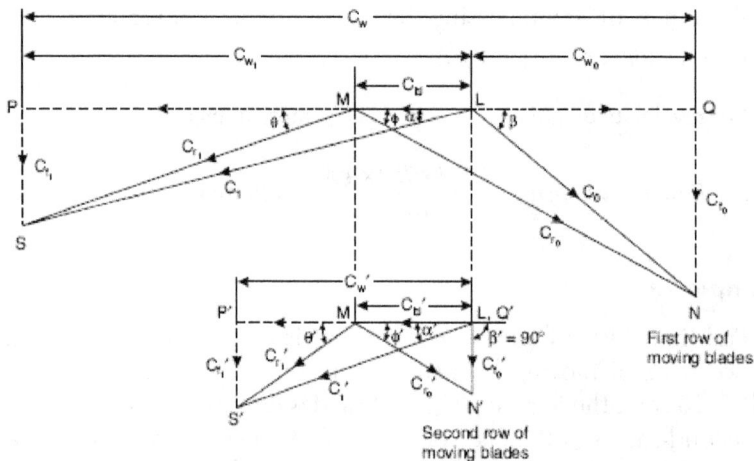

FIGURE 10.28. Sketch for Example 19.

Draw $\phi' = 30°$ and draw perpendicular through the point L (or Q') to LM as $\beta' = 90°$

This meets the line MN' at N'. This completes the outlet triangle LMN'.

Measure $C_{r_0}' = MN' = 3.5$ cm

$$C_{r_1}' = \frac{C_{r_0}'}{0.9} = \frac{3.5}{0.9} = 3.9 \text{ cm}.$$

Draw an $\angle\alpha' = 20°$ and draw an arc of radius 3.9 cm with center at M to cut the line LS' at S'. Join MS'. This completes the inlet velocity $\Delta LMS'$. Measure $LS' = C_1' = 6.5$ cm .

First row of moving blades:

The following steps are involved in drawing velocity triangle for the *first row of moving blades.*

Draw $LM = C_{bl} (= 3 \text{ cm})$

$$C_0 = \frac{C_1'}{0.9} = \frac{6.5}{0.9} = 7.22 \text{ cm}.$$

Draw $\angle\phi = 30°$,through the point M to the line LM.

Draw an arc of radius 7.22 cm with center L. This arc cuts the line LN at point N. Join MN. This completes the outlet ΔLMN.

Measure $\qquad\qquad C_{r_0} = MN = 9.7$ cm

$$MS = C_{r_1} = \frac{C_{r_0}}{0.9} = \frac{9.7}{0.9} = 10.8 \text{ cm}$$

— Draw an $\angle\alpha = 15°$. through a point L. Draw an arc of radius of 10.8 cm with the center at M. This arc cuts the line LS at S. Join MS. This completes the inlet velocity triangle.

Measure LS from the velocity triangle

$$LS = 13.8 \text{ cm} = C_1 = 540 \text{ m/s}.$$

The scale is now calculated from the previous steps.

$$\therefore \qquad \text{Scale 1 cm} = \frac{540}{13.8} = 39.1 \text{ m/s}$$

(*i*) Blade velocity, C_{bl}:

Measure the following distances from the velocity diagram and convert into velocities:

$$C_{bl} = LM = 3 \text{ cm} = 3 \times 39.1 = \textbf{117.3 m / s.} \quad \textbf{(Ans.)}$$

(*ii*) Blade efficiency, η_{bl} (*ii*) Blade velocity, η_{bl}:

$$C_w = PQ = 18.8 \text{ cm} = 18.8 \times 39.1 = 735.1 \text{ m/s}$$

$$C_w' = P'Q' = 6.2 \text{ cm} = 6.2 \times 39.1 = 242.4 \text{ m/s}$$

$$= \frac{2C_{bl}(C_w + C_w')}{C_1^2}$$

$$= \frac{2 \times 117.3(735\ 1 + 242.4)}{(540)^2} = \textbf{0.786 or 78.6\%. (Ans.)}$$

(*iii*) Specific steam consumption, m_s:

$$1 = \frac{m_s(C_w + C_w')C_{bl}}{3600 \times 1000} = \frac{m_s(735.1 + 242.4) \times 117.3}{3600 \times 1000}$$

$$\therefore \qquad m_s = \frac{3600 \times 1000}{(735.1 + 242.4) \times 117.3} = \textbf{31.39 kg / kWh. (Ans.)}$$

Example 20

The following particulars relate to a two-row velocity compounded impulse wheel:

Steam velocity at nozzle outlet	= 650 m / s
Mean blade velocity	= 125 m / s
Nozzle outlet angle	= 16°
Outlet angle, first row of moving blades	= 18°
Outlet angle, fixed guide blades	= 22°
Outlet angle, second row of moving blades	= 36°
Steam flow	= 2.5 kg / s

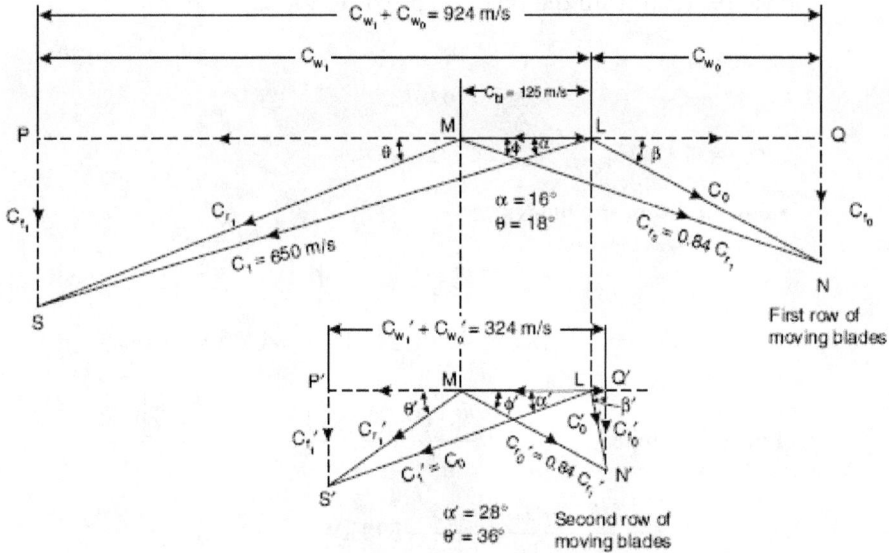

FIGURE 10.29. Sketch for Example 20.

The ratio of the relative velocity at outlet to that at inlet is 0.84 for all blades. Determine the following:

(i) The axial thrust on the blades

(ii) The power developed

(iii) The efficiency of the wheel

■ **Solution:**

With given data, $C_{bl} = 125 \text{ m/s}$, $C_1 = 650 \text{ m/s}$, $\alpha = 16°$,

The first row inlet velocity diagram is drawn.

Now with given, $\qquad\qquad\qquad C_{r_0} = 0.84 C_{r_1}$, $\phi = 18°$,

the first row exit diagram is drawn.

With $\qquad\qquad\qquad\qquad C_1' = C_0$; $\alpha' = 22°$,

the second row inlet velocity diagram is drawn.

With $\qquad\qquad\qquad\qquad C_{r_0}' = 0.84\, C_{r_1}$; $\phi' = 36°$,

the second row exit diagram is drawn.

The values read from the diagram are as follows:

$C_{f_1} = 180\,\text{m/s},$ $\qquad\qquad$ $C_{f_0} = 138\,\text{m/s},$

$C_{f_1}' = 122\,\text{m/s},$ $\qquad\qquad$ $C_{f_0}' = 107\,\text{m/s},$

$C_{w_1} + C_{w_0} = 924\,\text{m/s},$ $\qquad$ $C_{w_1}' + C_{w_0}' = 324\,\text{m/s}.$

(*i*)**Axial thrust on the blades**

$$= \dot{m}_s\left[\left(C_{f_1} - C_{f_0}\right) + \left(C_{f_1}' - C_{f_0}'\right)\right]$$

$$= 2.5\left[(180 - 138) + (122 - 107)\right] = \mathbf{1425\,N.} \quad \textbf{(Ans.)}$$

(ii) **Power developed**
$$= \frac{\dot{m}_s\left[\left(C_{w_1} + C_{w_0}'\right) + \left(C_{w_1}' + C_{w_0}'\right)\right]C_{bl}}{1000}$$

$$= \frac{2.5(924 + 324)\times 125}{1000} = \mathbf{390\,kW.} \quad \textbf{(Ans.)}$$

(*iii*) **Efficiency**
$$= \frac{(924 + 324)\times 125}{C_1^2/2} = \frac{(924 + 324)\times 125}{650^2/2}$$

$$= \mathbf{0.738\ or\ 73.8\%.} \quad \textbf{(Ans.)}$$

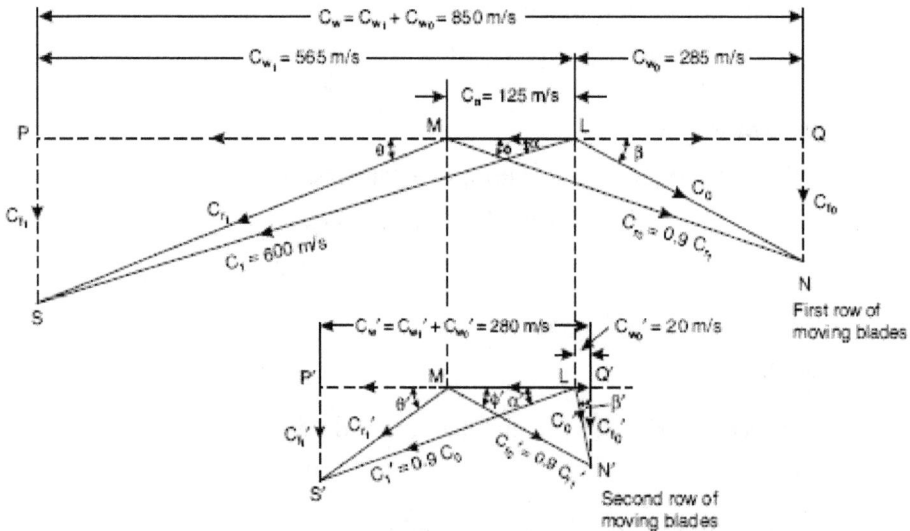

FIGURE 10.30. Sketch for Example 21.

Example 21

The first stage of an impulse turbine is compounded for velocity and has two rings of moving blades and one ring of fixed blades. The nozzle angle is 20° and the leaving angles of the blades are as follows:

First moving 20°, fixed 25° and second moving 30°. Velocity of steam leaving the nozzles is 600 m/sec and the steam velocity relative to the blade is reduced by 10% during the passage through each ring. Find the diagram efficiency and power developed for a steam flow of 4 kg per second. Blade speed may be taken as 125 m/sec.

■ **Solution:**

$C_{bl} = 125 \, \text{m/s}, \; C_1 = 600 \, \text{m/s}$

$$\alpha = 20°, \; \phi = 20°$$

$\alpha' = 25°, \; \phi' = 30°$

$$K = \left(1 - \frac{10}{100}\right) = 0.9$$

$\dot{m}_s = 4 \, \text{kg/s}$

With these values velocity triangles can be drawn.

From diagram (by measurement):

$$C_{w_1} = 565 \, \text{m/s}, \qquad C_{w_0} = 285 \, \text{m/s}$$

$$C_{w_1}' = 260 \, \text{m/s}, \qquad C_{w_0}' = 20 \, \text{m/s}$$

Now, $\quad C_w = C_{w_1} + C_{w_0} = 565 + 285 = 850 \, \text{m/s}$

$C_w' = C_{w_1}' + C_{w_0}' = 260 + 20 = 280 \, \text{m/s}$

Power developed
$$= \frac{\dot{m}_s\left(C_w + C_w'\right)C_{bl}}{1000}$$

$$= \frac{4 \times \left(850 + 280\right) \times 125}{1000} = \textbf{565 kW.} \qquad \textbf{(Ans.)}$$

Diagram efficiency
$$= \frac{C_{bl}\left(C_w + C'_w\right)}{C_1^2 / 2}$$

$$= \frac{2 \times 125 \left(850 + 280\right)}{600^2} = \textbf{0.7847 or 78.47\%. (Ans.)}$$

Example 22

The following data relate to a compound impulse turbine having two rows of moving blades and one row of fixed blades in between them.

The velocity of steam leaving the nozzle = 600 m / s

Blade speed	*= 125 m / s*
Nozzle angle	*= 20°*
First moving blade discharge angle	*= 20°*
First fixed blade discharge angle	*= 25°*
Second moving blade discharge angle	*= 30°*
Friction loss in eachring	*= 10% of relative velocity.*

Find: (i) Diagram efficiency;

(ii) Power developed for a steam flow of 6 kg/s.

■ Solution.

Refer to Figure 10.31.

First row of moving blades:

To draw velocity triangles for the first row of moving blades, the following procedure may be followed:

Select a suitable scale.

Draw LM = blade velocity (C_{bl}) = 125 m/s.

Make $\angle MLS$ = nozzle angle, $\alpha = 20°$.

Draw LS = velocity of steam leaving the nozzle = 600 m/s.

Join MS to complete the inlet triangle LMS.

Make $\angle LMN$ = outlet angle of first moving blades = 20°.

and   cut $MN = 0.9 \, MS$, since $K = 0.9$.

— Join LN to complete the outlet ΔLMN.

Second row of moving blades:

The velocity triangles for the second row of moving blades may be drawn as follows:

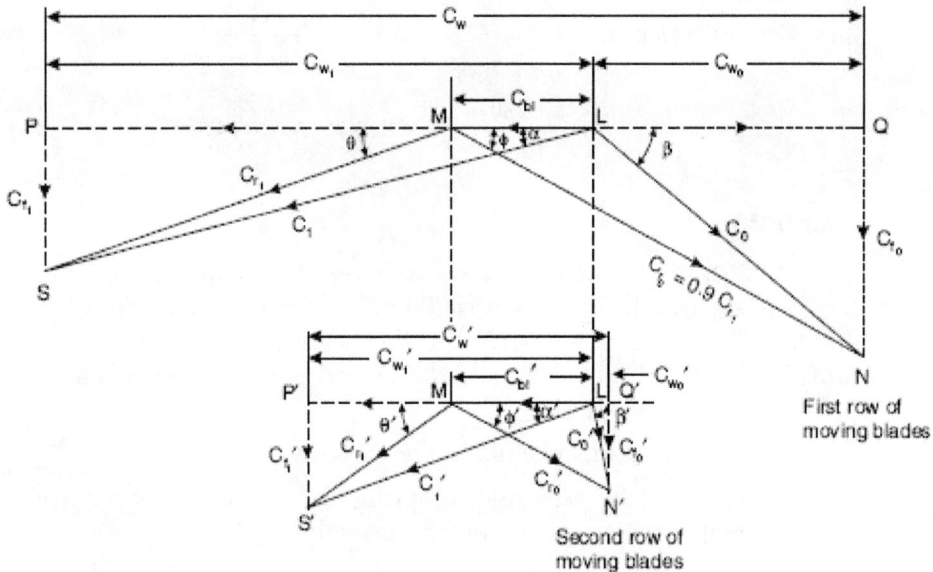

FIGURE 10.31. Sketch for Example 21.

Draw *LM* = blade velocity (C_{bl}) = 125 m/s.

Make $\angle MLS'$ = outlet angle of fixed blade = 25°

and cut *LS'* = 0.9 *LN*. $(\because K=0.9)$

Join *MS'*. The inlet velocity triangle *LMS'* is completed.

Make $\angle LMN'$ = outer angle of second moving blades = 30°

and   cut *MN'* = 0.9 *MS'* $(\because K = 0.9)$

Join *LN'*. The outlet velocity triangle is completed.

The following required data may now be scaled off from the diagram:

$$C_w = C_{w_1} + C_{w_0} = PQ = 845 \text{ m/s}$$

$C_w' = P'Q' = 280 \text{ m/s}.$

(*i*) **Diagram efficiency,** $\eta_{bl} = \dfrac{2C_{bl}\left(C_w + C_w'\right)}{C_1^2}$

$$= \frac{2 \times 125\left(845 + 280\right)}{\left(600\right)^2} = \textbf{0.781 or 78.1\%.}\quad \textbf{(Ans.)}$$

(*ii*) **Power developed,** $\qquad\qquad P = \dfrac{\dot{m}_s\left(C_w + C_w{}'\right)}{1000} C_{bl}$

$$= \dfrac{6\left(845 + 280\right) \times 125}{1000} = \textbf{843.75\,kW.} \quad \textbf{(Ans.)}$$

Example 23

The first stage of a turbine is a two-row velocity compounded impulse wheel. The steam velocity at inlet is 600 m/s and the mean blade velocity is 120 m/s. The nozzle angle is 16° and the exit angles for the first row of moving blades, the fixed blades, and the second row of moving blades are 18°, 21°, and 35° respectively.

(i) Calculate the blade inlet angles for each row.

(ii) Calculate also for each row of moving blades, the driving force and the axial thrust on the wheel for a mass flow of 1 kg/s.

(iii) Calculate the diagram efficiency for the wheel and the diagram power per kg/s steam flow.

(iv) What would be the maximum possible diagram efficiency for the given steam inlet velocity and nozzle angle?

Take the blade velocity coefficient as 0.9 for all blades.

■ Solution:

Refer to Figure 10.32.

$$\alpha = 16°, \ \phi = 18°, \ C_1 = 600 \text{ m/s}, \ C_{bl} = 120 \text{ m/s}$$

$$\alpha' = 21°, \ \phi' = 35°, \ \dot{m}_s = 1 \text{ kg/s},$$

Blade velocity coefficient, $K = 0.9$

With the above data velocity triangles can be drawn.

From the diagram (by measurement)

$$C_w = C_{w_1} + C_{w_0} = 875 \text{ m/s}; \ C_w{}' = C_{w_1}{}' + C_{w_0}{}' = 294 \text{ m/s}$$

$$C_{f_1} = 168 \text{ m/s}, \ C_{f_0} = 135 \text{ m/s}; \ C'_{f_1} = 106 \text{ m/s}, \ C'_{f_0} = 97 \text{ m/s}.$$

(*i*) Blade inlet angles:

First row: $\qquad\qquad\qquad \theta = 20°$ (moving blade)

$$\beta = 24.5° \text{ (fixed blade)}$$

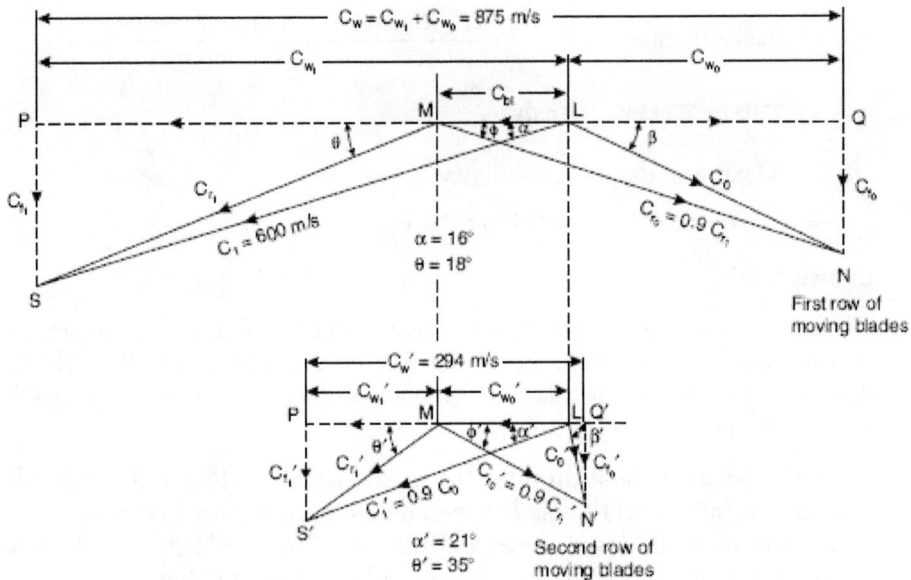

FIGURE 10.32. Sketch for Example 23.

Second row: $\theta' = 34.5°$ (moving blade).

(*ii*) Driving force:

First row of moving blades $= \dot{m}_s\left(C_{w_1} + C_{w_0}\right) = \dot{m}_s C_w = 1 \times 875 = \mathbf{875\,N}$. **(Ans)**

Second row of moving blades $= \dot{m}_s\left(C_{w_1}' + C_{w_0}'\right) = \dot{m}_s C_w' = 1 \times 294 = \mathbf{294\,N}$. **(Ans)**

Axial thrust:

First row of moving blades $= \dot{m}_s\left(C_{f_1}' + C_{f_0}'\right) = 1 \times (168 - 135) = 33\,N$

Second row of moving blades $= \dot{m}_s\left(C_{f_1}' - C_{f_0}'\right) = 1 \times (106 - 97) = 9\,N$

Total axial thrust $= 33 + 9 = \mathbf{42\,N\ per\ kg/s}$. **(Ans.)**

(*iii*) **Power developed** $= \dfrac{\dot{m}_s\left(C_w + C_w'\right) C_{bl}}{1000}$

$= \dfrac{1 \times (875 + 294) \times 120}{1000} = \mathbf{140.28\,kW\ per\ kg/s}$. **(Ans.)**

Diagram efficiency $$= \frac{2C_{bl}(C_w + C_w')}{C_1^2} = \frac{2 \times 120 \,(875+294)}{(600)^2}$$

= 0.7793 or 77.93%. (Ans.)

(iv) Maximum diagram efficiency

$$= \cos^2 \alpha = \cos^2 16° = 0.924 \text{ or } 92.4\%. \qquad \text{Ans}$$

Example 24.

An impulse stage of a turbine has two rows of moving blades separated by fixed blades. The steam leaves the nozzles at an angle of 20° with the direction of motion of the blades. The blade exit angles are: 1st moving 30°; fixed 22°; and 2nd moving 30°.

If the adiabatic heat drop for the nozzle is 186.2 kJ/kg and the nozzle efficiency is 90%, find the blade speed necessary if the final velocity of the steam is to be axial. Assume a loss of 15% in relative velocity for all blade passages. Find also the blade efficiency and the stage efficiency.

■ Solution:

Steam velocity, $C_1 = 44.72 \sqrt{\eta_n h_d} = 44.72 \sqrt{0.9 \times 186.2} = 579$ m/s.

The velocity diagram *for the axial discharge turbine is drawn in reverse direction* (see Figure 10.33).

The blade velocity (C_{bl}) *LM* is drawn to any convenient scale.

As discharge is axial *LN'* is drawn perpendicular to *LM*.

Knowing the outlet angle of the second moving ring $(\phi' = 30°)$, *N'* is located. *MN'* represents relative velocity at outlet $\left(C_{r_0}'\right)$.

Relative velocity at inlet to the second moving blade is

$$C_{r_0}' = MS' = \frac{MN'}{0.85} \left(\frac{C_{r_0}'}{K}\right)$$

The triangle at inlet to the second moving blades ring *LMS'* is obtained by drawing the discharge angle $\angle MLS'$ $(\alpha' = 22°)$, where *LS'* (C_1') is the exit velocity of the second blade ring.

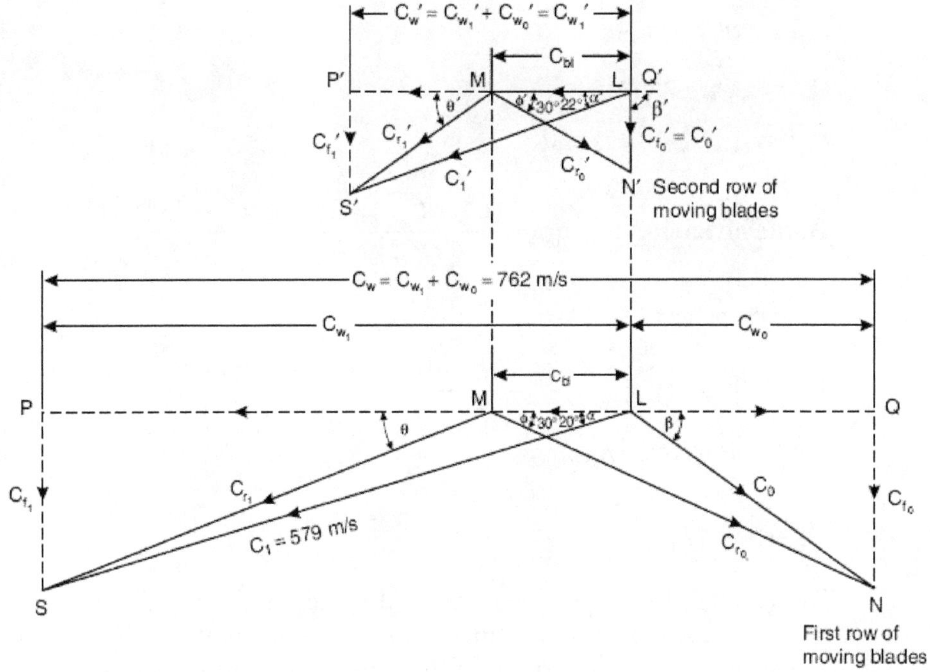

FIGURE 10.33. Sketch for Example 24.

Now, $C_0 = LN = \dfrac{C_1'}{0.85} = \dfrac{LS'}{0.85}$ assuming a loss of 15% (given).

With compass at center L and radius LN arc is drawn and the velocity triangle at the exit of the first moving blade ring LMN is completed, knowing the exit angle of the first moving blade ring ($\phi = 30°$)

With $\alpha = 20°$ and $C_{r_1} = MS = \dfrac{C_{r_0}}{0.85} = \dfrac{MN}{0.85}$ the inlet velocity triangle is completed.

Now LS is the absolute velocity from the nozzle. Since this velocity is known, scale can be calculated. It is

$$= \frac{\text{Absolute velocity at exit from the nozzle}}{\text{Length } LS}$$

and then the blade velocity, and so forth, can be calculated.

From the diagram:

$C_{bl}(= LM) = 117$ m/s

$C_{w_1} + C_{w_0} = 762$ m/s

$C_{w_1}' + C_{w_0}' = C_{w_1}' = 234$ m/s $\hspace{2cm}$ ($\because C_{w_0}' = 0$)

Blade efficiency, $\quad \eta_{bl} = \dfrac{(C_w + C_W')\times C_{bl}}{C_1^2 / 2} = \dfrac{(762 + 234)\times 117}{579^2 / 2}$

$= \mathbf{0.6952}$ **or** $\mathbf{69.52\%.}$ **(Ans.)**

Stage efficiency, $\eta_{\text{stage}} = \dfrac{(C_w + C_w')\times C_{bl}}{h_d} = \dfrac{(762 + 234)\times 117}{186.2 \times 1000}$

$= \mathbf{0.626}$ **or** $\mathbf{62.6\%.}$ **(Ans.)**

10.8. Reaction Turbines

The reaction turbines which are used these days are really impulse-reaction turbines. Pure reaction turbines are *not* in general use. The expansion of steam and heat drop occur both in fixed and moving blades.

Velocity Diagram for Reaction Turbine Blade

Figure 10.34 shows the velocity diagram for a reaction turbine blade. In the case of an impulse turbine blade, the relative velocity of steam either remains constant as the steam glides over the blades or is reduced slightly due

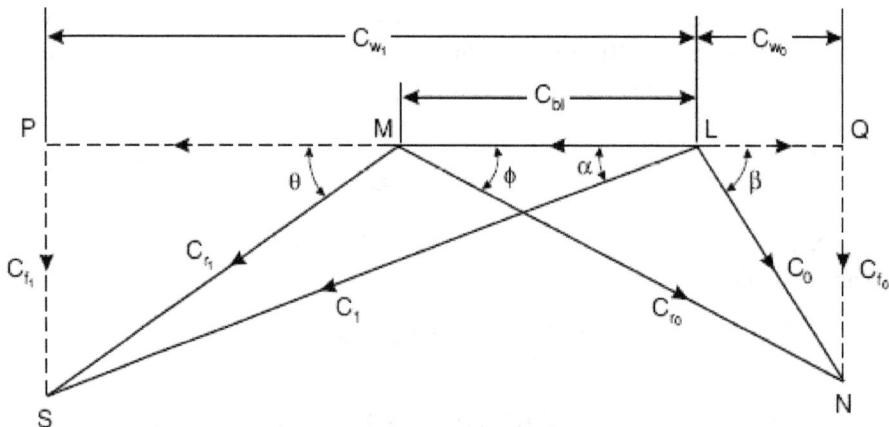

FIGURE 10.34. Velocity diagram for reaction turbine blade.

to friction. In reaction turbine blades, the steam continuously expands at it flows over the blades. The effect of the continuous expansion of steam during the flow over the blade is to increase the relative velocity of the steam.

$\therefore$ $C_{r_0} > C_{r_1}$ for *reaction turbine blade*.

($C_{r_0} \leq C_{r_1}$ for *impulse turbine blade*).

10.8.1. Degree of Reaction (R_d)

The degree of reaction of the reaction turbine stage is defined as the ratio of heat drop over the moving blades to the total heat drop in the stage.

Thus, the degree of reaction of the reaction turbine is given by,

$$R_d = \frac{\text{Heat drop in moving blades}}{\text{Heat drop in the stage}}$$

$$= \frac{\Delta h_m}{\Delta h_f + \Delta h_m} \text{ as shown in Fig. } \textbf{10.35}$$

The heat drop in the moving blades is equal to the increase in relative velocity of steam passing through the blade.

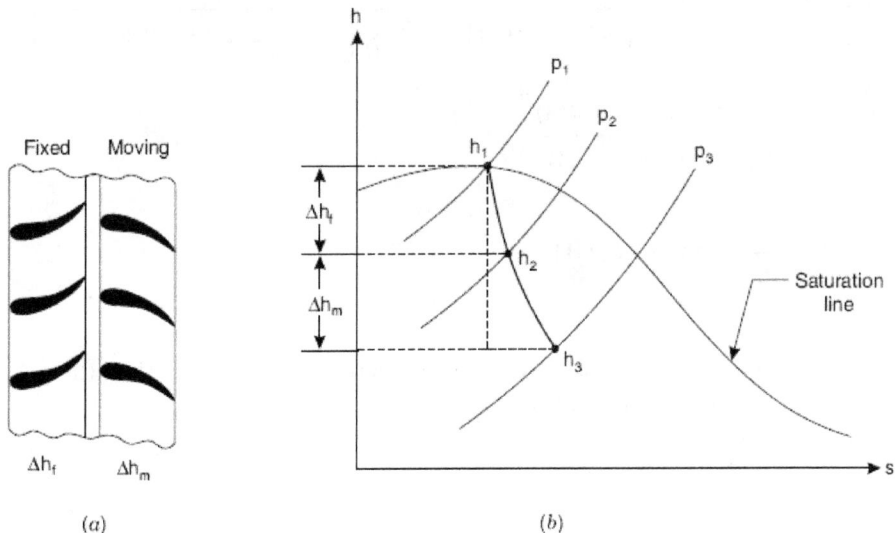

FIGURE 10.35. Blades configuration and h-S diagram.

$$\therefore \qquad \Delta h_m = \frac{C_{r_0}^{\,2} - C_{r_1}^{\,2}}{2}$$

The total heat drop in the stage $\Delta h_f + \Delta h_m$ is equal to the work done by the steam in the stage and it is given by

$$\Delta h_f + \Delta h_m = C_{bl}(C_{w_1} + C_{w_0})$$

$$\therefore \qquad R_d = \frac{C_{r_0}^{\,2} - C_{r_1}^{\,2}}{2C_{bl}(C_{w_1} + C_{w_0})}$$

Referring to Figure 10.36,

$C_{ro} = C_{fo} \operatorname{cosec} \phi$ and $C_{r1} = C_{f1} \operatorname{cosec} \theta$

and $\qquad\qquad C_{w1} = C_{wo} = \cot \theta + C_{fo} \cot \phi$

The velocity of flow generally remains constant through the blades.

$$\therefore \qquad C_{f_1} = C_{f_0} = C_f.$$

Substituting the values of C_{r_1}, C_{r_0} and $\left(C_{w_1} + C_{w_0}\right)$ in equation (22), we get

$$R_d = \frac{C_f^2 \left(\operatorname{cosec}^2\phi - \operatorname{cosec}^2\theta\right)}{2C_{bl}C_f \left(\cot\theta + \cot\phi\right)} = \frac{C_f}{2C_{bl}} \left[\frac{\left(\cot^2\phi + 1\right) - \left(\cot^2\theta + 1\right)}{\cot\theta + \cot\phi} \right]$$

$$= \frac{C_f}{2C_{bl}} \left[\frac{\cot^2\phi - \cot^2\theta}{\cot\phi + \cot\theta} \right]$$

$$= \frac{C_f}{2C_{bl}} (\cot\phi - \cot\theta)$$

If the turbine is designed for 50% reaction $\left(\Delta h_f = \Delta h_m\right)$, then the equation (23) can be written as

$$\frac{1}{2} = \frac{C_f}{2C_{bl}} (\cot\theta - \cot\phi)$$

$$\therefore \qquad C_{bl} = C_f (\cot\phi - \cot\theta)$$

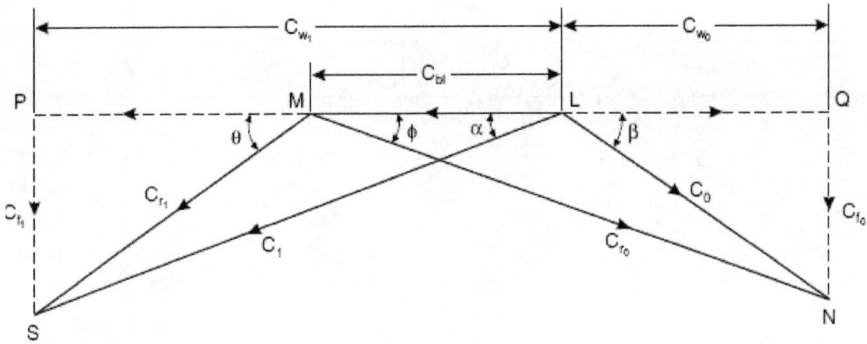

FIGURE 10.36. Velocity diagram for the blades of the Parson's reaction turbine.

Also, C_{bl} can be written as

$$C_{bl} = C_f \left(\cot \phi - \cot \beta \right)$$

and $$C_{bl} = C_f \left(\cot \alpha - \cot \theta \right)$$

$C_{f_i} = C_{f_o} = C_f$ is assumed in writing the above equations.

Comparing the equation (24), (25), and (26)

$\theta = \beta$ and $\phi = \alpha$

which *means that the moving blade and fixed blade* must have the same shape if the degree of reaction is 50%. This condition gives symmetrical velocity diagrams. This type of turbine is known as "Parson's reaction turbine." A velocity diagram for the blades of this turbine is given in Figure 10.36.

Example 25

Define the term "degree of reaction" as applied to a steam turbine. Show that for the Parson's reaction turbine, the degree of reaction is 50%.

■ **Solution:**

Refer to Figure 10.37.

The pressure drop in reaction turbines takes place in both fixed and moving blades. The division generally is given in terms of enthalpy drops. The criterion used is the *degree of reaction*. It is defined as

$$\frac{\text{Enthalpy drop in rotor blades}}{\text{Total enthalpy drop in stage}} = \frac{\Delta h_m}{\Delta h_f + \Delta h_m}$$

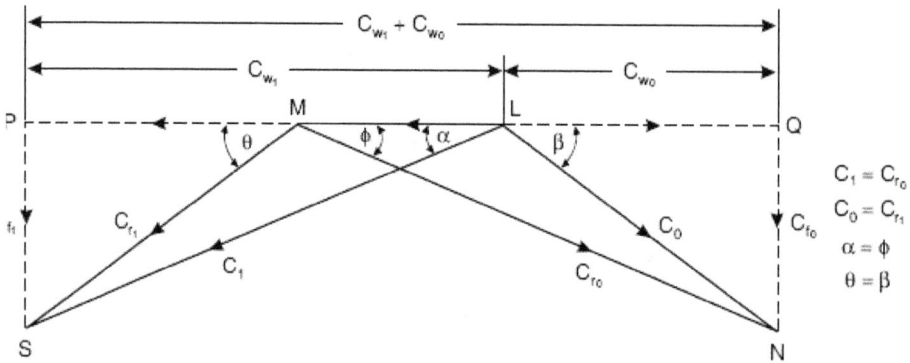

FIGURE 10.37. Sketch for Example 25.

A special case is when the degree of reaction is zero; it means no heat drop in the moving blades. This becomes a case of impulse stage. Another common case is of Parson's turbine, which has the same reason for both the fixed and moving blades. The blades are *symmetrical, that is*, the exit angle of the moving blade is equal to the exit angle of the fixed blade, and the inlet angle of the moving blade is equal to the inlet angle of the fixed blade. Since the blades are symmetrical the velocity diagram is also symmetrical. In such a case the degree of reaction is 50%.

Applying the steady flow energy equation to the fixed blades and assuming that the velocity of steam entering the fixed blade is equal to the absolute velocity of steam leaving the previous moving row, we have

$$\Delta h_f = \frac{C_1^{\,2} - C_0^{\,2}}{2}$$

Similarly, for the moving blades

$$\Delta h_m = \frac{C_{r_0}^{\,2} - C_{r_1}^{\,2}}{2}$$

But $\qquad\qquad\qquad\qquad\qquad\qquad C_1 = C_{r_0} \text{ and } C_0 = C_{r_1}$

$\therefore \qquad\qquad\qquad\qquad\qquad\qquad \Delta h_f = \Delta h_m$

Hence degree of reaction $\qquad\qquad\qquad = \dfrac{\Delta h_m}{\Delta h_m + \Delta h_m} = \dfrac{1}{2}$

This is a proof that Parson's reaction turbine is a 50% reaction turbine.

Example 26

(a) Explain the functions of the blading of a reaction turbine.

(b) A certain stage of a Parson's turbine consists of one row of fixed blades and one row of moving blades. The details of the turbine are as below:

The mean diameter of the blades	= 68 cm
R.P.M. of the turbine	= 3,000
The mass of steam passing per sec	= 13.5 kg
Steam velocity at exit from fixed blades	= 143.7 m / s
The blade outlet angle	= 20°

Calculate the power developed in the stage and gross efficiency, assuming carryover coefficient as 0.74 and the efficiency of conversion of heat energy into kinetic energy in the blade channel as 0.92.

■ Solution:

(*a*) The blades of the reaction turbine have to perform two functions:

They change the direction of the motion of steam, causing a change of momentum, which is responsible for motive force.

The blades also act as nozzles causing a pressure drop as steam moves in the blade passage.

(*b*) $D = 0.68$ m, $N = 3000$ r.p.m., $\dot{m} = 13.5$ kg/s

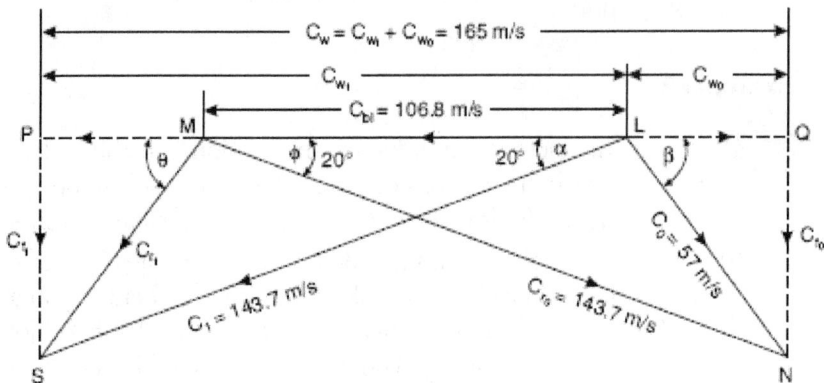

FIGURE 10.38. Sketch for Example 26.

$C_{r_0} = 143.7$ m/s, $\phi=20°$, $\psi = 0.74$, $\eta=0.92$

Blade velocity, $\quad C_{bl} = \dfrac{\pi DN}{60} = \dfrac{\pi \times 0.68 \times 3000}{60} = 106.8$ m/s

$C_{w_1} + C_{w_0} = 165$ m/s (Fig. 39)

Power developed $\quad \dot{m}_s \times \dfrac{(C_{w_1} + C_{w_0})C_{bl}}{1000}$

$= 13.5 \times \dfrac{165 \times 106.8}{1000} = 237.89$ kW. **(Ans.)**

In Parson turbine blades are symmetrical, that is,

$\alpha = \phi, \theta = \beta$

$C_1 = C_{r_0}, C_{r_1} = C_0$

Enthalpy drop $= 2 \times \dfrac{C_1^2 - \psi C_0^2}{2\eta} = 2 \times \left(\dfrac{143.7^2 - 0.74 \times 57^2}{2 \times 0.92 \times 1000} \right) = 19.83$ KJ/kg

$\therefore$ **Gross efficiency** $= \dfrac{\text{Work done /kg}}{\text{Enthalpy drop /kg}} = \dfrac{\left(C_{w_1} + C_{w_0}\right)C_{bl}}{19.83 \times 1000}$

$= \dfrac{165 \times 106.8}{19.83 \times 1000} = $ **0.888 or 88.8%. (Ans.)**

Example 27

(a) Discuss the factors that influence the erosion of turbine blades. On a sketch mark the portions of the blades more likely to be eroded. Sketch the methods used to prevent erosion of steam turbine blades.

(b) A reaction turbine running at 360 r.p.m. consumes 5 kg of steam per second. Tip leakage is 10%. The discharge blade tip angle for both moving and fixed blades is 20°. Axial velocity of flow is 0.75 times blade velocity. The power developed by a certain pair is 4.8 kW, where the pressure is 2 bar and the dryness fraction is 0.95. Find the drum diameter and blade height.

■ **Solution:**

(*a*) In the high pressure and intermediate pressure stages of the turbine, the pressures and temperatures are high and the blade material should be such that it withstands high pressures and temperatures. In the intermediate pressure stages steam is wet, therefore, the material should be able to withstand both corrosion and erosion due to the presence of water particles. In addition to corrosion and erosion the blades are also subjected to high centrifugal stresses as the low-pressure stages are longer, therefore, the blade material and its design should be such that it stands corrosion, erosion, and high centrifugal stresses.

When the speed is high and moisture exceeds 10% the effect of moisture is most prominent. The most effected portion is the back of the inlet edge of the blade, where either grooves are formed or even some portion breaks away. Due to centrifugal force the water particles tend to concentrate in the outer annulus and their tip speed is greater than the root speed, and hence the erosion effect occurs mostly on tips (see Figure 10.39).

Methods adopted to prevent erosion:

By raising the temperature of the steam at the inlet, so that at the exit of the turbine the wetness does not exceed 10%.

By adopting the reheat cycle, so that the wetness at the exit remains within limit.

Drainage belts are provided on the turbine, so that the water droplets which are on outer periphery due to centrifugal force are drained. The drained amount is about 25% of the total water particles present.

The leading edge of the turbine is provided with a shield of hard material.

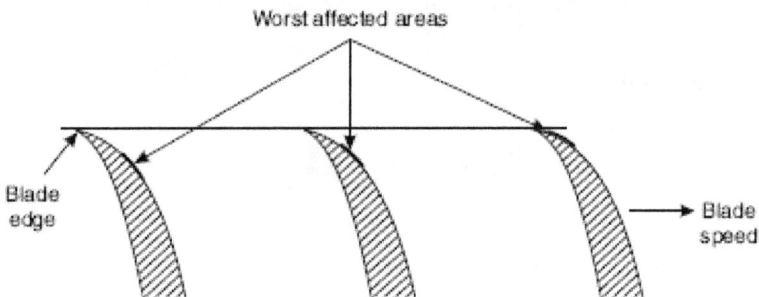

FIGURE 10.39. Water particles hit areas close to the tip of the blades.

In the method (*i*) difficulties are the limits of temperature a material can withstand.

The reheat cycle (method [*ii*]) has its own advantages and disadvantages.

Drainage belts (method [*iii*]) cause structural changes in the turbine casing design, however, bleeding may help.

The most satisfactory solution is *providing a tungsten shield. This prolongs the blade life; however, it does not remove the resistance which the water droplets impose on the rotation of the rotor.*

(*b*) Speed, $N = 360$ r.p.m.

$$\dot{m}_s = 5\left[1 - \frac{10}{100}(\text{tip leakage})\right] = 4.5 \text{ kg/s}$$

$$\alpha = \phi = 20°, C_f = 0.75\, C_{bl}$$

Power developed in a certain pair

= 4.8 kW at 2 bar (*x* = 0.95)

Blade velocity, $C_{bl}(LM) = \dfrac{\pi D N}{60} = \dfrac{\pi \times D \times 360}{60} = 18.85\, D$ m/s.

$C_{f_1} = 0.75 \times 18.85\, D = 14.138\, D$ m/s

From, $\Delta\, LSP$, $\qquad C_1 = \dfrac{C_{f_1}}{\sin 20°} = \dfrac{14.138\, D}{\sin 20°}$ m/s

and $\qquad LP = C_1 \cos 20° = 14.13\, D \times \dfrac{\cos 20°}{\sin 20°} = 14.138\, D \cot 20°$

$PQ = 2LP - LM = (2 \times 14.138\, D \cot 20° - 18.85\, D)$ m/s

Power developed $\qquad = \dfrac{\dot{m}_s \times PQ \times LM}{1000}$

$$4.8 = \frac{4.5 \times \left(2 \times 14.138\, D \cot 20° - 18.85 D\right) \times 18.85\, D}{1000}$$

Solving this equation, we get

Drum diameter, D = **0.98 m. (Ans.)**

From steam tables at 2 bar, $v_g = 0.885$ m³/kg

$$\text{Mass flow rate} = \frac{\text{Flow area} \times \text{flow velocity}}{\text{Specific volume}}$$

$$4.5 = \frac{\pi D h \times C_{f_1}}{x v_g}$$

$$4.5 = \frac{\pi \times 0.98 \times h \times (14.138 \times 0.98)}{0.95 \times 0.885}$$

$$h = \frac{4.5 \times 0.95 \times 0.885}{\pi \times 0.98 \times 14.138 \times 0.98} = 0.0887 \text{m}$$

∴ Blade height   **=0.0887m. (Ans.)**

Example 28

(a) List the advantages of steam turbines over gas turbines.

(b) Determine the isentropic enthalpy drop in the stage of the Parson's reaction turbine which has the following particulars:

Speed = 1500 rpm ; mean diameter of rotor   = 1 m ;
stage efficiency = 80% ; speed ratio   = 0.7 ;
blade outlet angle   = 20°.

■ **Solution:**

(a) Advantages of steam turbines over gas turbines:

1. The load control in steam turbines is easy, simply by throttle governing or cut-off governing. In gas turbines the air-fuel ratio becomes too high, 100 to 150 at part loads. This causes problems to sustain the flame.

2. The steam turbine works on a Rankine cycle. In this cycle most of the heat is supplied at constant temperature in the form of latent heat of evaporation. Also, the heat is rejected in the condenser isothermally. Hence the cycle is more efficient, and

its efficiency is close to that of a Carnot cycle. On the other hand, the gas turbine works on a Brayton cycle whose efficiency is much less than that of the Carnot cycle, working between the same maximum and minimum limits of temperatures.

3. The efficiency of the steam turbine at part load is not very much reduced. In gas turbines the maximum cycle temperature decreases considerably at part load; therefore, its part load efficiency is considerably low.

4. The blade material for steam turbines is cheap. For gas turbines the blade material is costly as it is required to sustain considerably high temperatures.

For a Parson's reaction turbine, the velocity triangles are symmetrical.

Given : $N = 1500$ r.p.m., $D = 1$ m, $\eta_{\text{stage}} = 80\%$;

Speed ratio, $\dfrac{C_{bl}}{C_1} = 0.7, \phi = \alpha = 20^{\circ}$

$C_1 = C_{r_0}$ and $C_{r_1} = C_0$

$Cbl = \dfrac{\pi DN}{60} = \dfrac{\pi \times 1 \times 1500}{60} = 78.54 \text{ m/s}$

Speed ratio $= 0.7 = \dfrac{C_{bl}}{C_1}$

$\therefore$ $\qquad C1 = \dfrac{78.54}{0.7} = 112.2 \text{ m/s}$

$Cr_1^2 = C_1^2 + C_{bl}^2 - 2C_1 C_{bl} \cos \alpha$

$= (112.2)^2 + (78.54)^2 - 2 \times 112.2 \times 78.54 \cos 20^{\circ} = 2195.84$

or $\qquad\qquad\qquad\qquad\qquad C_{r1} = 46.86 \text{ m/s}$

Δh = Actual enthalpy drop for the stage

$= \dfrac{1}{2}\left[\left(C_1^2 - C_0^2\right) + \left(C_{r_0}^2 - C_{r_1}^2\right)\right]$

$$= \frac{1}{2}\left[\left(C_1^2 - C_{r_1}^2\right) + \left(C_1^2 - C_{r_1}^2\right)\right] = C_1^2 - C_{r_1}^2$$

$(\because \quad C_0 = C_{r_1} ; C_{r_0} = C_1)$

or $\qquad \Delta h = [(112.2)^2 - (46.84)^2] \times 1/1000 \text{ kJ/kg} = 10.39 \text{ kJ/kg}$

Isentropic enthalpy drop, $\left(\Delta h'\right) = \dfrac{(\Delta h)}{\eta_{stage}} = \dfrac{10.39}{0.8} = \textbf{12.99 kJ / kg. (Ans.)}$

Condition for Maximum Efficiency

The condition for *maximum efficiency* is derived by making the following *assumptions*:

(*i*) The degree of reaction is 50%.

(*ii*) The moving and fixed blades are symmetrical.

(*iii*) The velocity of steam at the exit from the preceding stage is same as the velocity of the steam at the entrance to the succeeding stage.

Refer to Figure 10.37 (velocity diagram for reaction blade).

Work done per kg of steam,

$$W = C_{bl} (C_{w1} = C_{wo}) = C_{bl}[C_1 \cos \alpha + (C_{r0} \cos \phi - C_{b0})]$$

as $\qquad \phi = \alpha$ and $C_{r_0} = C_{r_1}$ as per the assumptions

$\therefore \qquad W = C_{bl} [2C_1 \cos \alpha - C_{bl}]$

or $\qquad W = C_1^2 \left[\dfrac{2C_{bl}C_1 \cos \alpha}{C_1^2} - \dfrac{C_{bl}^2}{C_1^2} \right]$

$$= C_1^2 [2\rho. \cos \alpha - \rho^2]$$

where $\rho = \dfrac{C_{bl}}{C_1}$.

The K.E. supplied to the fixed blade $= \dfrac{C_1^2}{2g}$.

The K.E. supplied to the moving blade $= \dfrac{C_{r_0}^{\,2} - C_{r_1}^{\,2}}{2}$.

∵ Total energy supplied to the stage,

$$\Delta h = \frac{C_1^2}{2} + \frac{C_{r_0}^{\,2} - C_{r_1}^{\,2}}{2}$$

as $\qquad\qquad\qquad\qquad C_{r_0} = C_1$ for symmetrical triangles.

∴ $\qquad\qquad\qquad\qquad \Delta h = \dfrac{C_1^2}{2} + \dfrac{C_1^2 - C_{r_1}^{\,2}}{2}$

$$= C_1^2 - \frac{C_{r_1}^{\,2}}{2}$$

Considering the ΔLMS (Figure 10.40)

$$C_{r_1}^{\,2} = C_1^2 + C_{bl}^{\,2} - 2C_1 \cdot C_{bl} \cdot \cos\alpha$$

Substituting this value of $C_{r_1}^{\,2}$ is equations (35), we have

Total energy supplied to the stage

$$\Delta h = C_1^2 - \left(C_1^2 + C_{bl}^{\,2} - 2C_1.C_{bl}.\cos\alpha \right)/2$$

$$= \left(C_1^2 + 2C_1\,C_{bl}\,\cos\alpha - C_{bl}^{\,2} \right)/2$$

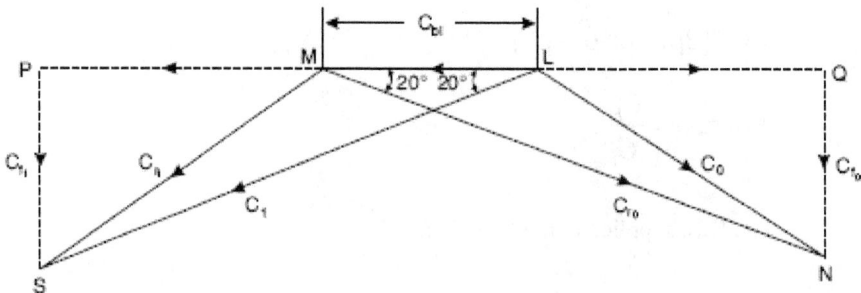

FIGURE 10.40. Sketch for Example 27.

$$= \frac{}{} - \left[1 + \frac{C_{bl}}{C_1} . \cos \alpha - \left(\frac{C_{bl}}{C_1} \right) \right]$$

$$= \frac{C_1^2}{2} - \left[1 + 2\rho \cos \alpha - \rho^2 \right]$$

The blade efficiency of the reaction turbine is given by,

$$\eta_{bl} = \frac{w}{\Delta h}$$

Substituting the value of W and Δh, we get

$$\eta_{bl} = \frac{C_1^2 \left[2\rho \cos \alpha - \rho^2 \right]}{\dfrac{C_1^2}{2} - \left[1 + 2\rho \cos \alpha - \rho^2 \right]}$$

$$= \frac{2 \left(2\rho \cos \alpha - \rho^2 \right)}{\left(1 + 2\rho \cos \alpha - \rho^2 \right)} = \frac{2\rho \left(2\rho \cos \alpha - \rho \right)}{\left(1 + 2\rho \cos \alpha - \rho^2 \right)} = \frac{2 \left(2\rho \cos \alpha - \rho^2 \right) - 2}{\left(1 + 2\rho \cos \alpha - \rho^2 \right)}$$

$$= 2 - \frac{2}{\left(1 + 2\rho \cos \alpha - \rho^2 \right)}$$

The η_{bl} becomes maximum when the value of $\left(1 + 2\rho \cos \alpha - \rho^2 \right)$ becomes maximum.

• ∴ The required equation is

$$\frac{d}{d\rho} \left(1 + 2\rho \cos \alpha - \rho^2 \right) = 0$$

$$2 \cos \alpha - 2\rho = 0$$

∴ $$\rho = \cos \alpha$$

Substituting the value of ρ the value of *maximum efficiency* is given by,

$$\left(\eta_{bl}\right)_{max} = 2 - \frac{2}{1 + 2\cos^2\alpha - \cos^2\alpha} = 2\left(1 - \frac{1}{1 + \cos^2\alpha}\right) = \frac{2\cos^2\alpha}{1 + \cos^2\alpha}$$

Hance
$$\left(\eta_{bl}\right)_{max} = \frac{2\cos^2\alpha}{1 + \cos^2\alpha}$$

The variation of η_{bl} with blade speed ratio $\left(\dfrac{C_{bl}}{C_1}\right)$ for the reaction stage is shown in Figure 10.40.

10.8.2. Turbine Efficiencies

1. Blade or diagram efficiency (η_{bl}). *It is the ratio of work done on the blade per second to the energy entering the blade per second.*

2. Stage efficiency (η_{stage}). The stage efficiency covers all the losses in the nozzles, blades, diaphragms, and discs that are associated with that stage.

$$\eta_{stage} = \frac{\text{Network done on shaft per stage per kg of steam flowing}}{\text{Adiabatic heat drop per stage}}$$
$$= \frac{\text{Network done on blades} - \text{Disc friction and windage}}{\text{Adiabatic heat drop per stage}}.$$

3. Internal efficiency ($\eta_{internal}$). This is equivalent to the stage efficiency when applied to the whole turbine, and is given by:

$$\eta_{internal} = \frac{\text{Heat coverted into useful work}}{\text{Total adiabatic heat drop}}.$$

4. Overall or turbine efficiency ($\eta_{overall}$). This efficiency covers internal and external losses; for example, bearings and steam friction, leakage, radiation, and so on.

$$\eta_{overall} = \frac{\text{Work delivered at the turbine coupling in heat units per kg of steam}}{\text{Total adiabatic heat drop}}.$$

5. Net efficiency or efficiency ratio (η_{net}). It is the ratio

$$\frac{\text{Brake thermal efficiency}}{\text{Thermal efficiency on the Rankine cycle}}$$

Also, the actual thermal efficiency

$$= \frac{\text{Heat converted into useful work per kg of steam}}{\text{Total heat in steam at stop valve} - \text{Water heat in exhaust}}.$$

Again, Rankine efficiency

$$= \frac{\text{Adiabatic heat drop}}{\text{Total heat in steam at stop valve} - \text{Water heat in exhaust}}.$$

$$\eta_{net} = \frac{\text{Heat converted into useful work}}{\text{Total adiabatic heat drop}}.$$

Hance $\qquad \eta_{net} = \eta_{overall.}$

It is the overall or net efficiency that is meant when the efficiency of a turbine is spoken of without qualification.

10.9. Types of Power in Steam Turbine Practice

In steam turbine performance the following types of power are generally used:

1. Adiabatic power (A.P.). It is the power based on the total internal steam flow and adiabatic heat drop.

2. Shaft power (S.P.). It is the actual power transmitted by the turbine.

3. Rim power (R.P.). It is the power developed at the rim. It is also called

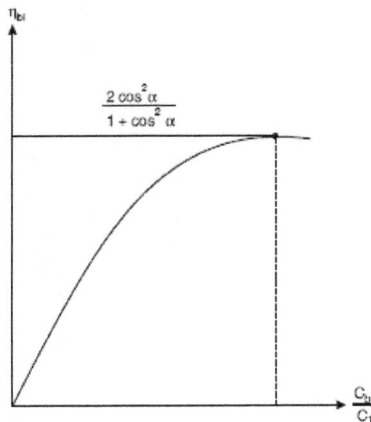

FIGURE 10.41. Sketch for Example 28.

blade power.

Power losses are usually expressed as follows:

(*i*) $(P)_D$ = Power lost in overcoming disc friction.

(*ii*) $(P)_{bw}$ = Power lost in blade windage losses.

Let us consider the case of an impulse turbine. Let $\dot{m}_s$ be the total internal steam flow in kg/s.

Refer to Figure 10.42. The line (1–2) represents the adiabatic or isentropic expansion of steam in the nozzle from pressure p_1 to p_2. But the actual path of the stage point during expansion in the nozzles is shown by (1–3) which takes into account the effect of "nozzle losses."

Then $\qquad$ A.P $= \dot{m}_s \ (h_1 - h_2)$kW

After expansion in the nozzle the steam enters the blades where the R.P. is developed. Due to blade friction the steam is somewhat reheated, and this reheating is shown by (3–4) along the constant pressure p_2 line just for convenience. But in practice the pressure at outlet of the blade is equal to that at the inlet, so the pressure in the blade channels is not constant. However, with this simplification:

$$\text{R.P} = \dot{m}_s \ (h_1 - h_4)\text{kW}$$

4–5 shows the further reheating due to friction and blade windage and these losses are given as

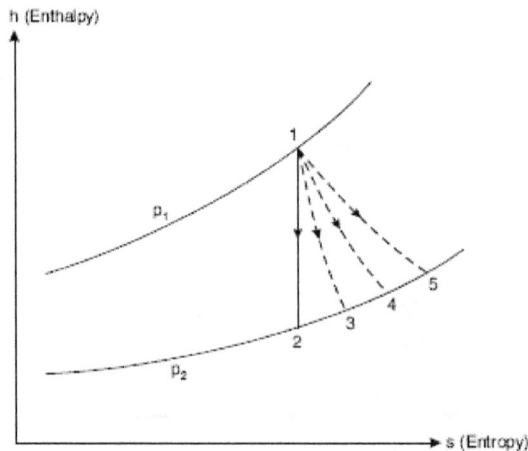

FIGURE 10.42. Adiabatic or isentropic expansion of steam in the nozzle from pressure p1 to p2.

$$\left(P\right)_{w.f.} = \dot{m}_s \ (h_5 - h_4)\text{kW}$$

Now points 1 and 5 are the initial and final stage points respectively for a single-stage impulse turbine. It, therefore, follows that

S.P. $= m_s (h_1 - h_5)$ kW.

REACTION TURBINES

Example 29

The following data refer to a particular stage of a Parson's reaction turbine:

Speed of theturbine $= 1500$ r.p.m.

Mean diameter of therotor $= 1$ metre

Stage efficiency $= 80$ per cent

Blade outlet angle $= 20°$

Speed ratio $= 0.7$

Determine the available isentropic enthalpy drop in the stage.

■ **Solution:**
Mean diameter of the rotor, $D = 1$ m

Turbine speed, $\quad N = 1500$ r.p.m.

Blade outlet angle, $\quad \phi = 20°$

Speed ratio, $\quad \rho = \dfrac{C_b}{C_1} = 0.7$

Stage efficiency, $\quad \eta_{stage} = 80\%$

Isentropic enthalpy drop:

Blade speed, $\quad C_{bl} = \dfrac{\pi DN}{60} = \dfrac{\pi \times 1 \times 1500}{60} = 78.54$ m/s

But $\qquad \rho = \dfrac{C_{bl}}{C_1} = 0.7 \text{ (given)}$

$\therefore \qquad C_1 = \dfrac{C_{bl}}{0.7} = \dfrac{75.54}{0.7} = 112.2 \text{ m/s}$

In Parson's turbine $\quad \alpha = \phi$.

With the above data known, the velocity diagram for the turbine can be drawn to a suitable scale as shown in Figure 10.43.

By measurement (from the diagram)

$C_{w_1} = 106.5 \text{ m/s} ; C_{w_0} = 27 \text{ m/s}$

$\eta_{\text{stage}} = \dfrac{C_{bl}\left(C_{w_1} + C_{w_0}\right)}{h_d}$, where h_d = isentropic enthalpy drop.

i.e., $\qquad 0.8 = \dfrac{78.54(106.25 + 27)}{h_d \times 1000}$

$\therefore \qquad h_d = \dfrac{78.54(106.25 + 27)}{0.8 \times 1000} = 13.08 \text{ kJ}$

Hence, *isentropic enthalpy drop* = 13.08 kJ/kg. (Ans.)

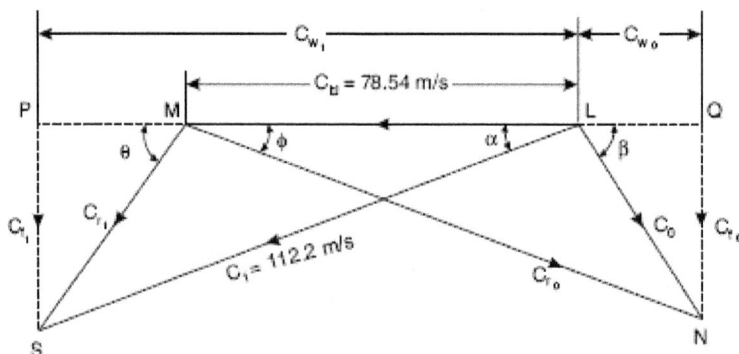

FIGURE 10.43. Sketch for Example 29.

Example 30

In a reaction turbine, the blade tips are inclined at 35° and 20° in the direction of motion. The guide blades are of the same shape as the moving blades, but reversed in direction. At a certain place in the turbine, the drum diameter is 1 meter and the blades are 10 cm high. At this place, the steam has a pressure of 1.75 bar and dryness 0.935. If the speed of this turbine is 250 r.p.m. and the steam passes through the blades without shock, find the mass of steam flow and power developed in the ring of moving blades.

■ **Solution:**

Refer to Figure 10.44.

Angles, $\alpha = \phi = 20^\circ$, and $\theta = \beta = 35^\circ$

Mean drum diameter, $D_m = 1 + 0.1 = 1.1\,\mathrm{m}$

Area of flow $= \pi D_w h$, where h is the height of blade

$$= \phi \times 1.1 \times 0.1 = 0.3456\,\mathrm{m}^2$$

Steam pressure = 1.75 bar
Dryness fraction of steam, $x = 0.935$
Speed of the turbine, $N = 250$ r.p.m.

Rate of steam flow, $\dot{m}_s$:

Blade speed, $\quad C_{bl} = \dfrac{\pi D N}{60} = \dfrac{\pi \times 1.1 \times 250}{60} = 14.4\,\mathrm{m/s}$

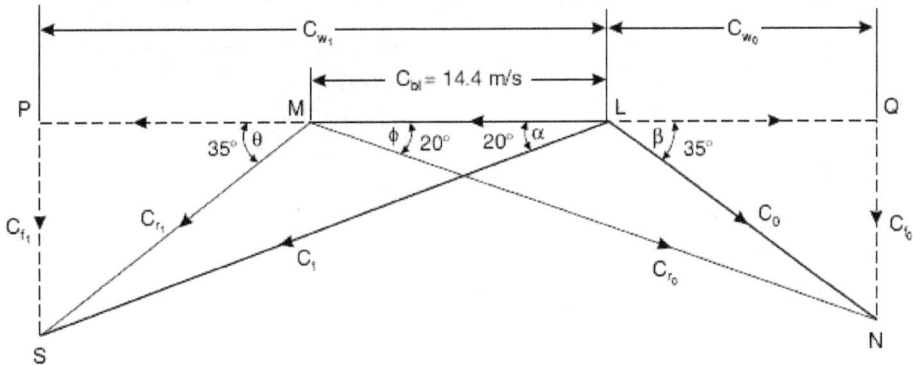

FIGURE 10.44. Sketch for Example 30.

With the given data the velocity diagram can be drawn to a suitable scale as shown in Figure 10.44.

By measurement (from diagram):

$$C_{w_1} = 30 \text{ m/s} \; ; \; C_{w_0} = 15.45 \text{ m/s} \; ; \; C_{f_1} = C_{f_0} = 10.8 \text{ m/s}$$

From steam tables corresponding to 1.75 bar pressure.

v_g = Specific volume of dry saturated steam

$= 1.004 \text{ m}^3 / \text{kg}$

$x = 0.935$ (given)

$\therefore$ Specific volume of wet steam $= x v_g = 0.935 \times 1.004 = 0.938 \text{ m}^3 / \text{kg}$

Mean flow rate is given by:

$$\dot{m}_s = \frac{\text{Area of flow} \times \text{Velocity of flow}}{\text{Specific volume of steam}} = \frac{0.3456 \times 10.8}{0.938} = 3.98 \text{ kg/s.}$$

Power developed, P:

$$P = \frac{\dot{m}_s (C_{w_1} + C_{w_0}) C_{bl}}{1000} \text{ kW}$$

$$= \frac{3.98(30 + 15.45) \times 14.4}{1000} = 2.6 \text{ kW.} \quad \textbf{(Ans.)}$$

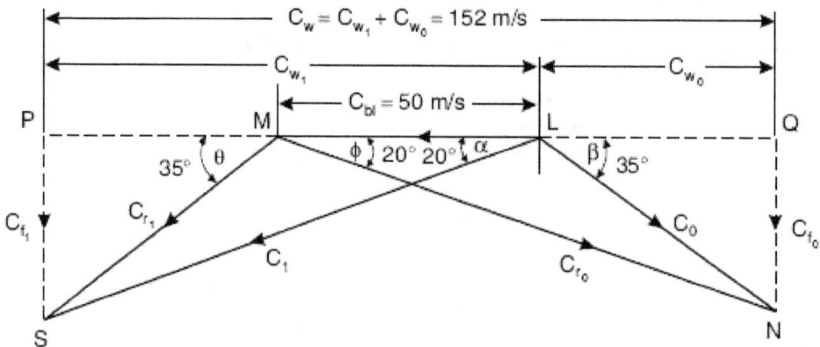

FIGURE 10.45. Sketch for Example 31.

Example 31

In a reaction turbine, the fixed blades and moving blades are of the same shape but reversed in direction. The angles of the receiving tips are 35° and of the discharging tips 20°. Find the power developed per pair of blades for a steam consumption of 2.5 kg/s, when the blade speed is 50 m/s. If the heat drop per pair is 10.04 kJ/kg, find the efficiency of the pair.

■ **Solution:**

Angles of receiving tips, $\theta = \beta = 35°$

Angles of discharging tips, $\alpha = \phi = 20°$

Steam consumption, $m_s = 2.5 \text{ kg / s}$

Blade speed, $c_{bl} = 50 \text{m / s}$

Heat drop per pair, $h_d = 10.04 \text{ kJ / kg}$
Power developed per pair of blades:

Refer to Figure 46.

$NS = PQ = 152 \text{ m / s}$

Work done per pair per kg of steam

$= (C_{w_1} + C_{w_0}) + C_{bl} = 152 \times 50 = 7600 \text{ Nm/kg of steam.}$

$$Power/pair = \frac{\dot{m}s(C_{w_1} + C_{w_0})C_{bl}}{1000} = \frac{2.5 \times 7600}{1000} = \textbf{19 kW. (Ans.)}$$

Efficiency of the pair:

$$Efficiency = \frac{\text{Work done per pair per kg of steam}}{h_d}$$

$$= \frac{7600}{10.04 \times 1000} = 0.757 = \textbf{75.7\%. (Ans.)}$$

Example 32

A stage of a turbine with Parson's blading delivers dry saturated steam at 2.7 bar from the fixed blades at 90 m/s. The mean blade height is 40 mm, and the moving blade exit angle is 20°. The axial velocity of steam

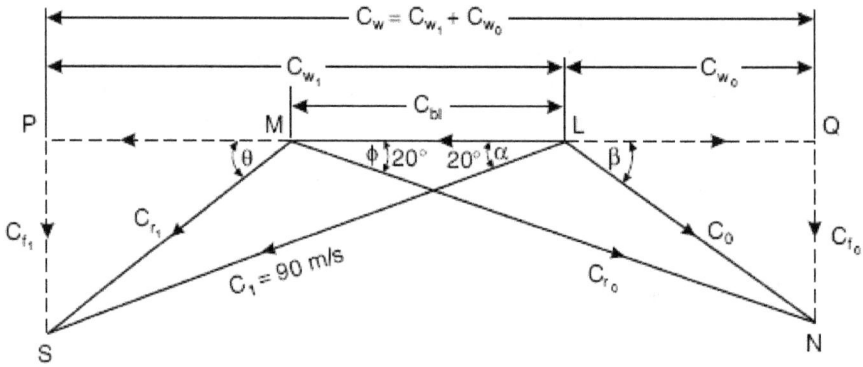

(a)

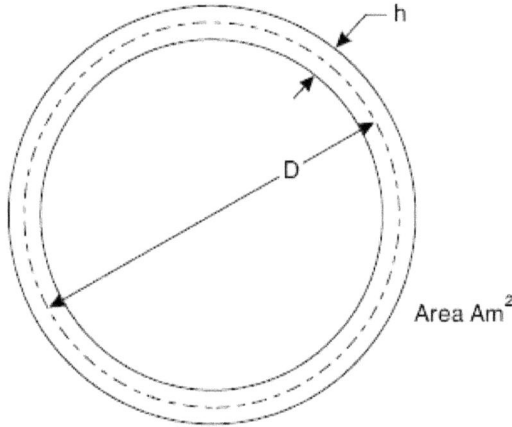

(b)

FIGURE 10.46. Sketch for Example 32.

is 3/4 of the blade velocity at the mean radius. Steam is supplied to the stage at the rate of 9000 kg/h. The effect of the blade tip thickness on the annulus area can be neglected. Calculate:

(i) The wheel speed in r.p.m.

(ii) The diagram power

(iii) The diagram efficiency

(iv) The enthalpy drop of the steam in this stage

■ **Solution.**

The velocity diagram is shown in Figure 10.46 (*a*) and the blade wheel annulus is represented in Figure 10.48(*b*).

Pressure = 2.7 bar, $x = 1$, $C_1 = 90$ m / s, $h = 40$ mm $= 0.04$ m

$\alpha = \phi = 20°$, $C_{f_1} = C_{f_o} = 3/4\ C_{bl}$

Rate of steam supply = 9000kg / h.

(*i*)   Wheel speed, N:

$C_f = 3/4\ C_{bl} = C_1\ \sin 20° = 90\ \sin 20° = 30.78$ m / s

∴ $C_{bl} = 30.78 \times 4/3 = 41.04$ m / s

The mass flow of steam is given by : $\dot{m}_s = \dfrac{C_f A}{v}$

(where *A* is the annulus area, and *v* is the specific volume of the steam).

In this case, $v = v_g$ at 2.7 bar $= 0.6686$ m^3 / kg

∴ $\dot{m}_s = \dfrac{9000}{3600} = \dfrac{30.78}{0.6686}$ or $A = \dfrac{9000 \times 0.6686}{3600 \times 30.78} = 0.054$ m^2

Now, annulus area, $A = \pi D h$

(where *D* is the mean diameter, and *h* is the mean blade height)

∴ $0.054 = \pi D \times 0.04$ or $D = \dfrac{0.054}{\pi \times 0.04} = 0.43$ m

Also, $C_{bl} = \dfrac{\pi D N}{60}$ or $41.04 = \dfrac{\pi \times 0.43 \times N}{60}$

or $N = \dfrac{41.04 \times 60}{\pi \times 0.43} = \mathbf{1823 r.p.m.\ (Ans.)}$

(*ii*) The diagram power:

Diagram power $= \dot{m}_s C_w C_{bl}$

Now, $C_w = 2C_1\ \cos \alpha - C_{bl}$

$$= 2 \times 90 \times \cos 20° - 41.04 = 128.1 \text{ m / s}$$

$$\therefore \textit{Diagram power} \quad = \frac{9000 \times 128.1 \times 41.04}{3600 \times 1000} = \textbf{13.14 kW. (Ans.)}$$

(*iii*) The diagram efficiency:

Rate of doing work per kg / s $= C_w \, C_{bl} = 128.1 \times 41.04$ N m / s

Also, energy input to the moving blades per stage

$$= \frac{C_1^2}{2} + \frac{C_{r_0}^2 - C_{r_1}^2}{2} = \frac{C_1^2}{2} + \frac{C_1^2 - C_{r_1}^2}{2} = C_1^2 - \frac{C_{r_1}^2}{2} \quad (\because C_{r_0} = C_1)$$

Referring to Figure 10.48(*a*), we have

$$C_{r_1}^2 = C_1^2 + C_{bl}^2 - 2 C_1 \, C_{bl} \cos \alpha$$

$$= 90^2 + 41.04^2 - 2 \times 90 \times 41.04 \times \cos 20°$$

$$= 8100 + 1684.28 - 6941.69$$

$$\therefore \qquad\qquad\qquad C_{r_1} = 53.3 \text{ m/s}$$

$$\text{Energy input} \qquad\qquad = 90^2 - \frac{53.3^2}{2} = 6679.5 \text{ Nm per kg / s}$$

$$\therefore \textit{Diagram efficiency} \quad = \frac{128.1 \times 41.04}{6679.5} = \textbf{0.787 or 78.7\%. (Ans.)}$$

(*iv*)Enthalpy drop in the stage:

Enthalpy drop in the moving blades

$$= \frac{C_{r_0}^2 - C_{r_1}^2}{2} = \frac{90^2 - 53.3^2}{2 \times 1000} = 2.63 \text{ kJ/kg} \qquad\qquad (\because C_{r_0} = C_{r_1})$$

$$\therefore \textit{Total enthalpy drop per stage} = 2 \times 2.63 = \textbf{5.26 kJ / kg. (Ans.)}$$

Example 33

The outlet angle of the blade of a Parson's turbine is 20° and the axial velocity of flow of steam is 0.5 times the mean blade velocity. If the diameter of the ring is 1.25 m and the rotational speed is 3000 r.p.m. determine:

(i) Inlet angles of blades.

(ii) Power developed if dry saturated steam at 5 bar passes through the blade with a height assumed as 6 cm. Neglect the effect of blade thickness.

■ **Solution.**

Refer to Figure 10.47.

Angles, $\alpha = \phi = 20°$

Axial velocity of flow of steam,

$C_{f_1} = C_{f_0} = 0.5\, C_{bl}$ (blade speed)

Diameter of the ring, 1.25 m

Rotational speed, $N = 3000$ r.p.m.

Blade speed, $C_{bl} = \dfrac{\pi DN}{60} = \dfrac{\pi \times 1.25 \times 3000}{60} = 196\,\text{m/s}$

$\therefore \qquad C_{f_1} = C_{f_0} = 0.5 \times 196 = 98\,\text{m/s}$

Velocity diagram is drawn as follows:

Takes $LM\,(C_{bl}) = 196\,\text{m/s}$, and $\alpha = \phi = 20°$.

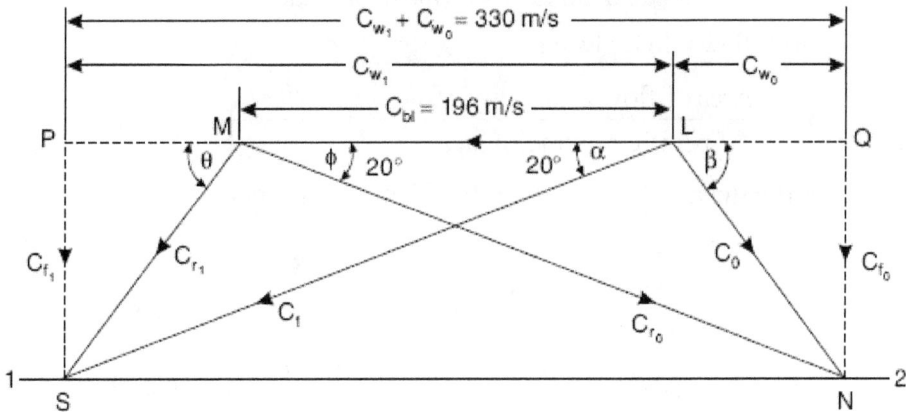

FIGURE 10.47. Sketch for Example 33.

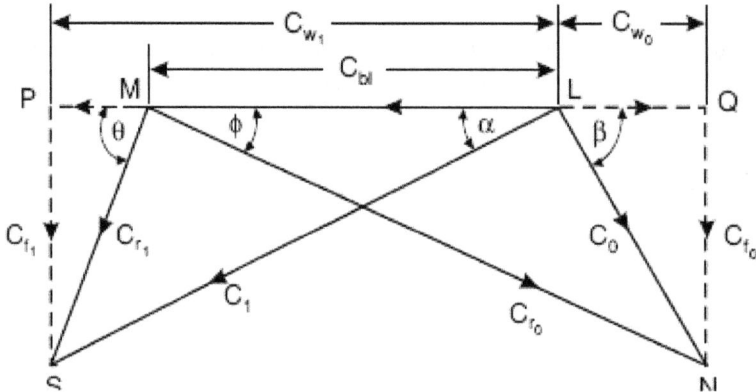

FIGURE 10.48. Sketch for Example 34.

Draw line 1–2 parallel to LM at a value of 98 m/s (according to scale). The points S and N are thus located on the line 1–2.

Complete the rest of the diagram as shown in Figure 48.

(i) Inlet angles of blades:

The *inlet angles* (by measurement) are:

$$\beta = \theta = 55° \quad \textbf{(Ans.)}$$

(ii) Power developed, P:

Area of flow is given by,

$A = \pi \times D$ (mean diameter) $\times h$ (mean of blade)

Mean flow rate is given by,

$$\dot{m}_s = \frac{\text{Area of flow} \times \text{Velocity of flow}}{\text{pecific volume of steam}} = \frac{\pi D h \times C_f}{v}$$

From steam tables, $vg = 0.375 \, \text{m}^3 / \text{kg at 5 bar}$

$$\therefore \quad \dot{m}_s = \frac{\pi \times 1.25 \left(\dfrac{6}{100}\right) \times 98}{0.375} = 61.57 \, \text{kg/s}$$

Power developed, $\quad P = \dfrac{\dot{m}_s \times C_w \times C_{bl}}{1000} = \dfrac{61.57 \times 330 \times 196}{1000} = \textbf{3982.3 kW.} \textbf{(Ans.)}$

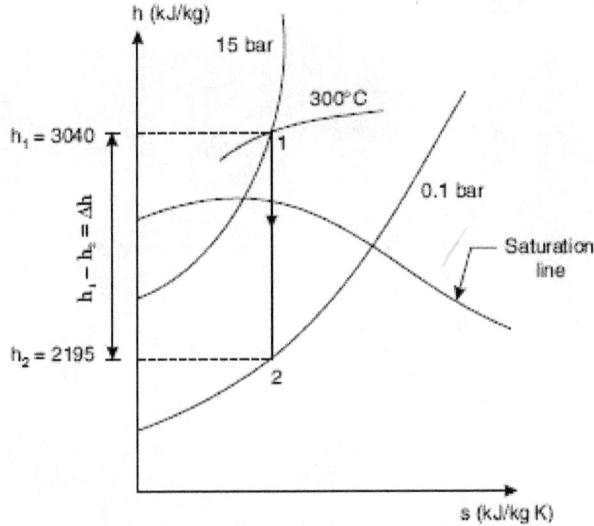

FIGURE 10.49. Sketch for Example 37, h-s diagram.

Example 34

A 50% reaction turbine (with symmetrical velocity triangles) running at 400 r.p.m. has the exit angle of the blades as 20° and the velocity of steam relative to the blades at the exit is 1.35 times the mean blade speed. The steam flow rate is 8.33 kg/s and at a particular stage the specific volume is 1.381 m3/kg. Calculate for this stage:

(i) A suitable blade height, assuming the rotor mean diameter 12 times the blade height,

(ii) The diagram work

■ **Solution:**

Speed, $\qquad N = 400$ r.p.m. ; $\alpha = 20°$

$C_{r_0} = C_1 = 1.35 \, C_{bl}$; $\dot{m}_s = 8.33$ kg/s

$v = 1.381 \text{ m}^3/\text{kg}$; $D = 12 \, h$

(i) Blade height, h:

Refer to Figure 10.50.

Axial flow velocity, $\qquad C_{f_1} = C_{f_0} = C_f = C_1 \sin \alpha$

$$= 1.35\, C_{bl}\, \sin 20°$$

$$= 0.4617\, C_{bl}$$

Area of flow,
$$A = \pi D h = \pi D \times \frac{D}{12} = \frac{\pi D^2}{12}$$

Mass flow rate,
$$\dot{m} = \frac{AC_f}{v} \text{ or } 8.33 = \frac{A \times 0.4617 C_{bl}}{1.381}$$

or
$$\frac{8.33 \times 1.381}{0.4917} = A \times C_{bl} = \frac{\pi D^2}{12} \times \frac{\pi D N}{60}$$

or
$$24.916 = \frac{\pi^2 D^3 \times 400}{720} \text{ or } D^3 = \frac{24.919 \times 720}{\pi^2 \times 400} \text{ or } D = 1.656\,\text{m}$$

∴ Blade height,
$$h = \frac{D}{12} = \frac{1.656}{12} = 0.138 \text{ m or } \mathbf{138\,mm. (Ans.)}$$

(*ii*) The diagram work:

Diagram work
$$= \dot{m} \times C_{bl}\, (C_{w_1} + C_{w_0})$$

$$= \dot{m} \times C_{bl}\, (2\, C_1 \cos \alpha - C_{bl})$$

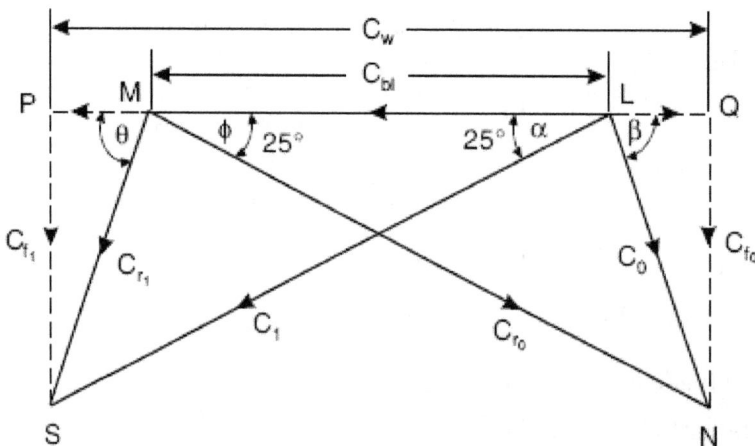

FIGURE 10.50. Sketch for Example 37.

$= 8.33 \times C_{bl} \; (2 \times 1.35 \, C_{bl} \, \cos 20° - C_{bl})$

$= 8.33 \times C_{bl}^{2} \; (2 \times 1.35 \cos 20° - 1)$

$= 8.33 \times \left(\dfrac{\pi DN}{60}\right)^{2} \times 1.537 = 8.33 \times \left(\dfrac{\pi \times 1656 \times 400}{60}\right)^{2} \times 1.537$

$= 15401.2$ W   or   **15.4 kW.**   **(Ans.)**

Example 35

300 kg/min of steam (2 bar, 0.98 dry) flows through a given stage of a reaction turbine. The exit angle of fixed blades as well as moving blades is 20° and 3.68 kW of power is developed. If the rotor speed is 360 r.p.m. and tip leakage is 5%, calculate the mean drum diameter and the blade height. The axial flow velocity is 0.8 times the blade velocity.

■ **Solution:**

Rate of flow of steam through the turbine, $\dot{m}_s = \dfrac{300}{60} = 5$ kg/s

Pressure and condition of steam, $p = 2$ bar, $x = 0.98$.

The exit angles of fixed blades as well as moving blades, $\alpha = \phi = 20°$

Power developed, $P = 3.68$ kW

Speed of the rotor, $N = 360$ r.p.m.

Tip leakage $= 5$ per cent

Axial flow velocity, $C_f = 0.8 \, C_{bl}$ (blade velocity)

Refer to Figure 10.50.

Mean drum diameter, D:

Mean blade velocity, $C_{bl} = \dfrac{\pi DN}{60} = \dfrac{\pi D \times 360}{60} = 18.85 \, D$ m/s

Power developed, $P = \dfrac{\dot{m}_s C_{bl} C_w}{1000}$

or

$$3.67 = \frac{(5 \times 0.95) \times 18.85 \, D \times C_w}{1000}$$

∴

$$C_w = \frac{3.67 \times 1000}{(5 \times 0.95) \times 18.85 D} = \frac{40.988}{D}$$

Assuming Parson's reaction turbine, we have

$$C_{f_1} = C_1 \sin \alpha \text{ or } C_1 = \frac{C_{f_1}}{\sin \alpha} = \frac{0.8 C_{bl}}{\sin \alpha} = \frac{0.8 \times 18.85 D}{\sin 20°} = 44.091 \, D$$

$$(C_{f_1} = C_{f_0} = C_f)$$

Also, $C_w = 2C_1 \cos \alpha - C_{bl}$ or $\dfrac{40.988}{D} = 2 \times 44.091 \, D \cos 20° - 18.85 \, D = 64.01 \, D$

or $40.988 = 64.01 \, D^2$ or $D = 0.8$ m or **800 mm.**   (**Ans.**)

Blade height, h:

Mean steam flow rate, $\dot{m}_s = \dfrac{\pi D h C_f}{x v_g}$

or $(5 \times 0.95) = \dfrac{\pi \times 0.8 \times h \times C_1 \sin \alpha}{0.98 \times 0.885} = \dfrac{\pi \times 0.8 \times h \times \left(44.091 D \times \sin 20°\right)}{0.98 \times 0.885}$

$\left(\text{At 2 bar: } v_g = 0.885 \, \text{kg/m}^3\right)$

or $h = \dfrac{(5 \times 0.95) \times 0.98 \times 0.885}{\pi \times 0.8 \times \left(44.091 \times 0.80 \times 0.3420\right)} = 0.1359$ m or **135.9 mm.** (**Ans.**)

Example 36

(a) Why is drum type construction preferred to disc type construction in reaction turbines?

(b) Why is partial admission of steam adopted for H.P. impulse stages while full admission is essential for any stage of a reaction turbine?

(c) In a 50% reaction turbine, the speed of rotation of a blade group is 3000 r.p.m. with a mean blade velocity of 120 m/s. The velocity ratio is 0.8 and the exit angle of the blades is 20°. If the mean blade height is 30 mm, calculate the total steam flow rate through the turbine. Neglect the effect of blade edge thickness of the annular area but consider 10% of the total steam flow rate as the tip leakage loss. The mean condition of steam in that blade group is found to be 2.7 bar and 0.95 dry.

(d) What do you mean by once-through boiler?

■ **Solution.**

(a) The rotor of the turbine can be of drum type or disc type. *Disc type construction is difficult (complicated) to make, but lighter in weight.* Hence the centrifugal stresses are lower at a particular speed. On the other hand, drum type construction is simple in construction, and it is easy to attach aerofoil shape blades. Further it is easier to design for tip leakage reduction which is a major problem in reaction turbines. Moreover, due to small pressure drop per stage (larger number of stages) in reaction turbines, their rotational speeds are lower and so the *centrifugal stresses are not very high* (even the reaction blades are lighter). Therefore, *drum type construction is preferred to disc type in reaction turbines.*

To accommodate an increase in specific volume at lower pressures the drum diameter is stepped up, which allows greater area without unduly increasing blade height. The *increased drum diameter also increases the torque due to steam pressure.*

(b) In impulse turbines there is no expansion of steam in moving blades, and the pressure of steam remains constant while flowing over the moving blades. The expansion takes place only in the nozzles at the inlet to the turbine in H.P. stages, or through the fixed blades in the subsequent stages. The nozzles need not occupy the complete circumference. Therefore, partial admission of steam is feasible and adopted for H.P. impulse stages.

In reaction turbines, *pressure drop is required in the moving blades also. This is not possible with partial admission. Hence full admission is essential for all stages of a reaction turbine.*

Given: $N = 3000$ r.p.m. ; $\phi = \alpha = 20°$; $C_{bl} = 120$ m/s ; $\dfrac{C_{bl}}{C_1} = 0.8$;

$\therefore$ $C_1 = \dfrac{C_{bl}}{0.8} = \dfrac{120}{0.8} = 150$ m/s

Also $C_{bl} = \dfrac{\pi DN}{60}$

or $120 = \dfrac{\pi DN}{60}$

$\therefore$ $D = \dfrac{120 \times 60}{\pi \times 3000} = 0.764$ m.

From steam tables, v_g (at 2.7 bar) $= 0.668$ m^3/kg

$v = 0.95 \times 0.668 = 0.6346$ m^3/kg

Flow area $A = \pi Dh = \pi \times 0.764 \times \dfrac{30}{1000} = 0.072$ m^2

Flow velocity $C_f = C_1 \sin \alpha = 150 \sin 20° = 51.3$ m/s ($C_{f_1} = C_{f_0} = C_f$)

Mass flow rate $\dot{m} = \dfrac{AC_f}{v} = \dfrac{0.072 \times 51.3}{0.6346} = 5.82$ kg/s

Accounting for 10% leakage (of total steam flow), the total steam flow rate is

$\dfrac{5.82}{0.9} = $ **6.467 kg/s. (Ans.)**

(c) A *once-through* boiler is a boiler which does not require any water or a steam drum. It is a monotube boiler using about 1.5 kg long tube arranged in the combustion chamber and the furnace. The economizer, boiler, and superheater are in series with no fixed surfaces as separators between the steam and water.

A Benson boiler is an example of once-through boiler, operating at supercritical pressure. The tube length to diameter ratio of such a boiler is about 2500. Due to large frictional resistance, the feed pressure should be about 1.4 times the boiler pressure.

Example 37

A twenty-stage Parson turbine receives steam at 15 bar at 300°C. The steam leaves the turbine at 0.1 bar pressure. The turbine has a stage efficiency of 80% and the reheat factor is 1.06. The total power developed by the turbine is 10665 kW. Find the steam flow rate through the turbine assuming all stages develop equal power.

The pressure of steam at certain stage of the turbine is 1 bar abs., and is dry and saturated. The blade exit angle is 25° and the blade speed ratio is 0.75. Find the mean diameter of the rotor of this stage and also the rotor speed. Take blade height as 1/12th of the mean diameter. The thickness of the blades may be neglected.

■ Solution:

Number of stage	= 20
Steam supply pressure	= 15 bar, 300°C
Exhaust pressure	= 0.1 bar
Stage efficiency of turbine, η_{stage}	= 80%
Reheat factor	= 1.06
Total power developed	= 10665 kW

Steam pressure at a certain stage = 1 bar abs., $x = 1$

Blade exit angle = 25°

Blade speed ratio, $\rho = \dfrac{C_{bl}}{C_1} = 0.75$

Height of the blade, $h = \dfrac{1}{12} D$ (mean dia. of rotor) (*i*) **Steam flow rate,** $\dot{m}_s$:

Refer to Figure 10.51.

Isentropic drop, $\qquad\qquad\left(\Delta h\right)_{\text{isentropic}} = h_1 - h_2 = 3040 - 2195 = 845 \text{ kJ} / \text{kg}$

$\eta_{\text{overall}} = \eta_{\text{stage}} \times \text{Reheat factor} = 0.8 \times 1.06 = 0.848$

Work done $\qquad\qquad\qquad\qquad$ = Actual enthalpy drop

$= \left(\Delta h\right)_{\text{isentropic}} \times \eta_{\text{overall}}$

$= 845 \times 0.848 = 716.56 \text{ kJ} / \text{kg}$

Work done per stage per kg $\qquad = \dfrac{716.56}{20} = 35.83 \text{ kJ}$ $\qquad\qquad$ (*i*)

Also, total power $\qquad\qquad$ = No. of stages $\times \dot{m}_s \times$ work done / kg stage

$\therefore \qquad\qquad\qquad\qquad 10665 = 20 \times \dot{m}_s \times 35.83$

$\therefore \qquad\qquad\qquad\qquad \dot{m}_s = \dfrac{10665}{20 \times 35.83} = \textbf{14.88 kg / s.} \quad \textbf{(Ans.)}$

(*ii*) **Mean diameter of rotor, D:**

Rotor speed, N:

Refer to Figure 10.50.

Work done per kg per stage $\qquad = C_{bl} \times C_w = C_{bl} \left(2 C_1 \cos 25° - C_{bl}\right)$

Also, $\qquad\qquad\qquad\qquad \dfrac{C_{bl}}{C_1} = 0.75$ $\qquad\qquad\qquad$ ...(Given)

$\therefore \qquad\qquad\qquad\qquad C_1 = \dfrac{C_{bl}}{0.75} = 1.33 C_{bl}$

i.e., $\quad$ Work done per kg per stage

$= C_{bl}(2 \times 1.33 C_{bl} \times 0.906 - C_{bl})$

$= 1.41 \ C_{bl}^{2} \text{ Nm}$ $\qquad\qquad\qquad\qquad\qquad\qquad\qquad$ (*ii*)

Equating (*i*) and (*ii*), we get

$1.41 \ C_{bl}^{2} = 35.83 \times 1000$

$\therefore$ $$C_{bl}^{2} = \frac{35.83 \times 1000}{1.41} \text{ or } C_{bl} = 159.41 \text{ m/s}$$

$\therefore$ $$C_1 = 1.33 \times 159.41 = 212 \text{ m/s}$$

From Figure 10.50

$$C_{f_1} = C_1 \sin \alpha = 212 \sin 25° = 89.59 \text{ m/s}$$

v_g = Specific volume at 1 bar when steam is dry and saturated

$= 1.694 \text{ m}^3 \text{/kg (from steam tables)}$

Mass flow rate, $$\dot{m}_s = \frac{\pi D h C_{f_1}}{v}$$

$\therefore$ $$14.88 = \frac{\pi \times D \times \left(\dfrac{D}{12}\right) 89.59}{1.694} \text{ or } D^2 = \frac{14.88 \times 1.694 \times 12}{\pi \times 89.59}$$

$\therefore$ $$D = \textbf{1.036 m. (Ans.)}$$

Now, $$h = \frac{D}{12} = \frac{1.036}{12} = 0.086 \text{ m} = \textbf{8.6 cm. (Ans.)}$$

Also, $$C_{bl} = \frac{\pi D N}{60}$$

$\therefore$ $$N = \frac{C_{bl} \times 60}{\pi D} = \frac{159.41 \times 60}{\pi \times 1.036} = \textbf{2938.7 r.p.m. (Ans.)}$$

Example 38

The following data relate to a stage of a reaction turbine:

Mean rotor diameter = 1.5 m; speed ratio = 0.72; blade outlet angle = 20°; rotor speed = 3000 r.p.m.

(i) Determine the diagram efficiency.

(ii) Determine the percentage increase in diagram efficiency and rotor speed if the rotor is designed to run at the best theoretical speed, the exit angle being 20°.

■ **Solution:**

Mean rotor diameter, $D = 1.5$ m

Speed ratio, $\rho = \dfrac{C_{bl}}{C_1} = 0.72$

Blade outlet angle $= 20°$

Rotor speed, $N = 3000$ r.p.m.

(This example solved purely by calculations (Figure 10.51) is not drawn to scale.)

(i) Diagram efficiency:

Blade velocity, $C_{bl} = \dfrac{\pi DN}{60} = \dfrac{\pi \times 1.5 \times 3000}{60} = 235.6$ m/s

Speed ratio, $\rho = \dfrac{C_{bl}}{C_1} = 0.72$

$\therefore$ $C_1 = \dfrac{C_{bl}}{0.72} = \dfrac{235.6}{0.72} = 327.2$ m/s

Assuming that velocity triangles are symmetrical

$\alpha = \phi = 20°$

From the velocity ΔLMS

$C_{r_1}^{~2} = C_1^{~2} + C_{bl}^{~2} - 2C_1 C_{bl} \cos \alpha$

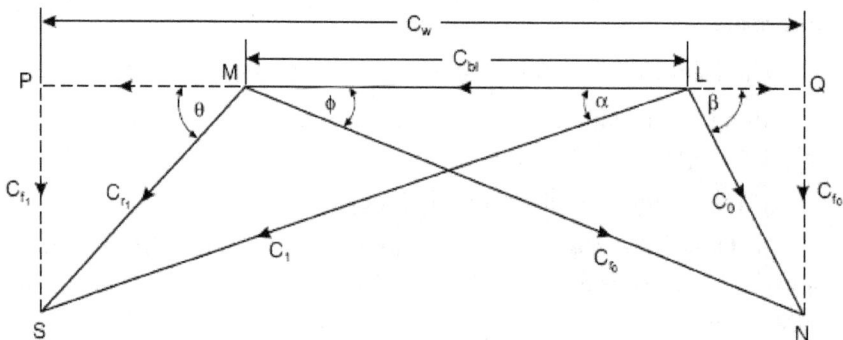

FIGURE 10.51. Sketch for Example 38.

$$C_{r_1} = \sqrt{(327.2)^2 + (235.6)^2 - 2 \times 327.2 \times 235.6 \cos 20°}$$

$$= 100\sqrt{10.7 + 5.55 - 14.48} = 133 \text{ m/s}$$

i.e., $\qquad\qquad\qquad\qquad C_{r_1} = 133 \text{ m/s}$

Work done per kg of steam $\qquad = C_{bl}C_w = C_{bl}(2C_1 \cos \alpha - C_{bl})$

$$= 235.6 \ (2 \times 327.2 \cos 20° - 235.6) = 89371.3 \text{ Nm.}$$

Energy supplied per kg of steam

$$= \frac{C_1^2 + C_{r_0}^2 - C_{r_1}^2}{2}$$

$$= \frac{2C_1^2 - C_{r_1}^2}{2} \qquad\qquad \left(\because C_1 = C_{r_0} \right)$$

$$= \frac{2 \times (327.2)^2 - (133)^2}{2} = 98215.3 \text{ Nm}$$

$\therefore$ *Diagram efficiency* $\qquad = \dfrac{89371.3}{98215.3} = 0.91 = \textbf{91\%. (Ans.)}$

(*ii*) Percentage increase in diagram efficiency:

For the best diagram efficiency (*maximum*), the required condition is

$$\rho = \frac{C_{bl}}{C_1} = \cos \alpha$$

$\therefore \qquad\qquad\qquad C_{bl} = C_1 \cos \alpha = 372.2 \cos 20° = 307.46 \text{ m/s}$

For this blade speed, the value of C_{r_1} is again calculated by using equation (*i*),

$$C_{r_1} = \sqrt{(327.2)^2 + (307.46)^2 - 2 \times 327.2 \times 307.46 \times \cos 20°}$$

$$= 100 \ \sqrt{10.7 + 9.45 - 18.906} = 111.5 \text{ m/s}$$

Diagram efficiency $\qquad\qquad = \dfrac{2C_{bl}\left(2C_1 \cos \alpha - C_{bl}\right)}{\left(C_1^2 + C_{r_0}^2 - C_{r_1}^2\right)}$

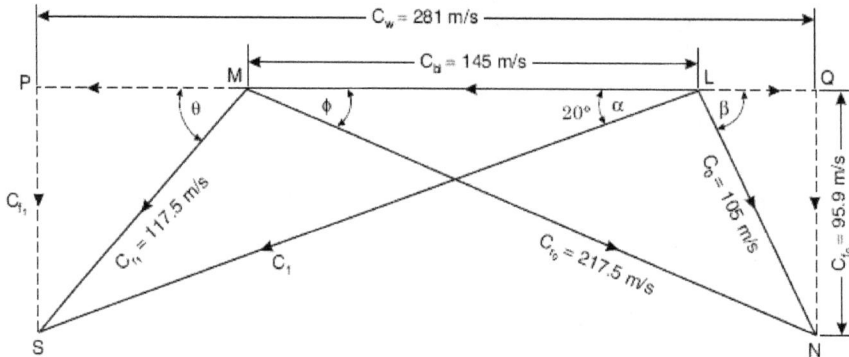

FIGURE 10.52. Sketch for Example 39.

$$= \frac{2 \times 307.46 \left(2 \times 327.2 \ \cos \ 20° - 307.46\right)}{\left(327.2\right)^2 + \left(327.2\right)^2 - \left(111.5\right)^2} = 0.937 \ \text{or} \ 93.7\%$$

Percentage increase in diagram efficiency

$$= \frac{0.937 - 0.91}{0.91} = \textbf{0.0296 or 2.96\%. (Ans.)}$$

> The diagram efficiency for the best speed can also be calculated by using relation
>
> $$\eta_{\text{blade}} = \frac{2 \cos^2 \alpha}{1 + \cos^2 \alpha} = \frac{2 \times \cos^2 20°}{1 + \cos^2 20°} = \frac{1.766}{1 + 0.883} = 0.937 \ \text{or} \ 93.7\%. \ \left(\textbf{Ans.}\right)$$

The best theoretical speed of the rotor is given by,

$$C_{bl} = \frac{\pi DN}{60} \ \text{or} \ N = \frac{60 C_{bl}}{\pi D} = \frac{60 \times 307.46}{\pi \times 1.5} = \textbf{3914.7 r.p.m.} \ \left(\textbf{Ans.}\right)$$

Impulse reaction turbine

Example 39

The following data relate to a stage of an impulse reaction turbine:

Steam velocity coming out of nozzle = 245 m/s; nozzle angle = 20°; blade mean speed = 145 m/s; speed of the rotor = 300 r.p.m.; blade height = 10 cm; specific volume of steam at nozzle outlet and blade outlet respectively = 3.45 m³/kg and 3.95 m³/kg; Power developed by the turbine = 287 kW; efficiency of nozzle and blades combined = 90%; carryover coefficient = 0.82.

Find:

(i) The heat drop in each stage

(ii) Degree of reaction

(iii) Stage efficiency

■ **Solution:**

Steam velocity coming out of nozzle, $C_1 = 245$ m/s

Nozzle angle, $\qquad\qquad \alpha = 20°$

Blade mean speed, $\qquad C_{bl} = 145$ m/s

Speed of the rotor, $\qquad N = 3000$ r.p.m.

Blade height, $\qquad\qquad h = 10$ cm $= 0.1$ m

Specific volume of steam at nozzle outlet or blade inlet, $v_1 = 3.45$ m^3/kg

Specific volume of steam at blade outlet, $v_0 = 3.95$ m^3/kg

Power developed by the turbine $= 287$ kW

Efficiency of nozzle and blades combinedly $= 90\%$

Carry over co-efficient, $\psi = 0.82$

Blade speed, $\qquad\qquad C_{bl} = \dfrac{\pi DN}{60}$

$\therefore \qquad\qquad\qquad D = \dfrac{60\,C_{bl}}{\pi N} = \dfrac{60 \times 145}{\pi \times 3000} = 0.923$ m

Mass flow rate, $\qquad \dot{m}_s = \dfrac{C_{f_1} \times \pi Dh}{v_1} = \dfrac{C_1 \sin \alpha \times \pi Dh}{v_1}$

$\qquad\qquad = \dfrac{245 \times \sin 20° \times \pi \times 0.923 \times 0.1}{3.45} = 7.04$ kg/s

Also $\qquad\qquad\qquad \dot{m}_s = \dfrac{C_{f_0} \pi Dh}{v_o}$

$$\therefore \qquad C_{f_0} = \frac{\dot{m}_s v_0}{\pi D h} = \frac{7.04 \times 3.95}{\pi \times 0.923 \times 0.1} = 95.9 \text{ m/s}$$

The power is given by, $\qquad P = \dfrac{\dot{m}_s \times C_{bl} \times C_w}{1000}$

$$287 = \frac{7.04 \times 145 \times C_w}{1000}$$

$$\therefore \qquad C_w = \frac{287 \times 1000}{7.04 \times 145} = 281 \text{ m/s}$$

Now draw velocity triangles as follows:

Select a suitable scale (say 1 cm = 25 m/s)

- Draw LM = blade velocity = 145 m/s; $\angle MLS$ = nozzle angle = 20°. Join MS to complete the inlet ΔLMS.

- Draw a perpendicular from S which cuts the line through LM at point P. Mark the point Q such that $PQ = C_w = 281$ m/s.

- Draw a perpendicular through point Q and the point N as $QN = 95.9$ m/s.

Join LN and MN to complete the *outlet* velocity triangle.

From the velocity triangles;

$C_{r_1} = 117.5$ m/s ; $C_{r_0} = 217.5$ m/s ; $C_0 = 105$ m/s.

(*i*) Heat drop in each stage, $(\Delta h)_{\text{stage}}$:

Heat drop in *fixed blades* (Δh_f)

$= \dfrac{C_1^2 - \psi C_0^2}{2 \times \eta_{\text{nozzle}}}$, where ψ is a carry over co-efficient

$= \dfrac{(245)^2 - 0.82 \times (105)^2}{2 \times 0.9 \times 1000} = 28.32$ kJ/kg

Heat drop in *moving blades* (Δh_m)

$= \dfrac{C_{r_0}^2 - C_{r_1}^2}{2 \times \eta_{\text{nozzle}}} = \dfrac{(217.5)^2 - (117.5)^2}{2 \times 0.9 \times 1000} = 18.61$ kJ/kg

Total heat drop in a stage,

$$\left(\Delta h\right)_{\text{stage}} = \Delta h_f + \Delta h_m = 28.32 + 18.61 = \textbf{46.93 kJ / kg. (Ans.)}$$

(*ii*) Degree of reaction, R_d:

$$R_d = \frac{\Delta h_m}{\Delta h_m + \Delta h_f} = \frac{18.61}{18.61 + 28.32} = \textbf{0.396.} \quad (\textbf{Ans.})$$

(*iii*) Stage efficiency, η_{stage}:

Work done per kg of steam

$$= \frac{C_{bl} \times C_w}{1000} = \frac{145 \times 281}{1000} = 40.74 \text{ kJ/kg of steam}$$

$$\therefore \quad \eta_{\text{stage}} = \frac{\text{Work done per kg of steam}}{\text{Total heat drop in a stage}} = \frac{40.74}{46.93} = \textbf{0.868 or 86.8\%. (Ans.)}$$

The terms *state point locus* and *reheat factor* are discussed in the following section:

10.10. State Point Locus and Reheat Factor

State Point Locus

The state point may be defined as that point on an h-s diagram which represents the condition of steam at that instant. Thus, knowing the initial condition of steam entering the nozzle of a turbine, the initial state point of Figure 10.53 may be located on the *h-s* diagram. If stage efficiency $\left[\eta_{\text{stage}} = \frac{\text{S.P.}}{\text{A.P.}} = \frac{h_1 - h_5}{h_1 - h_2}\right]$ is known or assumed, the position of the end state point 5 for a stage is readily obtained. The point 5 now becomes the initial state point for the succeeding stage of the turbine.

Let us now consider a *multistage turbine having four stages*.

The initial point 1 is located according to the given initial condition. (1-2) is adiabatic expansion in the first stage. h_2' may be calculated from the following relation, $h_1 - h_2' = \eta_{\text{stage}} \left(h_1 - h_2\right)$. Point $2'$ is then located with the value h_2' on the p_2 line. Then ($2'$ -3) is drawn showing adiabatic expansion. Point $3'$ may be located by finding h_3' from $h_2' - h_3' = \eta_{\text{stage}} \left(h_2' - h_3\right)$

. Proceeding in this way all stage points **3′**, **4′**, and **5′** may be fixed. The locus passing through these points is called the *"State point locus."* The sum of the adiabatic heat drops (1-2) + (2′ -3) + (3′ -4) + (4′ -5) is generally called the *"cumulative heat drop"* and is represented as h_{cum}. For the purpose of design the various quantities obtained from the Mollier diagram are set out in some curves form, and these curves are termed "condition curves."

Reheat Factor

Referring to Figure 10.53, the *adiabatic heat drop* (h_{adi}) from pressure p_1 to final pressure p_5, considering all the stages as one unit, is $(h_1 - h_6)$ It will be found that h_{adi} is less that h_{cum}. The ratio $\dfrac{h_{cum}}{h_{adi}}$ is termed as the *Reheat factor.*

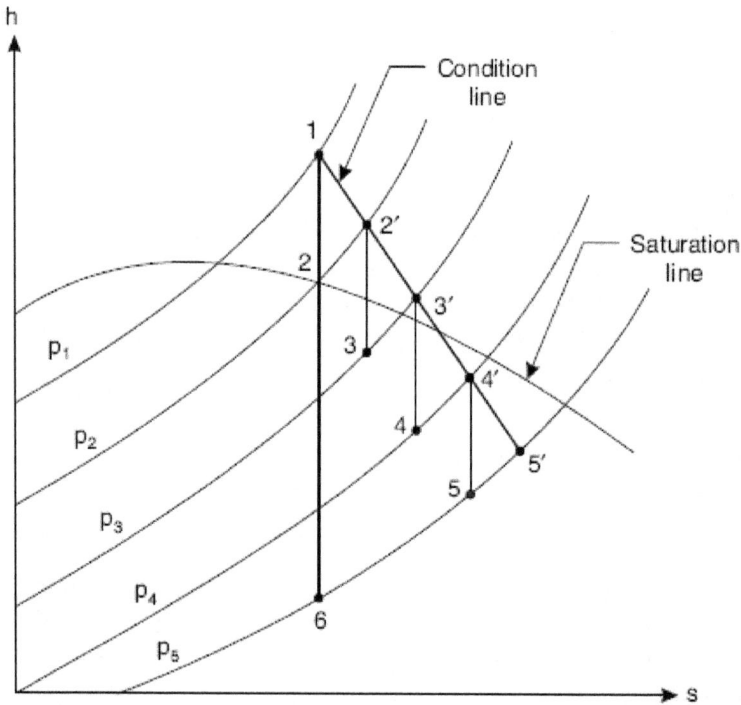

FIGURE 10.53. Initial condition of steam entering the nozzle of a turbine.

The value of the reheat factor depends on the following factors:

(*i*) Stage efficiency;

(*ii*) Initial pressure and condition of steam;

(*iii*) Final pressure.

Example 40

In a three-stage steam turbine, steam enters at 35 bar and 400°C and exhausts at 0.05 bar, 0.9 dry. If the work developed per stage is equal, determine:

(i) The condition of steam at the entry to each stage.

(ii) The stage efficiencies.

(iii) The reheat factor.

(iv) The Internal turbine efficiency.

Assume condition line to be straight.

■ Solution:

Initial condition of steam = 35 bar, 400°C

Exhaust condition of steam = 0.05 bar, 0.9 dry

(*i*) Condition of steam at entry to each stage:

Refer to Figure 10.54.

Locate points L_1 and L_4 corresponding to entry and exhaust conditions of steam.

Since the condition line is straight (given), points L_1 and L_4 are joined by a straight line.

Heat drop due to expansion from L_1 to L_4 = 3222 – 2315 = 907 kJ/kg.

Since the work developed per stage is equal,

$$\therefore \quad \text{Useful work/stage} \quad = \frac{907}{3} = 302.3 \text{ kJ/kg}$$

i.e., $\left(h_{L_1} - h_{N_1} \right) = 302.3 \text{ kJ/kg}$

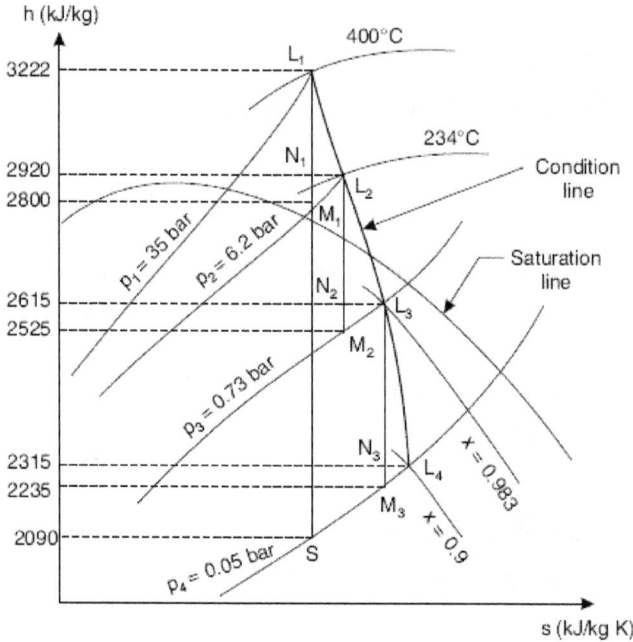

FIGURE 10.54. Condition of steam at entry to each stage.

Produce horizontal to cut condition line at L_2. L_1N_1 produced cuts the pressure line through L_2 at M_1. Thus L_2, N_1, and M_1 are located.

Proceeding for other stages likewise, we get the following results:

Points	Pressure	Temp. or quality of steam	
L_2	6.2 bar	234°C	(Ans.)
L_3	0.73 bar	0.98 dry	(Ans.)

$$h_{L_1 M_1} = h_{L_1} - h_{M_1} = 3222 - 2800 = 422 \text{ kJ/kg}$$

$$h_{L_2 M_2} = h_{L_2} - h_{M_2} = 2920 - 2525 - 395 \text{ kJ/kg}$$

$$h_{L_3 M_3} = h_{L_3} - h_{M_3} = 2615 - 2235 = 380 \text{ kJ/kg}$$

(*ii*) Stage efficiencies:

Efficiency of stage 1, $\eta_1 = \dfrac{h_{L_1} - h_{N_1}}{h_{L_1} - h_{M_1}} = \dfrac{302.3}{422} = 0.7163$ **or 71.63%. (Ans.)**

Efficiency of stage 2, $\eta_2 = \dfrac{h_{L_2} - h_{N_2}}{h_{L_2} - h_{M_2}} = \dfrac{302.3}{395} = 0.7653$ **or 76.53%. (Ans.)**

Efficiency of stage 3, $\eta_3 = \dfrac{h_{I_3} - h_{N_3}}{h_{I_3} - h_{M_3}} = \dfrac{302.3}{380} = 0.7955 = \mathbf{79.55\%.}$ **(Ans.)**

(*iii*) Reheat factor:

$$\text{Reheat factor} \quad = \frac{\text{Cumulative drop}}{\text{Isentropic enthalpy drop}} = \frac{h_{L_1M_1} + h_{L_2M_2} + h_{L_3M_3}}{h_{I_1} - h_s}$$

$$= \frac{422 + 395 + 380}{3222 - 2090} = \mathbf{1.057.} \quad \textbf{(Ans.)}$$

(*iv*) Internal turbine efficiency:

$$\text{Internal turbine efficiency} = \frac{h_{I_1} - h_{L_4}}{h_{I_1} - h_s} = \frac{3222 - 2315}{3222 - 2090} = \mathbf{0.801\ or\ 80.1\%.} \quad \textbf{(Ans.)}$$

10.10.1. Turbine Bleeding

Bleeding is the process of draining steam from the turbine at certain points during its expansion, and using this steam for heating the feed water supplied to the boiler. In this process a small quantity of steam, at certain sections of the turbine, is drained from the turbine and is then circulated around the feed water pipe leading from the hot well to the boiler. The steam is thus condensed due to relatively cold water, and the heat so lost by steam is transferred to the feed water. The condensed steam then finds its way to the hot well.

There is a usual practice in bleeding installations to allow the bled steam to mix with the feed water. The mixture of steam and water then proceeds to the boiler.

By the bleeding process hotter water is supplied to the boiler, though at the cost of the loss of a small amount of turbine work. *Due to this process efficiency is slightly increased, but at the same time power developed is also decreased.*

The bleeding process in steam turbines approximates to cascade heating and tends to modify the Rankine cycle to a reversible cycle, thus increasing the efficiency; but any increase in efficiency due to an approach to the condition of thermodynamic reversibility is accompanied by a decrease in power. Hence it follows that the thermodynamic benefits derived from the process of bleeding are of a *limited character. The ideal Rankine cycle, modified to take into account the effect of bleeding, is known as the regenerative cycle.*

10.10.2. Energy Losses in Steam Turbines

The increase in heat energy required for doing mechanical work in practice as compared to the theoretical value, in which the process of expansion takes place strictly according to the adiabatic process, is termed as energy loss in a steam turbine.

The losses which appear in an actual turbine may be divided into two following groups:

1. Internal losses. Losses directly connected with the steam conditions while in its flow through the turbine. They may be further classified as:

(*i*) Losses in regulating valves.

(*ii*) Losses in nozzles (guide blades).

(*iii*) Losses in moving blades:

(a) losses due to trailing edge wake;

(b) impingement losses;

(c) losses due to leakage of steam through the angular space;

(d) frictional losses;

(e) losses due to turning of the steam jet in the blades;

(f) losses due to shrouding;

(*iv*) Leaving velocity losses (exit velocity).

(*v*) Losses due to friction of disc carrying the blades and windage losses.

(*vi*) Losses due to clearance between the rotor and guide blade discs.

(*vii*) Losses due to wetness of steam.

(*viii*) Losses in exhaust piping, and so on.

2. External losses. Losses which do not influence the steam conditions. They may be further classified as:

(*i*) Mechanical losses.

(*ii*) Losses due to leakage of steam from the labyrinth gland seals.

10.11. Steam Turbine Governing and Control

The objective of governing is to keep the turbine speed fairly constant irrespective of load.

The principal methods of steam turbine governing are as follows:

1. Throttle governing

2. Nozzle governing

3. By-pass governing

4. Combination of 1 and 2 or 1 and 3.

10.11.1. Throttle Governing

Throttle governing is the most widely used, particularly on *small turbines,* because *its initial cost is less* and *the mechanism is simple.* The object of throttle governing is to *throttle the steam* whenever *there is a reduction of load* compared to economic or design load for maintaining speed and *vice versa.*

Figure 10.55 shows a simple throttle arrangement. To start the turbine for full load running, valve *A* is opened. The operation of double beat valve *B* is carried out by an oil servomotor which is controlled by a centrifugal governor. As the steam turbine gains speed, the valve *B* closes to throttle the steam and reduces the supply to the nozzle.

For a turbine governed by throttling, the relationship between steam consumption and load is given by the well-known *Willan's line* as shown in

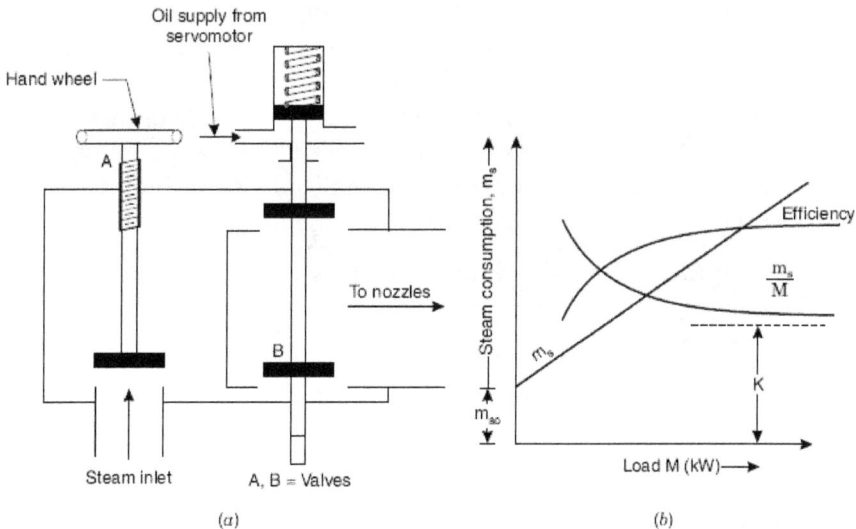

Figure 10.55 Throttle arrangement and steam consumption.

Figure 10.55 (b). Several tests have shown that when a turbine is governed by throttling, the Willan's line is straight. It is expressed as:

$$m_s = KM + m_{s_0}$$

where, $\dot{m}_s$ = Steam consumption in kg/h at any load M,

m_{s_0} = Steam consumption in kg/h at no load,

m_{s_1} = Steam consumption in kg/h at full load,

M = Any other load in kW,

M_1 = Full load in kW, and

K = Constant.

m_{s_0} varies from about 0.1 to 0.14 times the full load consumption. The equation (37) can also be written as:

$$\frac{m_s}{M} = K + \frac{m_{s_0}}{M}, \text{ where } \frac{m_s}{M} \text{ is called the steam consumption per kWh.}$$

10.11.2. Nozzle Governing

The efficiency of a steam turbine is considerably reduced if throttle governing is carried out at low loads. An alternative and more efficient form of governing is by means of nozzle control. Figure 10.56 shows a diagrammatic arrangement of typical nozzle control governing. In this method of governing, the nozzles are grouped together in 3 to 5 or more groups and the supply of steam to each group is controlled by regulating valves. Under full load conditions the valves remain fully open.

When the load on the turbine becomes more or less than the design value, the supply of steam to a group of nozzles may be varied accordingly so as to restore the original speed.

Nozzle control can only be applied to the first stage of a turbine. It is suitable for a *simple impulse turbine* and *larger units which have an impulse stage followed* by an *impulse-reaction turbine.* In pressure compounded impulse turbines, there will be some drop-in pressure at entry to the second stage when some of the first stage nozzles are cut out.

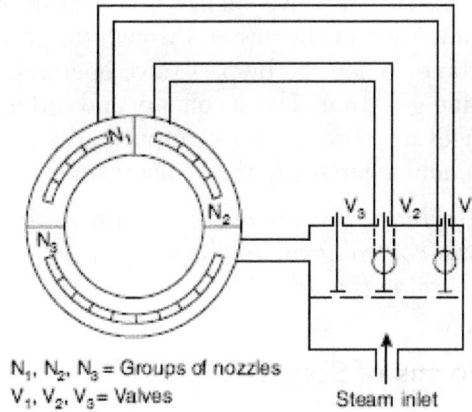

N₁, N₂, N₃ = Groups of nozzles
V₁, V₂, V₃ = Valves

Steam inlet

FIGURE 10.56. Diagrammatic arrangement of typical nozzle control governing.

10.11.3. Comparison of Throttle and Nozzle Control Governing

1.	*Throttling losses*	Severe	No throttling losses (Actually there is a little bit of throttling losses in nozzle valves which are partially open).
2.	*Partial admission losses*	Low	High
3.	*Heat drop available*	Lesser	Larger
4.	*Use*	Used in both impulse and reaction turbines.	Used in impulse and also in reaction (if initial stage impulse) turbines.
5.	*Suitability*	Small turbines	Medium and larger turbines

TABLE 10.2. Comparison of Throttle and Nozzle Control Governing

10.11.4. By-Pass Governing

With steam turbines that are designed to work at economic load, it is desirable to have full admission of steam in the high-pressure stages. At the maximum load, which is greater than the economic load, the additional steam required could not pass through the first stage since additional nozzles are not available. By-pass regulation allows for this in a turbine which is throttle governed, by means of a second by-pass valve in the first stage nozzle (Figure 10.57). This valve opens when the throttle valve has opened a definite amount. Steam is by-passed through the second valve to a lower stage in the turbine. When the by-pass valve operates, it is under the control of the turbine governor. The secondary and tertiary supplies of steam in the lower stages increase the work output in these stages, but there is a *loss in efficiency* and a curving of the Willian's line.

In reaction turbines, because of the pressure drop required in the moving blades, nozzle control governing is not possible, so *throttle governing plus by-pass governing is used.*

10.12. Special Forms of Steam Turbines

In many industries such as chemical, sugar refining, papermaking, textile, and so on, where combined use of power and heating and process work is required, it is wasteful to generate steam for power and process purposes separately, because about 70% of the heat supplied for power purposes will normally be carried away by the cooling water. On the other hand, if the engine or turbine is operated with a normal exhaust pressure, then the temperature of the exhaust steam is too low to be of any use for the heating process. If suitable modification of the initial steam pressure and exhaust pressure is made, it would be possible to generate the required power and

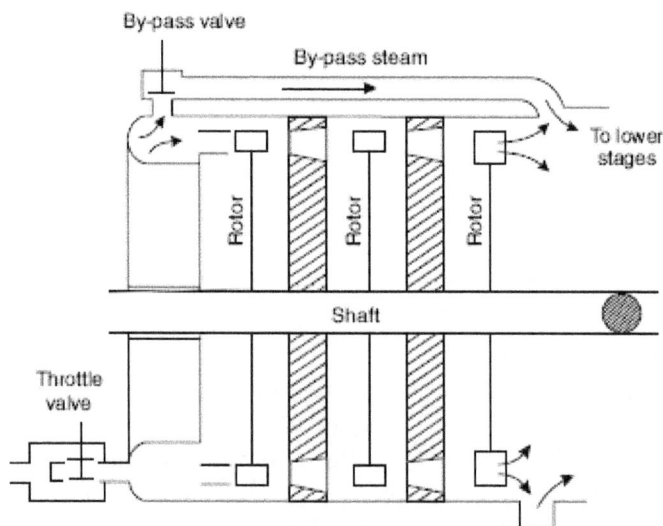

FIGURE 10.57. By-pass governing.

still have available for process work a large quantity of heat in the exhaust steam. Thus, in combined power and process plants, the following types of steam turbines are used: (a) *Back-pressure turbines* and (b) *Steam extraction or pass-out turbines*.

10.12.1. Back-Pressure Turbine

In this type of turbine steam at boiler pressure enters the turbine and is exhausted into a pipe. This pipe leads to a process plant or another turbine. The back-pressure turbine may be used in cases where the power generated (by expanding steam) from economical initial pressure down to the heating pressure is equal to, or greater than, the power requirements. The steam exhausted from the turbine is *usually superheated* and in most cases is not suitable for process work due to the following reasons:

(*i*) It is impossible to control its temperature, and

(*ii*) The rate of the heat transfer from superheated steam to the heating surface is lower than that of saturated steam. Consequently, a desuperheater is invariably used. To enhance the power capacity of the existing installation, a high-pressure boiler and a back-pressure turbine are added to it. This added high pressure boiler supplies steam to the back-pressure turbine, which exhausts into the old low-pressure turbine.

10.12.2. Extraction Pass-Out Turbine

It is found that in several cases the power available from a back-pressure turbine (through which the whole of the steam flows) is appreciably less than that required in the factory, and this may be due to the following reasons:

(*i*) Small heating or process requirements;

(*ii*) A relatively high exhaust pressure; and

(*iii*) A combination of both.

In such a case it would be possible to install a back-pressure turbine to provide the heating steam and a condensing turbine to generate extra power, but it is possible, and useful, to combine functions of both machines in a single turbine. Such a machine is called an extraction or pass-out turbine, and here at some intermediate point between inlet and exhaust, some steam is extracted or passed out for process or heating purposes. In this type of turbine, a sensitive governor is used, which controls the admission of steam to the high-pressure section so that regardless of power or process requirements, constant speed is maintained.

10.12.3. Exhaust or Low Pressure Turbine

If an uninterrupted supply of low pressure steam is available (such as from reciprocating steam engine exhaust), it is possible to improve the efficiency of the whole plant by fitting an exhaust or low-pressure turbine. The use of an exhaust turbine is chiefly made where there are several reciprocating steam engines that work intermittently and are non-condensing (*e.g.*, rolling mill and colliery engines). The exhaust steam from these engines is expanded in an exhaust turbine and then condensed. In this turbine some form of heat accumulator is needed to collect the more or less irregular supply of low pressure steam from the non-condensing steam engines and deliver it to the turbine at the rate required. In some cases when the supply of low pressure steam falls below the demand, live steam from the boiler, with its pressure and temperature reduced, is used to make up the deficiency.

The necessary drop in pressure may be obtained by the use of a reducing valve, or for large flows, more economically by expansion through another turbine. The high pressure and low-pressure turbines are sometimes combined on a common spindle, and because of two supply pressures this combined unit is known as a "mixed pressure turbine."

10.13. Exercises

Choose the Correct Answer:

1. In case of an impulse steam turbine

    (a) there is enthalpy drop in fixed and moving blades

    (b) there is enthalpy drop only in moving blades

    (c) there is enthalpy drop in nozzles

    (d) none of the above.

2. A De-Laval turbine is

    (a) pressure compounded impulse turbine

    (b) velocity compounded impulse turbine

    (c) simple single wheel impulse turbine

    (d) simple single wheel reaction turbine.

3. The pressure on the two sides of the impulse wheel of a steam turbine

 (a) is the same
 (b) is different
 (c) increases from one side to the other side
 (d) decreases from one side to the other side.

4. In a De Laval steam turbine

 (a) the pressure in the turbine rotor is approximately the same as in the condenser
 (b) the pressure in the turbine rotor is higher than pressure in the condenser
 (c) the pressure in the turbine rotor gradually decreases from inlet to exit to condenser
 (d) none of the above.

5. In the case of a reaction steam turbine

 (a) there is an enthalpy drop both in fixed and moving blades
 (b) there is an enthalpy drop only in fixed blades
 (c) there is an enthalpy drop only in moving blades
 (d) none of the above.

6. The Curtis turbine is

 (a) a reaction steam turbine
 (b) a pressure velocity compounded steam turbine
 (c) a pressure compounded impulse steam turbine
 (d) a velocity compounded impulse steam turbine.

7. The Rateau steam turbine is

 (a) a reaction steam turbine
 (b) a velocity compounded impulse steam turbine
 (c) a pressure compounded impulse steam turbine
 (d) a pressure velocity compounded steam turbine.

8. The Parson's turbine is

 (a) a pressure compounded steam turbine
 (b) a simple single wheel impulse steam turbine

(c) a simple single wheel reaction steam turbine

(d) a multiwheel reaction steam turbine.

9. Blade or diagram efficiency is given by

(a) $\dfrac{\left(C_{w_1} \pm C_{w_0}\right) C_{bl}}{C_1}$

(b) $\dfrac{2C_{bl}\left(C_{w_1} \pm C_{w_0}\right)}{C_1^2}$

(c) $\dfrac{C_{bl}^2}{C_1^2}$

(d) $\dfrac{C_1^2 - C_0^2}{C_1^2}$

10. Axial thrust on the rotor of a steam turbine is

(a) $\dot{m}_s\left(C_{f_1} - C_{f_0}\right)$

(b) $\dot{m}_s\left(C_{f_1} - 2C_{f_0}\right)$

(c) $\dot{m}_s\left(C_{f_1} + C_{f_0}\right)$

(d) $\dot{m}_s\left(2C_{f_1} - C_{f_0}\right)$.

11. Stage efficiency of a steam turbine is

(a) $\eta_{blade}/\eta_{nozzle}$

(b) $\eta_{nozzle}\,\eta_{blade}$

(c) $\eta_{nozzle} \times \eta_{blade}$

(d) none of the above.

12. For maximum blade efficiency for a single-stage impulse turbine

(a) $\rho\left(=\dfrac{C_{bl}}{C_1}\right) = \cos^2 \alpha$

(b) $\rho = \cos \alpha$

(c) $\rho = \dfrac{\cos \alpha}{2}$

(d) $\rho = \dfrac{\cos^2 \alpha}{2}$

13. Degree of reaction as referring to a steam turbine is defined as

(a) $\dfrac{\Delta h_f}{\Delta h_m}$

(b) $\dfrac{\Delta h_m}{\Delta h_f}$

(c) $\dfrac{\Delta h_m}{\Delta h_m + \Delta h_f}$

(d) $\dfrac{\Delta h_f}{\Delta h_f + \Delta h_m}$

14. For a Parson's reaction steam turbine, the degree of reaction is

(a) 75%

(b) 100%

(c) 50%

(d) 60%.

15. The maximum efficiency for a Parson's reaction turbine is given by

$\eta_{max} = \dfrac{\cos \alpha}{1 + \cos \alpha}$

$\eta_{max} = \dfrac{2 \cos \alpha}{1 + \cos \alpha}$

$\eta_{max} = \dfrac{2 \cos^2 \alpha}{1 + \cos^2 \alpha}$

$\eta_{max} = \dfrac{1 + \cos^2 \alpha}{2 \cos^2 \alpha}$

16. Reheat factor in steam turbines depends on

(a) exit pressure only

(b) stage efficiency only

(c) initial pressure and temperature only

(d) all of the above.

17. For multistage steam turbines the reheat factor is defined as

(a) stage efficiency × nozzle efficiency

(b) commulative enthalpy drop × η_{nozzle}

(c) $\dfrac{\text{commulative enthalpy drop}}{\text{isen tropic enthalpy drop}}$

(d) $\dfrac{\text{isentropic enthalpy drop}}{\text{cumulative actual enthalpy drop}}$.

18. The value of the reheat factor normally varies from

(a) 0.5 to 0.6

(b) 0.9 to 0.95

(c) 1.02 to 1.06

(d) 1.2 to 1.6.

19. Steam turbines are governed by the following methods

   (a) Throttle governing

   (b) Nozzle control governing

   (c) By-pass governing

   (d) All of the above.

20. In steam turbines the reheat factor

   (a) increases with the increase in the number of stages

   (b) decreases with the increase in the number of stages

   (c) remains same irrespective of the number of stages

   (d) none of the above.

21. Define a steam turbine and state its fields of application.

22. How are the steam turbines classified?

23. Discuss the advantages of a steam turbine over a steam engine.

24. Explain the difference between an impulse turbine and a reaction turbine.

   What do you mean by the compounding of steam turbines? Discuss various methods of compounding steam turbines.

25. What methods are used in reducing the speed of the turbine rotor?

26. Explain with the help of neat sketch a single-stage impulse turbine. Also explain the pressure and velocity variations along the axial direction.

27. Define the following as related to steam turbines:
   (i) Speed ratio
   (ii) Blade velocity coefficient
   (iii) Diagram efficiency
   (iv) Stage efficiency

28. In the case of steam turbines, derive expressions for the following:
   (i) Force
   (ii) Work done

(*iii*) Diagram efficiency
(*iv*) Stage efficiency
(*v*) Axial thrust.

29. Derive the expression for maximum blade efficiency in a single-stage impulse turbine.

30. Explain the pressure compounded impulse steam turbine showing pressure and velocity variations along the axis of the turbine.

31. Explain the velocity compounded impulse steam turbine showing pressure and velocity variations along the axis of the turbine.

32. Define the term "degree of reaction" used in reaction turbines and prove that it is given by

$$R_d = \frac{C_f}{2C_{bl}}\left(\cot\phi - \cot\theta\right) \text{ when } C_{f_1} = C_{f_0} = C_f.$$

Further prove that the moving and fixed blades should have the same shape for a 50% reaction.

33. Prove that the diagram or blade efficiency of a single-stage reaction turbine is given by

$$\eta_{bl} = 2 - \frac{2}{1 + 2\rho\cos\alpha - \rho^2} \text{ where } R_d = 50\% \text{ and } C_{f_1} = C_{f_0}$$

Further prove that maximum blade efficiency is given by

$$\left(\eta_{bl}\right)_{max} = \frac{2\cos^2\alpha}{1 + \cos^2\alpha}$$

34. Explain "reheat factor." Why is its magnitude always greater than unity?

35. Describe the process and purpose of reheating as applicable to steam flowing through a turbine.

36. State the advantages and disadvantages of reheating steam.

37. Write a short note on the "bleeding of steam turbines."

38. Enumerate the energy losses in steam turbines.

39. Describe briefly the various methods of "steam turbine governing."

IMPULSE TURBINES

40. A steam jet enters the row of blades with a velocity of 380 m/s at an angle of 22° with the direction of motion of the moving blades. If the blade speed is 180 m/s and there is no thrust on the blades, determine the inlet and outlet blade angles. The velocity of steam while passing over the blade is reduced by 10%. Also determine the power developed by the turbine when the rate of flow of steam is 1000 kg per minute.

41. In a simple impulse turbine, the nozzles are inclined at 20° to the direction of motion of the moving blades. The steam leaves the nozzles at 375 m/s. The blade speed is 165 m/s. Find suitable inlet and outlet angles for the blades in order that the axial thrust is zero. The relative velocity of steam as it flows over the blades is reduced by 15% by friction. Determine also the power developed for a flow rate of 10 kg/s.

42. In an impulse turbine the nozzles are inclined at 24° to the plane of rotation of the blades. The steam speed is 1000 m/s and blade speed is 400 m/s. Assuming equiangular blades, determine:
(*i*) Blade angles
(*ii*) Force on the blades in the direction of motion
(*iii*) Power developed for a flow rate of 1000 kg/h.

43. In a De Laval turbine, the steam issues from the nozzles with a velocity of 850 m/s. The nozzle angle is 20°. Mean blade velocity is 350 m/s. The blades are equiangular. The mass flow rate is 1000 kg/min. The friction factor is 0.8. Determine: (*i*) blade angles, (*ii*) axial thrust on the end bearing, (*iii*) power developed in kW, (iv) blade efficiency, and (*v*) stage efficiency, if nozzle efficiency is 93%.

44. In a single-stage impulse turbine, the nozzles discharge the steam onto the blades at an angle of 25° to the plane of rotation and the fluid leaves the blades with an absolute velocity of 300 m/s at an angle of 120° to the direction of motion of the blades. If the blades have equal inlet and outlet angles and there is no axial thrust, estimate:
(*i*) Blade angle
(*ii*) Power produced per kg/s flow of steam
(*iii*) Diagram efficiency.

45. Steam enters the blade row of an impulse turbine with a velocity of 600 m/s at an angle of 25° to the plane of rotation of the blades. The

mean blade speed is 255 m/s. The blade angle on the exit side is 30°.
The blade friction coefficient is 10%. Determine:
(*i*) Work done per kg of steam
(*ii*) Diagram efficiency
(*iii*) Axial thrust per kg of steam/s.

46. The nozzles of an impulse turbine are inclined at 22° to the plane of
rotation. The blade angles both at inlet and outlet are 36°. The mean
diameter of the blade ring is 1.25 m and the steam velocity is 680 m/s.
Assuming shockless entry determine: (*i*) the speed of the turbine rotor
in r.p.m., (*ii*) the absolute velocity of steam leaving the blades, and
(*iii*) the torque on the rotor for a flow rate of 2500 kg/h.

47. A single-stage steam turbine is provided with nozzles from which
steam is released at a velocity of 1000 m/s at an angle of 24° to the
direction of the motion of blades. The speed of the blades is 400 m/s.
The blade angles at inlet and outlet are equal. Find: (*i*) inlet blade
angle, (*ii*) force exerted on the blades in the direction of their motion,
and (*iii*) power developed in kW for a steam flow rate of 40000 kg/h.
Assume that the steam enters and leaves the blades without shock.

48. In a single-row impulse turbine the nozzle angle is 30° and the blade
speed is 215 m/s. The steam speed is 550 m/s. The blade friction
coefficient is 0.85. Assuming axial exit and a flow rate of 700 kg/h,
determine:
(*i*) Blade angles.
(*ii*) Absolute velocity of steam at exit.
(*iii*) The power output of the turbine.

49. In a steam turbine, steam expands from an inlet condition of 7 bar
and 300°C with an isentropic efficiency of 0.9. The nozzle angle is
20°. The stage operates at optimum blade speed ratio. The blade inlet
angle is equal to the outlet angle. Determine:
(*i*) Blade angles.

(*ii*) Power developed if the steam flow rate is 0.472 kg/s.

50. Steam at 7 bar and 300°C expands to 3 bar in an impulse stage. The
nozzle angle is 20°, the rotor blades have equal inlet and outlet angles
and the stage operates with the optimum blade speed ratio. Assuming
that isentropic efficiency of nozzles is 90% and velocity at entry to the

stage is negligible, deduce the blade angles used and the mass flow required for this stage to produce 50 kW.

51. In a two-stage velocity compounded steam turbine, the mean blade speed is 150 m/s while the steam velocity as it is issued from the nozzle is 675 m/s. The nozzle angle is 20°. The exit angle of the first row moving blade, fixed blade, and the second row moving blades are 25°, 25°, and 30°, respectively. The blade friction coefficient is 0.9. If the steam flow rate is 4.5 kg/s, determine:

    (*i*) Power output.
    (*ii*) Diagram efficiency.

REACTION TURBINES

52. At a particular stage of a reaction turbine, the mean blade speed is 60 m/s and the steam pressure is 3.5 bar with a temperature of 175°C. The identical fixed and moving blades have inlet angles of 30° and outlet angles of 20°. Determine:

    (*i*) The blade height if it is $\dfrac{1}{10}$ of the blade ring diameter, for a flow rate of 13.5 kg/s.

    (*ii*) The power developed by a pair.

    (*iii*) The specific enthalpy drop if the stage efficiency is 85%.

53. At a stage in a reaction turbine the pressure of steam is 0.34 bar and the dryness 0.95. For a flow rate of 36000 kg/h, the stage develops 950 kW. The turbine runs at 3600 r.p.m. and the velocity of flow is 0.72 times the blade velocity. The outlet angle of both stator and rotor blades is 20°. Determine at this stage:

    (i) Mean rotor diameter.
    (*ii*) Height of blades.

54. In a multi-stage reaction turbine at one of the stages the rotor diameter is 1250 mm and speed ratio 0.72. The speed of the rotor is 3000 r.p.m. Determine: (*i*) the blade inlet angle if the outlet blade angle is 22°, (*ii*) diagram efficiency, and (*iii*) the percentage increase in diagram efficiency and rotor speed if the turbine is designed to run at the best theoretical speed.

55. In a 50% reaction turbine stage running at 3000 r.p.m., the exit angles are 30° and the inlet angles are 50°. The mean diameter is 1 m. The steam flow rate is 10000 kg/min and the stage efficiency is 85%. Determine:

   (*i*) power output of the stage.

   (*ii*) the specific enthalpy drop in the stage.

   (*iii*) the percentage increase in the relative velocity of steam when it flows over the moving blades.

56. Twelve successive stages of a reaction turbine have blades with effective inlet and outlet angles of 80° and 20°, respectively. The mean diameter of the blade row is 1.2 m and the speed or rotation is 3000 r.p.m. Assuming constant velocity of flow throughout, estimate the enthalpy drop per stage.

For a steam inlet condition of 10 bar and 250°C and an outlet condition of 0.2 bar, estimate the stage efficiency.

Assuming a reheat factor of 1.04, determine the blade height at a stage where the specific volume is 1.02 m3/kg.

11

GAS POWER CYCLES

In This Chapter

- Overview
- Definition of a Cycle
- Air Standard Efficiency
- The Carnot Cycle
- Otto Cycle
- Exercises

11.1. Overview

The gas power cycle has the advantage of a lack of phase change compared to water/steam plants, but at the cost of lower power capacity. Overall, a gas power plant requires less equipment and is easier to start and stop in operation compared to steam power plants. In this chapter we discuss several gas power cycles.

11.2. Definition of a Cycle

A *cycle* is defined as a repeated series of thermodynamic operations or processes occurring in a certain order. The processes may be repeated in the same order. The cycle may be of an imaginary perfect engine or an actual engine. The former is called an *ideal cycle* and the latter an *actual cycle*. In an ideal cycle all accidental heat losses and irreversibility are absent, and the working substance is assumed to behave like a perfect working substance.

11.3. Air Standard Efficiency

To compare the effects of different cycles, it is of paramount importance that the effect of the calorific value of the fuel is altogether eliminated, and this can be achieved by considering air (which is assumed to behave as a perfect gas) as the working substance in the engine cylinder. The efficiency of an engine using air as the working medium is known as an *Air standard efficiency*. This efficiency is often called an ideal efficiency.

The actual efficiency of a cycle is always less than the air standard efficiency of that cycle under ideal conditions. This is taken into account by introducing a new term, *Relative efficiency*, which is defined as:

$$\eta_{relative} = \frac{\text{Actual thermal efficiency}}{\text{Air standard efficiency}}$$

The analysis of all air standard cycles is based upon the following assumptions:

1. The gas in the engine cylinder is a perfect gas, that is, it obeys the ideal gas laws and has constant specific heats.

2. The physical constants of the gas in the cylinder are the same as those of air at moderate temperatures, that is, the molecular weight (atomic mass) of a cylinder gas is 29, and

$$c_p = 1.005 \text{ kJ/kg K, } c_v = 0.718 \text{ kJ/kg K.}$$

3. The compression and expansion processes are adiabatic, and they take place without internal friction; that is, these processes are isentropic.

4. No chemical reaction takes place in the cylinder. Heat is supplied or rejected by bringing a hot body or a cold body in contact with the cylinder at appropriate points during the process.

5. The cycle is considered closed with the same air as working fluid, always remaining in the cylinder to repeat the cycle.

11.4. The Carnot Cycle

This cycle has the highest possible efficiency and consists of four simple operations, namely:

1. Isothermal expansion

2. Adiabatic expansion

3. Isothermal compression

4. Adiabatic compression

The condition of the Carnot cycle may be imagined to occur in the following way:

Refer to Figure 11.1(a). One kg of air is enclosed in the cylinder, which (except at the end) is made of a perfect non-conducting material. A source of heat H is supposed to provide an unlimited quantity of heat, non-conducting cover C, and a sump S which is of infinite capacity, so that its temperature remains unchanged despite how much heat is supplied to it. The temperature of source H is T_1, and is the same as that of the working substance. The working substance, while rejecting heat to sump "S," has the temperature T_2, that is, the same as that of sump S.

Following are the *four stages* of the Carnot cycle.

Stage (1): Line 1–2 represents the isothermal expansion which takes place at temperature T_1 when source of heat H is applied to the end of cylinder. Heat supplied in this case is given by

RT_1 in r, where r is the ratio of expansion.

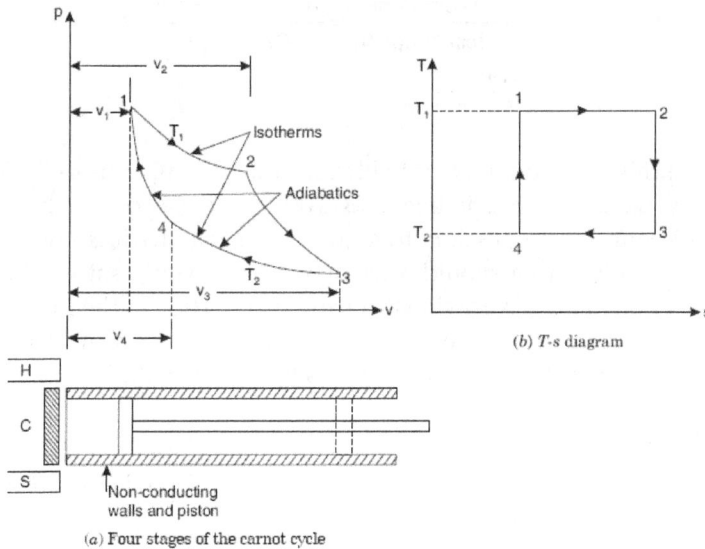

(b) T-s diagram

(a) Four stages of the carnot cycle

FIGURE 11.1 Carnot cycle.

Stage (2): Line 2–3 represents the application of non-conducting cover to the end of the cylinder. This is followed by the adiabatic expansion, and the temperature falls from T_1 to T_2.

Stage (3): Line 3–4 represents the isothermal compression which takes place when sump S is applied to the end of cylinder. Heat is rejected during this operation whose value is given by $RT_2 \ln r$ where in r is the ratio of compression.

Stage (4): Line 4–1 represents repeated application of non-conducting cover and adiabatic compression due to which the temperature increases from T_2 to T_1.

It may be noted that the ratio of expansion during isotherm 1–2 and the ratio of compression during isotherm 3–4 must be equal to get a closed cycle. Figure 11.1(*b*) represents the Carnot cycle on *T-s* coordinates.

According to the 1st law,

Heat supplied $\quad\quad\quad\quad = $ Work done + Heat rejected

Work done $\quad\quad\quad\quad\quad = $ Heat supplied – Heat rejected

$$= RT_1 . \log_e r - RT_2 . \log_e r$$

Efficiency of cycle $\quad\quad = \dfrac{\text{Work done}}{\text{Heat supplied}} = \dfrac{R \log_e r \left(T_1 - T_2\right)}{RT_1 . \log_e r}$

$$= \dfrac{T_1 - T_2}{T_1}$$

From this equation, it is quite obvious that if temperature T_2 decreases efficiency increases, and it becomes 100% if T_2 becomes absolute zero which, of course, is impossible to attain. Furthermore, it is not possible to produce an engine that should work on Carnot's cycle, as it would necessitate the piston to travel very slowly during first portion of the forward stroke (isothermal expansion) and to travel more quickly during the remainder of the stroke (adiabatic expansion) which is not practicable.

Example 1

A Carnot engine working between 400 °C and 40 °C produces 130 kJ of work.

Determine:

(*i*) The engine thermal efficiency.

(*ii*) The heat added.

(*iii*) The entropy changes during the heat rejection process.

■ **Solution.**

Temperature, $\quad T_1 = T_2 = 400 + 273 = 673 \text{ K}$

Temperature, $\quad\quad\quad\quad\quad T_3 = T_4 = 40 + 273 = 313 \text{ K}$

Work produced, $\quad\quad\quad W = 130 \text{ kJ.}$

(*i*) **Engine thermal efficiency, ηth:**

$$\eta_{th.} = \frac{673 - 313}{673} = \textbf{0.535 or 53.5\%.} \ \left(\textbf{Ans.}\right)$$

(*ii*) **Heat added:**

$$\eta_{th.} = \frac{\text{Work done}}{\text{Heat added}}$$

i.e., $\quad 0.535 = \dfrac{130}{\text{Heat added}}$

∴ Heat added $\quad\quad\quad\quad\quad = \dfrac{130}{0.535} = \textbf{243 kJ.} \ \left(\textbf{Ans.}\right)$

(*iii*) **Entropy change during the heat rejection process, $(S_3 - S_4)$:**

Heat rejected $\quad\quad\quad\quad$ = Heat added − Work done

$= 243 - 130 = 113 \text{ kJ}$

Heat rejected $\quad\quad\quad\quad\quad\quad\quad = T_3 \left(S_3 - S_4\right) = 113$

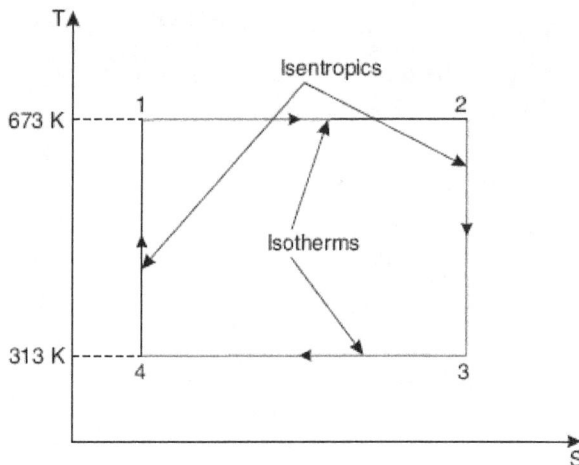

FIGURE 11.2 Sketch for Example 1.

$\therefore$ $\qquad\qquad\qquad\qquad\qquad (S_3 - S_4) = \dfrac{113}{T_3} = \dfrac{113}{313} = \mathbf{0.361\ kJ\ /\ k.}\ \ \mathbf{(Ans.)}$

Example 2

0.5 kg of air (ideal gas) executes a Carnot power cycle having a thermal efficiency of 50%. The heat transfer to the air during the isothermal expansion is 40 kJ. At the beginning of the isothermal expansion, the pressure is 7 bar and the volume is 0.12 m³. Determine:

(i)   The maximum and minimum temperatures for the cycle in K;

(ii) The volume at the end of the isothermal expansion in m3;

(iii) The heat transfer for each of the four processes in kJ.

For air c_v = 0.721 kJ/kg K, and c_p = 1.008 kJ/kg K.

■ Solution.

Refer to Figure 11.3. *Given:* m = 0.5 kg; η_{th} = 50%; Heat transferred during isothermal expansion = 40 kJ; p_1 = 7 bar, V_1 = 0.12 m³; c_v = 0.721 kJ/kg K; c_p = 1.008 kJ/kg K.

(i) The maximum and minimum temperatures, T1, T2:

$$p_1 V_1 = mRT_1$$

$$7 \times 10^5 \times 0.12 = 0.5 \times 287 \times T_1$$

$\therefore$ Maximum temperature, $\mathbf{T_1} = \dfrac{7 \times 10^5 \times 0.12}{0.5 \times 287} = \mathbf{585.4\ K.}\ \ \mathbf{(Ans.)}$

$$\eta_{cycle} = \dfrac{T_1 - T_2}{T_1} \Rightarrow 0.5 = \dfrac{585.4 - T_2}{585.4}$$

$\therefore$ Minimum temperature, $\mathbf{T_2}$ = 585.4 − 0.5 × 585.4 = **292.7 K.** **(Ans.)**

(ii) The volume at the end of isothermal expansion, V_2:

Heat transferred during isothermal expansion

$$= p_1 V_1 \ \text{In}\,(r) = mRT_1 \ \text{In}\left(\dfrac{V_2}{V_1}\right) = 40 \times 10^3 \ ...... \ \text{(Given)}$$

or $0.5 \times 287 \times 585.4 \ \text{In}\left(\dfrac{V_2}{0.12}\right) = 40 \times 10^3$

or $\text{In}\left(\dfrac{V_2}{0.12}\right) = \dfrac{40 \times 10^3}{0.5 \times 287 \times 585.4} = 0.476$

or $\qquad\qquad\qquad\qquad\qquad \mathbf{V_2} = 0.12 \times (e)^{0.476} = \mathbf{0.193\ m^3.}\ \ \mathbf{(Ans.)}$

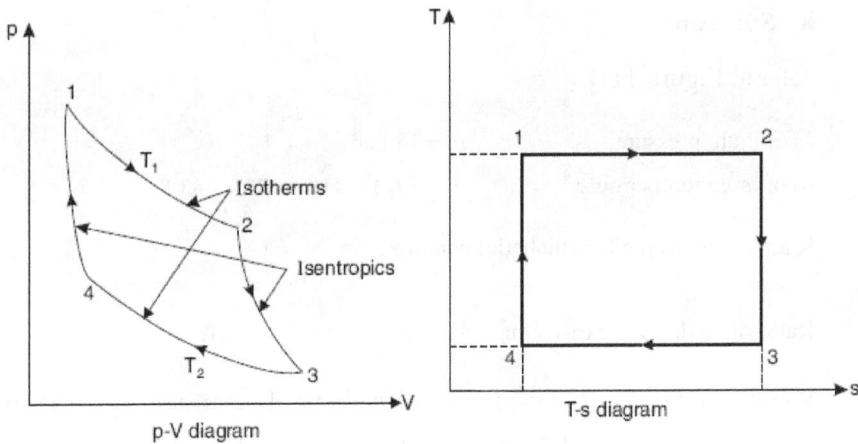

FIGURE 11.3 Sketch for Example 2.

(*iii*) The heat transfer for each of the four processes:

Process	Classification	Heat transfer
1–2	Isothermal expansion	40 kJ
2–3	Adiabatic reversible expansion	Zero
3–4	Isothermal compression	– 40 kJ
4–1	Adiabatic reversible compression	zero. (**Ans.**)

Example 3

In a Carnot cycle, the maximum pressure and temperature are limited to 18 bar and 410°C. The ratio of isentropic compression is 6 and isothermal expansion is 1.5. Assuming the volume of the air at the beginning of isothermal expansion as 0.18 m3, determine:

(*i*) *The temperature and pressures at main points in the cycle.*

(*ii*) *Change in entropy during isothermal expansion.*

(*iii*) *Mean thermal efficiency of the cycle.*

(*iv*) *Mean effective pressure of the cycle.*

(*v*) *The theoretical power if there are 210 working cycles per minute.*

■ Solution.

Refer to Figure 11.4.

Maximum pressure, $p_1 = 18$ bar

Maximum temperature, $T_1 = (T_2) = 410 + 273 = 683$ K

Ratio of isentropic (or adiabatic) compression, $\dfrac{V_4}{V_1} = 6$

Ratio of isothermal expansion, $\dfrac{V_2}{V_1} = 1.5$.

Volume of the air at the beginning of isothermal expansion, $V_1 = 0.18$ m³.

(*i*) Temperatures and pressures at the main points in the cycle:

For the *isentropic process* 4–1:

$$\frac{T_1}{T_4} = \left(\frac{V_4}{V_1}\right)^{\gamma-1} = (6)^{1.4-1} = (6)^{0.4} = 2.05$$

∴ $$T_4 = \frac{T_1}{2.05} = \frac{683}{2.05} = 333.2 \text{ K} = T_3$$

Also, $$\frac{p_1}{p_4} = \left(\frac{V_4}{V_1}\right)^{\gamma} = (6)^{1.4} = 12.29$$

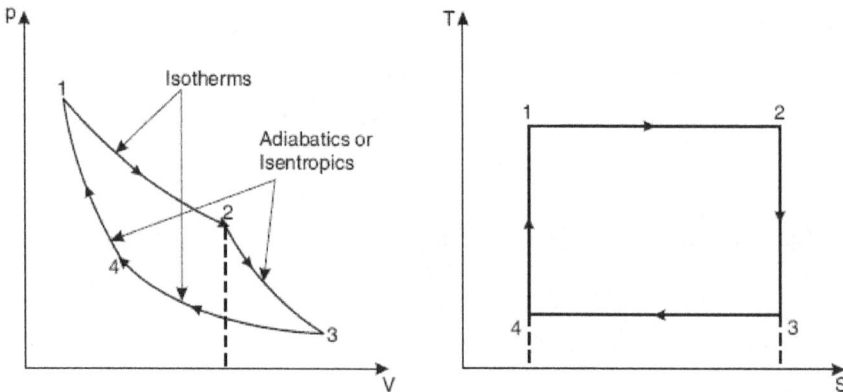

FIGURE 11.4 Sketch for Example 3.

$$\therefore \qquad p_4 = \frac{p_1}{12.29} = \frac{18}{12.29} = 1.46 \text{ bar}$$

For the *isothermal process* 1–2:

$$p_1 V_1 = p_2 V_2$$

$$p_2 = \frac{p_1 V_1}{V_2} = \frac{18}{1.5} = 12 \text{ bar}$$

For *isentropic process* 2–3, we have:

$$p_2 V_2^{\gamma} = p_3 V_3^{\gamma}$$

$$p_3 = p_2 \times \left(\frac{V_2}{V_3}\right)^{\gamma} = 12 \times \left(\frac{V_1}{V_4}\right)^{\gamma} \qquad\qquad \left[\because \ \frac{V_4}{V_1} = \frac{V_3}{V_2}\right]$$

$$= 12 \times \left(\frac{1}{6}\right)^{1.4} = \mathbf{0.97 \ bar. \ (Ans.)}$$

Hence
$$\left. \begin{array}{l} p_1 = \mathbf{18 \ bar} \quad T_1 = T_2 = 683 \text{ K} \\ p_2 = \mathbf{12 \ bar} \\ p_3 = \mathbf{0.97 \ bar} \ \ T_3 = T_4 = 333.2 \text{ K} \\ p_4 = \mathbf{1.46 \ bar} \end{array} \right\} \quad \mathbf{(Ans.)}$$

(*ii*) Change in entropy:

Change in entropy during isothermal expansion,

$$S_2 - S_1 = mR \log_e \left(\frac{V_2}{V_1}\right) = \frac{p_1 V_1}{T_1} \log_e \left(\frac{V_2}{V_1}\right) \qquad \left[\begin{array}{l} \therefore \ pV = mRT \\ or \ mR = \dfrac{pV}{T} \end{array} \right]$$

$$= \frac{18 \times 10^5 \times 0.18}{10^3 \times 683} \log_e (1.5) = \mathbf{0.192 \ kJ \, / \, K \ \ (Ans.)}$$

(*iii*) Mean thermal efficiency of the cycle:

Heat supplied, $\quad Q_s = p_1 V_1 \log_e \left(\dfrac{V_2}{V_1}\right)$

$$= T_1 (S_2 - S_1)$$
$$= 683 \times 0.192 = 131.1 \text{ kJ}$$

Heat rejected, $Q_r = p_4 V_4 \log_e\left(\dfrac{V_3}{V_4}\right)$

$= T_4(S_3 - S_4)$ because increase in entropy during heat addition
is equal to decrease in entropy during heat rejection.

$\therefore Q_r = 333.2 \times 0.192 = 63.97$ kJ

$\therefore$ Efficiency, $\eta = \dfrac{Q_s - Q_r}{Q_s} = 1 - \dfrac{Q_r}{Q_s}$

$= 1 - \dfrac{63.97}{131.1} = 0.512$ or **51.2%. (Ans.)**

(iv) Mean effective pressure of the cycle, p_m:

The mean effective pressure of the cycle is given by

$p_m = \dfrac{\text{Work done per cycle}}{\text{Stroke volume}}$

$\dfrac{V_3}{V_1} = 6 \times 1.5 = 9$

Stroke volume, $V_s = V_3 - V_1 = 9V_1 - V_1 = 8V_1 = 8 \times 0.18 = 1.44$ m³

$\therefore \qquad p_m = \dfrac{(Q_s - Q_r) \times J}{V_s} = \dfrac{(Q_s - Q_r) \times 1}{V_s}$ $\qquad (\because J = 1)$

$= \dfrac{(131.1 - 63.97) \times 10^3}{1.44 \times 10^5} = 0.466$ **bar. (Ans.)**

(v) Power of the engine, P:

Power of the engine working on this cycle is given by

$P = (131.1 - 63.97) \times (210/60) = $ **234.9 kW. (Ans.)**

Example 4

A reversible engine converts one-sixth of the heat input into work. When the temperature of the sink is reduced by 70°C, its efficiency is doubled. Find the temperature of the source and the sink.

■ **Solution.**

Let, T_1 = temperature of the source (K), and

T_2 = temperature of the sink (K)

First Case:

$$\frac{T_1 - T_2}{T_1} = \frac{1}{6}$$

i.e., $\qquad\qquad\qquad 6T_1 - 6T_2 = T_1$

or $\qquad\qquad\qquad 5T_1 = 6T_2 \text{ or } T_1 = 1.2T_2$

Second Case:

$$\frac{T_1 - \left[T_2 - (70 + 273)\right]}{T_1} = \frac{1}{3}$$

$$\frac{T_1 - T_2 + 343}{T_1} = \frac{1}{3}$$

$$3T_1 - 3T_2 + 1029 = T_1$$

$$2T_1 = 3T_2 - 1029$$

$$2 \times (1.2T_2) = 3T_2 - 1029 \qquad\qquad (\because \ T_1 = 1.2T_2)$$

$$2.4T_2 = 3T_2 - 1029$$

or $\quad 0.6T_2 = 1029$

$\therefore \qquad\qquad\qquad T_2 = \dfrac{1029}{0.6} = \textbf{1715 K or 1442°C.} \quad (\textbf{Ans.})$

and $\qquad\qquad\qquad T_1 = 1.2 \times 1715 = \textbf{2058 K or 1785°C.} \ (\textbf{Ans.})$

Example 5

An inventor claims that a new heat cycle will develop 0.4 kW for a heat addition of 32.5 kJ/min. The temperature of the heat source is 1990 K and that of the sink is 850 K. Is his claim possible?

■ **Solution.**

Temperature of heat source, $\quad T_1 = 1990$ K

Temperature of sink, $\qquad\qquad\qquad T_2 = 850$ K

Heat supplied, $\qquad\qquad\qquad\qquad = 32.5$ kJ/min

Power developed by the engine, $P = 0.4\,\text{kW}$

The *most efficient engine is one that works on the Carnot cycle*

$$\eta_{\text{carnot}} = \frac{T_1 - T_2}{T_1} = \frac{1990 - 850}{1990} = 0.573 \text{ or } 57.3\%$$

Also, thermal efficiency of the engine,

$$\eta_{\text{th}} = \frac{\text{Work done}}{\text{Heat supplied}} = \frac{0.4}{(32.5/60)} = \frac{0.4 \times 60}{32.5}$$

$$= 0.738 \text{ or } 73.8\%$$

which is not feasible as no engine can be more efficient than that working on the Carnot cycle.

Hence claims of the inventor are **not true.**

Example 6

An ideal engine operates on the Carnot cycle using a perfect gas as the working fluid. The ratio of the greatest to the least volume is fixed and is $x : 1$; the lower temperature of the cycle is also fixed, but the volume compression ratio "r" of the reversible adiabatic compression is variable. The ratio of the specific heats is γ.

Show that if the work done in the cycle is a maximum then,

$$(\gamma - 1)\log_e \frac{x}{r} + \frac{1}{r^{\gamma-1}} - 1 = 0.$$

■ **Solution.**

Refer to Figure 11.1.

$$\frac{V_3}{V_1} = x \; ; \; \frac{V_4}{V_1} = r$$

During isotherms, since compression ratio = expansion ratio

$$\therefore \quad \frac{V_3}{V_4} = \frac{V_2}{V_1}$$

Also $\dfrac{V_3}{V_4} = \dfrac{V_3}{V_1} \times \dfrac{V_1}{V_4} = x \times \dfrac{1}{r} = \dfrac{x}{r}$

Work done per kg of the gas

$$= \text{Heat supplied} - \text{Heat rejected} = RT_1 \log_e \frac{x}{r} - RT_2 \log_e \frac{x}{r}$$

$$= R(T_1 - T_2) \log_e \frac{x}{r} = RT_2 \left(\frac{T_1}{T_2} - 1 \right) \log_e \frac{x}{r}$$

But
$$\frac{T_1}{T_2} = \left(\frac{V_4}{V_1} \right)^{\gamma - 1} = (r)^{\gamma - 1}$$

$\therefore$ Work done per kg of the gas,

$$W = RT_2 \left(r^{\gamma - 1} - 1 \right) \log_e \frac{x}{r}$$

Differentiating W w.r.t. "r" and equating to zero

$$\frac{dW}{dr} = RT_2 \left[\left(r^{\gamma - 1} - 1 \right) \left\{ \frac{r}{x} \times \left(-xr^{-2} \right) \right\} + \log_e \frac{x}{r} \left\{ (\gamma - 1) r^{\gamma - 2} \right\} \right] = 0$$

or
$$\left(r^{\gamma - 1} - 1 \right) \left(-\frac{1}{r} \right) + (\gamma - 1) \times r^{\gamma - 2} \log_e \frac{x}{r} = 0$$

or
$$-r^{\gamma - 2} + \frac{1}{r} + r^{\gamma - 2} (\gamma - 1) \log_e \frac{x}{r} = 0$$

or
$$r^{\gamma - 2} \left\{ -1 + \frac{1}{r \cdot r^{\gamma - 2}} + (\gamma - 1) \log_e \frac{x}{r} \right\} = 0$$

or
$$-1 + \frac{1}{r \cdot r^{\gamma - 2}} + (\gamma - 1) \log_e \frac{x}{r} = 0$$

$$(\gamma - 1) \log_e \frac{x}{r} + \frac{1}{r^{\gamma - 1}} - 1 = 0. \quad \textbf{......Proved.}$$

11.5. Otto Cycle

This cycle is so named as it was conceived by Nikolaus Otto (1876). On this cycle, petrol, gas, and many types of oil engines work. It is the standard of comparison for internal combustion engines.

Figure 11.5 (*a*) and (*b*) shows the theoretical *p-V* diagram and *T-s* diagrams of this cycle, respectively.

- The point 1 represents that the cylinder is full of air with volume V_1, pressure p_1, and absolute temperature T_1.

- Line 1–2 represents the *adiabatic compression* of air due to which p_1, V_1, and T_1 change to p_2, V_2, and T_2, respectively.

- Line 2–3 shows the *supply of heat* to the air *at constant volume* so that p_2 and T_2 change to p_3 and T_3 (V_3 being the same as V_2).

- Line 3–4 represents the *adiabatic expansion* of the air. During expansion p_3, V_3, and T_3 change to a final value of p_4, V_4 or V_1, and T_4, respectively.

- Line 4–1 shows the *rejection of heat* by air at a *constant volume* till the original state (point 1) is reached.

Consider 1 kg of air (working substance):

Heat supplied at constant volume = $c_v(T_3 - T_2)$.

Heat rejected at constant volume = $c_v(T_4 - T_1)$.

But, work done   = Heat supplied – Heat rejected

$= c_v(T_3 - T_2) - c_v(T_4 - T_1)$

$$\therefore \ \text{Efficiency} = \frac{\text{Work done}}{\text{Heat supplied}} = \frac{c_v(T_3 - T_2) - c_v(T_4 - T_1)}{c_v(T_3 - T_2)}$$

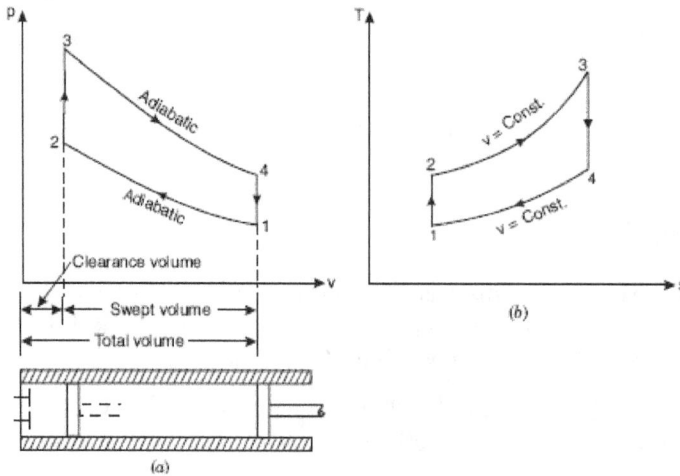

FIGURE 11.5 Theoretical p-V (a) and T-s (b) diagrams for Otto cycle.

$$= 1 - \frac{T_4 - T_1}{T_3 - T_2}$$

Let compression ratio, $\qquad\qquad r_c (= r) = \dfrac{v_1}{v_2}$

and expansion ratio, $\qquad\qquad r_e (= r) = \dfrac{v_4}{v_3}$

(These *two ratios are same* in this cycle)

As $\quad \dfrac{T_2}{T_1} = \left(\dfrac{v_1}{v_2}\right)^{\gamma-1}$

Then, $\qquad\qquad\qquad\qquad T_2 = T_1 \cdot (r)^{\gamma-1}$

Similarly, $\quad \dfrac{T_3}{T_4} = \left(\dfrac{v_4}{v_3}\right)^{\gamma-1}$

or, $\qquad\qquad\qquad\qquad T_3 = T_4 \cdot (r)^{\gamma-1}$

Inserting the values of T_2 and T_3 in equation (i), we get

$$\eta_{otto} = 1 - \frac{T_4 - T_1}{T_4 \cdot (r)^{\gamma-1} - T_1 \cdot (r)^{\gamma-1}} = 1 - \frac{T_4 - T_1}{r^{\gamma-1}(T_4 - T_1)}$$

$$= 1 - \frac{1}{(r)^{\gamma-1}}$$

This expression is known as the air standard efficiency of the Otto cycle.

It is clear from the above expression that efficiency increases with the increase in the value of r, which means we can have maximum efficiency by increasing r to a considerable extent, but *due to practical difficulties its value is limited to about 8.*

The *net work done per kg* in the Otto cycle can also be expressed in terms of p, v. If p is expressed in bar, that is, $10^5 \, N/m^2$, then work done

$$W = \left(\frac{p_3 v_3 - p_4 v_4}{\gamma - 1} - \frac{p_2 v_2 - p_1 v_1}{\gamma - 1}\right) \times 10^2 \ kJ$$

Also $\qquad\qquad \dfrac{p_3}{p_4} = r^\gamma = \dfrac{p_2}{p_1}$

$\therefore \qquad\qquad \dfrac{p_3}{p_2} = \dfrac{p_4}{p_1} = r_p$

where r_p stands for *pressure ratio*.

and $\quad v_1 = rv_2 = v_4 = rv_3 \quad \left[\therefore \dfrac{v_1}{v_2} = \dfrac{v_4}{v_3} = r \right]$

$$W = \frac{1}{\gamma-1}\left[p_4 v_4 \left(\frac{p_3 v_3}{p_4 v_4} - 1 \right) - p_1 v_1 \left(\frac{p_2 v_2}{p_1 v_1} - 1 \right) \right]$$

$$= \frac{1}{\gamma-1}\left[p_4 v_4 \left(\frac{p_3}{p_4 r} - 1 \right) - p_1 v_1 \left(\frac{p_2}{p_1 r} - 1 \right) \right]$$

$$= \frac{v_1}{\gamma-1}\left[p_4 \left(r^{\gamma-1} - 1 \right) - p_1 \left(r^{\gamma-1} - 1 \right) \right]$$

$$= \frac{v_1}{\gamma-1}\left[\left(r^{\gamma-1} - 1 \right)\left(p_4 - p_1 \right) \right]$$

$$= \frac{p_1 v_1}{\gamma-1}\left[\left(r^{\gamma-1} - 1 \right)\left(r_p - 1 \right) \right]$$

Mean effective pressure (p_m) is given by:

$$p_m = \left[\left(\frac{p_3 v_3 - p_4 v_4}{\gamma-1} - \frac{p_2 v_2 - p_1 v_1}{\gamma-1} \right) \div \left(v_1 - v_2 \right) \right] \text{bar}$$

Also $\quad p_m = \dfrac{\left[\dfrac{p_1 v_1}{\gamma-1}\left(r^{\gamma-1} - 1 \right)\left(r_p - 1 \right) \right]}{\left(v_1 - v_2 \right)}$

$$= \frac{\dfrac{p_1 v_1}{\gamma-1}\left[\left(r^{\gamma-1} - 1 \right)\left(r_p - 1 \right) \right]}{v_1 - \dfrac{v_1}{r}}$$

$$= \frac{\dfrac{p_1 v_1}{\gamma-1}\left[\left(r^{\gamma-1} - 1 \right)\left(r_p - 1 \right) \right]}{v_1 \left(\dfrac{r-1}{r} \right)}$$

i.e., $\qquad \mathbf{p_m} = \dfrac{p_1 r \left[\left(r^{\gamma-1} - 1 \right)\left(r_p - 1 \right) \right]}{\left(\gamma-1 \right)\left(r-1 \right)}$

Example 7

The efficiency of an Otto cycle is 60% and $\gamma = 1.5$. What is the compression ratio?

■ **Solution.**

Efficiency of Otto cycle, $\eta = 60\%$

Ratio of specific heats, $\quad\quad\quad \gamma = 1.5$

Compression ratio, $\quad\quad\quad\quad\quad r = ?$

Efficiency of Otto cycle is given by,

$$\eta_{\text{Otto}} = 1 - \frac{1}{(r)^{\gamma-1}}$$

$$0.6 = 1 - \frac{1}{(r)^{1.5-1}}$$

or $\quad \dfrac{1}{(r)^{0.5}} = 0.4 \quad$ or $\quad (r)^{0.5} = \dfrac{1}{0.4} = 2.5 \quad$ or $\quad r = 6.25$

Hence, *compression ratio* $\quad\quad\quad\quad\quad\quad = \mathbf{6.25}.$ (**Ans.**)

Example 8

An engine of 250 mm bore and 375 mm stroke works on an Otto cycle. The clearance volume is 0.00263 m³. The initial pressure and temperature are 1 bar and 50 °C. If the maximum pressure is limited to 25 bar, find the following:

(i) The air standard efficiency of the cycle.

(ii) The mean effective pressure for the cycle.

Assume the ideal conditions.

■ **Solution.**

Bore of the engine, $\quad D = 250$ mm $= 0.25$ m

Stroke of the engine, $\quad\quad\quad L = 375$ mm $= 0.375$ m

Clearance volume, $\quad\quad\quad\quad V_c = 0.00263$ m³

Initial pressure, $\quad\quad\quad\quad\quad p_1 = 1$ bar

Initial temperature, $\quad\quad\quad\quad T_1 = 50 + 273 = 323$ K

Maximum pressure, $\quad\quad\quad\quad p_3 = 25$ bar

Swept volume, $\quad\quad\quad\quad\quad V_s = \pi / 4 \, D^2 L = \pi / 4 \times 0.25^2 \times 0.375 = 0.0184$ m³

Compression ratio, $\quad\quad\quad\quad r = \dfrac{V_s + V_c}{V_c} = \dfrac{0.0184 + 0.00263}{0\ 00263} = 8.$

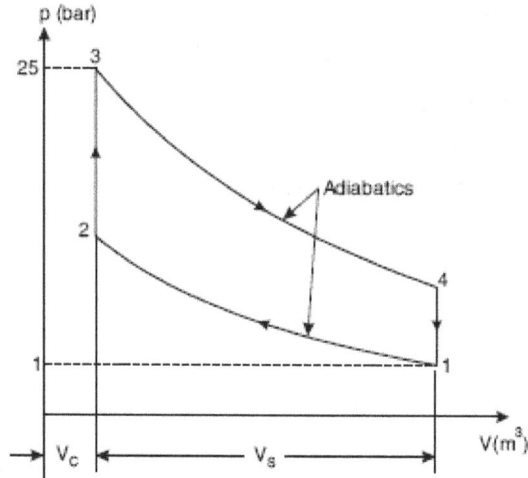

FIGURE 11.6 Sketch for Example 8.

(*i*) **Air standard efficiency:**

The air standard efficiency of the Otto cycle is given by

$$\eta_{Otto} = 1 - \frac{1}{(r)^{\gamma-1}} = 1 - \frac{1}{(8)^{1.4-1}} = 1 - \frac{1}{(8)^{0.4}}$$

$$= 1 - 0.435 = \mathbf{0.565} \quad \text{or} \quad \mathbf{56.5\%.} \quad \textbf{(Ans.)}$$

(*ii*) Mean effective pressure, pm:

For adiabatic (or isentropic) process 1–2

$$p_1 V_1^{\gamma} = p_2 V_2^{\gamma}$$

or

$$p_2 = p_1 \left(\frac{V_1}{V_2}\right)^{\gamma} = 1 \times (r)^{1.4} = 1 \times (8)^{1.4} = 18.38 \text{ bar}$$

∴ Pressure ratio,

$$r_p = \frac{p_3}{p_2} = \frac{25}{18.38} = 1.36$$

The mean effective pressure is given by

$$p_m = \frac{p_1 r \left[\left(r^{\gamma-1}-1\right)\left(r_p-1\right)\right]}{(\gamma-1)(r-1)} = \frac{1 \times 8 \left[\{(8)^{1.4-1}-1\} \, (1.36-1)\right]}{(1.4-1) \, (8-1)}$$

$$= \frac{8\,(2.297-1)(0.36)}{0.4 \times 7} = 1.334 \text{ bar}$$

Hence *mean effective pressure* = **1.334 bar. (Ans.)**

Example 9

The minimum pressure and temperature in an Otto cycle are 100 kPa and 27 °C. The amount of heat added to the air per cycle is 1500 kJ/kg.

(i) Determine the pressures and temperatures at all points of the air standard Otto cycle.

(ii) Also calculate the specific work and thermal efficiency of the cycle for a compression ratio of 8:1.

$$C_v = 0.72 \ \frac{kj}{kg.K} \ \text{and } \gamma = 1.4, \text{ for air.}$$

■ Solution.

Refer Fig. 7. *Given*: $p_1 = $ 100 kPa $= 10^5\,\text{N}/\text{m}^2$ or 1 bar ;

$T_1 = 27 + 273 = 300$ K ; Heat added $= 1500$ kJ / kg ;

$r = 8 : 1$; $c_v = 0.72$ kJ / kg ; $\gamma = 1.4$.

Consider 1 kg of air.

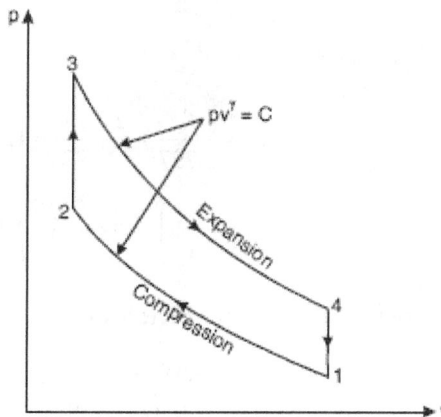

FIGURE 11.7. Sketch for Example 9.

(i) Pressures and temperatures at all points:

Adiabatic compression process 1–2:

$$\frac{T_2}{T_1} = \left(\frac{v_1}{v_2}\right)^{\gamma-1} = (r)^{\gamma-1} = (8)^{1.4-1} = 2.297$$

∴ $T_2 = 300 \times 2.297 = \textbf{689.1 K.}$ **(Ans.)**

Also $p_1 v_1^{\gamma} = p_2 v_2^{\gamma}$

or $\dfrac{p_2}{p_1} = \left(\dfrac{v_1}{v_2}\right)^{\gamma} = (8)^{1.4} = 18.379$

∴ $p_2 = 1 \times 18.379 = \textbf{18.379 bar.}$ **(Ans.)**

Constant volume process 2–3:

Heat added during the process,

$c_v(T_3 - T_2) = 1500$

or $0.72\,(T_3 - 689.1) = 1500$

or $T_3 = \dfrac{1500}{0.72} + 689.1 = \textbf{2772.4 K.}$ **(Ans.)**

Also, $\dfrac{p_2}{T_2} = \dfrac{p_3}{T_3}$ $\Rightarrow$ $p_3 = \dfrac{p_2 T_3}{T_2} = \dfrac{18.379 \times 2772.4}{689.1} = \textbf{73.94 bar.}$ **(Ans.)**

Adiabatic Expansion process 3–4:

$$\frac{T_3}{T_4} = \left(\frac{V_4}{V_3}\right)^{\gamma-1} = (r)^{\gamma-1} = (8)^{1.4-1} = 2.297$$

∴ $T_4 = \dfrac{T_3}{2.297} = \dfrac{2772.4}{2.297} = \textbf{1206.9 K.}$ **(Ans.)**

Also $p_3 v_3^{\gamma} = p_4 v_4^{\gamma}$ $\Rightarrow$ $p_4 = p_3 \times \left(\dfrac{v_3}{v_4}\right)^{\gamma} = 73.94 \times \left(\dfrac{1}{8}\right)^{1.4} = \textbf{4.023 bar.}$ **(Ans.)**

(ii) Specific work and thermal efficiency:

Specific work = Heat added – heat rejected

$= c_v(T_3 - T_2) - c_v(T_4 - T_1) = c_v[(T_3 - T_2) - (T_4 - T_1)]$

$= 0.72\left[(2772.4 - 689.1) - (1206.9 - 300)\right] = \textbf{847 kJ/kg.}$ **(Ans.)**

Thermal efficiency, $\eta_{th} = 1 - \dfrac{1}{(r)^{\gamma-1}}$

$= 1 - \dfrac{1}{(8)^{1.4-1}} = 0.5647$ or **56.47%.** (**Ans.**)

Example 10

An air standard Otto cycle has a volumetric compression ratio of 6, the lowest cycle pressure of 0.1 MPa, and operates between temperature limits of 27 °C and 1569 °C.

(i) Calculate the temperature and pressure after the isentropic expansion (*ratio of specific heats* = 1.4).

(ii) Since it is observed that values in (i) are well above the lowest cycle operating conditions, the expansion process was allowed to continue down to a pressure of 0.1 MPa. Which process is required to complete the cycle? Name the cycle so obtained.

(iii) Determine by what percentage the cycle efficiency has been improved.

■ **Solution.**

Solution. Refer Fig. 8. *Given* : $\dfrac{v_1}{v_2} = \dfrac{v_4}{v_3} = r = 6$; $p_1 = 0.1$ MPa $= 1$ bar ; $T_1 = 27 + 273 =$ 300 K ; $T_3 = 1569 + 273 = 1842$ K ; $\gamma = 1.4$.

(*i*) **Temperature and pressure after the isentropic expansion, T_4, p_4. Consider 1 kg of air:**

For *the compression process 1–2*:

$$p_1 v_1^{\gamma} = p_2 v_2^{\gamma} \Rightarrow p_2 = p_1 \times \left(\dfrac{v_1}{v_2}\right)^{\gamma} = 1 \times (6)^{1.4} = 12.3 \text{ bar}$$

Also $\dfrac{T_2}{T_1} = \left(\dfrac{v_1}{v_2}\right)^{\gamma-1} = (6)^{1.4-1} = 2.048$

∴ $T_2 = 300 \times 2.048 = 614.4$ K

For *the constant volume process 2–3*:

$\dfrac{p_2}{T_2} = \dfrac{p_3}{T_3} \Rightarrow p_3 = \dfrac{p_2 T_3}{T_2} = 12.3 \times \dfrac{1842}{614.4} = 36.9 \text{ bar}$

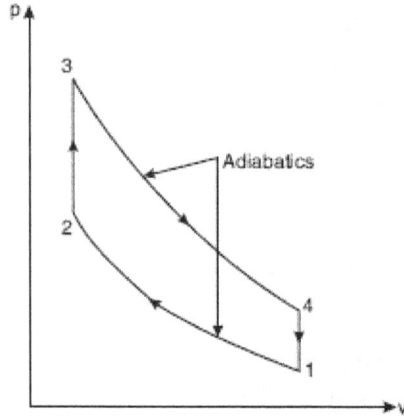

FIGURE 11.8. Sketch for Example 10, p-V diagram.

For *the expansion process 3–4:*

$$\frac{T_3}{T_4} = \left(\frac{v_4}{v_3}\right)^{\gamma-1} = \left(6\right)^{1.4-1} = 2.048$$

$$\therefore \qquad\qquad \mathbf{T_4} = \frac{T_3}{2.048} = \frac{1842}{2.048} = \mathbf{900\ K.}\quad (\mathbf{Ans.})$$

Also $p_3 v_3^{\gamma} = p_4 v_4^{\gamma} \;\Rightarrow\; p_4 = p_3 \times \left(\frac{v_3}{v_4}\right)^{\gamma}$

or $\qquad\qquad\qquad\qquad \mathbf{p_4} = 36.9 \times \left(\frac{1}{6}\right)^{1.4} = \mathbf{3\ bar.}\ (\mathbf{Ans.})$

(*ii*) Process required to complete the cycle:

Process required to complete the cycle is the *constant pressure scavenging.*

The cycle is called **Atkinson cycle** (Refer to Figure 11.9).

(*iii*) Percentage improvement/increase in efficiency:

$$\eta_{\text{Otto}} = 1 - \frac{1}{(r)^{\gamma-1}} = 1 - \frac{1}{(6)^{1.4-1}} = 0.5116 \text{ or } \mathbf{51.16\%.}\quad (\mathbf{Ans.})$$

$$\eta_{\text{Atkinson}} = \frac{\text{Work done}}{\text{Heat supplied}} = \frac{\text{Heat supplied} - \text{Heat rejected}}{\text{Heat supplied}}$$

$$= \frac{c_v(T_3 - T_2) - c_p(T_5 - T_1)}{c_v(T_3 - T_2)} = 1 - \frac{c_p(T_5 - T_1)}{c_v(T_3 - T_2)} = 1 - \frac{\gamma(T_5 - T_1)}{(T_3 - T_2)}$$

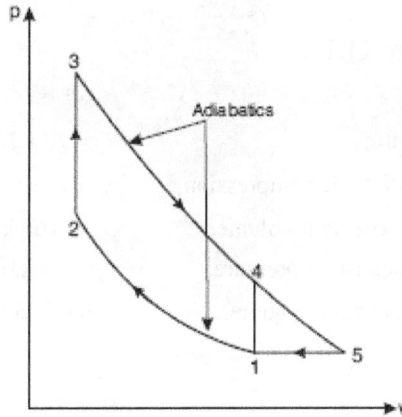

FIGURE 11.9. Atkinson cycle, Example 10.

Now,
$$\frac{T_5}{T_3} = \left(\frac{p_5}{p_3}\right)^{\frac{\gamma-1}{\gamma}} \quad \text{or} \quad T_5 = 1842 \times \left(\frac{1.0}{36.9}\right)^{\frac{1.4-1}{1.4}} = 657 \text{ K}$$

$$\therefore \quad \eta_{\text{Atkinson}} = 1 - \frac{1.4(657-300)}{(1842-614.4)} = \textbf{0.5929} \text{ or } \textbf{59.29\%.}$$

$\therefore$ **Improvement in efficiency** $= 59.29 - 51.16 = \textbf{8.13\%.}$ **(Ans.)**

Example 11

A certain quantity of air at a pressure of 1 bar and a temperature of 70 °C is compressed adiabatically until the pressure is 7 bar in an Otto cycle engine. 465 kJ of heat per kg of air is now added at a constant volume. Determine:

(i) Compression ratio of the engine.

(ii) Temperature at the end of compression.

(iii) Temperature at the end of heat addition.

Take for air $c_p = 1.0$ kJ/kg K, $c_v = 0.706$ kJ/kg K.

Show each operation on p-V and T-s diagrams.

■ **Solution.**

Refer to Figure 11.10.

Initial pressure,	$p_1 = 1$ bar
Initial temperature,	$T_1 = 70 + 273 = 343$ K
Pressure after adiabatic compression,	$p_2 = 7$ bar
Heat addition at constant volume,	$Q_s = 465$ kJ / kg of air
Specific heat at constant pressure,	$c_p = 1.0$ kJ / kg K
Specific heat at constant volume,	$c_v = 0.706$ kJ / kg K

$$\therefore \qquad \gamma = \frac{c_p}{c_v} = \frac{1.0}{0.706} = 1.41$$

(*i*) Compression ratio of engine, r:
According to *adiabatic compression 1–2*

$$p_1 V_1^\gamma = p_2 V_2^\gamma$$

or
$$\left(\frac{V_1}{V_2}\right)^\gamma = \frac{p_2}{p_1}$$

or
$$(r)^\gamma = \frac{p_2}{p_1} \qquad\qquad \left(\therefore \frac{V_1}{V_2} = r\right)$$

or
$$r = \left(\frac{p_2}{p_1}\right)^{\frac{1}{\gamma}} = \left(\frac{7}{1}\right)^{\frac{1}{1.41}} = (7)^{0.709} = 3.97$$

Hence *compression ratio of the engine* = **3.97**. (**Ans.**)

(*ii*) Temperature at the end of compression, T₂:
In case of *adiabatic compression 1–2*,

$$\frac{T_2}{T_1} = \left(\frac{V_1}{V_2}\right)^{\gamma-1} = (3.97)^{1.41-1} = 1.76$$

$$\therefore \qquad\qquad T_2 = 1.76\, T_1 = 1.76 \times 343 = 603.7 \text{ K or } 330.7°C$$

Hence *temperature at the end of compression* = **330.7°C**. (**Ans.**)

(*iii*) Temperature at the end of heat addition, T₃:

According to *constant volume heating operation 2–3*
$$Q_s = c_v(T_3 - T_2) = 465$$
$$0.706\,(T_3 - 603.7) = 465$$

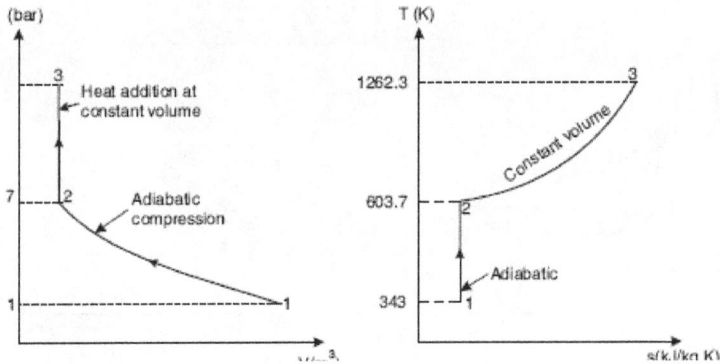

FIGURE 11.10. Sketch for Example 11.

or $T_3 - 603.7 = \dfrac{465}{0.706}$

or $T_3 = \dfrac{465}{0.706} + 603.7 = 1262.3$ K or $989.3°$C

Hence *temperature at the end of heat addition* = **989.3°C**.   (**Ans.**)

Example 12

In a constant volume "Otto cycle," the pressure at the end of compression is 15 times that at the start, the temperature of air at the beginning of compression is 38 °C and maximum temperature attained in the cycle is 1950 °C. Determine:

(i) Compression ratio.

(ii) Thermal efficiency of the cycle.

(iii) Work done.

Take γ for air = 1.4.

■ **Solution.**

Refer to Figure 11.11.

Initial temperature, $T_1 = 38 + 273 = 311$ K

Maximum temperature, $T_3 = 1950 + 273 = 2223$ K.

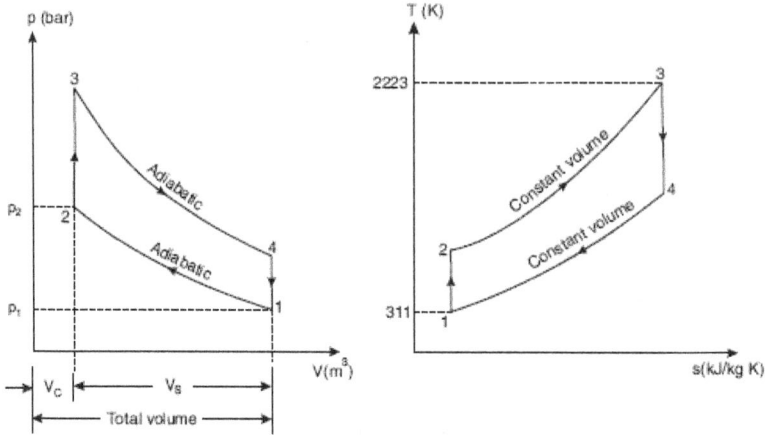

FIGURE 11.11. Sketch for Example 12.

(*i*) Compression ratio, r:

For *adiabatic compression 1–2,*

$$p_1 V_1^\gamma = p_2 V_2^\gamma$$

or

$$\left(\frac{V_1}{V_2}\right)^\gamma = \frac{p_2}{p_1}$$

But

$$\frac{p_2}{p_1} = 15 \qquad \qquad \qquad \ldots(\text{given})$$

$$\therefore \quad (r)^\gamma = 15 \qquad\qquad \left[\because \ \mathrm{r} = \frac{V_1}{V_2}\right]$$

or

$$(r)^{1.4} = 15$$

or

$$r = (15)^{\frac{1}{1.4}} = (15)^{0.714} = 6.9$$

Hence *compression ratio* = **6.9. (Ans.)**

(*ii*) Thermal efficiency:

Thermal efficiency, $\quad \eta_{th} = 1 - \dfrac{1}{(r)^{\gamma-1}} = 1 - \dfrac{1}{(6.9)^{1.4-1}} = \textbf{0.538 or 53.8\%. (Ans.)}$

(*iii*) **Work done:**

Again, for *adiabatic compression 1–2*,

$$\frac{T_2}{T_1} = \left(\frac{V_1}{V_2}\right)^{\gamma-1} = (r)^{\gamma-1} = (6.9)^{1.4-1} = (6.9)^{0.4} = 2.16$$

or
$$T_2 = T_1 \times 2.16 = 311 \times 2.16 = 671.7 \text{ K or } 398.7°C$$

For *adiabatic expansion process 3–4*

$$\frac{T_3}{T_4} = \left(\frac{V_4}{V_3}\right)^{\gamma-1} = (r)^{\gamma-1} = (6.9)^{0.4} = 2.16$$

or
$$T_4 = \frac{T_3}{2.16} = \frac{2223}{2.16} = 1029 \text{ K or } 756°C$$

Heat supplied *per kg of air*

$$= c_v (T_3 - T_2) = 0.717 (2223 - 671.7)$$

$$= 1112.3 \text{ kJ/kg or air}$$

$$\left[\begin{matrix} c_v = \dfrac{R}{\gamma - 1} = \dfrac{0.287}{1.4 - 1} \\ = 0.717 \text{ kJ/kg K} \end{matrix} \right]$$

Heat rejected per kg of air

$$= c_v (T_4 - T_1) = 0.717 (1029 - 311)$$

$$= 514.8 \text{ kJ/kg of air}$$

∴ *Work done per kg of air* = Heat supplied – heat rejected

$$= 1112.3 - 514.8$$

$$= \textbf{597.5 kJ or 597500 Nm. (Ans.)}$$

Example 13

An engine working on an Otto cycle has a volume of 0.45 m³, pressure of 1 bar, and a temperature of 30 °C at the beginning of the compression stroke. At the end of compression stroke, the pressure is 11 bar. 210 kJ of heat is added at a constant volume. Determine:

(i) Pressures, temperatures, and volumes at salient points in the cycle.

(ii) Percentage clearance.

(iii) Efficiency.

(iv) Mean effective pressure.

(v) Ideal power developed by the engine if the number of working cycles per minute is 210.

Assume the cycle is reversible.

■ **Solution.**

Refer to Figure 11.12.

Volume,	$V_1 = 0.45 \text{m}^3$
Initial pressure,	$p_1 = 1$ bar
Initial temperature,	$T_1 = 30 + 273 = 303$ K
Pressure at the end of compression stroke,	$p_2 = 11$ bar
Heat added at constant volume	$= 210$ kJ
Number of working cycles / min.	$= 210.$

(*i*) **Pressures, temperatures, and volumes at salient points:**

For *adiabatic compression 1–2*

$$p_1 V_1^\gamma = p_2 V_2^\gamma$$

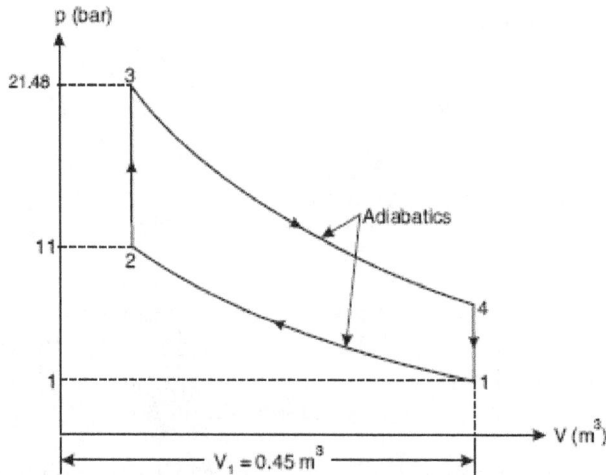

FIGURE 11.12. Sketch for Example 13.

or $\qquad \dfrac{p_2}{p_1} = \left(\dfrac{V_1}{V_2}\right)^{\gamma} = (r)^{\gamma}$ or $r = \left(\dfrac{p_2}{p_1}\right)^{\frac{1}{\gamma}} = \left(\dfrac{11}{1}\right)^{\frac{1}{1.4}} = (11)^{0.714} = 5.5$

Also $\qquad\qquad\qquad \dfrac{T_2}{T_1} = \left(\dfrac{V_1}{V_2}\right)^{\gamma-1} = (r)^{\gamma-1} = (5.5)^{1.4-1} = 1.977 \simeq 1.98$

$\therefore \qquad\qquad\qquad T_2 = T_1 \times 1.98 = 303 \times 1.98 = \mathbf{600\ K.}\ \ \mathbf{(Ans.)}$

Applying gas laws to points 1 and 2

$$\dfrac{p_1 V_1}{T_1} = \dfrac{p_2 V_2}{T_2}$$

$\therefore \qquad\qquad\qquad V_2 = \dfrac{T_2}{T_1} \times \dfrac{p_1}{p_2} \times V_1 = \dfrac{600 \times 1 \times 0.45}{303 \times 11} = \mathbf{0.081\ m^3.}\ \ \mathbf{(Ans.)}$

The heat supplied during the *process 2–3* is given by:

$$Q_s = m\ c_v\ (T_3 - T_2)$$

where $m = \dfrac{p_1 V_1}{RT_1} = \dfrac{1 \times 10^5 \times 0.45}{287 \times 303} = 0.517\ \text{kg}$

$\therefore \qquad\qquad\qquad 210 = 0.517 \times 0.71\ (T_3 - 600)$

or $\qquad\qquad\qquad T_3 = \dfrac{210}{0.517 \times 0.71} + 600 = \mathbf{1172\ K.}\ \ \mathbf{(Ans.)}$

For the *constant volume process 2–3*

$$\dfrac{p_3}{T_3} = \dfrac{p_2}{T_2}$$

$\therefore \qquad\qquad\qquad p_3 = \dfrac{T_3}{T_2} \times p_2 = \dfrac{1172}{600} \times 11 = \mathbf{21.48\ bar.}\ \ \mathbf{(Ans.)}$

$V_3 = V_2 = \mathbf{0.081\ m^3.}\ \ \mathbf{(Ans.)}$

For the *adiabatic (or isentropic) process 3–4*

$$p_3 V_3^{\gamma} = p_4 V_4^{\gamma}$$

$$p_4 = p_3 \times \left(\dfrac{V_3}{V_4}\right)^{\gamma} = p_3 \times \left(\dfrac{1}{r}\right)^{\gamma}$$

$$= 21.48 \times \left(\dfrac{1}{5.5}\right)^{1.4} = \mathbf{1.97\ bar.}\ \ \mathbf{(Ans.)}$

Also $\qquad\qquad\qquad \dfrac{T_4}{T_3} = \left(\dfrac{V_3}{V_4}\right)^{\gamma-1} = \left(\dfrac{1}{r}\right)^{\gamma-1} = \left(\dfrac{1}{5.5}\right)^{1.4-1} = 0.505$

$\therefore$ $\qquad\qquad\qquad\qquad\qquad$ $T_4 = 0.505\ T_3 = 0.505 \times 1172 = \textbf{591.8 K. (Ans.)}$

$V_4 = V_1 = \textbf{0.45 m}^3\textbf{. (Ans.)}$

(*ii*) **Percentage clearance:**

Percentage clearance

$$= \frac{V_c}{V_s} = \frac{V_2}{V_1 - V_2} \times 100 = \frac{0.081}{0.45 - 0.081} \times 100$$

$$= \textbf{21.95\%. (Ans.)}$$

(*iii*) **Efficiency:**

The heat rejected per cycle is given by

$Q_r = mc_v (T_4 - T_1)$

$= 0.517 \times 0.71\ (519.8 - 303) = 106\ \text{kJ}$

The air standard efficiency of the cycle is given by

$$\eta_{\text{otto}} = \frac{Q_s - Q_r}{Q_s} = \frac{210 - 106}{210} = \textbf{0.495 or 49.5\%. (Ans.)}$$

$$\left[\begin{array}{l} \text{Alternatively:} \\[2mm] \qquad\qquad \eta_{\text{otto}} = 1 - \dfrac{1}{(r)^{\gamma - 1}} = 1 - \dfrac{1}{(5.5)^{1.4-1}} = \textbf{0.495 or 49.5\%. (Ans.)} \end{array} \right]$$

(*iv*) **Mean effective pressure, p$_m$:**

The mean effective pressure is given by

$$p_m = \frac{W\ (\text{work done})}{V_s\ (\text{swept volume})} = \frac{Q_s - Q_r}{(V_1 - V_2)}$$

$$= \frac{(210 - 106) \times 10^3}{(0.45 - 0.081) \times 10^5} = \textbf{2.818 bar. (Ans.)}$$

(*v*) **Power developed, P:**

power developed, $\qquad\qquad\qquad$ P = Work done per second

= Work done per cycle $\times$ number of cycles per second

$= (210 - 106) \times (210/60) = \textbf{364 kW. (Ans.)}$

Example 14

(a) Show that the compression ratio for the maximum work to be done per kg of air in an Otto cycle between upper and lower limits of absolute temperatures T_3 and T_1 is given by

$$r = \left(\frac{T_3}{T_1}\right)^{1/2(\gamma-1)}$$

(b) Determine the air standard efficiency of the cycle when the cycle develops maximum work with the temperature limits of 310 K and 1220 K and the working fluid is air. What will be the percentage change in efficiency if helium is used as a working fluid instead of air? The cycle operates between the same temperature limits for maximum work development.

Consider that all conditions are ideal.

■ **Solution.**

Refer to Figure 11.13.

(a) The work done per kg of fluid in the cycle is given by
$$W = Q_s - Q_r = c_v\left(T_3 - T_2\right) - c_v\left(T_4 - T_1\right)$$

But
$$\frac{T_2}{T_1} = \left(\frac{v_1}{v_2}\right)^{\gamma-1} = \left(r\right)^{\gamma-1}$$

∴
$$T_2 = T_1 \cdot \left(r\right)^{\gamma-1}$$

Similarly,
$$T_3 = T_4 \cdot \left(r\right)^{\gamma-1}$$

∴
$$W = c_v\left[T_3 - T_1 \cdot \left(r\right)^{\gamma-1} - \frac{T_3}{\left(r\right)^{\gamma-1}} + T_1\right]$$

This expression is a function of r when T_3 and T_1 are fixed. The value of W will be maximum when,

$$\frac{dW}{dr} = 0.$$

∴
$$\frac{dW}{dr} = -T_1 \cdot \left(\gamma-1\right)\left(r\right)^{\gamma-2} - T_3\left(1-\gamma\right)\left(r\right)^{-\gamma} = 0$$

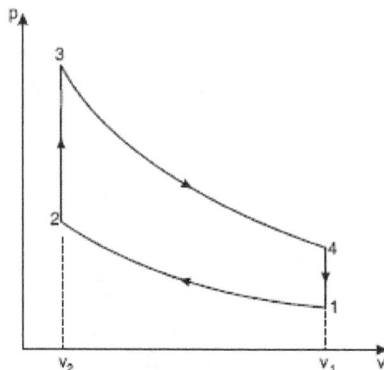

FIGURE 11.13. Sketch for Example 14.

or
$$T_3 \left(r\right)^{-\gamma} = T_1 \left(r\right)^{\gamma-2}$$

or
$$\frac{T_3}{T_1} = \left(r\right)^{2(\gamma-1)}$$

∴
$$r = \left(\frac{T_3}{T_1}\right)^{1/2(\gamma-1)} . \qquad \text{......Proved.}$$

(b) Change in efficiency:

For *air* $\gamma = 1.4$

∴ $r = \left(\dfrac{T_3}{T_1}\right)^{1/2(1.4-1)} = \left(\dfrac{1220}{310}\right)^{1/0.8} = 5.54$

The air-standard efficiency is given by

$$\eta_{\text{otto}} = 1 - \frac{1}{\left(r\right)^{\gamma-1}} = 1 - \frac{1}{\left(5.54\right)^{1.4-1}} = 0.495 \text{ or } 49.5\%. \quad \textbf{(Ans.)}$$

If *helium* is used, then the values of

$c_p = 5.22$ kJ/kg K and $c_v = 3.13$ kJ/kg K

∴
$$\gamma = \frac{c_p}{c_v} = \frac{5.22}{3.13} = 1.67$$

The compression ratio for maximum work for the temperature limits T_1 and T_3 is given by

$$r = \left(\frac{T_3}{T_1}\right)^{1/2(\gamma-1)} = \left(\frac{1220}{310}\right)^{1/2(1.67-1)} = 2.77$$

The air standard efficiency is given by

$$\eta_{\text{otto}} = 1 - \frac{1}{(r)^{\gamma-1}} = 1 - \frac{1}{(2.77)^{1.67-1}} = 0.495 \text{ or } \textbf{49.5\%}.$$

Hence change in efficiency is nil. (Ans.)

Example 15

(a) An engine working on Otto cycle, in which the salient points are 1, 2, 3, and 4, has upper and lower temperature limits T_3 and T_1. If the maximum work per kg of air is to be done, show that the intermediate temperature is given by

$$T_2 = T_4 = \sqrt{T_1 T_3}.$$

(b) If an engine works on an Otto cycle between temperature limits 1450 K and 310 K, find the maximum power developed by the engine assuming the circulation of air per minute as 0.38 kg.

■ Solution.

(a) Refer to Figure 11.13 (Example 14).

Using the equation (iii) of example 14.

$$W = c_v \left[T_3 - T_1 \cdot (r)^{\gamma-1} - \frac{T_3}{(r)^{\gamma-1}} + T_1 \right]$$

and differentiating W w.r.t. r and equating to zero

$$r = \left(\frac{T_3}{T_1} \right)^{1/2(\gamma-1)}$$

$$T_2 = T_1 (r)^{\gamma-1} \text{ and } T_4 = T_3 \big/ (r)^{\gamma-1}$$

Substituting the value of r in the above equation, we have

$$T_2 = T_1 \left[\left(\frac{T_3}{T_1} \right)^{1/2(\gamma-1)} \right]^{\gamma-1} = T_1 \left(\frac{T_3}{T_1} \right)^{1/2} = \sqrt{T_1 T_3}$$

Similarly,

$$T_4 = \frac{T_3}{\left[\left(\frac{T_3}{T_1} \right)^{1/2(\gamma-1)} \right]^{\gamma-1}} = \frac{T_3}{\left(\frac{T_3}{T_1} \right)^{1/2}} = \sqrt{T_3 T_1}$$

$$\therefore \quad T_2 = T_4 = \sqrt{T_1 T_3}. \quad \textbf{Proved.}$$

(*b*) **Power developed, P:**

$$\left.\begin{array}{l} T_1 = 310 \text{ K} \\ T_3 = 1450 \text{ K} \\ m = 0.38 \text{ kg} \end{array}\right\} \quad ....(\text{given})$$

Work done

$$W = c_v\left[(T_3 - T_2) - (T_4 - T_1)\right]$$

$$T_2 = T_4 = \sqrt{T_1 T_3} = \sqrt{310 \times 1450} = 670.4 \text{ K}$$

$$\therefore \qquad\qquad W = 0.71\left[(1450 - 670.4) - (670.4 - 310)\right]$$

$$= 0.71 \ (779.6 - 360.4) = 279.6 \ \text{kJ/kg}$$

Work done per second $= 297.6 \times (0.38/60) = 1.88 \text{ kJ/s}$

Hence **power developed, P = 1.88 kW. (Ans.)**

Example 16

For the same compression ratio, show that the efficiency of the Otto cycle is greater than that of the Diesel cycle.

■ **Solution.**

Refer to Figure 11.14.

We know that

$$\eta_{\text{otto}} = 1 - \frac{1}{(r)^{\gamma-1}}$$

and

$$\eta_{\text{diesel}} = 1 - \frac{1}{(r)^{\gamma-1}} \times \frac{1}{\gamma}\left\{\frac{\rho^\gamma - 1}{\rho - 1}\right\}$$

As the compression ratio is the same,

$$\frac{V_1}{V_2} = \frac{V_1'}{V_2'} = r$$

If $\dfrac{V_4'}{V_3'} = r_1$, then cut off ratio, $\rho = \dfrac{V_3'}{V_2'} = \dfrac{r}{r_1}$

Putting the value of ρ in η_{diesel}, we get

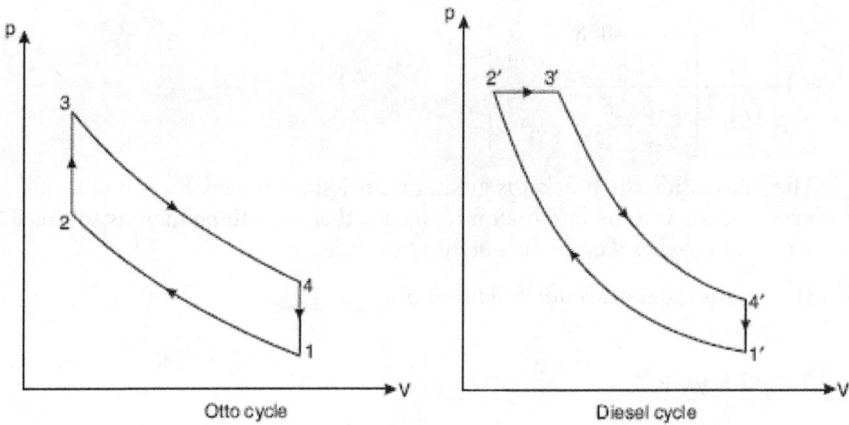

FIGURE 11.14. Sketch for Example 16.

$$\eta_{\text{diesel}} = 1 - \frac{1}{(r)^{\gamma-1}} \times \frac{1}{\gamma}\left[\frac{\left(\dfrac{r}{r_1}\right)^{\gamma} - 1}{\dfrac{r}{r_1} - 1}\right]$$

From above equation, we observe

$$\frac{r}{r_1} > 1$$

Let $r_1 = r - \delta$, where δ is a small quantity.

Then

$$\frac{r}{r_1} = \frac{r}{r-\delta} = \frac{r}{r\left(1-\dfrac{\delta}{r}\right)} = \left(1-\frac{\delta}{r}\right)^{-1} = 1 + \frac{\delta}{r} + \frac{\delta^2}{r^2} + \frac{\delta^3}{r^3} + \ldots\ldots$$

and

$$\left(\frac{r}{r_1}\right)^{\gamma} = \frac{r^{\gamma}}{r^{\gamma}\left(1-\dfrac{\delta}{r}\right)^{\gamma}} = \left(1-\frac{\delta}{r}\right)^{-\gamma} = 1 + \frac{\gamma\delta}{r} + \frac{\gamma(\gamma+1)}{2!}\cdot\frac{\delta^2}{r^2} + \ldots\ldots$$

$$\therefore \qquad \eta_{\text{diesel}} = 1 - \frac{1}{(r)^{\gamma-1}} \times \frac{1}{r}\left[\frac{\dfrac{\gamma.\delta}{r} + \dfrac{\gamma(\gamma+1)}{2!}\cdot\dfrac{\delta^2}{r^2} + \ldots\ldots}{\dfrac{\delta}{r} + \dfrac{\delta^2}{r^2} + \ldots\ldots}\right]$$

$$= 1 - \frac{1}{(r)^{\gamma-1}} \left[\frac{\dfrac{\delta}{r} + \dfrac{\gamma+1}{2} \cdot \dfrac{\delta^2}{r^2} + \ldots\ldots}{\dfrac{\delta}{r} + \dfrac{\delta^2}{r^2} + \ldots\ldots} \right]$$

The ratio inside the bracket is greater than 1 since the coefficient of terms δ^2/r^2 is greater than 1 in the numerator. Its means that something more is subtracted in the case of the diesel cycle than in the Otto cycle.

Hence, *for the same compression ratio* $\eta_{\text{otto}} > \eta_{\text{diesel}}$.

11.6. Diesel Cycle

This cycle was introduced by Rudolf Diesel in 1897. It differs from the Otto cycle in that heat is supplied at constant pressure instead of at constant volume. Figure 11.15(*a* and *b*) shows the *p-v* and *T-s* diagrams of this cycle respectively.

This cycle comprises the following operations:

(*i*)   1–2......*Adiabatic compression.*

(*ii*) 2–3......*Addition of heat at a constant pressure.*

(*iii*) 3–4......*Adiabatic expansion.*

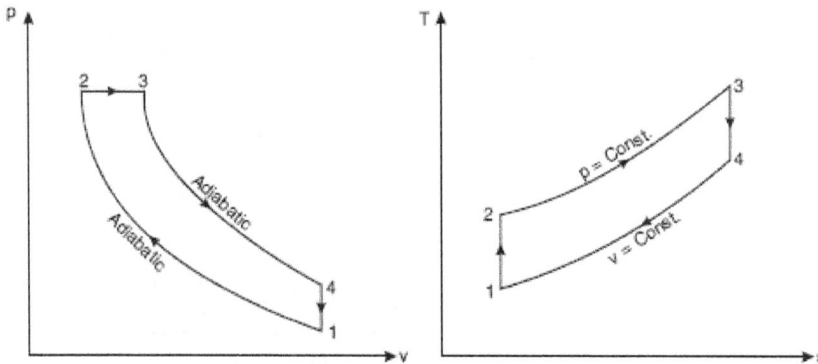

FIGURE 11.15. Diesel Cycle on p-v (left) and T-s (right) diagrams.

(iv) 4–1......*Rejection of heat at a constant volume.*

Point 1 represents that the cylinder is full of air. Let p_1, V_1, and T_1 be the corresponding pressure, volume, and absolute temperature. The piston then compresses the air adiabatically (*i.e., pVγ* = constant) till the values become p_2, V_2, and T_2, respectively (at the end of the stroke) at point 2. Heat is then added from a hot body at a constant pressure. During this addition of heat let volume increase from V_2 to V_3 and temperature T_2 to T_3, corresponding to point 3. This point (3) is called the *point of cut-off*. The air then expands adiabatically to the conditions p_4, V_4, and T_4, corresponding to point 4. Finally, the air rejects the heat to the cold body at a constant volume till point 1 where it returns to its original state.

Consider 1 kg of air.

Heat supplied at constant pressure $= c_p \left(T_3 - T_2 \right)$

Heat rejected at constant volume $= c_v \left(T_4 - T_1 \right)$

Work done $\qquad\qquad$ = Heat supplied − heat rejected

$$= c_p \left(T_3 - T_2 \right) - c_v \left(T_4 - T_1 \right)$$

$\therefore \qquad\qquad\qquad\qquad \eta_{\text{diesel}} = \dfrac{\text{Work done}}{\text{Heat supplied}}$

$$= \dfrac{c_p \left(T_3 - T_2 \right) - c_v \left(T_4 - T_1 \right)}{c_p \left(T_3 - T_2 \right)}$$

$$= 1 - \dfrac{\left(T_4 - T_1 \right)}{\gamma \left(T_3 - T_2 \right)} \qquad\qquad ...(i) \left[\because \ \dfrac{c_p}{c_v} = \gamma \right]$$

Let compression ratio, $r = \dfrac{v_1}{v_2}$, and cut-off ratio, $\qquad \rho = \dfrac{v_3}{v_2}$ *i.e.,* $\dfrac{\text{Volume at cut-off}}{\text{Clearance volume}}$

Now, during *adiabatic compression 1–2,*

$$\dfrac{T_2}{T_1} = \left(\dfrac{v_1}{v_2} \right)^{\gamma - 1} = \left(r \right)^{\gamma - 1} \ \text{ or } T_2 = T_1 . \left(r \right)^{\gamma - 1}$$

During *constant pressure process 2–3,*

$$\dfrac{T_3}{T_2} = \dfrac{v_3}{v_2} = \rho \ \text{ or } T_3 = \rho . T_2 = \rho . T_1 . \left(r \right)^{\gamma - 1}$$

During *adiabatic expansion 3–4*

$$\frac{T_3}{T_4} = \left(\frac{v_4}{v_3}\right)^{\gamma-1}$$

$$= \left(\frac{r}{\rho}\right)^{\gamma-1} \qquad\qquad \left(\because\ \frac{v_4}{v_3} = \frac{v_1}{v_3} = \frac{v_1}{v_2}\times\frac{v_2}{v_3} = \frac{r}{\rho}\right)$$

$$\therefore \qquad\qquad T_4 = \frac{T_3}{\left(\dfrac{r}{\rho}\right)^{\gamma-1}} = \frac{\rho.T_1\left(r\right)^{\gamma-1}}{\left(\dfrac{r}{\rho}\right)^{\gamma-1}} = T_1.\rho^{\gamma}$$

By inserting values of T_2, T_3, and T_4 in equation (i), we get

$$\eta_{\text{diesel}} = 1 - \frac{\left(T_1.\rho^{\gamma} - T_1\right)}{\gamma\left(\rho.T_1.(r)^{\gamma-1} - T_1.(r)^{\gamma-1}\right)} = 1 - \frac{\left(\rho^{\gamma} - 1\right)}{\gamma(r)^{\gamma-1}\left(\rho-1\right)}$$

or
$$\eta_{\text{diesel}} = 1 - \frac{1}{\gamma(r)^{\gamma-1}}\left[\frac{\rho^{\gamma} - 1}{\rho-1}\right]$$

It may be observed that efficiency equation of diesel cycle is different from that of the Otto cycle only in the bracketed factor. This factor is always greater than unity, because $\rho > 1$. Hence *for a given compression ratio, the Otto cycle is more efficient.*

The *net work* for a diesel cycle can be expressed in terms of pv as follows:

$$W = p_2\left(v_3 - v_2\right) + \frac{p_3 v_3 - p_4 v_4}{\gamma - 1} - \frac{p_2 v_2 - p_1 v_1}{\gamma - 1}$$

$$= p_2\left(\rho v_2 - v_2\right) + \frac{p_3 \rho v_2 - p_4 r v_2}{\gamma - 1} - \frac{p_2 v_2 - p_1 r v_2}{\gamma - 1}$$

$$\left[\begin{array}{l}\because\ \dfrac{v_3}{v_2} = \rho\ \therefore\ v_3 = \rho v_2\ \text{and}\ \dfrac{v_1}{v_2} = r\ \therefore\ v_1 = r v_2\\ \text{But}\ v_4 = v_1\ \therefore\ v_4 = r v_2\end{array}\right]$$

$$= p_2 v_2\left(\rho - 1\right) + \frac{p_3 \rho v_2 - p_4 r v_2}{\gamma - 1} - \frac{p_2 v_2 - p_1 r v_2}{\gamma - 1}$$

$$= \frac{v_2\left[p_2\left(\rho - 1\right)\left(\gamma - 1\right) + p_3 \rho - p_4 r - \left(p_2 - p_1 r\right)\right]}{\gamma - 1}$$

$$= \frac{v_2\left[p_2(\rho-1)(\gamma-1)+p_3\left(\rho-\frac{p_4 r}{p_3}\right)-p_2\left(1-\frac{p_1 r}{p_2}\right)\right]}{\gamma-1}$$

$$= \frac{p_2 v_2\left[(\rho-1)(\gamma-1)+\rho-\rho^\gamma . r^{1-\gamma}-\left(1-r^{1-\gamma}\right)\right]}{\gamma-1}$$

$$\left[\because \quad \frac{p_4}{p_3}=\left(\frac{v_3}{v_4}\right)^\gamma=\left(\frac{\rho}{r}\right)^\gamma=\rho^\gamma r^{-\gamma}\right]$$

$$= \frac{p_1 v_1 r^{\gamma-1}\left[(\rho-1)(\gamma-1)+\rho-\rho^\gamma r^{1-\gamma}-\left(1-r^{1-\gamma}\right)\right]}{\gamma-1}$$

$$\left[\because \quad \frac{p_2}{p_1}=\left(\frac{v_1}{v_2}\right)^\gamma \text{ or } p_2=p_1.r^\gamma \text{ and } \frac{v_1}{v_2}=r \text{ or } v_2=v_1 r^{-1}\right]$$

$$= \frac{p_1 v_1 r^{\gamma-1}\left[\gamma(\rho-1)-r^{1-\gamma}\left(\rho^\gamma-1\right)\right]}{(\gamma-1)}$$

Mean effective pressure p_m is given by:

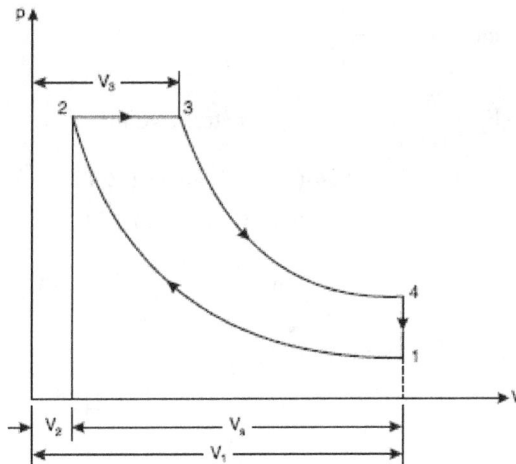

FIGURE 11.16. Sketch for Example 17.

$$p_m = \frac{p_1 v_1 r^{\gamma-1}\left[\gamma(\rho-1)-r^{1-\gamma}\left(\rho^{\gamma}-1\right)\right]}{(\gamma-1)v_1\left(\dfrac{r-1}{r}\right)}$$

or

$$\mathbf{p_m} = \frac{p_1 r^{\gamma}\left[\gamma(\rho-1)-r^{1-\gamma}\left(\rho^{\gamma}-1\right)\right]}{(\gamma-1)(r-1)}.$$

Example 17

A diesel engine has a compression ratio of 15 and heat addition at constant pressure takes place at 6% of stroke. Find the air standard efficiency of the engine.

Take γ for air as 1.4.

■ **Solution.**

Compression ratio, $r\left(=\dfrac{v_1}{v_2}\right) = 15$

γ for air = 1.4

Air standard efficiency of diesel cycle is given by

$$\eta_{\text{diesel}} = 1 - \frac{1}{\gamma(r)^{\gamma-1}}\left[\frac{\rho^{\gamma}-1}{\rho-1}\right]$$

where ρ = cut-off ratio = $\dfrac{V_3}{V_2}$

But $\quad V_3 - V_2 = \dfrac{6}{100} V_s \qquad (V_s = \text{stroke volume})$

$$= 0.06\left(V_1 - V_2\right) = 0.06\left(15\,V_2 - V_2\right)$$

$$= 0.84\,V_2 \text{ or } V_3 = 1.84\,V_2$$

∴

$$\rho = \frac{V_3}{V_2} = \frac{1.84\,V_2}{V_2} = 1.84$$

Putting the value in equation (i), we get

$$\eta_{\text{diesel}} = 1 - \frac{1}{1.4(15)^{1.4-1}}\left[\frac{(1.84)^{1.4}-1}{1.84-1}\right]$$

$= 1 - 0.2417 \times 1.605 = \mathbf{0.612} \text{ or } \mathbf{61.2\%}. \quad \textbf{(Ans.)}$

Example 18

The stroke and cylinder diameter of a compression ignition engine are 250 mm and 150 mm, respectively. If the clearance volume is 0.0004 m³ and fuel injection takes place at a constant pressure for 5% of the stroke, determine the efficiency of the engine. Assume the engine is working on the diesel cycle.

■ Solution.

Refer to Figure 11.16.

Length of stroke,	$L = 250 \text{ mm} = 0.25 \text{ m}$
Diameter of cylinder,	$D = 150 \text{ mm} = 0.15 \text{ m}$
Clearance volume,	$V_2 = 0.0004 \text{ m}^3$
Swept volume,	$V_s = \pi/4 \ D^2 L = \pi/4 \times 0.15^2 \times 0.25 = 0.004418 \text{ m}^3$
Total cylinder volume	= Swept volume + clearance volume
	$= 0.004418 + 0.0004 = 0.004818 \text{ m}^3$

Volume at point of cut-off,
$$V_3 = V_2 + \frac{5}{100} \ V_s$$

$$= 0.0004 + \frac{5}{100} \times 0.004418 = 0.000621 \text{ m}^3$$

∴ Cut-off ratio,
$$\rho = \frac{V_3}{V_2} = \frac{0.000621}{0.0004} = 1.55$$

Compression ratio, $r = \dfrac{V_1}{V_2} = \dfrac{V_s + V_2}{V_2} = \dfrac{0.004418 + 0.0004}{0.0004} = 12.04$

Hence, $\eta_{diesel} = 1 - \dfrac{1}{\gamma(r)^{\gamma-1}} \left[\dfrac{\rho^\gamma - 1}{\rho - 1} \right] = 1 - \dfrac{1}{1.4 \times (12.04)^{1.4-1}} \left[\dfrac{(1.55)^{1.4} - 1}{1.55 - 1} \right]$

$$= 1 - 0.264 \times 1.54 = \mathbf{0.593} \text{ or } \mathbf{59.3\%}. \text{ (Ans.)}$$

Example 19

Calculate the percentage loss in the ideal efficiency of a diesel engine with a compression ratio of 14 if the fuel cut-off is delayed from 5% to 8%.

■ Solution.

Let the clearance volume (V_2) be unity.

Then, compression ratio, $\qquad r = 14$

Now, when the fuel is cut off at 5%, we have

$$\frac{\rho-1}{r-1} = \frac{5}{100} \text{ or } \frac{\rho-1}{14-1} = 0.05 \text{ or } \rho-1 = 13 \times 0.05 = 0.65$$

$$\therefore \qquad \rho = 1.65$$

$$\eta_{\text{diesel}} = 1 - \frac{1}{\gamma(r)^{\gamma-1}} \left[\frac{\rho^\gamma-1}{\rho-1}\right] = 1 - \frac{1}{1.4 \times (14)^{1.4-1}} \left[\frac{(1.65)^{1.4}-1}{1.65-1}\right]$$

$$= 1 - 0.248 \times 1.563 = 0.612 \quad \text{or} \quad 61.2\%$$

When the fuel is cut off at 8%, we have

$$\frac{\rho-1}{r-1} = \frac{8}{100} \text{ or } \frac{\rho-1}{14-1} = \frac{8}{100} = 0.08$$

$$\therefore \qquad \rho = 1 + 1.04 = 2.04$$

$$\eta_{\text{diesel}} = 1 - \frac{1}{\gamma(r)^{\gamma-1}} \left[\frac{\rho^\gamma-1}{\rho-1}\right] = 1 - \frac{1}{1.4 \times (14)^{1.4-1}} \left[\frac{(2.04)^{1.4}-1}{2.04-1}\right]$$

$$= 1 - 0.248 \times 1.647 = 0.591 \quad \text{or} \quad 59.1\%$$

Hence percentage loss in efficiency is due to a delay in fuel cut-off

$$= 61.2 - 59.1 = \textbf{2.1\%.} \quad \textbf{(Ans.)}$$

Example 20

The mean effective pressure of a Diesel cycle is 7.5 bar and the compression ratio is 12.5. Find the percentage cut-off of the cycle if its initial pressure is 1 bar.

■ Solution.

Mean effective pressure, $p_m = 7.5$ bar

Compression ratio, $\qquad\qquad r = 12.5$

Initial pressure, $\qquad\qquad p_1 = 1$ bar

Refer Figure 15.

The mean effective pressure is given by

$$p_m = \frac{p_1 r^\gamma \left[\gamma(\rho-1) - r^{1-\gamma}(\rho^\gamma-1)\right]}{(\gamma-1)(r-1)}$$

$$7.5 = \frac{1 \times (12.5)^{1.4} \left[1.4(\rho-1) - (12.5)^{1-1.4}(\rho^{1.4}-1)\right]}{(1.4-1)(12.5-1)}$$

$$7.5 = \frac{34.33 \left[1.4\,\rho - 1.4 - 0.364\rho^{1.4} + 0.364\right]}{4.6}$$

$$7.5 = 7.46 \left(1.4 \, \rho - 1.036 - 0.364 \, \rho^{1.4} \right)$$

$$1.005 = 1.4 \, \rho - 1.036 - 0.364 \, \rho^{1.4}$$

or $\qquad 2.04 = 1.4 \, \rho - 0.364 \, \rho^{1.4}$ or $0.364 \, \rho^{1.4} - 1.4 \, \rho + 2.04 = 0$

Solving by trial and error method, we get

$$\rho = 2.24$$

∴ $\qquad$ % cut - off $= \dfrac{\rho - 1}{r - 1} \times 100 = \dfrac{2.24 - 1}{12.5 - 1} \times 100 = \mathbf{10.78\%.}$ **(Ans.)**

Example 21

An engine with a 200-mm cylinder diameter and a 300-mm stroke works on a theoretical Diesel cycle. The initial pressure and temperature of air used are 1 bar and 27 °C. The cut-off is 8% of the stroke. Determine:

(i) Pressures and temperatures at all salient points.

(ii) Theoretical air standard efficiency.

(iii) Mean effective pressure.

(iv) Power of the engine if the working cycles per minute are 380.

Assume that the compression ratio is 15 and the working fluid is air.

Consider all conditions to be ideal.

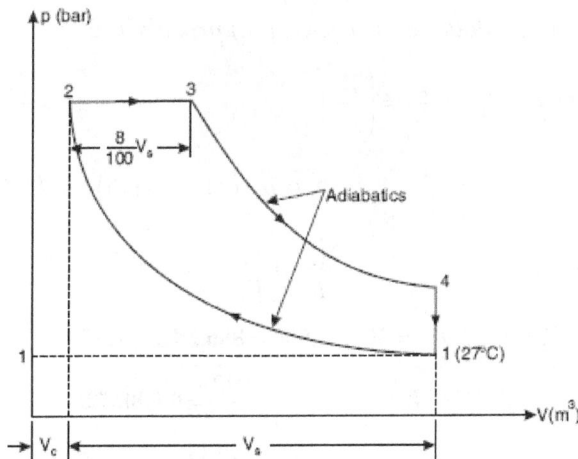

FIGURE 11.17. Sketch for Example 21.

■ **Solution.**

Cylinder diameter, $D = 200$ mm or 0.2 m

Stroke length, $L = 300$ mm or 0.3 m

Initial pressure, $p_1 = 1.0$ bar

Initial temperature, $T_1 = 27 + 273 = 300$ K

Cut-off $= \dfrac{8}{100} V_s = 0.08\, V_s$

(i) Pressures and temperatures at salient points:

Now, stroke volume, $V_s = \pi/4\ D^2 L = \pi/4 \times 0.2^2 \times 0.3 = 0.00942$ m³

$$V_1 = V_s + V_c = V_s + \frac{V_s}{r-1} \qquad \left[\because V_c = \frac{V_s}{r-1} \right]$$

$$= V_s \left(1 + \frac{1}{r-1} \right) = \frac{r}{r-1} \times V_s$$

i.e., $V_1 = \dfrac{15}{15-1} \times V_s = \dfrac{15}{14} \times 0.00942 = \mathbf{0.0101\ m^3}\,V$ **(Ans.)**

Mass of the air in the cylinder can be calculated by using the gas equation,

$$p_1 V_1 = m R T_1$$

$$m = \frac{p_1 V_1}{R T_1} = \frac{1 + 10^5 \times 0.0101}{287 \times 300} = 0.0117 \text{ kg/cycle}$$

For the *adiabatic (or isentropic) process 1–2*

$$p_1 V_1^{\gamma} = p_2 V_2^{\gamma} \text{ or } \frac{p_2}{p_1} = \left(\frac{V_1}{V_2} \right)^{\gamma} = (r)^{\gamma}$$

∴ $p_2 = p_1 \cdot (r)^{\gamma} = 1 \times (15)^{1.4} = \mathbf{44.31\ bar.\ (Ans.)}$

Also, $\dfrac{T_2}{T_1} = \left(\dfrac{V_1}{V_2} \right)^{\gamma - 1} = (r)^{\gamma - 1} = (15)^{1.4 - 1} = 2.954$

∴ $T_2 = T_1 \times 2.954 = 300 \times 2.954 = \mathbf{886.2\ K.\ (Ans.)}$

$$V_2 = V_c = \frac{V_s}{r-1} = \frac{0.00942}{15-1} = 0.0006728 \text{ m}^3. \qquad \textbf{(Ans.)}$$

$$p_2 = p_3 = \textbf{44.31 bar.} \qquad \textbf{(Ans.)}$$

% cut-off ratio $= \dfrac{\rho - 1}{r - 1}$

$$\dfrac{8}{100} = \dfrac{\rho - 1}{15 - 1}$$

i.e., $\rho = 0.08 \times 14 + 1 = 2.12$

∴ $V_3 = \rho \, V_2 = 2.12 \times 0.0006728 = \textbf{0.001426 m}^3.$ **(Ans.)**

$$\left[\begin{array}{l} V_3 \text{ can also be calculated as follows :} \\ V_3 = 0.08 V_s + V_c = 0.08 \times 0.00942 + 0.0006728 = 0.001426 \text{ m}^3 \end{array} \right]$$

For the *constant pressure process 2–3,*

$$\dfrac{V_3}{T_3} = \dfrac{V_2}{T_2}$$

∴ $T_3 = T_2 \times \dfrac{V_3}{V_2} = 886.2 \times \dfrac{0.001426}{0.0006728} = \textbf{1878.3 K.}$ **(A**

For the *isentropic process 3–4,*

$$p_3 V_3^{\gamma} = p_4 V_4^{\gamma}$$

$$p_4 = p_3 \times \left(\dfrac{V_3}{V_4} \right)^{\gamma} = p_3 \times \dfrac{1}{(7.07)^{1.4}} \quad \left[\begin{array}{l} \because \dfrac{V_4}{V_3} = \dfrac{V_4}{V_2} \times \dfrac{V_2}{V_3} = \dfrac{V_1}{V_2} \times \dfrac{V_2}{V_3} \\ = \dfrac{r}{\rho}, \; \because V_4 = V_1 = \dfrac{15}{2.12} = 7.07 \end{array} \right]$$

$$= \dfrac{44.31}{(7.07)^{1.4}} = \textbf{2.866 bar.} \text{ (Ans.)}$$

Also, $\dfrac{T_4}{T_3} = \left(\dfrac{V_3}{V_4} \right)^{\gamma - 1} = \left(\dfrac{1}{7.07} \right)^{1.4 - 1} = 0.457$

∴ $T_4 = T_3 \times 0.457 = 1878.3 \times 0.457 = \textbf{858.38 K.}$ **(Ans.)**

$$V_4 = V_1 = \textbf{0.0101 m}^3. \text{ (Ans.)}$$

(*ii*) Theoretical air standard efficiency:

$$\eta_{\text{diesel}} = 1 - \dfrac{1}{\gamma (r)^{\gamma - 1}} \left[\dfrac{\rho^{\gamma} - 1}{\rho - 1} \right] = 1 - \dfrac{1}{1.4 (15)^{1.4 - 1}} \left[\dfrac{(2.12)^{1.4} - 1}{2.12 - 1} \right]$$

$$= 1 - 0.2418 \times 1.663 = \textbf{0.598 or 59.8\%.} \text{ (Ans.)}$$

(*iii*) Mean effective pressure, p_m:
Mean effective pressure of Diesel cycle is given by

$$p_m = \frac{p_1\,(r)^{\gamma}\,[\gamma(\rho-1)-r^{1-\gamma}\left(\rho^{\gamma}-1\right)]}{(\gamma-1)(r-1)}$$

$$= \frac{1\times(15)^{1.4}\left[1.4(2.12-1)-(15)^{1-1.4}\left(2.12^{1.4}-1\right)\right]}{(1.4-1)(15-1)}$$

$$= \frac{44.31[1.568-0.338\times1.863]}{0.4\times14} = \textbf{7.424 bar.} \quad \textbf{(Ans.)}$$

(iv) Power of the engine, P:

Work done per cycle	$= p_m V_s = \dfrac{7424\times10^5 \times0.00942}{10^3} = 6.99 \text{ kJ/cycle}$
Work done per second	= Work done per cycle $\times$ no. of cycles per second

$$= 6.99\times380/60 = 44.27 \text{ kJ/s} = 44.27 \text{ kW}$$

Hence *power of the engine* = **44.27 kW.**   **(Ans.)**

Example 22

The volume ratios of compression and expansion for a diesel engine as measured from an indicator diagram are 15.3 and 7.5, respectively. The pressure and temperature at the beginning of the compression are 1 bar and 27 °C.

Assuming an ideal engine, determine the mean effective pressure, the ratio of maximum pressure to mean effective pressure, and cycle efficiency.

Also find the fuel consumption per kWh if the indicated thermal efficiency is 0.5 of ideal efficiency, mechanical efficiency is 0.8, and the calorific value of oil is 42000 kJ/kg.

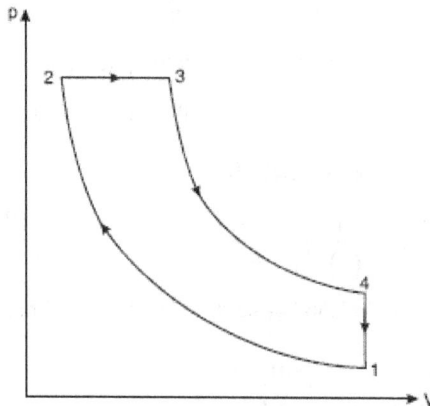

FIGURE 11.18. Sketch for Example 22, Diesel cycle.

■ **Solution.**

Given: $\dfrac{V_1}{V_2} = 15.3$; $\dfrac{V_4}{V_3} = 7.5$

$p_1 = 1$ bar; $T_1 = 27 + 273 = 300$ K; $\eta_{th(I)} = 0.5 \times \eta_{air-standard}$; $\eta_{mech.}$

$= 0.8$; $C = 42000$ kJ / kg. The cycle is shown in Figure 11.18; the subscripts denote the respective points in the cycle.

Mean effective pressure, p_m:

$$p_m = \frac{\text{Work done by the cycle}}{\text{Swept volume}}$$

Work done   = Heat added – heat rejected

Heat added $= mc_p \left(T_3 - T_2 \right)$, and

Heat rejected $= mc_v \left(T_4 - T_1 \right)$

Now assume air as a perfect gas and the mass of oil in the air-fuel mixture is negligible and is not taken into account.

Process 1–2 is an *adiabatic compression process*, thus

$$\frac{T_2}{T_1} = \left(\frac{V_1}{V_2} \right)^{\gamma-1} \quad \text{or} \quad T_2 = T_1 \times \left(\frac{V_1}{V_2} \right)^{1.4-1} \quad \left(\text{since } \gamma = 1.4 \right)$$

or $$T_2 = 300 \times \left(15.3 \right)^{0.4} = 893.3 \text{ K}$$

Also, $$p_1 V_1^{\gamma} = p_2 V_2^{\gamma} \ \Rightarrow \ p_2 = p_1 \times \left(\frac{V_1}{V_2} \right)^{\gamma} = 1 \times \left(15.3 \right)^{1.4} = 45.56 \text{ bar}$$

Process 2–3 is a *constant pressure process*, hence

$$\frac{V_2}{T_2} = \frac{V_3}{T_3} \ \Rightarrow \ T_3 = \frac{V_3 T_2}{V_2} = 2.04 \times 893.3 = 1822.3 \text{ K}$$

Assume that the volume at point 2 (V_2) is 1 m^3. Thus, the mass of air involved in the process,

$$m = \frac{p_2 V_2}{RT_2} = \frac{45.56 \times 10^5 \times 1}{287 \times 893.3} = 17.77 \text{ kg} \qquad \begin{bmatrix} \because \ \dfrac{V_4}{V_3} = \dfrac{V_1}{V_3} = \dfrac{V_1}{V_2} \times \dfrac{V_2}{V_3} \\[2mm] \text{or} \ \dfrac{V_3}{V_2} = \dfrac{V_1}{V_2} \times \dfrac{V_3}{V_4} = \dfrac{15.3}{7.5} = 2.04 \end{bmatrix}$$

Process 3–4 is an *adiabatic expansion process*, thus

$$\frac{T_4}{T_3} = \left(\frac{V_3}{V_4}\right)^{\gamma-1} = \left(\frac{1}{7.5}\right)^{1.4-1} = 0.4466$$

or $T_4 = 1822.3 \times 0.4466 = 813.8$ K

$\therefore$ Work done $= mc_p (T_3 - T_2) - mc_v (T_4 - T_1)$

$= 17.77 \left[1.005(1822.3 - 893.3) - 0.718(813.8 - 300)\right] = 10035$ kJ

$\therefore$ $\mathbf{P_m} = \dfrac{\text{Work done}}{\text{Swept volume}} = \dfrac{10035}{(V_1 - V_2)} = \dfrac{10035}{(15.3V_2 - V_2)} = \dfrac{10035}{14.3}$

$= 701.7$ kN/m^2 = **7.017 bar. (Ans.)**

$$\left(\because \ V_2 = 1 \text{ m}^3 \text{ assumed}\right)$$

Ratio of maximum pressure to mean effective pressure

$$= \frac{P_2}{P_m} = \frac{45.56}{7.017} = \mathbf{6.49.} \ \mathbf{(Ans.)}$$

Cycle efficiency, η_{cycle}:

$$\eta_{\text{cycle}} = \frac{\text{Work done}}{\text{Heat supplied}}$$

$$= \frac{10035}{mc_p (T_3 - T_2)} = \frac{10035}{17.77 \times 1.005(1822.3 - 897.3)} = \mathbf{0.6048 \text{ or } 60.48\%.} \ \mathbf{(Ans.)}$$

Fuel consumption per kWh; m_f:

$\eta_{\text{th(I)}} = 0.5 \ \eta_{\text{cycle}} = 0.5 \times 0.6048 = 0.3024$ or 30.24%

$\eta_{\text{th(B)}} = 0.3024 \times 0.8 = 0.242$

Also, $\eta_{\text{th(B)}} = \dfrac{\text{B.P}}{m_f \times C} = \dfrac{1}{\dfrac{m_f}{3600} \times 42000} = \dfrac{3600}{m_f \times 42000}$

or $0.242 = \dfrac{3600}{m_f \times 42000}$

or $\mathbf{m_f} = \dfrac{3600}{0.242 \times 42000} = \mathbf{0.354 \text{ kg/kWh.}} \ \mathbf{(Ans.)}$

11.7. Dual Combustion Cycle

This cycle (also called the *limited pressure cycle or mixed cycle*) is a combination of Otto and Diesel cycles, in the way that heat is added partly at constant volume and partly at constant pressure; the advantage is that more time is available to the fuel (which is injected into the engine cylinder before the end of compression stroke) for combustion. Because of the

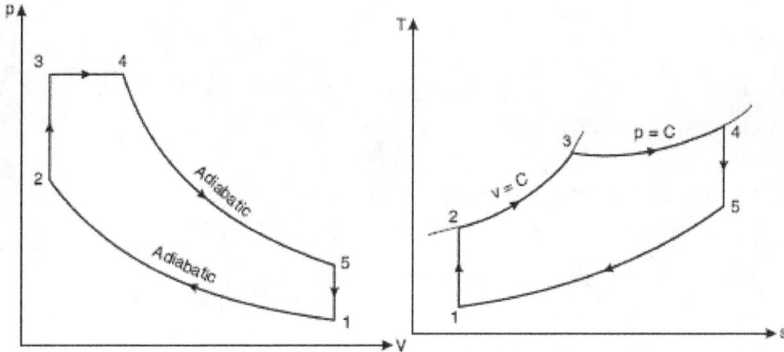

FIGURE 11.19. The dual combustion cycle, p-V and T-S diagrams.

lagging characteristics of the fuel, this cycle is invariably used for diesel and hot spot ignition engines.

The dual combustion cycle (Figure 11.19) consists of the following operations:

- 1–2—*Adiabatic compression*
- 2–3—*Addition of heat at constant volume*
- 3–4—*Addition of heat at constant pressure*
- 4–5—*Adiabatic expansion*
- 5–1—*Rejection of heat at constant volume.*

Consider 1 kg of air.

Total heat supplied = Heat supplied during the operation $2-3$
+ heat supplied during the operation $3-4$

$= c_v\left(T_3 - T_2\right) + c_p\left(T_4 - T_3\right)$

Heat rejected during operation $5 - 1 = c_v\left(T_5 - T_1\right)$

Work done = Heat supplied – heat rejected

$= c_v\left(T_3 - T_2\right) + c_p\left(T_4 - T_3\right) - c_v\left(T_5 - T_1\right)$

$$\eta_{dual} = \frac{\text{Work done}}{\text{Heat supplied}} = \frac{c_v\left(T_3 - T_2\right) + c_p\left(T_4 - T_3\right) - c_v\left(T_5 - T_1\right)}{c_v\left(T_3 - T_2\right) + c_p\left(T_4 - T_3\right)}$$

$$= 1 - \frac{c_v\left(T_5 - T_1\right)}{c_v\left(T_3 - T_2\right) + c_p\left(T_4 - T_3\right)}$$

$$= 1 - \frac{c_v \left(T_5 - T_1 \right)}{\left(T_3 - T_2 \right) + \gamma \left(T_4 - T_3 \right)} \qquad ...(i) \qquad \left(\because \gamma = \frac{c_p}{c_v} \right)$$

Compression ratio, $\qquad r = \dfrac{v_1}{v_2}$

During *adiabatic compression process 1–2*,

$$\frac{T_2}{T_1} = \left(\frac{v_1}{v_2} \right)^{\gamma - 1} = \left(r \right)^{\gamma - 1} \qquad\qquad ...(ii)$$

During *constant volume heating process,*

$$\frac{p_3}{T_3} = \frac{p_2}{T_2}$$

or $\quad \dfrac{T_3}{T_2} = \dfrac{p_3}{p_2} = \beta$, where β is known as **pressure or explosion ratio.**

or $\quad T_2 = \dfrac{T_3}{\beta} \qquad\qquad ...(iii)$

During *adiabatic expansion process,*

$$\frac{T_4}{T_5} = \left(\frac{v_5}{v_4} \right)^{\gamma - 1}$$

$$= \left(\frac{r}{\rho} \right)^{\gamma - 1} \qquad\qquad ...(iv)$$

$$\left(\because \quad \frac{v_5}{v_4} = \frac{v_1}{v_4} = \frac{v_1}{v_2} \times \frac{v_2}{v_4} = \frac{v_1}{v_2} \times \frac{v_3}{v_4} = \frac{r}{\rho}, \ \rho \text{ being the cut- off ratio} \right)$$

During *constant pressure heating process,*

$$\frac{v_3}{T_3} = \frac{v_4}{T_4}$$

$$T_4 = T_3 \frac{v_4}{v_3} = \rho \, T_3 \qquad\qquad ...(v)$$

Putting the value of T_4 in the equation (iv), we get

$$\frac{\rho T_3}{T_5} = \left(\frac{r}{\rho}\right)^{\gamma-1} \quad \text{or} \quad T_5 = \rho \cdot T_3 \cdot \left(\frac{\rho}{r}\right)^{\gamma-1}$$

Putting the value of T_2 in equation (ii), we get

$$\frac{\dfrac{T_3}{\beta}}{T_1} = (r)^{\gamma-1}$$

$$T_1 = \frac{T_3}{\beta} \cdot \frac{1}{(r)^{\gamma-1}}$$

Now inserting the values of T_1, T_2, T_4, and T_5 in equation (i), we get

$$\eta_{\text{dual}} = 1 - \frac{\left[\rho \cdot T_3 \left(\dfrac{\rho}{r}\right)^{\gamma-1} - \dfrac{T_3}{\beta} \cdot \dfrac{1}{(r)^{\gamma-1}}\right]}{\left[\left(T_3 - \dfrac{T_3}{\beta}\right) + \gamma(\rho T_3 - T_3)\right]} = 1 - \frac{\dfrac{1}{(r)^{\gamma-1}}\left(\rho^r - \dfrac{1}{\beta}\right)}{\left(1 - \dfrac{1}{\beta}\right) + \gamma(\rho-1)}$$

i.e.,
$$\eta_{\text{dual}} = 1 - \frac{1}{(r)^{\gamma-1}} \cdot \frac{(\beta \cdot \rho^\gamma - 1)}{\left[(\beta-1) + \beta\gamma(\rho-1)\right]}$$

Work done is given by,

$$W = p_3(v_4 - v_3) + \frac{p_4 v_4 - p_5 v_5}{\gamma-1} - \frac{p_2 v_2 - p_1 v_1}{\gamma-1}$$

$$= p_3 v_3(\rho-1) + \frac{(p_4\rho v_3 - p_5 r v_3) - (p_2 v_3 - p_1 r v_3)}{\gamma-1}$$

$$= \frac{p_3 v_3(\rho-1)(\gamma-1) + p_4 v_3\left(\rho - \dfrac{p_5}{p_4}r\right) - p_2 v_3\left(1 - \dfrac{p_1}{p_2}r\right)}{\gamma-1}$$

Also
$$\frac{p_5}{p_4} = \left(\frac{v_4}{v_5}\right)^\gamma = \left(\frac{\rho}{r}\right)^\gamma \quad \text{and} \quad \frac{p_2}{p_1} = \left(\frac{v_1}{v_2}\right)^\gamma = r^\gamma$$

also,
$$p_3 = p_4, \quad v_2 = v_3, \quad v_5 = v_1$$

∴
$$W = \frac{v_3\left[p_3(\rho-1)(\gamma-1) + p_3\left(\rho - \rho^\gamma r^{1-\gamma}\right) - p_2\left(1 - r^{1-\gamma}\right)\right]}{(\gamma-1)}$$

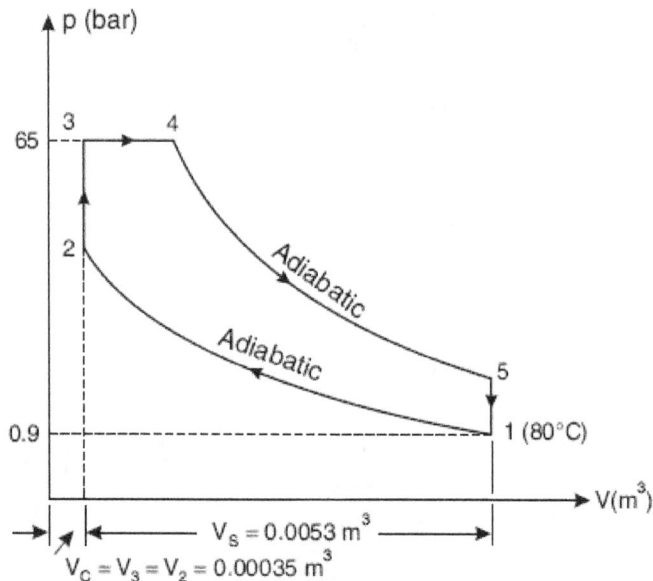

FIGURE 11.20. Sketch for Example 23.

$$= \frac{p_2 v_2 \left[\beta(\rho-1)(\gamma-1) + \beta(\rho-\rho^\gamma r^{1-\gamma}) - (1-r^{1-\gamma})\right]}{(\gamma-1)}$$

$$= \frac{p_1 (r)^\gamma v_1 / r \left[\beta\gamma(\rho-1) + (\beta-1) - r^{1-\gamma}(\beta\rho^\gamma - 1)\right]}{\gamma-1}$$

$$= \frac{p_1 v_1 r^{\gamma-1} \left[\beta\gamma(\rho-1) + (\beta-1) - r^{1-\gamma}(\beta\rho^\gamma - 1)\right]}{\gamma-1}$$

Mean effective pressure (p_m) is given by,

$$p_m = \frac{W}{v_1 - v_2} = \frac{W}{v_1\left(\dfrac{r-1}{r}\right)} = \frac{p_1 v_1 \left[r^{1-\gamma}\beta\gamma(\rho-1) + (\beta-1) - r^{1-\gamma}(\beta\rho^\gamma - 1)\right]}{(\gamma-1)v_1\left(\dfrac{r-1}{r}\right)}$$

$$\mathbf{p_m} = \frac{p_1 (r)^\gamma \left[\beta(\rho-1) + (\beta-1) - r^{1-\gamma}(\beta\rho^\gamma - 1)\right]}{(\gamma-1)(r-1)}$$

Example 23

The swept volume of a diesel engine working on dual cycle is 0.0053 m³ and clearance volume is 0.00035 m3. The maximum pressure is 65 bar. Fuel injection ends at 5% of the stroke. The temperature and pressure at the start of the compression are 80 °C and 0.9 bar. Determine the air standard efficiency of the cycle. Take γ for air = 1.4.

■ **Solution.**

Swept volume, $V_s = 0.0053 \text{ m}^3$

Clearance volume, $V_c = V_3 = V_2 = 0.00035 \text{ m}^3$

Maximum pressure, $p_3 = p_4 = 65 \text{ bar}$

Initial temperature, $T_1 = 80 + 273 = 353 \text{ K}$

Initial pressure, $p_1 = 0.9 \text{ bar}$

$\eta_{\text{dual}} = ?$

The efficiency of a dual combustion cycle is given by

$$\eta_{\text{dual}} = 1 - \frac{1}{(r)^{\gamma-1}}\left[\frac{\beta.\rho^{\gamma}-1}{(\beta-1)+\beta\gamma(\rho-1)}\right] \qquad ...(i)$$

Compression ratio, $r = \dfrac{V_1}{V_2} = \dfrac{V_s + V_c}{V_c} = \dfrac{0.0053 + 0.00035}{0.00035} = 16.14$

$[\because \ V_2 = V_c = \text{Clearance volume}]$

Cut-off ratio, $\rho = \dfrac{V_4}{V_3} = \dfrac{\dfrac{5}{100}V_s + V_3}{V_3} = \dfrac{0.05V_s + V_c}{V_c} \ (\because \ V_2 = V_3 = V_c)$

$= \dfrac{0.05 \times 0.00053 + 0.00035}{0.00035} = 1.757 \text{ say } 1.76$

Also during the *compression operation 1–2,*

$p_1 V_1^{\gamma} = p_2 V_2^{\gamma}$

or $\dfrac{p_2}{p_1} = \left(\dfrac{V_1}{V_2}\right)^{\gamma} = (16.14)^{1.4} = 49.14$

or $p_2 = p_1 \times 49.14 = 0.9 \times 49.14 = 44.22 \text{ bar}$

Pressure or explosion ratio, $\beta = \dfrac{p_3}{p_2} = \dfrac{65}{44.22} = 1.47$

Putting the value of r, ρ and β in equation (i), we get

$$\eta_{\text{dual}} = 1 - \frac{1}{(16.14)^{1.4-1}} \left[\frac{1.47 \times (1.76)^{1.4} - 1}{(1.47 - 1) + 1.47 \times 1.4 (1.76 - 1)} \right]$$

$$= 7 - 0.328 \left[\frac{3.243 - 1}{0.47 + 1.564} \right] = \mathbf{0.6383} \text{ or } \mathbf{63.83\%.} \text{ (Ans.)}$$

Example 24

An oil engine working on the dual combustion cycle has a compression ratio of 14 and the explosion ratio obtained from an indicator card is 1.4. If the cut-off occurs at 6% of stroke, find the ideal efficiency. Take γ for air = 1.4.

■ Solution.

Refer to Figure 11.19.

Compression ratio, $r = 14$

Explosion ratio, $\beta = 1.4$

If ρ is the cut – off ratio, then $\dfrac{\rho - 1}{r - 1} = \dfrac{6}{100}$ or $\dfrac{\rho - 1}{14 - 1} = 0.06$

$\therefore \rho = 1.78$

Ideal efficiency is given by

$$\eta_{\text{ideal or dual}} = 1 - \frac{1}{(r)^{\gamma - 1}} \left[\frac{(\beta \rho^{\gamma} - 1)}{(\beta - 1) + \beta \gamma (\rho - 1)} \right]$$

$$= 1 - \frac{1}{(14)^{1.4-1}} \left[\frac{1.4 \times (1.78)^{1.4} - 1}{(1.4 - 1) + 1.4 \times 1.4 (1.78 - 1)} \right]$$

$$= 1 - 0.348 \left[\frac{3.138 - 1}{0.4 + 1.528} \right] = \mathbf{0.614} \text{ or } \mathbf{61.4\%.} \text{ (Ans.)}$$

Example 25

The compression ratio for a single-cylinder engine operating on a dual cycle is 9. The maximum pressure in the cylinder is limited to 60 bar. The pressure and temperature of the air at the beginning of the cycle are 1 bar and 30 °C. Heat is added during a constant pressure process up to 4% of the stroke. Assuming the cylinder diameter and stroke length as 250 mm and 300 mm respectively, determine:

(i) The air standard efficiency of the cycle.

(ii) The power developed if the number of working cycles are 3 per second.

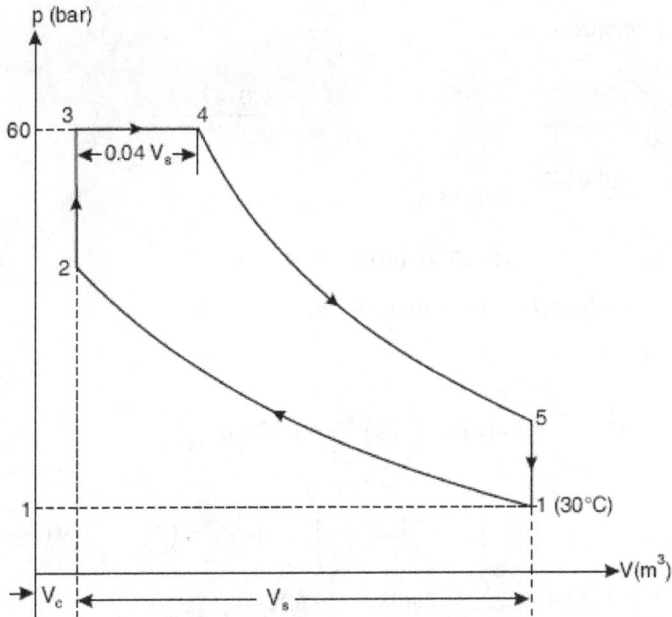

FIGURE 11.21. Sketch for Example 25 p-V diagram.

Take for air $c_v = 0.71$ kJ/kg K and: $c_p = 1.0$. kJ/kg K

■ **Solution.**

Cylinder diameter, $D = 250$ mm $= 0.25$ m

Compression ratio, $r = 9$

Stroke length, $L = 300$ mm $= 0.3$ m

Initial pressure, $p_1 = 1$ bar

Initial temperature, $T_1 = 30 + 273 = 303$ K

Maximum pressure, $p_3 = p_4 = 60$ bar

Cut-off $= 4\%$ of stroke volume

Number of working cycles/sec. $= 3$.

(i) Air standard efficiency:

Now, swept volume, $V_s = \pi / 4\, D^2 L = \pi / 4 \times 0.25^2 \times 0.3$

$= 0.0147$ m^3

Also, compression ratio, $\quad r = \dfrac{V_s + V_c}{V_c}$

i.e., $\qquad\qquad\qquad\qquad\qquad 9 = \dfrac{0.0147 + V_c}{V_c}$

$\therefore \quad V_c = \dfrac{0.0147}{8} = 0.0018 \text{ m}^3$

$\therefore \quad V_1 = V_s + V_c = 0.0147 + 0.0018 = 0.0165 \text{ m}^3$

For the *adiabatic (or isentropic) process 1–2,*

$p_1 V_1^{\gamma} = p_2 V_2^{\gamma}$

$p_2 = p_1 \times \left(\dfrac{V_1}{V_2}\right)^{\gamma} = 1 \times (r)^{\gamma} = 1 \times (9)^{1.4} = 21.67 \text{ bar}$

Also $\qquad\qquad\qquad \dfrac{T_2}{T_1} = \left(\dfrac{V_1}{V_2}\right)^{\gamma-1} = (r)^{\gamma-1} = (9)^{1.4-1} = (9)^{0.4} = 2.408$

$\therefore \quad T_2 = T_1 \times 2.408 = 303 \times 2.408 = 729.6 \text{ K}$

For the *constant volume process 2–3,*

$\dfrac{T_3}{p_3} = \dfrac{T_2}{p_2}$

$\therefore \quad T_3 = T_2 \cdot \dfrac{p_3}{p_2} = 729.6 \times \dfrac{60}{21.67} = 2020 \text{ K}$

Also, $\quad \dfrac{\rho - 1}{r - 1} = \dfrac{4}{100} \text{ or } 0.04$

$\therefore \quad \dfrac{\rho - 1}{9 - 1} = 0.04 \text{ or } \rho = 1.32$

For the *constant pressure process 3–4,*

$\dfrac{V_4}{T_4} = \dfrac{V_3}{T_3} \text{ or } \dfrac{T_4}{T_3} = \dfrac{V_4}{V_3} = \rho$

$\therefore \quad T_4 = T_3 \times \rho = 2020 \times 1.32 = 2666.4 \text{ K}$

Also expansion ratio, $\qquad \dfrac{V_5}{V_4} = \dfrac{V_5}{V_2} \times \dfrac{V_2}{V_4} = \dfrac{V_1}{V_2} \times \dfrac{V_3}{V_4} = \dfrac{r}{\rho} \quad \left[\because V_5 = V_1 \text{ and } V_2 = V_3\right]$

For *adiabatic process 4–5,*

$\dfrac{T_5}{T_4} = \left(\dfrac{V_4}{V_5}\right)^{\gamma-1} = \left(\dfrac{\rho}{r}\right)^{\gamma-1}$

$$\therefore \quad T_5 = T_4 \times \left(\frac{\rho}{r}\right)^{\gamma-1} = 2666.4 \times \left(\frac{1.32}{9}\right)^{1.4-1} = 1237 \text{ K}$$

Also $\quad p_4 V_4^\gamma = p_5 V_5^\gamma$

$$p_5 = p_4 \cdot \left(\frac{V_4}{V_5}\right)^\gamma = 60 \times \left(\frac{r}{\rho}\right)^\gamma = 60 \times \left(\frac{1.32}{9}\right)^{1.4} = 4.08 \text{ bar}$$

Heat supplied, $\qquad\qquad Q_s = c_v(T_3 - T_2) + c_p(T_4 - T_3)$

$= 0.71\ (2020 - 729.6) + 1.0\ (2666.4 - 2020) = 1562.58 \text{ kJ/kg}$

Heat rejected, $\qquad\qquad Q_r = c_v(T_5 - T_1)$

$= 0.71\ (1237 - 303) = 663.14 \text{ kJ/kg}$

$$\eta_{\text{air-standard}} = \frac{Q_s - Q_r}{Q_s} = \frac{1562.85 - 663.14}{1562.58} = \textbf{0.5756 or 57.56\%. (Ans.)}$$

(*ii*) Power developed by the engine, P:

Mass of air in the cycle is given by

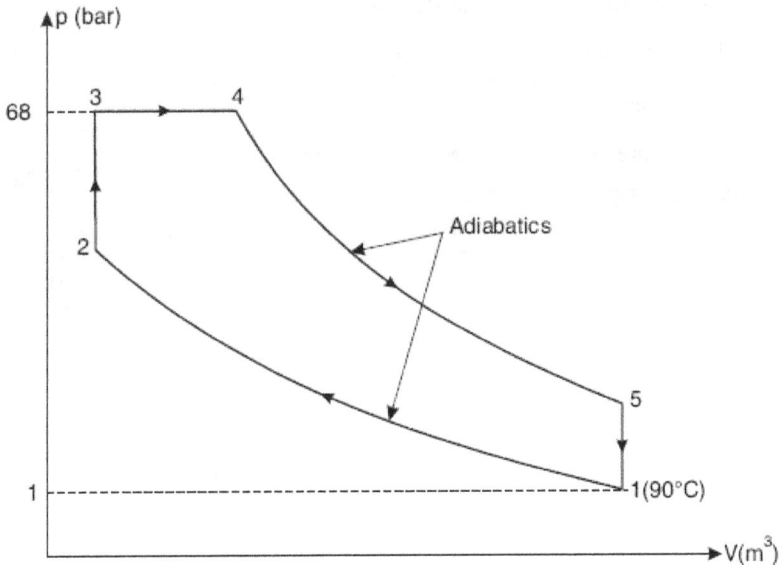

FIGURE 11.22. Sketch for Example 26.

$$m = \frac{p_1 V_1}{RT_1} = \frac{1 \times 10^5 \times 0.0165}{287 \times 303} = 0.0189 \text{ kg}$$

∴ Work done per cycle $= m(Q_s - Q_r)$

$= 0.0189\,(1562.58 - 663.14) = 16.999 \text{ kJ}$

Power developed = Work done per cycle × no. of cycles per second

$= 16.999 \times 3 = \textbf{50.99 say 51 kW. (Ans.)}$

Example 26

In an engine working on a Dual cycle, the temperature and pressure at the beginning of the cycle are 90 °C and 1 bar. The compression ratio is 9. The maximum pressure is limited to 68 bar and the total heat supplied per kg of air is 1750 kJ. Determine:

(i) Pressure and temperatures at all salient points

(ii) Air standard efficiency

(iii) Mean effective pressure

■ **Solution.**

Refer to Figure 11.22.

Initial pressure,	$p_1 = 1$ bar
Initial temperature,	$T_1 = 90 + 273 = 363$ K
Compression ratio,	$r = 9$
Maximum pressure,	$p_3 = p_4 = 68$ bar
Total heat supplied	$= 1750$ kJ / kg

(*i*) Pressures and temperatures at salient points:

$$p_1 V_1^\gamma = p_2 V_2^\gamma$$

$$p_2 = p_1 \times \left(\frac{V_1}{V_2}\right)^\gamma = 1 \times (r)^\gamma = 1 \times (9)^{1.4} = \textbf{21.67 bar. (Ans.)}$$

Also, $$\frac{T_2}{T_1} = \left(\frac{V_1}{V_2}\right)^{\gamma-1} = (r)^{\gamma-1} = (9)^{1.4-1} = 2.408$$

∴ $T_2 = T_1 \times 2.408 = 363 \times 2.408 = \textbf{874.1 K. (Ans.)}$

$p_3 = p_4 = \textbf{68 bar. (Ans.)}$

For the *constant volume process* 2–3,

$$\frac{p_2}{T_2} = \frac{p_3}{T_3}$$

$$\therefore \quad T_3 = T_2 \times \frac{p_3}{p_2} = 874.1 \times \frac{68}{21.67} = \mathbf{2742.9 \ K. \ (Ans.)}$$

Heat added at constant volume

$$= c_v \ (T_3 - T_2) = 0.71 \ (2742.9 - 874.1) = 1326.8 \ \text{kJ/kg}$$

$\therefore$ Heat added at constant pressure

= Total heat added – heat added at constant volume

$$= 1750 - 1326.8 = 423.2 \ \text{kJ/kg}$$

$$\therefore \quad c_p(T_4 - T_3) = 423.2$$

or $\quad 1.0(T_4 - 2742.9) = 423.2$

$$\therefore \quad \boldsymbol{T_4 = 3166 \ K. \ (Ans.)}$$

For *constant pressure process* 3–4,

$$\rho = \frac{V_4}{V_3} = \frac{T_4}{T_3} = \frac{3199}{2742.9} = 1.15$$

For *adiabatic (or isentropic) process* 4–5,

$$\frac{V_5}{V_4} = \frac{V_5}{V_2} \times \frac{V_2}{V_4} = \frac{V_1}{V_2} \times \frac{V_3}{V_4} = \frac{r}{\rho} \qquad \left(\because \ \rho = \frac{V_4}{V_3} \right)$$

Also $\quad p_4 V_4^{\gamma} = p_5 V_5^{\gamma}$

$$\therefore \quad p_5 = p_4 \times \left(\frac{V_4}{V_5} \right)^{\gamma} = 68 \times \left(\frac{\rho}{r} \right)^{\gamma} = 68 \times \left(\frac{1.15}{9} \right)^{1.4} = \mathbf{3.81 \ bar. \ (Ans.)}$$

Again, $\quad \dfrac{T_5}{T_4} = \left(\dfrac{V_4}{V_5} \right)^{\gamma-1} = \left(\dfrac{\rho}{r} \right)^{\gamma-1} = \left(\dfrac{1.15}{9} \right)^{14-1} = 0.439$

$$\therefore \qquad\qquad\qquad T_5 = T_4 \times 0.439 = 3166 \times 0.439 = \mathbf{1389.8 \ K.} \quad \mathbf{(Ans.)}$$

(*ii*) Air standard efficiency:

Heat rejected during constant volume process 5–1,

$$Q_r = c_v \ (T_5 - T_1) = 0.71(1389.8 - 363) = 729 \ \text{kJ/kg}$$

$$\therefore \qquad\qquad \eta_{\text{air-standard}} = \frac{\text{Work done}}{\text{Heat supplied}} = \frac{Q_s - Q_r}{Q_s}$$

$$= \frac{1750 - 729}{1750} = 0.5834 \text{ or } \textbf{58.34\%. (Ans.)}$$

(*iii*) Mean effective pressure, p_m:

Mean effective pressure is given by

$$p_m = \frac{\text{Work done per cycle}}{\text{Stroke volume}}$$

or $$p_m = \frac{1}{V_s}\left[p_3\left(V_4 - V_3\right) + \frac{p_4 V_4 - p_5 V_5}{\gamma - 1} - \frac{p_2 V_2 - p_1 V_1}{\gamma - 1}\right]$$

$$V_1 = V_5 = rV_c, V_2 = V_3 = V_c, V_4 = \rho V_c, \quad \left[\therefore r = \frac{V_s + V_c}{V_c} = 1 + \frac{V_s}{V_c}\right]$$
$$V_s = (r-1)V_c \qquad\qquad \left[\therefore V_s = (r-1)V_c\right]$$

$$\therefore \quad p_m = \frac{1}{(r-1)V_c}\left[p_3(\rho V_c - V_c) + \frac{p_4\rho V_c - p_5 \times rV_c}{\gamma - 1} - \frac{p_2 V_c - p_1 rV_c}{\gamma - 1}\right]$$

r = 9, ρ = 1.15, γ = 1.4

$p_1 = 1$ bar, $p_2 = 21.67$ bar, $p_3 = p_4 = 68$ bar, $p_5 = 3.81$ bar

Substituting these values in the previous equation, we get

$$p_m = \frac{1}{(9-1)}\left[68(1.15-1) + \frac{68 \times 1.15 - 3.81 \times 9}{1.4-1} - \frac{21.67-9}{1.4-1}\right]$$

$$= \frac{1}{8}(10.2 + 109.77 - 31.67) = 11.04$$

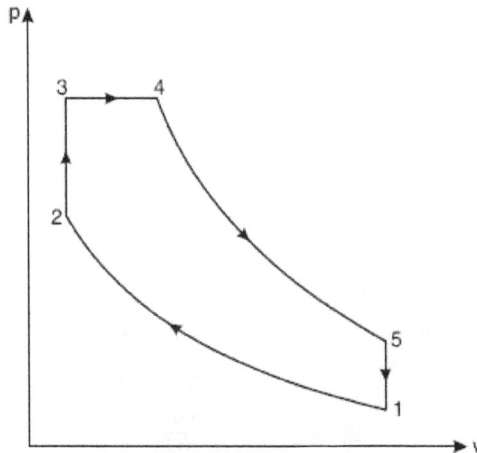

FIGURE 11.23. Sketch for Example 27, Dual cycle.

Hence, *mean effective pressure* = 11.04 bar. (Ans.)

Example 27

In an I.C. engine operating on the dual cycle (limited pressure cycle), the temperature of the working fluid (air) at the beginning of compression is 27 °C. The ratio of the maximum and minimum pressures of the cycle is 70 and the compression ratio is 15. The amounts of heat added at constant volume and at constant pressure are equal. Compute the air standard thermal efficiency of the cycle. State three main reasons why the actual thermal efficiency is different from the theoretical value Take γ for air = 1.4.

■ Solution.

Refer Fig.23. *Given* : $T_1 = 27 + 273 = 300$ K; $\dfrac{p_3}{p_1} = 70, \dfrac{v_1}{v_2} = \dfrac{v_1}{v_3} = 15$

Air standard efficiency, $\eta_{\text{air-standard}}$:

Consider 1 kg of air.

Adiabatic compression process 1–2:

$$\frac{T_2}{T_1} = \left(\frac{v_1}{v_2}\right)^{\gamma-1} = (15)^{1.4-1} = 2.954$$

$$\therefore \ T_2 = 300 \times 2.954 = 886.2 \text{ K}$$

$$\frac{p_2}{p_1} = \left(\frac{v_1}{v_2}\right)^{\gamma} = (15)^{1.4} \Rightarrow p_2 = 44.3 \, p_1$$

Constant pressure process 2–3:

$$\frac{p_2}{T_2} = \frac{p_3}{T_3}$$

or $\ T_3 = T_2 \times \dfrac{p_3}{p_2} = 886.2 \times \dfrac{70 p_1}{44.3 p_1} = 1400$ K

Also, Heat added at constant volume = Heat added at constant pressure ...(Given)

or $\qquad\qquad c_v \left(T_3 - T_2\right) = c_p \left(T_4 - T_3\right)$

or $\qquad\qquad T_3 - T_2 = \gamma \left(T_4 - T_3\right)$

or $\qquad\qquad T_4 = T_3 + \dfrac{T_3 - T_2}{\gamma} = 1400 + \dfrac{1400 - 886.2}{1.4} = 1767$ K

Constant volume process 3–4:

$$\frac{v_3}{T_3} = \frac{v_4}{T_4} \quad \Rightarrow \quad \frac{v_4}{v_3} = \frac{T_4}{T_3} = \frac{1767}{1400} = 1.26$$

Also,
$$\frac{v_4}{v_3} = \frac{v_4}{(v_1/15)} = 1.26 \text{ or } v_4 = 0.084 \, v_1$$

Also,
$$v_5 = v_1$$

Adiabatic expansion process 4–5:

$$\frac{T_4}{T_5} = \left(\frac{v_5}{v_4}\right)^{\gamma-1} = \left(\frac{v_1}{0.084 v_1}\right)^{14-1} = 2.69$$

$$\therefore \quad T_5 = \frac{T_4}{2.69} = \frac{1767}{2.69} = 656.9 \text{ K}$$

$$\therefore \quad \eta_{\text{air-standard}} = \frac{\text{Work done}}{\text{Heat supplied}} = \frac{\text{Heat supplied} - \text{Heat rejected}}{\text{Heat supplied}}$$

$$= 1 - \frac{\text{Heat rejected}}{\text{Heat supplied}}$$

$$= 1 - \frac{c_v (T_5 - T_1)}{c_v (T_3 - T_2) + c_p (T_4 - T_3)}$$

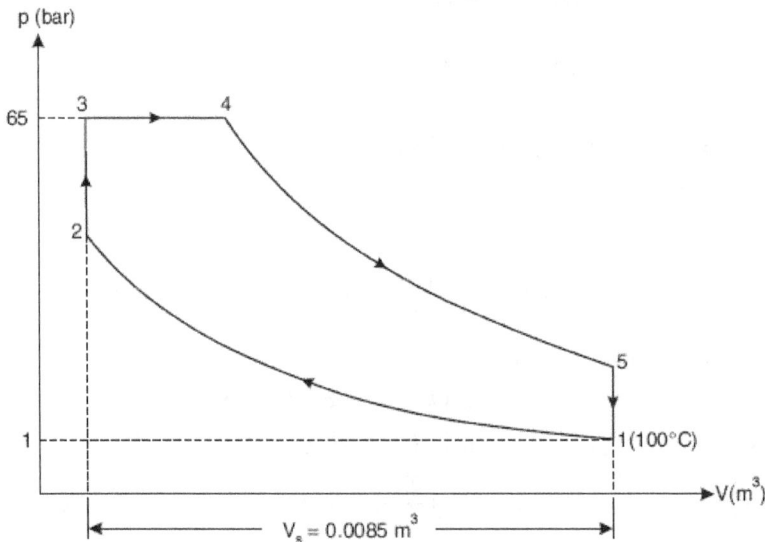

FIGURE 11.24. Sketch for Example 28, p-V diagram.

$$= 1 - \frac{(T_5 - T_1)}{(T_3 - T_2) + \gamma(T_4 - T_3)}$$

$$= 1 - \frac{(656.9 - 300)}{(1400 - 886.2) + 1.4(1767 - 1400)} = \textbf{0.653 or 65.3\%. (Ans.)}$$

Reasons for actual thermal efficiency being different from the theoretical value:

1. In a theoretical cycle the working substance is taken *air*, whereas in an actual cycle *air with fuel acts as the working substance.*

2. The fuel combustion phenomenon and associated problems like dissociation of gases, dilution of charge during suction stroke, and so on, have *not* been taken into account.

3. The effect of variable specific heat, heat loss through cylinder walls, inlet and exhaust velocities of air/gas, and so on, have not been taken into account.

Example 28

A Diesel engine working on a dual combustion cycle has a stroke volume of 0.0085 m³ and a compression ratio of 15 : 1. The fuel has a calorific value of 43890 kJ/kg. At the end of suction, the air is at 1 bar and 100 °C. The maximum pressure in the cycle is 65 bar and the air-fuel ratio is 21 : 1. Find for the ideal cycle the thermal efficiency. Assume c_p = 1.0 and c_v = 0.71.

■ **Solution.**

Refer to Figure 11.24.

Initial temperature,	$T_1 = 100 + 273 = 373$ K
Initial pressure,	$p_1 = 1$ bar
Maximum pressure in the cycle,	$p_3 = p_4 = 65$ bar
Stroke volume,	$Vs = 0.0085$ m³
Air-fuel ratio	$= 21 : 1$
Compression ratio,	$r = 15 : 1$
Calorific value of fuel,	$C = 43890$ kJ / kg
$c_p = 1.0, c_v = 0.71$	

Thermal efficiency:

$$V_s = V_1 - V_2 = 0.0085$$

and as

$$r = \frac{V_1}{V_2} = 15, \text{ then } V_1 = 15V_2$$

$\therefore$

$$15V_2 - V_2 = 0.0085$$

or

$$14V_2 = 0.0085$$

or

$$V_2 = V_3 = V_c = \frac{0.0085}{14} = 0.0006 \text{ m}^3$$

or

$$V_1 = 15V_2 = 15 \times 0.0006 = 0.009 \text{ m}^3$$

For the *adiabatic compression process 1–2*,

$$p_1 V_1^\gamma = p_2 V_2^\gamma$$

or

$$p_2 = p_1 \cdot \left(\frac{V_1}{V_2}\right)^\gamma = 1 \times (15)^{1.41} \qquad \left[\gamma = \frac{c_p}{c_v} = \frac{1.0}{0.71} = 1.41\right]$$

$$= 45.5 \text{ bar}$$

Also,

$$\frac{T_2}{T_1} = \left(\frac{V_1}{V_2}\right)^{\gamma-1} = (r)^{\gamma-1} = (15)^{1.41-1} = 3.04$$

$\therefore$

$$T_2 = T_1 \times 3.04 = 373 \times 3.04 = 1134 \text{ K or } 861°C$$

For the *constant volume process 2–3*,

$$\frac{p_2}{T_2} = \frac{p_3}{T_3}$$

or

$$T_3 = T_2 \times \frac{p_3}{p_2} = 1134 \times \frac{65}{45.5} = 1620 \text{ K or } 1347°C$$

According to the characteristic equation of gas,

$$p_1 V_1 = mRT_1$$

$\therefore$

$$m \frac{p_1 V_1}{RT_1} = \frac{1 \times 10^5 \times 0.009}{287 \times 373} = 0.0084 \text{ kg (air)}$$

Heat added during the *constant volume process 2–3*,

$$= m \times c_v (T_3 - T_2)$$

$$= 0.0084 \times 0.71 (1620 - 1134)$$

$$= 2.898 \text{ kJ}$$

Amount of fuel added during the constant volume process 2–3,

$$= \frac{2.898}{43890} = 0.000066 \text{ Kg}$$

Also as air-fuel ratio is 21 : 1.

$$\therefore \text{Total amount of fuel added} \quad = \frac{0.0084}{21} = 0.0004 \text{ kg}$$

Quantity of fuel added during the process 3–4,

$$= 0.0004 - 0.000066 = 0.000334 \text{ kg}$$

∴ *Heat added during the constant pressure operation* $3-4$

$$= 0.000334 \times 43890 = 14.66 \text{ kJ}$$

But $(0.0084 + 0.0004) c_p (T_4 - T_3) = 14.66$

or $\quad 0.0088 \times 1.0 \ (T_4 - 1620) \ = \ 14.66$

$$\therefore \quad T_4 = \frac{14.66}{0.0088} + 1620 = 3286 \text{ K or } 3013^\circ\text{C}$$

Again for *process 3–4,*

$$\frac{V_3}{T_3} = \frac{V_4}{T_4} \text{ or } V_4 = \frac{V_3 T_4}{T_3} = \frac{0.0006 \times 3286}{1620} = 0.001217 \text{ m}^3$$

For the *adiabatic expansion operation 4–5,*

$$\frac{T_4}{T_5} = \left(\frac{V_5}{V_4}\right)^{\gamma-1} = \left(\frac{0.009}{0.001217}\right)^{1.41-1} = 2.27$$

or $\quad T_5 = \dfrac{T_4}{2.27} = \dfrac{3286}{2.27} = 1447.5 \text{ K or } 1174.5^\circ\text{C}$

Heat rejected during the constant volume process 5–1,

$$= m \ c_v \ (T_5 - T_1)$$

$$= (0.00854 + 0.0004) \times 0.71 \ (1447.5 - 373) = 6.713 \text{ kJ}$$

Work done $\qquad\qquad\qquad$ = Heat supplied – Heat rejected

$$= (2.898 + 14.66) - 6.713 = 10.845 \text{ kJ}$$

∴ *Thermal efficiency,*

$$\eta_{th}. = \frac{\text{Work done}}{\text{Heat supplied}} = \frac{10.845}{(2.898 + 14.66)} = \textbf{0.6176 or 61.76\%. (Ans.)}$$

Example 29

The compression ratio and expansion ratio of an oil engine working on the dual cycle are 9 and 5, respectively. The initial pressure and temperature

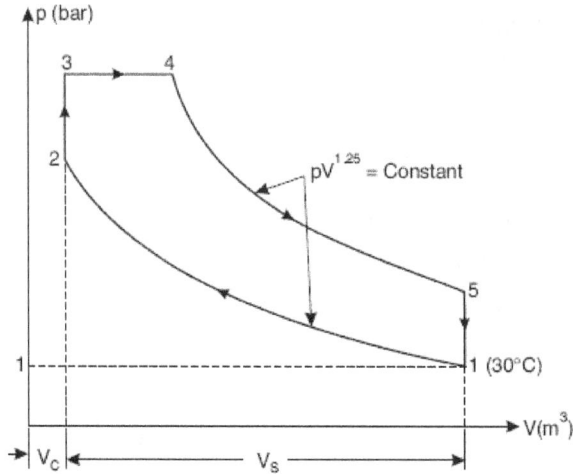

FIGURE 11.25. Sketch for Example 29, p-V diagram.

of the air are 1 bar and 30 °C. The heat liberated at constant pressure is twice the heat liberated at constant volume. The expansion and compression follow the law $pV^{1.25}$ = constant. Determine:

(i) Pressures and temperatures at all salient points.

(ii) Mean effective pressure of the cycle.

(iii) Efficiency of the cycle.

(iv) Power of the engine if working cycles per second are 8.

Assume: Cylinder bore = 250 mm and stroke length = 400 mm.

■ **Solution.**

Refer to Figure 11.25.

Initial temperature,	$T_1 = 30 + 273 = 303$ K
Initial pressure,	$p_1 = 1$ bar
Compression and expansion law,	$pV^{1.25}$ = Constant
Compression ratio,	$r_c = 9$
Expansion ratio,	$r_e = 5$ $_V$
Number of cycles / sec.	$= 8$
Cylinder diameter,	$D = 250\,\text{mm} = 0.25\,\text{m}$
Stroke length,	$L = 400\text{ mm} = 0.4\text{ m}$

Heat liberated at constant pressure

$$= 2 \times \text{heat liberated at constant volume}$$

(*i*) Pressures and temperatures at all salient points:

For *compression process 1–2,*

$$p_1 V_1^n = p_2 V_2^n$$

$$\therefore \quad p_2 = p_1 \times \left(\frac{V_1}{V_2}\right)^n = 1 \times (9)^{1.25} = \mathbf{15.59\ bar.\ (Ans.)}$$

Also, $\dfrac{T_2}{T_1} = \left(\dfrac{V_1}{V_2}\right)^{n-1} = (9)^{1.25-1} = 1.732$

$$\therefore \quad \boldsymbol{T_2} = T_1 \times 1.732 = 303 \times 1.732$$

Also, $c_p (T_4 - T_3) = 2 \times c_v (T_3 - T_2)$ ...... (given)

For *constant pressure process 3–4,*

$$\frac{T_4}{T_3} = \frac{V_4}{V_3} = \rho = \frac{\text{Compression ratio}\,(r_c)}{\text{Expansion ratio}\,(r_e)}$$

$$= \frac{9}{5} = 1.8$$

$$T_4 = 1.8 T_3$$

$$\left[\begin{array}{l} \dfrac{V_5}{V_4}\,(i.e., r_e) = \dfrac{V_5}{V_3} \times \dfrac{V_3}{V_4} \\[2.5ex] \qquad = \dfrac{V_1}{V_3} \times \dfrac{1}{\rho} \\[2.5ex] \qquad = \dfrac{V_1}{V_2} \times \dfrac{1}{\rho} = \dfrac{r_c}{\rho} \\[2.5ex] \therefore\, \rho = \dfrac{r_c}{\dfrac{V_5}{V_4}} = \dfrac{r_c}{r_e} \end{array}\right]$$

Substituting the values of T_2 and T_4 in the eqn. (*i*), we get

$$1.0\left(1.8 T_3 - T_3\right) = 2 \times 0.71\left(T_3 - 524.8\right)$$

$$0.8 T_3 = 1.42\left(T_3 - 524.8\right)$$

$$0.8 T_3 = 1.42 T_3 - 745.2$$

$$\therefore \qquad\qquad\qquad 0.62 T_3 = 745.2$$

$$T_3 = \mathbf{1201.9\ K}\ \text{or}\ \mathbf{928.9°C.\ (Ans.)}$$

Also, $\dfrac{p_3}{T_3} = \dfrac{p_2}{T_2}$ ...... for process $2-3$

$$\therefore \quad \boldsymbol{p_3} = p_2 \times \frac{T_3}{T_2} = 15.59 \times \frac{1201.9}{524.8} = \mathbf{35.7\ bar.}\quad \mathbf{(Ans.)}$$

$p_4 = p_3 = \textbf{35.7 bar. (Ans.)}$

$T_4 = 1.8 T_3 = 1.8 \times 1201.9 = \textbf{2163.4 K}$ or $\textbf{1890.4°C. (Ans.)}$

For expansion process 4–5,

$p_4 V_4^n = p_5 V_5^n$

$p_5 = p_4 \times \left(\dfrac{V_4}{V_5}\right)^n = p_4 \times \dfrac{1}{(r_e)^n} = \dfrac{35.7}{(5)^{1.25}} = \textbf{4.77 bar. (Ans.)}$

Also

$\dfrac{T_5}{T_4} = \left(\dfrac{V_4}{V_5}\right)^{n-1} = \dfrac{1}{(r_e)^{n-1}} = \dfrac{1}{(5)^{1.25-1}} = 0668$

$\therefore \quad T_5 = T_4 \times 0.668 = 2163.4 \times 0.668 = \textbf{1445 K}$ or $\textbf{1172°C.}$ **(Ans.)**

(*ii*) Mean effective pressure, p_m:

Mean effective pressure is given by

$p_m = \dfrac{1}{V_s}\left[p_3(V_4 - V_3) + \dfrac{p_4 V_4 - p_5 V_5}{n-1} - \dfrac{p_2 V_2 - p_1 V_1}{n-1}\right]$

$= \dfrac{1}{(r_c - 1)}\left[p_3(\rho - 1) + \dfrac{p_4 \rho - p_5 r_c}{n-1} - \dfrac{p_2 - p_1 r_c}{n-1}\right]$

Now, $r_c = \rho$, $\rho = 1.8$, $n = 1.25$, $p_1 = 1$ bar, $p_2 = 15.59$ bar, $p_3 = 35.7$ bar, $p_4 = 35.7$ bar, $p_5 = 4.77$ bar

$\therefore \quad p_m = \dfrac{1}{(9-1)}\left[35.7(1.8-1) + \dfrac{35.7 \times 1.8 - 4.77 \times 9}{1.25-1} - \dfrac{15.59 - 1 \times 9}{1.25-1}\right]$

$= \dfrac{1}{8}\left[28.56 + 85.32 - 26.36\right] = 10.94$ bar

Hence *mean effective pressure* = 10.94 bar. (Ans.)

(*iii*) Efficiency of the cycle:

Work done per cycle is given by $W = p_m V_s$

Here, $\qquad\qquad V_s = \pi/4 D^2 L = \pi/4 \times 0.25^2 \times 0.4 = 0.0196$ m^3

$\therefore \qquad\qquad W = \dfrac{10.94 \times 10^5 \times 0.0196}{1000}$ kJ/cycle $= 21.44$ kJ/cycle

Heat supplied per cycle $\qquad\qquad = m Q_s$,

where m is the mass of air per cycle which is given by

$$m = \frac{p_1 V_1}{RT_1} \text{ where } V_1 = V_s + V_c = \frac{r_c}{r_c - 1} V_s$$

$$\left[r_c = \frac{V_s + V_c}{V_c} = 1 + \frac{V_s}{V_c} \text{ or } V_c = \frac{V_s}{r_c - 1} \right.$$

$$\left. \therefore \ V_1 = V_s + \frac{V_s}{r_c - 1} = V_s \left(1 + \frac{1}{r_c - 1} \right) = \frac{r_c}{r_c - 1} V_s \right]$$

$$= \frac{9}{9 - 1} \times 0.0196 = 0.02205 \text{ m}^3$$

$$\therefore \quad m = \frac{1 \times 10^5 \times 0.02205}{287 \times 303} = 0.02535 \text{ kg/cycle}$$

∴ Heat supplied per cycle

$$= mQ_s = 0.02535 \left[c_v (T_3 - T_2) + c_p (T_4 - T_3) \right]$$

$$= 0.02535 \left[0.71 (1201.9 - 524.8) + 1.0 (2163.4 - 1201.9) \right]$$

$$= 36.56 \text{ kJ / cycle}$$

$$\text{Efficiency} \quad = \frac{\text{Work done per cycle}}{\text{Heat supplied per cycle}} = \frac{21.44}{36.56}$$

$$= \textbf{0.5864} \text{ or } \textbf{58.64\%. (Ans.)}$$

(*iv*) Power of the engine, P:

Power of the engine, P = Work done per second

= Work done per cycle × no. of cycles/sec.

= 21.44 × 8 = **171.52 kW. (Ans.)**

11.8. Comparison of Otto, Diesel, and Dual Combustion Cycles

The following are the important variable factors which are used as a basis for comparison of the cycles:

- Compression ratio
- Maximum pressure
- Heat supplied
- Heat rejected
- Network

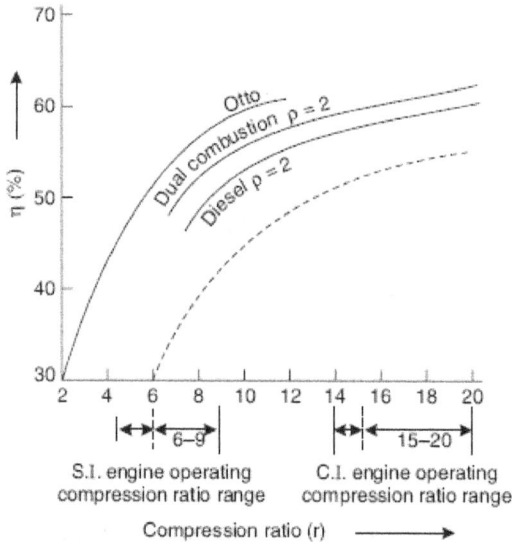

FIGURE 11.26. Comparison of efficiency at various compression ratios.

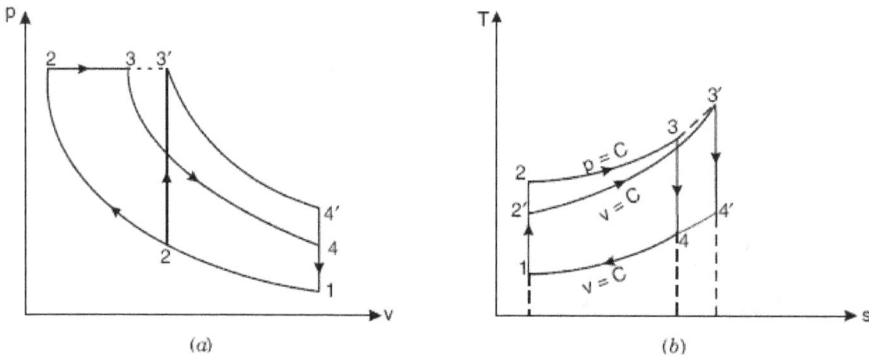

FIGURE 11.27. (A) p-v diagram, (b) T-s diagram.

Some of the above-mentioned variables are fixed when the performance of Otto, Diesel, and dual combustion cycles is to be compared.

11.8.1. Efficiency Versus Compression Ratio

Figure 11.26 shows the comparison for the air standard efficiencies of the Otto, Diesel, and Dual combustion cycles at various compression ratios and with a given cut-off ratio for the Diesel and Dual combustion cycles. It is evident from Figure 11.26 that the air standard efficiencies increase with the increase in the compression ratio. For a given compression ratio the

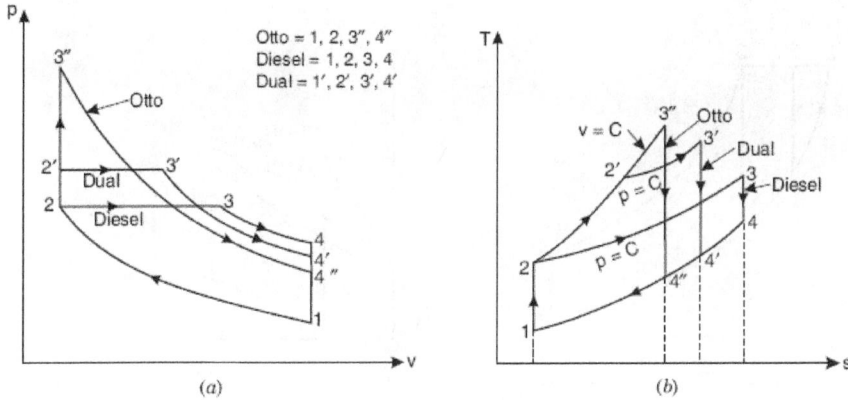

FIGURE 11.28. (A) p-v diagram, (b) T-s diagram.

Otto cycle is the most efficient while the Diesel cycle is the least efficient $(\eta_{otto} > \eta_{dual} > \eta_{diesel})$.

The maximum compression ratio for the gasoline engine is limited by detonation. In their respective ratio ranges, the Diesel cycle is more efficient than the Otto cycle.

A comparison of the cycles (Otto, Diesel, and Dual) on the p-v and T-s diagrams for the *same compression ratio and heat supplied* is shown in Figure 11.27.

We know that, $$\eta = 1 - \frac{\text{Heat rejected}}{\text{Heat supplied}} \qquad ...(13)$$

Since all the cycles reject their heat at the same specific volume, shown by the process line from state 4 to 1, the quantity of *heat rejected from each cycle is represented by the appropriate area under the line 4 to 1 on the T-s diagram*. As is evident from the equation (13), the cycle which has the least heat rejected will have the highest efficiency. Thus, the Otto cycle is the most efficient and the Diesel cycle is the least efficient of the three cycles.

i.e., $$\eta_{otto} > \eta_{dual} > \eta_{diesel}.$$

Figure 11.28 shows the Otto and Diesel cycles on p-v and T-s diagrams for constant maximum pressure and heat input, respectively.

— For the maximum pressure the points 3 and 3' must lie on a constant pressure line.

— On a T-s diagram the heat rejected from the Diesel cycle is represented by the area under the lines 4 to 1, and this area is less than the

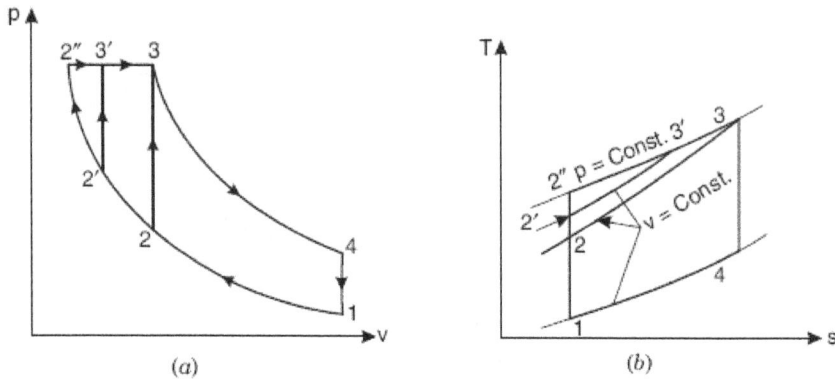

FIGURE 11.29. Sketch for Example 30, p-V and T-S diagrams.

Otto cycle area under the curve 4′ to 1; hence the Diesel cycle is more efficient than the Otto cycle for the condition of maximum pressure and heat supplied.

Example 30

With the help of p-v and T-s diagrams compare the cold air standard otto, diesel, and dual combustion cycles for the same maximum pressure and maximum temperature.

■ **Solution.**

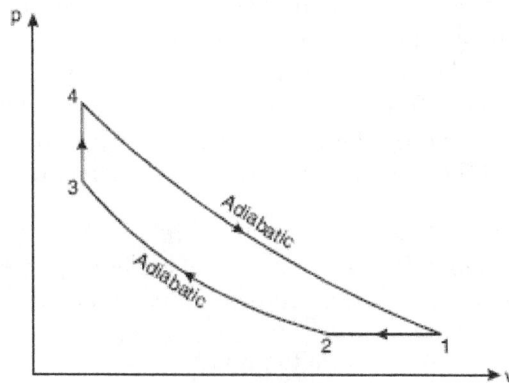

FIGURE 11.30. Atkinson cycle shown on a p-V diagram.

The air-standard Otto, Dual, and Diesel cycles are drawn on common p-v and T-s diagrams for the same maximum pressure and maximum temperature, for the purpose of comparison.

Otto 1–2–3–4–1, Dual 1–2′–3–4–1, Diesel 1–2″–3–4–1 (Figure 11.29 (a)).

The slope of the constant volume lines on the T-s diagram is higher than that of the constant pressure lines. (Figure 11.29 (b)).

Here the Otto cycle must be limited to a low compression ratio (r) to fulfill the condition that point 3 (same maximum pressure and temperature) is to be a common state for all three cycles.

The construction of cycles on the T-s diagram proves that for the given conditions, the heat rejected is the same for all three cycles (area under process line 4–1). Since, by definition,

$$\eta = 1 - \frac{\text{Heat rejected, } Q_r}{\text{Heat supplied, } Q_s} = 1 - \frac{\text{Const.}}{Q_s}$$

the cycle, with greater heat addition, will be more efficient. From the T-s diagram,

$$Q_{s(\text{diesel})} = \text{Area under } 2'' - 3$$

$$Q_{s(\text{dual})} = \text{Area under } 2' - 3' - 3$$

$$Q_{s(\text{otto})} = \text{Area under } 2 - 3$$

It can be seen that $Q_{s(\text{diesel})} > Q_{s(\text{dual})} > Q_{s(\text{otto})}$

and thus, $\eta_{\text{diesel}} > \eta_{\text{dual}} > \eta_{\text{otto}}$.

11.9. Atkinson Cycle

This cycle consists of two adiabatics, a constant volume, and a constant pressure process. A p-V diagram of this cycle is shown in Figure 11.30. It consists of the following *four operations*:

- 1–2—Heat rejection at a constant pressure

- 2–3—Adiabatic compression

- 3–4—Addition of heat at a constant volume

- 4–1—Adiabatic expansion.

  Considering 1 kg of air.

Compression ratio $\qquad = \dfrac{v_2}{v_3} = \alpha$

Expansion ratio $\qquad = \dfrac{v_1}{v_4} = r$

Heat supplied at constant volume $= c_v \left(T_4 - T_3 \right)$

Heat rejected $\qquad\qquad\qquad = c_V (T_1 - T_2)$

Work done $\qquad\qquad\qquad$ = Heat supplied – heat rejected

$\qquad\qquad\qquad\qquad = c_V (T_4 - T_3) - c_V (T_1 - T_2)$

$$\eta = \dfrac{\text{Work done}}{\text{Heat supplied}} = \dfrac{c_V \left(T_4 - T_3 \right) - c_p \left(T_1 - T_2 \right)}{c_v \left(T_4 - T_3 \right)}$$

$$= 1 - \gamma \cdot \dfrac{\left(T_1 - T_2 \right)}{\left(T_4 - T_3 \right)} \qquad\qquad\qquad (i)$$

During *adiabatic compression 2–3,*

or $\qquad\qquad\qquad\qquad\qquad T_3 = T_2 \left(\alpha \right)^{\gamma - 1} \qquad\qquad (ii)$

During *constant pressure operation 1–2,*

$$\dfrac{v_1}{T_1} = \dfrac{v_2}{T_2}$$

or $\qquad \dfrac{T_2}{T_1} = \dfrac{v_2}{v_1} = \dfrac{\alpha}{r} \qquad ...(iii) \qquad \left(\dfrac{v_2}{v_1} = \dfrac{v_2}{v_3} \times \dfrac{v_3}{v_1} = \dfrac{v_2}{v_3} \times \dfrac{v_4}{v_1} = \dfrac{\alpha}{r} \right)$

During *adiabatic expansion 4–1,*

$$\dfrac{T_4}{T_1} = \left(\dfrac{v_1}{v_4} \right)^{\gamma - 1} = \left(r \right)^{\gamma - 1}$$

$$T_1 = \dfrac{T_4}{\left(r \right)^{\gamma - 1}} \qquad\qquad\qquad\qquad (iv)$$

Putting the value of T_1 in equation (*iii*), we get

$$T_2 = \dfrac{T_4}{\left(r \right)^{\gamma - 1}} \cdot \dfrac{\alpha}{r}$$

$$= \dfrac{\alpha T_4}{r^{\gamma}} \qquad\qquad\qquad\qquad (v)$$

Substituting the value of T_2 in equation (*ii*), we get

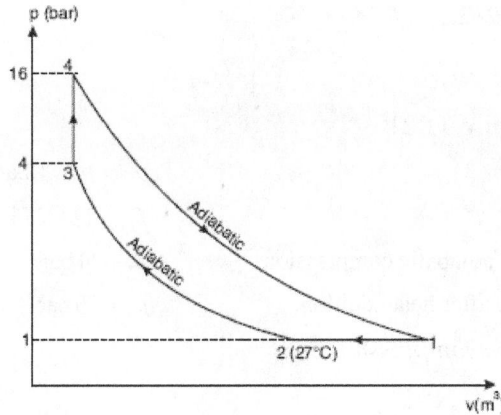

FIGURE 11.31. Sketch for Example 31.

$$T_3 = \frac{\alpha T_4}{r^\gamma}(\alpha)^{\gamma-1} = \left(\frac{\alpha}{r}\right)^\gamma . T_4$$

Finally putting the values of T_1, T_2, and T_3 in equation (i), we get

$$\eta = 1 - \gamma \left(\frac{\dfrac{T_4}{r^{\gamma-1}} - \dfrac{\alpha T_4}{(r)^\gamma}}{T_4 - \left(\dfrac{\alpha}{r}\right)^\gamma . T_4} \right) = 1 - \gamma \left(\frac{r - \alpha}{r^\gamma - \alpha^\gamma} \right)$$

Hence, air standard efficiency $= 1 - \gamma . \left(\dfrac{r - \alpha}{r^\gamma - \alpha^\gamma} \right)$

Example 31

A perfect gas undergoes a cycle which consists of the following processes taken in order:

(a) Heat rejection at a constant pressure.

(b) Adiabatic compression from 1 bar and 27 °C to 4 bar.

(c) Heat addition at a constant volume to a final pressure of 16 bar.

(d) Adiabatic expansion to 1 bar.

Calculate: (i) Work done/kg of gas.

(ii) Efficiency of the cycle.

Take : $c_p = 0.92, c_v = 0.75$.

■ **Solution.**

Refer to Figure 11.31.

Pressure,	$p_2 = p_1 = 1$ bar
Temperature,	$T_2 = 27 + 273 = 300$ K
Pressure after adiabatic compression,	$p_3 = 4$ bar
Final pressure after heat addition,	$p_4 = 16$ bar

For *adiabatic compression 2–3*,

$$\frac{T_3}{T_2} = \left(\frac{p_3}{p_2}\right)^{\frac{\gamma-1}{\gamma}} = \left(\frac{4}{1}\right)^{\frac{1.22-1}{1.22}} = 1.284 \qquad \left[\gamma = \frac{c_p}{c_v} = \frac{0.92}{0.75} = 1.22\right]$$

$\therefore \qquad T_3 = T_2 \times 1.284 = 300 \times 1.284 = 385.2$ K or $112.2°$C

For *constant volume process 3–4*,

$$\frac{p_4}{T_4} = \frac{p_3}{T_3}$$

$$T_4 = \frac{p_4 T_3}{p_3} = \frac{16 \times 385.2}{4} = 1540.8 \text{ K or } 1267.8°\text{C}$$

For *adiabatic expansion process 4–1*,

$$\frac{T_4}{T_1} = \left(\frac{p_4}{p_1}\right)^{\frac{\gamma-1}{\gamma}} = \left(\frac{16}{1}\right)^{\frac{1.22-1}{1.22}} = 1.648$$

or $\quad T_1 = \frac{T_4}{1.648} = \frac{1540.8}{1.648} = 934.9$ K or $661.9°$C.

(*i*) Work done per kg of gas, W:

Heat supplied	$= c_v (T_4 - T_3)$
	$= 0.75 \ (1540.8 - 85.2) = 866.7$ kJ / kg
Heat rejected	$= c_p (T_1 - T_2) = 0.92 (934.9 - 300) = 584.1$ kJ / kg

Work done / kg of gas, $\qquad W$ = Heat supplied – heat rejected

$= 866.7 - 584.1 = 282.6$ kJ / kg $= \textbf{282600 Nm / kg. (Ans.)}$

(*ii*) Efficiency of the cycle:

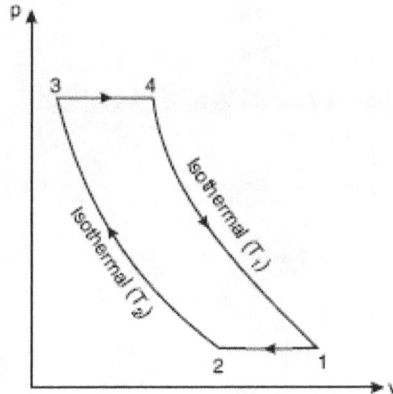

FIGURE 11.32. Ericsson Cycle shown on a p-V diagram.

$$Efficiency, \quad \eta = \frac{\text{Work done}}{\text{Heat supplied}} = \frac{282.6}{866.7} = \textbf{0.326 or 32.6\%.} \quad \textbf{(Ans.)}$$

11.10. Ericsson Cycle

It is so named as it was invented by John Ericsson (1833). Figure 11.32 shows a *p-v* diagram of this cycle.

It comprises the following operations:

- 1–2—*Rejection of heat at a constant pressure*
- 2–3—*Isothermal compression*
- 3–4—*Addition of heat at a constant pressure*
- 4–1—*Isothermal expansion.*

Considering 1 kg of air.

Volume ratio, $\quad r = \dfrac{v_2}{v_3} = \dfrac{v_1}{v_4}$

Heat supplied to air from an external source

= Heat supplied during the isothermal expansion $4-1$

= $RT_1 \log_e r$

Heated rejected by air to an external source

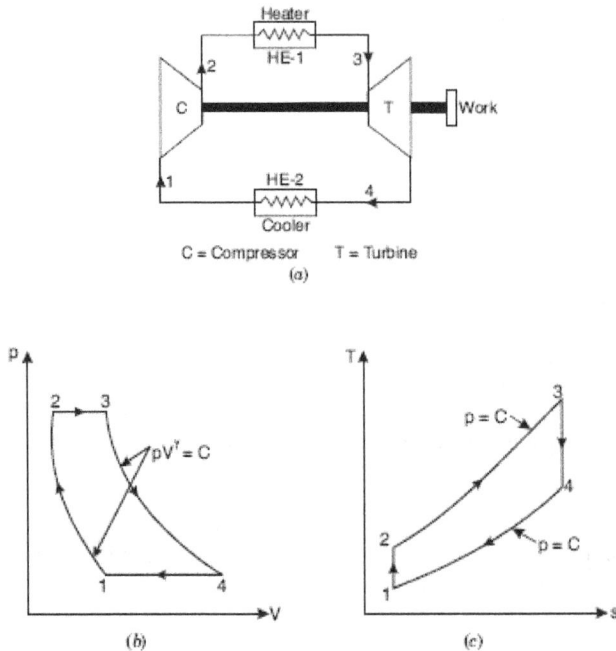

FIGURE 11.33. Brayton cycle: (A) Basic components of a gas turbine power plant, (B) p-v diagram, (C) T-S diagram.

$$= RT_2 \cdot \log_e r$$

Work done $\qquad$ = Heat supplied – heat rejected

$$= RT_1 \cdot \log_e r - RT_2 \cdot \log_e r = R \log_e r (T_1 - T_2)$$

$$\eta = \frac{\text{Work done}}{\text{Heat supplied}} = \frac{R \log_e r (T_1 - T_2)}{RT_1 . \log_e r} 1$$

$$= \frac{T_1 - T_2}{T_1}$$

which is the same as a Carnot cycle.

11.11. Brayton Cycle

The Brayton cycle is a constant pressure cycle for a perfect gas. The heat transfers are achieved in reversible constant pressure heat exchangers. An ideal gas turbine plant would perform the processes that make up a Brayton cycle. The cycle is shown in the Figure 11.33 (*a*) and it is represented on *p-v* and *T-s* diagrams as shown in Figure 11.33 (*b*, *c*).

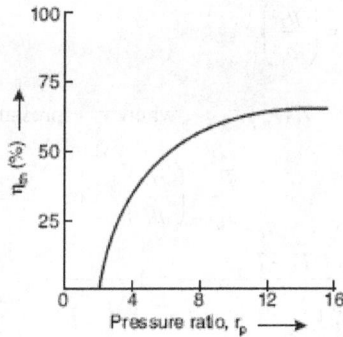

FIGURE 11.34. Effect of the pressure ratio on the efficiency of the Brayton cycle.

The various operations are as follows:

Operation 1–2. The air is compressed isentropically from the lower pressure p_1 to the upper pressure p_2, the temperature rising from T_1 to T_2. No heat flow occurs.

Operation 2–3. Heat flows into the system increasing the volume from V_2 to V_3 and the temperature from T_2 to T_3 while the pressure remains constant at p_2. Heat received = $mc_p\,(T_3 - T_2)$.

Operation 3–4. The air is expanded isentropically from p_2 to p_1, the temperature falling from T_3 to T_4. No heat flow occurs.

Operation 4–1. Heat is rejected from the system as the volume decreases from V_4 to V_1 and the temperature from T_4 to T_1 while the pressure remains constant at p_1. Heat rejected= $mc_p\,(T_4 - T_1)$.

$$\eta_{\text{air-standard}} = \frac{\text{Work done}}{\text{Heat received}}$$

$$= \frac{\text{Heat received/cycle} - \text{Heat rejected/cycle}}{\text{Heat received/cycle}}$$

$$= \frac{mc_p\,(T_3 - T_2) - mc_p\,(T_4 - T_1)}{mc_p\,(T_3 - T_2)} = 1 - \frac{T_4 - T_1}{T_3 - T_2}$$

Now, from isentropic expansion,

$$\frac{T_2}{T_1} = \left(\frac{p_2}{p_1}\right)^{\frac{\gamma-1}{\gamma}}$$

$$T_2 = T_1\left(r_p\right)^{\frac{\gamma-1}{\gamma}} \quad , \text{ where } r_p = \text{pressure ratio.}$$

Similarly
$$\frac{T_3}{T_4} = \left(\frac{p_2}{p_1}\right)^{\frac{\gamma-1}{\gamma}} \quad \text{or} \quad T_3 = T_4\left(r_p\right)^{\frac{\gamma-1}{\gamma}}$$

$$\therefore \ \eta_{\text{air-standard}} = 1 - \frac{T_4 - T_1}{T_4\left(r_p\right)^{\frac{\gamma-1}{\gamma}} - T_1\left(r_p\right)^{\frac{\gamma-1}{\gamma}}} = 1 - \frac{1}{\left(r_p\right)^{\frac{\gamma-1}{\gamma}}}$$

This equation shows that the efficiency of the ideal joule cycle increases with the pressure ratio. The absolute limit of upper pressure is determined by the limiting temperature of the material of the turbine; at the point at which this temperature is reached by the compression process alone, no further heating of the gas in the combustion chamber would be permissible and the work of expansion would ideally just balance the work of compression so that no excess work would be available for external use.

Pressure Ratio for maximum work:

Now we shall prove that the *pressure ratio for maximum work is a function of the limiting temperature ratio.*

Work output during the cycle

= Heat received / cycle – heat rejected / cycle

$$= mc_p\left(T_3 - T_2\right) - mc_p\left(T_4 - T_1\right)$$

$$= mc_p\left(T_3 - T_4\right) - mc_p\left(T_2 - T_1\right)$$

$$= mc_p T_3\left(1 - \frac{T_4}{T_3}\right) - T_1\left(\frac{T_2}{T_1} - 1\right)$$

In the case of a given turbine the minimum temperature T_1 and the maximum temperature T_3 are prescribed, T_1 being the temperature of the atmosphere and T_3 the maximum temperature which the metals of turbine would withstand. Consider the specific heat at constant pressure c_p to be constant. Then,

Since,
$$\frac{T_3}{T_4} = \left(r_p\right)^{\frac{\gamma-1}{\gamma}} = \frac{T_2}{T_1}$$

Using the constant
$$\text{'}z\text{'} = \frac{\gamma-1}{\gamma},$$

we have, work output/cycle

$$W = K\left[T_3\left(1 - \frac{1}{r_p{}^z}\right) - T_1\left(r_p{}^z - 1\right)\right]$$

Differentiating with respect to r_p

$$\frac{dW}{dr_p} = K\left[T_3 \times \frac{z}{r_p{}^{(z+1)}} - T_1 z r_p{}^{(z-1)}\right] = 0 \text{ for a maximum}$$

$\therefore$
$$\frac{zT_3}{r_p{}^{(z+1)}} = T_1 z\left(r_p\right)^{(z-1)}$$

$\therefore$
$$r_p{}^{2z} = \frac{T_3}{T_1}$$

$\therefore$
$$r_p = \left(T_3 / T_1\right)^{1/2z} \quad i.e., \quad r_p = \left(T_3 / T_1\right)^{\frac{\gamma}{2(\gamma-1)}}$$

Thus, the *pressure ratio for maximum work is a function of the limiting temperature ratio.*

Work Ratio

Work ratio is defined as the *ratio of net work output to the work done by the turbine.*

$\therefore$ Work ratio $= \dfrac{W_T - W_C}{W_T}$

$$\left[\begin{array}{l}\text{where,} \quad W_T = \text{Work obtained from this turbine,}\\ \text{and} \qquad W_C = \text{Work supplied to the compressor.}\end{array}\right]$$

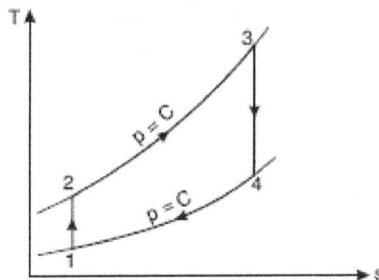

FIGURE 11.35. Sketch for Example 32.

$$= \frac{mc_p\left(T_3 - T_4\right) - mc_p\left(T_2 - T_1\right)}{mc_p\left(T_3 - T_4\right)} = 1 - \frac{T_2 - T_1}{T_3 - T_4}$$

$$= 1 - \frac{T_1}{T_3}\left[\frac{\left(r_p\right)^{\frac{\gamma-1}{\gamma}} - 1}{1 - \frac{1}{\left(r_p\right)^{\frac{\gamma-1}{\gamma}}}}\right] = 1 - \frac{T_1}{T_3}\left(r_p\right)^{\frac{\gamma-1}{\gamma}}$$

Example 32

Air enters the compressor of a gas turbine plant operating on a Brayton cycle at 101.325 kPa, 27 °C. The pressure ratio in the cycle is 6. Calculate the maximum temperature in the cycle and the cycle efficiency. Assume $W_T = 2.5\, W_C$, where W_T and W_C are the turbine and the compressor work, respectively. Take $\gamma = 1.4$.

■ Solution.

Pressure of intake air, $p_1 = 101.325$ kPa

Temperature of intake air, $T_1 = 27 + 273 = 300$ K

The pressure ratio in the cycle, $r_p = 6$

(i) Maximum temperature in the cycle, T_3:

Refer to Figure 11.35.

$$\frac{T_2}{T_1} = \left(\frac{p_2}{p_1}\right)^{\frac{\gamma-1}{\gamma}} = \left(r_p\right)^{\frac{\gamma-1}{\gamma}} = (6)^{\frac{1.4-1}{1.4}} = 1.668$$

$$\therefore \quad T_2 = 1.668\, T_1 = 1.668 \times 300 = 500.4\,\text{K}$$

Also
$$\frac{T_3}{T_4} = \left(r_p\right)^{\frac{\gamma-1}{\gamma}} = (6)^{\frac{1.4-1}{1.4}} = 1.668$$

$$\therefore \quad T_4 = \frac{T_3}{1.668}$$

But $\quad W_T = 2.5\, W_C \qquad\qquad\qquad\qquad\qquad\qquad\qquad\qquad\text{(given)}$

$$\therefore \qquad\qquad mc_p\left(T_3 - T_4\right) = 2.5\, mc_p\left(T_2 - T_1\right)$$

or $\quad T_3 - \dfrac{T_3}{1.668} = 2.5\,(500.4 - 300) = 501 \quad$ or $\quad T_3\left(1 - \dfrac{1}{1.668}\right) = 501$

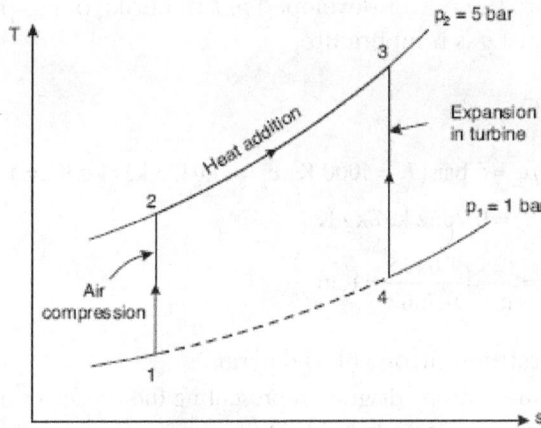

FIGURE 11.36. Sketch for Example 33.

$$\therefore \quad T_3 = \frac{501}{\left(1 - \dfrac{1}{1.668}\right)} = \textbf{1251 K} \quad \text{or} \quad \textbf{978°C.} \qquad \textbf{(Ans.)}$$

(ii) **Cycle efficiency**, η_{cycle} :

Now, $T_4 = \dfrac{T_3}{1.668} = \dfrac{1251}{1.668} = 750\ \text{K}$

$$\therefore \qquad \eta_{\text{cycle}} = \frac{\text{Net work}}{\text{Head added}} = \frac{mc_p\left(T_3 - T_4\right) - mc_p\left(T_2 - T_1\right)}{mc_p\left(T_3 - T_2\right)}$$

$$= \frac{\left(1251 - 750\right) - \left(500.4 - 300\right)}{1251 - 500.4} = \textbf{0.4} \quad \text{or} \quad \textbf{40\%.} \quad \textbf{(Ans.)}$$

$$\left[\ \textbf{Check}\ ;\ \eta_{\text{cycle}} = 1 - \frac{1}{\left(r_p\right)^{\left(\frac{\gamma-1}{\gamma}\right)}} = 1 - \frac{1}{\left(6\right)^{\frac{1.4-1}{1.4}}} = \textbf{0.4} \quad \text{or} \quad \textbf{40\%. (Ans.)}\ \right]$$

Example 33

A gas turbine is supplied with gas at 5 bar and 1000 K and expands it adiabatically to 1 bar. The mean specific heat at constant pressure and constant volume are 1.0425 kJ/kg K and 0.7662 kJ/kg K, respectively.

(i) Draw the temperature-entropy diagram to represent the processes of the simple gas turbine system.

(ii) Calculate the power developed in kW per kg of gas per second and the exhaust gas temperature.

■ **Solution.**

Given:

$p_1 = 1$ bar ; $p_2 = 5$ bar ; $T_3 = 1000$ K ; $c_p = 1.0425$ kJ / kg K ; $= 1.0425$ kJ/kg K ;

$$c_v = 0.7662 \text{ kJ / kg K}$$

$$\gamma = \frac{c_p}{c_v} = \frac{1.0425}{0.7662} = 1.36$$

(*i*) Temperature-entropy (T-s) diagram:

A temperature-entropy diagram representing the processes of the simple gas turbine system is shown in Figure 11.36.

(*ii*) Power required:

$$\frac{T_4}{T_3} = \left(\frac{p_1}{p_2}\right)^{\frac{\gamma-1}{\gamma}} = \left(\frac{1}{5}\right)^{\frac{1.36-1}{1.36}} = 0.653$$

$$\therefore \quad T_4 = 1000 \times 0.653 = 653 \text{ K}$$

Power developed per kg of gas per second

$$= c_p (T_3 - T_4)$$

$$= 1.0425 (1000 - 653) = \textbf{361.7 kW.} \quad \textbf{(Ans.)}$$

Example 34

An isentropic air turbine is used to supply 0.1 kg/s of air at 0.1 MN/m² and at 285 K to a cabin. The pressure at the inlet to the turbine is 0.4 MN/m². Determine the temperature at the turbine inlet and the power developed by the turbine. Assume $c_p = 1.0$ kJ/kg K.

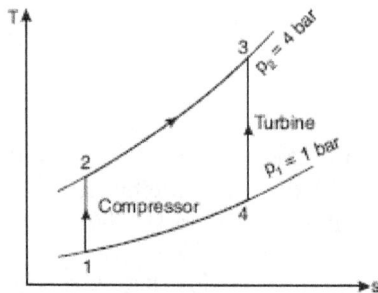

FIGURE 11.37. Sketch for Example 34.

■ **Solution.**

Given: $\dot{m}_a = 0.1$ kg/s ; $p_1 = 0.1$ MN/m$^2 = 1$ bar, $T_4 = 285$ K ; p_2

$= 0.4$ MN/m$^2 = 4$ bar ; $c_p = 1.0$ kJ/kg K

Temperature at turbine inlet, T$_3$:

$$\frac{T_3}{T_4} = \left(\frac{P_2}{P_1}\right)^{\frac{\gamma-1}{\gamma}} = \left(\frac{4}{1}\right)^{\frac{1.4-1}{1.4}} = 1.486$$

$\therefore$ T$_3 = 285 \times 1.486 = $ **423.5 K.**   **(Ans.)**

Power developed, P:

$P = \dot{m}_a c_p (T_3 - T_4)$

$= 0.1 \times 1.0 \, (423.5 - 285)$

$= $ **13.85 kW. (Ans.)**

Example 35

Consider an air standard cycle in which the air enters the compressor at 1.0 bar and 20 °C. The pressure of air leaving the compressor is 3.5 bar and the temperature at the turbine inlet is 600 °C. Determine per kg of air:

(i) Efficiency of the cycle,   *(ii) Heat supplied to air,*

(iii) Work available at the shaft,   *(iv) Heat rejected in the cooler, and*

(v) Temperature of air leaving the turbine.

For air $\gamma = 1.4$ and $c_p = 1.005 \, kJ / kg \, K.$

■ **Solution.**

Refer to Figure 11.38

Pressure of air entering the compressor, $p_1 = 1.0$ bar

Temperature at the inlet of compressor, $T_1 = 20 + 273 = 293$ K

Pressure of air leaving the compressor, $p_2 = 3.5$ bar

Temperature of air at turbine inlet, $T_3 = 600 + 273 = 873$ K

(i) **Efficiency of the cycle, η_{cycle} :**

$$\eta_{cycle} = 1 - \frac{1}{\left(r_p\right)^{\frac{\gamma-1}{\gamma}}} = 1 - \frac{1}{(3.5)^{\frac{1.4-1}{1.4}}} = 0.30 \text{ or } 30\%. \quad \textbf{(Ans.)} \quad \left(\because r_p = \frac{p_2}{p_1} = \frac{3.5}{1.0} = 3.5\right)$$

(*ii*) Heat supplied to the air:

For the compression process 1–2, we have

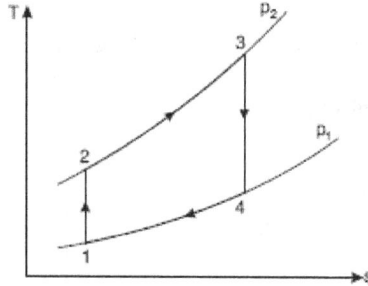

FIGURE 11.38. Sketch for Example 35.

$$\frac{T_2}{T_1} = \left(\frac{P_2}{P_1}\right)^{\frac{\gamma-1}{\gamma}} = \left(\frac{3.5}{1}\right)^{\frac{1.4-1}{1.4}} = 1.43$$

$$\therefore \quad T_2 = T_1 \times 1.43 = 293 \times 1.43 \simeq 419 \text{ K}$$

$\therefore$ Heat supplied to air, $Q_1 = c_p \ (T_3 - T_2) = 1.005 \ (873 - 419) = \mathbf{456.27 \ kJ / kg. \ (Ans.)}$

(*iii*) Work available at the shaft, W:

We know that, $\qquad \eta_{\text{cycle}} = \dfrac{\text{Work output } (W)}{\text{Heat input } (Q_1)}$

or $\quad 0.30 = \dfrac{W}{456.27} \quad$ or $\quad W = 0.3 \times 456.27 = 136.88 \text{KJ/Kg}$

(*iv*) Heat rejected in the cooler, Q_2:

Work output (W) = Heat supplied (Q_1) – heat rejected (Q_2)

$\therefore \quad Q_2 = Q_1 - W = 456.27 - 136.88 = \mathbf{319.39 \ kJ / kg.} \quad \mathbf{(Ans.)}$

(*v*) Temperature of air leaving the turbine, T_4:

For the *expansion (isentropic) process 3–4*, we have

$$\frac{T_3}{T_4} = \left(r_p\right)^{\frac{\gamma-1}{\gamma}} = (3.5)^{\frac{1.4-1}{1.4}} = 1.43$$

$$\therefore \quad T_4 = \frac{T_3}{1.43} = \frac{873}{1.43} = \mathbf{610.5 \ K.} \quad \mathbf{(Ans.)}$$

[Check: Heat rejected in the air cooler at a constant pressure during the process 4–1 can also be calculated as: Heat rejected $= m \times c_p \ (T_4 - T_1) = 1 \times 1.005 \times (610.5 - 293) = 319.1 \text{ kJ/kg}]$

Example 36

Air enters the compressor of a gas turbine plant operating on a Brayton cycle at 1 bar, 27 °C. The pressure ratio in the cycle is 6. If $W_T = 2.5\ W_C$, where W_T and W_C are the turbine and compressor work, respectively, calculate the maximum temperature and the cycle efficiency.

■ Solution.

Given : $p_1 = 1$ bar ; $T_1 = 27 + 273 = 300$ K ; $\dfrac{p_2}{p_1} = 6$; $W_T = 2.5\ W_C$

Maximum temperature, T_3 :

Now, $\qquad \dfrac{T_2}{T_1} = \left(\dfrac{P_2}{P_1}\right)^{\frac{\gamma-1}{\gamma}} = (6)^{\frac{1.4-1}{1.4}} = (6)^{\frac{0.4}{1.4}} = 1.688$

$\therefore \qquad T_2 = 300 \times 1.668 = 500.4$ K

Also, $\qquad \dfrac{T_3}{T_4} = \left(\dfrac{P_2}{P_1}\right)^{\frac{\gamma-1}{\gamma}} = (6)^{\frac{0.4}{1.4}} = 1.688$

or $\qquad T_3 = 1.668\ T_4$

Now, compressor work

$W_C = mc_p\ (T_2 - T_1),$

and turbine work,

$W_T = mc_p\ (T_3 - T_4)$

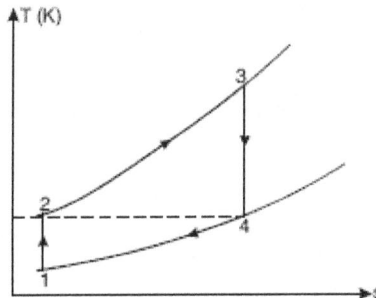

FIGURE 11.39. Sketch for Example 36.

Since $W_T = 2.5\,W_C$ (Given)

$\therefore$ $mc_p(T_3 - T_4) = 2.5 \times mc_p(T_2 - T_1)$

$\left(T_3 - \dfrac{T_3}{1.668}\right) = 2.5(500.4 - 300)$

or $T_3\left(1 - \dfrac{1}{1.668}\right) = 501$

$\therefore$ **$T_3 = 1251$ K.   (Ans.)**

Cycle efficiency, η_{cycle} :

$$\eta_{cycle} = \frac{W_T - W_C}{mc_p(T_3 - T_2)} = \frac{mc_p(T_3 - T_4) - mc_p(T_2 - T_1)}{mc_p(T_3 - T_2)}$$

$$= \frac{\left(1251 - \dfrac{1251}{1.668}\right) - (500.4 - 300)}{1251 - 500.4}$$

$$= \frac{501 - 200.4}{750.6} = 0.40 \text{ or } 40\%. \text{ (Ans.)}$$

Example 37

A closed cycle ideal gas turbine plant operates between temperature limits of 800 °C and 30 °C and produces a power of 100 kW. The plant is designed such that there is no need for a regenerator. A fuel of calorific of 45000 kJ/kg is used. Calculate the mass flow rate of air through the plant and the rate of fuel consumption.

Assume $c_p = 1$ kJ / kg K and $\gamma = 1.4$.

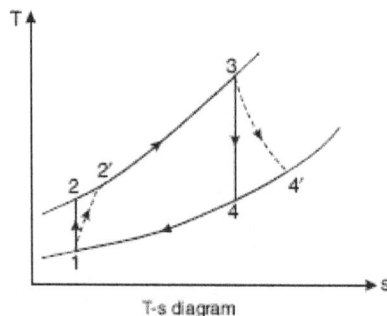

FIGURE 11.40. Sketch for Example 37.

■ **Solution.**

Given : $T_1 = 30 + 273 = 303$ K ; $T_3 = 800 + 273 = 1073$ K ;

$C = 45000$ kJ/kg ; $c_p = 1$ kJ/kg K ; $\gamma = 1.4$; $W_{turbine} - W_{compressor} = 100$ kW.

$\dot{m}_a, \dot{m}_{f:}$

Since no regenerator is used we can assume the turbine expands the gases up to T_4 in such a way that the exhaust gas temperature from the turbine is equal to the temperature of air coming out of the compressor, that is, $T_2 = T_4$

$$\frac{P_2}{P_1} = \frac{P_3}{P_4}, \frac{P_2}{P_1} = \left(\frac{T_2}{T_1}\right)^{\frac{\gamma}{\gamma-1}} \text{ and } \frac{P_3}{P_4} = \left(\frac{T_3}{T_4}\right)^{\frac{\gamma}{\gamma-1}}$$

$$\therefore \quad \frac{T_2}{T_1} = \frac{T_3}{T_4} = \frac{T_3}{T_2}$$

($\because \ T_2 = T_4 \ $ ......assumed)

or $\quad T_2^2 = T_1 T_3 \ $ or $\ T_2 = \sqrt{T_1 T_3}$

or $\quad T_2 = \sqrt{303 \times 1073} = 570.2$ K

Now, $\quad W_{turbine} - W_{compressor} = \dot{m}_f \times C \times \eta$

or $\quad 100 = \dot{m}_f \times 45000 \times \left[1 - \frac{T_4 - T_1}{T_3 - T_2}\right]$

$$= \dot{m}_f \times 45000 \times \left[1 - \frac{570.2 - 303}{1073 - 570.2}\right]$$

$$= \dot{m}_f \times 21085.9$$

or $\quad \dot{m}_f = \dfrac{100}{21085.9} = \mathbf{4.74 \times 10^{-3}}$ **kg / s. (Ans.)**

Again, $\quad W_{turbine} - W_{compressor} = 100$ kW

$(\dot{m}_a + \dot{m}_f)(T_3 - T_4) - \dot{m}_a \times 1 \times (T_2 - T_1) = 100$

or $\ (\dot{m}_a + 0.00474)(1073 - 570.2) - \dot{m}_a (570.2 - 303) = 100$

or $\ (\dot{m}_a + 0.00474) \times 502.8 - 267.2 \ \dot{m}_a = 100$

or $\ 502.8 \ \dot{m}_a + 2.383 - 267.2 \ \dot{m}_a = 100$

or $\ 235.6 \ \dot{m}_a = 97.617$

$\therefore \quad \dot{m}_a = \mathbf{0.414 \, kg / s. \ (Ans.)}$

Example 38

In a gas turbine plant working on a Brayton cycle, the air at the inlet is 27 °C, 0.1 MP$_a$. The pressure ratio is 6.25 and the maximum temperature is 800 °C. The turbine and compressor efficiencies are each 80%. Find compressor work, turbine work, heat supplied, cycle efficiency, and turbine exhaust temperature. The mass of air may be considered as 1 kg. Draw a T-s diagram.

■ **Solution.**

Given : $T_1 = 27 + 273 = 300$ K ; $p_1 = 0.1$ MP$_a$;

$r_p = 6.25$, $T_3 = 800 + 273 = 1073$ K ; $\eta_{comp.} = \eta_{turbine} = 0.8$.

For the *compression process 1–2*, we have

$$\frac{T_2}{T_1} = \left(\frac{P_2}{P_1}\right)^{\frac{\gamma-1}{\gamma}} = \left(r_p\right)^{\frac{\gamma-1}{\gamma}} = (6.25)^{\frac{14-1}{14}} = 1.688$$

or $T_2 = 300 \times 1.688 = 506.4$ K

Also, $\eta_{comp.} = \dfrac{T_2 - T_1}{T_2' - T_1}$ or $0.8 = \dfrac{506,4 - 300}{T_2' - 300}$

or $T_2' = \dfrac{506.4 - 300}{0.8} + 300 = 558$ K

∴ **Compressor work,** $W_{comp.} = 1 \times c_p \times (T_2' - T_1)$

$= 1 \times 1.005 (558 - 300) = \textbf{259.29 kJ / kg. (Ans.)}$

For *expansion process 3–4*, we have

$$\frac{T_3}{T_4} = \left(\frac{P_3}{P_4}\right)^{\frac{\gamma-1}{\gamma}} = \left(r_p\right)^{\frac{\gamma-1}{\gamma}} (6.25)^{\frac{1.4-1}{1.4}} = 1.688$$

or $T_4 = \dfrac{T_3}{1.688} = \dfrac{1073}{1.688} = 635.66$ K

Also, $\eta_{turbine} = \dfrac{T_3 - T_4'}{T_3 - T_4}$ or $0.8 = \dfrac{1073 - T_4'}{1073 - 635.66}$

or $T_4' = 1073 - 0.8 (1073 - 635.66) = 723.13$ K

∴ **Turbine work,** $W_{turbine} = 1 \times c_p \times (T_3 - T_4')$ (neglecting fuel mass)

$= 1 \times 1.005 (1073 - 723.13) = 351.6$ kJ/kg. (Ans.)

Net work output, $W_{net} = W_{turbine} - W_{comp.} = 351.6 - 259.29 = 92.31$ kJ/kg

Heat supplied, $Q_s = 1 \times c_p \times (T_3 - T_2')$

$= 1 \times 1.005 \times (1073 - 558) = \textbf{517.57 kJ / kg.} \quad \textbf{(Ans.)}$

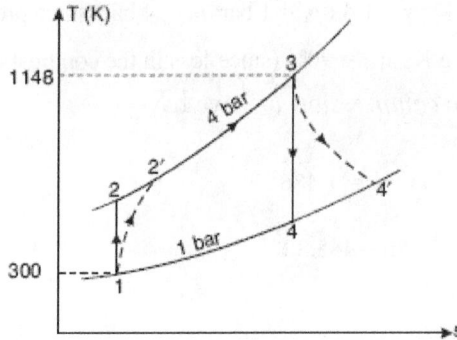

FIGURE 11.41. Sketch for Example 38.

Cycle efficiency, $\eta_{cycle} = \dfrac{W_{net}}{Q_s} = \dfrac{92.31}{517.57} = 0.1783$ or **17.83%.** (Ans.)

Turbine exhaust temperature, $T_4' = $ **723.13 K** or **450.13°C.** (Ans.)

The *T-s* diagram is shown in Figure 11.40.

Example 39

Find the required air-fuel ratio in a gas turbine whose turbine and compressor efficiencies are 85% and 80%, respectively. The maximum cycle temperature is 875 °C. The working fluid can be taken as air $(c_p = 1.0\ kJ\ /\ kg\ K, \gamma = 1.4)$ which enters the compressor at 1 bar and 27 °C. The pressure ratio is 4. The fuel used has calorific value of 42000 kJ/kg. There is a loss of 10% of calorific value in the combustion chamber.

■ Solution.

Given : $\eta_{turbine} = 85\%$; $\eta_{compressor} = 80\%$; $T_3 = 273 + 875 = 1148\ K$; $T_1 = 27 + 273 = 300$ K ;

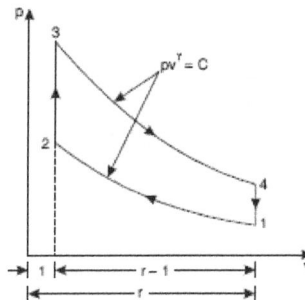

FIGURE 11.42. Sketch for Example 39.

$c_p = 1.0$ kJ/kg K ; $\gamma = 1.4$; $p_1 = 1$ bar; $p_2 = 4$ bar (since pressure ratio is 4) ;

$C = 42000$ kJ/kg K, $\eta_{cc} = 90\%$ (since loss in the combustion chamber is 10%)

For isentropic compression 1–2, we have

$$\frac{T_2}{T_1} = \left(\frac{P_2}{P_1}\right)^{\frac{\gamma-1}{\gamma}} = (4)^{\frac{1.4-1}{1.4}} = 1.486$$

$\therefore$ $T_2 = 300 \times 1.486 = 445.8$ K

$$\eta_{compressor} = \frac{T_2 - T_1}{T_2' - T_1}$$

or $0.8 = \dfrac{445.8 - 300}{T_2' - 300}$

or $T_2' = \dfrac{445.8 - 300}{0.8} + 300 = 482.2$ K

Now, heat supplied by the fuel = heat taken by the burning gases

$$0.9 \times m_f \times C = (m_a + m_f) \times c_p \times (T_3 - T_2')$$

$\therefore$ $$C = \left(\frac{m_a + m_f}{m_f}\right) \times \frac{c_p (T_3 - T_2')}{0.9} = \left(\frac{m_a}{m_f} + 1\right) \times \frac{c_p (T_3 - T_2')}{0.9}$$

or $$42000 = \left(\frac{m_a}{m_f} + 1\right) \times \frac{1.00(1148 - 482.27)}{0.9} = 739.78\left(\frac{m_a}{m_f} + 1\right)$$

$\therefore$ $\dfrac{m_a}{m_f} = \dfrac{42000}{739.78} - 1 = 55.77$ say 56

$\therefore$ **A / F ratio = 56 : 1. (Ans.)**

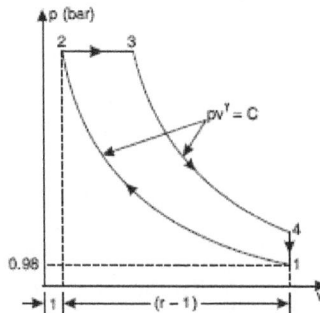

FIGURE 11.43. Sketch for Example 40.

Example 40

In an engine working on the Otto cycle between given lower and upper limits of absolute temperature, T_1 and T_2, respectively, show that for maximum work to be done per kg, the ratio of compression is given by:

$$r = \left(\frac{T_3}{T_1}\right)^{1.25}$$

where, γ = Ratio of specific heats = 1.4

■ **Solution.**

Figure 11.43 shows the cycle for the Otto cycle.

$$T_2 = T_1 \times (r)^{\gamma-1} ;\ T_4 = T_3 \times \left(\frac{1}{r}\right)^{\gamma-1}$$

Work done, $W = c_v(T_3 - T_2) - c_v(T_4 - T_1)$

$$= c_v(T_3 - T_1 \times r^{\gamma-1}) - c_v\left\{T_3\left(\frac{1}{r}\right)^{r-1} - T_1\right\}$$

For maximum work, W is differentiated with the variable r and equated to 0.

i.e., $\qquad \dfrac{dW}{dr} = \left[0 - (\gamma-1)T_1(r)^{\gamma-2}\right]$

$$-\left[-T_3 \times (\gamma-1)r^{-\gamma} - 0\right] = 0$$

or, $\qquad (\gamma-1)\ T_1\ (r)^{\gamma-2} = (\gamma-1)T_3\ r^{-\gamma}$

or, $\qquad \dfrac{T_3}{T_1} = \dfrac{(r)^{\gamma-2}}{(r)^{-\gamma}} = (r)^{2(\gamma-1)}$

∴ $\qquad r = \left(\dfrac{T_3}{T_1}\right)^{\frac{1}{2(\gamma-1)}} = \left(\dfrac{T_3}{T_1}\right)^{\frac{1}{2(1.4-1)}} = \left(\dfrac{T_3}{T_1}\right)^{1.25}$ (Proved)

Example 41

A diesel engine contains 0.1 m³ of air at 0.98 bar and 30 °C at the beginning of compression. The compression ratio is 15 and the volume at cut-off is 0.0125 m³. Determine for the corresponding air standard cycle:

(i) The cut-off ratio; (ii) The percent clearance;

(iii) The work done; (iv) The air standard efficiency.

Take c_p = 1.005 kJ/kg K and γ = 1.4.

■ **Solution.**

Figure 11.43 shows the diesel cycle on a *p-v* diagram.

Given : V_1 = 0.1 m³ ; T_1 = 30 + 273 = 303 K ;

p_1 = 0.98 bar.

Compression ratio, $r = \dfrac{V_1}{V_2} = 15$

∴ Clearance volume, $V_2 = \dfrac{V_1}{15} = \dfrac{0.1}{15} = 0.006667 \text{ m}^3$

(*i*) **Cut-off ratio,** $\rho = \dfrac{V_3}{V_2}$

$= \dfrac{0.0125}{0.006667} = \mathbf{1.875. \ (Ans.)}$

(*ii*) **The per cent clearance** $= \dfrac{1}{15} = 0.06667$

or **6.667%. (Ans.)**

(*iii*) The work done, W:

$T_2 = T_1 \times (r)^{\gamma-1} = 303 \times (15)^{1.4-1} = 895.1 \text{ K}$

$T_3 = \rho \times T_2 = 1.875 \times 895.1 = 1678.3 \text{ K}$

$T_4 = T_3 \times \left(\dfrac{\rho}{r}\right)^{\gamma-1} = 1678.3 \times \left(\dfrac{1.875}{15}\right)^{1.4-1} = 730.5 \text{ K}$

Heat supplied, $Q_{2-3} = c_p (T_3 - T_2) = 1.005(1678.3 - 895.1) = 787.1 \text{ kJ/kg}$

Heat rejected, $Q_{4-1} = c_v (T_4 - T_1)$

$= 0.718 (730.5 - 303) = 306.9 \text{ kJ/kg}$ $\left(\because c_v = \dfrac{c_p}{\gamma} = \dfrac{1.005}{1.4} = 0.718 \right)$

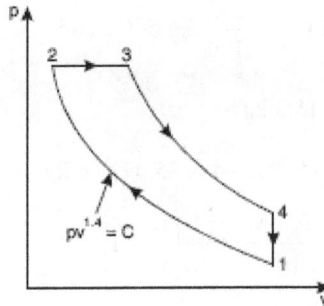

FIGURE 11.44. Sketch for Example 42.

∴ Work done, $W = Q_{2-3} - Q_{4-1} = 787.1 - 306.9 = \textbf{480.2 kJ / kg.}$ **(Ans.)**

(*iv*) The air standard efficiency, $\eta_{air\text{-}standard}$:

∴ $\quad \eta_{air\text{-}standard} = \dfrac{W}{Q_{2-3}} = \dfrac{480.2}{787.1} = \textbf{0.61}$ or **61%.** **(Ans.)**

$$\left[\text{Check:} \eta_{diesel} = 1 - \frac{\rho^{\gamma} - 1}{\gamma(r)^{\gamma-1}(\rho - 1)} = 1 - \frac{(1.875)^{1.4} - 1}{1.4 \times (15)^{1.4-1} \times (1.875 - 1)} = 0.61 \text{ or } 61\% \right]$$

Example 42

A diesel engine having a compression ratio of 16 uses the fuel C_7H_{16}. The compression follows the law $pv^{1.4}$ = constant. If the engine uses 64% excess air and the temperature at the beginning of compression is 325 K, find the percentage of stroke at which combustion is completed.

Assume $c_p = 1.0 + 21 \times 10^{-5} T$ kJ / kg K and T is in degrees' kelvin. C.V. = 44000 kJ / kg. Assume that air contains 23% by weight of O_2.

■ Solution.

Given : $r = 16$; Excess air used = 64% ; $T_1 = 325$ K ; C.V. = 44000 kJ/kg.

Figure 11.44 shows the cycle of operation. The minimum amount of air required to burn 1 kg of fuel can be calculated by using the following chemical equation:

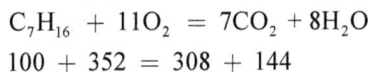

$C_7H_{16} + 11O_2 = 7CO_2 + 8H_2O$

$100 + 352 = 308 + 144$

$\therefore$ 1 kg of fuel requires $\dfrac{352}{100}$ kg of O_2 and

$\dfrac{352}{100} \times \dfrac{100}{23}$ kg of air $= 15.3$ kg

Actual air supplied $= 15.3 \times 1.64 = 25.092$ kg/kg of fuel.

$$\% \text{ cut-off} = \frac{v_3 - v_2}{v_1 - v_2} = \frac{v_3 - v_2}{16 v_2} = \frac{1}{16}\left(\frac{v_3}{v_2} - 1\right) \qquad \qquad ...(i)$$

From *compression process 1–2*, we have $\qquad\qquad\left(\because \ \dfrac{v_1}{v_2} = r = 16\right)$

$$\frac{T_2}{T_1} = \left(\frac{v_1}{v_2}\right)^{\gamma - 1} = (16)^{1.4 - 1} \ , \text{or}, \ T_2 = 325 \times (16)^{0.4} = 985 \text{ K}$$

The heat supplied per kg of fuel = Heat given to the gases during the process 2–3.

$$\therefore \qquad\qquad 44000 = \int_{T_2}^{T_3}\left(\frac{m_a}{m_f} + 1\right) c_p \ dT$$

$$44000 = \int_{T_2}^{T_3}(25.092 + 1)(1.0 + 21 \times 10^{-5} T) \ dT$$

or, $$\frac{44000}{26.092} = \int_{985}^{T_3}(1.0 + 21 \times 10^{-5} T) \ dT$$

or, $$1686.34 = (T_3 - 985) + \frac{21 \times 10^{-5}}{2}[T_3^2 - (985)^2]$$

or, $$1686.34 = T_3 - 985 + 10.5 \times 10^{-5} \ T_3^2 - 101.87$$

or, $10.5 \times 10^{-5} \ T_3^2 + T_3 - 2762.7 = 0$

or, $$T_3 = \frac{-1 \pm \sqrt{1^2 + 4 \times 10.5 \times 10^{-5} \times 2762.7}}{2 \times 10.5 \times 10^{-5}}$$

$$= \frac{-1 + 1.47}{2 \times 10.5 \times 10^{-5}} = 2238 \text{ K}.$$

From *constant pressure process 2–3*, we have

$$\frac{v_2}{T_2} = \frac{v_3}{T_3} \ , \text{or}, \ \frac{v_3}{v_2} = \frac{T_3}{T_2} = \frac{2238}{985} = 2.272$$

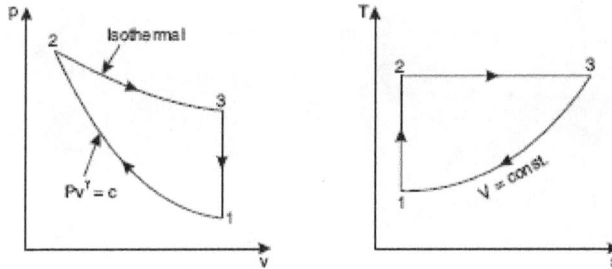

FIGURE 11.45. Sketch for Example 43.

Substituting this in equation (i), we have

$$\therefore \qquad \% \text{ cut-off} = \frac{1}{16}(2.272 - 1) = \textbf{0.0795} \text{ or } \textbf{7.95\%. (Ans.)}$$

Example 43

In a hypothetical air cycle consisting of three processes, an adiabatic compression is followed by an isothermal expansion to the initial volume of compression. Finally, a heat rejection process completes the cycle.

(i) *Draw the cycle on p-v and T-s diagrams and derive an expression for thermal efficiency of the cycle in terms of compression ratio r.*

(ii) *Also find the efficiency and m.e.p. of the cycle if the compression ratio is 14 and the induction conditions are 1 bar and 27 °C. Take for air, $c_p = 1.005$ kJ/kg K and $c_v = 0.718$ kJ/kg K.*

■ **Solution.**

Given : $r = 14$; $p_1 = 1$ bar ; $T_1 = 27 + 273 = 300$ K ; $c_p = 1.005$ kJ / kg K ; $c_v = 0.718$ kJ / kg K

(i) The *p-v* and *T-s* diagrams of the cycle are shown in Figure 11.45. Consider 1 kg of air.

Consider *adiabatic process 1–2:*

$$\frac{T_2}{T_1} = \left(\frac{v_1}{v_2}\right)^{\gamma-1} = (r)^{\gamma-1} \qquad \therefore \ T_2 = T_1(r)^{\gamma-1}; \ Q_{1-2} = 0$$

Consider *isothermal process 2–3*:

$$T_2 = T_3$$

$$Q_{2-3} = W = RT_2 \ ln(r)$$

Consider *heat-rejection (constant volume) process 3–1*:

$$Q_{3-1} = c_v(T_3 - T_1)$$

Efficiency, $\qquad \eta = \dfrac{\text{Work done}}{\text{Heat added}} = \dfrac{Q_{2-3} - Q_{3-1}}{Q_{2-3}} = 1 - \dfrac{Q_{3-1}}{Q_{2-3}}$

$$= 1 - \frac{c_v(T_3 - T_1)}{RT_2 \ \ln(r)} = 1 - \frac{c_v[T_1(r)^{\gamma-1} - T_1]}{RT_1(r)^{\gamma-1} \ln(r)}$$

$$= 1 - \frac{\mathbf{c_v}}{\mathbf{R \ ln(r)}}\left[1 - \frac{1}{(r)^{\gamma-1}}\right]. \quad \textbf{(Ans.)}$$

(*ii*) $\qquad\qquad Q_{added} = Q_{2-3} = RT_2 \ \ln(r)$

where, $\qquad\qquad\qquad R = c_p - c_v = 1.005 - 0.718 = 0.287 \text{ kJ/kg K},$

$T_2 = T_1(r)^{\gamma-1} = 300 \times (14)^{1.4-1} = 300 \times 2.874 = 862 \text{ K}$

$$\left(\because \gamma = \frac{c_p}{c_v} = \frac{1.005}{0.718} = 1.4\right)$$

$\therefore \qquad\qquad Q_{added} = 0.287 \times 862 \times \ln(14) = 652.88 \text{ kJ/kg}$

$\therefore \qquad\qquad Q_{rejected} = c_v(T_3 - T_1) = 0.718(862 - 300) = 403.52 \text{ kJ/kg}$

$$\eta = 1 - \frac{Q_{rejected}}{Q_{added}} = 1 - \frac{403.52}{652.88} = \textbf{0.3819} \text{ or } \textbf{38.19\%} \quad \textbf{(Ans.)}$$

$$\left[\textit{Alternatively}: \ \eta = 1 - \frac{0.718}{0.278 \times \ln(14)}\left\{1 - \frac{1}{(14)^{1.4-1}}\right\}\right.$$

$$= 1 - 0.9479 \times 0.652 = 0.3819 \text{ or } 38.19\%\Big]$$

$W_{net} = Q_{added} - Q_{rejected}$

$= 652.88 - 403.52 = 249.36 \text{ kJ/kg}$

Swept volume, $\qquad v_s = v_1 - v_2 = v_1 - \dfrac{v_1}{14} = \dfrac{13}{14}v_1$

or, $\dfrac{13}{14}\left(\dfrac{RT_1}{P_1}\right)=\dfrac{13}{14}\times\dfrac{(0.287\times1000)}{1\times10^5}\times300=0.7995\ \mathrm{m^3/kg}$

∴ **m.e.p.** $=\dfrac{W_{net}}{v_s}=\dfrac{249.36}{0.7995}=\mathbf{311.89\ kN/m^2}$ or **3.1189 bar.** (Ans.)

11.12. Exercises

1. The air standard Otto cycle comprises
    (*a*) two constant pressure processes and two constant volume processes
    (*b*) two constant pressure processes and two constant entropy processes
    (*c*) two constant volume processes and two constant entropy processes
    (*d*) none of the above.

2. The air standard efficiency of the Otto cycle is given by
    $(a)\ \eta=1+\dfrac{1}{(r)^{\gamma+1}}$ $(b)\ \eta=1-\dfrac{1}{(r)^{\gamma-1}}$
    $(c)\ \eta=1-\dfrac{1}{(r)^{\gamma+1}}$ $(d)\ \eta=2-\dfrac{1}{(r)^{\gamma-1}}.$

3. The thermal efficiency of a theoretical Otto cycle
    (*a*) increases with increase in compression ratio
    (*b*) increases with increase in isentropic index γ
    (*c*) does not depend upon the pressure ratio
    (*d*) follows all the above.

4. The work output of a theoretical Otto cycle
    (*a*) increases with increase in compression ratio
    (*b*) increases with increase in pressure ratio
    (*c*) increases with increase in adiabatic index γ
    (*d*) follows all the above.

5. For the same compression ratio

   (*a*) thermal efficiency of an Otto cycle is greater than that of a Diesel cycle

   (*b*) thermal efficiency of an Otto cycle is less than that of a Diesel cycle

   (*c*) thermal efficiency of an Otto cycle is same as that for Diesel cycle

   (*d*) thermal efficiency of an Otto cycle cannot be predicted.

6. In an air standard Diesel cycle, at a fixed compression ratio and fixed value of adiabatic index (γ)

   (*a*) thermal efficiency increases with an increase in heat addition cut-off ratio

   (*b*) thermal efficiency decreases with an increase in heat addition cut-off ratio

   (*c*) thermal efficiency remains the same with an increase in heat addition cut-off ratio

   (*d*) none of the above.

7. What is a cycle? What is the difference between an ideal and an actual cycle?

8. What is an air-standard efficiency?

9. What is relative efficiency?

10. Derive expressions of efficiency in the following cases:

    (*i*) Carnot cycle   (*ii*) Diesel cycle   (*iii*) Dual combustion cycle.

11. Explain "Air standard analysis" which has been adopted for I.C. engine cycles. State the assumptions made for air standard cycles.

12. Derive an expression for the "Atkinson cycle."

13. A Carnot engine working between 377 °C and 37 °C produces 120 kJ of work. Determine:

    (*i*) The heat added in kJ.   (*ii*) The entropy change during the heat rejection process.

    (*iii*) The engine thermal efficiency.

14. Find the thermal efficiency of a Carnot engine whose hot and cold bodies have temperatures of 154 °C and 15 °C, respectively.

15. Derive an expression for change in efficiency for a change in compression ratio. If the compression ratio is increased from 6 to 8, what will be the percentage increase in efficiency?

16. The efficiency of an Otto cycle is 50% and γ is 1.5. What is the compression ratio?

17. An engine working on an Otto cycle has a volume of 0.5 m³, pressure of 1 bar, and a temperature 27 °C at the commencement of the compression stroke. At the end of the compression stroke, the pressure is 10 bar. Heat added during the constant volume process is 200 kJ. Determine:

(*i*) Percentage clearance   (*ii*) Air standard efficiency

(*iii*) Mean effective pressure

(*iv*) Ideal power developed by the engine if the engine runs at 400 r.p.m. so that there are 200 complete cycles per minute

18. The compression ratio in an air-standard Otto cycle is 8. At the beginning of the compression process, the pressure is 1 bar and the temperature is 300 K. The heat transfer to the air per cycle is 1900 kJ/ kg of air.

19. Calculate:

(*i*) Thermal efficiency   (*ii*) The mean effective pressure.

20. An engine of 200 mm bore and 300 mm stroke works on an Otto cycle. The clearance volume is 0.0016 m³. The initial pressure and temperature are 1 bar and 60 °C. If the maximum pressure is limited to 24 bar, find:

(*i*) The air-standard efficiency   (*ii*) The mean effective pressure for
     of the cycle   the cycle.

     Assume ideal conditions.

21. Calculate the air standard efficiency of a four stroke Otto cycle engine with the following data:

22. Piston diameter (bore) = 137 mm; Length of stroke = 130 mm;

23. Clearance volume 0.00028 m³.

Express clearance as a percentage of swept volume.

24. In an ideal Diesel cycle, the temperatures at the beginning of compression, at the end of compression, and at the end of the heat addition are 97°C, 789°C, and 1839°C. Find the efficiency of the cycle.

25. An air-standard Diesel cycle has a compression ratio of 18, and the heat transferred to the working fluid per cycle is 1800 kJ/kg. At the beginning of the compression stroke, the pressure is 1 bar and the temperature is 300 K. Calculate: (*i*) Thermal efficiency, (*ii*) The mean effective pressure.

26. 1 kg of air is taken through a Diesel cycle. Initially the air is at 15 °C and 1 atma. The compression ratio is 15 and the heat added is 1850 kJ. Calculate: (*i*) The ideal cycle efficiency, (*ii*) The mean effective pressure.

27. What will be loss in the ideal efficiency of a Diesel engine with a compression ratio of 14 if the fuel cut-off is delayed from 6% to 9%?

28. The pressures on the compression curve of a diesel engine are at $\frac{1}{8}$ stroke 1.4 bar and at $\frac{7}{8}$ stroke 14 bar. Estimate the compression ratio. Calculate the air standard efficiency of the engine if the cut-off occurs at $\frac{1}{15}$ of the stroke.

29. A compression ignition engine has a stroke 270 mm, and a cylinder diameter of 165 mm. The clearance volume is 0.000434 m³ and the fuel ignition takes place at constant pressure for 4.5% of the stroke. Find the efficiency of the engine assuming it works on the Diesel cycle.

30. The following data belong to a Diesel cycle:
Compression ratio = 16 : 1; Heat added = 2500 kJ/kg; Lowest pressure in the cycle = 1 bar; Lowest temperature in the cycle = 27 °C.
Determine:

(*i*) Thermal efficiency of the cycle. (*ii*) Mean effective pressure.

31. The compression ratio of an air-standard Dual cycle is 12 and the maximum pressure in the cycle is limited to 70 bar. The pressure and temperature of the cycle at the beginning of compression process are 1 bar and 300 K. Calculate: (*i*) Thermal efficiency, (*ii*) Mean effective pressure.

Assume: cylinder bore = 250 mm, stroke length = 300 mm, c_p = 1.005, c_v = 0.718 and γ = 1.4.

32. The compression ratio of a Dual cycle is 10. The temperature and pressure at the beginning of the cycle are 1 bar and 27 °C. The maximum pressure of the cycle is limited to 70 bar and heat supplied is limited to 675 kJ/kg of air. Find the thermal efficiency of the cycle.

33. An air standard Dual cycle has a compression ratio of 16, and compression begins at 1 bar, 50 °C. The maximum pressure is 70 bar. The heat transferred to air at constant pressure is equal to that at constant volume. Determine:

(*i*) The cycle efficiency. (*ii*) The mean effective pressure of the cycle.

Take: c_p = 1.005 kJ/kg K, c_v = 0.718 kJ/kg K.

34. Compute the air standard efficiency of a Brayton cycle operating between a pressure of 1 bar and a final pressure of 12 bar. Take γ = 1.4.

12

GAS TURBINES AND JET PROPULSION

In This Chapter

- Overview
- Classification of Gas Turbines
- Merits of Gas Turbines
- Demerits of Gas Turbines
- Constant Pressure Combustion Gas Turbines
- Constant Volume Combustion Turbines
- GasTurbineFuels
- Jet Propulsion
- Rocket Engines
- Exercises

12.1. Overview

Probably a windmill was the first turbine to produce useful work, wherein there is no pre-compression and no combustion. The characteristic features of a gas turbine as we think of the name today include a *compression process* and a *heat addition* (or combustion) *process*. The gas turbine represents perhaps the most satisfactory way of producing very large quantities of power in a self-contained and compact unit. The gas turbine may have a future use in conjunction with the oil engine. For smaller gas turbine units, the inefficiencies in compression and expansion processes become greater, and to improve the thermal efficiency it is necessary to use a heat exchanger. In order that a small gas turbine may compete for economy with the small oil engine or petrol engine, it is necessary that a compact effective heat exchanger be used in the gas turbine cycle. The thermal efficiency of the gas turbine alone is still quite modest—20 to 30%—compared with that of a modern steam turbine plant—38 to 40%. It is possible to construct combined plants whose efficiencies are of the order of 45% or more. Higher efficiencies might be attained in the future.

The following *are* the major fields of application of gas turbines:

- Aviation

- Power generation

- Oil and gas industry

- Marine propulsion

The efficiency of a gas turbine is not the criteria for the choice of this plant. A gas turbine is used in aviation and marine fields because it is self-contained, lightweight, does not require cooling water, and generally fits into the overall shape of the structure. It is selected for power generation because of its simplicity, lack of cooling water, quick installation, and quick starting. It is used in the oil and gas industry because of a cheaper supply of fuel and low installation cost.

The gas turbines have the following limitations: (a) They are not self-starting; (b) low efficiencies at part loads; (c) non-reversibility; (d) higher rotor speeds and (e) overall efficiency of the plant low.

12.2. Classification of Gas Turbines

The gas turbines are mainly divided into two groups:

1. Constant pressure combustion gas turbine

(a) Open cycle constant pressure gas turbine
(b) Closed cycle constant pressure gas turbine

2. Constant volume combustion gas turbine

In almost all the fields open cycle gas turbine plants are used. Closed cycle plants were introduced at one stage because of their ability to burn cheap fuel. In between their progress remained slow because of availability of cheap oil and natural gas. Because of rising oil prices, now again, the attention is being paid to closed cycle plants.

12.3. Merits of Gas Turbines

(*i*) Merits over Internal Combustion(I.C.) engines:

1. The mechanical efficiency of a gas turbine (95%) is quite high as compared with an I.C. engine (85%) since the I.C. engine has a large number of sliding parts.

2. A gas turbine does not require a flywheel as the torque on the shaft is continuous and uniform, whereas a flywheel is a must in the case of an I.C. engine.

3. The weight of gas turbine per H.P. developed is less than that of an I.C. engine.

4. The gas turbine can be driven at a very high speed (40000 r.p.m.) whereas this is not possible with I.C. engines.

5. The work developed by a gas turbine per kg of air is more as compared to an I.C. engine. This is due to the fact that gases can be expanded up to atmospheric pressure in the case of a gas turbine whereas in an I.C. engine expansion up to atmospheric pressure is not possible.

6. The components of the gas turbine can be made lighter since the pressures used in it are very low, say 5 bar, compared with I.C. engine, say 60 bar.

7. In the gas turbine the ignition and lubrication systems are much simpler as compared with I.C. engines.

8. Cheaper fuels such as paraffin type, residue oils, or powdered coal can be used, whereas special grade fuels are employed in petrol engines to check knocking or pinking.

9. The exhaust from a gas turbine is less polluting comparatively since excess air is used for combustion.

10. Because of low specific weight the gas turbines are particularly suitable for use in aircrafts.

(*ii*) Merits over steam turbines:

The gas turbine entails the following *advantages over steam turbines*:

1. Capital and running cost less.

2. For the same output the space required is far less.

3. Starting is more easy and quick.

4. Weight per H.P. is far less.

5. Can be installed anywhere.

6. Control of a gas turbine is much easier.

7. A boiler along with accessories is not required.

12.4. Demerits of Gas Turbines

1. The thermal efficiency of a simple turbine cycle is low (15 to 20%) as compared with I.C. engines (25 to 30%).

2. With wide operating speeds the fuel control is comparatively difficult.

3. Due to higher operating speeds of the turbine, it is imperative to have a speed reduction device.

4. It is difficult to start a gas turbine as compared to an I.C. engine.

5. The gas turbine blades need a special cooling system.

6. One of the main demerits of a gas turbine is its *very poor thermal efficiency at part loads,* as the quantity of air remains the same irrespective of load, and output is reduced by reducing the quantity of fuel supplied.

7. Owing to the use of nickel-chromium alloy, the manufacture of the blades is difficult and costly.

8. For the same output the gas turbine produces five times the exhaust gases of an I.C. engine.

9. Because of the prevalence of high temperatures (1000 K for blades and 2500 K for the combustion chamber) and centrifugal force, the life of the combustion chamber and blades is short/small.

12.5. Constant Pressure Combustion Gas Turbines

12.5.1. Open Cycle Gas Turbines

The fundamental gas turbine unit is one operating on the open cycle, in which a rotary compressor and a turbine are mounted on a common shaft. As shown in Figure 12.1, air is drawn into the compressor and after

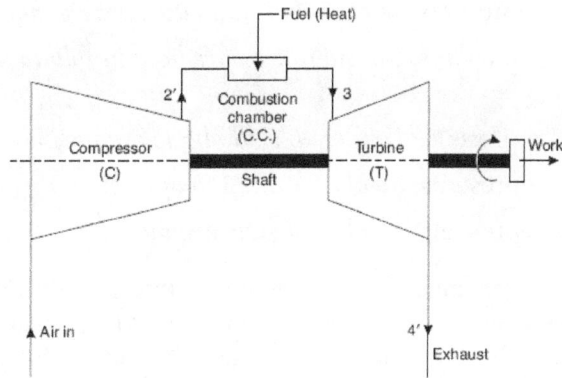

FIGURE 12.1 Open cycle gas turbine.

compression passes to a combustion chamber. Energy is supplied in the combustion chamber by spraying fuel into the airstream, and the resulting hot gases expand through the turbine to the atmosphere. In order to achieve network output from the unit, the turbine must develop more gross work output than is required to drive the compressor and to overcome mechanical losses in the drive. The products of combustion coming out from the turbine are exhausted to the atmosphere as they cannot be used anymore. The working fluids (air and fuel) must be replaced continuously as they are exhausted into the atmosphere.

If pressure loss in the combustion chamber is neglected, this cycle may be drawn on a *T-s* diagram as shown in Figure 12.2.

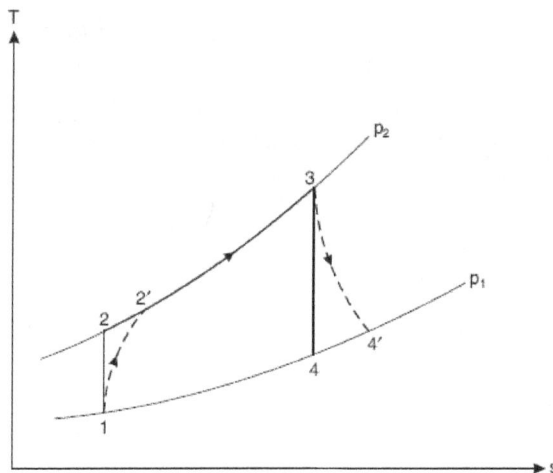

FIGURE 12.2 Gas turbine cycle on a T-S diagram.

- 1-2′ represents: *Irreversible adiabatic compression.*
- 2′-3 represents: *Constant pressure heat supply in the combustion chamber.*
- 3-4′ represents: *Irreversible adiabatic expansion.*
- 1–2 represents: *Ideal isentropic compression.*
- 3–4 represents: *Ideal isentropic expansion.*

Assuming change in kinetic energy between the various points in the cycle to be negligibly small compared with enthalpy changes and then applying the flow equation to each part of the cycle, for unit mass, we have

Work input (compressor) $\qquad = c_p \left(T_2' - T_1 \right)$

Heat supplied (combustion chamber) $\qquad = c_p \left(T_3 - T_2' \right)$

Work output (turbine) $\qquad = c_p \left(T_3 - T_4' \right)$

∴ Network output $\qquad$ = Work output − Work input

$$= c_p \left(T_3 - T_4' \right) - c_p \left(T_2' - T_1 \right)$$

and $\eta_{thermal} = \dfrac{\text{Network output}}{\text{Heat supplied}}$

$$= \dfrac{c_p \left(T_3 - T_4' \right) - c_p \left(T_2' - T_1 \right)}{c_p \left(T_3 - T_2' \right)}$$

Compressor isentropic efficiency, η_{comp}

$$= \dfrac{\text{Work input required in isentropic compression}}{\text{Actual work required}}$$

$$= \dfrac{c_p (T_2 - T_1)}{c_p (T_2' - T_1)} = \dfrac{T_2 - T_1}{T_2' - T_1}$$

Turbine isentropic efficiency, $\eta_{turbine}$

$$= \dfrac{\text{Actual work output}}{\text{Isentropic work output}}$$

$$= \dfrac{c_p (T_3 - T_4')}{c_p (T_3 - T_4)} = \dfrac{T_3 - T_4'}{T_3 - T_4}$$

With the variation in temperature, the value of the specific heat of a real gas varies, and in the open cycle, the specific heat of the gases in the combustion chamber and in turbine is different from that in the compressor because fuel has been added and a chemical change has taken place. Curves showing the variation of c_p with a temperature and air/fuel ratio can be used, and a suitable mean value of c_v and hence γ can be found out. It is usual in gas turbine practice to assume a fixed mean value of c_p and γ for the expansion process, and fixed mean values of c_p and γ for the compression process. In an open cycle gas turbine unit, the mass flow of gases in the turbine is greater than that in the compressor due to the mass of fuel burned, but it is possible to neglect mass of fuel, since the air/fuel ratios used are large. Also, in many cases, air is bled from the compressor for cooling purposes, or in the case of aircraft at high altitudes, bled air is used for de-icing and cabin air-conditioning. This amount of air bled is approximately the same as the mass of fuel injected therein.

12.5.2. Methods for Improvement of Thermal Efficiency of Open Cycle Gas Turbine Plants

The following methods are employed to increase the specific output and thermal efficiency of the plant:

1. Intercooling   2. Reheating   3. Regeneration

1. **Intercooling.** A compressor in a gas turbine cycle utilizes the major percentage of power developed by the gas turbine. The work required by the compressor can be reduced by compressing the air in two stages and *incorporating an intercooler* between the two as shown in Figure 12.3. Turbine plant with intercooler. The corresponding *T-s* diagram for the unit is shown in Figure 12.4. The actual processes take place as follows:

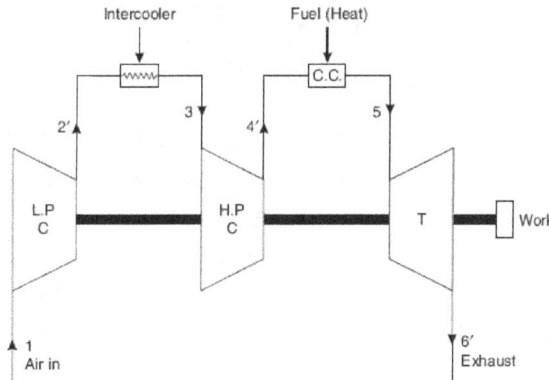

FIGURE 12.3 Turbine plant with intercooler.

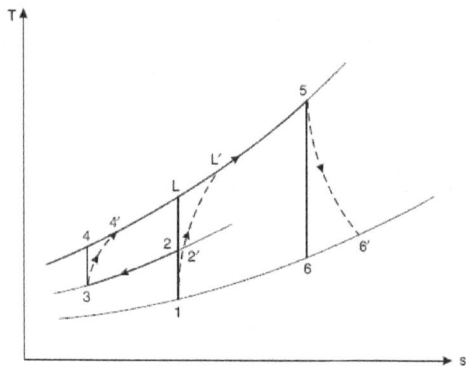

FIGURE 12.4 T-s diagram for the unit.

1-2'	...	L.P. (Low pressure) compression
2'-3	...	Intercooling
3-4'	...	H.P. (High pressure) compression
4'-5	...	C.C. (Combustion chamber)-heating
5-6'	...	T (Turbine)-expansion

The ideal cycle for this arrangement is 1-2-3-4-5-6; the compression process without intercooling is shown as 1-L in the actual case, and 1-L in the ideal isentropic case.

Now,

Work input (*with intercooling*)

$$= c_p \left(T_2' - T_1 \right) + c_p \left(T_4' - T_3 \right) \tag{12.1}$$

Work input (*without intercooling*)

$$= c_p \left(T_L' - T_1 \right) = c_p \left(T_2' - T_1 \right) + c_p \left(T_L' - T_2' \right) \tag{12.2}$$

By comparing equation (12.2) with equation (12.1) it can be observed that the *work input with intercooling is less than the work input with no intercooling*, when $c_p \left(T_4' - T_3 \right)$ is *less* than $c_p \left(T_L' - T_2' \right)$. This is so if it is assumed that the isentropic efficiencies of the two compressors, operating separately, are each equal to the isentropic efficiency of the single compressor, which would be required if no intercooling were used. Then $\left(T_4' - T_3 \right) < \left(T_L' - T_2' \right)$ since the pressure lines diverge on the *T-s* diagram from left to right.

Again, work ratio $\quad = \dfrac{\text{Network output}}{\text{Gross work output}}$

$$= \dfrac{\text{Work of expansion } - \text{Work of compression}}{\text{Work of expansion}}$$

From this we may conclude that *when the compressor work input is reduced, then the work ratio is increased.*

However, the heat supplied in the combustion chamber when intercooling is used in the cycle is given by

Heat supplied *with intercooling* $= c_p \left(T_5 - T_4' \right)$

Also the heat supplied when intercooling is not used, with the same maximum cycle temperature T_5, is given by

Heat supplied *without intercooling* $= c_p \left(T_5 - T_L' \right)$

Thus, the *heat supplied when intercooling is used is greater than with no intercooling. Although the network output is increased by intercooling it is found in general that the increase in heat to be supplied causes the thermal efficiency to decrease.* When intercooling is used a supply of cooling water must be readily available. The additional bulk of the unit may offset the advantage to be gained by increasing the work ratio.

2. Reheating. The output of a gas turbine can be amply improved by expanding the gases in two stages with a *reheater* between the two as shown in Figure 12.5. The H.P. turbine drives the compressor and the L.P. turbine provides the useful power output. The corresponding *T-s* diagram is shown in Figure 12.6. The line 4'-L' represents the expansion in the L.P. turbine if reheating is *not* employed.

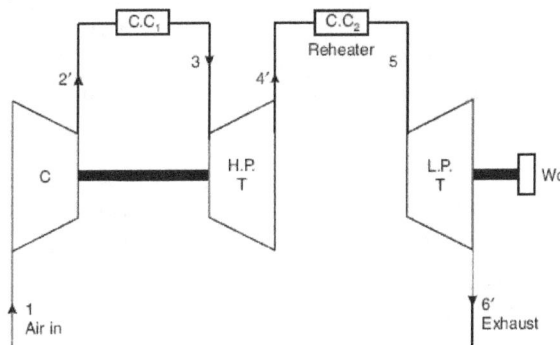

FIGURE 12.5 Gas turbine with reheater.

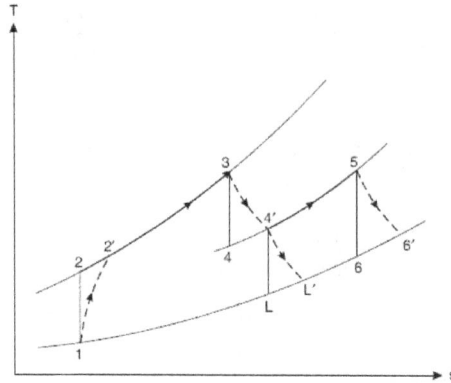

FIGURE 12.6 T-s diagram for the unit.

Neglecting mechanical losses the work output of the H.P. turbine must be exactly equal to the work input required for the compressor, that is,

$$c_{pa}\left(T_2' - T_1\right) = c_{pg}\left(T_3 - T_4'\right)$$

The work output (net output) of the L.P. turbine is given by

$$\text{Network output (with reheating)} = c_{pg}\left(T_5 - T_6'\right)$$

$$\text{and Network output (without reheating)} = c_{pg}\left(T_4' - T_L'\right)$$

Since the pressure lines diverge to the right on the T-s diagram, it can be seen that the temperature difference $(T_5 - T_6')$ is always greater than $(T_4' - T_L')$, so that reheating increases the network output.

Although the network is increased by reheating, the heat to be supplied is also increased, and the net effect can be to reduce the thermal efficiency

$$\text{Heat supplied } = c_{pg}\left(T_3 - T_2'\right) + c_{pg}\left(T_5 - T_4'\right).$$

c_{pa} and c_{pg} stand for specific heats of air and gas, respectively, at constant pressure.

3. Regeneration. The exhaust gases from a gas turbine carry a large quantity of heat with them since their temperature is far above the ambient temperature. They can be used to heat the air coming from the compressor, thereby reducing the mass of fuel supplied in the combustion chamber. Figure 12.7 shows a gas turbine plant with a regenerator. The corresponding T-s diagram is shown in Figure 12.8. 2′–3 represents the heat flow into

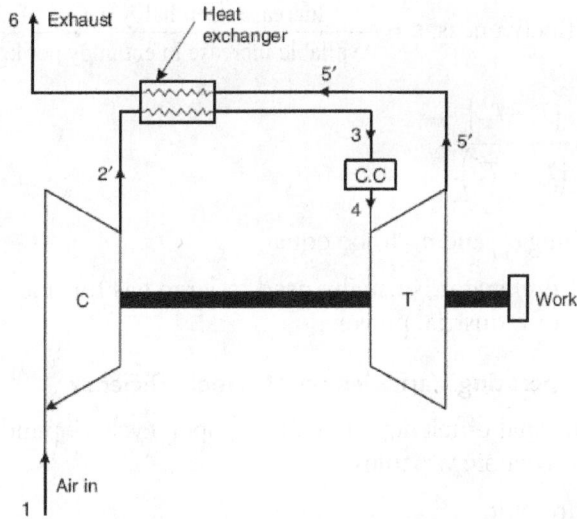

FIGURE 12.7 Gas turbine with regenerator.

the compressed air during its passage through the heat exchanger, and 3–4 represents the heat taken in from the combustion of fuel. Point 6 represents the temperature of exhaust gases at discharge from the heat exchanger. The maximum temperature to which the air could be heated in the heat exchanger is ideally that of the exhaust gases, but less than this is obtained in practice because a temperature gradient must exist for an unassisted transfer of energy. The effectiveness of the heat exchanger is given by:

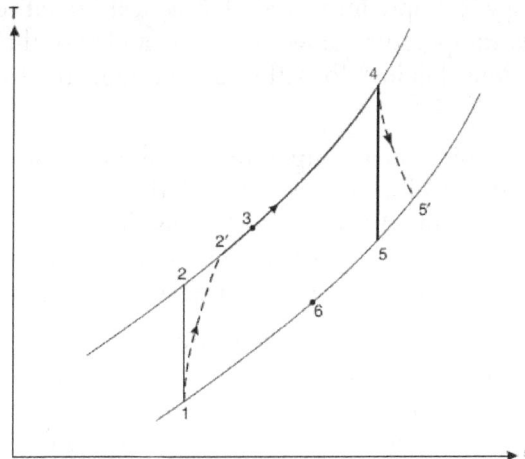

FIGURE 12.8 T-s diagram for the unit.

$$\text{Effectiveness, } \varepsilon = \frac{\text{Increase in enthalpy per kg of air}}{\text{Available increase in enthalpy per kg of air}}$$

$$= \frac{\left(T_3 - T_2'\right)}{\left(T_5' - T_2'\right)}$$

(assuming c_{pa} and c_{pg} to be equal)

A heat exchanger is usually used in large gas turbine units for marine propulsion or industrial power.

Effect of Operating Variables on Thermal Efficiency

The thermal efficiency of an actual open cycle depends on the following thermodynamic variables:

Pressure ratio

Turbine inlet temperature (T_3)

Compressor inlet temperature (T_1)

Efficiency of the turbine (η_{turbine})

Efficiency of the compressor (η_{comp}).

Effect of turbine inlet temperature and pressure ratio:

If the permissible turbine inlet temperature (with the other variables being constant) of an open cycle gas turbine power plant is increased, its thermal efficiency is amply improved. A practical limitation to increasing the turbine inlet temperature, however, is the ability of the material available for the turbine blading to withstand the high rotative and thermal stresses.

For a given turbine inlet temperature, as the pressure ratio increases, the heat supplied as well as the heat rejected are reduced. But the ratio of change of heat supplied is not the same as the ratio of change of heat rejected. Consequently, there exists an optimum pressure ratio to produce maximum thermal efficiency for a given turbine inlet temperature.

As the pressure ratio increases, the thermal efficiency also increases until it becomes maximum and then it drops off with a further increase in the pressure ratio (Figure 12.9). Further, as the turbine inlet temperature increases, the peaks of the curves flatten out, giving a greater range of ratios of pressure optimum efficiency.

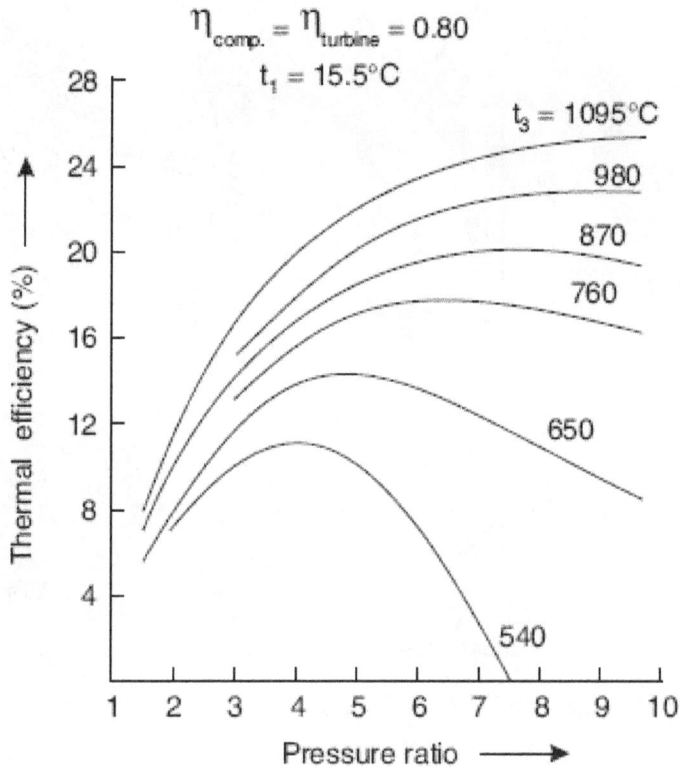

$$\eta_{comp.} = \eta_{turbine} = 0.80$$
$$t_1 = 15.5°C$$

FIGURE 12.9 Effect of the pressure ratio and turbine inlet temperature.

The following particulars are worth noting:

Gas temperatures	Efficiency (gas turbine)
550 to 600 °C	20 to 22%
900 to 1000 °C	32 to 35%
Above 1300 °C	more than 50%

The effect of turbine and compressor efficiencies:

As shown in Figure 12.10, the thermal efficiency of the actual gas turbine cycle is very sensitive to variations in the efficiencies of the compressor and turbine. There is a particular pressure ratio at which maximum efficiencies occur. For lower efficiencies, the peak of the thermal efficiency occurs at lower pressure ratios and vice versa.

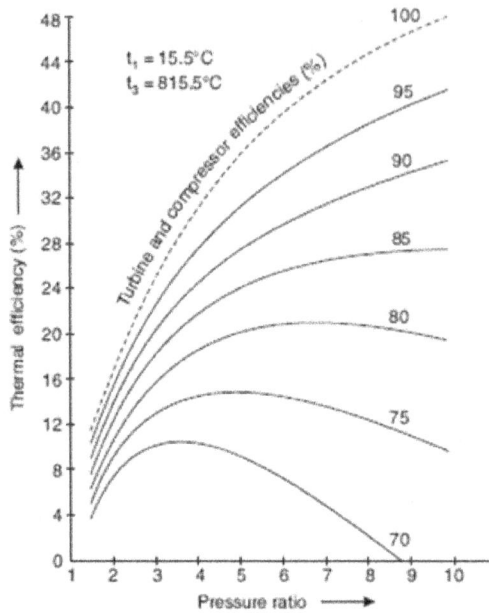

FIGURE 12.10 Effect of component efficiency.

Effect of compressor inlet temperature:

As shown in Figure 12.11, with the decrease in the compressor inlet temperature there is an increase in thermal efficiency of the plant. Also the peaks of thermal efficiency occur at high pressure ratios and the curves become flatter, giving thermal efficiency over a wider pressure ratio range.

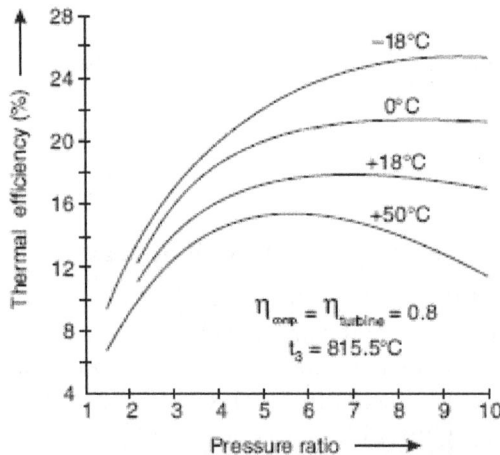

FIGURE 12.11 Effect of compressor inlet temperature.

12.5.3. Closed Cycle Gas Turbine

Figure 12.12 shows a gas turbine operating on a constant pressure cycle in which the closed system consists of air behaving as an ideal gas. The various operations are as follows: Refer to Figure 12.13 and Figure 12.14.

Operation 1–2: The air is compressed isentropically from the lower pressure p_1 to the upper pressure p_2, the temperature rising from T_1 to T_2. No heat flow occurs.

Operation 2–3: Heat flows into the system increasing the volume from V_2 to V_3 and temperature from T_2 to T_3, while the pressure remains constant at p_2. Heat received = $mc_p\left(T_3 - T_2\right)$.

Operation 3–4: The air is expanded isentropically from p_2 to p_1, the temperature falling from T_3 to T_4. No heat flow occurs.

Operation 4–1: Heat is rejected from the system as the volume decreases from V_4 to V_1 and the temperature from T_4 to T_1, while the pressure remains constant at p_1. Heat rejected = $mc_p\left(T_4 - T_1\right)$.

$$\eta_{air\text{-}standard} = \frac{\text{Work done}}{\text{Heat received}}$$

$$= \frac{\text{Heat received/cycle} - \text{Heat rejected/cycle}}{\text{Heat received/cycle}}$$

$$= \frac{mc_p\left(T_3 - T_2\right) - mc_p\left(T_4 - T_1\right)}{mc_p\left(T_3 - T_2\right)} = 1 - \frac{T_4 - T_1}{T_3 - T_2}$$

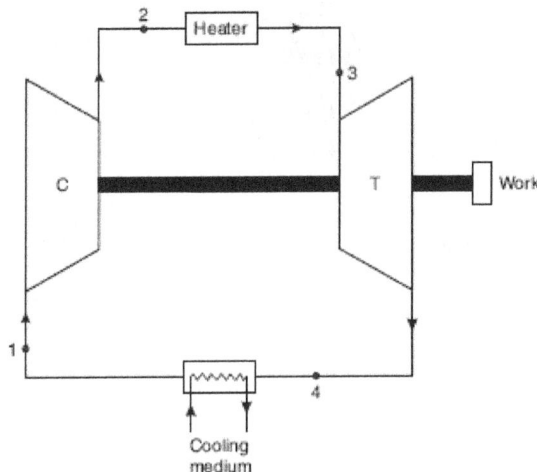

FIGURE 12.12 Closed cycle gas turbine.

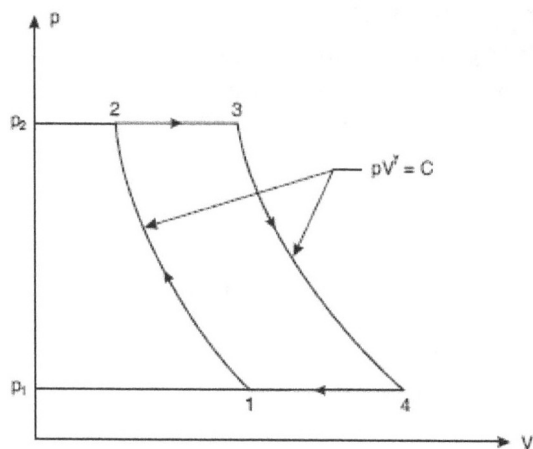

FIGURE 12.13 p-V diagram.

Now, from isentropic expansion

$$\frac{T_2}{T_1} = \left(\frac{p_2}{p_1}\right)^{\frac{\gamma-1}{\gamma}}$$

$$T_2 = T_1(r_p)^{\frac{\gamma-1}{\gamma}}, \text{ where } r_p = \text{Pressure ratio}$$

Similarly, $\dfrac{T_3}{T_4} = \left(\dfrac{p_2}{p_1}\right)^{\frac{\gamma-1}{\gamma}}$ or $T_3 = T_4(r_p)^{\frac{\gamma-1}{\gamma}}$

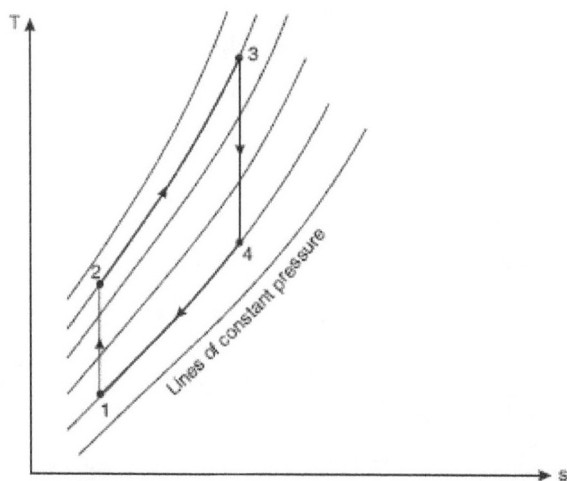

FIGURE 12.14 T-s diagram.

$$\therefore \quad \eta_{air\text{-}standard} = 1 - \frac{T_4 - T_1}{T_4\left(r_p\right)^{\frac{\gamma-1}{\gamma}} - T_1\left(r_p\right)^{\frac{\gamma-1}{\gamma}}} = 1 - \frac{1}{\left(r_p\right)^{\frac{\gamma-1}{\gamma}}}$$

The expression shows that the efficiency of the ideal joule cycle increases with the pressure ratio. The absolute limit of pressure is determined by the limiting temperature of the material of the turbine at the point at which this temperature is reached, so by the compression process alone, no further heating of the gas in the combustion chamber would be permissible and the work of expansion would ideally just balance the work of compression so that no excess work would be available for external use.

Now we shall prove that the pressure ratio for maximum work is a function of the limiting temperature ratio.

Work output during the cycle

= Heat received / cycle – Heat rejected / cycle

$$= mc_p\left(T_3 - T_2\right) - mc_p\left(T_4 - T_1\right) = mc_p\left(T_3 - T_4\right) - mc_p\left(T_2 - T_1\right)$$

$$= mc_pT_3\left(1 - \frac{T_4}{T_3}\right) - T_1\left(\frac{T_2}{T_1} - 1\right)$$

In the case of a given turbine the minimum temperature T_1 and the maximum temperature T_3 are prescribed, T_1 being the temperature of the atmosphere and T_3 the maximum temperature which the metals of the turbine would withstand. Consider the specific heat at constant pressure c_p to be constant. Then,

Since, $$\frac{T_3}{T_4} = \left(r_p\right)^{\frac{\gamma-1}{\gamma}} = \frac{T_2}{T_1}$$

Using the constant $$\text{'}z\text{'} = \frac{\gamma-1}{\gamma},$$

we have, work output/cycle $W = K\left[T_3\left(1 - \frac{1}{r_p^{\,z}}\right) - T_1\left(r_p^{\,z} - 1\right)\right]$

Differentiating with respect to r_p

$$\frac{dW}{dr_p} = K\left[T_3 \times \frac{z}{r_p^{\,(z+1)}} - T_1 z r_p^{\,(z-1)}\right] = 0 \text{ for a maximum}$$

$$\therefore \quad \frac{zT_3}{r_p^{(z+1)}} = T_1 z \left(r_p \right)^{(z-1)}$$

$$\therefore \quad r_p^{2z} = \frac{T_3}{T_1}$$

$$\therefore \qquad r_p = \left(T_3 / T_1 \right)^{1/2z} \text{ i.e., } r_p = \left(T_3 / T_1 \right)^{\frac{r}{2(\gamma-1)}}$$

Thus, the pressure ratio for maximum work is a function of the limiting temperature ratio.

Figure 12.15 shows an arrangement of a closed cycle stationary gas turbine plant in which air is continuously circulated. This ensures that the air is not polluted by the addition of combustion waste product, since the heating of air is carried out in the form of the heat exchanger shown in the diagram as an air heater. The air exhausted from the power turbine is cooled before readmission to the L.P. compressor. The various operations as indicated on the T-s diagram (Figure 12.16) are as follows:

Operation 1-2′:   Air is compressed from p_1 to p_x in the L.P. compressor.

Operation 2′-3:   Air is cooled in the intercooler at constant pressure p_x.

Operation 3–4′:   Air is compressed in the H.P. compressor from p_x to p_2.

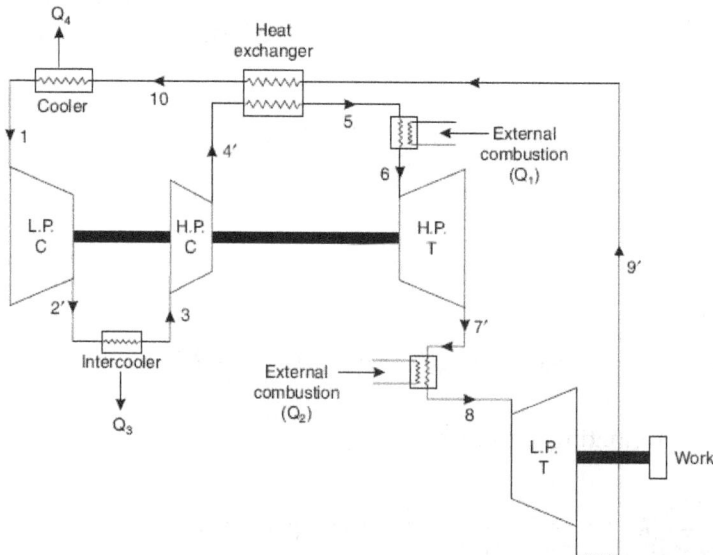

FIGURE 12.15 Closed cycle gas turbine plant.

Operation 4′–5: High pressure air is heated at constant pressure by exhaust gases from the power turbine in the heat exchanger to T_5.

Operation 5–6: High pressure air is further heated at constant pressure to the maximum temperature T_6 by an air heater (through external combustion).

Operation 6–7′: The air is expanded in the H.P. turbine from p_2 to p_x, producing work to drive the compressor.

Operation 7′–8: Exhaust air from the H.P. turbine is heated at constant pressure in the air heater (through external combustion) to the maximum temperature $T_8(= T_6)$.

Operation 8–9′: The air is expanded in the L.P. turbine from p_x to p_1, producing energy for a flow of work externally.

Operation 9′–10: Air the from L.P. turbine is passed to the heat exchanger, where energy is transferred to the air delivered from the H.P. compressor. The temperature of air leaving the heat exchanger and entering the cooler is T_{10}.

Operation 10-11: Air is cooled to T_1 by the cooler before entering the L.P. compressor.

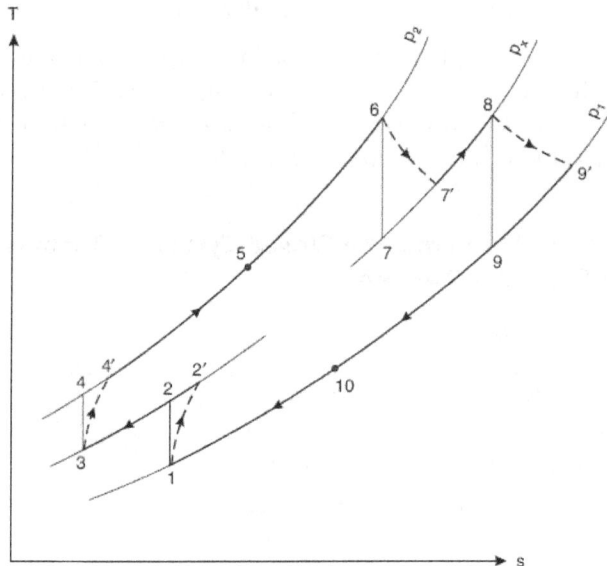

FIGURE 12.16 T-s diagram for the plant.

The energy balance for the whole plant is as follows:

$$Q_1 + Q_2 - Q_3 - Q_4 = W$$

In a closed cycle plant, in practice, the control of power output is achieved by varying the mass flow by the use of a reservoir in the circuit. The *reservoir maintains the design pressure and temperature and therefore achieves an approximately constant level of efficiency for varying* loads. In this cycle since it is closed, *gases other than air with favorable properties can be used;* furthermore, it is possible to burn solid fuels in the combustion heaters. *The major factor responsible for inefficiency in this cycle is the large irreversible temperature drop which occurs in the air heaters between the furnace and circulating gas.*

In closed cycle gas turbines, although air has been extensively used, the use of helium, though of a lower density, has been inviting the attention of manufacturers for its use for large output gas turbine units. *The specific heat of helium at a constant pressure is about five times that of air, therefore for each kg mass flow the heat drop and hence energy dealt within helium machines is nearly five times of those using air.* The *surface area of the heat exchanger for helium can be kept as low as 1/3 of that required for a gas turbine plant using air as a working medium.* For the same temperature ratio and for the plants of the same output, the *cross-sectional area required for helium is much less than that for air.* It may therefore be concluded that the *size of a helium unit is considerably small comparatively.*

Some gas turbine plants work on a combination of two cycles: the open cycle and the closed cycle. Such a combination is called the *semi-closed cycle. Here a part of the working fluid is confined within the plant and another part flows from and to the atmosphere.*

12.5.4. Merits and Demerits of a Closed Cycle Gas Turbine over an Open Cycle Gas Turbine

Merits of closed cycle:

1. Higher thermal efficiency	2. Reduced size
3. No contamination	4. Improved heat transmission
5. Improved part load efficiency	6. Lesser fluid friction
7. No loss of working medium	8. Greater output
9. Inexpensive fuel.	

Demerits of closed cycle:

Complexity.

A large amount of cooling water is required. This limits its use to stationary installation or marine use where water is available in abundance.

Dependent system.

The weight of the system per H.P. developed is high comparatively, therefore it is not economical for moving vehicles.

It requires the use of a very large air heater.

12.6. Constant Volume Combustion Turbines

Refer to Figure 12.17. In a constant volume combustion turbine, the compressed air from an air compressor C is admitted into the combustion chamber D through the valve A. When the valve A is closed, the fuel is admitted into the chamber by means of a fuel pump P. Then the mixture is ignited by means of a spark plug S. The combustion takes place at a constant volume with an increase of pressure. The valve B opens and the hot gases flow to the turbine T, and, finally, they are discharged into the atmosphere. The energy of the hot gases is thereby converted into mechanical energy. For continuous running of the turbine these operations are repeated.

The main demerit associated with this type of turbine is that the *pressure difference and velocities of hot gases are not constant, so the turbine speed fluctuates.*

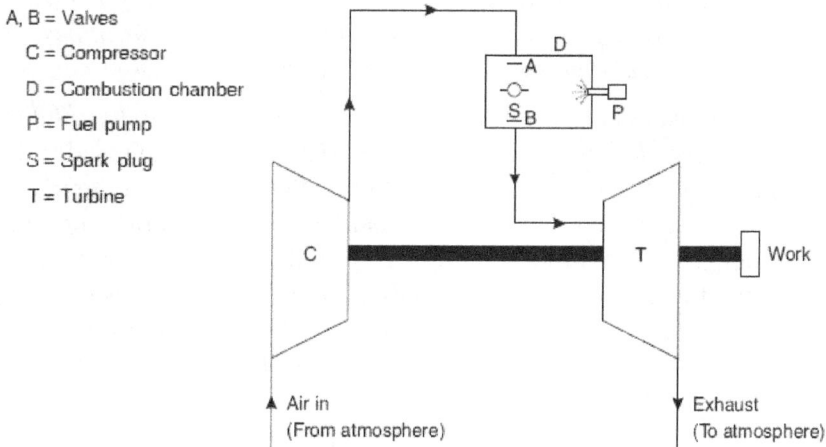

A, B = Valves
C = Compressor
D = Combustion chamber
P = Fuel pump
S = Spark plug
T = Turbine

C

T

Work

Air in
(From atmosphere)

Exhaust
(To atmosphere)

FIGURE 12.17 Constant volume combustion gas turbine.

USES OF GAS TURBINES

Gas turbines find wide *applications* in the following *fields*:

1. Supercharging	2. Turbo-jet and turbo-propeller engines
3. Marine field	4. Railway
5. Road transport	6. Electric power generation
7. Industry.	

12.7. Gas Turbine Fuels

The various fuels used in gas turbines are enumerated and discussed as follows:

1. Gaseous fuels

2. Liquid fuels

3. Solid fuels

1. **Gaseous fuels.** *Natural gas is the ideal fuel for gas turbines*, but this is not available everywhere.

   Blast furnace and *producer gases* may also be used for gas turbine power plants.

2. **Liquid fuels.** Liquid fuels of petroleum origin such as distillate oils or residual oils are most commonly used for gas turbine plants. The essential qualities of these fuels include *proper volatility, viscosity, and calorific value*. At the same time it *should be* free *from any contents of moisture* and *suspended impurities that would clog the small passages of the nozzles and damage valves and plungers of the fuel pumps*.

   Minerals like *sodium, vanadium,* and *calcium* prove *very harmful* for the *turbine blading* as these build deposits or corrode the blades. The sodium in ash should be less than 30% of the vanadium content as otherwise the ratio tends to be critical. The actual sodium content may be between 5 ppm to 10 ppm (part per million). If the vanadium is over 2 ppm, the magnesium in ash tends to become critical. *It is necessary that the magnesium in ash is at least three times the quantity of vanadium.* The content of calcium and lead should not be over 10 ppm and 5 ppm, respectively.

   Sodium is removed from residual oils by mixing with 5% water and then double centrifuging when sodium leaves with water. Magnesium is added to the washed oil in the form of Epsom salts, before the oil is sent into the combustor. This checks the corrosive action of vanadium. Residual oils burn with less ease than distillate oils and the latter are often used to start the unit from cold, after which the residual oils are fed in the combustor. In cold conditions residual oils need to be preheated.

3. **Solid fuels.** The use of solid fuels such as coal in pulverized form in gas turbines presents several difficulties, most of which have been only partially overcome yet. The pulverizing plant for coal in gas turbine applications is much lighter and small than its counterpart in steam generators. The *introduction of fuel in the combustion chamber of a gas turbine is required to be done against a high pressure, whereas the pressure in the furnace of a steam plant is atmospheric.* Furthermore, the *degree of completeness of combustion in gas turbine applications has to be very high, as otherwise soot and dust in the gas would deposit on the turbine blading.*

Some practical applications of solid fuel burning in turbine combustors have been made commercially available in recent years. In one such design finely crushed coal is used instead of pulverized fuel. This fuel is carried in a stream of air tangentially into one end of a cylindrical furnace while gas comes out at the center of opposite end. As the fuel particles roll around the circumference of the furnace, they are burnt, and a high temperature of about 1650 °C is maintained, which causes the mineral matter of the fuel to be converted into a liquid slag. The slag covers the walls of the furnace and runs out through a top hole in the bottom. The result is that fly-ash is reduced to a very small content in the gases. In *another design* a regenerator is used to transfer the heat to the air, the combustion chamber being located on the outlet of the turbine, and the combustion is carried out in the turbine exhaust stream. The advantage is that only clean air is handled by the turbine.

Example 1

The air enters the compressor of an open cycle constant pressure gas turbine at a pressure of 1 bar and a temperature of 20 °C. The pressure of the air after compression is 4 bar. The isentropic efficiencies of the compressor and the turbine are 80% and 85%, respectively. The air-fuel ratio used is 90 : 1. If the flow rate of air is 3.0 kg/s, find:

(i) Power developed.

(ii) Thermal efficiency of the cycle.

Assume $c_p = 1.0$ kJ/kg K and $\gamma = 1.4$ of air and gases

Calorific value of fuel = 41800 kJ/kg.

■ **Solution.**

$p_1 = 1$ bar ; $T_1 = 20 + 273 = 293$ K

$p_2 = 4$ bar; $\eta_{compressor} = 80\%$; $\eta_{turbine} = 85\%$

Air-fuel ratio = 90 : 1 ; Air flow rate, $m_a = 3.0$ kg/s

(i) **Power developed, P:**

Refer to Figure 12.18 (b)

$$\frac{T_2}{T_1} = \left(\frac{p_2}{p_1}\right)^{\frac{\gamma-1}{\gamma}} = \left(\frac{4}{1}\right)^{\frac{1.4-1}{1.4}} = 1.486$$

$\therefore$ $T_2 = (20 + 273) \times 1.486 = 435.4$ K

$$\eta_{compressor} = \frac{T_2 - T_1}{T_2' - T_1}$$

$$0.8 = \frac{435.4 - 293}{T_2' - 293}$$

$\therefore$ $T_2' = \dfrac{435.4 - 293}{0.8} + 293 = 471$ K

Heat supplied by fuel = Heat taken by burning gases

$$m_f \times C = (m_a + m_f) c_p \left(T_3 - T_2'\right)$$

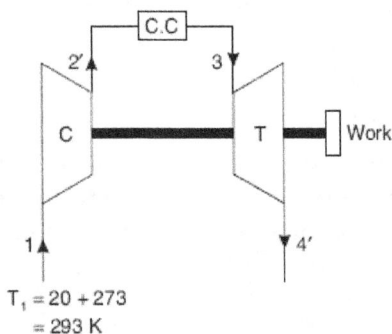

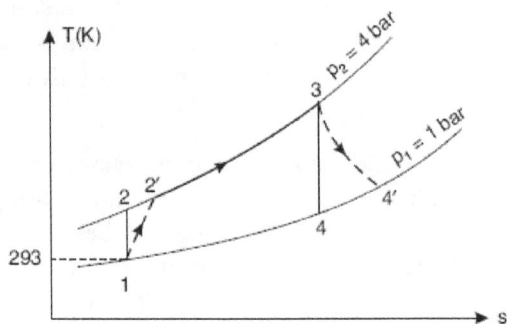

FIGURE 12.18 Sketch for Example 1.

(where m_a = mass of air, m_f = mass of fuel)

$$\therefore \quad C = \left(\frac{m_a}{m_f}+1\right)c_p\left(T_3 - T_2'\right)$$

$$\therefore \quad 41800 = (90+1)\times 1.0 \times (T_3 - 471)$$

i.e., $T_3 = \dfrac{41800}{91} + 471 = 930$ K

Again, $\qquad \dfrac{T_4}{T_3} = \left(\dfrac{p_4}{p_3}\right)^{\frac{\gamma-1}{\gamma}} = \left(\dfrac{1}{4}\right)^{0.4/1.4} = 0.672$

$$\therefore \quad T_4 = 930 \times 0.672 = 624.9 \text{ K}$$

$$\eta_{turbine} = \frac{T_3 - T_4'}{T_3 - T_4}$$

$$0.85 = \frac{930 - T_4'}{930 - 824.9}$$

$$\therefore T_4' = 930 - 0.85(930 - 624.9) = 670.6 \text{ K}$$

$$W_{turbine} = m_g \times c_p \times \left(T_3 - T_4'\right)$$

(where m_g is the mass of hot gases formed per kg of air)

$$\therefore W_{turbine} = \left(\frac{90+1}{90}\right)\times 1.0 \times (930 - 670.6)$$

$$= 262.28 \text{ kJ/kg of air.}$$

$$W_{compressor} = m_a \times c_p \times (T_2' - T_1) = 1 \times 1.0 \times (471 - 293)$$

$$= 178 \text{ kJ / kg of air}$$

$$W_{net} = W_{turbine} - W_{compressor}$$

$$= 262.28 - 178 = 84.28 \text{ kJ / kg of air.}$$

Hence, *power developed,* $P = 84.28 \times 3 = \textbf{252.84 kW / kg of air.}$ **(Ans.)**

(*ii*) **Thermal efficiency of cycle,** $\eta_{thermal}$**:**
Heat supplied per kg of air passing through combustion chamber

$$= \frac{1}{90} \times 41800 = 464.44 \text{ kJ / kg of air}$$

$$\therefore \quad \eta_{thermal} = \frac{\text{Work output}}{\text{Heat supplied}} = \frac{84.28}{464.44} = \textbf{0.1814 or 18.14\%. (Ans.)}$$

Example 2

A gas turbine unit has a pressure ratio of 6 : 1 and maximum cycle temperature of 610 °C. The isentropic efficiencies of the compressor and turbine are 0.80 and 0.82, respectively. Calculate the power output in kilowatts of an electric generator geared to the turbine when the air enters the compressor at 15 °C at the rate of 16 kg/s.

Take c_p = 1.005 kJ/kg K and γ = 1.4 for the compression process, and take c_p = 1.11 kJ/kg K and γ = 1.333 for the expansion process.

■ **Solution.**

$$T_1 = 15 + 273 = 288 \text{ K} ; \ T_3 = 610 + 273 = 883 \text{ K} ; \ \frac{p_2}{p_1} = 6,$$

$\eta_{compressor}$ = 0.80 ; $\eta_{turbine}$ = 0.82 ; Air flow rate = 16 kg/s

For compression process : c_p = 1.005 kJ / kg K, γ = 1.4

For expansion process : c_p = 1.11 kJ / kg K, γ = 1.333

In order to evaluate the network output, it is necessary to calculate temperatures T_2' and T_4'. To calculate T_2' we must first calculate T_2 and then use the isentropic efficiency.

For an isentropic process, $\dfrac{T_2}{T_1} = \left(\dfrac{p_2}{p_1}\right)^{\frac{\gamma-1}{\gamma}} = (6)^{\frac{1.4-1}{1.4}} = 1.67$

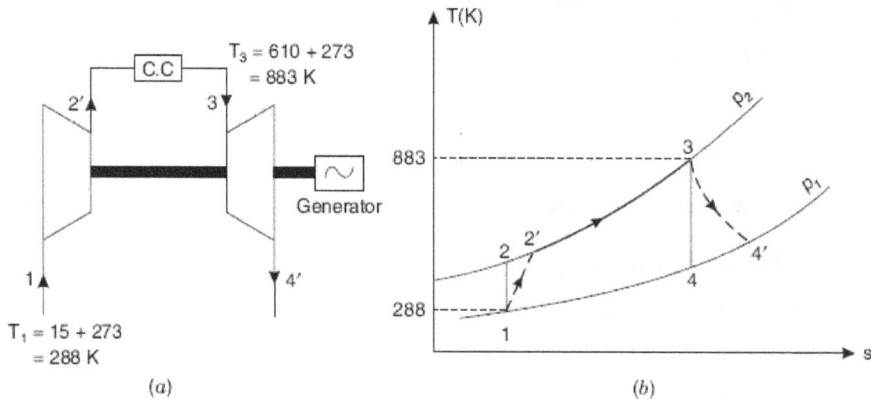

FIGURE 12.19 Sketch for Example 2.

$$\therefore \ T_2 = 288 \times 1.67 = 481 \ K$$

Also, $\eta_{compressor} = \dfrac{T_2 - T_1}{T_2' - T_1}$

$$0.8 = \dfrac{481 - 288}{T_2' - T_1}$$

$$\therefore \ T_2' = \dfrac{481 - 288}{0.8} + 288 = 529 \ K$$

Similarly for the turbine, $\dfrac{T_3}{T_4} = \left(\dfrac{p_3}{p_4}\right)^{\frac{\gamma-1}{\gamma}} = \left(\dfrac{p_2}{p_1}\right)^{\frac{\gamma-1}{\gamma}} = (6)^{\frac{1.333-1}{1.333}} = 1.565$

$$\therefore \ T_4 = \dfrac{T_3}{1.565} = \dfrac{883}{1.565} = 564 \ K$$

Also, $\eta_{turbine} = \dfrac{T_3 - T_4'}{T_3 - T_4} = \dfrac{883 - T_4'}{883 - 564}$

$$\therefore \ 0.82 = \dfrac{883 - T_4'}{883 - 564}$$

$$\therefore \ T_4' = 883 - 0.82\,(883 - 564) = 621.4 \ K$$

Hence,

Compressor work input, $W_{compressor} = c_p\left(T_2' - T_1\right)$

$$= 1.005\,(529 - 288) = 242.2 \ kJ/kg$$

Turbine work output, $\qquad W_{turbine} = c_p\left(T_3 - T_4'\right)$

$$= 1.11\,(883 - 621.4) = 290.4 \ kJ/kg$$

$\therefore$ Network output, $\qquad W_{net} = W_{turbine} - W_{compressor}$

$$= 290.4 - 242.2 = 48.2 \ kJ/kg$$

Power in kilowatts $= 48.2 \times 16 = $ **771.2 kW.** **(Ans.)**

Example 3

A gas turbine unit receives air at 1 bar and 300 K and compresses it adiabatically to 6.2 bar. The compressor efficiency is 88%. The fuel has a heating valve of 44186 kJ/kg and the fuel-air ratio is 0.017 kJ/kg of air.

The turbine internal efficiency is 90%. Calculate the work of the turbine and compressor per kg of air compressed and thermal efficiency. $Cp = 1.147$ kJ/kg.K and, for products.

■ **Solution.**

Given: $p_1 \left(= p_4\right) = 1$ bar, $T_1 = 300$ K ; $p_2 \left(= p_3\right) = 6.2$ bar ; $\eta_{compressor} = 88\%$;

$C = 44186$ kJ/kg ; Fuel-air ratio = 0.017 kJ/kg of air, $\eta_{turbine} = 90\%$;

$c_p = 1.147$ kJ/kg K ; $\gamma = 1.333$.

For isentropic compression process 1–2:

$$\frac{T_2}{T_1} = \left(\frac{p_2}{p_1}\right)^{\frac{\gamma-1}{\gamma}} = \left(\frac{6.2}{1}\right)^{\frac{1.4-1}{1.4}} = 1.684$$

$\therefore T_2 = 300 \times 1.684 = 505.2$ K

Now, $\eta_{compressor} = \dfrac{T_2 - T_1}{T_2' - T_1}$

$0.88 = \dfrac{505.2 - 300}{T_2' - 300}$

$T_2' = \left(\dfrac{505.2 - 300}{0.88} + 300\right) = 533.2$ K

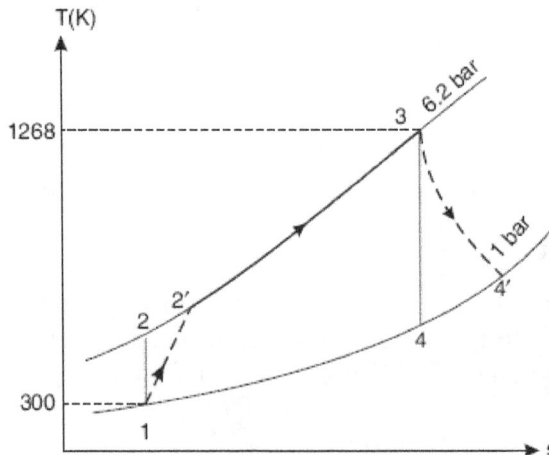

FIGURE 12.20 Sketch for Example 3.

Heat supplied $= \left(m_a + m_f\right) \times c_p (T_3 - T_2') = m_f \times C$

or $\left(1 + \dfrac{m_f}{m_a}\right) \times c_p \left(T_3 - T_2'\right) = \dfrac{m_f}{m_a} \times C$

or $(1 + 0.017) \times 1.005(T_3 - 533.2) = 0.017 \times 44186$

$\therefore T_3 = \dfrac{0.017 \times 44186}{(1 + 0.017) \times 1.005} + 533.2 = 1268 \text{ K}$

For isentropic expression process 3–4:

$\dfrac{T_4}{T_3} = \left(\dfrac{p_4}{p_3}\right)^{\frac{\gamma-1}{\gamma}} = \left(\dfrac{1}{6.2}\right)^{\frac{1.333-1}{1.333}} = 0.634$

$\therefore T_4 = 1268 \times 0.634 = 803.9 \text{ K}$ $\qquad (\because \gamma_g = 1.333 \ \ ......\text{Given})$

Now, $\eta_{\text{turbine}} = \dfrac{T_3 - T_4'}{T_3 - T_4}$

$0.9 = \dfrac{1268 - T_4'}{1268 - 803.9}$

$\therefore T_4' = 1268 - 0.9(1268 - 803.9) = 850.3 \text{ K}$

$W_{\text{compressor}} = c_p \left(T_2' - T_1\right) = 1.005(533.2 - 300) = 234.4 \text{ kJ / kg}$

$W_{\text{turbine}} = c_{pg} \left(T_3 - T_4'\right) = 1.147(1268 - 850.3) = 479.1 \text{ kJ/kg}$

Network $= W_{\text{turbine}} - W_{\text{compressor}}$

$= 479.1 - 234.4 = 244.7 \text{ kJ/kg}$

Heat supplied per kg of air

$= 0.017 \times 44186 = 751.2 \text{ kJ / kg}$

$\therefore$ Thermal efficiency, $\eta_{\text{th.}} = \dfrac{\text{Network}}{\text{Heat supplied}}$

$= \dfrac{244.7}{751.2} = 0.3257 \text{ or } \textbf{32.57\%. (Ans.)}$

Example 4

Find the required air-fuel ratio in a gas turbine whose turbine and compressor efficiencies are 85% and 80%, respectively. The maximum cycle temperature is 875 °C. The working fluid can be taken as air ($c_p = 1.0\ kJ\ /\ kg\ K, \gamma = 1.4$) which enters the compressor at 1 bar and 27 °C. The pressure ratio is 4. The fuel used has a calorific value of 42000 kJ/kg. There is a loss of 10% of calorific value in the combustion chamber.

■ **Solution.**

Given : $\eta_{turbine} = 85\%$; $\eta_{compressor} = 80\%$; $T_3 = 273 + 875 = 1148$ K,

$T_1 = 27 + 273 = 300$ K ; $c_p = 1.0$ kJ / kg K ; $\gamma = 1.4, p_1 = 1$ bar, $p_2 = 4$ bar (Since pressure ratio is 4); $C = 42000$ kJ / kg K, $\eta_{cc} = 90\%$ (since loss in the combustion chamber is 10%)

For isentropic compression 1–2:

$$\frac{T_2}{T_1} = \left(\frac{p_2}{p_1}\right)^{\frac{\gamma-1}{\gamma}} = (4)^{\frac{1.4-1}{1.4}} = 1.486$$

$$\therefore T_2 = 300 \times 1.486 = 445.8\ K$$

$$\eta_{compressor} = \frac{T_2 - T_1}{T_2' - T_1}$$

$$\text{or } 0.8 = \frac{445.8 - 300}{T_2' - 300}$$

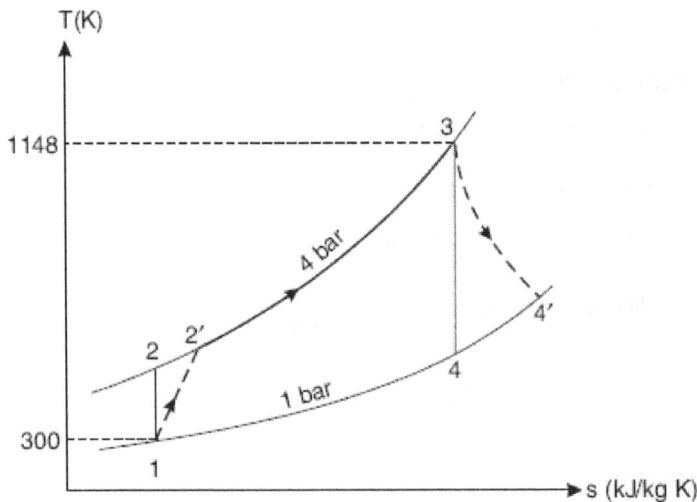

FIGURE 12.21 Sketch for Example 4.

or $T_2' = \dfrac{445.8 - 300}{0.8} + 300 = 482.2$ K

Now, heat supplied by the fuel = heat taken by the burning gases

$$0.9 \times m_f \times C = \left(m_a + m_f\right) \times c_p \times \left(T_3 - T_2'\right)$$

$$\therefore \quad C = \left(\dfrac{m_a + m_f}{m_f}\right) \times \dfrac{c_p \left(T_3 - T_2'\right)}{0.9} = \left(\dfrac{m_a}{m_f} + 1\right) \times \dfrac{c_p \left(T_3 - T_2'\right)}{0.9}$$

or $\quad 42000 = \left(\dfrac{m_a}{m_f} + 1\right) \times \dfrac{1.00\left(1148 - 482.2\right)}{0.9} = 739.78\left(\dfrac{m_a}{m_f} + 1\right)$

$\therefore \quad \dfrac{m_a}{m_f} = \dfrac{42000}{739.78} - 1 = 55.77$ say 56

$\therefore$ **A/F ratio = 56 : 1. (Ans.)**

Example 5

Calculate the thermal efficiency and work ratio of the plant in example 4, assuming that c_p for the combustion process is 1.11 kJ/kg K.

■ **Solution.**

Heat supplied $= c_p \left(T_3 - T_2'\right)$

$= 1.11 \left(883 - 529\right) = 392.9$ kJ / kg

$\eta_{thermal} = \dfrac{\text{Network output}}{\text{Heat supplied}} = \dfrac{48.2}{392.9} = $ **0.1226 or 12.26%. (Ans.)**

Now, *Work ratio* $= \dfrac{\text{Network output}}{\text{Gross work output}} = \dfrac{48.2}{W_{turbine}} = \dfrac{48.2}{290.4} = $ **0.166. (Ans.)**

Example 6

In a constant pressure open cycle gas turbine, air enters at 1 bar and 20 °C and leaves the compressor at 5 bar. Using the following data: temperature of gases entering the turbine = 680°C, pressure loss in the combustion chamber

$= 0.1 \ bar, \eta_{compressor} = 85\%, \eta_{turbine} = 80\%, \eta_{combustion} = 85\%, \ \gamma = 1.4 \ and$ $c_p = 1.024 \ kJ / kg \ K \ for \ air \ and \ gas, \ find$

(i) *The quantity of air circulation if the plant develops 1065 kW.*

(ii) *Heat supplied per kg of air circulation.*

(iii) The thermal efficiency of the cycle.

Mass of the fuel may be neglected.

■ **Solution.**

Given : $p_1 = 1$ bar, $p_2 = 5$ bar, $p_3 = 5 - 0.1 = 4.9$ bar, $p_4 = 1$ bar,

$T_1 = 20 + 273 = 293$ K, $T_3 = 680 + 273 = 953$ K,

$\eta_{compressor} = 85\%$, $\eta_{turbine} = 80\%$, $\eta_{combustion} = 85\%$,

For air and gases : $c_p = 1.024$ kJ / kg K, $\gamma = 1.4$

Power developed by the plant, $P = 1065$ kW.

(i) **The quantity of air circulation, ma:**

For *isentropic compression 1-2,*

$$\frac{T_2}{T_1} = \left(\frac{p_2}{p_1}\right)^{\frac{\gamma-1}{\gamma}} = \left(\frac{5}{1}\right)^{\frac{1.4-1}{1.4}} = 1.584$$

$\therefore \quad T_2 = 293 \times 1.584 = 464$ K

$$\text{Now, } \eta_{compressor} = \frac{T_2 - T_1}{T_2' - T_1} \quad i.e., 0.85 = \frac{464 - 293}{T_2' - 293}$$

$\therefore \quad T_2' = \dfrac{464 - 293}{0.85} + 293 = 494$ K

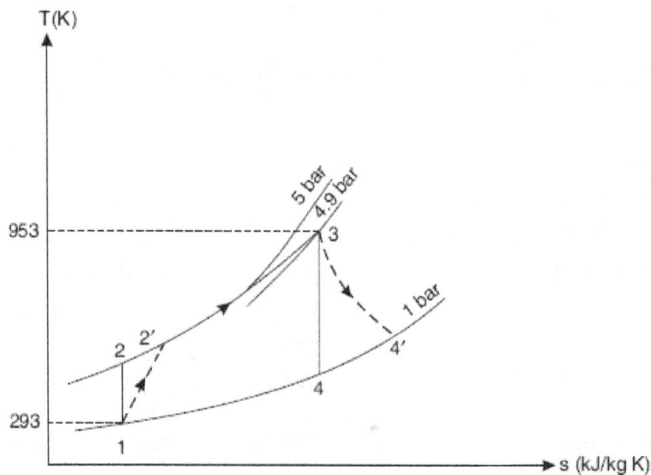

FIGURE 12.22 Sketch for Example 6.

For *isentropic expansion process 3-4,*

$$\frac{T_4}{T_3} = \left(\frac{p_4}{p_3}\right)^{\frac{\gamma-1}{\gamma}} = \left(\frac{1}{4.9}\right)^{\frac{1.4-1}{1.4}} = 0.635$$

$$\therefore \ T_4 = 953 \times 0.635 = 605 \text{ K}$$

$$\text{Now, } \eta_{turbine} = \frac{T_3 - T_4'}{T_3 - T_4}$$

$$0.8 = \frac{953 - T_4'}{953 - 605}$$

$$\therefore \ T_4' = 953 - 0.8(953 - 605) = 674.6 \text{ K}$$

$$W_{compressor} = c_p (T_2' - T_1) = 1.024 \ (494 - 293) = 205.8 \text{ kJ/kg}$$

$$W_{turbine} = c_p (T_3 - T_4') = 1.024 \ (953 - 674.6) = 285.1 \text{ kJ/kg.}$$

$$\therefore \ W_{net} = W_{turbine} - W_{compressor} = 285.1 - 205.8 = 79.3 \text{ kJ/kg of air}$$

If the mass of air is flowing is m_a kg/s, the power developed by the plant is given by

$$P = m_a \times W_{net} \text{ kW}$$

$$1065 = m_a \times 79.3$$

$$\therefore \ m_a = \frac{1065}{79.3} = 13.43 \text{ kg.}$$

i.e., **Quantity of air circulation = 13.43 kg. (Ans.)**

(*ii*) **Heat supplied per kg of air circulation:**

Actual heat supplied per kg of air circulation

$$= \frac{c_p \left(T_3 - T_2'\right)}{\eta_{combustion}} = \frac{1.024(953 - 494)}{0.85} = 552.9 \text{ kJ/kg.}$$

(*iii*) **Thermal efficiency of the cycle, $\eta_{thermal}$:**

$$\eta_{thermal} = \frac{\text{Work output}}{\text{Heat supplied}} = \frac{79.3}{552.9} = 0.1434 \text{ or } \textbf{14.34\%.} \quad \textbf{(Ans.)}$$

Example 7

In a gas turbine the compressor is driven by the high-pressure turbine. The exhaust from the high-pressure turbine goes to a free low-pressure turbine which runs the load. The air flow rate is 20 kg/s and the minimum and maximum temperatures are 300 K and 1000 K, respectively. The compressor pressure ratio is 4. Calculate the pressure ratio of the low-pressure turbine and the temperature of exhaust gases from the unit. The compressor and turbine are isentropic. c_p of air and exhaust gases = 1 kJ/kg K and $\gamma = 1.4$.

■ **Solution.**

Given: $\dot{m}_a = 20$ kg/s ; $T_1 = 300$ K ; $T_3 = 1000$ K, $\dfrac{p_2}{p_1} = 4$; $c_p = 1$ kJ/kg K ; $\gamma = 1.4$,

Pressure ratio of low pressure turbine, $\dfrac{\mathbf{p_4}}{\mathbf{p_5}}$:

Since the compressor is driven by a high-pressure turbine,

$$\therefore \quad \frac{T_2}{T_1} = \left(\frac{p_2}{p_1}\right)^{\frac{\gamma-1}{\gamma}} = (4)^{\frac{0.4}{1.4}} = 1.486$$

or $T_2 = 300 \times 1.486 = 445.8$ K

Also, $\dot{m}_a c_p (T_2 - T_1) = \dot{m}_a c_p (T_3 - T_4)$ (neglecting mass of fuel)

or $T_2 - T_1 = T_3 - T_4$

$445.8 - 300 = 1000 - T_4$, or $T_4 = 854.2$ K

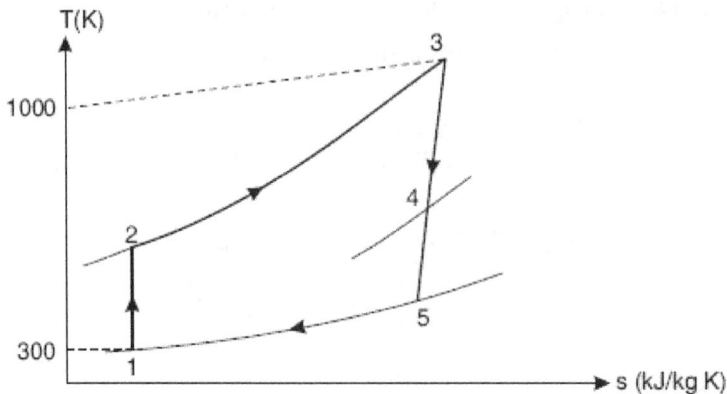

FIGURE 12.23 Sketch for Example 7.

For process 3–4:

$$\frac{T_3}{T_4} = \left(\frac{p_3}{p_4}\right)^{\frac{\gamma-1}{\gamma}}, \text{ or } \frac{p_3}{p_4} = \left(\frac{T_3}{T_4}\right)^{\frac{1.4}{0.4}}$$

or $\dfrac{p_3}{p_4} = \left(\dfrac{1000}{854.2}\right)^{3.5} = 1.736$

Now, $\dfrac{p_3}{p_4} = \dfrac{p_3}{p_5} \times \dfrac{p_5}{p_4} = 4 \times \dfrac{p_5}{p_4} \qquad \left(\because \dfrac{p_3}{p_5} = \dfrac{p_2}{p_1} = 4\right)$

$\therefore \quad \dfrac{p_5}{p_4} = \dfrac{1}{4}\left(\dfrac{p_3}{p_4}\right) = \dfrac{1}{4} \times 1.736 = 0.434$

Hence pressure ratio of low pressure turbine $= \dfrac{p_4}{p_5} = \dfrac{1}{0.434} = \mathbf{2.3.}$ **(Ans.)**

Temperature of the exhaust from the unit T_5:

$$\frac{T_4}{T_5} = \left(\frac{p_4}{p_5}\right)^{\frac{\gamma-1}{\gamma}} = (2.3)^{\frac{1.4-1}{1.4}} = 1.269$$

$\therefore \quad \mathbf{T_5} = \dfrac{T_4}{1.269} = \dfrac{854.2}{1.269} = \mathbf{673\ K.}$

Example 8

In an air-standard regenerative gas turbine cycle the pressure ratio is 5. Air enters the compressor at 1 bar, 300 K and leaves at 490 K. The maximum temperature in the cycle is 1000 K. Calculate the cycle efficiency, given that the efficiency of the regenerator and the adiabatic efficiency of

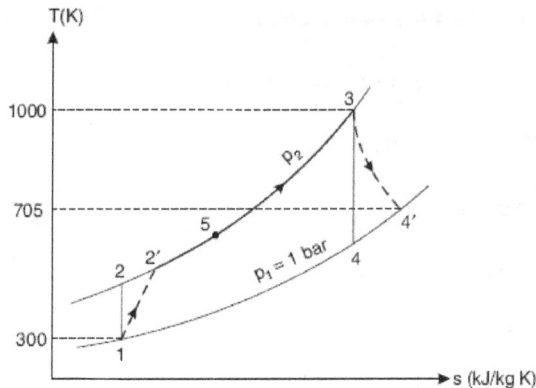

FIGURE 12.24 Sketch for Example 8.

the turbine are each 80%. Assume for air, the ratio of specific heats is 1.4. Also, show the cycle on a T-s diagram.

■ **Solution.**

Given: $p_1 = 1$ bar ; $T_1 = 300$ K, $T_2' = 490$ K ; $T_3 = 1000$ K

$\dfrac{p_2}{p_1} = 5, \eta_{turbine} = 80\%, \varepsilon = 80\% = 0.8 \; ; \gamma = 1.4$

Now, $\dfrac{T_3}{T_4} = \left(\dfrac{p_2}{p_1}\right)^{\frac{\gamma-1}{\gamma}} = (5)^{\frac{1.4-1}{1.4}} = 1.5838$

$\therefore \; T_4 = \dfrac{T_3}{1.5838} = \dfrac{1000}{1.5838} = 631.4$ K

Also, $\eta_{turbine} = \dfrac{T_3 - T_4'}{T_3 - T_4}$

or $\;\; 0.8 = \dfrac{1000 - T_4'}{1000 - 631.4}$

$\therefore \; T_4' = 1000 - 0.8 \, (1000 - 631.4) = 705$ K

Effectiveness of heat exchanger, $\varepsilon = \dfrac{T_5 - T_2'}{T_4' - T_2'}$

or $\;\; 0.8 = \dfrac{T_5 - 490}{705 - 490}$

$\therefore \; T_5 = 0.8(705 - 490) + 490 = 662$ K

Work consumed by compressor $= c_p \left(T_2' - T_1\right)$

$= 1.005(490 - 300) = 190.9$ kJ/kg

Work done by turbine $= c_p \left(T_3 = T_4'\right)$

$= 1.005 \; (1000 - 705) = 296.5$ kJ/kg

Heat supplied $= c_p \left(T_3 - T_5\right)$

$= 1.005 \; (1000 - 662) = 339.7$ kJ/kg

$$\therefore \quad \text{Cycle efficiency,} \eta_{cycle} = \frac{\text{Network}}{\text{Heat supplied}}$$

$$= \frac{\text{Turbine work} - \text{Compressor work}}{\text{Heat supplied}}$$

$$= \frac{296.5 - 190.9}{339.7} = 0.31 \text{ or } \mathbf{31\%}. \quad (\mathbf{Ans.})$$

Example 9

A gas turbine plant consists of two turbines. One is a compressor turbine to drive the compressor and other a power turbine to develop power output; both have their own combustion chambers, which are served by air directly from the compressor. Air enters the compressor at 1 bar and 288 K and is compressed to 8 bar with an isentropic efficiency of 76%. Due to heat added in the combustion chamber, the inlet temperature of gas to both turbines is 900 °C. The isentropic efficiency of the turbines is 86% and the mass flow rate of air at the compressor is 23 kg/s. The calorific value of fuel is 4200 kJ/kg. Calculate the output of the plant and the thermal efficiency if the mechanical efficiency is 95% and the generator efficiency is 96%. Take $c_p = 1.005$ kJ/kg K and $\gamma = 1.4$ for air and $c_{pg} = 1.128$ kJ/kg K and $\gamma = 1.34$ for gases.

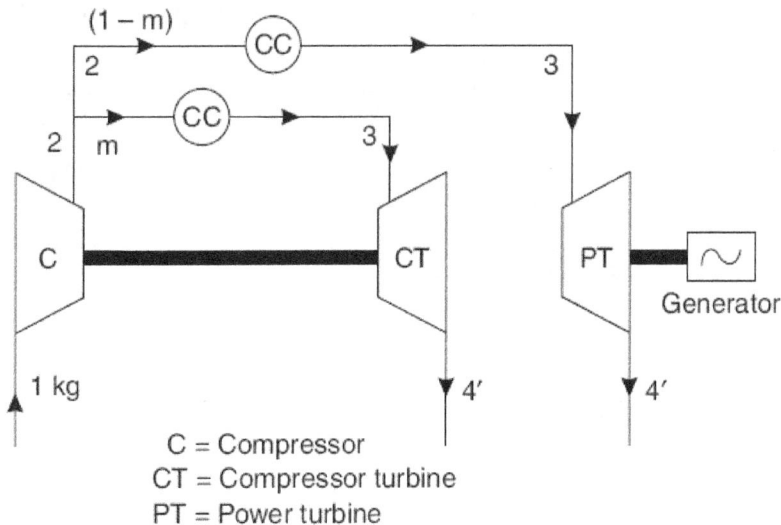

C = Compressor
CT = Compressor turbine
PT = Power turbine

(a)

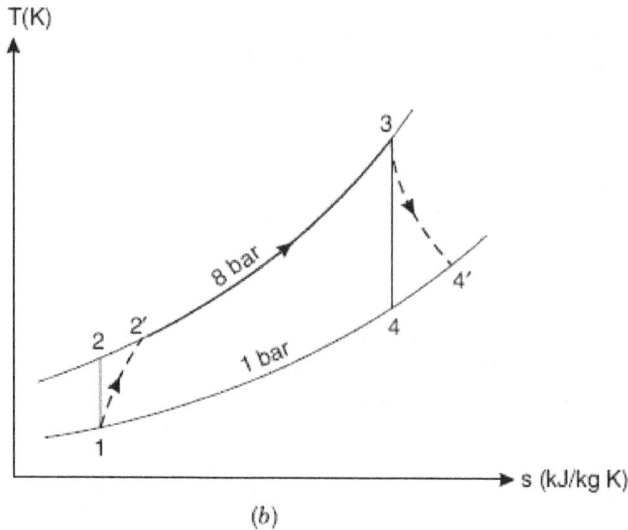

FIGURE 12.25 Sketch for Example 9.

■ **Solution.**

Given : $p_1 = 1$ bar ; $T_1 = 288$ K ; $p_2 = 8$ bar, $\eta_{(isen)} = 76\%$; $T_3 = 900°C$ or 1173 K, $\eta_{T(isen.)} = 86\%$, $m_a = 23$ kg/s ; C.V. = 4200 kJ/kg ; $\eta_{mech} = 95\%$; $\eta_{gen.} = 96\%$; $c_p = 1.005$ kJ/kg ; $\gamma_a = 1.4$; $c_{pg} = 1.128$ kJ/kg K ; $\gamma_g = 1.34$.

The arrangement of the plant and the corresponding *T-s* diagram is shown in Figure 12.25 (a), (b).

Considering *isentropic compression process 1–2*, we have

$$\frac{T_2}{T_1} = \left(\frac{p_2}{p_1}\right)^{\frac{\gamma-1}{\gamma}} = \left(\frac{8}{1}\right)^{\frac{1.4-1}{14}} = 1.811$$

$$\therefore \quad T_2 = 288 \times 1.811 = 521.6 \text{ K}$$

Also, $\eta_{C(isen.)} = \dfrac{T_2 - T_1}{T_2' - T_1}$

or $0.76 = \dfrac{521.6 - 288}{T_2' - 288}$

or $T_2' = \dfrac{521.6 - 288}{0.76} + 288 = 595.4 \text{ K}$

Considering *isentropic expansion process 3–4*, we have

$$\frac{T_4}{T_3} = \left(\frac{p_4}{p_3}\right)^{\frac{\gamma-1}{\gamma}} = \left(\frac{1}{8}\right)^{\frac{1.34-1}{1.34}} = 0.59$$

$$\therefore \quad T_4 = 1173 \times 0.59 = 692.1 \text{ K}$$

Also, $\eta_{T(\text{isen})} = \dfrac{T_3 - T_4'}{T_3 - T_4}$

or $\quad 0.86 = \dfrac{1173 - T_4'}{1173 - 692.1}$

$$\therefore \quad T_4' = 1173 - 0.86\,(1173 - 692.1) = 759.4 \text{ K}$$

Consider 1 kg of air flow through the compressor

$$W_{compressor} = c_p\left(T_2' - T_1'\right) = 1.005(595.4 - 288) = 308.9 \text{ kJ}$$

This is equal to work of the compressor turbine.

$$\therefore 308.9 = m_1 \times c_{pg}\left(T_3 - T_4'\right), \quad \text{neglecting fuel mass}$$

or $\quad m_1 = \dfrac{308.9}{1.128\,(1173 - 759.4)} = 0.662 \text{ kg}$

and flow through the power turbine $= 1 - m = 1 - 0.662 = 0.338$ kg

$$\therefore \quad W_{PT} = (1-m) \times c_{pg}\left(T_3 - T_4'\right)$$

$$= 0.338 \times 1.128\,(1173 - 759.4) = 157.7 \text{ kJ}$$

$\therefore$ **Power Output** $= 23 \times 157.7 \times \eta_{\text{mech.}} \times \eta_{\text{gen.}}$

$$= 23 \times 157.7 \times 0.95 \times 0.96 = \textbf{3307.9 kJ. (Ans.)}$$

$$Q_{\text{input}} = c_{pg}T_3 - c_{pa}T_2'$$

$$= 1.128 \times 1173 - 1.005 \times 595.4 = 724.7 \text{ kJ/kg of air}$$

Thermal efficiency, $\eta_{\text{th}} = \dfrac{157.7}{724.7} \times 100 = \textbf{21.76\%. (Ans.)}$

Example 10

Air is drawn in a gas turbine unit at 15 °C and 1.01 bar and pressure ratio is 7 : 1. The compressor is driven by the H.P. turbine and the L.P. turbine drives a separate power shaft. The isentropic efficiencies of the compressor, and the H.P. and L.P. turbines, are 0.82, 0.85, and 0.85, respectively. If the maximum cycle temperature is 610 °C, calculate:

(i) The pressure and temperature of the gases entering the power turbine.

(ii) The net power developed by the unit per kg/s mass flow.

(iii) The work ratio.

(iv) The thermal efficiency of the unit.

Neglect the mass of fuel and assume the following:

For the compression process c_{pa} = 1.005 kJ/kg K and γ = 1.4

For the combustion and expansion processes; c_{pg} = 1.15 kJ/kg and γ = 1.333.

■ **Solution.**

Given : $T_1 = 15 + 273 = 288$ K, $p_1 = 1.01$ bar, Pressure ratio $= \dfrac{p_2}{p_1} = 7,$

$\eta_{compressor} = 0.82$, $\eta_{turbine\ (H.P.)} = 0.85$, $\eta_{turbine\ (L.P.)} = 0.85,$

Maximum cycle temperature, $T_3 = 610 + 273 = 883$ K

(*i*) **Pressure and temperature of the gases entering the power turbine, p4′ and T4 ′:**

Considering *isentropic compression 1–2*, we have

$$\frac{T_2}{T_1} = \left(\frac{p_{2\gamma}}{p_1}\right)^{\frac{\gamma-1}{\gamma}} = (7)^{\frac{1.4-1}{1.4}} = 1.745$$

$\therefore\quad T_2 = 288 \times 1.745 = 502.5$ K

Also $\eta_{compressor} = \dfrac{T_2 - T_1}{T_2' - T_1}$

$0.82 = \dfrac{502.5 - 288}{T_2' - 288}$

$\therefore\quad T_2' = \dfrac{502.5 - 288}{0.82} + 288 = 549.6$ K

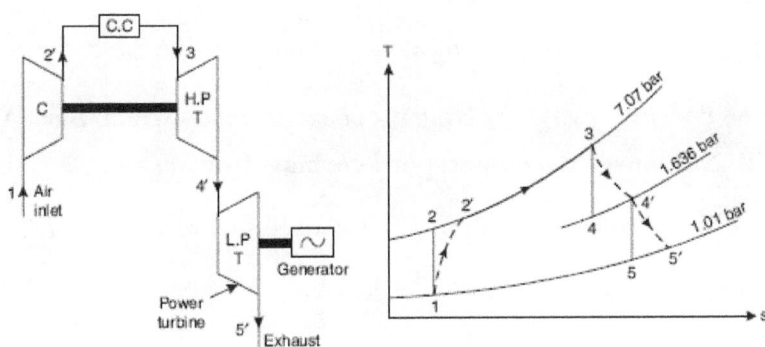

FIGURE 12.26 Sketch for Example 10.

$$W_{compressor} = c_{pa}\left(T_2' - T_1\right) = 1.005 \times \left(549.6 - 288\right) = 262.9 \text{ kJ/kg}$$

Now, the *work output of H.P. turbine = Work input to compressor*

$$\therefore \quad c_{pg}\left(T_3 - T_4'\right) = 262.9$$

i.e., $\quad 1.15\left(883 - T_4'\right) = 262.9\,T$

$$\therefore \quad T_4' = 883 - \frac{262.9}{1.15} = 654.4 \text{ K}$$

i.e., Temperature of gases entering the power turbine = **654.4 K. (Ans.)**

Again, for H.P. turbine:

$$\eta_{turbine} = \frac{T_3 - T_4'}{T_3 - T_4} \text{ i.e., } 0.85 = \frac{883 - 654.4}{883 - T_4}$$

$$\therefore \quad T_4 = 883 - \left(\frac{883 - 654.4}{0.85}\right) = 614 \text{ K}$$

Now, considering the *isentropic expansion process 3-4,* we have

$$\frac{T_3}{T_4} = \left(\frac{p_3}{p_4}\right)^{\frac{\gamma-1}{\gamma}}$$

or $\quad \dfrac{p_3}{p_4} = \left(\dfrac{T_3}{T_4}\right)^{\frac{\gamma-1}{\gamma}} = \left(\dfrac{883}{614}\right)^{\frac{1.33}{0.33}} = 4.32$

i.e.,
$$p_4 = \frac{p_3}{4.32} = \frac{7.07}{4.32} = 1.636 \text{ bar}$$

i.e., Pressure of gases entering the power turbine = **1.636 bar. (Ans.)**

(*ii*) **Net power developed per kg/s mass flow, P:**

To find the power output it is now necessary to calculate T_5'.

The pressure ratio, $\dfrac{p_4}{p_5}$, is given by $\dfrac{p_4}{p_3} \times \dfrac{p_3}{p_5}$

i.e.,
$$\frac{p_4}{p_5} = \frac{p_4}{p_3} \times \frac{p_2}{p_1} = \frac{7}{4.32} = 1.62 \qquad (\because \; p_2 = p_3 \text{ and } p_5 = p_1)$$

Then,
$$\frac{T_4'}{T_5} = \left(\frac{p_4}{p_5}\right)^{\frac{\gamma-1}{\gamma}} = (1.62)^{\frac{0.33}{1.33}} = 1.127$$

$\therefore \quad T_5 = \dfrac{T_4'}{1.127} = \dfrac{654.4}{1.127} = 580.6 \text{ K.}$

Again, for the L.P. turbine

$$\eta_{turbine} = \frac{T_4' - T_5'}{T_4' - T_5'}$$

i.e.,
$$0.85 = \frac{654.4 - T_5'}{654.4 - 580.6}$$

$\therefore \quad T_5' = 654.4 - 0.85\,(654.4 - 580.6) = 591.7 \text{ K}$

$W_{\text{L.P. } turbine} = c_{pg}\left(T_4' - T_5'\right) = 1.15\,(654.4 - 591.7) = 72.1 \text{ kJ/kg}$

Hence *net power output* (per kg/s mass flow) = **72.1 kW.** **(Ans.)**

(*iii*) **Work ratio:**

$$Work\ ratio = \frac{\text{Network output}}{\text{Gross work output}} = \frac{72.1}{72.1 + 262.9} = \textbf{0.215. (Ans.)}$$

(*iv*) **Thermal efficiency of the unit,** η_{thermal}:

Heat supplied $= c_{pg}\left(T_3 - T_2'\right) = 1.15\,(883 - 549.6) = 383.4 \text{ kJ/kg}$

$$\therefore \quad \eta_{thermal} = \frac{\text{Network output}}{\text{Heat suppiled}} = \frac{72.1}{383.4} = \textbf{0.188 or 18.8\%} \quad (\textbf{Ans.})$$

Example 11

The pressure ratio of an open-cycle gas turbine power plant is 5.6. Air is taken at 30 °C and 1 bar. The compression is carried out in two stages with perfect intercooling in between. The maximum temperature of the cycle is limited to 700 °C. Assuming the isentropic efficiency of each compressor stage as 85% and that of the turbine as 90%, determine the power developed and efficiency of the power plant, if the air flow is 1.2 kg/s. The mass of fuel may be neglected, and it may be assumed that Cp=1.02 kJ/kg.K and γ = 1.41.

■ **Solution.**

Pressure ratio of the open-cycle gas turbine = 5.6

Temperature of intake air, T_1 = 30 + 273 = 303 K

Pressure of intake air, p_1 = 1 bar

Maximum temperature of the cycle, T_5 = 700 + 273 = 973 K

Isentropic efficiency of each compressor, $\eta_{comp.}$ =85%

Isentropic efficiency of turbine. $\eta_{turbine}$ =90%

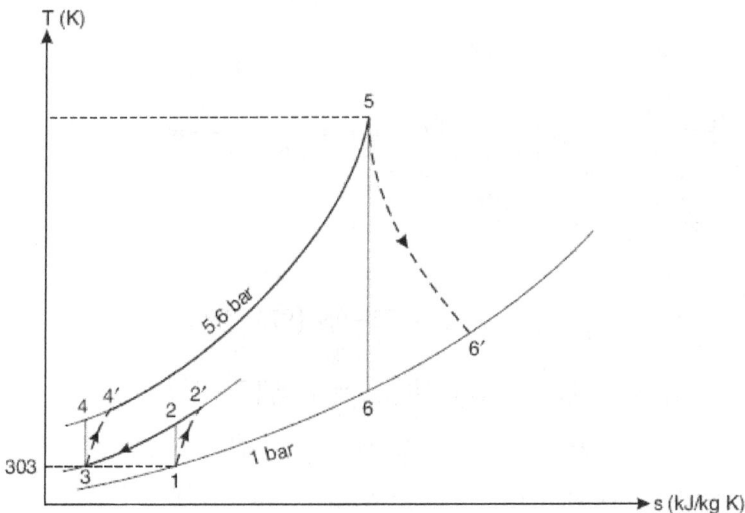

FIGURE 12.27 Sketch for Example 11.

Rate of air-flow $\dot{m}_a = 1.2\,\text{kg/s}$

$c_p = 1.02$ kJ/kg K and $\gamma = 1.41$.

Power developed and efficiency of the power plant:

Assuming that the pressure ratio in each stage is the same, we have

$$\frac{p_2}{p_1} = \frac{p_4}{p_3} = \sqrt{\frac{p_4}{p_1}} = \sqrt{5.6} = 2.366$$

Since the pressure ratio and the isentropic efficiency of each compressor is the same, then the *work input required for each compressor is the same,* since both the compressors have the same inlet temperature (*perfect intercooling*), *that is, $T_1 = T_3$ and $T_2' = T_4'$:*

Now, $\dfrac{T_2}{T_1} = \left(\dfrac{p_2}{p_1}\right)^{\frac{\gamma-1}{\gamma}} = (2.366)^{\frac{1.41-1}{1.41}} = 1.2846$ or $T_2 = 303 \times 1.2846 = 389.23$ K

Also. $\eta_{comp.} = \dfrac{T_2 - T_1}{T_2' - T_1}$ or $0.85 = \dfrac{389.23 - 303}{T_2' - 303}$

or $\quad T_2' = \dfrac{389.23 - 303}{0.85} + 303 = 404.44$ K

Work input to 2-stage compressor, $W_{comp.} = 2 \times m \times c_p \left(T_2' - T_1\right)$

$= 2 \times 1.2 \times 1.02 (404.44 - 303) = 248.32$ kJ/s

For the turbine, we have

$\dfrac{T_5}{T_6} = \left(\dfrac{p_5}{p_6}\right)^{\frac{\gamma-1}{\gamma}} = (5.6)^{\frac{1.41-1}{1.41}} = 1.65$ or $T_6 = \dfrac{T_5}{1.65} = \dfrac{973}{1.65} = 589.7$ K

Also, $\eta_{turbine} = \dfrac{T_5 - T_6'}{T_5 - T_6}$

or $\quad 0.9 = \dfrac{973 - T_6'}{973 - 589.7}$ or $T_6' = 973 - 0.9\,(973 - 589.7) = 628$ K

∴ *Work output of thrbin,* $W_{turbine} = m \times c_p \left(T_5 - T_6'\right)$

$= 1.2 \times 1.02 (973 - 628) = 422.28$ kJ/s

Network output, $W_{net} = W_{turbine} - W_{comp.}$

$= 422.28 - 248.32 = 173.96$ kJ/s or kW

Heat *power developed* $= 173.96\,\text{kW}.$ (**Ans.**)

Heat supplied, $\quad Q_s = m \times c_p \times \left(T_5 - T_4' \right)$

$= 1.2 \times 1.02 \times (973 - 404.44) = 695.92\ \text{kJ/s}$

$\therefore$ Power plant efficiency, $\quad \eta_{th} = \dfrac{W_{net}}{Q_s} = \dfrac{173.96}{695.92} = 0.25$ or **25%**. (**Ans.**)

Example 12

(a) Why are the back work ratios relatively high in gas turbine plants compared to those of steam power plants?

(b) In a gas turbine plant compression is carried out in two stages with perfect intercooling and expansion in one stage of the turbine. If the maximum temperature $(T_{max}\ \text{K})$ and minimum temperature $(T_{min}\ \text{K})$ in the cycle remain constant, show that for maximum specific output of the plant, the optimum overall pressure ratio is given by

$$r_{opt} = \left(\eta_T . \eta_c . \frac{T_{max}}{T_{max}} \right)^{\frac{2\gamma}{3(\gamma-1)}}$$

where γ = *Adiabatic index;* η_T = *Isentropic efficiency of the turbine.* η_c is isentropic efficiency of the compressor.

■ Solution.

(a) **Back work ratio** may be defined as *the ratio of negative work to the turbine work in a power plant.* In gas turbine plants, air is compressed from the turbine exhaust pressure to the combustion chamber pressure. This

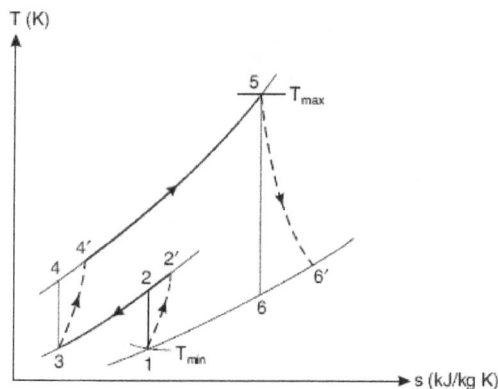

FIGURE 12.28 Sketch for Example 12.

work is given by $-\int vdp$. As the specific volume of air is very high (even in closed cycle gas turbine plants), the compressor work required is very high, and a *bulky compressor is required*. In steam power plants, the turbine exhaust is changed to a liquid phase in the condenser. The pressure of condensate is raised to boiler pressure by a condensate extraction pump and a boiler feed pump in series; since the specific volume of water is very small as compared to that of the air, the pump work $(-\int vdp)$ is also very small. From the above reasons, the back work ratio

$$= \frac{-\int vdp}{\text{Turbine work}}$$

for gas *turbine plants is relatively high compared to that for steam power plants.*

(*b*) Assuming an optimuvm pressure ratio in each stage of the compressors $\sqrt{r}$,

$$\frac{T_2}{T_1} = \left(\frac{p_2}{p_1}\right)^{\frac{\gamma-1}{\gamma}}$$

or
$$T_2 = T_{\min} \times (r)^{\frac{\gamma-1}{2\gamma}}$$

$$W_{compressor} = 2\left[c_p\left(T_2' - T_1\right)\right] \text{ for both compressors}$$

$$= 2c_p \frac{T_2 - T_1}{\eta c} = \frac{2c_p}{\eta c} T_{\min} \left[(r)^{\frac{\gamma-1}{2\gamma}} - 1\right], \text{ as } T_1 = T_{\min}$$

Also,
$$\frac{T_5}{T_6} = \left(\frac{p_2}{p_1}\right)^{\frac{\gamma-1}{\gamma}} = (r)^{\frac{\gamma-1}{\gamma}}$$

$$\therefore \qquad T_6 = \frac{T_5}{(r)^{\gamma-1/\gamma}} = \frac{T_{\max}}{(r)^{\gamma-1/\gamma}}, \text{ as } T_5 = T_{\max}$$

$$W_{turbine} = c_p\left(T_5 - T_6'\right) = c_p\left[T_{\max} - \frac{T_{\max}}{(r)^{\gamma-1/\gamma}}\right]\eta_T, \text{ as } \eta_T = \frac{T_5 - T_6'}{T_5 - T_6}$$

$$= c_p T_{\max}\left[1 - \frac{1}{(r)^{\gamma-1/\gamma}}\right]\eta_T$$

$$W_{net} = W_{turbine} - W_{compressor}$$

$$= c_p \, \eta_T \, T_{max} \left[1 - \frac{1}{(r)^{\gamma-1/\gamma}} \right] - \frac{2c_p}{\eta c} T_{min} \left[(r)^{\frac{\gamma-1}{2\gamma}} - 1 \right]$$

For maximum work output,

$$\frac{dW_{net}}{dr} = 0$$

or $\quad -c_p \, \eta_T \, T_{max} \left(-\frac{\gamma-1}{\gamma} \right) (r)^{-\left(\frac{\gamma-1}{\gamma}\right)-1} - \frac{2c_p}{\eta c} T_{min} \left(\frac{\gamma-1}{2\gamma} \right) (r)^{\frac{\gamma-1}{2\gamma}-1} = 0$

or $\quad \eta_T \, \eta c \, \dfrac{T_{max}}{T_{min}} = (r)^{3(\gamma-1)/2\gamma}$, on simplification.

Hence, the optimum pressure ratio is

$$r_{opt} = \left[\eta_T . \eta c . \frac{T_{max}}{T_{min}} \right]^{\frac{2\gamma}{3(\gamma-1)}} ......\textbf{Proved.}$$

Example 13

In a gas turbine the compressor takes in air at a temperature of 15 °C and compresses it to four times the initial pressure with an isentropic efficiency of 82%. The air is then passed through a heat exchanger heated by the turbine exhaust before reaching the combustion chamber. In the heat exchanger 78% of the available heat is given to the air. The maximum temperature after constant pressure combustion is 600 °C, and the efficiency of the turbine is 70%. Neglecting all losses except those mentioned, and

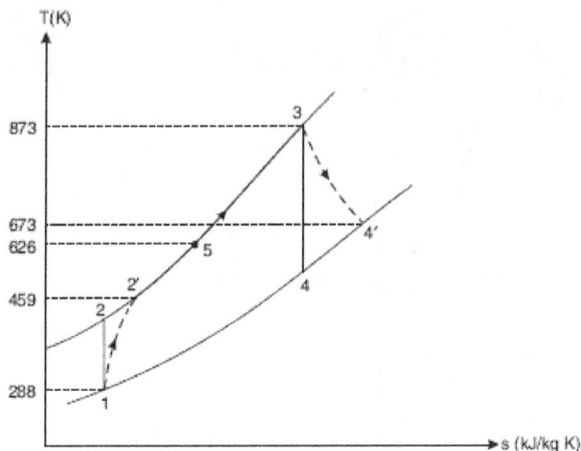

FIGURE 12.29 Sketch for Example 13.

assuming the working fluid throughout the cycle to have the characteristic of air, find the efficiency of the cycle.

Assume $R = 0.287$ kJ/kg K and $\gamma = 1.4$ for air and constant specific heats throughout.

■ **Solution.**

Given: $T_1 = 15 + 273 = 288$ K, Pressure ratio, $\dfrac{p_2}{p_1} = \dfrac{p_3}{p_4} = 4$, $\eta_{compressor} = 82\%$.

Effectiveness of the heat exchanger, $\varepsilon = 0.78$,

$\eta_{turbine} = 70\%$, Maximum temperature, $T_3 = 600 + 273 = 873$ K.

Efficiency of the cycle η cycle:

Considering the *isentropic compression 1–2*, we have

$$\frac{T_2}{T_1} = \left(\frac{p_2}{p_1}\right)^{\frac{\gamma-1}{\gamma}} = (4)^{\frac{1.4-1}{1.4}} = 1.486$$

$\therefore \quad T_2 = 288 \times 1.486 = 428$ K

Now, $\quad \eta_{compressor} = \dfrac{T_2 - T_1}{T_2' - T_1}$

i.e., $\quad 0.82 = \dfrac{428 - 288}{T_2' - 288}$

$\therefore \quad T_2' = \dfrac{428 - 288}{0.82} + 288 = 459$ K

Considering the *isentropic expansion process 3–4*, we have

$$\frac{T_3}{T_4} = \left(\frac{p_3}{p_4}\right)^{\frac{\gamma-1}{\gamma}} = (4)^{\frac{1.4-1}{1.4}} = 1.486$$

$\therefore \quad T_4 = \dfrac{T_3}{1.486} = \dfrac{873}{1.486} = 587.5$ K.

Again, $\quad \eta_{turbine} = \dfrac{T_3 - T_4'}{T_3 - T_4} = \dfrac{873 - T_4'}{873 - 587.5}$

i.e., $\quad 0.70 = \dfrac{873 - T_4'}{873 - 587.5}$

$\therefore \quad T_4' = 873 - 0.7\,(873 - 587.5) = 673$ K

$W_{compressor} = c_p\left(T_2' - T_1\right)$

But $c_p = R \times \dfrac{\gamma}{\gamma-1} = 0.287 \times \dfrac{1.4}{1.4-1} = 1.0045$ kJ/kg K

$\therefore \quad W_{compressor} = 1.0045\,(459-288) = 171.7$ kJ/kg

$W_{turbine} = c_p\left(T_3 - T_4'\right) = 1.0045\,(873-673) = 200.9$ kJ/kg.

$\therefore \quad$ Network $= W_{turbine} - W_{compressor} = 200.9 - 171.7 = 29.2$ kJ/kg.

Effectiveness for heat exchanger, $\varepsilon = \dfrac{T_5 - T_2'}{T_4' - T_2'}$

i.e., $\quad 0.78 = \dfrac{T_5 - 459}{673 - 459}$

$\therefore \quad T_5 = (673-459)\times 0.78 + 459 = 626$ K

$\therefore \quad$ Heat supplied by fuel per kg

$= c_p\,(T_3 - T_5) = 1.0045\,(873-626) = 248.1$ kJ/kg

$\therefore \quad \eta_{cycle} = \dfrac{\text{Network done}}{\text{Heat supplied by the fuel}} = \dfrac{29.2}{248.1} = \mathbf{0.117 \text{ or } 11.7\% \text{ (Ans.)}}$

Example 14

A gas turbine employs a heat exchanger with a thermal ratio of 72%. The turbine operates between the pressures of 1.01 bar and 4.04 bar and the ambient temperature is 20 °C. Isentropic efficiencies of the compressor and the turbine are 80% and 85% respectively. The pressure drop on each side of the heat exchanger is 0.05 bar and in the combustion chamber 0.14 bar. Assume combustion efficiency to be unity and calorific value of the fuel to be 41800 kJ/kg.

Calculate the increase in efficiency due to the heat exchanger over that for a simple cycle.

Assume c_p is constant throughout and is equal to 1.024 kJ/kg K, and assume $\gamma = 1.4$.

For the simple cycle the air-fuel ratio is 90 : 1, and for the heat exchange cycle the turbine entry temperature is the same as for a simple cycle.

■ **Solution.**

Simple Cycle.

$$\frac{T_2}{T_1} = \left(\frac{p_2}{p_1}\right)^{\frac{\gamma-1}{\gamma}} = \left(\frac{4.40}{1.10}\right)^{\frac{1.4-1}{1.4}} = 1.486$$

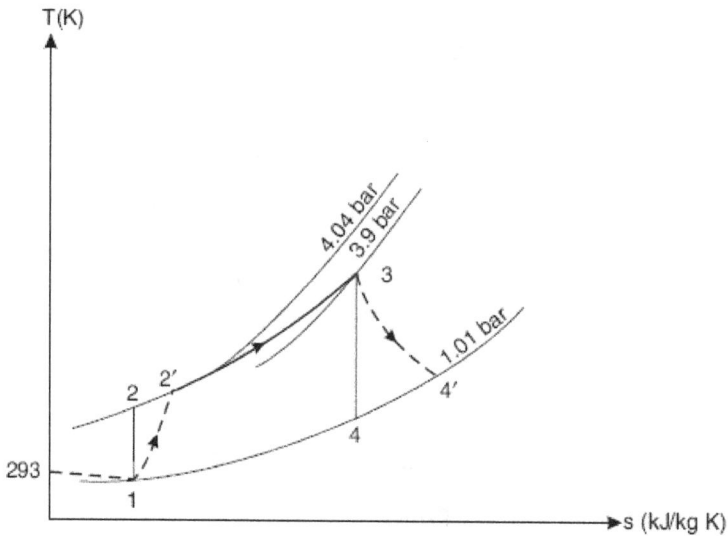

FIGURE 12.30 Sketch for Example 14.

$$\therefore \quad T_2 = 293 \times 1.486 = 435.4$$

$$\text{Also, } \eta_{compressor} = \frac{T_2 - T_1}{T_2' - T_1}$$

$$0.8 = \frac{435.4 - 293}{T_2' - 293}$$

$$\therefore \quad T_2' = \frac{435.4 - 293}{0.8} + 293 = 471 \text{ K}$$

$$\text{Now, } \quad m_f \times C = \left(m_a + m_f\right) \times c_p \times \left(T_3 - T_2'\right)$$

$$\left[m_a = \text{mass of air}, \ m_f = \text{mass of fuel}\right]$$

$$\therefore \quad T_3 = \frac{m_f \times C}{c_p\left(m_a + m_f\right)} + T_2' = \frac{1 \times 41800}{1.024\left(90+1\right)} + 471 = 919.5 \text{ K}$$

$$\text{Also,} \qquad \frac{T_4}{T_3} = \left(\frac{p_4}{p_3}\right)^{\frac{\gamma-1}{\gamma}}$$

$$\text{or} \qquad T_4 = T_3 \times \left(\frac{p_4}{p_3}\right)^{\frac{\gamma-1}{\gamma}} = 919.5 \times \left(\frac{1.01}{3.9}\right)^{\frac{1.4-1}{1.4}} = 625 \text{ K}$$

Again, $\eta_{turbine} = \dfrac{T_3 - T_4'}{T_3 - T_4}$

$\therefore \quad 0.85 = \dfrac{919.5 - T_4'}{919.5 - 625}$

$\therefore \quad T_4' = 919.5 - 0.85\left(919.5 - 625\right) = 669 \text{ K}$

$\eta_{thermal} = \dfrac{\left(T_3 - T_4'\right) - \left(T_2' - T_1\right)}{\left(T_3 - T_2'\right)}$.

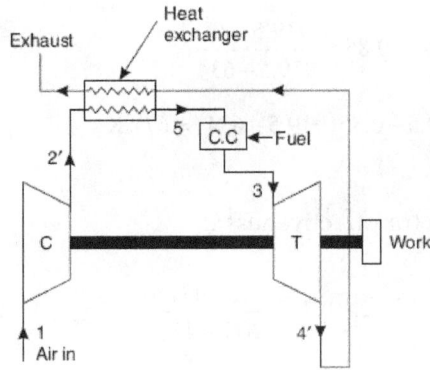

(a)

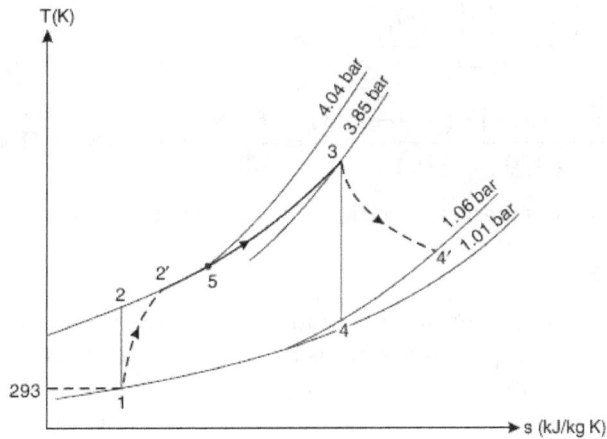

(b)

FIGURE 12.31 Sketch for Example 14, heat exchanger cycle.

$$= \frac{(919.5-669)-(471-293)}{(919.5-471)} = \frac{72.5}{448.5} = 0.1616 \text{ or } \mathbf{16.16\%}. \quad (\mathbf{Ans.})$$

Heat Exchanger Cycle.

$T_2' = 471 \text{ K} \left(\text{as for simple cycle}\right); T_3 = 919.5 \text{ K}\left(\text{as for simple cycle}\right)$

To find T$_4$':

$p_3 = 4.04 - 0.14 - 0.05 = 3.85 \text{ bar}; p_4 = 1.01 + 0.05 = 1.06 \text{ bar}$

$$\therefore \quad \frac{T_4}{T_3} = \left(\frac{p_4}{p_3}\right)^{\frac{\gamma-1}{\gamma}} = \left(\frac{1.06}{3.85}\right)^{\frac{1.4-1}{1.4}} = 0.69$$

i.e., $\quad T_4 = 919.5 \times 0.69 = 634 \text{ K}$

$$\eta_{turbine} = \frac{T_3 - T_4'}{T_3 - T_4}; \quad 0.85 = \frac{919.5 - T_4'}{919.5 - 634}$$

$$\therefore \quad T_4' = 919.5 - 0.85(919.5 - 634) = 677 \text{ K}$$

To find T$_5$:

Thermal ratio (or effectiveness),

$$\varepsilon = \frac{T_5 - T_2'}{T_4' - T_2'} \qquad \therefore \quad 0.72 = \frac{T_5 - 471}{677 - 471}$$

$$\therefore \quad T_5 = 0.72(677 - 471) + 471 = 619 \text{ K}$$

$$\eta_{thermal} = \frac{\left(T_3 - T_4'\right)-\left(T_2' - T_1\right)}{(T_3 - T_5)}$$

$$= \frac{(919.5 - 677)-(471 - 293)}{(919.5 - 619)} = \frac{64.5}{300.5} = 0.2146 \text{ or } \mathbf{21.46\%}$$

$\therefore$ **Increase in thermal efficiency** $= 21.46 - 16.16 = \mathbf{5.3\%}. \quad (\mathbf{Ans.})$

Example 15

A 4500 kW gas turbine generating set operates with two compressor stages; the overall pressure ratio is 9 : 1. A high pressure turbine is used to drive the compressors, and a low-pressure turbine drives the generator. The temperature of the gases at entry to the high-pressure turbine is 625 °C and the gases are reheated to 625 °C after expansion in the first turbine. The exhaust gases leaving the low-pressure turbine are passed through a heat exchanger to heat air leaving the high-pressure stage compressor.

The compressors have equal pressure ratios and intercooling is complete between the stages. The air inlet temperature to the unit is 20 °C. The isentropic efficiency of each compressor stage is 0.8, the isentropic efficiency of each turbine stage is 0.85, and the heat exchanger thermal ratio is 0.8. A mechanical efficiency of 95% can be assumed for both the power shaft and the compressor turbine shaft. Neglecting all pressure losses and changes in kinetic energy calculate:

(i) The thermal efficiency;   (ii) Work ratio of the plant;

(iii) The mass flow in kg/s.

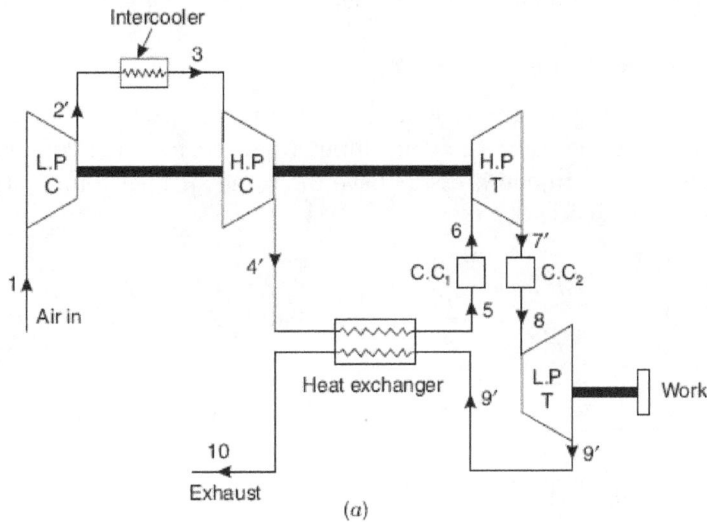

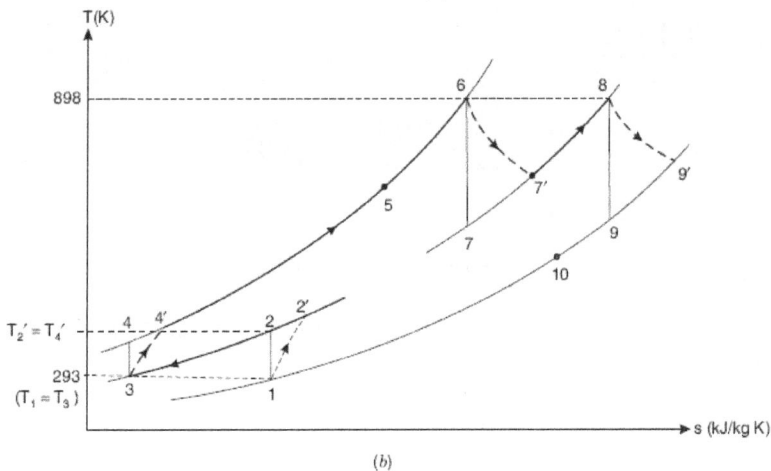

FIGURE 12.32 Sketch for Example 15.

Neglect the mass of the fuel and assume the following:

For air : c_{pa} = 1.005 kJ/kg K and γ = 1.4

For gases in the combustion chamber and in turbines and the heat exchanger, c_{pg} = 1.15 kJ/kg K and and γ = 1.333

■ **Solution.**

Given : $T_1 = 20 + 273 + 293$ K, $T_6 = T_8 = 625 + 273 = 898$ K

Efficiency of each compressor stage $= 0.8$

Efficiency of each turbine stage $= 0.85$, $\eta_{mech} = 0.95$, $\varepsilon = 0.8$

(*i*) **Thermal efficiency, η_{thermal}:**

Since the pressure ratio and the isentropic efficiency of each compressor is the same, then the work input required for each compressor is the same, since both compressors have the same air inlet temperature, that is, $T_1 = T_3$ and $T_2' = T_4'$.

Also,
$$\frac{T_2}{T_1} = \left(\frac{p_2}{p_1}\right)^{\frac{\gamma-1}{\gamma}} \text{ and } \frac{p_2}{p_1} = \sqrt{9} = 3$$

$\therefore$
$$T_2 = (20 + 273) \times (3)^{\frac{1.4-1}{1.4}} = 401 \text{ K}$$

Now, $\eta_{compressor}$ (L.P.) $= \dfrac{T_2 - T_1}{T_2' - T_1}$

$$0.8 = \frac{401 - 293}{T_2' - 293}$$

i.e.,
$$T_2' = \frac{401 - 298}{0.8} + 293 = 428 \text{ K}$$

Work input per compressor stage

$$= c_{pa}\left(T_2' - T_1\right) = 1.005(428 - 293) = 135.6 \; kJ \, / \, kg$$

The H.P. turbine is required to drive both compressors and to overcome mechanical friction.

i.e., work output of H.P. turbine $= \dfrac{2 \times 135.6}{0.95} = 285.5 \text{ kJ/kg}$

$\therefore$
$$c_{pa}\left(T_6 - T_7'\right) = 285.5$$

i.e., $1.15(898 - T_7{}') = 285.5$

$\therefore \qquad\qquad T_7{}' = 898 - \dfrac{285.5}{1.15} = 650 \text{ K}$

Now, $\qquad \eta_{turbine\,(\text{H.P.})} = \dfrac{T_6 - T_7{}'}{T_6 - T_7} ; \; 0.85 = \dfrac{898 - 650}{898 - T_7}$

$\therefore \qquad\qquad T_7 = 898 - \left(\dfrac{898 - 650}{0.85}\right) = 606 \text{ K}$

Also, $\qquad\qquad \dfrac{T_6}{T_7} = \left(\dfrac{p_6}{p_7}\right)^{\frac{\gamma-1}{\gamma}}$

or $\qquad\qquad \dfrac{p_6}{p_7} = \left(\dfrac{T_6}{T_7}\right)^{\frac{\gamma}{\gamma-1}} = \left(\dfrac{898}{606}\right)^{\frac{1.333}{0.333}} = 4.82$

Then, $\qquad \dfrac{p_8}{p_9} = \dfrac{9}{4.82} = 1.86$

Again, $\qquad\qquad \dfrac{T_8}{T_9} = \left(\dfrac{p_8}{p_9}\right)^{\frac{\gamma-1}{\gamma}} = (1.86)^{\frac{1.333-1}{1.333}} = 1.16$

$\therefore \qquad\qquad T_9 = \dfrac{T_8}{1.16} = \dfrac{898}{1.16} = 774 \text{ K}$

Also, $\qquad \eta_{turbine(\text{L.P.})} = \dfrac{T_8 - T_9{}'}{T_8 - T_9} ; \; 0.85 = \dfrac{898 - T_9{}'}{898 - 774}$

$\therefore \qquad\qquad T_9{}' = 898 - 0.85(898 - 774) = 792.6 \text{ K}$

$\therefore$ Network output $= c_{pg}(T_8 - T_9{}') \times 0.95$

$= 1.15\,(898 - 792.6) \times 0.95 = 115.15 \text{ kJ/kg}$

Thermal ratio or effectiveness of heat exchanger,

$\varepsilon = \dfrac{T_5 - T_4{}'}{T_9{}' - T_4{}'} = \dfrac{T_5 - 428}{792.6 - 428}$

i.e., $\quad 0.8 = \dfrac{T_5 - 428}{792.6 - 428}$

$\therefore \quad T_5 = 0.8\,(792.6 - 428) + 428 = 719.7 \text{ K}$

Now, Heat supplied $= c_{pg}(T_6 - T_5) + c_{pg}(T_8 - T_7{}')$

$$= 1.15 \, (898 - 719.7) + 1.15 \, (898 - 650) = 490.2 \text{ kJ/kg}$$

$$\therefore \ \eta_{thermal} = \frac{\text{Network output}}{\text{Heat supplied}} = \frac{115.15}{490.2}$$

$$= 0.235 \text{ or } 23.5\%. \quad \textbf{(Ans.)}$$

(*ii*) **Work ratio:**

Gross work of the plant $= W_{turbine \, (H.P.)} + W_{turbine \, (L.P.)}$

$$= 285.5 + \frac{115.15}{0.95} = 406.7 \text{ kJ/kg}$$

$$\therefore \ \textbf{Work ratio} \quad = \frac{\text{Network output}}{\text{Gross work output}} = \frac{115.15}{406.7} = \textbf{0.283.} \quad \textbf{(Ans.)}$$

(*iii*) **Mass flow in $\dot{m}$:**

Let the mass flow be $\dot{m}$, then

$$\dot{m} \times 115.15 = 4500$$

$$\therefore \qquad \dot{m} = \frac{4500}{115.15} = 39.08 \text{ kg/s}$$

i.e., **Mass flow** $= \textbf{39.08 kg / s. (Ans.)}$

Example 16

In a closed cycle gas turbine, there is a two-stage compressor and a two-stage turbine. All the components are mounted on the same shaft. The pressure and temperature at the inlet of the first-stage compressor are 1.5 bar and 20 °C. The maximum cycle temperature and pressure are limited to 750 °C and 6 bar. A perfect intercooler is used between the two-stage compressors and a reheater is used between the two turbines. Gases

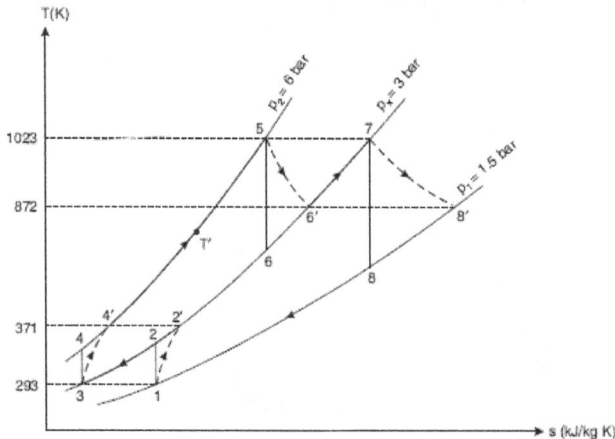

FIGURE 12.33 Sketch for Example 16.

are heated in the reheater to 750 °C before entering into the L.P. turbine. Assuming the compressor and turbine efficiencies as 0.82, calculate:

(i) The efficiency of the cycle without regenerator.

(ii) The efficiency of the cycle with a regenerator whose effectiveness is 0.70.

(iii) The mass of the fluid circulated if the power developed by the plant is 350 kW.

The working fluid used in the cycle is air. For air: $\gamma = 1.4$ *and* $c_p = 1.005$ kJ/kg K.

■ **Solution.**

Given : $T_1 = 20 + 273 = 293$ K, $T_5 = T_7 = 750 + 273 = 1023$ K, $p_1 = 1.5$ bar, $p_2 = 6$ bar, $\eta_{compressor} = \eta_{turbine} = 0.82$.

Effectiveness of regenerator, $\varepsilon = 0.70$, Power developed, $P = 350$ kW.
For air : $c_p = 1.005$ kJ/kg, K, $\gamma = 1.4$
As per given conditions : $T_1 = T_3$, $T_2' = T_4'$

$$\frac{T_2}{T_1} = \left(\frac{p_2}{p_1}\right)^{\frac{\gamma-1}{\gamma}} \text{ and } p_x = \sqrt{p_1 p_2} = \sqrt{1.5 \times 6} = 3 \text{ bar}$$

Now, $$T_2 = T_1 \times \left(\frac{p_2}{T_1}\right)^{\frac{\gamma-1}{\gamma}} = 293 \times \left(\frac{3}{1.5}\right)^{\frac{1.4-1}{1.4}} = 357 \text{ K}$$

$$\eta_{compressor\,(L.P.)} = \frac{T_2 - T_1}{T_2' - T_1}$$

$$0.82 = \frac{357 - 293}{T_2' - 293}$$

∴ $$T_2' = \frac{357 - 293}{0.82} + 293 = 371 \text{ K } i.e., \ T_2' = T_4' = 371 \text{ K}$$

Now, $$\frac{T_5}{T_6} = \left(\frac{T_5}{T_6}\right)^{\frac{\gamma-1}{\gamma}} = \left(\frac{p_2}{p_x}\right)^{\frac{1.4-1}{1.4}} \qquad \left[\begin{array}{l} \because p_5 = p_2 \\ p_6 = p_x \end{array}\right]$$

$$\frac{1023}{T_6} = \left(\frac{6}{3}\right)^{0.286} = 1.219$$

$$\therefore \qquad T_6 = \frac{1023}{1.219} = 839 \text{ K}$$

$$\eta_{turbine \text{ (H.P.)}} = \frac{T_5 - T_6'}{T_5 - T_6}$$

$$0.82 = \frac{1023 - T_6'}{1023 - 839}$$

$$\therefore \qquad T_6' = 1023 - 0.82 \,(1023 - 839) = 872 \text{ K}$$

$$T_8' = T_6' = 872 \text{ K as } \eta_{turbine(\text{H.P.})} = \eta_{turbine(\text{L.P.})}$$

and $\quad T_7 = T_5 = 1023$ K

Effectiveness of regenerator, $\varepsilon = \dfrac{T' - T_4'}{T_8' - T_4'}$

where T' is the temperature of air coming out of regenerator

$$\therefore \qquad 0.70 = \frac{T' - 371}{872 - 371} \quad i.e., \ T' = 0.70 \,(872 - 371) + 371 = 722 \text{ K}$$

Network available, $W_{net} = \left[W_{\text{T(L.P.)}} + W_{\text{T(L.P.)}} \right] - \left[W_{\text{C(H.P.)}} + W_{\text{C(L.P.)}} \right]$

$= 2 \left[W_{\text{T(L.P.)}} - W_{\text{C(L.P.)}} \right]$ *as the work developed by each turbine is the same and work absorbed by each compressor is the same.*

$$\therefore \quad W_{net} = 2c_p \,[(T_5 - T_6') - (T_2' - T_1)]$$

$$= 2 \times 1.005 \,[(1023 - 872) - (371 - 293)] = 146.73 \text{ kJ/kg of air}$$

Heat supplied per kg of air *without regenerator*

$$= c_p (T_5 - T_4') + c_p (T_7 - T_6')$$

$$= 1.005 \,[(1023 - 371) + (1023 - 872)] = 807 \text{ kJ/kg of air}$$

Heat supplied per kg of air *with regenerator*

$$= c_p (T_5 - T') + c_p (T_7 - T_6')$$

$$= 1.005 \,[(1023 - 722) + (1023 - 872)]$$

$$= 454.3 \text{ kJ/kg}$$

(*i*) $\eta_{thermal(\text{without regenerator})} = \dfrac{146.73}{807} = \mathbf{0.182}$ **or 18.2%.** (**Ans.**)

(*ii*) $\eta_{thermal(\text{with regenerator})} \qquad = \dfrac{146.73}{454.3} = \mathbf{0.323}$ **or 32.3%.** (**Ans.**)

(*iii*) **Mass of fluid circulated, $\dot{m}$:**

Power developed, $\quad P = 146.73 \times \dot{m}$ kW

$\therefore \qquad 350 = 146.73 \times \dot{m}$

i.e., $\qquad \dot{m} = \dfrac{350}{146.73} = 2.38 kgsm$

i.e., *Mass of fluid circulated* = **2.38 kg/s.** (**Ans.**)

Example 17

The air in a gas turbine plant is taken in a L.P. compressor at 293 K and 1.05 bar and after compression it is passed through an intercooler where its temperature is reduced to 300 K. The cooled air is further compressed in an H.P. unit and then passed into the combustion chamber, where its temperature is increased to 750 °C by burning the fuel. The combustion products expand in the H.P. turbine which runs the compressors, and further expansion is continued in the L.P. turbine which runs the alternator. The gases coming out from the L.P. turbine are used for heating the incoming air from the H.P. compressor and are then expanded to atmosphere.

The pressure ratio of each compressor = 2, isentropic efficiency of each compressor stage = 82%, isentropic efficiency of each turbine stage = 82%, effectiveness of heat exchanger = 0.72, airflow = 16 kg/s, calorific value of fuel = 42000 kJ/kg, c_v (for gas) = 1.0 kJ/kg K, c_p (gas) = 1.15 kJ/kg K, γ (for air) = 1.4, γ (for gas) = 1.33.

Neglecting the mechanical, pressure, and heat losses of the system and fuel mass, also determine the following:

(i) The power output.   (ii) Thermal efficiency.

(iii) Specific fuel consumption.

■ **Solution.**

Given : $T_1 = 293$ K, $T_3 = 300$ K, $\dfrac{p_2}{p_1} = \dfrac{p_4}{p_3} = 2$, $T_6 = 750 + 273 = 1023$ K

$h_{compressor} = 82\%$, $\eta_{turbine} = 82\%$, $\varepsilon = 0.72$, $\dot{m}_a = 16$ kg/s, C = 42000 kJ/kg, $c_{pa} = 1.0$ kJ/kg K, cpg = 1.15 kJ/kg K, γ (for air) = 1.4, γ (for gas) 1.33.

$$\frac{T_2}{T_1} = \left(\frac{p_2}{p_1}\right)^{\frac{\gamma-1}{\gamma}} = (2)^{\frac{1.4-1}{1.4}} = 1.219$$

$\therefore \qquad T_2 = 293 \times 1.219 = 357$ K

Also, $\qquad \eta_{compressor} = \dfrac{T_2 - T_1}{T_2{}' - T_1}$

$\therefore \qquad 0.82 = \dfrac{357 - 293}{T_2{}' - 293}$

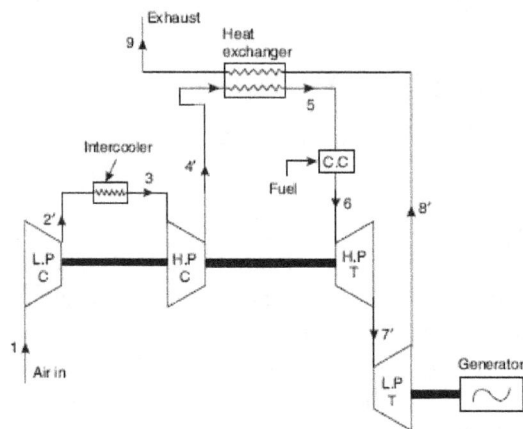

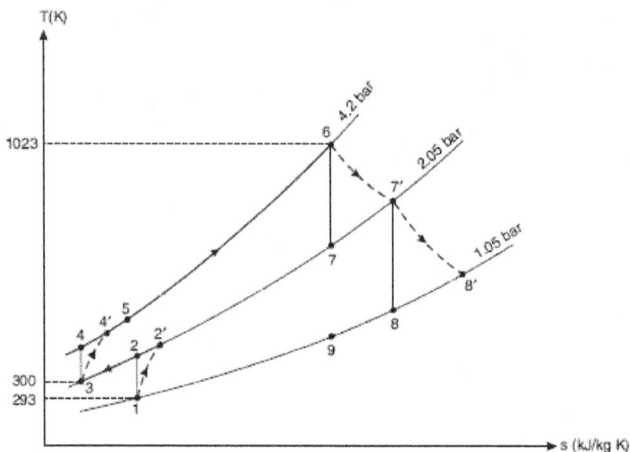

FIGURE 12.34 Sketch for Example 17.

$$\therefore \qquad T_2' = \left(\frac{357 - 293}{0.82}\right) + 293 = 371 \text{ K}$$

Similarly, $\qquad \dfrac{T_4}{T_3} = \left(\dfrac{p_4}{p_3}\right)^{\frac{\gamma-1}{\gamma}} = (2)^{\frac{1.4-1}{1.4}} = 1.219$

$$\therefore \qquad T_4 = 300 \times 1.219 = 365.7 \text{ K and } \eta_{compressor} = \frac{T_4 - T_3}{T_4' - T_3}$$

$$\therefore \qquad 0.82 = \frac{365.7 - 300}{T_4' - 300}$$

i.e., $T_4' = \left(\dfrac{365.7 - 300}{0.82} \right) + 300 = 380\,\text{K}$

Work output of H.P. turbine = Work input to compressor.

Neglecting mass of fuel we can write

$c_{pg}(T_6 - T_7') = c_{pa}[(T_2' - T_1) + (T_4' - T_3)]$

$1.15\,(1023 - T_7') = 1.0\,[(371 - 293) + (380 - 300)]$

or $1.15\,(1023 - T_7') = 158$

∴ $T_7' = 1023 - \dfrac{15.8}{1.15} = 886\,\text{K}$

Also, $\eta_{turbine\ (\text{H.P.})} = \dfrac{T_6 - T_7'}{T_6 - T_7}$

i.e., $0.82 = \dfrac{1023 - 886}{1023 - T_7}$

∴ $T_7 = 1023 - \left(\dfrac{1023 - 886}{0.82} \right) = 856\,\text{K}$

Now, $\dfrac{T_6}{T_7} = \left(\dfrac{p_6}{p_7} \right)^{\frac{\gamma-1}{\gamma}}$

∴ $\dfrac{p_6}{p_7} = \left(\dfrac{T_6}{T_7} \right)^{\frac{\gamma-1}{\gamma}} = \left(\dfrac{1023}{856} \right)^{\frac{1.33}{1.33-1}} = 2.05$

i.e., $p_7 = \dfrac{p_6}{2.05} = \dfrac{4.2}{2.05} = 2.05\,\text{bar}$ [∵ $p_6 = 1.05 \times 4 = 4.2\,\text{bar}$]

$\dfrac{T_7'}{T_8} = \left(\dfrac{p_7}{p_8} \right)^{\frac{\gamma-1}{\gamma}} = \left(\dfrac{2.05}{1.05} \right)^{\frac{1.33-1}{1.33}} = 1.18$

$T_8 = \dfrac{T_7'}{1.18} = \dfrac{886}{1.18} = 751\,\text{K}$

Again, $\eta_{turbine\ (\text{L.P.})} = \dfrac{T_7' - T_8'}{T_7' - T_8}$

$0.82 = \dfrac{886 - T_8'}{886 - 751}$

∴ $T_8' = 886 - 0.82\,(886 - 751) = 775\,\text{K}$

(*i*) **Power output:**

Net power output $= c_{pg}(T_7' - T_8')$

$= 1.15\,(886 - 775) = 127.6$ kJ/kg

∴ Net output per second $= \dot{m} \times 127.6$

$= 16 \times 127.6 = 2041.6$ kJ/s $= \textbf{2041.6 kW. (Ans.)}$

(*ii*) **Thermal efficiency:**

Effectiveness of heat exchanger, $\varepsilon = \dfrac{T_5 - T_4'}{T_8' - T_4'}$

i.e., $\quad 0.72 = \dfrac{T_5 - 380}{778 - 380}$

∴ $\quad T_5 = 0.72\,(775 - 380) + 380 = 664$ K

Heat supplied in combustion chamber per second
$= \dot{m}_a\,c_{pg}(T_6 - T_5)$

$= 16 \times 1.15\,(1023 - 664) = 6605.6$ kJ/s

∴ $\eta_{thermal} = \dfrac{2041.6}{6605.6} = \textbf{0.309} \text{or } \textbf{30.9\%. (Ans.)}$

(*iii*) **Specific fuel consumption :**

If m_f is the mass of fuel supplied per kg of air, then
$m_f \times 42000 = 1.15\,(1023 - 664)$

∴ $\quad \dfrac{1}{m_f} = \dfrac{42000}{1.15\,(1023 - 664)} = \dfrac{101.7}{1}$

∴ Air-fuel ratio $= 101.7 : 1$

∴ Fuel supplied per hour $= \dfrac{16 \times 3600}{101.7} = 566.37$ kg/h

∴ *Specific fuel consumption* $= \dfrac{566.37}{2041.6} = \textbf{0.277 kg / kWh. (Ans.)}$

Example 18

Air is taken in a gas turbine plant at 1.1 bar 20°C. The plant comprises L.P. and H.P. compressors and L.P. and H.P. turbines. The compression in the L.P. stage is up to 3.3 bar followed by intercooling to 27°C. The pressure of air after the H.P. compressor is 9.45 bar. Loss in pressure during intercooling is 0.15 bar. Air from the H.P. compressor is transferred to the heat exchanger of effectiveness 0.65, where it is heated by the gases from the L.P. turbine. After the heat exchanger the air passes through the combustion chamber. The temperature of gases supplied to the H.P. turbine is 700 °C. The gases expand in the H.P. turbine to 3.62 bar and air is then reheated to 670 °C before expanding in the L.P. turbine. The loss of pressure in the reheater is 0.12 bar. Determine:

(i) The overall efficiency   (ii) The work ratio

(iii) Mass flow rate when the power generated is 6000 kW.

Assume : Isentropic efficiency of compression in both stages =0.82. Isentropic efficiency of expansion in turbines =0.85.

For air : c =1.005 kJ / kg K, γ =1.4.

For gases : c =1.15 kJ / kg K, γ =1.33.

Neglect the mass of fuel.

■ **Solution.**

Given : $T_1 = 20 + 273 = 293$ K, $p_1 = 1.1$ bar, $p_2 = 3.3$ bar,

$T_3 = 27 + 273 = 300$ K,

$p_3 = 3.3 - 0.15 = 3.15$ bar, $p_4 = p_6 = 9.45$ bar, $T_6 = 973$ K,

$T_8 = 670 + 273 = 943$ K, $p_8 = 3.5$ bar,

$\eta_{compressors}$ = 82 %, $\eta_{turbines}$ = 85 %, Power generated = 6000 kW,

Effectiveness, ε = 0.65, c_{pa} = 1.005 kJ/kg K, γ_{air} = 1.44, c_{pg} = 1.15 kJ/kg K and γ_{gases} = 1.33

Refer to Figure 12.35.

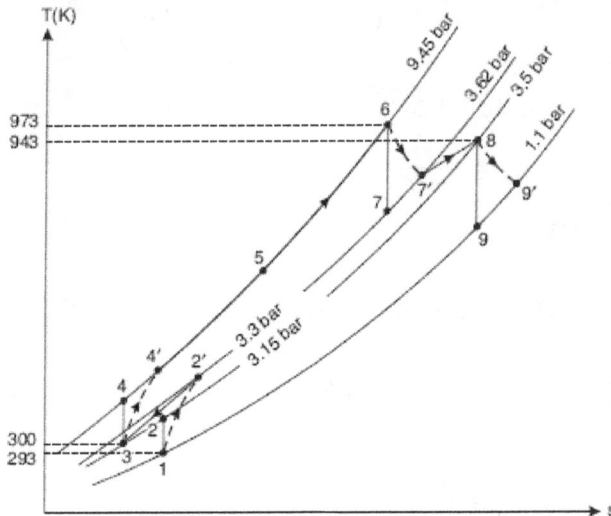

FIGURE 12.35 Sketch for Example 18.

Now, $\dfrac{T_2}{T_1} = \left(\dfrac{p_2}{p_1}\right)^{\frac{\gamma-1}{\gamma}} = \left(\dfrac{3.3}{1.1}\right)^{\frac{1.4-1}{1.4}} = 1.369$

$\therefore$ $T_2 = 293 \times 1.369 = 401\ \text{K}$

$\eta_{compressor\ (L.P.)} = 0.82 = \dfrac{T_2 - T_1}{T_2' - T_1} = \dfrac{401 - 293}{T_2' - 293}$

$\therefore$ $T_2' = \left(\dfrac{401 - 293}{0.82}\right) + 293 = 425\ \text{K}$

Again, $\dfrac{T_4}{T_3} = \left(\dfrac{p_4}{p_3}\right)^{\frac{\gamma-1}{\gamma}} = \left(\dfrac{9.45}{3.15}\right)^{\frac{1.4-1}{1.4}} = 1.369$

$\therefore$ $T_4 = 300 \times 1.369 = 411\ \text{K}$

Now $\eta_{compressor\ (H.P.)} = \dfrac{T_4 - T_3}{T_4' - T_3}$

$0.82 = \dfrac{411 - 300}{T_4' - 300}$

$\therefore$ $T_4' = \left(\dfrac{411 - 300}{0.82}\right) + 300 = 435\ \text{K}$

Similarly, $\dfrac{T_6}{T_7} = \left(\dfrac{p_6}{p_7}\right)^{\frac{\gamma-1}{\gamma}} = \left(\dfrac{9.45}{3.62}\right)^{\frac{1.33-1}{1.33}} = 1.268$

$\therefore$ $T_7 = \dfrac{T_6}{1.268} = \dfrac{973}{1.268} = 767\ \text{K}$

Also, $\eta_{turbine\ (H.P.)} = \dfrac{T_6 - T_7'}{T_6 - T_7}$

$0.85 = \dfrac{973 - T_7'}{973 - 767}$

$\therefore$ $T_7' = 973 - 0.85\ (973 - 767) = 798\ \text{K}$

Again, $\dfrac{T_8}{T_9} = \left(\dfrac{p_8}{p_9}\right)^{\frac{\gamma-1}{\gamma}} = \left(\dfrac{3.5}{1.1}\right)^{\frac{1.33-1}{1.33}} = 1.332$

$\therefore$ $T_9 = \dfrac{T_8}{1.332} = \dfrac{943}{1.332} = 708\ \text{K}$

$$\eta_{turbine\,(L.P.)} = \frac{T_8 - T_9'}{T_8 - T_9}$$

$$0.85 = \frac{943 - T_9'}{943 - 708}$$

$$\therefore \quad T_9' = 943 - 0.85\,(943 - 708) = 743\,\text{K}.$$

Effectiveness of heat exchanger,

$$\varepsilon = 0.65 = \frac{T_5 - T_4'}{T_9' - T_4'}$$

i.e., $\quad 0.65 = \dfrac{T_5 - 435}{743 - 435}$

$$\therefore \quad T_5 = 0.65\,(743 - 435) + 435 = 635\,\text{K}$$

$$W_{turbine(H.P.)} = c_{pg}\,(T_6 - T_7')$$

$$= 1.15\,(973 - 798) = 201.25\;\text{kJ / kg of gas}$$

$$W_{turbine\,(L.P.)} = c_{pg}\,(T_8 - T_9')$$

$$= 1.15\,(943 - 743) = 230\;\text{kJ / kg of gas}$$

$$W_{compressor(L.P.)} = c_{pa}\,(T_2' - T_1)$$

$$= 1.005\,(425 - 293) = 132.66\;\text{kJ / kg of air}$$

$$W_{compressor\,(H.P.)} = c_{pa}\,(T_4' - T_3)$$

$$= 1.005\,(435 - 300) = 135.67\;\text{kJ / kg of air}$$

Heat supplied $\quad = c_{pg}\,(T_6 - T_5) + c_{pg}\,(T_8 - T_7')$

$$= 1.15\,(973 - 635) + 1.15\,(943 - 798) = 555.45\;\text{kJ / kg of gas}$$

(*i*) **Overall efficiency** η_{overall}**:**

$$\eta_{overall} = \frac{\text{Network done}}{\text{Heat supplied}}$$

$$= \frac{[W_{turbine\,(H.P.)} + W_{turbine\,(L.P.)}] - [W_{comp.\,(L.P.)} + W_{comp.\,(H.P.)}]}{\text{Heat supplied}}$$

$$= \frac{(201.25 + 230) - (132.66 + 135.67)}{555.45}$$

$$= \frac{162.92}{555.45} = 0.293 \text{ or } \mathbf{29.3\%}. \quad (\mathbf{Ans.})$$

(*ii*) **Work ratio:**

$$Work\,ratio = \frac{\text{Network done}}{\text{Turbine work}}$$

$$= \frac{\left[W_{turbine(\text{H.P.})} + W_{turbine(\text{L.P.})} \right] - \left[W_{comp.(\text{L.P.})} + W_{comp.(\text{H.P.})} \right]}{\left[W_{turbine(\text{H.P.})} + W_{turbine(\text{L.P.})} \right]}$$

$$= \frac{(201.25 + 230) - (132.66 + 135.67)}{(201.25 + 230)} = \frac{162.92}{431.25} = 0.377.$$

i.e.,   *Work ratio* = **0.377.** $\left(\mathbf{Ans.} \right)$

(*iii*) **Mass flow rate, ṁ :**

Network done   = 162.92 kJ / kg.

Since mass of fuel is neglected, for 6000 kW, mass flow rate,

$$\dot{m} = \frac{6000}{162.92} = 36.83 \text{ kg / s}$$

i.e.,   *Mass flow rate* = **36.83 kg / s.** $(\mathbf{Ans.})$

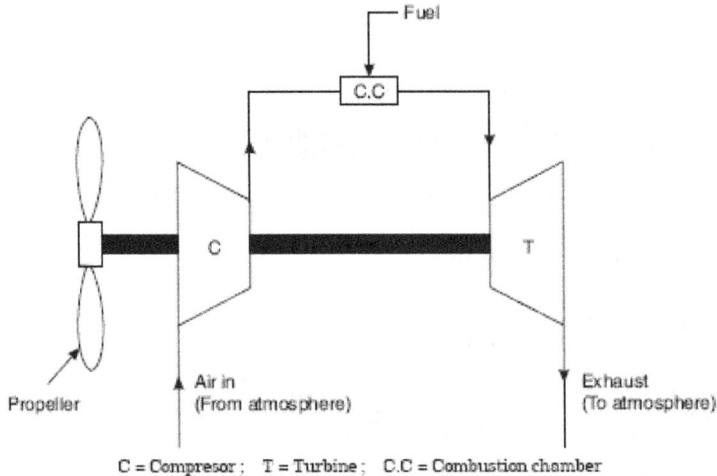

FIGURE 12.36 Power plant for the screw propeller.

12.8. Jet Propulsion

The principle of jet propulsion involves imparting momentum to a mass of fluid in such a manner that the reaction of the imparted momentum provides a propulsive force. It may be achieved by expanding the gas, which is at high temperature and pressure, through a nozzle, due to which a high velocity jet of hot gases is produced (in the atmosphere) that gives a propulsive force (in the opposite direction due to its reaction). For jet propulsion the open cycle gas turbine is most suitable.

The propulsion system may be classified as follows:

1. **Airstream jet engines**, (Air-breathing engines)

   (a) Steady combustion systems; continuous air flow

   (i) Turbo-jet   (ii) Turbo-prop

   (iii) Ram jet

   (b) Intermittent combustion system; intermittent flow

   (i) Pulse jet or flying bomb

2. **Self-contained rocket engines** (Non-air breathing engines)

   (i) Liquid propellant   (ii) Solid propellant

In airstream jet engines the oxygen necessary for the combustion is taken from the surrounding atmosphere, whereas in a rocket engine the fuel and the oxidizer are contained in the body of the unit which is to be propelled.

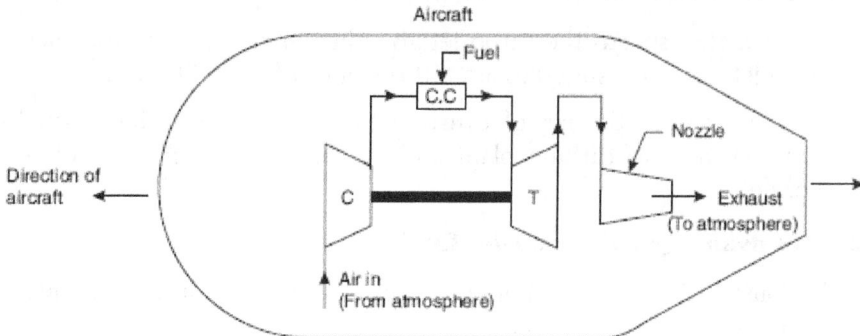

FIGURE 12.37 turbine plant for a turbo-jet.

Note that the turbo-jet and turbo-prop are modified forms of a simple open cycle gas turbine. The ram jet and pulse jet are athodyd's (aero-thermo-dynamic ducts), that is, straight duct types of jet engines having no compressor and turbine wheels.

In the past air propulsion was achieved by a "Screw propeller." In this system the total power developed by the turbine (full expansion) is used to drive the compressor and propeller. Figure 12.36 shows the power plant for the screw propeller. By controlling the supply of fuel in the combustion chamber, the power supplied to the propeller can be controlled. The rate of increase of efficiency of the screw propeller is higher at lower speeds but its efficiency falls rapidly at higher speeds above the sonic velocity.

As shown in Figure 12.37, a turbo-jet unit consists of a *diffuser* at the entrance which slows down the air (entering at a velocity equal to the plane speed), and part of the kinetic energy of the airstream is converted into pressure; this type of compression is called *ram compression*.

The air is further compressed to a pressure of 3 to 4 bar in a rotary compressor (usually of an axial flow type).

The compressed air then enters the combustion chamber (C.C.) where fuel is added. The combustion of fuel takes place at a sensibly constant pressure and subsequently temperature rises rapidly.

The hot gases then enter the gas turbine where *partial expansion* takes place. The *power produced is just sufficient to drive the compressor, fuel pump, and other auxiliaries*.

The exhaust gases from the gas turbine, which are at a higher pressure than the atmosphere, are expended in a nozzle, and a very high velocity jet is produced which provides a forward motion to the aircraft by the jet reaction (Newton's third law of motion).

At higher speeds the turbo-jet gives higher propulsion efficiency. The turbo-jets are most suited to aircraft traveling *above 800 km/h*.

The overall efficiency of a turbo-jet is the product of the thermal efficiency of the gas turbine plant and the propulsive efficiency of the jet (nozzle).

12.8.1. Advantages of Turbo-Jet Engines

1. Construction is much simpler (as compared to a multi-cylinder piston engine of comparable power).

2. Engine vibrations are absent.

3. Much higher speeds are possible (more than 3000 km/h achieved).

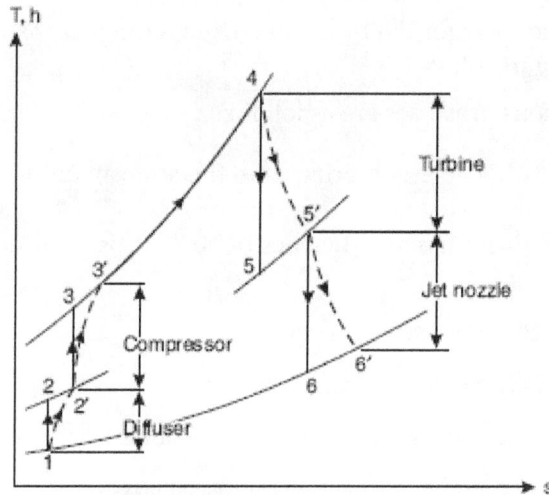

FIGURE 12.38 T-s diagram of turbo-jet.

4. The power supply is uninterrupted and smooth.

5. Weight to power ratios are superior (as compared to that of the reciprocating type of aero-engine).

6. The rate of climb is higher.

7. The requirement of major overhauls is less frequent.

8. Radio interference is much less.

9. The maximum altitude ceiling as compared to turbo-prop and conventional piston type engines.

10. The frontal area is smaller.

11. Fuel can be burnt over a large range of mixture strength.

12.8.2. Disadvantages of Turbo-Jet Engines

1. Less efficient.

2. Life of the unit is comparatively shorter.

3. The turbo-jet becomes rapidly inefficient below 550 km/h.

4. More noisy (than a reciprocating engine).

5. Materials required are quite expensive.

6. Require a longer strip since the length of take-off is too much.

7. At take-off the thrust is low; this effect is overcome by boosting.

12.8.3. Basic Cycle for a Turbo-Jet Engine

The basic cycle for the turbo-jet engine is the *Joule or Brayton cycle* as shown in Figure 12.38.

The various processes are as follows:

Process 1-2: The air entering from the atmosphere is *diffused isentropically* from velocity C_1 down to zero (*i.e.*, $C_2 = 0$). This indicates that the diffuser has an efficiency of 100%; this is termed as *ram compression*.

Process 1-2′ is the *actual process*.

Process 2-3: *Isentropic* compression of air.

Process 2′-3′ shows the *actual* compression of air.

Process 3-4: *Ideal* addition of heat at constant pressure $p_3 = p_4$.

Process 3′-4 shows the *actual* addition of heat at constant process $p_3 = p_4$.

Process 4-5: *Isentropic expansion* of gas in the *turbine*.

Process 4-5′ shows the *actual* expansion in the turbine.

Process 5-6: *Isentropic expansion* of gas in the *nozzle*.

Process 5′-6′ shows the *actual expansion* of gas in the *nozzle*.

Consider *1 kg of working fluid flowing through the system*.

Diffuser:

Between states 1 and 2, the energy equation is given by:

$$\frac{C_a^2}{2} + h_1 + Q_{1-2} = \frac{C_2^2}{2} + h_2 + W_{1-2}$$

where $C_a (= C_1)$ = Velocity of entering air from atmosphere.

In an ideal diffuser $C_2 = 0$, $Q_{1-2} = 0$ and $W_{1-2} = 0$.

∴   Enthalpy at state 2 is, $h_2 = h_1 + \dfrac{C_a^2}{2}$ kJ/kg

or $T_2 = T_1 + \dfrac{C_a^2}{2.c_p}$

($\because \ h = c_p.T$)

Process 1-2' shows actual process in diffuser.

Diffuser efficiency, $\eta_d = \dfrac{h_2 - h_1}{h_2' - h_1} = \dfrac{T_2 - T_1}{T_2' - T_1}$

or $h_2' = h_1 + \dfrac{C_a^2}{2\eta_d}$

or $T_2' = T_1 + \dfrac{C_a^2}{2.c_p.\eta_d}$

Compressor:

Energy equation between states 2 and 3 gives

$$h_2 + \dfrac{C_2^2}{2} + Q_{2-3} + W_c = h_3 + \dfrac{C_3^2}{2}$$

Assuming changes in potential and kinetic energies to be negligible, the ideal work expended in running the compressor is given as,

$$W_c = h_3 - h_2 = c_p\left(T_3 - T_2\right)$$

The actual compressor work (to be supplied by the turbine)

$$= h_3' - h_2 = \dfrac{h_3 - h_2}{\eta_c} = \dfrac{c_p\left(T_3 - T_2\right)}{\eta_c}$$

$\left(\text{where } \eta_c = \text{Isentropic efficiency of compressor}\right)$

Combustion chamber:

Ideal heat supplied per kg, $Q = h_4 - h_3 = c\left(T_4 - T_3\right)$

Actual heat supplied $= \left(1 + \dfrac{m_f}{m_a}\right)h_4 - h_3'$

or $Q_a = c_{pg}\left(1 + \dfrac{m_f}{m_a}\right)T_4 - c_{pg} . T_3'$

(where c_{pg} and c_{pa} are specific heats of gases and air at constant pressure respectively)

TURBINE:

Between states 4 and 5, the energy equation is given by:

$$h_4 + \dfrac{C_4^2}{2} + Q_{4-5} = h_5 + \dfrac{C_5^2}{2} + W_t$$

If $Q_{4-5} = 0$, then turbine work,

$$W_t = (h_4 - h_5) + \frac{\left(C_4^2 - C_5^2\right)}{2}$$

If the change in kinetic energy is neglected, we have

$$W_t = (h_4 - h_5) = c_p(T_4 - T_5)$$

Actual turbine work $= h_4 - h_5' = c_p(T_4 - T_5') = c_p(T_4 - T_5) \times \eta_t$

(where η_t = Isentropic efficiency of turbine)

For the simplification, turbine work = compressor work

or $\quad c_p(T_4 - T_5') = c_p(T_4 - T_5)\eta_t = \dfrac{c_p(T_3 - T_2)}{\eta_c}$

or $\quad T_5' = T_4 - (T_4 - T_5)\eta_t = T_4 - \dfrac{c_p(T_3 - T_2)}{\eta_c}$

Jet nozzle:

Energy equation between states 5 and 6 gives

$$h_5 + \frac{C_5^2}{2} = h_6 + \frac{C_6^2}{2} \qquad \text{...Ideal case}$$

$$h_5' + \frac{C_5'^2}{2} = h_6' + \frac{C_6'^2}{2} \qquad \text{...Actual case}$$

If $C_5'^2$ is much less as compared to $C_6'^2$, we have

$$h_5' = h_6' + \frac{C_6'^2}{2}$$

or $\quad C_6' = \sqrt{2(h_5' - h_6')} = \sqrt{2\,\eta_n(h_5' - h_6)}$

or $\quad C_6' = \sqrt{2\eta_n\, c_p(T_5' - T_6)}$

(where η_n = Nozzle efficiency)

Thermal efficiency (η_{th}) is given by :

$$\eta_{th} = \frac{(h_4 - h_6') - (h_3' - h_1)}{(h_4 - h_3')}$$

$$= \frac{(T_4 - T_6') - (T_3' - T_1)}{(T_4 - T_3')}$$

Thrust, Thrust-power, Propulsive Efficiency, and Thermal Efficiency Thrust (T)

Let C_a = Forward velocity of aircraft through air, m / s. Assuming the atmospheric air to be still, the velocity of air, relative to the aircraft, at entry to the aircraft will be C_a. It is called the *velocity of approach of air.*

C_j = Velocity of jet (gases) relative to the exit nozzle / aircraft ; m / s.

$$\left[1 + \frac{\text{fuel } (m_f)}{\text{air } (m_a)}\right] = \text{Mass of products leaving the nozzle for 1 kg of air.}$$

Thrust is the force *produced due to a change of momentum.*

Now, absolute velocity of gases leaving aircraft $= \left(C_j - C_a\right)$

Absolute velocity of air entering the aircraft = 0

$\therefore$ Change of momentum $= \left(1 + \dfrac{m_f}{m_a}\right) \left(C_j - C_a\right)$

Hence, thrust, $T = \left(1 + \dfrac{m_f}{m_a}\right) \left(C_j - C_a\right)$ N / kg of air / s

$T = \left(C_j - C_a\right)$ N / kg of air / s, *neglecting mass of fuel*

Thrust power (T.P.):

It is defined as the rate at which work must be developed by the engine if the aircraft is to be kept moving at a constant velocity C_a against friction force or drag.

$\therefore$ Thrust power = Forward thrust $\times$ speed of aircraft

or T.P. $= \left[\left(1 + \dfrac{m_f}{m_a}\right)\left(C_j - C_a\right)\right] C_a$ W/kg of air

$= \left(C_j - C_a\right) C_a$ W/kg of air if *mass of fuel is neglected*

$= \dfrac{\left(C_j - C_a\right) C_a}{1000}$ kW/kg of air

Propulsive power (P.P.):

The energy required to change the momentum of the mass flow of gas represents the propulsive power. It is expressed as the difference between the rate of kinetic energies of the entering air and exit gases.

Mathematically, $\text{P.P.} = \Delta\text{K.E.} = \dfrac{\left(1+\dfrac{m_f}{m_a}\right)C_j^2}{2} - \dfrac{C_a^2}{2}$ W/kg

$= \dfrac{\left(C_j^2 - C_a^2\right)}{2}$ W/kg, *neglecting mass of fuel*

$= \dfrac{C_j^2 - C_a^2}{2\times1000}$ kW/kg of air

Propulsive efficiency (H$_{PROP}$):

The *ratio of thrust power to propulsive power is called the propulsive efficiency of the propulsive unit.*

$$\eta_{prop.} = \frac{\text{Thrust power}}{\text{Propulsive power}} = \frac{\left[\left(1+\dfrac{m_f}{m_a}\right)(C_j-C_a)\right]C_a}{\left[\dfrac{\left(1+\dfrac{m_f}{m_a}\right)C_j^2}{2} - \dfrac{C_a^2}{2}\right]} = \frac{2\left[\left(1+\dfrac{m_f}{m_a}\right)(C_j-C_a)\right]C_a}{\left[\left(1+\dfrac{m_f}{m_a}\right)(C_j^2-C_a^2)\right]}$$

Neglecting the mass of fuel,

$$\eta_{prop.} = \frac{2(C_j-C_a)C_a}{C_j^2-C_a^2} = \frac{2(C_j-C_a)C_a}{(C_j+C_a)(C_j-C_a)}$$

or $\eta_{prop.} = \dfrac{2C_a}{C_j+C_a}$ or $\dfrac{2}{\left(\dfrac{C_j}{C_a}+1\right)}$

From the last equation it is evident that the *propulsive efficiency increases with an increase in aircraft velocity C_a.* $\eta_{prop.}$ becomes 100% when C_a approaches C_j; thrust reduces to zero (Equation 12).

Thermal efficiency, (η_{th}):

It is defined as the ratio of propulsive work and the energy released by the combustion of fuel.

or $\eta_{th} = \dfrac{\text{Propulsive work}}{\text{Heat released by the combustion of fuel}}$

$= \dfrac{\text{Increase in kinetic energy of the gases}}{\text{Heat released by the combustion of fuel}}$

or $\qquad \eta_{th} = \dfrac{\left(1+\dfrac{m_f}{m_a}\right)\left(C_j^{\,2} - C_a^{\,2}\right)}{2\left[\dfrac{m_f}{m_a} \times \text{calorific value}\right]}$

$\approx \dfrac{\left(C_j^{\,2} - C_a^{\,2}\right)}{2\times\left(\dfrac{m_f}{m_a}\right)\times \text{calorific value}}$

Overall efficiency (η_0) is given by:

$\eta_0 = \eta_{th} \times \eta_{prop.} = \dfrac{\left(C_j^{\,2} - C_a^{\,2}\right)}{2\times\left(\dfrac{m_f}{m_a}\right)\times \text{calorific value}} \times \dfrac{2C_a}{C_j + C_a}$

$= \dfrac{\left(C_j - C_a\right)C_a}{\left(\dfrac{m_f}{m_a}\right)\times \text{calorific value}}$

For maximum overall efficiency the aircraft velocity C_a is one half of the jet velocity C_j. The jet efficiency (η_{jet}) is defined as:

$\eta_{jet} = \dfrac{\text{Final kinetic energy in the jet}}{\text{Isentropic heat drop in the jet pipe} + \text{Carry over from the turbine.}}$

Example 19

A turbo-jet engine consumes air at the rate of 60.2 kg/s when flying at a speed of 1000 km/h. Calculate:

(i) Exit velocity of the jet when the enthalpy change for the nozzle is 230 kJ/kg and velocity coefficient is 0.96

(ii) Fuel flow rate in kg/s when air-fuel ratio is 70 : 1

(iii) Thrust specific fuel consumption

(iv) Thermal efficiency of the plant when the combustion efficiency is 92% and the calorific value of the fuel used is 42000 kJ/kg

(v) Propulsive power

(vi) Propulsive efficiency

(vii) Overall efficiency.

■ **Solution.**

Rate of air consumption, $\dot{m}_a = 60.2$ kg/s

Enthalpy change for nozzle, $\Delta h = 230$ kJ / kg

Velocity coefficient, $z = 0.96$

Air-fuel ratio $= 70 : 1$

Combustion efficiency, $\eta_{combustion} = 92\%$

Calorific value offuel, C.V. $= 42000$ kJ / kg

Aircraft velocity, $C_a = \dfrac{1000 \times 1000}{60 \times 60} = 277.8$ m/s

(*i*) **Exit velocity of jet, C_j:**

$C_j = z\sqrt{2\,\Delta h \times 1000}$, where Δh is in kJ

$= 0.96\sqrt{2 \times 230 \times 1000} = 651$ m/s.

i.e.,   **Exit velocity of jet = 651 m/s. (Ans.)**

(*ii*) **Fuel flow rate:**

Rate of fuel consumption, $\dot{m}_f = \dfrac{\text{Rate of air consumption}}{\text{Air-fuel ratio}}$

$= \dfrac{60.2}{70} = \mathbf{0.86\ kg / s.\ (Ans.)}$

(*iii*) **Thrust specific fuel consumption:**

Thrust is the force produced due to change of momentum.

Thrust produced $= \dot{m}_a\left(C_j - C_a\right)$, neglecting mass of fuel.

$= 60.2\left(651 - 277.8\right) = 22466.6$ N.

∴ Thrust specific fuel consumption

$= \dfrac{\text{Fuel consumption}}{\text{Thrust}} = \dfrac{0.86}{22466.6}$

$= \mathbf{3.828 \times 10^{-5}\ kg / N\ of\ thrust / s.}$ **(Ans.)**

(*iv*) **Thermal efficiency, η_{thermal}:**

$\eta_{thermal} = \dfrac{\text{Work output}}{\text{Heat supplied}}$

$= \dfrac{\text{Gain in kinetic energy per kg of air}}{\text{Heat supplied by fuel per kg of air}}$

$$= \frac{\left(C_j^{\,2} - C_a^{\,2} \right)}{\left(\dfrac{m_f}{m_a} \right) \times C.V. \times \eta_{combustion} \times 1000}$$

$$= \frac{(651^2 - 277.8^2)}{2 \times \dfrac{1}{70} \times 42000 \times 0.92 \times 1000} = 0.3139 \text{ or } 31.39\%$$

i.e.,   *Thermal efficiency*   = **31.39%. (Ans.)**

(*v*) **Propulsive power:**

Propulsive power $= \dot{m}_a \times \left(\dfrac{C_j^{\,2} - C_a^{\,2}}{2} \right) = \dfrac{60.2}{1000} \times \left(\dfrac{651^2 - 277.8^2}{2} \right)$ kW

= **10433.5 kW. (Ans.)**

(*vi*) **Propulsive efficiency,** $\eta_{\text{prop.}}$**:**

$$\eta_{prop.} = \frac{\text{Thrust power}}{\text{Propulsive power}} = \frac{2C_a}{C_j + C_a}$$

$$= \frac{2 \times 277.8}{651 + 277.8} = 0.598 \text{ or } \textbf{59.8\%. (Ans.)}$$

(*vii*) **Overall efficiency,** η_0**:**

$$\eta_0 = \frac{\text{Thrust work}}{\text{Heat supplied by fuel}} = \frac{\left(C_j - C_a \right) C_a}{\left(\dfrac{m_f}{m_a} \right) \times C.V. \times \eta_{combustion}}$$

$$= \frac{(651 - 277.8) \times 277.8}{\dfrac{1}{70} \times 42000 \times 0.92 \times 1000} = \textbf{0.1878 or 18.78\%. (Ans.)}$$

Example 20

The following data pertain to a turbo-jet flying at an altitude of 9500 m:

Speed of the turbo-jet	= 800 km/ h
Propulsive efficiency	= 55%
Overall efficiency of the turbine plant	= 17%
Density of air at 9500 m altitude	= 0.17 kg /m3
Drag on the plane	= 6100 N

Assuming calorific value of the fuels used as 46000 kJ / kg,

Calculate:

(i) Absolute velocity of the jet. (ii) Volume of air compressed per min.

(iii) Diameter of the jet. (iv) Power output of the unit.

(v) Air-fuel ratio.

■ **Solution.**

Given : Altitude = 9500 m, $C_a = \dfrac{800 \times 1000}{60 \times 60} = 222.2$ m/s,

$\eta_{propulsive}$ = 55%, $\eta_{overall}$ = 17%; density of air at 9500 m altitude = 0.17 kg/m³; drag on the plane = 6100 N.
drag on the plane = 6100 N.

(*i*) **Absolute velocity of the jet, ($C_J - C_A$):**

$\eta_{propulsive} = 0.55 = \dfrac{2C_a}{C_j + C_a}$

where, c_j = Velocity of gases at nonnle exit relative to the aircraft, and

C_A = VELOCITY OF THE TURBO-JET/AIRCRAFT.

∴ $0.55 = \dfrac{2 \times 222.2}{C_j + 222.2}$

i.e., $C_j = \dfrac{2 \times 222.2}{0.55} - 222.2 = 585.8$ m/s

∴ Absolute velocity of jet $= C_j - C_a = 585.8 - 222.2 = 363.6$ m / s.

(*ii*) **Volume of air compressed/min.:**

Propulsiveforce $= \dot{m}_a \left(C_j - C_a \right)$

$6100 = \dot{m}_a (585.8 - 222.2)$

∴ $\dot{m}_a = 16.77$ kg / s

∴ Volume of air compressed / min. $= \dfrac{16.77}{0.17} \times 60 = $ **5918.8 kg / min.** (**Ans.**)

(*iii*) **Diameter of the jet, d:**

Now, $\dfrac{\pi}{4} d^2 \times C_j = 5918.8$

i.e., $\dfrac{\pi}{4} d^2 \times 585.8 = (5918.8 / 60)$

∴ $d = \left(\dfrac{5918.8 \times 8 \times 4}{60 \times \pi \times 585.8} \right)^{1/2} = 0.463$ m = 463 mm *i.e.,* **Diameter of the jet = 463 mm.** **(Ans.)**

(*iv*) Power output of the unit:

Thrust power = Drag force × velocity of turbo-jet

$$= 6100 \times 222.2 \text{ N-m/s}$$

$$= \frac{6100 \times 222.2}{1000} = 1355.4 \text{ kW}$$

Turbine output $= \dfrac{\text{Thrust power}}{\text{Propulsive efficiency}} = \dfrac{1355.4}{0.55} = $ **2464.4 kW.** **(Ans.)**

(*v*) Overall efficiency, η_0:

$$\eta_0 = \frac{\text{Heat equivalent of output}}{\dot{m}_f \times \text{C.V.}}$$

i.e., $\qquad 0.17 = \dfrac{2464.4}{\dot{m}_f \times 46000}$

$\therefore \qquad \dot{m}_f = \dfrac{2464.4}{0.17 \times 46000} = 0.315 \text{ kg/s}$

$\therefore \quad$ Air-fuel ratio $= \dfrac{\text{Air used (in kg/s)}}{\text{Fuel used (in kg / s)}} = \dfrac{16.77}{0.315} = 53.24$

i.e., $\quad$ *Air - fuel ratio* = **53.24 : 1.** **(Ans.)**

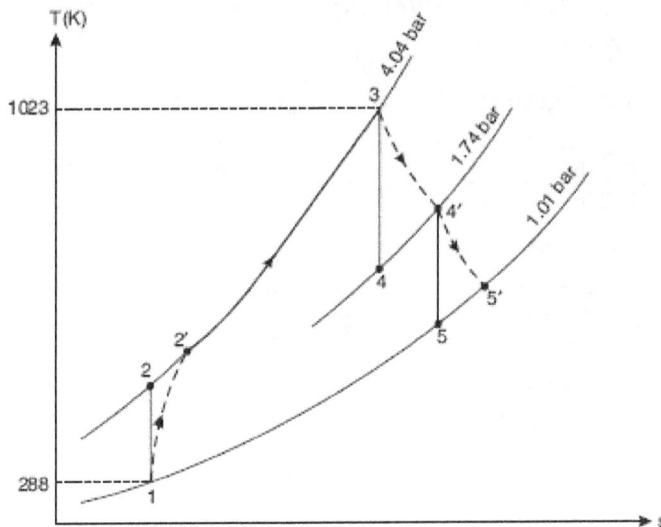

FIGURE 12.39 Sketch for Example 21.

Example 21

In a jet propulsion unit air is drawn into the rotary compressor at 15 °C and 1.01 bar and delivered at 4.04 bar. The isentropic efficiency of compression is 82% and the compression is uncooled. After delivery the air is heated at constant pressure until the temperature reaches 750 °C. The air then passes through a turbine unit which drives the compressor only and has an isentropic efficiency of 78% before passing through the nozzle and expanding to an atmospheric pressure of 1.01 bar with an efficiency of 88%. Neglecting any mass increase due to the weight of the fuel and assuming that R and γ are unchanged by combustion, determine:

(i) The power required to drive the compressor.

(ii) The air-fuel ratio if the fuel has a calorific value of 42000 kJ/kg.

(iii) The pressure of the gases leaving the turbine.

(iv) The thrust per kg of air per second.
   Neglect any effect of the velocity of approach.
   Assume for air: R = 0.287 kJ/kg K, γ = 1.4.

■ **Solution.**

Given: $T1 = 15 + 273 = 288$ K,

$p_1 = 1.01$ bar, $p_2 = 4.04$ bar, $T_3 = 750 + 273 = 1023$ K,

$\eta_{compressor} = 82\%$, $\eta_{turbine} = 78\%$, $\eta_{nozzle} = 88\%$,

$R_{air} = 0.287$ kJ/kg K, $\gamma_{air} = 1.4$.

Refer to Figure 12.39.

$$\frac{T_2}{T_1} = \left(\frac{P_2}{P_1}\right)^{\frac{\gamma-1}{\gamma}} = \left(\frac{4.04}{1.01}\right)^{\frac{1.4-1}{1.4}} = 1.486$$

$\therefore \quad T_2 = 2.88 \times 1.486 = 428$ K

$$\eta_{compressor} = \frac{T_2 - T_1}{T_2' - T_1} \quad i.e., \ 0.82 = \frac{428 - 288}{T_2' - 288}$$

$\therefore \quad T_2' = \left(\frac{428 - 288}{0.82}\right) + 288 = 458.7$ K

$$c_p = R \times \left(\frac{\gamma}{\gamma - 1}\right) = 0.287 \times \frac{14}{1.4 - 1} = 1.004 \text{ kJ/kg K}$$

(*i*) **Power required to drive the compressor:**

Power required to drive the compressor (per kg of air/sec.)

$$= c_p(T_2' - T_1) = 1.004(458.7 - 288) = \textbf{171.38 kW.} \quad \textbf{(Ans.)}$$

(*ii*) **Air-fuel ratio:**

$$m_f \times C = (m_a + m_f) \times c_p \times (T_3 - T_2')$$

where, m_a = Mass of air per kg of fuel, and

= Air-fuel ratio.

$$\therefore \quad m_a = \frac{m_f \times C}{c_p(T_3 - T_2')} - m_f$$

$$\therefore \quad \frac{m_a}{m_f} = \frac{C}{c_p(T_3 - T_2')} - 1$$

$$= \frac{42000}{1.004(1023 - 458.7)} - 1 = 73.1$$

i.e.,   **Air - fuel ratio**   **= 73.1 : 1.**   **(Ans.)**

(*iii*) **Pressure of the gases leaving the turbine, p₄:**

Neglecting effect of fuel on mass flow,

Actual turbine work = actual compressor work

i.e., $c_p(T_2' - T_1) = c_p(T_3 - T_4')$

or $T_2' - T_1 = T_3 - T_4'$

$\therefore$ $458.7 - 288 = 1023 - T_4'$

$\therefore$ $T_4' = 852.3$ K

Also, $\eta_{turbine} = \dfrac{T_3 - T_4'}{T_3 - T_4}$

$$0.78 = \frac{1023 - 852.3}{1023 - T_4}$$

$$\therefore \quad T_4 = 1023 - \left(\frac{1023 - 852.3}{0.78}\right) = 804 \text{ K}$$

$$\therefore \quad \frac{T_4}{T_3} = \left(\frac{P_4}{P_3}\right)^{\frac{\gamma - 1}{\gamma}}$$

or $\dfrac{p_4}{p_3} = \left(\dfrac{T_4}{T_3}\right)^{\frac{\gamma}{\gamma-1}} = \left(\dfrac{804}{1023}\right)^{\frac{1.4}{1.4-1}} = 0.43$

or $p_4 = 4.04 \times 0.43 = \textbf{1.74 bar.}$ **(Ans.)**

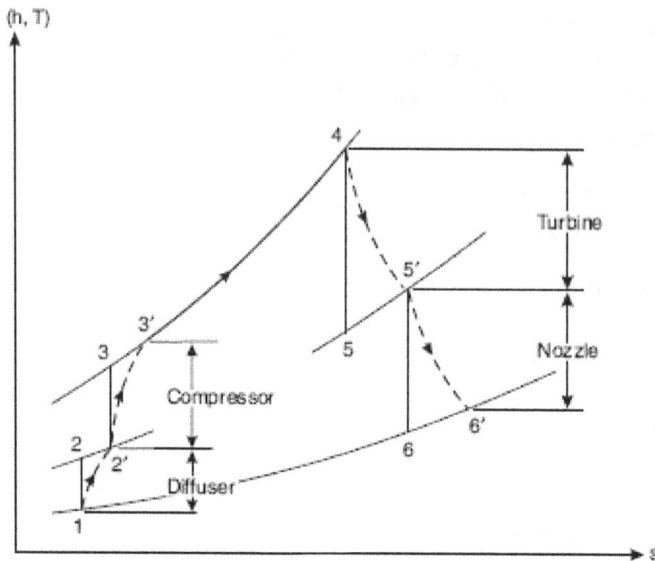

FIGURE 12.40 Sketch for Example 22.

(iv) **Thrust per kg of air per second:**

$$\frac{T_4'}{T_5} = \left(\frac{P_4}{P_5}\right)^{\frac{\gamma-1}{\gamma}} = \left(\frac{1.74}{1.01}\right)^{\frac{1.4-1}{1.4}} = 1.168$$

$$\therefore \quad T_5 = \frac{T_4'}{1.168} = \frac{852.3}{1.168} = 729.7 \text{ K}$$

$$\eta_{nozzle} = \frac{T_4' - T_5'}{T_4' - T_5}$$

$$0.88 = \frac{852 - T_5'}{852.3 - 729.7}$$

$$\therefore \quad T_5' = 852.3 - 0.88 \,(852.3 - 729.7) = 744.4 \text{ K}$$

If C_j is the jet velocity, then

$$\frac{C_j^2}{2} = c_p \left(T_4' - T_5'\right)$$

$$\therefore \quad C_j = \sqrt{2 \times c_p \left(T_4' - T_5'\right)}.$$

$$= \sqrt{2 \times 1.004 \,(852.3 - 744.4) \times 1000} = 465.5 \text{ m/s}$$

$\therefore$ Thrust per kg per second $= 1 \times 465.5 = $ **465.5 N.**   **(Ans.)**

Example 22

A turbo-jet engine travels at 216 m/s in air at 0.78 bar and – 7.2 °C. Air first enters the diffuser in which it is brought to rest relative to the unit, and it is then compressed in a compressor through a pressure ratio of 5.8 and fed to a turbine at 1110 °C. The gases expand through the turbine and then through the nozzle to atmospheric pressure (i.e., 0.78 bar). The efficiencies of diffuser, nozzle, and compressor are each 90%. The efficiency of the turbine is 80%. Pressure drop in the combustion chamber is 0.168 bar. Determine:

(i) Air-fuel ratio;

(ii) Specific thrust of the unit;

(iii) Total thrust, if the inlet cross-section of the diffuser is 0.12 m².

Assume calorific value of fuel as 44150 kJ/kg of fuel.

■ **Solution.**

Speed of the aircraft, $C_a = 216$ m/s

Intake air temperature, $T_1 = -7.2 + 273 = 265.8$ K

Intake air pressure, $p_1 = 0.78$ bar

Pressure ratio in the compressor, $r_p = 5.8$

Temperature of gases entering the gas turbine, $T_4 = 1110 + 273 = 1383$ K

Pressure drop in combustion chamber = 0.168 bar

$\eta_d = \eta_n; \eta_c = 90\%; \eta_t = 80\%$

Calorific value of fuel, $C.V. = 44150$ kJ/kg of coal

(*i*) **Air-fuel ratio:**

For an ideal diffuser (*i.e.*, process 1-2) the energy equation is given by:

$$h_2 = h_1 + \frac{C_a^2}{2} \text{ or } h_2 - h_1 = \frac{C_a^2}{2} \text{ or } T_2 - T_1 = \frac{C_a^2}{2c_p}$$

or $\quad T_2 = T_1 + \dfrac{C_a^2}{2c_p} = 265.8 + \dfrac{216^2}{2 \times 1.005 \times 1000} = 289$ K

For actual diffuser (*i.e.*, process 1-2′),

$$\eta_d = \left(\frac{h_2 - h_1}{h_2' - h_1} \right) \text{ or } h_2' - h_1 = \frac{h_2 - h_1}{\eta_d}$$

or $\quad h_2' = h_1 + \dfrac{h_2 - h_1}{\eta_d} = h_1 + \dfrac{C_a^2}{\eta_d}$

or $\quad T_2' = T_1 + \dfrac{C_a^2}{2c_p \eta_d} = 265.8 + \dfrac{216^2}{2 \times 1.005 \times 1000 \times 0.9} = 291.6$ K

Now, $\quad \dfrac{T_2}{T_1} = \left(\dfrac{P_2}{P_1}\right)^{\frac{\gamma-1}{\gamma}}$ or $\dfrac{289}{265.8} = \left(\dfrac{p_2}{0.78}\right)^{\frac{1.4-1}{1.4}}$ or $(1.087)^{3.5} = \left(\dfrac{p_2}{0.78}\right)$

or $\quad p_2 = 0.78 \times (1.087)^{3.5} = 1.044$ bar

Again, $\quad \dfrac{T_3}{T_2'} = (r_p)^{\frac{\gamma-1}{\gamma}} = (5.8)^{\frac{1.4-1}{1.4}} = 1.652$ or $T_3 = 291.6 \times 1.652 = 481.7$ K

Also, $\eta_c = \dfrac{T_3 - T_2'}{T_3' - T_2'}$ or $T_3' = T_2' + \dfrac{T_3 - T_2'}{\eta_c} = 291.6 + \dfrac{481.7 - 291.6}{0.9} = 502.8$ K

Assume $\quad c_{pg} = c_{pa} = c_p$

Heat supplied $\quad = (m_a + m_f) c_p T_4 - m_a c_p T_3' = m_f \times C$

or $\quad m_a c_p T_4 + m_f c_p T_4 - m_a c_p T_3' = m_f \times C$

or $\quad m_a c_p (T_4 - T_3') = m_f (C - c_p T_4)$

or $\quad \dfrac{m_a}{m_f} = \dfrac{C - c_p T_4}{c_p (T_4 - T_3')} = \dfrac{44150 - 1.005 \times 1383}{1.005(1383 - 502.8)} = 48.34$

$\therefore$ Air-fuel ratio = **48.34. (Ans.)**

(ii) Specific thrust of the unit:

$p_4 = p_3 - 0.168 = 5.8 \times 1.044 - 0.168 = 5.88$ bar

Assume that the turbine drives the *compressor only* (and not accessories also as is the usual case)

$\therefore \quad c_p(T_3' - T_2') = c_p(T_4 - T_5')$

or $\quad T_3' - T_2' = T_4 - T_5'$ or $T_5' = T_4 - (T_3' - T_2')$

$= 1383 - (502.8 - 291.6) = 1171.8$ K

Also, $\eta_t = \dfrac{T_4 - T_5'}{T_4 - T_5}$

or $\quad T_5 = T_4 - \dfrac{T_4 - T_5'}{\eta_t} = 1383 - \dfrac{1383 - 1171.8}{0.8} = 1119$ K

Now, $\dfrac{T_4}{T_5} = \left(\dfrac{P_4}{P_5}\right)^{\frac{\gamma-1}{\gamma}} = \left(\dfrac{5.88}{P_5}\right)^{\frac{1.4-1}{1.4}}$

or $\quad \left(\dfrac{1383}{1119}\right)^{3.5} = \dfrac{5.88}{p_5}$ or $p_5 = 2.8$ bar

Again, $\quad \dfrac{T_5'}{T_6} = \left(\dfrac{P_5}{P_6}\right)^{\frac{\gamma-1}{\gamma}} = \left(\dfrac{2.8}{0.78}\right)^{\frac{1.4-1}{1.4}} = 1.44$

or $\quad T_6 = \dfrac{T_5'}{1.44} = \dfrac{1171.8}{1.44} = 813.75$ K

and $\quad \eta_n = \dfrac{T_5' - T_6'}{T_5' - T_6}$ or $T_6' = T_5' - \eta_n(T_5' - T_6)$

$= 1171.8 - 0.9\,(1171.8 - 813.75) = 849.5$ K

Velocity at the exit of the nozzle,

$C_j = 44.72 \sqrt{h_5' - h_6'} = 44.72 \sqrt{c_p\,(T_5' - T_6')}$

$= 44.72 \sqrt{1.005(1171.8 - 849.5)} = 804.8$ m/s

Specific thrust $\quad = (1 + m_f) \times C_j = \left(1 + \dfrac{1}{48.34}\right) \times 804.8$

$= \mathbf{821.45\,N/kg\ of\ air/s.}$ **(Ans.)**

(*iii*) **Total Thrust:**

Volume of flowing air, $\quad V_1 = 0.12 \times 216 = 92$ m³/s

Mass flow, $\quad m_a = \dfrac{p_1 V_1}{RT_1} = \dfrac{0.78 \times 10^5 \times 25.92}{(0.287 \times 1000) \times 265.8} = 26.5$ kg/s

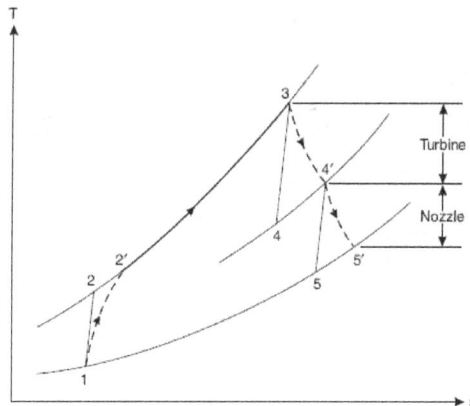

FIGURE 12.41 Sketch for Example 23.

$\therefore$ Total thrust $= 26.5 \times 821.45 = \mathbf{21768.4\ N.}$ **(Ans.)**

Example 23

The following data pertain to a jet engine flying at an altitude of 9000 meters with a speed of 215 m/s.

Thrust power developed	750 kW
Inlet pressure and temperature	0.32 bar, – 42 °C
Temperature of gases leaving the combustion chamber	690 °C
Pressure ratio 5.2	5.2
Calorific value of fuel	42500 kJ/kg
Velocity in ducts (constant)	195 m/s
Internal efficiency of turbine	86%
Efficiency of compressor	86%
Efficiency of jet tube	90%

For air : c_p = 1.005, γ = 1.4, R = 0.287
For combustion gases, c = 1.087
For gases during expansion, γ = 1.33.
Calculate the following:

(i) Overall thermal efficiency of the unit;

(ii) Rate of air consumption;

(iii) Power developed by the turbine;

(iv) The outlet area of the jet tube;

(v) Specific fuel consumption is kg per kg of thrust.

> ■ **Solution.**
> Refer to Figure 12.41.
> *Given*: T.P. = 750 kW ; p_1 = 0.32 bar, T_1 = – 42 + 273 = 231 K ;
> T_3 = 690 + 273 = 963 K ; r_{pc} = 5.2 ; C = 42500 kJ/kg ; C_a = 215 m/s,
> C'_4 = 19.5 m/s, η_c = 0.86; η_t = 0.86; η_{jt} = 0.9.
> Refer to Figure 12.41.
> Let m_f = kg of fuel required per kg of air
> Then, heat supplied per kg of air
> $$= 42500\ m_f = (1 + m_f) \times 1.087(T_3 - T_2') \tag{i}$$

Now, $\dfrac{T_2}{T_1} = \left(\dfrac{p_2}{p_1}\right)^{\frac{\gamma-1}{\gamma}} = (5.2)^{\frac{1.4-1}{1.4}} = (5.2)^{0.2857} = 1.60$

or $\quad T_2 = 231 \times 1.60 = 369.6$ K

Also, $\quad \eta_c = \dfrac{T_2 - T_1}{T_2' - T_1}$ or $T_2' = T_1 + \dfrac{T_2 - T_1}{\eta_c} = 231 + \dfrac{369.6 - 231}{0.86} = 392.2$ K

Substituting the value of T_2' in equation (i), we get

$42500\,m_f = \left(1 + m_f\right) \times 1.087(963 - 392.2) = 620.46\left(1 + m_f\right)$

or $\quad 42500\,m_f = 620.46 + 620.46\,m_f$

or $\quad m_f = \dfrac{620.46}{(42500 - 620.46)} = 0.0148 =$ fuel-air ratio

$\therefore$ Air-fuel ratio $= \dfrac{1}{0.0148} = 67.56 : 1$

The discharge velocity $C_j = C_5'$ cannot be determined from the thrust equation because the rate of air flow is not known. It may be determined from the expression of jet efficiency.

Jet efficiency,

$$\eta_{jet} = \dfrac{\text{Final kinetic energy in the jet}}{\text{Isentropic heat drop in the jet pipe} + \text{Carry-over from the turbine}}$$

or $\quad \eta_{jet} = \dfrac{c_j^2 / 2}{c_{pg}\left(T_4' - T_5\right) + C_4'^2 / 2} \quad \left(\text{where } C_4' = 195\text{m/s}\right)$ $\qquad (ii)$

Since the turbine's work is to drive the compressor only, therefore,

$$c_{pa}\left(T_2' - T_1\right) = c_{pg}\left(1 + \dfrac{m_f}{m_a}\right)\left(T_3 - T_4'\right)$$

or $\quad 1.005\,(392.2 - 231) = 1.087(1 + 0.0148)\,(963 - T_4')$

or $\quad T_4' = 963 - \dfrac{1.005(392.2 - 231)}{1.087(1 + 0.0148)} = 816.13$ K

Let $\quad r_{pt}$ = expansion pressure ratio in turbine $i.e.,\ r_{pt} = \dfrac{p_3}{p_4}$

r_{pj} = expansion pressure ratio in jet tube $i.e.,\ r_{pj} = \dfrac{p_4}{p_5}$

$\therefore \quad r_{pt} \times r_{pj} = \dfrac{p_3}{p_4} \times \dfrac{p_4}{p_5} \approx 5.2$

Now, $\eta_t = \dfrac{T_3 - T_4{}'}{T_3 - T_4}$ or $T_4 = T_3 - \dfrac{T_3 - T_4{}'}{\eta_t}$

$$= 963 - \dfrac{963 - 816.13}{0.86} = 792.2\,\text{K}$$

Also, $\dfrac{T_3}{T_4} = \left(\dfrac{p_3}{p_4}\right)^{\frac{\gamma-1}{\gamma}} = \left(\dfrac{p_3}{p_4}\right)^{\frac{1.33-1}{1.33}}$ or $\dfrac{963}{792.2} = \left(r_{pt}\right)^{0.248}$

or $\quad r_{pt} = \left(\dfrac{963}{792.2}\right)^{\frac{1}{0.248}} = 2.197$

$\therefore \quad r_{pj} = \dfrac{p_4}{p_5} = \dfrac{5.2}{2.197} = 2.366$

Thus, $\dfrac{T_4{}'}{T_5} = \left(r_{pj}\right)^{\frac{\gamma-1}{\gamma}} = (2.366)^{\frac{1.33-1}{1.33}} = 1.238$

or $\quad T_5 = \dfrac{T_4{}'}{1.238} = \dfrac{816.13}{1.238} = 659.23\,\text{K}$

Substituting the values in equation (ii), we get

$$0.9 = \dfrac{C_j^2/2}{1.087 \times 1000\,(816.13 - 659.23) + 195^2/2} = \dfrac{C_j^2/2}{189562.8}$$

$\therefore \quad C_j = \sqrt{0.9 \times 189562.8 \times 2} = 584.13\ \text{m/s}$

(i) **Overall efficiency, η_0:**

$$\eta_0 = \dfrac{\left[\left(1 + \dfrac{m_f}{m_a}\right)C_j - C_a\right]C_a}{\left(\dfrac{m_f}{m_a}\right) \times C} = \dfrac{\left[(1 + 0.0148) \times 584.13 - 215\right]215}{1000 \times 0.0148 \times 42500}$$

$= \mathbf{0.1291}$ or **12.91%. (Ans.)**

(ii) **Rate of air consumption, $\dot{m}_a$:**

Thrust power $=$ Thrust $\times$ Velocity of the unit

$$750 = \left[\dfrac{\left\{\left(1 + \dfrac{m_f}{m_a}\right)C_j - C_a\right\}\dot{m}_a}{1000}\right]C_a$$

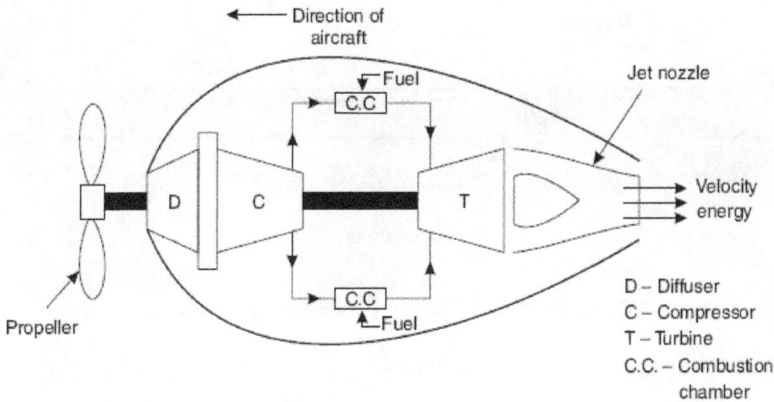

FIGURE 12.42 Turbo-prop.

or $\quad 750 = \dfrac{\{(1+0.0148)\times 584.13 - 215\}\times \dot{m}_a}{1000}\times 215 = 81.22\ \dot{m}_a$

or $\quad \dot{m}_a = \dfrac{750}{81.22} = \textbf{9.234 kg / s. (Ans.)}$

(*iii*) **Power developed by the turbine, P$_t$:**

$$p_t = \dot{m}_a\left(1+\dfrac{m_f}{m_a}\right)c_{pg}\left(T_3 - T_4'\right)$$

$$= 9.234(1+0.0148)\times 1.087(963-816.13) = \textbf{1496 kW. (Ans.)}$$

(*iv*) **The outlet area of the jet tube, A$_{jt}$:**

Now, $\quad \dfrac{C_j^2 - C_4'^2}{2} = c_{pg}\left(T_4' - T_5'\right)$

or $\quad T_5' = T_4' - \dfrac{C_j^2 - C_4'^2}{2\times c_{pg}}$

$$= 816.13 - \dfrac{\left(584.13^2 - 195^2\right)}{2\times 1.087\times 1000} = 676.67\ \text{K}$$

Assume the exit pressure of the gases is equal to the atmospheric pressure, that is, 0.32 bar.

Density of exhaust gases, $\quad \rho = \dfrac{p_5'}{RT_5'} = \dfrac{0.32\times 10^5}{0.29\times 1000\times 676.67} = 0.163\ \text{m}^3 / \text{kg}$

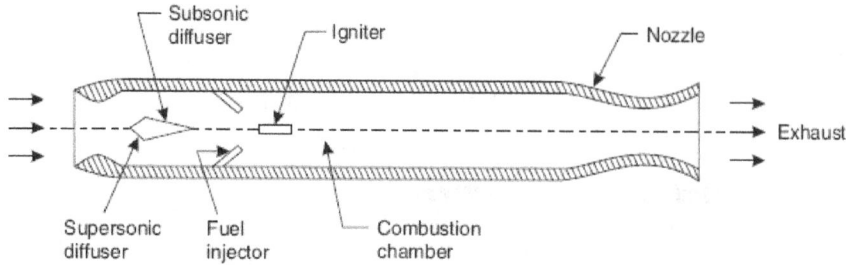

FIGURE 12.43 Schematic diagram of a Ram-jet propulsion unit.

(Assuming $R = 0.29$ for the gases)

Also, discharge of jet area $= A_{jt} \times C_j \times \rho = \dot{m}_a \left(1 + \dfrac{m_f}{m_a} \right)$

or $A_{jt} \times 584.13 \times 0.163 = 9.234 (1 + 0.0148)$

or $A_{jt} = \mathbf{0.0984\, m^2}.$ (**Ans.**)

(*v*) **Specific fuel consumption in kg per kg of thrust:**

Specific fuel consumption $= \dfrac{0.0148 \times 9.234 \times 3600}{1000 \times (750 / 215)}$

$= \mathbf{0.141\, kg\, /\, thrust\text{-}hour.}$ (**Ans.**)

12.8.4. Turbo-Propulsion

Figure 12.42 shows a turbo-prop system employed in aircrafts. Here the expansion of gases takes place *partly in the turbine* (80%) and *partly* (20%) *in the nozzle*. The power developed by the turbine is consumed in running the compressor and the propeller. The propeller and jet produced by the nozzle give forward motion to the aircraft. The turbo-prop entails the advantages of the turbo-jet (*i.e., low specific weight and simplicity in design*) and propeller (*i.e., high power for take-off and high propulsion efficiency at speeds below 600 km/h*). The overall efficiency of the turbo-prop is improved by providing the diffuser before the compressor as shown. The pressure rise takes place in the diffuser. This pressure rise takes place due to *conversion of kinetic energy of the incoming air* (equal to aircraft velocity) *into pressure energy by the diffuser*. This type of compression is known as the *"ram effect"*.

12.8.5. Ram-Jet

The ram-jet is also called *athodyd, Lorin tube,* or *flying stovepipe.* Ram-jet engines have the capability to fly at *supersonic speeds.* Figure 12.43 shows a schematic diagram of a ram-jet engine (a compressor and turbine are not necessary as the entire compression depends only on the ram compression).

The ram-jet engine consists of a *diffuser* (used for compression), *combustion chamber,* and *nozzle.*

The air enters the ram-jet plant with *supersonic speed* and is slowed down to *sonic velocity* in the *supersonic diffuser,* and consequently the pressure suddenly increases in the supersonic diffuser to the formation of a shock wave. The pressure of air is further increased in the subsonic diffuser *increasing the temperature of the air above the ignition temperature.*

In the combustion chamber, the fuel is injected through injection nozzles. The fuel air mixture is then ignited by means of a spark plug, and combustion temperatures of the order of 2000 K are attained. The expansion of gases toward the diffuser entrance is restricted by a pressure barrier at the other end of the diffuser, and as a result the hot gases are constrained to move toward the nozzle and undergo expansion; the pressure energy is converted into kinetic energy. The high velocity gases leaving the nozzle provide forward thrust to the unit.

The best performance of a ram-jet engine is obtained at flight speeds of 1700 km/h to 2000 km/h.

Advantages of A ram-jet engine

The ram-jet engine possesses the following *advantages* over other types of jet engines:

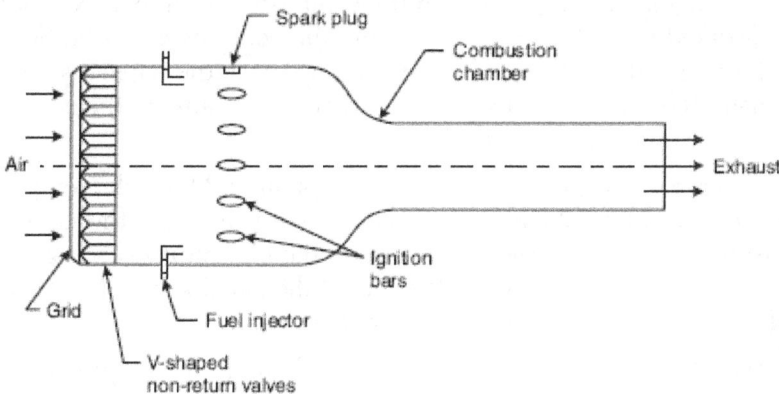

FIGURE 12.44 Pulse-Jet Engine.

No moving parts.

Light in weight.

A wide variety of fuels may be used.

Shortcomings/Limitations

It cannot be started on its own. It has to be accelerated to a certain flight velocity by some launching device. A ram-jet is always equipped with a small turbo-jet which starts the ram-jet.

The fuel consumption is too large at low and moderate speeds.

For successful operation, the diffuser needs to be designed carefully so that kinetic energy associated with high entrance velocities is efficiently converted into pressure.

To obtain steady combustion, certain elaborate devices in the form of flame holders or pilot flames are required.

12.8.6. Pulse-Jet Engine

A *pulse-jet engine* is an intermittent combustion engine and it operates on a cycle similar to a reciprocating engine, whereas the turbo-jet and ram-jet engines are continuous in operation and are based on a Brayton cycle. A pulse-jet engine, like an athodyd, develops thrust by a high velocity jet of exhaust gases without the aid of a compressor or turbine. Its development is primarily due to the inability of the ram-jet to be self-starting. Figure 12.44 shows a schematic arrangement of a pulse-jet propulsion unit.

The incoming air is compressed by ram effect in the diffuser section and the grid passages which are opened and closed by V-shaped non-return valves.

The fuel is then injected into the combustion chamber by fuel injectors (worked from the air pressure from the compressed air bottles). The combustion is then initiated by a spark plug (once the engine is operating normally, the spark is turned off and the residual flame in the combustion chamber is used for combustion).

As a result of combustion (of mixture of air and fuel) the temperature and pressure of combustion products increase. Because the combustion pressure is higher than the ram pressure, the non-return valves get closed and consequently the hot gases flow out of the tail pipe with a high velocity and in doing so give a forward thrust to the unit.

With the escape of gases to the atmosphere, the static pressure in the chamber falls and the high-pressure air in the diffuser forces the valves to open, and fresh air is admitted for combustion during a new cycle.

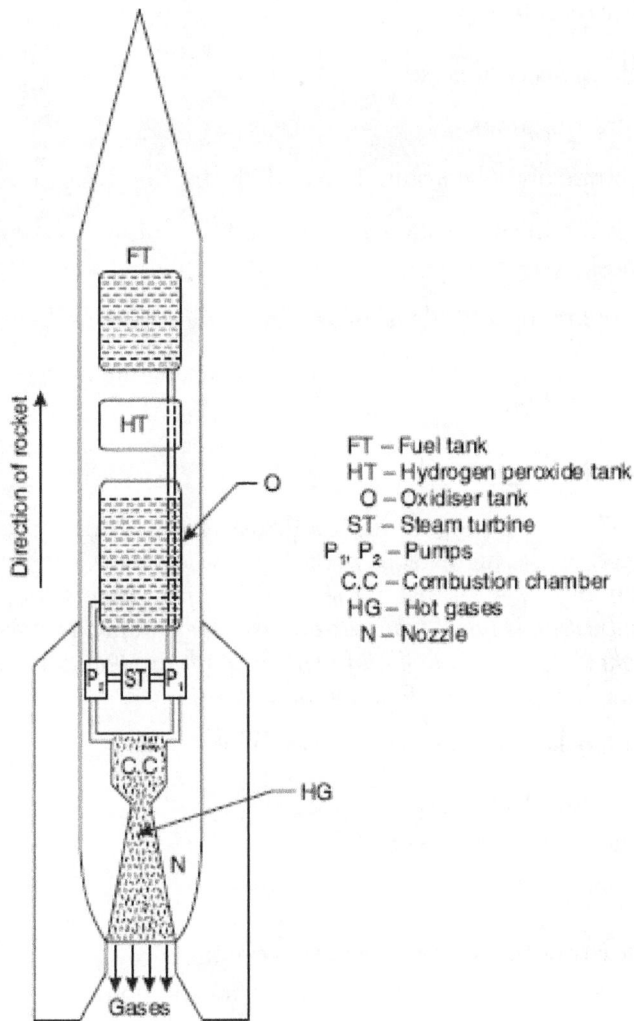

FIGURE 12.45 A simple type single-stage liquid propellant rocket

FT – Fuel tank
HT – Hydrogen peroxide tank
O – Oxidiser tank
ST – Steam turbine
P_1, P_2 – Pumps
C.C – Combustion chamber
HG – Hot gases
N – Nozzle

Advantages:

1. Simple in construction and very inexpensive as compared to a turbo-jet engine. Well adapted to pilotless aircraft.

2. Capable of producing static thrust and thrust in excess of drag at much lower speeds.

Shortcomings:

1. High intensity of noise.

2. Severe vibrations.

3. High rate of fuel consumption and low thermodynamic efficiency.

4. Intermittent combustion as compared to continuous combustion in a turbo-jet engine.

5. The operating altitude is limited by air density considerations.

6. Serious limitation to mechanical valve arrangement.

12.9. Rocket Engines

Similar to jet propulsion, the thrust required for rocket propulsion is produced by the high velocity jet of gases passing through the nozzle. But the main difference is that in the case of *jet propulsion the oxygen required for combustion is taken from the atmosphere and fuel is stored, whereas for the rocket engine, the fuel and oxidizer both are contained in a propelling body, and as such it can function in a vacuum.*

The rockets may be classified as follows:

1. **According to the type of propellants:**
   Solid propellant rocket

   Liquid propellant rocket

2. **According to the number of motors:**
   Single-stage rocket (consists of one rocket motor)

   Multi-stage rocket (consists of more than one rocket motor)

Figure 12.45 shows a simple type single-stage liquid propellant (the fuel and the oxidizer are commonly known as propellants) rocket. It consists of a fuel tank *FT*, an oxidizer tank *O*, two pumps P_1, P_2, a steam turbine *ST*, and a combustion chamber C.C. The fuel tank contains alcohol and the oxidizer tank contains liquid oxygen. The fuel and the oxidizer are supplied by the pumps to the combustion chamber where the fuel is ignited by electrical means. The pumps are driven with the help of a steam turbine. Here the steam is produced by mixing a very concentrated hydrogen-peroxide with potassium permanganate. The products of combustion are discharged from the combustion chamber through the nozzle N. So, the rocket moves in the opposite direction. In some modified form, this type of rocket may be used in missiles.

Requirements of an ideal rocket propellant

An ideal rocket propellant should have the following *characteristics/ properties:*

1. High heat value

2. Reliable smooth ignition

3. Stability and ease of handling and storing

4. Low toxicity and corrosiveness

5. Highest possible density so that it occupies less space

Applications of rockets

The fields of application of rockets are as follows:

1. Long range artillery

2. Lethal weapons

3. Signaling and firework display

4. Jet assisted take-off

5. For satellites

6. For spaceships

7. Research

12.9.1. Thrust Work, Propulsive Work, and Propulsive Efficiency

In rocket propulsion, since air is self-contained, the entry velocity relative to the aircraft is zero.

Neglecting the friction and other losses, we have the following formulae.

Thrust work $= C_j C_a$

Propulsive work $= C_j C_a + \dfrac{\left(C_j - C_a\right)^2}{2} = \dfrac{C_j^2 + C_a^2}{2}$

$$Rocket\ propulsive\ efficiency = \frac{C_j C_a}{(C_j^2 + C_a^2)/2} = \frac{2C_j C_a}{C_j^2 + C_a^2} = \frac{2\left(\dfrac{C_a}{C_j}\right)}{1 + \left(\dfrac{C_a}{C_j}\right)^2}$$

12.10 Exercises

1. Thermal efficiency of a gas turbine plant as compared to a Diesel engine plant is

   (a) higher   (b) lower
   (c) same   (d) may be higher or lower.

2. Mechanical efficiency of a gas turbine as compared to an internal combustion reciprocating engine is

   (a) higher   (b) lower
   (c) same   d) unpredictable.

3. For a gas turbine the pressure ratio may be in the range

   (a) 2 to 3   (b) 3 to 5
   (c) 16 to 18   (d) 18 to 22.

4. The air standard efficiency of a closed gas turbine cycle is given by $(r_p$ = pressure ratio for the compressor and turbine)

   (a) $\eta = 1 - \dfrac{1}{(r_p)^{\gamma-1}}$ (b) $\eta = 1 - (r_p)^{\gamma-1}$

   (c) $\eta = 1 - \left(\dfrac{1}{r_p}\right)^{\frac{\gamma-1}{\gamma}}$ (d) $\eta = (r_p)^{\frac{\gamma-1}{\gamma}} - 1.$

5. The work ratio of a closed cycle gas turbine plant depends upon

   (a) pressure ratio of the cycle and specific heat ratio

   (b) temperature ratio of the cycle and specific heat ratio

   (c) pressure ratio, temperature ratio, and specific heat ratio

   (d) only on pressure ratio.

6. Thermal efficiency of a closed cycle gas turbine plant increases by

   (*a*) reheating   (*b*) intercooling

   (*c*) regenerator   (*d*) all of the above.

7. With an increase in pressure ratio, thermal efficiency of a simple gas turbine plant with fixed turbine inlet temperature

   (*a*) decreases   (*b*) increases

   (*c*) first increases and then decreases   (*d*) first decreases and then increases.

8. The thermal efficiency of a gas turbine cycle with an ideal regenerative heat exchanger is

   (*a*) equal to work ratio   (*b*) less than work ratio

   (*c*) more than work ratio   (*d*) unpredictable.

9. In a two-stage gas turbine plant, reheating after the first stage

   (*a*) decreases thermal efficiency   (*b*) increases thermal efficiency

   (*c*) does not affect thermal efficiency   (*d*) none of the above.

10. In a two-stage gas turbine plant, reheating after the first stage

    (*a*) increases work ratio   (*b*) decreases work ratio

    (*c*) does not affect work ratio   (*d*) none of the above.

11. In a two-stage gas turbine plant, with intercooling and reheating

    (*a*) both work ratio and thermal efficiency improve

    (*b*) work ratio improves but thermal efficiency decreases

    (*c*) thermal efficiency improves but work ratio decreases

    (*d*) both work ratio and thermal efficiency decrease.

12. For a jet-propulsion unit, ideally the compressor work and turbine work are

    (*a*) equal   (*b*) unequal

    (*c*) not related to each other   (*d*) unpredictable.

13. The greater the difference between jet velocity and airplane velocity

    (*a*) the greater the propulsive   (*b*) the less the propulsive efficiency
        efficiency

    (*c*) the propulsive efficiency is   (*d*) none of the above.
        unaffected

14. What do you mean by the term "gas turbine"? How are gas turbines classified?

15. State the merits of gas turbines over I.C. engines and steam turbines. Discuss also the demerits over gas turbines.

16. Describe with neat sketches the working of a simple constant pressure open cycle gas turbine.

17. Discuss briefly the methods employed for improvement of thermal efficiency of open cycle gas turbine plants.

18. Describe with a neat diagram a closed cycle gas turbine. State also its merits and demerits.

19. Explain with a neat sketch the working of a constant volume combustion turbine.

20. Enumerate the various uses of gas turbines.

21. Write a short note on fuels used for gas turbines.

22. Explain the working difference between propeller-jet, turbo-jet, and turbo-prop.

23. State the fundamental differences between jet propulsion and rocket propulsion.

24. In an air standard gas turbine engine, air at a temperature of 15 °C and a pressure of 1.01 bar enters the compressor, where it is compressed through a pressure ratio of 5. Air enters the turbine at a temperature of 815 °C and expands to original pressure of 1.01 bar. Determine the ratio of turbine work to compressor work and the thermal efficiency when the engine operates on an ideal Brayton cycle.

25. In an open-cycle, constant-pressure gas turbine, air enters the compressor at 1 bar and 300 K. The pressure of air after the

compression is 4 bar. The isentropic efficiencies of the compressor and turbine are 78% and 85%, respectively. The air-fuel ratio is 80 : 1. Calculate the power developed and thermal efficiency of the cycle if the flow rate of air is 2.5 kg/s.

Take c_p = 1.005 kJ/kg K and γ = 1.4 for air and c_{pg} = 1.147 kJ/kg K and γ = 1.33 for gases. R = 0.287 kJ/kg K.
Calorific value of fuel = 42000 kJ / kg.

26. A gas turbine has a pressure ratio of 6/1 and a maximum cycle temperature of 600 °C. The isentropic efficiencies of the compressor and turbine are 0.82 and 0.85, respectively. Calculate the power output in kilowatts of an electric generator geared to the turbine when the air enters the compressor at 15 °C at the rate of 15 kg/s.

   Take c_p = 1.005 kJ/kg K and γ = 1.4 for the compression process, and take c_p = 1.111 kJ/kg K and γ = 1.333 for the expansion process

27. Calculate the thermal efficiency and the work ratio of the plant in example 3 (above), assuming that cp for the combustion process is 1.11 kJ / kg K.

28. The gas turbine has an overall pressure ratio of 5 : 1 and a maximum cycle temperature of 550 °C. The turbine drives the compressor and an electric generator, the mechanical efficiency of the drive being 97%. The ambient temperature is 20 °C and the isentropic efficiencies for the compressor and turbine are 0.8 and 0.83, respectively. Calculate the power output in kilowatts for an air flow of 15 kg/s. Calculate also the thermal efficiency and the work ratio.
   Neglect changes in kinetic energy and the loss of pressure in the combustion chamber.

29. Air is drawn in a gas turbine unit at 17 °C and 1.01 bar and the pressure ratio is 8 : 1. The compressor is driven by the H.P. turbine, and the L.P. turbine drives a separate power shaft. The isentropic efficiencies of the compressor and the H.P. and L.P. turbines are 0.8, 0.85, and 0.83, respectively. Calculate the pressure and temperature of the gases entering the power turbine, the net power developed by the unit per kg/s of mass flow, the work ratio, and the thermal efficiency of the unit. The maximum cycle temperature is 650 °C. For the compression process take c_p = 1.005 kJ/kg K and γ = 1.4

For the combustion process and expansion process, take
c_p = 1.15 kJ/kg K and γ = 1.333
Neglect the mass of fuel.

30. In a gas turbine plant, air is compressed through a pressure ratio of 6 : 1 from 15 °C. It is then heated to the maximum permissible temperature of 750 °C and expanded in two stages each of expansion ratio $\sqrt{6}$, the air being reheated between the stages to 750 °C. A heat exchanger allows the heating of the compressed gases through 75% of the maximum range possible. Calculate: (i) the cycle efficiency (ii), the work ratio (iii), and the work per kg of air.
The isentropic efficiencies of the compressor and turbine are 0.8 and 0.85, respectively.

31. At the design speed the following data apply to a gas turbine set employing the heat exchanger: isentropic efficiency of compressor = 75%, isentropic efficiency of the turbine = 85%, mechanical transmission efficiency = 99%, combustion efficiency = 98%,mass flow = 22.7 kg/s, pressure ratio = 6:1, heat exchanger effectiveness = 75%, and maximum cycle temperature = 1000 K.
The ambient air temperature and pressure are 15 °C and 1.013 bar. Calculate:

(i) The net power output   (ii) Specific fuel consumption

(iii) Thermal efficiency of the cycle.

Take the lower calorific value of fuel as 43125 kJ/kg and assume no pressure-loss in the heat exchanger and combustion chamber.

32. In a gas turbine plant air at 10 °C and 1.01 bar is compressed through a pressure ratio of 4 : 1. In a heat exchanger and combustion chamber the air is heated to 700 °C while its pressure drops 0.14 bar. After expansion through the turbine the air passes through a heat exchanger which cools the air through 75% of the maximum range possible, while the pressure drops 0.14 bar, and the air is finally exhausted to atmosphere. The isentropic efficiency of the compressor is 0.80 and that of the turbine is 0.85.
Calculate the efficiency of the plant.

33. In a marine gas turbine unit, a high-pressure stage turbine drives the compressor, and a low-pressure stage turbine drives the propeller through suitable gearing. The overall pressure ratio is 4 : 1, and the maximum temperature is 650 °C. The isentropic efficiencies

of the compressor, H.P. turbine, and L.P. turbine are 0.8, 0.83, and 0.85, respectively, and the mechanical efficiency of both shafts is 98%. Calculate the pressure between turbine stages when the air intake conditions are 1.01 bar and 25 °C. Calculate also the thermal efficiency and the shaft power when the mass flow is 60 kg/s. Neglect kinetic energy changes and pressure loss in combustion.

34. In a gas turbine unit comprising L.P. and H.P. compressors, air is taken at 1.01 bar 27 °C. Compression in the L.P. stage is up to 3.03 bar followed by intercooling to 30 °C. The pressure of air after the H.P. compressor is 58.7 bar. Loss in pressure during intercooling is 0.13 bar. Air from the H.P. compressor is transferred to a heat exchanger of effectiveness 0.60 where it is heated by gases from the L.P. turbine. The temperature of gases supplied to the H.P. turbine is 750 °C. The gases expand in the H.P. turbine to 3.25 bar and are then reheated to 700 °C before expanding in the L.P. turbine. The loss of pressure in the reheater is 0.1 bar. If isentropic efficiency of compression in both stages is 0.80 and isentropic efficiency of expansion in the turbine is 0.85, calculate:
    (*i*) Overall efficiency, (*ii*) Work ratio, and (*iii*) Mass flow rate when the gas power generated is 6500 kW. Neglect the mass of fuel.
    Take, for air : c_p = 1.005 kJ/kg K, γ = 1.4
    for gases c_{pg} = 1.115 kJ/kg K and γ = 1.3

35. In a gas turbine installation, air is taken in a L.P. compressor at 15 °C 1.1 bar and after compression it is passed through an intercooler where its temperature is reduced to 22 °C. The cooled air is further compressed in an H.P. unit and is then passed in the combustion chamber where its temperature is increased to 677 °C by burning the fuel. The combustion products expand in the H.P. turbine which runs the compressor and further expansion is continued in the L.P. turbine which runs the alternator. The gases coming out from the L.P. turbine are used for heating the incoming air from the H.P. compressor and are then exhausted to atmosphere. Taking the following data determine: (*i*) power output, (*ii*) specific fuel consumption, and (*iii*) Thermal efficiency:
    Pressure ratio of each compressor = 2, isentropic efficiency of each compressor stage = 85%, isentropic efficiency of each turbine stage = 85%, effectiveness of heat exchanger = 0.75, air flow = 15 kg/sec.,

calorific value of fuel = 45000 kJ / kg, c_p (for gas) = 1 kJ / kg K, c_p (for gas) = 1.15 kJ/kg K and γ (for air) = 1.4, γ (for gas) = 1.33.
Neglect the mechanical, pressure, and heat losses of the system and fuel mass also.

36. A turbo-jet engine flying at a speed of 960 km/h consumes air at the rate of 54.5 kg/s. Calculate: (*i*) Exit velocity of the jet when the enthalpy change for the nozzle is 200 kJ/kg and the velocity coefficient is 0.97, (*ii*) fuel flow rate in kg/s when air-fuel ratio is 75 : 1, (*iii*) Thrust specific fuel consumption, (*iv*) Thermal efficiency of the plant when the combustion efficiency is 93% and calorific value of the fuel is 45000 kJ/kg, (*v*) Propulsive power, (*vi*) Propulsive efficiency, and (*vii*) Overall efficiency.

37. A turbo-jet has a speed of 750 km/h while flying at an altitude of 10000 m. The propulsive efficiency of the jet is 50% and overall efficiency of the turbine plant is 16%. The density of air at 10000 m altitude is 0.173 kg/m3. The drag on the plant is 6250 N. The calorific value of the fuel is 48000 kJ/kg. Calculate: (*i*) Absolute velocity of the jet, (*ii*) Volume of air compressed per minute, (*iii*) Diameter of the jet, (*iv*) Power output of the unit in kW, and (*v*) Air-fuel ratio.

CHAPTER 13

REFRIGERATION

In This Chapter

- Overview
- Fundamventals of Refrigeration
- Air Refrigeration System
- Reversed Carnot Cycle
- Reversed Brayton Cycle
- Simple Vapor Compression System
- Factors Affecting the Performance of a Vapor Compression System
- Vapor Absorption System
- Refrigerants

13.1. Overview

Refrigeration is the science of producing and maintaining temperatures below that of the surrounding atmosphere. This means the removing of heat from a substance to be cooled. According to the 2nd law heat always passes downhill, from a warm body to a cooler one, until both bodies are at the same temperature. Maintaining perishables at their required temperatures is done by refrigeration. In addition, today many human work spaces in offices and factory buildings are air-conditioned, and a refrigeration unit is the heart of the system.

Before the advent of mechanical refrigeration water was kept cool by storing it in semi-porous jugs so that the water could seep through and evaporate. The evaporation carried away heat and cooled the water. This system was used by ancient civilizations for cooling water. Natural ice from lakes and rivers was often cut during winter and stored in caves, straw-lined pits, and later in sawdust-insulated buildings to be used as required. The

Romans carried pack trains of snow from the Alps to Rome for cooling the emperor's drinks. Though these methods of cooling all make use of natural phenomena, they were used to maintain a lower temperature in a space or product and may properly be called refrigeration.

13.2. Fundamventals of Refrigeration

In simple terms, refrigeration means the cooling of or removal of heat from a system. The equipment employed to maintain the system at a low temperature is termed as the refrigerating system, and the system which is kept at a lower temperature is called the refrigerated system. Refrigeration is generally produced in one of the following three ways:

(*i*)   By melting of a solid.

(ii) By sublimation of a solid.

(iii) By evaporation of a liquid.

Most of the commercial refrigeration is produced by the evaporation of a liquid called *refrigerant. Mechanical refrigeration* depends upon the evaporation of liquid refrigerant, and its circuit includes the equipment named the *evaporator, compressor, condenser,* and *expansion valve.* It is used for preservation of food, manufacture of ice, solid carbon dioxide, and control of air temperature and humidity in the air-conditioning system.

Important refrigeration applications:

1.   Ice making

2.   Transportation of foods above and below freezing

3.   Industrial air-conditioning

4.   Comfort air-conditioning

5.   Chemical and related industries

6.   Medical and surgical aids

7.   Processing food products and beverages

8.   Oil refining and synthetic rubber manufacturing

9.   Manufacturing and treatment of metals

10. Freezing food products

11. Miscellaneous applications:

13.2.1. Elements of Refrigeration Systems

All refrigeration systems must include at least four basic units as follows:

(*i*) *A low temperature thermal "sink" to which heat will flow from the space to be cooled.*

(*ii*) *Means of extracting energy from the sink, raising the temperature level of this energy, and delivering it to a heat receiver.*

(*iii*) *A receiver to which heat will be transferred from the high temperature, high-pressure refrigerant.*

(*iv*) *Means of reducing of pressure and temperature of the refrigerant as it returns from the receiver to the "sink."*

13.2.2. Refrigeration Systems

The various refrigeration systems may be enumerated as below:

1. Ice refrigeration

2. Air refrigeration system

3. Vapor compression refrigeration system

4. Vapor absorption refrigeration system

5. Special refrigeration systems

    (i)   Adsorption refrigeration system

    (*ii*)   Cascade refrigeration system

    (*iii*)   Mixed refrigeration system

    (*iv*)   Vortex tube refrigeration system

    (v)   Thermoelectric refrigeration

    (*vi*)   Steam jet refrigeration system.

13.2.3. Coefficient of Performance (C.O.P.)

The performance of a refrigeration system is expressed by a term known as the "coefficient of performance," which is defined as the ratio of heat absorbed by the refrigerant while passing through the evaporator to the work input required to compress the refrigerant in the compressor; in short it is the ratio between heat extracted and work done.

If, R_n = Net refrigerating effect,

W = Work expanded in by the machine during the same interval of time,

Then,   C.O.P. $= \dfrac{R_n}{W}$

and,   Relative C.O.P. $= \dfrac{\text{Actual C.O.P.}}{\text{Theoretical C.O.P.}}$

where,   Actual C.O.P. = Ratio of R_n and W

actually measured during a test and, Theoretical C.O.P. = Ratio of the theoretical values of R_n and W obtained by applying laws of thermodynamics to the refrigeration cycle

13.2.4. Standard Rating of a Refrigeration Machine

The rating of a refrigeration machine is obtained by a refrigerating effect or the amount of heat extracted in a given time from a body. The rating of the refrigeration machine is given by a unit of refrigeration known as *standard commercial metric ton of refrigeration*, which is defined as the refrigerating effect produced by the melting of 1 metric ton of ice from and at 0 °C in 24 hours. Since the latent heat of fusion of ice is 336 kJ/kg, the refrigerating effect of 336 × 1000 kJ in 24 hours is rated as one *metric ton*, *that is,*

$$1 \text{ metric ton of refrigeration } (\text{TR}) = \frac{336 \times 1000}{24} = 14000 \text{ kJ/h.}$$

Note that one Ton of Refrigeration is basically an American unit of refrigerating effect (R.E.). It originated from the rate at which heat is required to be removed to freeze one ton of water from and at 0 °C. Using American units this is equal to removal of 200 BTU of heat per minute, and in MKS units it is adopted as 50 kcal/min or 3000 kcal/hour. In S.I. units its conversion is rounded off to 3.5 kJ/s (kW) or 210 kJ/min.

In this book we shall be adopting,

1 metric ton of refrigeration = 14000 kJ/h (1 ton = 0.9 metric ton).

13.3. Air Refrigeration System

Air cycle refrigeration is one of the earliest methods of cooling developed. It became obsolete for several years because of its low coefficient of performance (C.O.P.) and high operating costs. It has, however, been applied to aircraft refrigeration systems, where with low equipment weight, it can utilize a portion of the cabin air according to the supercharger capacity. The main characteristic of an air refrigeration system is that throughout the cycle the refrigerant remains in a gaseous state.

The air refrigeration system can be divided in two systems:

(*i*) Closed system   (*ii*) Open system.

In a closed (or dense air) system the air refrigerant is contained within the piping or components of the system at all times and refrigerator pressure is usually above atmospheric pressure.

In the open system the refrigerator is replaced by the actual space to be cooled with the air expanded to atmospheric pressure, circulated through the cold room and then compressed to the cooler pressure. The pressure of operation in this system is inherently limited to operation at atmospheric pressure in the refrigerator.

A closed system claims the following *advantages* over an open system: (*i*) In a closed system the suction to the compressor may be at high pressure. The sizes of the expander and compressor can be kept within reasonable limits by using dense air; (*ii*) In open air system, the air picks up moisture from the products kept in the refrigerated chamber; the moisture may freeze during expansion and is likely to choke the valves, whereas it does not happen in closed system; and (*iii*) In an open system, the expansion of the refrigerant can be carried only up to atmospheric pressure prevailing in the cold chamber, but for a closed system there is no such restriction.

13.4. Reversed Carnot Cycle

If a machine working on a reversed Carnot cycle is driven from an external source, it will work or function as a refrigerator. The production of such a machine has not been possible practically because the adiabatic portion of the stroke would need a high speed, while during the isothermal portion of the stroke a very low speed would be necessary. This variation of speed during the stroke, however, is not practicable.

p-V and T-s diagrams of reversed Carnot cycles are shown in Figure 13.1 and Figure 13.2. Starting from point l, the clearance space of the cylinder is full of air, the air is then expanded adiabatically to point p during which its temperature falls from T_1 to T_2, and the cylinder is put in contact with a cold body at temperature T_2. The air is then expanded isothermally to the point n, as a result of which heat is extracted from the cold body at temperature T_2.

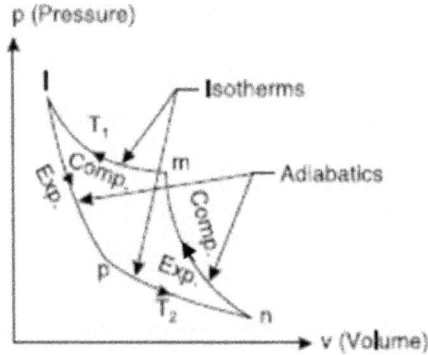

FIGURE 13.1 T-s diagram for a reversed Carnot cycle.

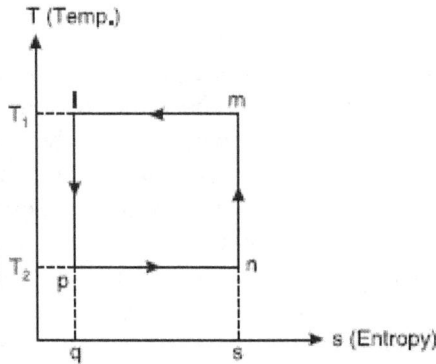

FIGURE 13.2 p-V diagram for reversed Carnot cycle.

Now the cold body is removed; from n to m the air undergoes adiabatic compression with the assistance of some external power, and the temperature rises to T_1. A hot body at temperature T_1 is put in contact with the cylinder. Finally, the air is compressed isothermally, during which process heat is rejected to the hot body. With reference to Figure 13.2, we have

Heat abstracted from the cold body $\quad$ = Area '$npqs$' = $T_2 \times pn$

Work done per cycle $\quad$ = Area '$lpnm$'

$= \left(T_1 - T_2 \right) \times pn$

Coefficient of performance,

$$\text{C.O.P.} = \frac{\text{Heat extracted from the cold body}}{\text{Work done per cycle}}$$

$$= \frac{\text{Area '}npqs\text{'}}{\text{Area '}lpnm\text{'}} = \frac{T_2 \times pn}{\left(T_1 - T_2 \right) \times pn} = \frac{T_2}{T_1 - T_2}$$

$$(13.1)$$

Since the coefficient of performance (C.O.P.) means the ratio of the desired effect in kJ/kg to the energy supplied in kJ/kg, therefore C.O.P. in the case of a Carnot cycle run either as a refrigerating machine, a heat pump, or as a heat engine will be as given as follows:

(*i*) **For a reversed Carnot cycle "refrigerating machine":**

$$\text{C.O.P.}_{\text{(ref.)}} = \frac{\text{Heat extracted from the cold body /cycle}}{\text{Work done per cycle}}$$

$$= \frac{T_2 \times pn}{(T_1 - T_2) \times pn} = \frac{T_2}{T_1 - T_2}$$

(*ii*) **For a Carnot cycle "heat pump":**

$$\text{C.O.P.}_{\text{(heat pump)}} = \frac{\text{Heat rejected to the hot body/cycle}}{\text{Work done per cycle}} = \frac{T_1 \times lm}{(T_1 - T_2) \times pn}$$

$$= \frac{T_1 \times pn}{(T_1 - T_2) \times pn} \qquad (13.2)$$

$$= \frac{T_1}{T_1 - T_2} \qquad (13.3)$$

$$= 1 + \frac{T_2}{T_1 - T_2} \qquad (13.4)$$

This indicates that the C.O.P. of the heat pump is greater than that of a refrigerator working on a reversed Carnot cycle between the same temperature limits T_1 and T_2 by unity.

(*iii*) **For a Carnot cycle "heat engine":**

$$\text{C.O.P.}_{\text{(heat engine)}} = \frac{\text{Work obtained/cycle}}{\text{Heat supplied/cycle}} = \frac{(T_1 - T_2) \times pn}{T_1 \times lm} = \frac{(T_1 - T_2) \times pn}{T_1 \times pn}$$

$$= \frac{T_1 - T_2}{T_1} \qquad (\because lm = pn) \qquad (13.5)$$

Example 1

A Carnot refrigerator requires 1.3 kW per metric ton of refrigeration to maintain a region at a low temperature of – 38 °C. Determine:

(*i*) *C.O.P. of Carnot refrigerator*

(*ii*) *Higher temperature of the cycle*

(*iii*) *The heat delivered and the C.O.P. when this device is used as a heat pump.*

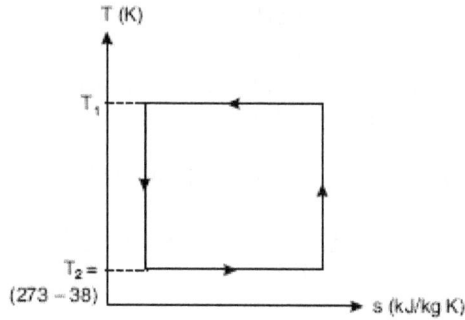

■ **Solution.**

$T_2 = 273 - 38 = 235$ K

Power required per metric ton of refrigeration = 1.3 kW

(i) C.O.P. of Carnot refrigerator:

$$\text{C.O.P.}_{\text{(Carnot ref.)}} = \frac{\text{Heat absorbed}}{\text{Work done}}$$

$$= \frac{1 \text{ metric ton}}{1.3} = \frac{14000 \text{ kJ/h}}{1.3 \times 60 \times 60 \text{ kJ/h}} = \textbf{2.99. (Ans.)}$$

(ii) Higher temperature of the cycle, T_1:

Figure 13.2

$$\text{C.O.P.}_{\text{(Carnot ref.)}} = \frac{T_2}{T_1 - T_2}$$

i.e.,

$$2.99 = \frac{235}{T_1 - 235}$$

∴

$$T_1 = \frac{235}{2.99} + 235 = 313.6 \text{ K}$$

$$= 313.6 - 273 = \textbf{40.6° C.} \quad \textbf{(Ans.)}$$

(iii) Heat delivered as heat pump

= Heat absorbed + Work done

$$= \frac{14000}{60} + 1.3 \times 60 = \textbf{311.3 kJ / min.} \quad \textbf{(Ans.)}$$

$$\textbf{C.O.P.}_{\text{(heat pump)}} = \frac{\text{Heat delivered}}{\text{Work done}} = \frac{311.3}{1.3 \times 60} = \textbf{3.99.} \quad \textbf{(Ans.)}$$

Example 2

A refrigerating system operates on the reversed Carnot cycle. The higher temperature of the refrigerant in the system is 35 °C and the lower temperature is – 15 °C. The capacity is to be 12 metric tons. Neglect all losses. Determine:

(*i*) *Coefficient of performance.*

(*ii*) *Heat rejected from the system per hour.*

(*iii*) *Power required.*

■ **Solution.**

(*i*)
$$T_1 = 273 + 35 = 308 \text{ K}$$
$$T_2 = 273 - 15 = 258 \text{ K}$$

Capacity $= 12$ metric ton

C.O.P. $= \dfrac{T_2}{T_1 - T_2} = \dfrac{258}{308 - 258} =$ **5.16. (Ans.)**

(*ii*) **Heat rejected from the system per hour:**

C.O.P. $= \dfrac{\text{Refrigerating effect}}{\text{Work input}}$

$5.16 = \dfrac{12 \times 14000 \text{ kJ/h}}{\text{Work input}}$

∴ Work input $= \dfrac{12 \times 1400}{5.16}$

Thus, *heat rejected / hour* = Refrigerating effect/hour + Work input/hour

$= 12 \times 14000 + 32558 =$ **200558 kJ / h. (Ans.)**

(*iii*) Power required:

Power required $= \dfrac{\text{Work input/hour}}{60 \times 60} = \dfrac{32558}{60 \times 60} =$ **9.04 kW. (Ans.)**

Example 3

A cold storage is to be maintained at – 5 °C while the surroundings are at 35 °C. The heat leakage from the surroundings into the cold storage is estimated to be 29 kW. The actual C.O.P. of the refrigeration plant used is one third that of an ideal plant working between the same temperatures. Find the power required to drive the plant.

■ **solution.**

$T_2 = -5 + 273 = 268 \text{ K} \; ; \; T_1 = 35 + 273 = 308 \text{ K}$

Heat leakage from the surroundings into the cold storage = 29 kW

Ideal C.O.P. $= \dfrac{T_2}{T_1 - T_2} = \dfrac{268}{308 - 268} = 6.7$

Actual C.O.P. $= \dfrac{1}{3} \times 6.7 = 2.233 = \dfrac{R_n}{W}$

where R_n = net refrigerating effect, and W = work done

or $2.233 = \dfrac{29}{W}$ or $W = \dfrac{29}{2.233} = 12.98$ kJ/s

Hence power required to drive the plant = **12.98 kW. (Ans.)**

Example 4

Ice is formed at 0 °C from water at 20 °C. The temperature of the brine is – 8 °C. Find out the kg of ice formed per kWh. Assume that the refrigeration cycle used is a perfect reversed Carnot cycle. Take latent heat of ice as 335 kJ/kg.

■ **Solution.**

Latent heat of ice = 335 kJ/kg

$T_1 = 20 + 273 = 293$ K

$T_2 = -8 + 273 = 265$ K

C.O.P. $= \dfrac{T_2}{T_1 - T_2} = \dfrac{265}{293 - 265} = 9.46$

Heat to be extracted per kg of water (to form ice at 0 °C, i.e., 273 K), R_n

$= 1 \times c_{pw} \times (293 - 273) +$ latent heat of ice

$= 1 \times 4.18 \times 20 + 335 = 418.6$ kJ/kg

Also, 1 kWh $= 1 \times 3600 = 3600$ kJ

Also, C.O.P. $= \dfrac{R_n}{W} = \dfrac{\text{Refrigerating effect in kJ/kg}}{\text{Work done in kJ}}$

∴ $9.46 = \dfrac{m_{ice} \times 418.6}{3600}$

i.e., $m_{ice} = \dfrac{9.46 \times 3600}{418.6} = 81.35$ kg

Hence ice formed per kWh = 81.35 kg. (Ans.)

Example 5

Find the least power of a perfect reversed heat engine that makes 400 kg of ice per hour at – 8 °C from feed water at 18 °C. Assume specific heat of ice as 2.09 kJ/kg K and latent heat as 334 kJ/kg.

■ **Solution.**

$T_1 = 18 + 273 = 291$ K

$T_2 = -8 + 273 = 265$ K

$$\text{C.O.P.} = \frac{T_2}{T_1 - T_2} = \frac{265}{291 - 265} = 10.19$$

Heat absorbed per kg of water (to form ice at – 8 °C, i.e., 265 K), R_n

$= 1 \times 4.18 (291 - 273) + 334 + 1 \times 2.09 \times (273 - 265) = 425.96$ kJ/kg

Also, $\text{C.O.P.} = \dfrac{\text{Net refrigerating effect}}{\text{Work done}} = \dfrac{R_n}{W}$

i.e., $10.19 = \dfrac{425.96 \times 400}{W}$

$\therefore \quad W = \dfrac{425.96 \times 400}{10.19} = 16720.7$ kJ/h

$= 4.64$ kJ/s or 4.64 kW

Hence, least power $= \textbf{4.64 kW. (Ans.)}$

Example 6

The capacity of the refrigerator (working on a reversed Carnot cycle) is 280 metric tons when operating between – 10 °C and 25 °C. Determine:

(i) Quantity of ice produced within 24 hours when water is supplied at 20 °C.

(ii) Minimum power (in kW) required.

■ **Solution.**

(i) Quantity of ice produced:

Heat to be extracted per kg of water (to form ice at 0 °C)
$= 4.18 \times 20 + 335 = 418.6$ kJ/kg

Heat extraction capacity of the refrigerator
= 280 metric tons

$= 280 \times 14000 = 3920000$ kJ/h

$\therefore$ Quantity of ice produced in 24 hours,

$$m_{ice} = \frac{3920000 \times 24}{418.6 \times 1000} = \textbf{224.75} \text{ metric tons.} \quad (\textbf{Ans.})$$

(ii) Minimum power required:

$$T_1 = 25 + 273 = 298 \text{ K}$$

$$T_2 = -10 + 273 = 263 \text{ K}$$

$$\text{C.O.P.} = \frac{T_2}{T_1 - T_2} = \frac{263}{298 - 263} = 7.51$$

Also, $\text{C.O.P.} = \dfrac{\text{Net refrigerating effect}}{\text{Work done /min}} = \dfrac{R_n}{W}$

i.e., $7.51 = \dfrac{3920000}{W}$

$\therefore \quad W = \dfrac{3920000}{7.51} \text{ kJ/h} = 145 \text{ kJ/s}$

$\therefore$ **Power required** $= \textbf{145 kW.} \quad (\textbf{Ans.})$

Example 7

A cold storage plant is required to store 20 metric tons of fish. The temperature of the fish when supplied = 25 °C; storage temperature of fish required = – 8 °C; specific heat of fish above freezing point = 2.93 kJ/kg °C; specific heat of fish below freezing point = 1.25 kJ/kg °C; freezing point of fish = – 3 °C. Latent heat of fish = 232 kJ/kg.

If the cooling is achieved within 8 hours, find out:

(i) Capacity of the refrigerating plant.
(ii) Carnot cycle C.O.P. between this temperature range.

(iii) If the actual C.O.P. is $\dfrac{1}{3}$ of the Carnot C.O.P. find out the power required to run the plant.

■ Solution.

Heat removed in 8 hours from each kg of fish
$$= 1 \times 2.93 \times \left[25 - (-3)\right] + 232 + 1 \times 1.25 \left[-3 - (-8)\right]$$

$$= 82.04 + 232 + 6.25 = 320.29 \text{ kJ / kg}$$

Heat removed by the plant /min
$$= \frac{320.29 \times 20 \times 1000}{8} = 800725 \text{ kJ/h}$$

(*i*) **Capacity of the refrigerating plant** $= \dfrac{800725}{14000} = \mathbf{57.19}$ metric tons. **(Ans.)**

(*ii*) $\qquad\qquad\qquad\qquad T_1 = 25 + 273 = 298 \text{ K}$

$T_2 = -8 + 273 = 265 \text{ K}$

∴ **C.O.P. of reversed Carnot cycle**

$= \dfrac{T_2}{T_1 - T_2} = \dfrac{265}{298 - 265} = \mathbf{8.03.}$ **(Ans.)**

(*iii*) **Power required:**

Actual C.O.P. $= \dfrac{1}{3} \times$ carnot C.O.P. $= \dfrac{1}{3} \times 8.03 = 2.67$

But actual C.O.P. $= \dfrac{\text{Net refrigerating effect/min}}{\text{Work done /min}} = \dfrac{R_n}{W}$

$2.67 = \dfrac{800725}{W} \text{ kJ/h}$

∴ $\quad W = \dfrac{800725}{2.67} = 299897 \text{ kJ/h} = 83.3 \text{ kJ/s}$

∴ *Power required to run the plant* = **83.3 kW. (Ans.)**

Example 8

A heat pump is used for heating the interior of a house in a cold climate. The ambient temperature is – 5 °C and the desired interior temperature is 25 °C. The compressor of the heat pump is to be driven by a heat engine working between 1000 °C and 25 °C. Treating both cycles as reversible, calculate the ratio in which the heat pump and heat engine share the heating load.

■ **Solution.**

Refer to Figure 13.3. *Given* : $T_1 = 1000 + 273 = 1273 \text{ K}$; $T_2 = 25 + 273 = 298 \text{ K}$

$T_3 = -5 + 273 = 268 \text{ K}$; $T_4 = 25 + 273 = 298 \text{ K}$

The ratio in which the heat pump and heat engine share the heating load, $\dfrac{Q_4}{Q_1}$:

Since both the cycles are *reversible*, therefore,

$\dfrac{Q_3}{Q_4} = \dfrac{T_3}{T_4}$ and $\dfrac{Q_2}{Q_1} = \dfrac{T_2}{T_1}$

or $\quad \dfrac{Q_3}{Q_4} = \dfrac{268}{298}$ or $Q_3 = \dfrac{268}{298} Q_4$ and $\dfrac{Q_2}{Q_1} = \dfrac{298}{1273}$

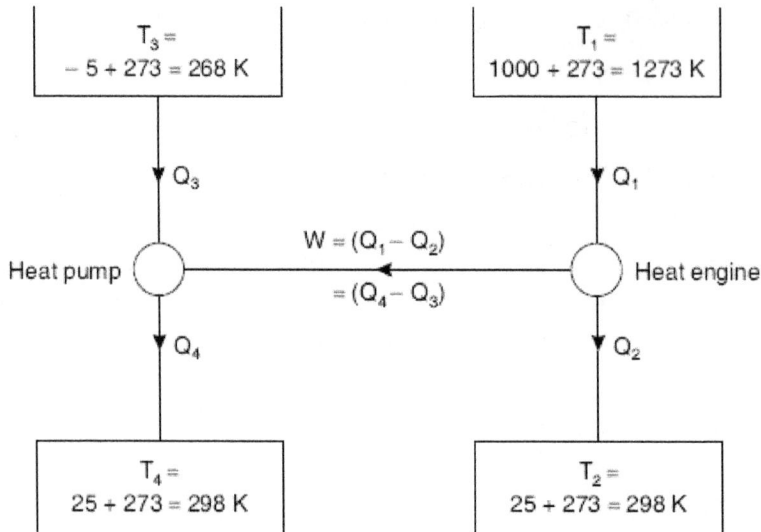

FIGURE 13.3 Sketch for Example 8.

Heat engine drives the heat pump,

$$\therefore \quad W = (Q_1 - Q_2) = Q_4 - Q_3$$

Dividing both sides by Q_1, we have

$$1 - \frac{Q_2}{Q_1} = \frac{Q_4 - Q_3}{Q_1}$$

$$1 - \frac{298}{1273} = \frac{Q_4 - \dfrac{268}{298} Q_4}{Q_1}$$

$$\frac{975}{1273} = \frac{30}{298} \times \frac{Q_4}{Q_1}$$

$$\therefore \quad \frac{Q_4}{Q_1} = \frac{975}{1273} \times \frac{298}{30} = 7.608. \quad (\textbf{Ans.})$$

13.5. Reversed Brayton Cycle

Figure 13.4 shows a schematic diagram of an air refrigeration system working on a reversed Brayton cycle. *Elements* of this systems are:

1. Compressor 2. Cooler (Heat exchanger)
3. Expander 4. Refrigerator.

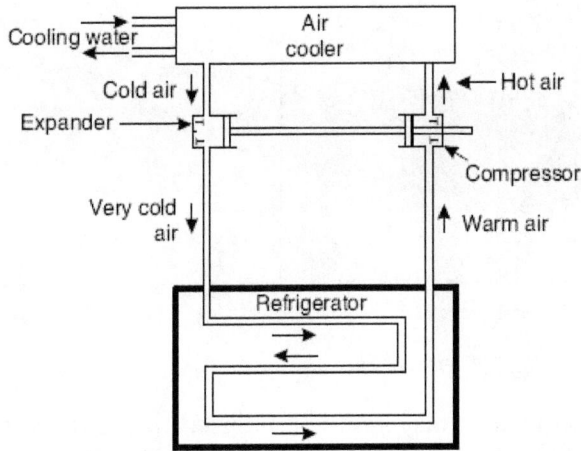

FIGURE 13.4 Air refrigeration system.

In this system, work gained from an expander is employed for compression of air, and consequently less external work is needed for operation of the system. In practice it may or may not be done; for example, in some aircraft refrigeration systems which employ an air refrigeration cycle, the expansion work may be used for driving other devices.

This system uses a reversed *Brayton cycle*, which is described below:

Figure 13.5 (a) and (b) shows p-V and T-s diagrams for a reversed Brayton cycle. Here it is assumed that (i) absorption and rejection of heat are constant pressure processes and (ii) compression and expansion are isentropic processes.

Considering m kg of air:

Heat absorbed in refrigerator, $Q_{added} = m \times c_p \times (T_3 - T_2)$

Heat rejected in cooler, $Q_{rejected} = m \times c_p \times (T_4 - T_1)$

If the process is considered to be polytropic, the *steady flow work of compression is* given by,

$$W_{comp} = \frac{n}{n-1} (p_4 V_4 - p_3 V_3)$$

(13.6)

Similarly work of expansion is given by,

$$W_{exp.} = \frac{n}{n-1} (p_1 V_1 - p_2 V_2)$$

(13.7)

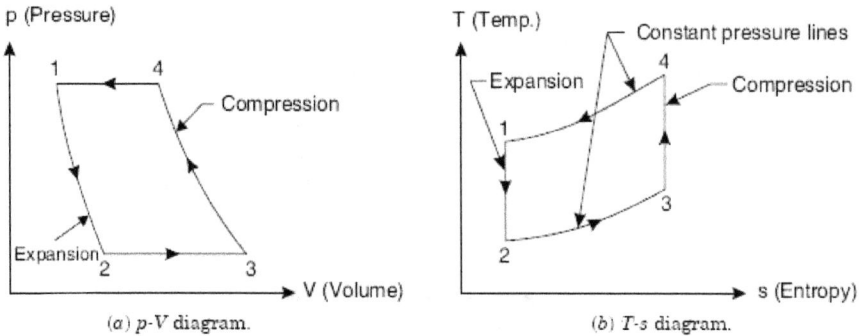

(a) p-V diagram.   (b) T-s diagram.

FIGURE 13.5 Reversed Brayton cycle.

Above equations may easily be reduced to the theoretical isentropic process shown in Figure 13.5 (b) by substituting $\gamma = n$ and the known relationship.

$$R = c_p \left(\frac{\gamma - 1}{\gamma} \right)$$

The net external work required for operation of the cycle

= Steady flow work of compression – Steady flow work of expansion

$$= W_{comp.} - W_{exp.}$$

$$= \left(\frac{n}{n-1} \right) \left(p_4 V_4 - p_3 V_3 - p_1 V_1 + p_2 V_2 \right) \qquad \left[\because \begin{array}{l} p_1 V_1 = mRT_1 \\ p_2 V_2 = mRT_2 \\ p_3 V_3 = mRT_3 \\ p_4 V_4 = mRT_4 \end{array} \right]$$

$$= \left(\frac{n}{n-1} \right) mR \left(T_4 - T_3 - T_1 + T_2 \right)$$

$$= \left(\frac{n}{n-1} \right) \frac{mR}{J} \left(T_4 - T_3 - T_1 + T_2 \right)$$

in heat units.

But $$R = c_p \left(\frac{\gamma - 1}{\gamma} \right)$$

(J = 1 in S.I. units)

$$\therefore \quad W_{comp.} - W_{exp.} = \left(\frac{n}{n-1} \right) \left(\frac{\gamma - 1}{\gamma} \right) mc_p \left(T_4 - T_3 + T_2 - T_1 \right)$$

(13.8)

For *isentropic compression* and *expansion*,

$$W_{net} = mc_p \left(T_4 - T_3 + T_2 - T_1 \right)$$

Now according to the law of conservation of energy, the net work on the gas must be equivalent to the net heat rejected.

Now, $\text{C.O.P.} = \dfrac{W_{added}}{Q_{rejected} - Q_{added}} = \dfrac{Q_{added}}{W_{net}}$

For the air cycle, assuming polytropic compression and expansion, the coefficient of performance is:

$$\text{C.O.P.} = \dfrac{m \times c_p \times (T_3 - T_2)}{\left(\dfrac{n}{n-1}\right)\left(\dfrac{\gamma-1}{\gamma}\right) m \times c_p \times (T_4 - T_3 + T_2 - T_1)}$$

$$= \dfrac{(T_3 - T_2)}{\left(\dfrac{n}{n-1}\right)\left(\dfrac{\gamma-1}{\gamma}\right)(T_4 - T_3 + T_2 - T_1)} \qquad (13.9)$$

The reversed Brayton cycle is same as the Bell-Coleman cycle. Conventionally the Bell-Coleman cycle refers to a closed cycle with expansion and compression taking place in a reciprocating expander and compressor, and heat rejection and heat absorption taking place in a condenser and evaporator, respectively.

With the development of efficient centrifugal compressors and gas turbines, the processes of compression and expansion can be carried out in centrifugal compressors and gas turbines, respectively. Thus, the shortcomings encountered with conventional reciprocating expanders and compressors is overcome. The reversed Brayton cycle finds its application for air-conditioning of airplanes where air is used as refrigerant.

13.5.1. Merits and Demerits of an Air-Refrigeration System

Merits

1. Since air is non-flammable, there is no risk of fire as in the machine using NH3 as the refrigerant.

2. It is cheaper as air is easily available as compared to the other refrigerants.

3. As compared to the other refrigeration systems, the weight of an air refrigeration system per metric ton of refrigeration is quite low; for this reason this system is employed in aircrafts.

Demerits

1. The C.O.P. of this system is very low in comparison to other systems.

2. The weight of air required to be circulated is more compared with refrigerants used in other systems. This is due to the fact that heat is carried by air in the form of *sensible heat*.

Example 9

A Bell-Coleman refrigerator operates between pressure limits of 1 bar and 8 bar. Air is drawn from the cold chamber at 9 °C, compressed, and then it is cooled to 29 °C before entering the expansion cylinder. Expansion and compression follow the law pv1.35 = constant. Calculate the theoretical C.O.P. of the system.

For air take $\gamma = 1.4$, $c_p = 1.003$ *kJ/kg K.*

■ **Solution.**

Given : $\qquad\qquad p_2 = 1.0$ bar ;

$p_1 = 8.0$ bar ;

$T_3 = 9 + 273 = 282$ K ;

$T_4 = 29 + 273 = 302$ K.

Considering *polytropic compression 3–4*, we have

$$\frac{T_4}{T_3} = \left(\frac{p_1}{p_2}\right)^{\frac{n-1}{n}} = \left(\frac{8}{1}\right)^{\frac{1.35-1}{1.35}} = (8)^{0.259} = 1.71$$

or $T_4 = T_3 \times 1.71 = 282 \times 1.71 = 482.2$ K

Again, considering *polytropic expansion 1–2*, we have

$$\frac{T_1}{T_2} = \left(\frac{p_1}{p_2}\right)^{\frac{n-1}{n}} = \left(\frac{8}{1}\right)^{\frac{1.35-1}{1.35}} = 1.71$$

$$T_2 = \frac{T_1}{1.71} = \frac{302}{1.71} = 176.6 \text{ K}$$

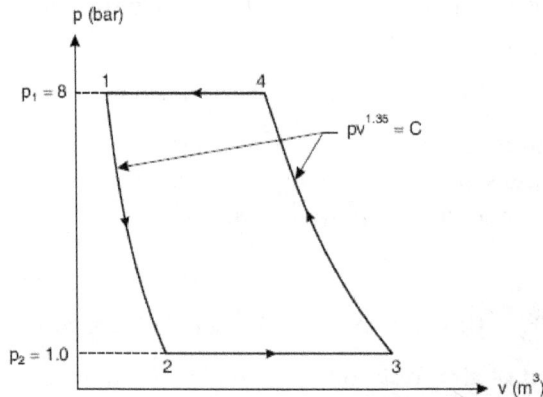

FIGURE 13.6 Sketch for Example 9.

Heat extracted from the cold chamber per kg of air

$$= c_p\left(T_3 - T_2\right) = 1.003\left(282 - 176.6\right) = 105.7 \text{ kJ/kg}.$$

Heat rejected in the cooling chamber per kg of air

$$= c_p\left(T_4 - T_1\right) = 1.003\left(482.2 - 302\right) = 180.7 \text{ kJ/kg}.$$

Since the compression and expansion are not isentropic, the difference between heat rejected and heat absorbed is not equal to the work done because there are heat transfers to the surroundings and from the surroundings during compression and expansion.

To find the work done, the area of the diagram '1-2-3-4' is to be considered:

$$\text{Work done} \ = \frac{n}{n-1}\left(p_4 V_4 - p_3 V_3\right) - \frac{n}{n-1}\left(p_1 V_1 - p_2 V_2\right)$$

$$= \frac{n}{n-1} R\left[\left(T_4 - T_3\right) - \left(T_1 - T_2\right)\right]$$

The value of R can be calculated as follows

$$\frac{c_p}{c_v} = \gamma$$

$$\therefore \ c_v = \frac{c_p}{\gamma} = \frac{1.003}{1.4} = 0.716$$

$$R = \left(c_p - c_v\right) = 1.003 - 0.716 = 0.287 \text{ kJ / kg K}.$$

$$\therefore \quad \text{Work done} \ = \frac{1.35}{0.35} \times 0.287 \left[\left(482.2 - 282\right) - \left(302 - 176.6\right)\right] = 82.8 \text{ kJ / kg}.$$

$$\therefore \ \textbf{C.O.P.} = \frac{\text{Heat abstracted}}{\text{Work done}} = \frac{105.7}{82.4} = \textbf{1.27. (Ans.)}$$

Example 10

An air refrigeration open system operating between 1 MPa and 100 kPa is required to produce a cooling effect of 2000 kJ/min. The temperature of the air leaving the cold chamber is – 5 °C and at leaving the cooler is 30 °C. Neglect losses and clearance in the compressor and expander. Determine:

(i) Mass of air circulated per min.;

(ii) Compressor work, expander work, cycle work;

(iii) C.O.P. and power in kW required.

■ **Solution.**

Pressure, $p_1 = 1$ MPa $= 1000$ kPa ; $p_2 = 100$ kPa

Refrigerating effect produced $= 2000$ kJ/min
Temperature of air leaving the cold chamber, $T_3 = -5 + 273 = 268$ K

Temperature of air leaving the cooler, $T_1 = 30 + 273 = 303$ K

(i) **Mass of air circulated per minute, m:**
For the *expansion process 1–2*, we have

$$\frac{T_1}{T_2} = \left(\frac{p_1}{p_2}\right)^{\frac{\gamma-1}{\gamma}} = \left(\frac{1000}{100}\right)^{\frac{1.4-1}{1.4}} = 1.9306$$

or $\qquad\qquad\qquad T_2 = \dfrac{T_1}{1.9306} = \dfrac{303}{1.9306} = 156.9$ K

Refrigerating effect per kg $= 1 \times c_p\left(T_3 - T_2\right) = 1.005\ \left(268 - 156.9\right) = 111.66$ kJ / kg

$\therefore$ Mass of air circulated per minute $= \dfrac{\text{Refrigerating effect}}{\text{Refrigerating effect per kg}}$

$$= \frac{2000}{111.66} = \textbf{17.91 kg / min. (Ans.)}$$

(ii) **Compressor work (W$_{comp.}$), expander work (W$_{exp.}$), and cycle work (W$_{cycle}$):**

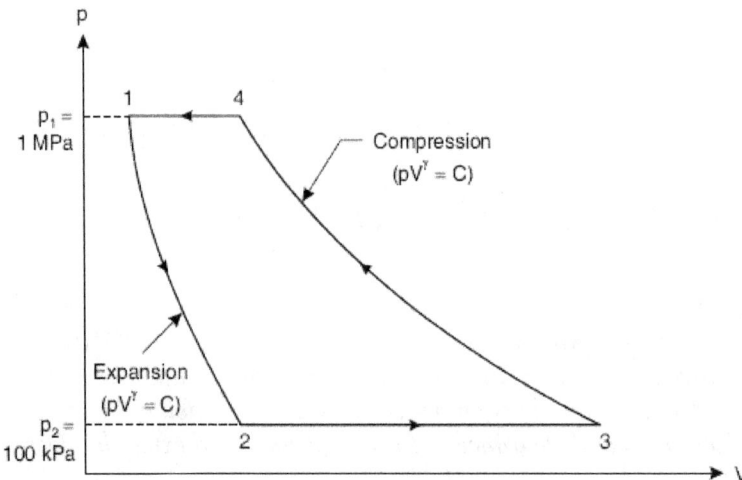

FIGURE 13.7 Sketch for Example 10.

For *compression process 3–4*, we have

$$\frac{T_4}{T_3} = \left(\frac{p_4}{p_3}\right)^{\frac{\gamma-1}{\gamma}} = \left(\frac{1000}{10}\right)^{\frac{1.4-1}{1.4}} = 1.9306$$

or $T_4 = 268 \times 1.9306 = \textbf{517.4 K. (Ans.)}$

Compressor work, W$_{\text{comp.}}$ $: \dfrac{\gamma}{\gamma-1} \, mR(T_4 - T_3)$

$$= \frac{1.4}{1.4-1} \times 17.91 \times 0.287 \, (517.4 - 268)$$

$= \textbf{4486.85 kJ / min. (Ans.)}$

Expander work, W$_{\text{exp.}}$ $: \dfrac{\gamma}{\gamma-1} \, mR \, (T_1 - T_2)$

$$= \frac{1.4}{1.4-1} \times 17.91 \times 0.287 \, (303 - 156.9)$$

$= \textbf{2628.42 kJ / min. (Ans.)}$

Cycle work, W$_{\text{cycle}}$**: W**$_{\text{comp.}}$ **– W**$_{\text{exp.}}$

$= 4486.85 - 2628.42 = \textbf{1858.43 kJ / min. (Ans.)}$

(*iii*) **C.O.P. and power required (P):**

$$\text{C.O.P.} = \frac{\text{Refrigerating effect}}{\text{Work required}} = \frac{2000}{1858.43} = \textbf{1.076 \ (Ans.)}$$

Power required,

$$P = \text{Work per second} = \frac{1858.43}{60} \text{ kJ/s or kW} = \textbf{30.97 kW. \ (Ans.)}$$

Example 11

A refrigerating machine of 6 metric tons capacity working on a Bell-Coleman cycle has an upper limit of pressure of 5.2 bar. The pressure and temperature at the start of the compression are 1.0 bar and 16 °C . The compressed air cooled at a constant pressure to a temperature of 41 °C enters the expansion cylinder. Assuming both expansion and compression processes to be adiabatic with γ = 1.4, calculate:

Coefficient of performance.

Quantity of air in circulation per minute.

Piston displacement of compressor and expander.

Bore of compressor and expansion cylinders. The unit runs at 240 r.p.m. and is double-acting. Stroke length = 200 mm.

Power required to drive the unit.

For air take $\gamma = 1.4$ *and* $c_p = 1.003$ *kJ / kg K.*

■ Solution.

$T_3 = 16 + 273 = 289$ K ; $T_1 = 41 + 273 = 314$ K

$p_1 = 5.2$ bar ; $p_2 = 1.0$ bar.

Considering the *adiabatic compression 3–4*, we have

$$\frac{T_4}{T_3} = \left(\frac{p_1}{p_2}\right)^{\frac{\gamma-1}{\gamma}} = \left(\frac{5.2}{1}\right)^{\frac{1.4-1}{1.4}} = (5.2)^{0.286} = 1.6$$

$\therefore$ $T_4 = 1.6$; $T_3 = 1.6 \times 289 = 462.4$ K

Considering the *adiabatic expansion 1–2*, we have

$$\frac{T_1}{T_2} = \left(\frac{p_1}{p_2}\right)^{\frac{\gamma-1}{\gamma}}$$

$$\frac{314}{T_2} = \left(\frac{5.2}{1}\right)^{\frac{0.4}{1.4}} = 1.6 \text{ or } T_2 = \frac{314}{1.6} = 196.25 \text{ K.}$$

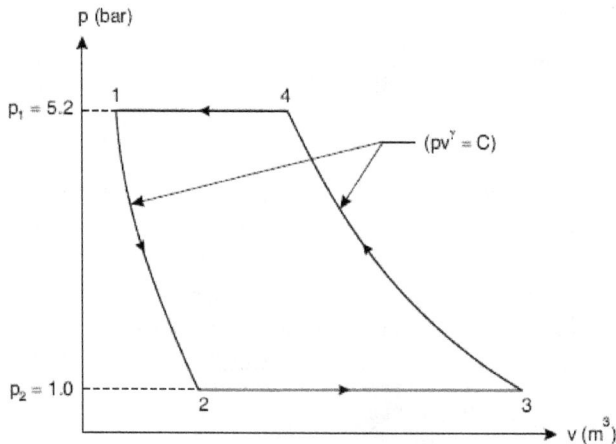

FIGURE 13.8 Sketch for Example 11.

(i) **C.O.P.:**

Since both the compression and expansion processes are isentropic/adiabatic reversible,

$$\therefore \quad C.O.P. \text{ of the machine} = \frac{T_2}{T_1 - T_2} = \frac{196.25}{314 - 196.25} = 1.67. \quad \textbf{(Ans.)}$$

(ii) **Mass of air in circulation:**

Refrigerating effect per kg of air

$$= c_p \left(T_3 - T_2 \right) = 1.003 \ (289 - 196.25) = 93.03 \text{ kJ/kg}.$$

Refrigerating effect produced by the refrigerating machine

$$= 6 \times 14000 = 84000 \text{ kJ/h}.$$

Hence mass of air in circulation

$$= \frac{84000}{93.03 \times 60} = \textbf{15.05 kg / min. (Ans.)}$$

(iii), *(iv)* **Piston displacement of compressor**

= Volume corresponding to point 3 *i.e.,* V_3

$$\therefore \ V_3 = \frac{mRT_3}{p_2} = \frac{15.05 \times 0.287 \times 1000 \times 289}{1.0 \times 10^5} = \textbf{12.48 m}^3 \textbf{ / min. (Ans.)}$$

$\therefore$ Swept volume per stroke

$$= \frac{12.48}{2 \times 240} = 0.026 \text{ m}^3$$

If, d_c = Dia. of compressor cylinder, and

l = Length of stroke,

then $\frac{\pi}{4} d_c^2 \times l = 0.026$

or $\frac{\pi}{4} d_c^2 \times \left(\frac{200}{1000} \right) = 0.026$

$$\therefore \ d_c = \left(\frac{0.026 \times 1000 \times 4}{\pi \times 200} \right)^{1/2} = 0.407 \text{ m or 407 mm}$$

i.e., Diameter or bore of the compressor cylinder = **407 mm. (Ans.)**

Piston displacement of expander

= Volume corresponding to point 2 *i.e.,* V_2

$$\therefore \ V_2 = \frac{mRT_2}{p_2} = \frac{15.05 \times 0.287 \times 1000 \times 196.25}{1 \times 10^5} = \textbf{8.476 m}^3 \textbf{ / min. (Ans.)}$$

∴ Swept volume per stroke

$$= \frac{8.476}{2 \times 240} = 0.0176 \text{ m}^3.$$

If d_e = dia. of the expander, and

l = length of stroke,

then $\frac{\pi}{4} d_c^2 \times l = 0.0176$

or $\frac{\pi}{4} d_c^2 \times \left(\frac{200}{1000} \right) = 0.0176$

∴ $d_e = \left(\frac{0.0176 \times 1000 \times 4}{\pi \times 200} \right)^{1/2} = 0.335 \text{ m or } 335 \text{ mm}$

i.e., Diameter or bore of the expander cylinder = **335 mm.** (**Ans.**)

(v) **Power required to drive the unit:**

$$\text{C.O.P.} = \frac{\text{Refrigerating effect}}{\text{Work done}} = \frac{R_n}{W}$$

$$1.67 = \frac{6 \times 14000}{W}$$

$$W = \frac{6 \times 14000}{1.67} = 50299.4 \text{ kJ/h} = 13.97 \text{ kJ/s}.$$

Hence power required = **13.97 kW.** (**Ans.**)

13.6. Simple Vapor Compression System

Out of all refrigeration systems, the vapor compression system is the most important system from the viewpoint of *commercial* and *domestic utility*. It is the most practical form of refrigeration. In this system the *working fluid is a vapor*. It readily evaporates and condenses or changes alternately between the vapor and liquid phases without leaving the refrigerating plant. During evaporation, it absorbs heat from the cold body. This heat is used as its latent heat for converting it from the liquid to vapor. In condensing or cooling or liquifying, it rejects heat to the external body, thus creating a cooling effect in the working fluid. This refrigeration system thus acts as a latent heat pump since it pumps its latent heat from the cold body or brine and rejects it or delivers it to the external hot body or cooling medium. The principle upon which the vapor compression system works applies to all the vapors for which tables of thermodynamic properties are available.

13.6.1. Simple Vapor Compression Cycle

In a simple vapor compression system, fundamental processes are completed in one cycle.

These are:

1. Compression

2. Condensation

3. Expansion

4. Vaporization.

The flow diagram of such a cycle is shown in Figure 13.9.

The *vapor at low temperature and pressure* (state '2') enters the *"compressor"* where it is compressed isentropically, and subsequently its temperature and pressure increase considerably (state '3'). This vapor after leaving the compressor enters the *"condenser"* where it is condensed into a *high-pressure liquid* (state '4') and is collected in a *"receiver tank."* From the receiver tank it passes through the *"expansion valve,"* and here it is *throttled down to a lower pressure* and has a low temperature (state '1'). After finding its way through the expansion "valve," it finally passes on to an *"evaporator"* where it *extracts heat from the surroundings* or *circulating fluid being refrigerated and vaporizes to a low-pressure vapor (state '2').*

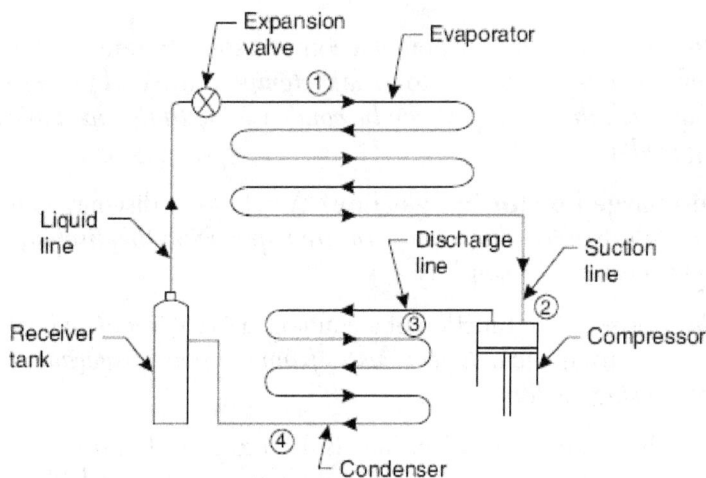

FIGURE 13.9 Vapor compression system.

Merits and demerits of a vapor compression system over an air refrigeration system:

Merits:

1. C.O.P. is quite high as the working of the cycle is very near to that of a reversed Carnot cycle.

2. When used on ground level the running cost of a vapor-compression refrigeration system is only 1/5 the running cost of an air refrigeration system.

3. For the same refrigerating effect the size of the evaporator is smaller.

4. The required temperature of the evaporator can be achieved simply by adjusting the throttle valve of the same unit.

Demerits:

1. Initial cost is high.

2. The major disadvantages are *inflammability*, *leakage of vapors,* and *toxicity.* These have been overcome to a great extent by improvement in design.

13.6.2. Functions of Parts of a Simple Vapor Compression System

Here follows the brief description of various parts of a simple vapor compression system, which is shown in Figure 13.9.

1. **Compressor.** The function of a compressor is to remove the *vapor* from the evaporator, and to *raise its temperature and pressure to a point such that it* (vapor) *can be condensed with the available condensing media.*

2. **Discharge line (or hot gas line).** A hot gas or discharge line *delivers the high-pressure, high-temperature vapor from the discharge of the compressor to the condenser.*

3. **Condenser.** The function of a condenser is to *provide a heat transfer surface through which heat passes from the hot refrigerant vapor to the condensing medium.*

4. **Receiver tank.** A receiver tank is used to provide *storage for a condensed liquid* so that a constant supply of liquid is available to the evaporator as required.

5. Liquid line. A liquid line carries the liquid refrigerant from the receiver tank to the refrigerant flow control.

6. Expansion valve (refrigerant flow control). Its function is to *meter the proper amount of refrigerant to the evaporator and to reduce the pressure of liquid entering the evaporator so that liquid will vaporize in the evaporator at the desired low temperature and take out a sufficient amount of heat.*

7. Evaporator. An evaporator *provides a heat transfer surface through which heat can pass from the refrigerated space into the vaporizing refrigerant.*

8. Suction line. The suction line *conveys the low-pressure vapor from the evaporator to the suction inlet of the compressor.*

13.6.3. Vapor Compression Cycle on a Temperature-Entropy (T-s) Diagram

We shall consider the following three cases:

1. When the vapor is dry and saturated at the end of compression. Figure 13.10 represents the vapor compression cycle on a *T-s* diagram; the points 1, 2, 3, and 4 correspond to the state points 1, 2, 3, and 4 in Figure 13.10.

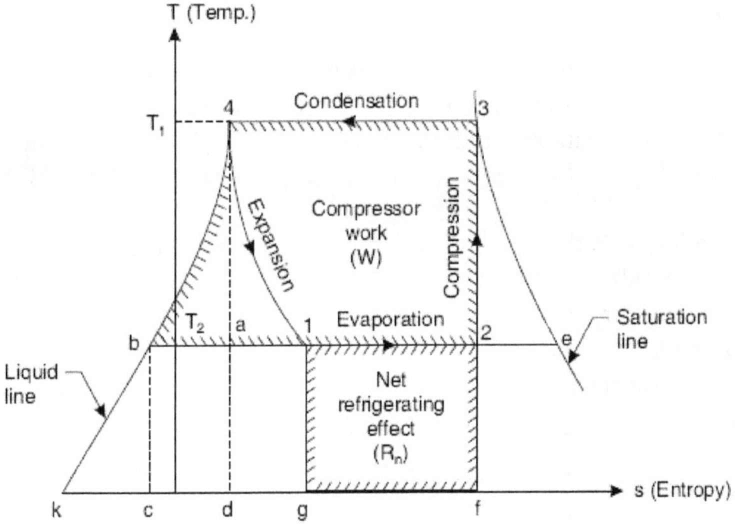

FIGURE 13.10 Vapor compression cycle on a T-s diagram.

At point '2' the vapor which is at low temperature (T_2) and low pressure enters the compressor's cylinder and is compressed adiabatically to '3' when its temperature increases to the temperature T_1. It is then condensed in the condenser (line 3–4) where it gives up its latent heat to the condensing medium. It then undergoes throttling expansion while passing through the expansion valve and it again reduces to T_2, represented by the line 4–1. From the T-s diagram it may be noted that due to this expansion the liquid partially evaporates, as its dryness fraction is represented by the ratio $\dfrac{b_1}{b_2}$. At '1' it enters the evaporator where it is further evaporated at constant pressure and constant temperature to the point '2' and the cycle is completed.

Work done by the compressor = W = Area '2-3-4-b-2'

Heat absorbed = Area '2-1-g-f-2'

$\therefore$ C.O.P. = $\dfrac{\text{Heat extracted or refrigerating effect}}{\text{Work done}} = \dfrac{\text{Area '2-1-}g\text{-}f\text{-2'}}{\text{Area '2-3-4-}b\text{-2'}}$

or C.O.P. = $\dfrac{h_2 - h_1}{h_3 - h_2}$

$$\text{[13.10(a)]}$$

$$= \dfrac{h_2 - h_4}{h_3 - h_2}$$

$$\text{[13.10(b)]}$$

($\because$ $h_1 = h_4$, *since during the throttling expansion $4-1$ the total heat content remains unchanged*)

2. **When the vapor is superheated after compression.** If the compression of the vapor is continued after it has become dry, the vapor will be superheated; its effect on the T-s diagram is shown in Figure 13.11. The vapor enters the compressor at condition '2' and is compressed to '3' where it is superheated to temperature T_{sup}. Then it enters the condenser. Here firstly superheated vapor cools to temperature T_1 (represented by line 3-3') and then it condenses at constant temperature along the line 3'-4; the remainder of the cycle, however, is the same as before.

Now, work done = Area '2-3-3'-4-b-2'

and Heat extracted/absorbed = Area '2-1-g-f-2'

$\therefore$ C.O.P = $\dfrac{\text{Heat extracted}}{\text{Work done}} = \dfrac{\text{Area '2-1-}g\text{-}f\text{-2'}}{\text{Area '2-3-3'-4-}b\text{-2'}} = \dfrac{h_2 - h_1}{h_3 - h_2}$

$$\text{[13.10(c)]}$$

In this case $h_3 = h_3' + c_p (T_{sup.} - T_{sat.})$ and h_3' = total heat of dry and saturated vapor at the point '3''.

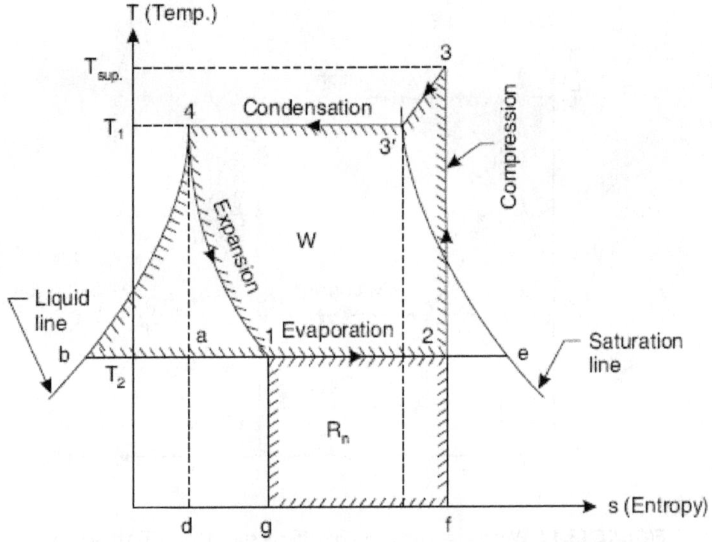

FIGURE 13.11 Superheated dry-vapor compression effects shown on a T-s diagram.

3. When the vapor is wet after compression. Refer to Figure 13.12.

Work done by the compressor = Area '2-3-4-*b*-2'

Heat extracted = Area '2-1-*g*-*f*-2'

$$\therefore \quad \text{C.O.P.} = \frac{\text{Heat extracted}}{\text{Work done}} = \frac{\text{Area '2-1-}g\text{-}f\text{-2'}}{\text{Area '2-3-4-}b\text{-2'}} = \frac{h_2 - h_1}{h_3 - h_2} \qquad [13.10(\text{d})]$$

If the vapor is not superheated after compression, the operation is called "WET COMPRESSION," and if the vapor is superheated at the end of compression, it is known as "DRY COMPRESSION." Dry compression in practice is always preferred, as it gives *higher volumetric efficiency* and *mechanical efficiency* and there are *less chances of compressor damage*.

13.6.4. Vapor Compression Cycles on a Pressure-Enthalpy (p-h) Chart

The diagrams commonly used in the analysis of the refrigeration cycle are:

(*i*) Pressure-enthalpy (*p-h*) chart

(*ii*) Temperature-entropy (*T-s*) chart.

Of the two, the pressure-enthalpy diagram seems to be more useful.

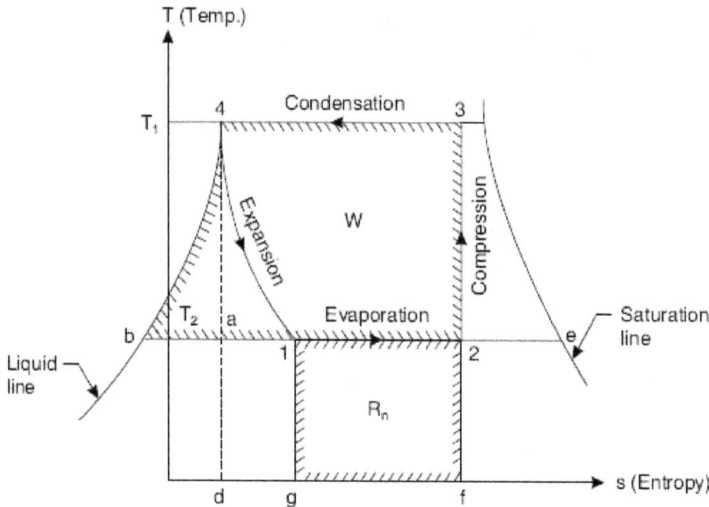

FIGURE 13.12 Wet-vapor compression effects shown on a T-s diagram.

The condition of the refrigerant in any thermodynamic state can be represented as a point on the p-h chart. This point may be located if any two properties of the refrigerant for that state are known; the other properties of the refrigerant for that state can be determined directly from the chart for studying the performance of the machines.

With reference to Figure 13.13, the chart is divided into three areas that are separated from each other by the saturated liquid and saturated vapor lines. The region on the chart to the *left* of the saturated liquid line is called the *sub-cooled region*. At any point in the sub-cooled region the refrigerant is in the liquid phase and its temperature is below the saturation temperature corresponding to its pressure. The area to the *right* of the saturated vapor line is a superheated region and the refrigerant is in the form of a *superheated vapor*. The section of the chart between the saturated liquid and saturated vapor lines is the two-phase region and represents the change in phase of the refrigerant between liquid and vapor phases. At any point between two saturation lines the refrigerant is in the form of a liquid vapor mixture. *The distance between the two lines along any constant pressure line, as read on the enthalpy scale at the bottom of the chart, is the latent heat of vaporization of the refrigerant at that pressure.*

The horizontal lines extending across the chart are lines of "constant pressure" and the vertical lines are lines of constant enthalpy. The lines of "constant temperature" in the sub-cooled region are almost vertical on the chart and parallel to the lines of constant enthalpy. In the center section,

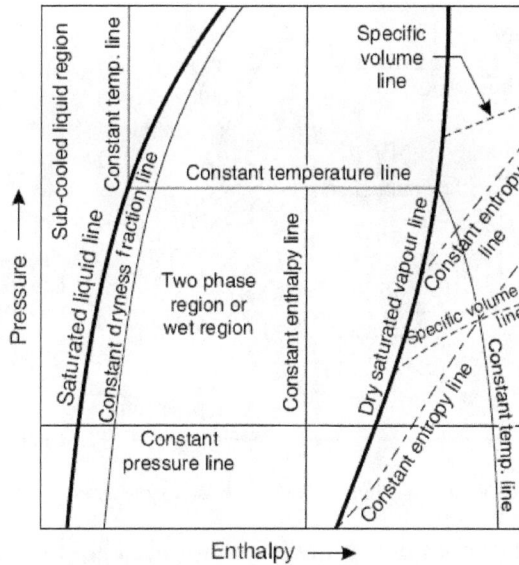

FIGURE 13.13 Pressure enthalpy (p-h) chart.

since the refrigerant changes state at a constant temperature and pressure, the lines of constant temperature are parallel to and coincide with the lines of constant pressure. At the saturated vapor line the lines of constant temperature change direction again and, in the superheated vapor region, fall of sharply toward the bottom of the chart.

The straight lines which extend diagonally and almost vertically across the superheated vapor region are lines of constant entropy. The curved, nearly horizontal lines crossing the superheated vapor region are lines of constant volume.

The p-h chart gives directly the changes in enthalpy and pressure during a process for thermodynamic analysis.

13.6.5. Simple Vapor Compression Cycle on a p-h Chart

Figure 13.14 shows a simple vapor compression cycle on a *p-h* chart. The points 1, 2, 3, and 4 correspond to the points marked in Figure 13.9.

The dry saturated vapor (at state 2) is drawn by the compressor from the evaporator at lower pressure p_1 and then it (vapor) is compressed isentropically to the upper pressure p_2. The isentropic compression is shown by the line 2-3. Since the vapor is dry and saturated at the start of compression, it becomes superheated at the end of compression as given by point 3. The process of *condensation which takes place at constant pressure* is given by

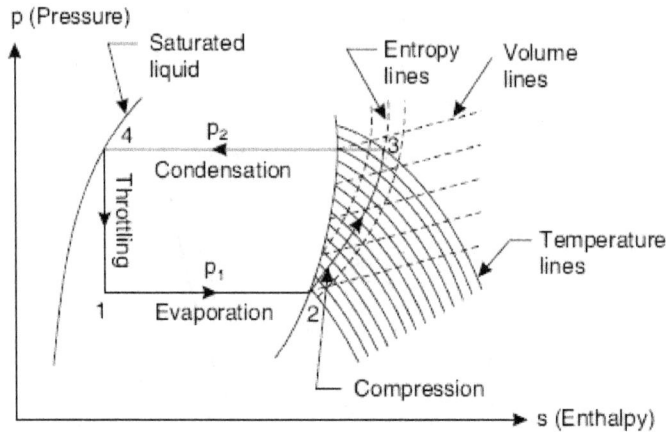

FIGURE 13.14 Simple vapor compression cycle on p-h chart.

the line 3-4. The vapor now reduced to saturated liquid is throttled through the expansion valve and the process is shown by the line 4-1. At the point 1 a mixture of vapor and liquid enters the evaporator where it gets dry saturated as shown by point 2. The cycle is thus completed.

Heat extracted (or refrigerating effect produced),

$$R_n = h_2 - h_1$$

Work done, $W = h_3 - h_2$

$$\therefore \text{ C.O.P.} = \frac{R_n}{W} = \frac{h_2 - h_1}{h_3 - h_2}$$

The values of h_1, h_2, and h_3 can be directly read from the p-h chart.

13.7. Factors Affecting the Performance of a Vapor Compression System

The factors which affect the performance of a vapor compression system are given as follows:

1. **Effect of suction pressure.** The effect of *decrease* in suction pressure is shown in Figure 13.15.

The C.O.P. of the original cycle,

$$\text{C.O.P.} = \frac{h_2 - h_1}{h_3 - h_2}$$

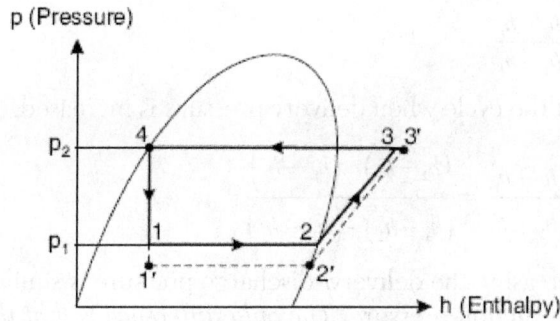

FIGURE 13.15 Effect of decrease in suction pressure.

The C.O.P. of the cycle when suction pressure is decreased,

$$C.O.P. = \frac{h_2' - h_1'}{h_3' - h_2'}$$

$$= \frac{(h_2 - h_1) - (h_2 - h_2')}{(h_3 - h_2) + (h_2 - h_2') + (h_3' - h_3)}$$

$$\left(\therefore \quad h_1 = h_1' \right)$$

This shows that the *refrigerating effect is decreased and work required is increased. The net effect is to reduce the refrigerating capacity of the system (with the same amount of refrigerant flow) and the C.O.P.*

2. Effect of delivery pressure. Figure 13.16 shows the effect of *increase in delivery pressure.*

C.O.P. of the original cycle,

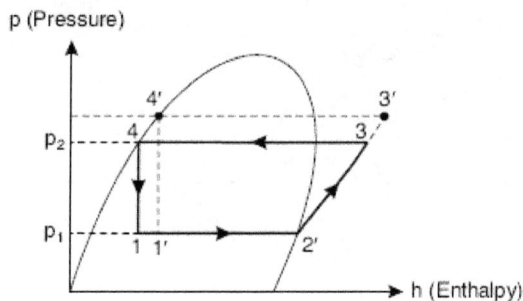

FIGURE 13.16 Effect of increase in delivery pressure.

$$\text{C.O.P.} = \frac{h_2 - h_1}{h_3 - h_2}$$

C.O.P. of the cycle when delivery pressure is increased,

$$\text{C.O.P.} = \frac{h_2 - h_1'}{h_3' - h_2} = \frac{\left(h_2 - h_1\right) - \left(h_1' - h_1\right)}{\left(h_3 - h_2\right) + \left(h_3' - h_3\right)}$$

The effect of increasing the delivery/ discharge pressure is similar to the effect of decreasing the suction pressure. *The only difference is that the effect of decreasing the suction pressure is more predominant than the effect of increasing the discharge pressure.*

The following points may be noted:

(*i*) *As the discharge temperature required in the summer is more as compared with winter, the same machine will give less refrigerating effect (load capacity decreased) at a higher cost.*

(*ii*) *The increase in discharge pressure is necessary for high condensing temperatures and a decrease in suction pressure is necessary to maintain low temperature in the evaporator.*

3. **Effect of superheating**. As may be seen from Figure 13.17 the effect of superheating is to increase the refrigerating effect, but this increase in refrigerating effect is at the cost of an increase in the amount of work spent to attain the upper pressure limit. *Since the increase in work is more as compared to the increase in the refrigerating effect, the overall effect of superheating is to give a low value of C.O.P.*

4. **Effect of sub-cooling of liquid.** *"Sub-cooling"* is the process of cooling the liquid refrigerant below the condensing temperature for a given pressure. In Figure 13.17 the process of sub-cooling is shown by 4-4′.

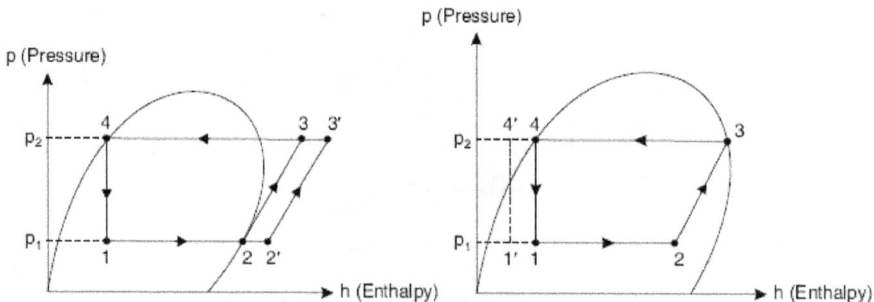

FIGURE 13.17 Effect of superheating.

As is evident from the figure the *effect of sub-cooling is to increase the refrigerating effect. Thus, sub-cooling results in an increase of C.O.P. provided that no further energy has to be spent to obtain the extra-cold coolant required.*

The *sub-cooling* or *undercooling* may be done by any of the following methods:

Inserting a special coil between the condenser and the expansion valve.

Circulating a greater quantity of cooling water through the condenser.

Using water cooler than the main circulating water.

5. Effect of suction temperature and condenser temperature. The performance of the vapor compression refrigerating cycle varies considerably with both vaporizing and condensing temperatures. Of the two, the *vaporizing temperature has the far greater effect.* It is seen that *the capacity and performance of the refrigerating system improve as the vaporizing temperature increases and the condensing temperature decreases.* Thus, the refrigerating system *should always be designed to operate at the highest possible vaporizing temperature and lowest possible condensing temperature*, of course, keeping in view the requirements of the application.

Actual Vapor Compression Cycle

The actual vapor compression cycle differs from the theoretical cycle in several ways because of the following *reasons*:

(*i*) Frequently the liquid refrigerant is sub-cooled before it is allowed to enter the expansion valve, and usually the gas leaving the evaporator is superheated a few degrees before it enters the compressor. This superheating may occur as a result of the type of expansion control used or through a pickup of heat in the suction line between the evaporator and compressor.

(*ii*) Compression, although usually assumed to be isentropic, may actually prove to be neither isentropic nor polytropic.

(*iii*) Both the compressor suction and discharge valves are actuated by pressure difference, and this process requires the actual suction pressure inside the compressor to be slightly below that of the evaporator and the discharge pressure to be above that of the condenser.

(*iv*) Although isentropic compression assumes no transfer of heat between the refrigerant and the cylinder walls, actually the cylinder walls are hotter than the incoming gases from the evaporator and colder than the compressed gases discharged to the condenser.

(*v*) A pressure drop in long suction and liquid line piping and any vertical differences in the head created by locating the evaporator and condenser at different elevations.

Figure 13.18 shows the actual vapor compression cycle on a *T-s* diagram. The various processes are discussed as follows:

Process 1-2-3. This process represents the passage of refrigerant through the evaporator, with *1-2* indicating the *gain of latent heat of vaporization, and 2-3, the gain of superheat before the entrance to the compressor*. Both of these processes approach very closely to the constant pressure conditions (assumed in theory).

Process 3-4-5-6-7-8. This path/process represents the passage of the vapor refrigerant from entrance to the discharge of the compressor. Path 3-4 represents the throttling action that occurs during passage through the suction valves, and path 7-8 represents the throttling during passage through the exhaust valves. Both of these actions are accompanied by an entropy increase and a slight drop in temperature.

Compression of the refrigerant occurs along path 5-6, which is actually neither isentropic nor polytropic. The heat transfers indicated by path 4-5 and 6-7 occur essentially at a constant pressure.

Process 8-9-10-11. This process represents the passage of refrigerant through the condenser with 8-9 indicating removal of superheat, 9-10 the removal of latent heat, and 10-11 removal of heat of liquid or sub-cooling.

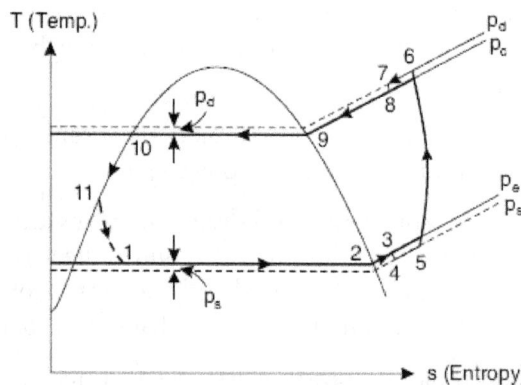

FIGURE 13.18 Actual vapor compression cycle (T-s diagram).

Process 11-1. This process represents passage of the refrigerant through the expansion valve, both theoretically and practically an irreversible adiabatic path.

Volumetric Efficiency

A compressor which is theoretically perfect would have neither clearance nor losses of any type and would pump on each stroke a quantity of refrigerant equal to piston displacement. No actual compressor is able to do this, since it is impossible to construct a compressor without clearance or one that will have no wire drawing through the suction and discharge valves, no superheating of the suction gases upon contact with the cylinder walls, or no leakage of gas past the piston or the valves. All these factors affect the volume of gas pumped or the capacity of the compressor, and some of them affect the H.P. requirements per metric ton of refrigeration developed.

"Volumetric efficiency" is defined as the *ratio of actual volume of gas drawn into the compressor* (at evaporator temperature and pressure) on *each stroke to the piston displacement*. If the effect of *clearance alone* is considered, the resulting expression may be termed *clearance volumetric efficiency*. The expression used for grouping into one constant all the factors affecting efficiency may be termed *total volumetric efficiency*.

Clearance volumetric efficiency. *"Clearance volume"* is the volume of space between the end of the cylinder and the piston when the latter is in a dead center position. The clearance volume is expressed as a percentage of piston displacement. In Figure 13.19 the piston displacement is shown as 4'-1.

During the suction stroke 4'-1, the vapor filled in clearance space at a discharge pressure p_d expands along 3-4 and the suction valve opens only

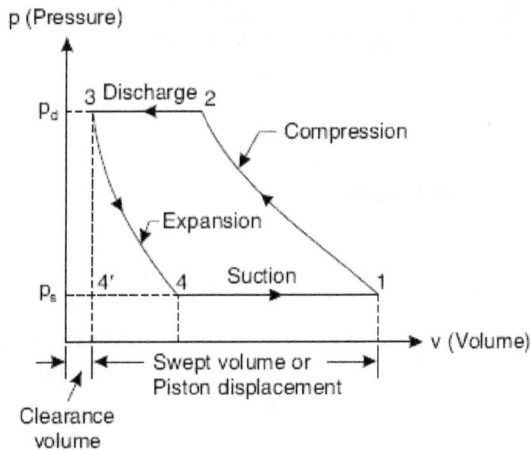

FIGURE 13.19 Compressor p-V cycle with piston displacement.

when pressure has dropped to suction pressure p_s, therefore actual volume sucked will be $(v_1 - v_4)$ while the swept volume is $(v_1 - v_4')$. The *ratio of actual volume of vapor sucked to the piston displacement is defined as clearance volumetric efficiency*.

Thus,

Clearance volumetric efficiency, $\quad \eta_{cv} = \dfrac{v_1 - v_4}{v_1 - v_4'} = \dfrac{v_1 - v_4}{v_1 - v_3} \left(\because \ v_4' = v_3 \right)$

Considering *polytropic expansion process 3-4*, we have

$$p_s v_4{}^n = p_d v_3{}^n$$

or, $\qquad \dfrac{p_d}{p_s} = \left(\dfrac{v_4}{v_3} \right)^n \qquad$ or $\quad v_4 = v_3 = \left(\dfrac{p_d}{p_s} \right)^{1/n}$

If the clearance ratio,

$$C = \frac{v_3}{v_1 - v_3} = \frac{\text{Clearance volume}}{\text{Swept volume}}$$

Thus, $\quad \eta_{cv} = \dfrac{v_1 - v_4}{v_1 - v_3} = \dfrac{(v_1 - v_4') - (v_4 - v_4')}{(v_1 - v_3)}$

$$= \frac{(v_1 - v_3) - (v_4 - v_3)}{(v_1 - v_3)} \quad (\because \ v_4' = v_3)$$

$$= 1 - \frac{v_4 - v_3}{v_1 - v_3}$$

$$= 1 - \frac{v_3 \left(\dfrac{p_d}{p_s} \right)^{1/n} - v_3}{v_1 - v_3} = 1 + \frac{v_3}{v_1 - v_3} \left[1 - \left(\frac{p_d}{p_s} \right)^{1/n} \right]$$

$$= 1 + C - C \left(\frac{p_d}{p_s} \right)^{1/n}$$

Hence clearance volumetric efficiency,

$$\eta_{cv} = 1 + C - C \left(\frac{p_d}{p_s} \right)^{1/n} \qquad\qquad (13.11)$$

Total volumetric efficiency. The total volumetric efficiency (η_{tv}) of a compressor is best obtained by actual *laboratory measurements of the amount of refrigerant compressed and delivered to the condenser*. It is very difficult to predict the effects of wire-drawing, cylinder wall heating, and piston leakage to allow any degree of accuracy in most cases. If the pressure drop through the suction valves and the temperature of the gases at the end of the suction stroke are known and if it is assumed that there is no leakage past the piston during compression, the total volumetric efficiency can be calculated by using the following equation:

$$\eta_{tv} = \left[1 + C - C \left(\frac{p_d}{p_s} \right)^{1/n} \right] \times \frac{p_c}{p_s} \times \frac{T_s}{T_c} \qquad (13.12)$$

where the subscript "c" refers to compressor cylinder and "s" refers to the evaporator or the suction line just adjacent to the compressor.

Mathematical Analysis of Vapor Compression Refrigeration

(*i*) **Refrigerating effect.** *Refrigerating effect is the amount of heat absorbed by the refrigerant in its travel through the evaporator.* In Figure 13.10 this effect is represented by the expression.

$$Q_{evap.} = \left(h_2 - h_1 \right) \text{ kJ/kg} \qquad (13.13)$$

In addition to the latent heat of vaporization it may include any heat of superheat absorbed in the evaporator.

(*ii*) **Mass of refrigerant.** Mass of refrigerant circulated (per second per metric ton of refrigeration) may be calculated by *dividing the amount of heat by the refrigerating effect.*

∴ Mass of refrigerant circulated,

$$m = \frac{14000}{3600 \left(h_2 - h_1 \right)} \text{ kg/s-metric ton} \qquad (13.14)$$

because one metric ton of refrigeration means a cooling effect of 14000 kJ/h.

(*iii*) **Theoretical piston displacement.** Theoretical piston displacement (per metric ton of refrigeration per minute) may be found by *multiplying the mass of refrigerant to be circulated (per metric ton of refrigeration per sec.) by the specific volume of the refrigerant gas,* $(v_g)_2$, at the entrance of the compressor. Thus,

$$\text{Piston displacement}_{(Theoretical)} = \frac{14000}{3600 (h_2 - h_1)} (v_g)_2 \text{ m}^3\text{/s-metric ton} \qquad (13.15)$$

(*iv*) **Power (Theoretical) required.** Theoretical power per metric ton of refrigeration is the power *theoretically required to compress the* *refrigerant.* Here volumetric and mechanical efficiencies are not taken into consideration. Power required may be calculated as follows:

(a) *When compression is isentropic:*

Work of compression $= h_3 - h_2$ (13.16)

Power required $= m\left(h_3 - h_2\right)$ kW

where m = Mass of refrigerant circulated in kg/s.

(b) *When compression follows the general law* $pV^n = constant$:

Work of compression $= \dfrac{n}{n-1}\ (p_3 v_3 - p_2 v_2)$ Nm/kg

Power required $= m \times \dfrac{n}{n-1}(p_3 v_3 - p_2 v_2) \times \dfrac{1}{10^3}$ kW (p is in N/m^2)   (13.17)

(v) **Heat rejected to compressor cooling water.** If the compressor cylinders are jacketed, an appreciable amount of heat may be rejected to the cooling water during compression. If the suction and discharge compression conditions are known, this heat can be determined as follows:

Heat rejected to compressor cooling water

$$= \left[\frac{n}{(n-1)}\left(\frac{p_3 v_3 - p_2 v_2}{1000}\right) - (h_3 - h_2)\right] \text{ kJ/kg} \qquad (p \text{ is in N/m}^2) \qquad (13.18)$$

(vi) **Heat removed through condenser.** Heat removed through the condenser includes all heat removed through the condenser, either as latent heat, heat of superheat, or heat of liquid. This is *equivalent to the heat absorbed in the evaporator plus the work of compression.*

∴ Heat removed through condenser

$= m\left(h_3 - h_4\right)$ kJ/s $\left(m = \text{mass of refrigerant circulated in kg/s}\right)$ (13.19)

13.8. Vapor Absorption System

In a vapor absorption system, the refrigerant is absorbed on leaving the evaporator, the absorbing medium being a solid or liquid. In order for the sequence of events to be continuous, it is necessary for the refrigerant to be separated from the absorbent and subsequently condensed before being returned to the evaporator. The separation is accomplished by the application of direct heat in a "generator." The solubility of the refrigerant and absorbent must be suitable and the plant which uses ammonia as the refrigerant and water as the absorbent will be described.

13.8.1. Simple Vapor Absorption System

Refer to Figure 13.20 for a simple absorption system. The solubility of ammonia in water at low temperatures and pressures is higher than it is at higher temperatures and pressures. The ammonia vapor leaving the evaporator at point 2 is readily absorbed in the low temperature hot solution in the absorber. This process is accompanied by the rejection of heat. The ammonia in the water solution is pumped to the higher pressure and is heated in the generator. *Due to the reduced solubility of ammonia in water at the higher pressure and temperature, the vapor is removed from the*

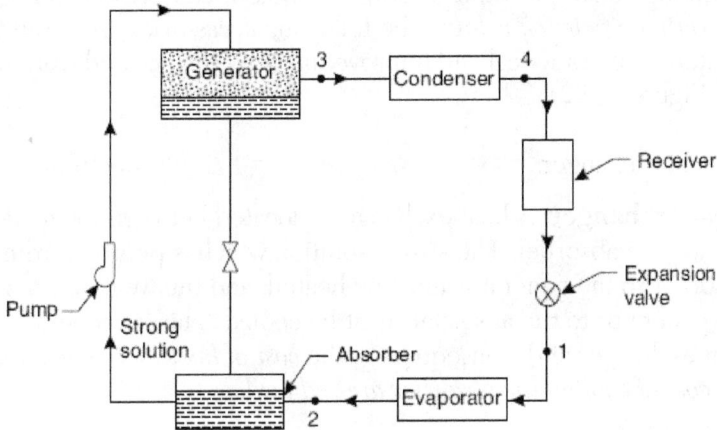

FIGURE 13.20 Simple vapor absorption system.

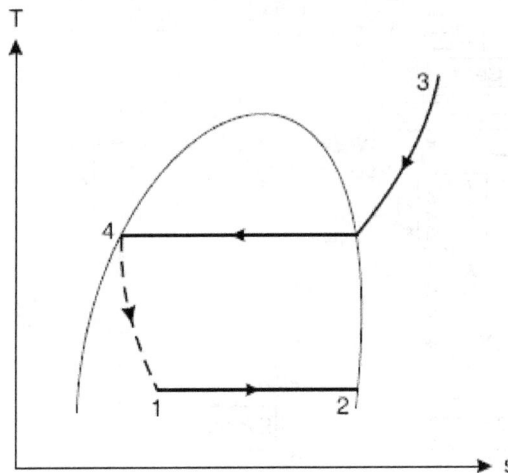

FIGURE 13.21 Simple vapor absorption system—T-s diagram.

solution. The vapor then passes to the condenser and the weakened ammonia in the water solution is returned to the absorber.

In this system the *work done on compression is less than in a vapor compression cycle* (since pumping a liquid requires much less work than compressing a vapor between the same pressures), but a heat input to the generator is required. The heat may be supplied by any convenient form, for example, steam or gas heating.

13.8.2. Practical Vapor Absorption System

Although a simple vapor absorption system can provide refrigeration, *its operating efficiency is low.* The following *accessories* are fitted to make the system more practical and improve the performance and working of the plant (Figure 13.22).

   1. Heat exchanger.   2. Analyzer.   3. Rectifier.

1. **Heat exchanger.** A heat exchanger is located between the generator and the absorber. The strong solution which is pumped from the absorber to the generator must be heated, and the weak solution from the generator to the absorber must be cooled. This is accomplished by a heat exchanger and, consequently, the *cost of heating the generator and the cost of cooling the absorber are reduced.*

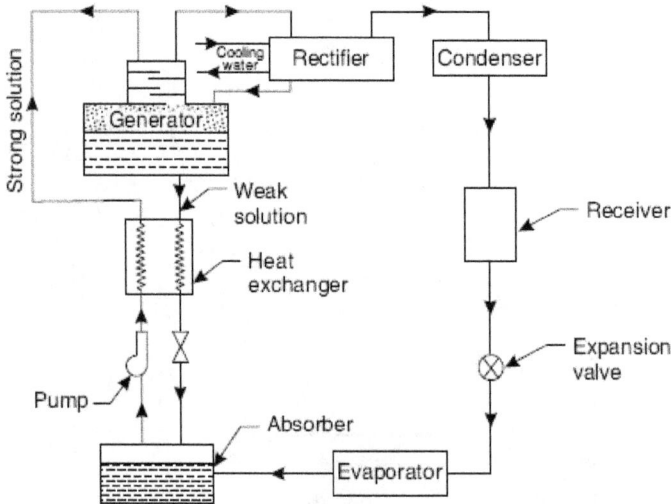

FIGURE 13.22 Practical Vapor Absorption system and components.

2. **Analyzer**. An analyzer consists of a series of trays mounted above the generator. Its main function is to partly remove some of the unwanted water particles associated with ammonia vapor going to the condenser. If these water vapors are permitted to enter condenser, they may enter the expansion valve and freeze; as a result the pipe line may get choked.

3. **Rectifier**. A rectifier is a water-cooled heat exchanger *which condenses water vapor and some ammonia and sends it back to the generator.* Thus, *final reduction or elimination of the percentage of water vapor takes place in a rectifier.*

The coefficient of performance (C.O.P.) of this system is given by:

$$\text{C.O.P.} = \frac{\text{Heat extracted from the evaporator}}{\text{Heat supplied in the generator} + \text{Work done by the liquid pump}}.$$

Comparison between Vapor Compression and Vapor Absorption Systems

Particulars	Vapor compression system	Vapor absorption system
Type of energy supplied	Mechanical—a high grade energy	Mainly heat—a low grade energy
Energy supply	Low	High
Wear and tear	More	Less
Performance at part loads	Poor	System not affected by variations of loads.
Suitability	Used where high grade mechanical energy is available	Can also be used at remote places as it can work even with a simple kerosene lamp (of course in small capacities)
Charging of refrigerant	Simple	Difficult
Leakage of refrigerant	More chances	No chance as there is no compressor or any reciprocating component to cause leakage.
Damage	Liquid traces in suction line may damage the compressor	Liquid traces of refrigerant present in piping at the exit of evaporator constitute no danger.

Example 12

A refrigeration machine is required to produce i.e., at 0 °C from water at 20 °C. The machine has a condenser temperature of 298 K while the evaporator temperature is 268 K. *The relative efficiency of the machine is 50%, and 6 kg of Freon-12 refrigerant is circulated through the system per minute. The refrigerant enters the compressor with a dryness fraction of 0.6. The specific heat of water is 4.187 kJ/kg K and the latent heat of ice is 335 kJ/kg. Calculate the amount of ice produced in 24 hours. The table of properties of Freon-12 is given below:*

Temperature K	Liquid heat kJ/kg	Latent heat kJ/g	Entropy of liquid kJ/kg
298	59.7	138.0	0.2232
268	31.4	154.0	0.1251

■ **Solution.**

Given : $m = 6$ kg/min. ; $\eta_{\text{relative}} = 50\%$; $x_2 = 0.6$; $c_{pw} = 4.187$ kJ/kg K ;

Latent heat of ice = 335 kJ/kg.

$h_{f_2} = 31.4$ kJ/kg ; $h_{fg_2} = 154.0$ kJ/kg ; $h_{f_3} = 59.7$ kJ/kg ;

$h_{fg_3} = 138$ kJ/kg ; $h_{f_4} = 59.7$ kJ/kg   ...From the table given above

$h_2 = h_{f_2} + x_2 h_{fg_2}$

$= 31.4 + 0.6 \times 154 = 123.8$ kJ/kg

For *isentropic compression 2-3*, we have
$s_3 = s_2$

$$s_{f_3} + x_3 \frac{h_{fg_3}}{T_3} = s_{f_2} + x_2 \frac{h_{fg_2}}{T_2}$$

$$0.2232 + x_3 \frac{138}{298} = 0.1251 + 0.6 \frac{154}{268}$$

$= 0.4698$

$$\therefore \quad x_3 = (0.4698 - 0.2232) \times \frac{298}{138} = 0.5325$$

Now, $h_3 = h_{f_3} + x_3 \, h_{fg_3} = 59.7 + 0.5325 \times 138 = 133.2$ kJ/kg

Also, $h_1 = h_{f_4} = 59.7$ kJ/kg

Theoretical   C.O.P. $= \dfrac{R_n}{W} = \dfrac{h_2 - h_1}{h_3 - h_2} = \dfrac{123.8 - 59.7}{133.2 - 123.8} = 6.82$

Actual   C.O.P. $= \eta_{\text{relative}} \times (\text{C.O.P.})_{\text{theoretical}} = 0.5 \times 6.82 = 3.41$

Heat extracted from 1 kg of water at 20 °C for the formation of 1 kg of ice at 0 °C

$= 1 \times 4.187 \times (20 - 0) + 335 = 418.74$ kJ / kg

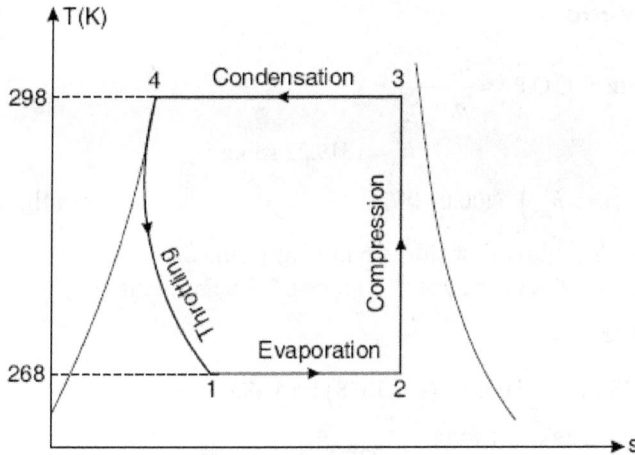

FIGURE 13.23 Sketch for Example 12.

Let m_{ice} = Mass of ice formed in kg/min.

$$(C.O.P.)_{actual} = 3.41 = \frac{R_n \text{(actual)}}{W} = \frac{m_{ice} \times 418.74}{m(h_3 - h_2)} = \frac{m_{ice} \times 418.74 \text{ (kJ/min)}}{6(133.2 - 123.8) \text{ (kJ/min)}}$$

$$\therefore \quad m_{ice} = \frac{6(133.2 - 123.8) \times 3.41}{418.74} = 0.459 \text{ kg/min}$$

$$= \frac{0.459 \times 60 \times 24}{1000} \text{ metric tons (in 24 hours)} = \textbf{0.661 metric ton. (Ans.)}$$

Example 13

Twenty-eight metric tons of ice from and at 0 °C is produced per day in an ammonia refrigerator. The temperature range in the compressor is from 25 °C to – 15 °C. The vapor is dry and saturated at the end of compression and an expansion valve is used. Assuming a coefficient of performance of 62% of the theoretical, calculate the power required to drive the compressor.

Temp. °C	Enthalpy (kJ/kg)		Entropy of liquid (kJ/kg K)	Entropy of vapor kJ/kg K
	Liquid	Vapor		
25	100.04	1319.22	0.3473	4.4852
– 15	– 54.56	1304.99	– 2.1338	5.0585

Take latent heat of ice = 335 kJ / kg.

■ **Solution.**

Theoretical C.O.P. $= \dfrac{h_2 - h_1}{h_3 - h_2}$

Here, $h_3 = 1319.22$ kJ/kg ;

$h_1 = h_4 \left(i.e., \ h_{f4}\right) = 100.04$ kJ/kg   ...From the table above

To find h_2, let us first find dryness at point 2.
Entropy at '2' = Entropy at '3' $\left(\text{Process 2-3 being isentropic}\right)$

$s_{f_2} + x_2 \ s_{fg_2} = s_{g_3}$

$-2.1338 + x_2 \times \left[5.0585 - \left(-2.1338\right)\right] = 4.4852$

$\therefore \quad x_2 = \dfrac{4.4852 + 2.1338}{5.0585 + 2.1338} = 0.92$

$\therefore \quad h_2 = h_{f_2} + x_2 \ h_{fg_2} = -54.56 + 0.92 \times [1304.99 - (-54.56)]$

$= 1196.23$ kJ/kg.

$\therefore \quad \text{C.O.P.}_{\text{(theoretical)}} = \dfrac{1196.23 - 100.04}{1319.22 - 1196.23} = 8.91.$

$\therefore \quad \text{C.O.P.}_{\text{(actual)}} = 0.62 \times \text{C.O.P.}_{\text{(theoretical)}}$ ...Given

i.e., $\text{C.O.P.}_{\text{(actual)}} = 0.62 \times 8.91 = 5.52$

Actual refrigerating effect per kg
$= \text{C.O.P.}_{\text{(actual)}} \times \text{work done}$

$= 5.52 \times \left(h_3 - h_2\right) = 5.52 \times \left(1319.22 - 1196.23\right)$

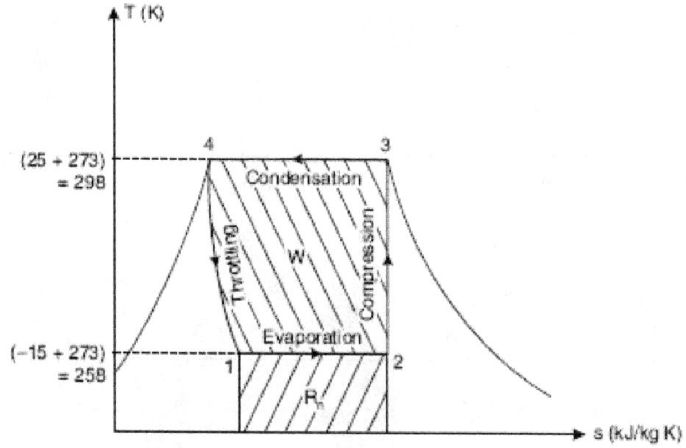

FIGURE 13.24 Sketch for Example 13.

$= 678.9$ kJ/kg

Heat to be extracted per hour

$$= \frac{28 \times 1000 \times 335}{24} = 390833.33 \text{ kJ}$$

Heat to be extracted per second $= \dfrac{390833.33}{3600} = 108.56$ kJ/s.

$\therefore$ Mass of refrigerant circulated per second $= \dfrac{108.56}{678.9} = 0.1599$ kg

Total work done by the compressor per second

$= 0.1599 \times (h_3 - h_2) = 0.1599 \, (1319.22 - 1196.23)$

$= 19.67$ kJ/s

i.e., **Power required to drive the compressor $= 19.67$ kW.** (**Ans.**)

Example 14

A refrigerating plant works between temperature limits of − 5 °C and 25 °C. The working fluid ammonia has a dryness fraction of 0.62 at entry to the compressor. If the machine has a relative efficiency of 55%, calculate the amount of ice formed during a period of 24 hours. The ice is to be formed at 0 °C from water at 15 °C and 6.4 kg of ammonia is circulated per minute. The specific heat of water is 4.187 kJ/kg and the latent heat of ice is 335 kJ/kg.

Properties of NH_3 *(datum – 40 °C).*

Temp .°C	Liquid heat kJ/kg	Latent heat kJ/kg	Entropy of liquid kJ/kg K
25	298.9	1167.1	1.124
− 5	158.2	1280.8	0.630

■ Solution.

Enthalpy at point '2', $h_2 = h_{f_2} + x_2 \, h_{fg_2} = 158.2 + 0.62 \times 1280.8 = 952.3$ kJ/kg

Enthalpy at point '1', $h_1 = h_{f_4} = 298.9$ kJ/kg

Also, entropy at point '2' = entropy at point '3'

i.e., $s_2 = s_3$

$s_{f_2} + x_2 \, s_{fg_2} = s_{f_3} + x_3 \, s_{fg_3}$

$$0.630 + 0.62 \times \frac{1280.8}{(-5 + 273)} = 1.124 + x_2 \times \frac{1167.1}{(25 + 273)}$$

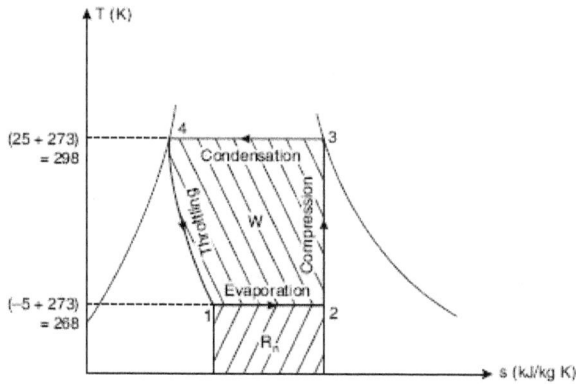

FIGURE 13.25 Sketch for Example 14.

i.e., $x_3 = 0.63$

∴   Enthalpy at point '3', $h_3 = h_{f_3} + x_3\, h_{fg_3}$

$= 298.9 + 0.63 \times 1167.1 = 1034.17 \text{ kJ/kg}$

$$\text{C.O.P.}_{\text{(theoretical)}} = \frac{h_2 - h_1}{h_3 - h_2} = \frac{952.3 - 298.9}{1034.17 - 952.3} = \frac{653.4}{81.87} = 7.98.$$

$\text{C.O.P.}_{\text{(actual)}} = 0.55 \times 7.98 = 4.39$

Work done per kg of refrigerant $= h_3 - h_2 = 1034.17 - 952.3 = 81.87 \text{ kJ/kg}$

Refrigerant in circulation, $m = 6.4$ kg/min.

∴ Work done per second $= 81.87 \times \dfrac{6.4}{60} = 8.73 \text{ kJ/s}$

Heat extracted per kg of ice formed $= 15 \times 4.187 + 335 = 397.8 \text{ kJ}.$

Amount of ice formed in 24 hours,

$$m_{\text{ice}} = \frac{8.73 \times 3600 \times 24}{397.8} = \textbf{1896.1 kg. (Ans.)}$$

Example 15

A simple vapor compression plant produces 5 metric tons of refrigeration. The enthalpy values at inlet to compressor, at exit from the compressor, and at exit from the condenser are 183.19, 209.41, and 74.59 kJ/kg, respectively. Estimate:

(i) The refrigerant flow rate,   *(ii) The C.O.P.,*

(iii) The power required to drive the compressor, and

(iv) The rate of heat rejection to the condenser.

■ **Solution.**

Total refrigeration effect produced = 5 TR(metric tons of refrigeration)

$= 5 \times 14000 = 70000$ kJ/h or 19.44 kJ/s $(\because 1\ TR = 14000$ kJ/h$)$

Given: $h_2 = 183.19$ kJ/kg ; $h_3 = 209.41$ kJ/kg ;

$h_4 (= h_1) = 74.59$ kJ/kg $($Throttling process$)$

(i) The refrigerant flow rate, ṁ:

Net refrigerating effect produced per kg = $h_2 - h_1$

$= 183.19 - 74.59 = 108.6$ kJ/kg

$\therefore$ Refrigerant flow rate, $\dot{m} = \dfrac{19.44}{108.6} = \mathbf{0.179\ kg/s.}$ $(\mathbf{Ans.})$

(ii) The C.O.P.:

$$\text{C.O.P.} = \frac{R_n}{W} = \frac{h_2 - h_1}{h_3 - h_2} = \frac{183.19 - 74.59}{209.41 - 183.19} = \mathbf{4.142.}\ (\mathbf{Ans.})$$

(iii) The power required to drive the compressor, P:

$P = \dot{m}\ (h_3 - h_2) = 0.179\ (209.41 - 183.19) = \mathbf{4.69\ kW.}$ $(\mathbf{Ans.})$

(iv) The rate of heat rejection to the condenser:

The rate of heat rejection to the condenser

$= \dot{m}\ (h_3 - h_4) = 0.179\ (209.41 - 74.59) = \mathbf{24.13\ kW.}$ $(\mathbf{Ans.})$

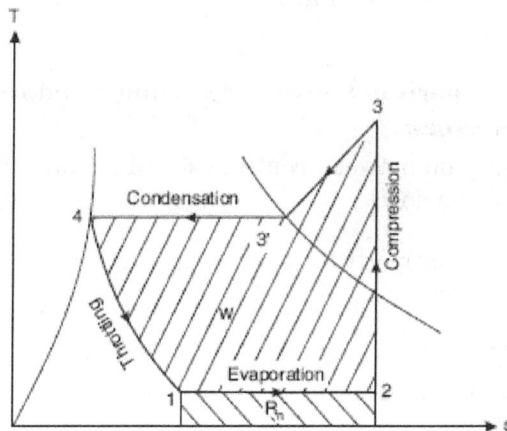

FIGURE 13.26 Sketch for Example 15.

Example 16

(*i*) *What are the advantages of using an expansion valve instead of an expander in a vapor compression refrigeration cycle?*

(*ii*) *Give a comparison between centrifugal and reciprocating compressors.*

(*iii*) *An ice-making machine operates on an ideal vapor compression refrigeration cycle using refrigerant R-12. The refrigerant enters the compressor as dry saturated vapor at – 15 °C and leaves the condenser as saturated liquid at 30 °C. Water enters the machine at 15 °C and leaves as ice at – 5 °C. For an ice production rate of 2400 kg in a day, determine the power required to run the unit. Find also the C.O.P. of the machine. Use the refrigerant table only to solve the problem. Take the latent heat of fusion for water as 335 kJ/kg.*

■ Solution.

(*i*) If an expansion cylinder is used in a vapor compression system, *the work recovered will be extremely small, in fact not even sufficient to overcome the mechanical friction*. It will not be possible to gain any work. Further, the *expansion cylinder is bulky*. On the other hand, the *expansion valve is a very simple and handy device, much cheaper than the expansion cylinder*. It *does not need installation, lubrication, or maintenance*.

The expansion valve *also controls the refrigerant flow rate according to the requirement, in addition to serving the function of reducting the pressure of the refrigerant*.

(*ii*) **The comparison between centrifugal and reciprocating compressors:**

The comparison between centrifugal and reciprocating compressors is given in the table below:

(*iii*) Using the property table of R-12:

$h_2 = 344.927$ kJ/kg

$h_4 = h_1 = 228.538$ kJ/kg

$\left(c_p\right)_v = 0.611$ kJ/kg°C

$s_2 = s_3$

or $1.56323 = 1.5434 + 0.611 \log_e \left[\dfrac{t_3 + 273}{30 + 273}\right]$

or $t_3 = 39.995$°C

Particulars	Centrifugal compressor	Reciprocating compressor
Suitability	Suitable for handling large volumes of air at low pressures	Suitable for low discharges of air at high pressure.
Operational speeds	Usually high	Low
Air supply	Continuous	Pulsating
Balancing	Less vibrations	Cyclic vibrations occur
Lubrication system	Generally simple lubrication systems are required.	Generally complicated
Quality of air delivered	Air delivered is relatively more clean	Generally contaminated with oil.
Air compressor size	Small for given discharge	Large for same discharge
Free air handled	2000–3000 m^3/min	250–300 m^3/min
Delivery pressure	Normally below 10 bar	500 to 800 bar
Usual standard of compression	Isentropic compression	Isothermal compression
Action of compressor	Dynamic action	Positive displacement.

$$h_3 = 363.575 + 0.611(39.995 - 30)$$
$$= 369.68 \text{ kJ/kg.}$$
$$R_n/\text{kg} = h_2 - h_1 = 344.927 - 228.538$$
$$= 116.389 \text{ kJ/kg}$$
$$W/\text{kg} = h_3 - h_2 = 369.68 - 344.927 = 24.753$$
$$\textbf{C.O.P.} = \frac{R_n}{W} = \frac{116.389}{24.753} = \textbf{4.702.} \quad \textbf{(Ans.)}$$

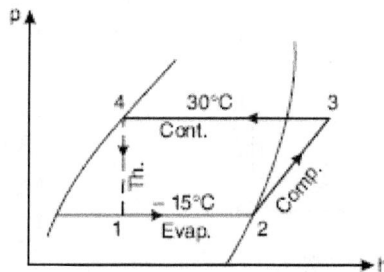

FIGURE 13.27 Sketch for Example 16.

Assuming c_p for ice $= 2.0935$ kJ/kg°C

Heat to be removed to produce ice

$$= \frac{2400}{24 \times 3600} \left[4.187\left(15-0\right) + 335 + 2.0935\left(0 - \left(-5\right)\right) \right]$$

$$= 11.3409 \text{ kJ/s} = \text{Work required, kJ/s } \left(\text{kW}\right) \times \text{C.O.P.}$$

$$\therefore \text{ Work required } \left(\text{Power}\right) = \frac{11.3409}{4.702} = \textbf{2.4 kW.} \quad \textbf{(Ans.)}$$

Example 17

A R-12 refrigerator works between the temperature limits of – 10 °C and + 30 °C. The compressor employed is a 20 cm × 15 cm, twin cylinder, single-acting compressor having a volumetric efficiency of 85%. The compressor runs at 500 r.p.m. The refrigerant is subcooled and it enters at 22 °C in the expansion valve. The vapor is superheated and enters the compressor at – 2 °C. Work out the following:

(*i*) *show the process on T-s and p-h diagrams;* (*ii*) *the amount of refrigerant circulated per minute;* (*iii*) *the metric tons of refrigeration; and* (*iv*) *the C.O.P. of the system.*

■ **Solution.**

(*i*) **Process on T-s and p-h diagrams:**

The processes on *T-s* and *p-h* diagrams are shown in Figure 13.28.

(*ii*) **Mass of refrigerant circulated per minute:**

The value of enthalpies and specific volume read from the *p-h* diagram are as follows:

$h_2 = 352$ kJ/kg ; $h_3 = 374$ kJ/kg

$h_4 = h_1 = 221$ kJ/kg ; $v_2 = 0.08$ m³/kg

Refrigerants effect per kg $= h_2 - h_1 = 352 - 221 = 131$ kJ/kg

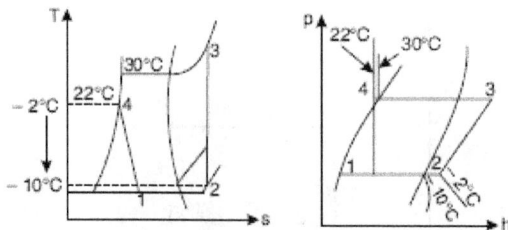

FIGURE 13.28 Sketch for Example 17.

Volume of refrigerant admitted per min.

$$= \frac{\pi}{4} = D^2 L \times r.p.m. \times 2 \times \eta_{vol}, \text{ for twin cylinder, single acting}$$

$$= \frac{\pi}{4}(0.2)^2 \times 0.15 \times 500 \times 2 \times 0.85 = 4 \text{ m}^3 / \text{min}$$

Mass of refrigerant per min $= \dfrac{4}{0.08} =$ **50 kg / min.**   **(Ans.)**

(*iii*) **Cooling capacity in metric tons of refrigeration:**

Cooling capacity $\qquad = 50(h_2 - h_1) = 50 \times 131$

$= 6550$ kJ/min or 393000 kJ/h

or $\qquad = \dfrac{393000}{14000} =$ **28.07 TR.**   **(Ans.)**

($\because \qquad$ 1 metric ton of refrigeration TR $= 14000$ kJ / h)

(*iv*) **The C.O.P. of the system:**

Work per kg $= (h_2 - h_1) = 374 - 352 = 22$

C.O.P. $= \dfrac{131}{22} =$ **5.95.**   **(Ans.)**

Example 18

In a standard vapor compression refrigeration cycle, operating between an evaporator temperature of – 10 °C and a condenser temperature of 40 °C, the enthalpy of the refrigerant, Freon-12, at the end of compression is 220 kJ/kg. Show the cycle diagram on a T-s plane. Calculate:

(i) The C.O.P. of the cycle.

(ii) The refrigerating capacity and the compressor power assuming a refrigerant flow rate of 1 kg/min. You may use the following extract of the Freon-12 property table:

t(°C)	p(MPa)	h_f(kJ/kg)	h_g(kJ/kg)
– 10	0.2191	26.85	183.1
40	0.9607	74.53	203.1

■ **Solution.**

The cycle is shown on the *T-s* diagram in Figure 13.29.

Given : Evaporator temperature $\qquad = -10°C$

Condenser temperature $\qquad = 40°C$

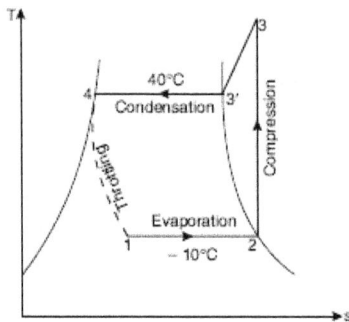

FIGURE 13.29 Sketch for Example 18.

Enthalpy at the end of compression, $h_3 = 220$ kJ/kg

From the table given, we have
$h_2 = 183.1$ kJ/kg ; $h_1 = h_{f_4} = 26.85$ kJ/kg

(i) The C.O.P. of the cycle:

$$\text{C.O.P.} = \frac{R_n}{W} = \frac{h_2 - h_1}{h_3 - h_2}$$

$$= \frac{183.1 - 74.53}{220 - 183.1} = \textbf{2.94.} \quad \textbf{(Ans.)}$$

(ii) Refrigerating capacity

Refrigerating capacity $\quad = m(h_2 - h_1)$

[where m = mass flow rate of refrigerant = 1 kg/min $\quad$ ...[(Given)]

$= 1 \times (183.1 - 74.53) = \textbf{108.57 kJ / min.} \quad \textbf{(Ans.)}$

Compressor power:

Compressor power $\quad = m(h_3 - h_2)$

$= 1 \times (220 - 183.1) = 36.9$ kJ/min $\quad$ or $\quad$ 0.615 kJ/s

$= \textbf{0.615 kW.} \quad \textbf{(Ans.)}$

Example 19

A Freon-12 refrigerator producing a cooling effect of 20 kJ/s operates on a simple cycle with pressure limits of 1.509 bar and 9.607 bar. The vapor leaves the evaporator dry saturated and there is no undercooling. Determine the power required by the machine.

If the compressor operates at 300 r.p.m. and has a clearance volume of 3% of stroke volume, determine the piston displacement of the compressor. For the compressor assume that the expansion following the law $pv^{1.13}$ = constant.

Given:

Temperature °C	Ps bar	Vg m³/kg	Enthalpy h_f	kJ/kg h_g	Entropy s_f	kJ/kg K s_g	Specific heat kJ/kg K
−20	1.509	0.1088	17.8	178.61	0.073	0.7082	—
40	9.607	—	74.53	203.05	0.2716	0.682	0.747

■ **Solution.**

Given: From the previous table:

$h_2 = 178.61$ kJ/kg ; $h_3' = 203.05$ kJ/kg ; $h_{f_4} = 74.53$ kJ/kg $= h_1$

Now, cooling effect $= \dot{m}\,(h_2 - h_1)$

$20 = \dot{m}\,(178.61 - 74.53)$

$\therefore \quad \dot{m} = \dfrac{20}{(178.60 - 74.53)} = 0.192$ kg/s

Also, $s_3 = s_2$

$s_3' + c_p \ln\left(\dfrac{T_3}{T_3'}\right) = 0.7082$

$0.682 + 0.747 \ln\left(\dfrac{T_3}{313}\right) = 0.7082$

or $\quad \ln\left(\dfrac{T_3}{313}\right) = \dfrac{0.7082 - 0.682}{0.747} = 0.03507$

or $\quad \dfrac{T_3}{313} = e^{0.03507} = 1.0357$

$\therefore \quad T_3 = 313 \times 1.0357 = 324.2$ K

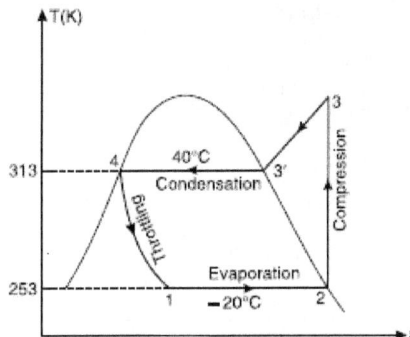

FIGURE 13.30 Sketch for Example 19.

Now, $$h_3 = h_3' + c_p(324.2 - 303)$$

$$= 203.05 + 0.747 (324.2 - 313) = 211.4 \text{ kJ/kg}$$

Power required:

Power required by the machine $= \dot{m}(h_3 - h_2)$

$$= 0.192 \, (211.4 - 178.61) = \mathbf{6.29 \text{ kW. (Ans.)}}$$

Piston displacement, V:

Volumetric efficiency, $$\eta_{\text{vol.}} = 1 + C - C\left(\frac{p_d}{p_s}\right)^{1/n}$$

$$= 1 + 0.03 - 0.03\left(\frac{9.607}{1.509}\right)^{\frac{1}{1.13}} = 0.876 \quad \text{or} \quad 87.6\%$$

The volume of the refrigerant at the intake condition is

$$\dot{m} \times v_g = 0.192 \times 0.1088 = 0.02089 \text{ m}^3/\text{s}$$

Hence the swept volume $$= \frac{0.02089}{\eta_{\text{vol.}}} = \frac{0.02089}{0.876} = 0.02385 \text{ m}^3/\text{s}$$

$$\therefore \quad \mathbf{V} = \frac{0.02385 \times 60}{300} = \mathbf{0.00477 \text{ m}^3}. \quad \textbf{(Ans.)}$$

Example 20

A food storage locker requires a refrigeration capacity of 50 kW. It works between a condenser temperature of 35 °C and an evaporator temperature of − 10 °C. The refrigerant is ammonia. It is sub-cooled by 5 °C before entering the expansion valve by the dry saturated vapor leaving the evaporator. Assume a single cylinder, single-acting compressor operating at 1000 r.p.m. with stroke equal to 1.2 times the bore.

Determine: (i) The power required, and

(ii) The cylinder dimensions.

Properties of ammonia are:

Saturation temperature, °C	Pressure bar	Enthalpy (kJ/kg)		Entropy (kJ/kg K)		Specific volume (m³/kg)		Specific heat (kJ/kg K)	
		Liquid	Vapor	Liquid	Vapor	Liquid	Vapor	Liquid	Vapor
− 10	2.9157	154.056	1450.22	0.82965	5.7550	—	0.417477	—	2.492
35	13.522	366.072	1488.57	1.56605	5.2086	1.7023	0.095629	4.556	2.903

■ **Solution.**

Given: From the previous table:

$h_2 = 1450.22$ kJ/kg ; $h_3' = 1488.57$ kJ/kg ; $h_{f_4} = 366.072$ kJ/kg ;

$h_{f_{4'}} = h_1 = h_{f_4} - 4.556\,(308 - 303)$

$= 366.07 - 4.556\,(308 - 303) = 343.29$ kJ/kg

Also $\qquad s_3 = s_2$

or $\quad s_3' + c_p \ln\left(\dfrac{T_3}{T_3'}\right) = 5.755$

$5.2086 + 2.903 \ln\left(\dfrac{T_3}{308}\right) = 5.755$

or $\qquad \ln\left(\dfrac{T_3}{308}\right) = \dfrac{5.755 - 5.2086}{2.903}$

$\dfrac{T_3}{308} = e^{0.1882} = 371.8$ K

Now, $\qquad\qquad h_3 = h_3' + c_p\,(T_3 - T_3')$

$= 1488.57 + 2.903\,(371.8 - 308) = 1673.8$ kJ/kg

Mass of refrigerant, $\qquad \dot{m} = \dfrac{50}{h_2 - h_1} = \dfrac{50}{1450.22 - 343.29}$

$= 0.04517$ kJ/s

(*i*) **Power required:**

Power required $\qquad\qquad = \dot{m}\,(h_3 - h_2)$

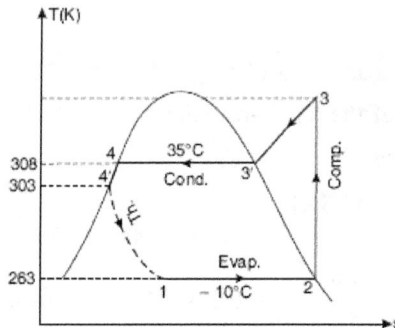

FIGURE 13.31 Sketch for Example 20.

= 0.04517 (1673.8 − 1450.22) = **10.1 kW. (Ans.)**

(ii) Cylinder dimensions:

$$\dot{m} = \frac{\pi}{4} D^2 \times L \times \frac{N}{60} \times 0.417477 = 0.04517 \text{ (calculated above)}$$

or $\qquad \dfrac{\pi}{4} D^2 \times 1.2 D \times \dfrac{1000}{60} \times 0.417477 = 0.04517$

or $\qquad D^3 = \dfrac{0.04517 \times 4 \times 60}{\pi \times 1.2 \times 1000 \times 0.417477} = 0.006888$

∴ Diameter of cylinder, $\quad$ **D** = $(0.006888)^{1/3}$ = **0.19 m.** $\quad$ **(Ans.)**

and, Length of the cylinder, $\quad$ L = 1.2D = 1.2 × 0.19 = **0.228 m.** $\quad$ **(Ans.)**

Example 21

A refrigeration cycle uses Freon-12 as the working fluid. The temperature of the refrigerant in the evaporator is − 10 °C. The condensing temperature is 40 °C. The cooling load is 150 W and the volumetric efficiency of the compressor is 80%. The speed of the compressor is 720 r.p.m. Calculate the mass flow rate of the refrigerant and the displacement volume of the compressor.

Properties of Freon-12

Temperature (°C)	Saturation pressure (MPa)	Enthalpy (kJ/kg)		Specific volume (m³/kg) Saturated vapor
		Liquid	Vapor	
− 10	0.22	26.8	183.0	0.08
40	0.96	74.5	203.1	0.02

■ Solution.

Given : Cooling load = 150 W ; $\eta_{vol.}$ = 0.8 ; N = 720 r.p.m.

Mass flow rate of the refrigerant ṁ :

Refrigerating effect $\quad = h_2 - h_1$

= 183 − 74.5 = 108.5 kJ/kg

Cooling load $\qquad = \dot{m} \times (108.5 \times 1000) = 150$

or $\qquad \dot{m} = \dfrac{150}{108.5 \times 1000} = $ **0.001382 kJ / s. (Ans.)**

Displacement volume of the compressor:

Specific volume at entry to the compressor,

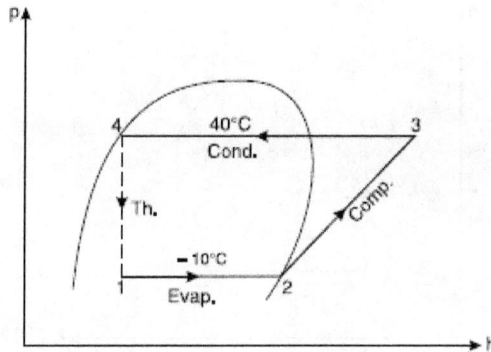

FIGURE 13.32 Sketch for Example 21.

$v_2 = 0.08 \text{ m}^3/\text{kg}$

(From table)

∴ Displacement volume of compressor $= \dfrac{\dot{m}v_2}{\eta_{\text{vol.}}} = \dfrac{0.001382 \times 0.08}{0.8}$

$= 0.0001382 \text{ m}^3/\text{s}$. **(Ans.)**

Example 22

In a simple vapor compression cycle, the following are the properties of the refrigerant R-12 at various points:

Compressor inlet : $h_2 = 183.2 \text{ kJ}/\text{kg}$ $v_2 = 0.0767 \text{ m}^3/\text{kg}$

Compressor discharge : $h_3 = 222.6 \text{ kJ}/\text{kg}$ $v_3 = 0.0164 \text{ m}^3/\text{kg}$

Compressor exit : $h_4 = 84.9 \text{ kJ}/\text{kg}$ $v_4 = 0.00083 \text{ m}^3/\text{kg}$

The piston displacement volume for compressor is 1.5 liters per stroke and its volumetric efficiency is 80%. The speed of the compressor is 1600 r.p.m.

Find: (i) Power rating of the compressor (kW);

(ii) Refrigerating effect (kW).

■ Solution.

Piston displacement volume $= \dfrac{\pi}{4} d^2 \times l = 1.5$ litres

$= 1.5 \times 1000 \times 10^{-6} \text{ m}^3/\text{stroke} = 0.0015 \text{ m}^3/\text{revolution}$.

(i) Power rating of the compressor (kW):

Compressor discharge

$= 0.0015 \times 1600 \times 0.8 \, (\eta_{\text{vol.}}) = 1.92 \text{ m}^3/\text{min}$.

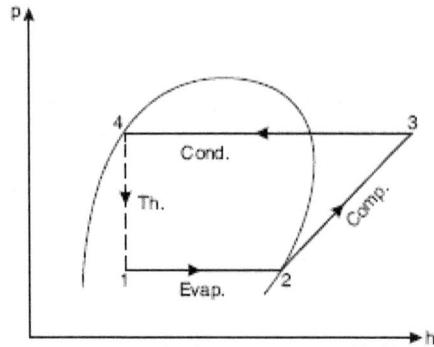

FIGURE 13.33 Sketch for Example 22.

Mass flow rate of compressor,

$$m = \frac{\text{Compressor discharge}}{v_2}$$

$$= \frac{1.92}{0.0767} = 25.03 \text{ kg/min.}$$

Power rating of the compressor

$$= \dot{m}(h_3 - h_2)$$

$$= \frac{25.03}{60} \left(222.6 - 183.2 \right)$$

$$= \mathbf{16.44 \text{ kW. (Ans.)}}$$

(ii) Refrigerating effect (kW):

Refrigerating effect $= \dot{m}\,(h_2 - h_1) = \dot{m}\,(h_2 - h_4)$ $(\because h_1 = h_4)$

$$= \frac{25.03}{60} \left(183.2 - 84.9 \right)$$

$$= \mathbf{41 \text{ kW. (Ans.)}}$$

Example 23

A refrigerator operating on a standard vapor compression cycle has a coefficiency performance of 6.5 and is driven by a 50 kW compressor. The enthalpies of saturated liquid and saturated vapor refrigerant at the operating condensing temperature of 35 °C are 62.55 kJ/kg and 201.45 kJ/kg, respectively. The saturated refrigerant vapor leaving evaporator has an enthalpy of 187.53 kJ/kg. Find the refrigerant temperature at the compressor discharge. The c_p of the refrigerant vapor may be taken to be 0.6155 kJ/kg°C.

■ **Solution,**

Given : C.O.P. = 6.5 ; W = 50 kW, h_3' = 201.45 kJ/kg,

$h_{f_4} = h_1$ = 69.55 kJ/kg ; h_2 = 187.53 kJ/kg

c_p = 0.6155 kJ/kg K

Temperature, t₃ :

Refrigerating capacity   = 50 × C.O.P.

= 50 × 6.5 = 325 kW

Heat extracted per kg of refrigerant

= 187.53 – 69.55 = 117.98 kJ/kg

Refrigerant flow rate $= \dfrac{325}{117.98} = 2.755$ kg/s

Compressor power   = 50 kW

∴ Heat input per kg $= \dfrac{50}{2.755} = 18.15$ kJ/kg

Enthalpy of vapor after compression

$= h_2 + 18.15 = 187.53 + 18.15$

$= 205.68$ kJ/kg

Superheat $= 205.68 - h_3' = 205.68 - 201.45$

$= 4.23$ kJ / kg

But $4.23 = 1 \times c_p \ (t_3 - t_3') = 1 \times 0.6155 \times (t_3 - 35)$

∴ $\mathbf{t_3} = \dfrac{4.23}{0.6155} + 35 = \mathbf{41.87°C.}$ **(Ans.)**

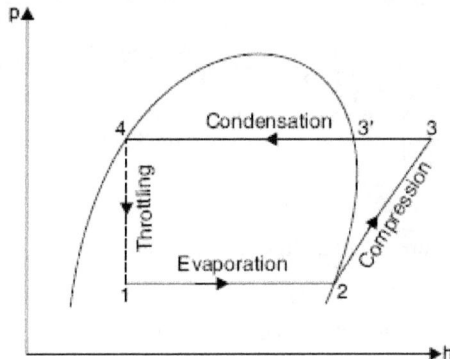

FIGURE 13.34 Sketch for Example 23.

Note that the compressor rating of 50 kW is assumed to be the enthalpy of compression, in the absence of any data on the efficiency of the compressor.

Example 24

A vapor compression heat pump is driven by a power cycle having a thermal efficiency of 25%. For the heat pump, refrigerant-12 is compressed from saturated vapor at 2.0 bar to the condenser pressure of 12 bar. The isentropic efficiency of the compressor is 80%. Saturated liquid enters the expansion valve at 12 bar. For the power cycle 80% of the heat rejected by it is transferred to the heated space, which has a total heating requirement of 500 kJ/min. Determine the power input to the heat pump compressor. The following data for refrigerant-12 may be used:

Pressure, bar	Temperature, °C	Enthalpy, kJ/kg		Entropy, kJ/kg K	
	°C	Liquid	Vapor	Liquid	Vapor
2.0	– 12.53	24.57	182.07	0.0992	0.7035
12.0	49.31	84.21	206.24	0.3015	0.6799

Vapour specific heat at constant pressure = 0.7 kJ / kg K.

■ Solution.

Heat rejected by the cycle $= \dfrac{500}{0.8} = 625$ kJ/min.

Assuming isentropic compression of the refrigerant, we have

Entropy of dry saturated vapor at 2 bar

= Entropy of superheated vapor at 12 bar

$$0.7035 = 0.6799 + c_p \ \ln \frac{T}{(49.31+273)} = 0.6799 + 0.7 \times \ln\left(\frac{T}{322.31}\right)$$

or $\ln\left(\dfrac{T}{322.31}\right) = \dfrac{0.7035 - 0.6799}{0.7} = 0.03371$

or $T = 322.31 \ (e)^{0.03371} = 333.4$ K

∴ Enthalpy of superheated vapour at 12 bar

$= 206.24 + 0.7 \left(333.4 - 322.31\right) = 214 \, \text{kJ/kg}$

Heat rejected per cycle $= 214 - 84.21 = 129.88$ kJ/kg

Mass flow rate of refrigerant $= \dfrac{625}{129.88} = 4.812$ kg/min

Work done on compressor $= 4.812 \ (214 - 182.07)$

$= 153.65$ kJ/min $= 2.56$ kW

Actual work of compression $= \dfrac{2.56}{\eta_{compressor}} = \dfrac{2.56}{0.8} = 3.2$ kW

Hence power input to the heat pump compressor = **3.2 kW. (Ans.)**

Example 25

A food storage locker requires a refrigeration system of 2400 kJ/min. capacity at an evaporator temperature of 263 K and a condenser temperature of 303 K. The refrigerant used is freon-12 and is subcooled by 6 °C before entering the expansion valve and vapor is superheated by 7 °C before leaving the evaporator coil. The compression of the refrigerant is reversible adiabatic. The refrigeration compressor is two-cylinder single-acting with stroke equal to 1.25 times the bore and operates at 1000 r.p.m.

Properties of freon-12

Satura-tion temp, K	Absolute pressure, bar bar	Specific volume of vapor, m³/ kg	Enthalpy, kJ/ kg Liq-uid	Va-por	Entropy, kJ/kg K Liquid	Vapor
263	2.19	0.0767	26.9	183.2	0.1080	0.7020
303	7.45	0.0235	64.6	199.6	0.2399	0.6854

Take: Liquid specific heat = 1.235 kJ/kg K; Vapor specific heat = 0.733 kJ/kg K. Determine:

(i) Refrigerating effect per kg.

(ii) Mass of refrigerant to be circulated per minute.

(iii) Theoretical piston displacement per minute.

(iv) Theoretical power required to run the compressor, in kW.

(v) Heat removed through condenser per min.

(vi) Theoretical bore and stroke of compressor.

■ **Solution.**

The cycle of refrigeration is represented on the *T-s* diagram in Figure 13.35.

Enthalpy at '2', $\quad h_2 = h_2' + c_p\left(T_2 - T_2'\right)$

From the given table:

$h_2' = 183.2$ kJ/kg

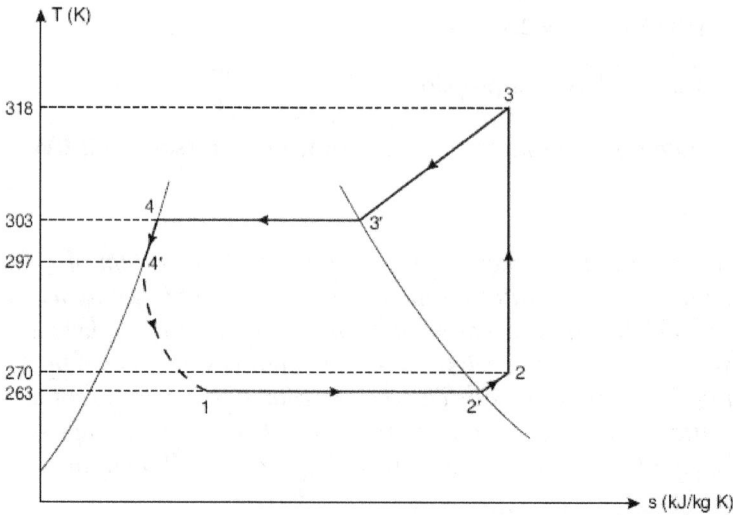

FIGURE 13.35 Sketch for Example 25.

$(T_2 - T_2') =$ Degree of superheat as the vapor enters the compressor
$= 7°C$

$\therefore \quad h_2 = 183.2 + 0.733 \times 7 = 188.33$ kJ/kg

Also, entropy at '2', $\quad s_2 = s_2' + c_p \, \log_e \, \dfrac{T_2}{T_2'}$

$= 0.7020 + 0.733 \, \log_e \left(\dfrac{270}{263}\right) = 0.7212$ kJ/kg K

For isentropic process 2-3

Entropy at '2' $\qquad\qquad$ = Entropy at '3'

$0.7212 = s_3' + c_p \, \log_e \left(\dfrac{T_3}{T_3'}\right)$

$= 0.6854 + 0.733 \, \log_e \left(\dfrac{T_3}{303}\right)$

$\therefore \quad \log_e \left(\dfrac{T_3}{303}\right) = 0.0488$

i.e., $\quad T_3 = 318$ K

Now, enthalpy at '3', $\quad h_3 = h_3' + c_p \left(T_3 - T_3'\right)$

$$= 199.6 + 0.733\,(318 - 303) = 210.6 \text{ kJ/kg}.$$

Also, enthalpy at 4′,

$$h_{f_4}' = h_{f_4} - \left(c_p\right)_{\text{liquid}} \left(T_4 - T_4'\right) = 64.6 - 1.235 \times 6 = 57.19 \text{ kJ/kg}$$

For the *process 4′-1,*

Enthalpy at 4′ = enthalpy at 1 = 57.19 kJ/kg

For specific volume at 2,

$$\frac{v_2'}{T_2'} = \frac{v_2}{T_2}$$

$$\therefore \quad v_2 = \frac{v_2'}{T_2'} \times T_2 = 0.0767 \times \frac{270}{263} = 0.07874 \text{ m}^3$$

(*i*) **Refrigerating effect per kg**

$$= h_2 - h_1 = 188.33 - 57.19 = \textbf{131.14 kJ / kg. (Ans.)}$$

(*ii*) **Mass of refrigerant to be circulated per minute** for producing effect of 2400 kJ/min.

$$= \frac{2400}{131.14} = \textbf{18.3 kg / min. (Ans.)}$$

(*iii*) **Theoretical piston displacement per minute**

= Mass flow/min. × specific volume at suction

$$= 18.3 \times 0.07874 = \textbf{1.441 m}^3 \textbf{/ min. (Ans.)}$$

(*iv*) **Theoretical power required to run the compressor**

= Mass flow of refrigerant per sec. × compressor work/kg

$$= \frac{18.3}{60} \times \left(h_3 - h_2\right) = \frac{18.3}{60}\,(210.6 - 188.33) \text{ kJ/s}$$

$$= \textbf{6.79 kJ / s or 6.79 kW. (Ans.)}$$

(*v*) **Heat removed through the condenser per min.**

= Mass flow of refrigerant × heat removed per kg of refrigerant

$$= 18.3\,(h_3 - h_{f_4'}) = 18.3\,(210.6 - 57.19) = \textbf{2807.4 kJ / min. (Ans.)}$$

(*vi*) **Theoretical bore (d) and stroke (l):**

Theoretical piston displacement per cylinder

$$= \frac{\text{Total displacement per minute}}{\text{Number of cylinder}} = \frac{1.441}{2} = 0.7205 \text{ m}^3/\text{min}.$$

Also,   length of stroke $= 1.25 \times$ diameter of piston

Hence, $0.7205 = \pi/4 \ d^2 \times (1.25 \ d) \times 1000$

i.e., $d = \mathbf{0.09}$ **m** or **90 mm.   (Ans.)**

and $l = 1.25 \ d = 1.25 \times 90 = \mathbf{112.5 \, mm.}$ **(Ans.)**

Example 26

A refrigeration system of 10.5 metric tons capacity at an evaporator temperature of –12 °C and a condenser temperature of 27 °C is needed in a food storage locker. The refrigerant ammonia is sub-cooled by 6 °C before entering the expansion valve. The vapor is 0.95 dry as it leaves the evaporator coil. The compression in the compressor is of adiabatic type.

Using a p-h chart find:

(i)   Condition of volume at outlet of the compressor

(ii) Condition of vapor at entrance to the evaporator

(iii) C.O.P.

(iv) Power required, in kW.

Neglect valve throttling and clearance effect.

■ **Solution.**

Using the *p-h* chart for ammonia,

• Locate point '2' where – 12 °C cuts the 0.95 dryness fraction line.

• From point '2' move along the constant entropy line and locate point '3' where it cuts the constant pressure line corresponding to + 27 °C temperature.

• From point '3' follow the constant pressure line till it cuts the + 21 °C temperature line to get to point '4'.

• From point '4' drop a vertical line to cut the constant pressure line corresponding to – 12°C and get to point '5'.

The values as read from the chart are:

$h_2 = 1597$ kJ/kg

$h_3 = 1790$ kJ/kg

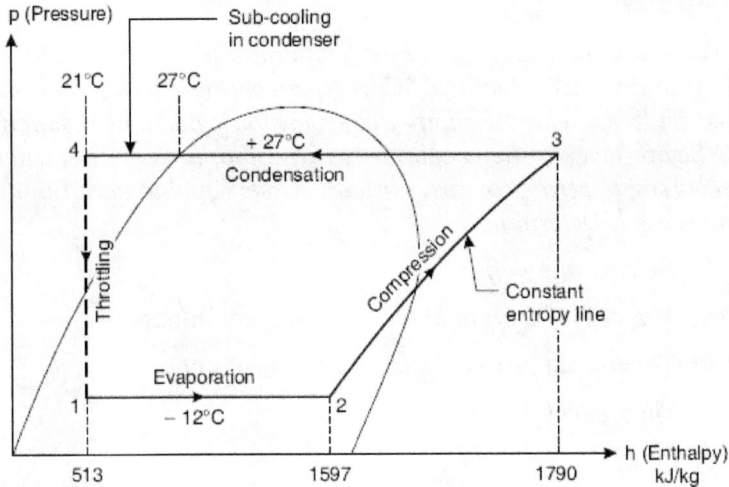

FIGURE 13.36 Sketch for Example 26.

$h_4 = h_1 = 513 \text{ kJ/kg}$

$t_3 = 58°C$

$x_1 = 0.13.$

(*i*) **Condition of the vapor at the outlet of the compressor**

$= 58 - 27 = \textbf{31°C superheat. (Ans.)}$

(*ii*) **Condition of vapor at the entrance to the evaporator,**

$x_1 = \textbf{0.13. (Ans.)}$

(*iii*) **C.O.P.** $= \dfrac{h_2 - h_1}{h_3 - h_2} = \dfrac{1597 - 513}{1790 - 1597} = \textbf{5.6. (Ans.)}$

(*iv*) **Power required:**

$$\text{C.O.P.} = \frac{\text{Net refrigerating effect}}{\text{Work done}} = \frac{R_n}{W}$$

$$5.6 = \frac{10.5 \times 14000}{W \times 60}$$

$$\therefore \quad W = \frac{10.5 \times 14000}{5.6 \times 60} \text{ kJ/min} = 437.5 \text{ kJ/min.}$$

$= 7.29 \text{ kJ/s.}$

i.e., **Power required = 7.29 kW. (Ans.)**

Example 27

The evaporator and condenser temperatures of a 20-metric ton capacity freezer are – 28 °C and 23 °C, respectively. The refrigerant – 22 is subcooled by 3 °C before it enters the expansion valve and is superheated to 8 °C before leaving the evaporator. The compression is isentropic. A six-cylinder single-acting compressor with stroke equal to bore running at 250 r.p.m. is used. Determine:

(i) Refrigerating effect/kg.

(ii) Mass of refrigerant to be circulated per minute.

(iii) Theoretical piston displacement per minute.

(iv) Theoretical power.

(v) C.O.P.

(vi) Heat removed through the condenser.

(vii) Theoretical bore and stroke of the compressor.

Neglect valve throttling and clearance effect.

■ Solution.

Following the procedure as given in the previous example, plot the points 1, 2, 3, and 4 on a *p-h* chart for freon-22. The following values are obtained:

$h_2 = 615$ kJ/kg

$h_3 = 664$ kJ/kg

$h_4 = h_1 = 446$ kJ/kg

$v_2 = 0.14$ m^3/kg.

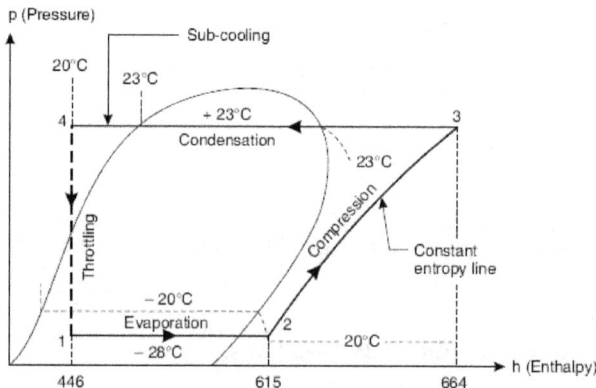

FIGURE 13.37 Sketch for Example 27.

(i) **Refrigerating effect per kg** $= h_2 - h_1 = 615 - 446 = $ **169 kJ / kg. (Ans.)**

(ii) **Mass of refrigerant to be circulated per minute,**

$$m = \frac{20 \times 14000}{169 \times 60} = 27.6 \text{ kg / min. (Ans.)}$$

(iii) **Theoretical piston displacement**

= Specific volume at suction × Mass of refrigerant used/min

$= 0.14 \times 27.6 = $ **3.864 m^3 / min. (Ans.)**

(iv) **Theoretical power**

$$= m \times (h_3 - h_2) = \frac{27.6}{60} (664 - 615) = 22.54 \text{ kJ/s}$$

$= $ **22.54 kW. (Ans.)**

(v) **C.O.P.** $= \dfrac{h_2 - h_1}{h_3 - h_2} = \dfrac{615 - 446}{664 - 615} = $ **3.45. (Ans.)**

(vi) **Heat removed through the condenser**

$= m(h_3 - h_4) = 27.6 (664 - 446) = $ **6016.8 kJ / min. (Ans.)**

(vii) **Theoretical displacement per minute per cylinder**

$$= \frac{\text{Total displacement / min.}}{\text{Number of cylinders}} = \frac{3.864}{6} = 0.644 \text{ m}^3 \text{/min}$$

Let diameter of the cylinder = d

Then, stroke length, $l = d$

Now, $\dfrac{\pi}{4} d^2 \times l = \dfrac{0.644}{950}$

or $\dfrac{\pi}{4} d^2 \times d = \dfrac{0.644}{950}$

i.e., **d = 0.0952 m or 95.2 mm. (Ans.)**

and **l = 95.2 mm. (Ans.)**

13.9. Refrigerants

A refrigerant is defined as any substance that absorbs heat through expansion or vaporization and loses it through condensation in a refrigeration system. The term refrigerant in the broadest sense is also applied to such secondary cooling mediums as cold water or brine solutions. Usually refrigerants include only those working mediums which pass through the

cycle of evaporation, recovery, compression, condensation, and liquifica-
tion. These substances absorb heat at one place at a low temperature level
and reject the same at some other place having a higher temperature and
pressure. The rejection of heat takes place at the cost of some mechanical
work. Thus, circulating cold mediums and cooling mediums (such as ice
and solid carbon dioxide) are not primary refrigerants. In the early days
only four refrigerants, air, ammonia (NH_3), carbon dioxide (CO_2), and
sulfur dioxide (SO_2), possessing chemical, physical, and thermodynamic
properties permitting their efficient application and service in the practical
design of refrigeration equipment, were used. All the refrigerants change
from liquid state to vapor state during the process.

Classification of Refrigerants

The refrigerants are classified as follows:

1. Primary refrigerants.

2. Secondary refrigerants.

1. **Primary refrigerants** *are those working mediums or heat carriers
which directly take part in the refrigeration system and cool the substance by the absorption of latent heat, for example, ammonia, carbon
dioxide, sulfur dioxide, methyl chloride, methylene chloride, ethyl chloride, and the Freon group.*

2. **Secondary refrigerants** *are those circulating substances which are
first cooled with the help of the primary refrigerants and are then
employed for cooling purposes, for example, ice and solid carbon dioxide.*
These refrigerants cool substances by absorption of their sensible heat.

The primary refrigerants are grouped as follows:

(*i*) **Halocarbon compounds.** In 1928, Charles Kettening and Dr. Thomas Mighey invented and developed this group of refrigerant. In this
group are included refrigerants which contain one or more of three
halogens, chlorine and bromine and they are sold in the market under
the names as *Freon, Genetron, Isotron*, and *Areton*. Since the refrigerants belonging to this group have outstanding merits over the other
group's refrigerants, therefore they find a wide field of application in
domestic, commercial, and industrial purposes.

The list of the halocarbon refrigerants commonly used is given as follows:

R-10 — Carbon tetrachloride (CCl_4)

R-11 — Trichloro-monofluoro methane (CCl_3F)

R - 12 — Dichloro - difluoro methane $\left(CCl_2F_2\right)$

R - 13 — Mono - bromotrifluoro methane $\left(CBrF_3\right)$

R - 21 — Dichloro monofluoro methane $\left(CHCl_2F\right)$

R - 22 — Mono chloro difluoro methane $\left(CHClF_2\right)$

R - 30 — Methylene - chloride $\left(CH_2Cl_2\right)$

R - 40 — Methyle chloride $\left(CH_3Cl\right)$

R - 41 — Methyle fluoride $\left(CH_3F\right)$

R - 100 — Ethyl chloride $\left(C_2H_5Cl\right)$

R - 113 — Trichloro trifluoroethane $\left(C_2F_3Cl_3\right)$

R - 114 — Tetra - fluoro dichloroethane $\left(Cl_2F_4Cl_2\right)$

R - 152 — Difluoro - ethane $\left(C_2H_6F_2\right)$

(*ii*) **Azeotropes.** The refrigerants belonging to this group consists of mixtures of different substances. These substances cannot be separated into components by distillations. They possess fixed thermodynamic properties and do not undergo any separation with changes in temperature and pressure. An azeotrope behaves like a simple substance.

Example. R-500. It contains 73.8% of (R-12) and 26.2% of (R-152).

(*iii*) Hydrocarbons. Most of the refrigerants of this group are organic compounds. Several hydrocarbons are used successfully in commercial and industrial installations. Most of them possess satisfactory thermodynamic properties but are highly inflammable. Some of the important refrigerants of this group are:

R - 50   — Methane $\left(CH_4\right)$

R - 170 — Ethane $\left(C_2H_6\right)$

R - 290 — Propane $\left(C_2H_8\right)$

R - 600 — Butane $\left(C_4H_{10}\right)$

R - 601 — Isobutane $\left[CH\left(CH_3\right)_3\right]$

(*iv*) **Inorganic compounds.** Before the introduction of the hydrocarbon group, these refrigerants were most commonly used for all purposes. The important refrigerants of this group are:

R - 717 — Ammonia $\left(NH_3\right)$

R - 718 — Water $\left(H_2O\right)$

R - 729 — Air (mixture of O_2, N_2, CO_2 etc.)

R - 744 — Carbon dioxide (CO_2)

R - 764 — Sulphur dioxide (SO_2)

(v) **Unsaturated organic compound.** The refrigerants belonging to this group possess ethylene or propylene as their constituents. They are:

R -1120 — Trichloroethylene ($C_3H_4Cl_3$)

R -1130 — Dichloroethylene ($C_2H_4Cl_2$)

R -1150 — Ethylene (C_3H_6)

R -1270 — Propylene.

Desirable Properties of an Ideal Refrigerant

An ideal refrigerant should possess the following properties:

1. Thermodynamic properties:

(i) Low boiling point

(ii) Low freezing point

(iii) Positive pressures (but not very high) in condenser and evaporator

(iv) High saturation temperature

(v) High latent heat of vaporization

2. Chemical properties:

(i) Non-toxicity

(ii) Non-flammable and non-explosive

(iii) Non-corrosiveness

(iv) Chemical stability in reacting

(v) No effect on the quality of stored (food and other) products like flowers, with other materials such as furs and fabrics

(vi) Non-irritating and odorless

3. Physical properties:

(i) Low specific volume of vapor

(ii) Low specific heat

(*iii*) High thermal conductivity

(*iv*) Low viscosity

(*v*) High electrical insulation

4. Other properties:

(*i*) Ease of leakage location

(*ii*) Availability and low cost

(*iii*) Ease of handling

(*iv*) High C.O.P.

(*v*) Low power consumption per metric ton of refrigeration

(*vi*) Low pressure ratio and pressure difference

Some important properties (mentioned above) are discussed as follows:

Freezing point. As the refrigerant must operate in the cycle above its freezing point, it is evident that the same for the refrigerant *must be lower than system temperatures.* It is found that except in the case of water for which the freezing point is 0°C, other refrigerants have reasonably low values. Water, therefore, can be used only in air-conditioning applications which are above 0°C.

Condenser and evaporator pressures. The evaporating pressure should be as near atmospheric as possible. If it is *too low,* it would result in *a large volume of the suction vapor.* If it is *too high, overall high pressures including condenser pressure would result necessitating stronger equipment and consequently higher cost. A positive pressure is required in order to eliminate the possibility of the entry of air and moisture into the system. The normal boiling point of the refrigerant should, therefore, be lower than the refrigerant temperature.*

Critical temperature and pressure. Generally, for high C.O.P. the *critical temperature should be very high so that the condenser temperature line on the p-h diagram is far removed from the critical point.* This ensures a reasonable refrigerating effect, as it is very small with the state of the liquid before expansion near the critical point.

The *critical pressure should be low so as to give low condensing pressure.*

Latent heat of vaporization. It should be *as large as possible to reduce the weight of the refrigerant to be circulated in the system. This* *reduces initial cost of the refrigerant. The size of the system will also be small and hence there will be low initial cost.*

Toxicity. Taking into consideration the comparative hazard to life due to gases and vapors, Underwriters Laboratories have divided the compounds into *six groups*. Group six contains compounds with a very low degree of toxicity. It includes R_{12}, R_{114}, R_{13}, and so on. Group one, at the other end of the scale, includes the most toxic substances such as SO_2.

Ammonia is not used in comfort air-conditioning and in domestic refrigeration because of inflammability and toxicity.

Inflammability. Hydrocarbons (*e.g.*, methane and ethane) are highly explosive and inflammable. Fluorocarbons are neither explosive nor inflammable. *Ammonia is explosive* in a mixture with air in a concentration of 16 to 25% by volume of ammonia.

Volume of suction vapor. The *size of the compressor depends on the volume of suction vapor per unit* (say per metric ton) *of refrigeration. Reciprocating compressors are used with refrigerants with high pressures and small volumes of the suction vapor. Centrifugal or turbo-compressors are used with refrigerants with low pressures and large volumes of the suction vapor.* A high-volume flow rate for a given capacity is required for centrifugal compressors to permit flow passages of sufficient width to minimize drag and obtain high efficiency.

Thermal conductivity. *For a high heat transfer coefficient, a high thermal conductivity is desirable.* R_{22} has better heat transfer characteristics than R_{12}; R_{21} is still better, and R_{13} has poor heat transfer characteristics.

Viscosity. For *a high heat transfer coefficient, a low viscosity is desirable*.

Leak tendency. The *refrigerants should have low leak tendency*. The greatest drawback of *fluorocarbons* is the fact that they are *odorless*. This, at times, results in a complete loss of costly gas from leaks without being detected. An ammonia leak can be very easily detected by its pungent odor.

Refrigerant cost. The cost factor is only relevant to the extent of the price of the initial charge of the refrigerant, which is very small compared to the total cost of the plant and its installation. The cost of losses due to leakage is also important. In small-capacity units requiring only a small charge of the refrigerant, the cost of the refrigerant is immaterial.

The *cheapest refrigerant is ammonia. R_{12} is slightly cheaper than R_{22}.* R_{12} and R_{22} have replaced ammonia in the dairy and frozen food industry (and even in cold storages) *because of the tendency of ammonia to attack some food products.*

Coefficient of performance and power per metric ton. Practically all common refrigerants have approximately the same C.O.P. and power requirement.

Table 13.1 gives the values of C.O.P. for some important refrigerants.

Action with oil. No chemical reaction between refrigerant and lubricating oil of the compressor should take place. Miscibility of the oil is quite important as some oil should be carried out of the compressor crankcase with the hot refrigerant vapor to lubricate the pistons and discharge valves properly.

Reaction with materials of construction. While selecting a material to contain the refrigerant, this material should be given due consideration. Some metals are attacked by the refrigerants; for example, *ammonia reacts with copper, brass, or other cuprous alloys in the presence of water*, therefore, in *ammonia systems the common metals used are iron and steel. The Freon group* does not react with steel, copper, brass, zinc, tin, and aluminum, but it is corrosive to magnesium and aluminum having magnesium more than 2%. Freon group refrigerants tend to dissolve natural rubber in packing and gaskets, but synthetic rubber such as neoprene is entirely suitable. The hydrogenated hydrocarbons may react with zinc but not with copper, aluminum, iron, and steel.

TABLE 13.1 C.O.P. of Some Important Refrigerants

Refrigerant	C.O.P.
Carnot value	5.74
R_{11}	5.09
R_{113}	4.92
Ammonia	4.76
R_{12}	4.70
R_{22}	4.66
R_{144}	4.49
CO_2	2.56

13.9.1. Properties and Uses of Commonly Used Refrigerants

1. Air
Properties:

(*i*) No cost involved; easily available.

(*ii*) Completely non-toxic.

(*iii*) Completely safe.

(*iv*) The C.O.P. of the air cycle operating between temperatures of 80 °C and – 15 °C is 1.67.

Uses:

(*i*) Air is one of the earliest refrigerants and was widely used even as late as World War I wherever a completely non-toxic medium was needed.

(*ii*) Because of low C.O.P., it is used only where *operating efficiency is secondary* as in *aircraft refrigeration.*

2. Ammonia (NH$_3$)
Properties:

(*i*) It is highly toxic and flammable.

(*ii*) It has excellent thermal properties.

(*iii*) It has the *highest refrigerating effect per kg of refrigerant.*

(*iv*) Low volumetric displacement.

(*v*) Low cost.

(*vi*) Low weight of liquid circulated per metric ton of refrigeration.

(*vii*) High efficiency.

(*viii*) The evaporator and condenser pressures are 3.5 bar abs. and 13 bar abs. (app.), respectively at standard conditions of – 15 °C and 30 °C.

Uses:

(*i*) It is widely used in large industrial and commercial reciprocating compression systems where high toxicity is secondary.

It is extensively used in *ice plants, packing plants, large cold storages, skating rinks,* and so on.

(*ii*) It is widely used as the refrigerant in *absorption systems.*

The following points are worth noting:

Ammonia *should never be used with copper, brass, and other copper alloys; iron and steel should be* used in *ammonia systems instead.*

In ammonia systems, to *detect leakage a sulfur candle is used, which gives off a dense white smoke when ammonia vapor is present.*

3. Sulphur dioxide (SO_2)
Properties:

(i) It is a colorless gas or liquid.

(ii) It is extremely toxic and has a pungent irritating odor.

(iii) It is non-explosive and non-flammable.

(iv) It has a liquid specific gravity of 1.36.

(v) Works at low pressures.

(vi) Possesses small latent heat of vaporization.

Uses:

It finds little use these days. However, it was used in small machines in the early days.

The leakage of sulfur dioxide may be detected by bringing aqueous ammonia near the leak, which gives off a white smoke.

4. Carbon dioxide (CO_2)
Properties:

(i) It is a colorless and odorless gas, and is heavier than air.

(ii) It has liquid specific gravity of 1.56.

(iii) It is non-toxic and non-flammable.

(iv) It is non-explosive and non-corrosive.

(v) It has *extremely high operating pressures.*

(vi) It gives *very low refrigerating effect.*

Uses:

This refrigerant has received only limited use because of the high power requirements per metric ton of refrigeration and the high operating pressures. In *former years* it was selected for *marine refrigeration,* for *theater air-conditioning systems,* and for *hotel and institutional refrigeration* instead of ammonia because it is non-toxic.

At the present time its use is limited primarily to the *manufacture of dry ice* (solid carbon dioxide).

The *leak detection of CO_2* is done by a *soap solution.*

5. Methyl chloride (CH_3Cl)
Properties:

(*i*) It is a colorless liquid with a faint, sweet, non-irritating odor.

(*ii*) It has liquid specific gravity of 1.002 at atmospheric pressure.

(*iii*) It is neither flammable nor toxic.

Uses:

It has been used in the past in both domestic and commercial applications. It should never be used with aluminum.

6. R-11 (Trichloro monofluoro methane)
Properties:

(*i*) It is composed of one carbon, one fluorine, and three chlorine atoms (or parts by weight) and is *non-corrosive, non-toxic,* and *non-flammable.*

(*ii*) It dissolves natural rubber.

(*iii*) It has a boiling point of – 24 °C.

(*iv*) It mixes completely with mineral lubricating oil under all conditions.

Uses:

It is employed for 50 metric tons capacity and over in small office buildings and factories. A centrifugal compressor is used in the plants employing this refrigerant.

Its leakage is detected by a halide torch.

7. R-12 (Dichloro-difluoro methane) or Freon-12
Properties:

(*i*) It is *non-toxic, non-flammable*, and *non-explosive*, therefore it is a *most suitable refrigerant*.

(*ii*) It is fully oil miscible therefore it simplifies the problem of oil return.

(*iii*) The operating pressures of R-12 in an evaporator and condenser under a *standard metric ton of refrigeration* are 1.9 bar abs. and 7.6 bar abs. (app.).

(*iv*) Its latent heat at – 15°C is 161.6 kJ/kg.

(*v*) C.O.P. = 4.61.

(*vi*) It does not break even under extreme operating conditions.

(*vii*) It condenses at moderate pressure and under atmospheric conditions.
   Uses:

1. It is suitable for high, medium, and low temperature applications.

2. It is used for domestic applications.

3. It is an *excellent electric insulator; therefore, it is universally used in sealed type compressors.*

8. R-22 (Monochloro-difluro methane) or Freon-22

The R-22 refrigerant is superior to R-12 in many respects. It has the following properties and uses:
   Properties:

(*i*) The compressor displacement per metric ton of refrigeration with R-22 is 60% less than the compressor displacement with R-12 as the refrigerant.

(*ii*) R-22 is miscible with oil at the condenser temperature but tries to separate at evaporator temperature when the system is used for very low temperature applications (– 90 °C). Oil separators must be incorporated to return the oil from the evaporator when the system is used for such low temperature applications.

(*iii*) The pressures in the evaporator and condenser at a standard metric ton of refrigeration are 2.9 bar abs. and 11.9 bar abs. (app.).

(*iv*) The latent heat at – 15 °C is low and is 89 kJ/kg.

The *major disadvantage of R-22 compared with R-12 is the high discharge temperature, which requires water cooling of the compressor head and cylinder.*

   Uses:

R-22 is universally used in commercial and industrial low temperature systems.

13.10 Exercises

Fill in the blanks:

1. _____ means the cooling of or removal of heat from a system.

2. Most commercial refrigeration is produced by the evaporation of a liquid _____

3. _____ is the ratio between the heat extracted and the work done.

4. _____ $= \dfrac{\text{Actual C.O.P}}{\text{Theoretical C.O.P.}}$.

5. The C.O.P. for a Carnot refrigerator is equal to _____

6. The C.O.P. for a Carnot heat pump is equal to _____

7. The C.O.P. for a Carnot refrigerator is _____ than that of a Carnot heat pump.

8. The C.O.P. of an air refrigeration system is _____ than a vapor compression system.

9. In a refrigeration system the heat rejected at higher temperatures = _____ + _____

10. Out of all the refrigeration systems, the _____ system is the most important system from the standpoint of commercial and domestic utility.

11. The function of a _____ is to remove the vapor from the evaporator and to raise its temperature and pressure to a point such that it (vapor) can be condensed with normally available condensing media.

12. The function of a _____ is to provide a heat transfer surface through which heat passes from the hot refrigerant vapor to the condensing medium.

13. The function of _____ is to meter the proper amount of refrigerant to the evaporator and to reduce the pressure of liquid entering the evaporator so that liquid will vaporize in the evaporator at the desired low temperature.

14. _____ provides a heat transfer surface through which heat can pass from the refrigerated space or product into the vaporizing refrigerant.

15. If the vapor is not superheated after compression, the operation is called _____

16. If the vapor is superheated at the end of compression, the operation is called _____

17. When the suction pressure decreases, the refrigerating effect and C.O.P. are _____

18. _____ results in the increase of C.O.P. provided that no further energy has to be spent to obtain the extra-cold coolant required.

19. _____ efficiency is defined as the ratio of actual volume of gas drawn into the compressor (at evaporator temperature and pressure) on each stroke to the piston displacement.

20. Define the following:

    (*i*) Refrigeration   (*ii*) Refrigerating system

    (*iii*) Refrigerated system.

21. Enumerate different ways of producing refrigeration.

22. Enumerate important refrigeration applications.

23. State elements of refrigeration systems.

24. Enumerate systems of refrigeration.

25. Define the following:

    (*i*) Actual C.O.P.   (*ii*) Theoretical C.O.P.

    (*iii*) Relative C.O.P.

26. What is a standard rating of a refrigeration machine?

27. What is main characteristic of an air refrigeration system?

28. Differentiate clearly between open and closed air refrigeration systems.

29. Explain briefly an air refrigerator working on a reversed Carnot cycle. Derive an expression for its C.O.P.

30. Derive an expression for C.O.P. for an air refrigeration system working on a reversed Brayton cycle.

31. State merits and demerits of an air refrigeration system.

32. Describe a simple vapor compression cycle, giving clearly its flow diagram.

33. State merits and demerits of a "Vapor compression system" over an "Air refrigeration system."

34. State the functions of the following parts of a simple vapor compression system:

    (*i*) Compressor   (*ii*) Condenser

    (*iii*) Expansion valve and   (*iv*) Evaporator.

35. Show the vapor compression cycle on a "Temperature-Entropy" (*T-s*) diagram for the following cases:

    (*i*) When the vapor is dry and saturated at the end of compression.

    (*ii*) When the vapor is superheated after compression.

    (*iii*) When the vapor is wet after compression.

36. What is the difference between "wet compression" and "dry compression"?

37. Write a short note on a "Pressure Enthalpy (*p-h*) chart."

38. Show the simple vapor compression cycle on a *p-h* chart.

39. Discuss the effect of the following on the performance of a vapor compression system:

    (*i*) Effect of suction pressure   (*ii*) Effect of delivery pressure

    (*iii*) Effect of superheating   (*iv*) Effect of sub-cooling of liquid

    (*v*) Effect of suction temperature and condenser temperature.

40. Show, with the help of diagrams, the difference between theoretical and actual vapor compression cycles.

41. Define the terms "Volumetric efficiency" and "Clearance volumetric efficiency."

42. Derive an expression for "Clearance volumetric efficiency."

43. Explain briefly the term "Total volumetric efficiency."

44. Explain briefly a simple vapor absorption system.

45. Give the comparison between a vapor compression system and a vapor absorption system.

46. The coefficient of performance of a Carnot refrigerator, when it extracts 8350 kJ/min from a heat source, is 5. Find the power required to run the compressor.

47. A reversed cycle has a refrigerating C.O.P. of 4.

    (*i*) Determine the ratio T_1/T_2; and

(*ii*) If this cycle is used as a heat pump, determine the C.O.P. and heat delivered.

48. An ice plant produces 10 metric tons of ice per day at 0 °C, using water at room temperature of 20 °C. Estimate the power rating of the compressor motor if the C.O.P. of the plant is 2.5 and overall electromechanical efficiency is 0.9.

Take latent heat of freezing for water = 335 kJ/kg

Specific heat of water = 4.18 kJ/kg.

49. An air refrigeration system operating on a Bell Coleman cycle takes in air from a cold room at 268 K and compresses it from 1.0 bar to 5.5 bar. The index of compression is 1.25. The compressed air is cooled to 300 K. The ambient temperature is 20 °C. Air expands in an expander where the index of expansion is 1.35.

Calculate: (*i*) C.O.P. of the system, (*ii*) Quantity of air circulated per minute for production of 1500 kg of ice per day at 0 °C from water at 20 °C, and (*iii*) Capacity of the plant in terms of kJ/s.

Take c_p = 4.18 kJ/kg K for water, c_p = 1.005 kJ/kg K for air

Latent heat of ice = 335 kJ/kg.

50. The temperature in a refrigerator coil is 267 K and that of the condenser coil is 295 K. Assuming that the machine operates on the reversed Carnot cycle, calculate:

(*i*) C.O.P.$_{(ref.)}$

(*ii*) The refrigerating effect per kW of input work.

(*iii*)The heat rejected to the condenser.

51. An ammonia vapor-compression refrigerator operates between an evaporator pressure of 2.077 bar and a condenser pressure of 12.37 bar. The following cycles are to be compared, and in each case there is no undercooling in the condenser, and isentropic compression may be assumed:

(*i*) The vapor has a dryness fraction of 0.9 at the entry to the compressor.

(*ii*) The vapor is dry saturated at the entry to the compressor.

(*iii*) The vapor has 5 K of superheat at the entry to the compressor.

In each case calculate the C.O.P.$_{(ref.)}$ and the refrigerating effect per kg.

What would be the C.O.P.$_{(ref.)}$ of a reversed Carnot cycle operating between the same saturation temperatures?

52. A refrigerator using Freon-12 operates between saturation temperatures of – 10 °C and 60 °C, at which temperatures the latent heats are 156.32 kJ/kg and 113.52 kJ/kg, respectively. The refrigerant is dry saturated at the entry to the compressor, and the liquid is not undercooled in the condenser. The specific heat of liquid Freon is 0.970 kJ/kg K and that of the superheated Freon vapor is 0.865 kJ/kg K. The vapor is compressed isentropically in the compressor. Using no other information than that given, calculate the temperature at the compressor delivery, and the refrigerating effect per kg of Freon.

53. A heat pump using ammonia as the refrigerant operates between saturation temperatures of 6°C and 38°C. The refrigerant is compressed isentropically from dry saturation and there is 6 K of undercooling in the condenser. Calculate:

(*i*) C.O.P.$_{(heat\ pump)}$ (*ii*) The mass flow of the refrigerant
(*iii*) The heat available per kilowatt input.

54. An ammonia vapor-compression refrigerator has a single-stage, single-acting reciprocating compressor which has a bore of 127 mm, a stroke of 152 mm, and a speed of 240 r.p.m. The pressure in the evaporator is 1.588 bar and that of the condenser is 13.89 bar. The volumetric efficiency of the compressor is 80% and its mechanical efficiency is 90%. The vapor is dry saturated on leaving the evaporator and the liquid leaves the condenser at 32 °C. Calculate the mass flow of refrigerant, the refrigerating effect, and the power ideally required to drive the compressor.

55. An ammonia refrigerator operates between evaporating and condensing temperatures of – 16 °C and 50 °C, respectively. The vapor is dry saturated at the compressor inlet, the compression process is isentropic, and there is no undercooling of the condensate. Calculate:

(*i*) The refrigerating effect per kg,
(*ii*) The mass flow and power input per kW of refrigeration, and
(*iii*) The C.O.P.$_{(ref.)}$

56. Metric tons of ice from and at 0 °C is produced in a day of 24 hours by an ammonia refrigerator. The temperature range in the compressor is from 298 K to 258 K. The vapor is dry saturated at the end of compression and an expansion valve is used. Assume a coefficient of performance of 60% of the theoretical and calculate the power in kW required to drive the compressor. The latent heat of ice is 334.72 kJ/kg.

Temp. K	Enthalpy (kJ/kg)		Entropy of liquid (kJ/kg K)	Entropy of vapor (kJ/kg)
	Liquid	Vapor		
298	100.04	1319.22	0.3473	4.4852
258	– 54.56	1304.99	– 2.1338	5.0585

57. A refrigerant plant works between temperature limits of – 5 °C (in the evaporator) and 25 °C (in the condenser). The working fluid ammonia has a dryness fraction of 0.6 at the entry to the compressor. If the machine has a relative efficiency of 50%, calculate the amount of ice formed during a period of 24 hours. The ice is to be formed at 0 °C from water at 20 °C and 6 kg of ammonia is circulated per minute. The specific heat of the water is 4.187 kJ/kg, and the latent heat of ice is 335 kJ/kg.

Properties of ammonia (datum – 40 °C):

Temp. K	Liquid heat kJ/kg	Latent heat kJ/kg	Entropy of liquid kJ/kg°C
298	298.9	1167.1	1.124
268	158.2	1280.8	0.630

58. A food storage locker requires a refrigeration system of 2500 kJ/min capacity at an evaporator temperature of – 10 °C and a condenser temperature of 30 °C. The refrigerant used is Freon-12 and sub-cooled by 5 °C before entering the expansion valve, and the vapor is superheated by 6 °C before leaving the evaporator coil. The compression of the refrigerant is reversible adiabatic. The refrigeration compressor is two-cylinder single-acting with stroke equal to 1.3 times the bore and operates at 975 r.p.m. Determine (using thermodynamic tables of properties for Freon-12):

(*i*) Refrigerating effect per kg.

(*ii*) Mass of refrigerant to be circulated per minute.

(*iii*) Theoretical piston displacement per minute.

(*iv*) Theoretical power required to run the compressor, in kW.

(*v*) Heat removed through the condenser per minute.

(*vi*) Theoretical bore and stroke of the compressor.

Properties of Freon-12:

Saturation temp. °C	Absolute pressure	Specific volume of vapor m³/kg	Enthalpy		Entropy	
			Liquid kJ/kg	Vapor kJ/kg	Liquid kJ/kg K	Vapor kJ/kg K
– 10°C	2.19	0.0767	26.9	183.2	0.1080	0.7020
30°C	7.45	0.0235	64.6	199.6	0.2399	0.6854

Take: Liquid specific heat = 1.235 kJ/kg K

Vapor specific heat = 0.735 kJ/kg K.

59. A vapor compression refrigerator uses methyl chloride and works in the pressure range of 11.9 bar and 5.67 bar. At the beginning of the compression, the refrigerant is 0.96 dry and at the end of isentropic compression, it has a temperature of 55 °C. The refrigerant liquid leaving the condenser is saturated. If the mass flow of the refrigerant is 1.8 kg/min, determine:

(*i*) Coefficient of performance.

(*ii*) The rise in temperature of condenser cooling water if the water flow rate is 16 kg/min.

(*iii*) The ice produced in the evaporator in kg/hour from water at 15 °C and ice at 0°C.

Properties of methyl chloride:

Saturation temp. (°C)	Pressure (bar)	Enthalpy (kJ/kg)		Entropy (kJ/kg K)	
		h_f	h_g	s_f	s_g
– 20	11.9	30.1	455.2	0.124	1.803
25	5.67	100.5	476.8	0.379	1.642

Take: Specific enthalpy of fusion of ice = 336 kJ/kg
specific heat of water = 4.187 kJ/kg.

60. A vapor compression refrigerator circulates 4.5 kg of NH_3 per hour. Condensation takes place at 30 °C and evaporation at – 15°C. There is no under-cooling of the refrigerant. The temperature after isentropic compression is 75 °C and the specific heat of the superheated vapor is 2.82 kJ/kg K. Determine:

(*i*) Coefficient of performance.

(*ii*) Ice produced in kg per hour in the evaporator from water at 20°C and ice at 0 °C. Take: Enthalpy of fusion of ice = 336 kJ/kg, specific heat of water = 4.187 kJ/kg.

(*iii*) The effective swept volume of the compressor in m³/min.

Properties of ammonia:

Sat. temp. (K)	Enthalpy (kJ/kg)		Entropy (kJ/kg K)		Volume (m³/kg)	
	h_f	h_g	s_f	s_g	v_f	v_g
303	323.1	1469	1.204	4.984	0.00168	0.111
258	112.3	1426	0.457	5.549	0.00152	0.509

CHAPTER 14

THERMODYNAMICS OF ENDOREVERSIBLE ENGINES

In This Chapter

- Overview
- Endoreversible Cycle Definition and Types
- Analysis of the Curzon-Ahlborn Engine
- Analysis of the Novikov Engine
- Müser Solar Thermal Engine
- Exercises

14.1 Overview

In classical thermodynamics we deal with systems at their equilibrium state and possible system limits as being reversible, mainly to maximize their efficiency. For example, Carnot's efficiency limit for an ideal engine is the maximum level that any heat engine may reach. However, real systems are far from the ideal case and the reversibility limit breaks down mainly due to irreversibility caused by friction, dissipation, and so on. The implicit assumption, or requirement, of having a reversible system is that of its slowness going through changes, through processes, and hence the need of having a very long time at hand for a reversible process/cycle to be completed. This is also referred to as a quasi-steady change/process. For example: in a Carnot engine, the piston should move very slowly in order to eliminate (or at least negligibly minimize) the friction between the piston and the cylinder wall; in a boiler we should have zero resistance for heat transfer; and an automotive factory should run its manufacturing line very slowly to have the maximum efficiency. These requirements put tough (or impractical) constraints on industries for meaningful power production and manufacturing, in general. In addition, for a real system, we would like to have practical and useful power production close to maximum capacity of the system, which requires having the process changes happening in a finite time. In other words, we can have energy conversion at an extremely slow pace (i.e., during a very long time) in order to maximize the efficiency or, rather, compromise the efficiency to an acceptable limit with having the energy conversion at a finite rate (i.e., finite time) in order to maximize the power production. A simple example could be as follows: after having a meal we can work at a very slow pace to have our performance efficiency at the maximum possible level or work at a reasonable pace to get things done within a finite time, but of course at the expense of having relatively lower efficiency than in the ideal case.

14.2. Endoreversible Cycle Definition and Types

The focus of classical thermodynamics, mainly developed in the 19th century, was the study of reversible systems at their equilibrium states. The practical requirements and necessities mentioned above encouraged the researchers to study irreversible processes further and develop the so-called *finite-time thermodynamics* in the 20th century [116], [117], [118]. Among the approaches developed [117] is one with the assumption of having a real system to be composed of ideal reversible equilibrium subsystems with irreversible interfaces or points of contact where entropy is generated. This approach, referred to as *endoreversible thermodynamics* (a sub-section of finite-time thermodynamics), takes advantage of powerful classical thermodynamics methods of analysis for reversible systems and produces results that are closer to what we measure and observe in practice, with taking irreversibility into consideration as well [119], [120], [121], [122]. For example, a Curzon-Ahlborn (CA) engine [123] is an endoreversible engine composed of a reversible Carnot's cycle operating between two heat reservoirs. The heat transfer mechanisms happen in a finite time with consideration of thermal resistance between the reservoirs and the engine, as sketched in Figure 14.1 in comparison with an ideal Carnot's engine.

14.3. Analysis of the Curzon-Ahlborn Engine

Historically [118], the analysis of endoreversible engines, such as the CA engine, goes back to R. B. Reitlinger in 1929 [124]. However, the work by Curzon-Ahlborn in 1975 [123] is the most cited one. We consider a Carnot engine (i.e., reversible) which receives heat $\dot{Q}_H$ at temperature T_1 from a

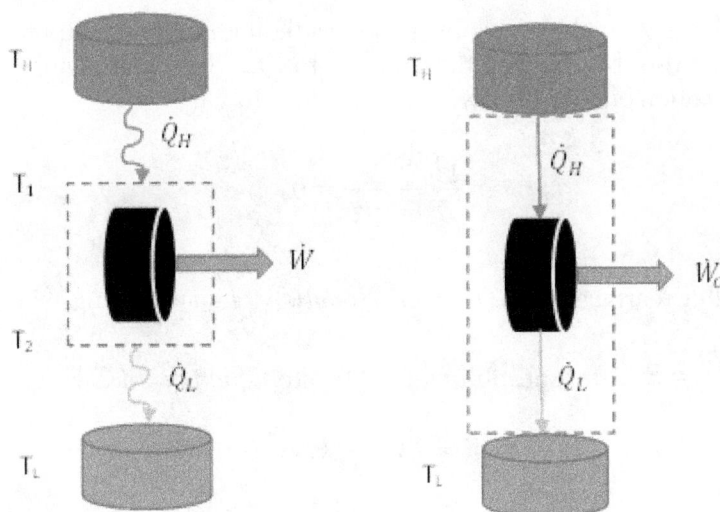

FIGURE 14.1 Sketch for an endoreversible system (left) and an ideal reversible system (right).

heat reservoir with constant temperature of T_H and rejects heat $\dot{Q}_L$ at temperature T_2 to a heat sink with constant temperature of T_L (see Figure 14.1). The differences in temperatures of the heat transfer mechanism represent the irreversibility of the system due to thermal resistance at the interfaces of the engine and the reservoirs. Note that for a fully reversible system (i.e., heat engine and the heat reservoirs) we would have $T_H = T_1$ and $T_L = T_2$ (see Figure 14.1). However, for our endoreversible system we have $T_H > T_1$ and $T_2 > T_L$. We use a Newtonian model for heat transfer with heat conductance (i.e., product of heat transfer coefficient and heat transfer surface area) K_H and K_L for the equipment delivering and rejecting heat, respectively. Therefore, we can write

$$\dot{Q}_H = K_H \left(T_H - T_1 \right) \tag{14.1}$$

and

$$\dot{Q}_L = K_L \left(T_2 - T_L \right) \tag{14.2}$$

By applying the 1st law, we have

$$\dot{W} = \dot{Q}_H - \dot{Q}_L \tag{14.3}$$

where $\dot{W}$ is the power expressed as the rate of work output from the engine. Also, by applying the 2nd law to the reversible engine, we have conservation of entropies as

$$\frac{\dot{Q}_H}{T_1} - \frac{\dot{Q}_L}{T_2} = 0 \tag{14.4}$$

After rearranging the terms in *Equation 14.4* and letting $\Theta = \dfrac{T_2}{T_1}$, we get $\dfrac{\dot{Q}_L}{\dot{Q}_H} = \Theta$. After substituting for $\dot{Q}_L$ into *Equation 14.3*, we get

$$\dot{W} = \dot{Q}_H (1 - \Theta) \tag{14.5}$$

We can rewrite *Equation 14.2* to get $T_2 = \dfrac{\dot{Q}_L}{K_L} + T_L$. Then we have

$T_1 = \dfrac{T_2}{\Theta} = \dfrac{1}{\Theta}\left(\dfrac{\dot{Q}_L}{K_L} + T_L \right) = \dfrac{\dot{Q}_H}{K_L} + \dfrac{T_L}{\Theta}$. After substituting for T_1 in *Equation 14.1* and solving for $\dot{Q}_H$ we have

$$\dot{Q}_H = \frac{K_H T_H}{\Theta}\left(\frac{\Theta - \dfrac{T_L}{T_H}}{1 + \dfrac{K_H}{K_L}} \right) \tag{14.6}$$

Writing Equation 14.6 in terms of engine efficiency $\eta = 1 - \theta$, gives

$$\dot{Q}_H = \frac{K_L K_H}{K_L + K_H}\left(\frac{T_H - T_L - \eta T_H}{1 - \eta} \right) \tag{14.7}$$

Therefore, the power produced is (using Equation 14.5), after rearranging the terms,

$$\dot{W} = \frac{K_L K_H \eta}{K_L + K_H}\left(\frac{T_H - T_L - \eta T_H}{1 - \eta} \right) \tag{14.8}$$

Equation 14.8 is a function of single variable η, assuming system design parameters K_H, T_H, K_L, and T_L as givens. To find the maximum power $\dot{W}_{max}$ we calculate the solutions of $\dfrac{d\dot{W}}{d\eta} = 0$ using Equation 14.8 , which yields $T_H\eta^2 - 2T_H\eta + T_H - T_L = 0$. Acceptable solution of this equation (i.e., $\eta < 1$) gives

$$\eta = 1 - \sqrt{T_L / T_H}$$
(14.9)

Equation 14.9 is the CA engine efficiency at maximum power production point. We can write Equation 14.9 in terms of Carnot efficiency, $\eta_c = 1 - \dfrac{T_L}{T_H}$, as

$$\eta = 1 - \sqrt{1 - \eta_c}$$
(14.10)

It is useful to compare, for example, $\eta_c = 60\%$. with $\eta = 1 - \sqrt{1 - 0.6} = 37\%$.

Back substituting for η, using Equation 14.9, into Equation 14.8 we get the maximum power as

$$\dot{W}_{max} = \frac{K_L K_H}{K_L + K_H}\left(\sqrt{T_H} - \sqrt{T_L}\right)^2$$
(14.11)

At the maximum power condition, the amount of thermal power required can be calculated by plugging Equation 14.9 into Equation 14.7 (or equivalently, divide Equation 14.11 by the efficiency, given by Equation 14.9) which gives, after some manipulation,

$$\dot{Q}_H\Big|_{max-power} = \frac{K_L K_H}{K_L + K_H}\left[\sqrt{T_H}\left(\sqrt{T_H} - \sqrt{T_L}\right)\right]$$
(14.12)

Similarly, it would be useful to calculate the total rate of generated entropy at maximum power for the CA engine, which is not zero due to irreversibility. Therefore, we can write $\sum \dot{S}_{gen} = -\dfrac{\dot{Q}_H}{T_H} + \dfrac{\dot{Q}_L}{T_L} = -\dfrac{\dot{Q}_H}{T_H} + \dfrac{(1-\eta)\dot{Q}_H}{T_L}$. and after substituting for $\dot{Q}_H$ from Equation 14.12 and some manipulation, we get

$$\dot{S}_{gen}\Big|_{max-power} = \frac{K_L K_H}{K_L + K_H} \frac{\left(\sqrt{T_H} - \sqrt{T_L}\right)^2}{\sqrt{T_L T_H}}$$

(14.13)

Now we investigate the variation of engine power versus its efficiency, as given by Equation 14.8.

The power in terms of efficiency can be written using Equation 14.8. We rewrite this equation in terms of η as the variable. Therefore, we get,

$$\frac{\dot{W}}{\left(\dfrac{K_L K_H}{K_L + K_H}\right)} = \eta T_H \left(\frac{1 - {T_L}/{T_H} - \eta}{1 - \eta}\right) = \eta T_H \left(\frac{\eta_c - \eta}{1 - \eta}\right) = \eta T_H \left(1 - \frac{1 - \eta_c}{1 - \eta}\right).$$

Therefore, we have

$$P_{CA} = \eta \left(1 - \frac{1 - \eta_c}{1 - \eta}\right)$$

(14.14)

where $P_{CA} = \dfrac{\dot{W}}{\left(\dfrac{K_L K_H}{K_L + K_H}\right) T_H}$ is the dimensionless power of the CA

engine. The variations of the P_{CA} versus η are shown in Figure 14.2. As shown in this figure, power is maximum at an efficiency equal to that of CA, η_{CA} and is zero at $\eta = 0$ and Carnot efficiency $\eta = \eta_c$. To find

the roots, we let $\eta \left(1 - \dfrac{1 - \eta_c}{1 - \eta}\right) = 0$. ThereforeTherefore, we get the

two roots; $\eta_1 = 0$ and $\eta_2 = \eta_c$ (as if no irreversibility existed in the system). To find the maximum, we differentiate Equation 14.14, hence

$\dfrac{\partial}{\partial \eta}(P_{CA})\Big|_{\eta_c} = 1 - (1 - \eta_c)(1 - \eta)^{-2} = 0$. This relation gives a quadratic

equations in terms of η, or $\eta^2 - 2\eta + \eta_c = 0$ which has two roots; $\eta_1 = 1 + \sqrt{1 - \eta_c}$ and $\eta_2 = \eta_1 = 1 - \sqrt{1 - \eta_c}$. Obviously the former (i.e.,

$\eta_1 = 1 + \sqrt{1 - \eta_c}$ is not acceptable, since efficiency is always less than 1) Therefore, the acceptable result is $\eta_2 = \eta_{CA} = 1 - \sqrt{1 - \eta_c}$ which is equal to CA engine efficiency (see Equation 14.10).

It is noteworthy to mention that for any value of efficiency lower than η_{CA} the corresponding power will decrease with decreasing efficiency (see Figure 14.2). This is not desirable for the CA engine, or in general any engine. But for any value of efficiency higher than η_{CA} the corresponding power will increase (decrease) with decreasing (increasing) efficiency (see Figure 14.2). This engine behavior is desirable, since we exchange power with efficiency or vice versa.

Also it is clearly seen from Figure 14.2 that η_{CA} is not the maximum efficiency possible for the CA engine;, rather the Carnot efficicny efficiency η_c is the maximum. But at η_{CA} we get the maximum power wheras at η_c power is null.

As mentioned, Equation 14.9 gives the efficiency of the CA engine, independent from thermal conductance, at its maximum power. Note that this efficiency (i.e., $1 - \sqrt{T_L / T_H}$) is not the maximum available efficiency for the CA engine and, therefore, not an upper bound limit, but it is still less than that of Carnot's, or $\eta < \eta_c$. It is observed that actual efficiency of a typical power plant is usually between these two efficiencies. Reported in

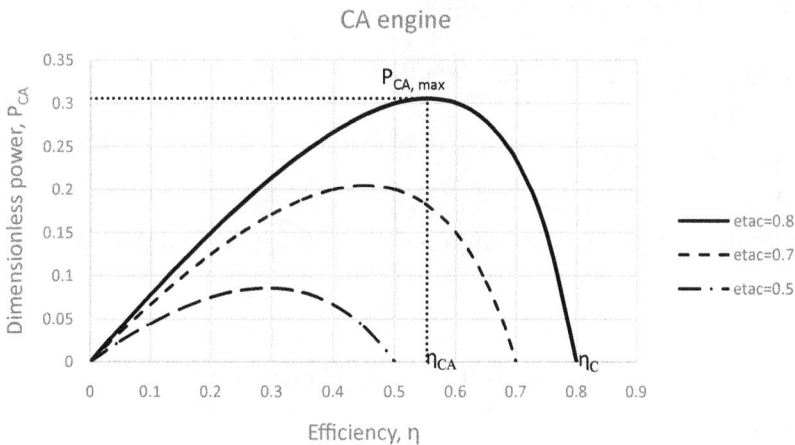

FIGURE 14.2. Typical Curzon-Ahlborn power vs. efficiency.

				η_c	η (Curzon-Ahlborn)	η
Power plant	Type	T_H [K]	T_L [K]	(Carnot)		(actual)
West Thurock (UK) [126]	Coal	838	298	64%	40%	36%
CANDU (Canada) [127]	Nuclear	573	298	48%	28%	30%
Landerello (Italy) [128]	Geothermal	523	353	32%	17%	16%
Doel 4 (Belgium) [122]	Nuclear	566	283	50%	29%	35%

TABLE 14.1. Actual Efficiency of Some Power Plants vs. That of Curzon-Ahlborn's and Carnot's

[117], [123], and [125], the actual efficiencies of some typical power plants are closer to that of the CA engine than Carnot's; see Table 14.1.

Now we discuss the system heat transfer properties in order to have the maximum power. In other words, we would like to know the thermal resistances distribution in order to have the maximum power from the CA engine, given by Equation 14.8, with conductance as parameters of the heat transfer equipment. In order to find the proportion of total conductance $K = K_H + K_L$ between heating equipment, attached to the hot reservoir, and the cooling equipment, attached to the cold reservoir, we rewrite Equation 14.8 in terms of K. Therefore, we have

$\dot{W} = \dfrac{K_L K_H \eta}{K_L + K_H}\left(\dfrac{T_H - T_L - \varsigma T_H}{1-\eta}\right) = \dfrac{K_L(K - K_L)}{K} f(\eta).$ This function

is then maximized by differentiating it with respect to K_L. Therefore,

$\dfrac{\partial \dot{W}}{\partial(K_L)} = \dfrac{K - 2K_L}{K} f(\eta) = 0,$ which gives $K_L = \dfrac{K}{2}.$ This result shows

that the conductance of the cooling equipment and heating equipment should be equal for maximum power production from the CA engine, or

$$K_L = K_H = \frac{K}{2} \tag{14.15}$$

Substituting Equation 14.15 into Equation 14.11 gives the maximum power as

$$\dot{W}_{max} = \frac{K}{4}\left(\sqrt{T_H} - \sqrt{T_L}\right)^2 \tag{14.16}$$

Also the thermal power input and entropy generated rate at maximum power are, using Equation 14.12 and Equation 14.13,

$$\dot{Q}_{max} = \frac{K}{4}\left[\sqrt{T_H}\left(\sqrt{T_H} - \sqrt{T_L}\right)\right]$$

(14.17)

and

$$\dot{S}_{gen} = \frac{K}{4}\frac{\left(\sqrt{T_H} - \sqrt{T_L}\right)^2}{\sqrt{T_L T_H}}$$

(14.18)

This concludes the analysis of the CA engine.

14.4. Analysis of the Novikov Engine

The Novikov engine [129], is a special case of the CA engine, for which the thermal resistance exists only at the interface of the reversible Carnot engine and the hot reservoir, as shown in Figure 14.3. Therefore, we have $T_2 = T_L$ (or infinite conductance, i.e., $K_L \to \infty$). Substituting for T_L into Equation 14.9, $\theta_N = \sqrt{\dfrac{T_2}{T_H}}$ gives , at which we get the maximum power. But $\theta_N = \dfrac{T_2}{T_1}$, by definition, hence we have $\dfrac{T_2}{T_1} = \sqrt{\dfrac{T_2}{T_H}}$ and after solving for T_1 we get (using $T_2 = T_L$)

$$T_1 = \sqrt{T_L T_H}$$

(14.19)

The efficiency of the Novikov engine is then calculated as $\eta_N = 1 - \dfrac{T_L}{T_1} = 1 - \dfrac{T_L}{\sqrt{T_L T_H}} = 1 - \sqrt{T_L/T_H}$ which is the same as that of the CA engine (see Equation 14.9). The maximum power for the Novikov engine can be calculated by rearranging Equation 14.11 as $\dot{W}_{max} = \dfrac{K_H}{1 + K_H/K_L}\left(\sqrt{T_H} - \sqrt{T_L}\right)^2$ and letting $K_L \to \infty$. Therefore, we have $K_H/K_L \to 0$ and the maximum power is given as

$$\dot{W}_{max}\Big|_{Novicov} = K_H\left(\sqrt{T_H} - \sqrt{T_L}\right)^2$$

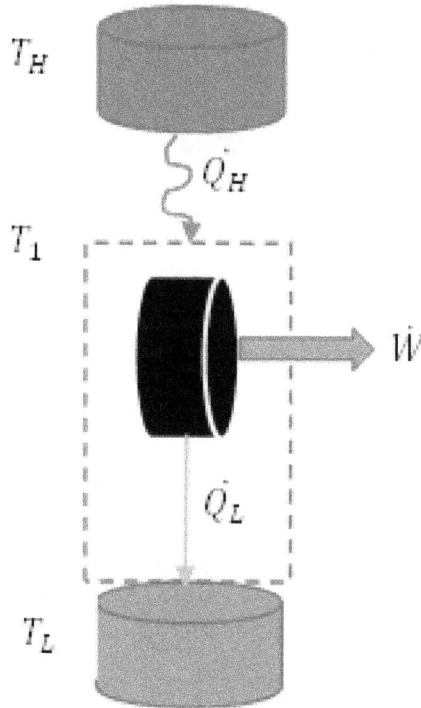

Figure 14.3 Sketch for the Novikov engine.

Similar analyses to those given above, can be carried out for common endoreversible heat engines including Otto, Diesel, Rankin, Brayton, Stirling cycles, and solar thermal systems [130], [131], [122], [120], [132] , [133]. A detailed discussion for a solar cell analysis is given in the reference [134].

As mentioned, Newtonian heat transfer mechanisms are common for CA engines. It is also possible to use different heat transfer mechanisms in an endoreversible system. For example, a radiation type (i.e., proportional to T^4) is used for solar thermal systems, [122], as described in a further section. These heat transfer mechanisms are unified by following Dulong-Petit law [135], [136] , or

$$\dot{Q} = K\left(T_1^n - T_2^n\right)^m$$

For example: for $n = m = 1$ the Newtonian linear heat transfer is recovered, and for $n = 4$, $m = 1$ the Stefan-Boltzmann radiative heat transfer is recovered.

14.5. Müser Solar Thermal Engine

The Müser engine is similar to the Novikov one, except the heat transfer is from a solar source through radiation considering irreversible thermal resistance. The heat sink is the ground with perfect conductance, hence no irreversibility is considered at the interface of the Carnot engine and the heat sink. A possible sketch of the engine is similar to Figure 14.3. The thermal power supplied is written as, according to the Stefan-Boltzmann radiation law,

$$\dot{Q}_H = K_R\left(T_H{}^4 - T_1^4\right)$$

(14.20)

where K_R is the conductance coefficient. By applying the 1st and 2nd laws (similar to Equation 14.3, Equation 14.4) and the definition of efficiency $\eta = \dfrac{\dot{W}}{\dot{Q}_H}$ we arrive at the relation, $T_1 = \dfrac{T_L}{1-\eta}$. After substituting for T_1 into Equation 14.20, we get

$$\dot{Q}_H = K_R\left[T_H{}^4 - \frac{T_L{}^4}{\left(1-\eta\right)^4}\right]$$

(14.21)

Therefore, the thermal power is (using, $\dot{W} = \eta\dot{Q}_H$)

$$\dot{W} = K_R\left[\eta T_H{}^4 - \frac{\eta T_L{}^4}{\left(1-\eta\right)^4}\right]$$

(14.22)

Equation 14.22 gives the power versus efficiency for the Müser engine. It is easy to check that power is zero at $\eta = 0$ and $\eta_C = 1 - \dfrac{T_L}{T_H}$, that is, Carnot efficiency. The efficiency at maximum power can be found by solving $\dfrac{\partial \dot{W}}{\partial \eta} = 0$, or $T_H{}^4\left(1-\eta\right)^8 - T_L{}^4\left(1-\eta\right)^4 - 4\eta\left(1-\eta\right)^3 T_L{}^4 = 0$. We divide both sides by $\left(1-\eta\right)^3$ (this operation is legitimate, since $\eta < 1$), to get $T_H{}^4\left(1-\eta\right)^5 - T_L{}^4\left(1-\eta\right) - 4\eta T_L{}^4 = 0$. Now we rewrite this expression, as a polynomial involving terms like $\left(1-\eta\right)$, or $T_H{}^4\left(1-\eta\right)^5 - T_L{}^4\left(1-\eta\right) - \underbrace{4\eta T_L^4 + 4T_L^4 - 4T_L^4}_{=0} = 0$, which gives, after rearrangement of the terms

$$T_H{}^4 \left(1-\eta\right)^5 + 3T_L{}^4 \left(1-\eta\right) - 4T_L{}^4 = 0 \tag{14.23}$$

The real root of Equation 14.23 is the efficiency at maximum power. However, this equation cannot be solved analytically (i.e., solution of a 5th order polynomial), and one should find its solution using numerical methods.

In order to find the real root of Equation 14.23, we use the Earth surface temperature $T_L = 273 + 15 = 288\text{K}$, and an approximation for effective sky temperature $T_H \approx 5^{0.25} T_L \cong 431\text{K}$ (see [122]). Therefore, Equation 14.23 reads (after dividing by $T_L{}^4$)

$$5\left(1-\eta\right)^5 - 3\eta - 1 = 0 \tag{14.24}$$

The only real solution for Equation 14.24 is, $\eta = 0.2029167$ (readers may use online equation solvers, like *www.wolframalpha.com*). Substituting for η into Equation 14.22, we get the maximum power available from the Müser engine, or as a scaled dimensionless quantity, $\dot{w}$

$$\dot{w} = \frac{\dot{W}}{K_R T_L{}^4} = \left[5\eta - \frac{\eta}{\left(1-\eta\right)^4}\right] = 0.51189$$

De Vos [122], gives an argument for the quantity $\dfrac{\dot{W}}{K_R T_L{}^4} = \dfrac{4\dot{W}}{f\sigma T_s{}^4}$, where $f = 2.16 \times 10^{-5}$ is the sunlight dilution factor for Earth, $\sigma = 5.67 \times 10^{-8}$ W/m².K⁴, $T_S = 5762$K sun temperature.

14.6. Exercises

1. What is the fill factor for the Curzon-Ahlborn and Novacon engines?

2. Derive the equation for CA engine power versus its efficiency.

3. Derive the equation for Novikov engine power versus its efficiency.

4. Derive the dimensionless power vs. efficiency for a Novikov engine. Make a graph and discuss the results for different values of power and efficiency.

CHAPTER 15

STEAM TABLES

S.I. Units

1. Saturated Water and Steam (Temperature) Tables

2. Saturated Water and Steam (Pressure) Tables

Symbols and Units Used in the Tables

t = Temperature, °C

t_s = Saturation temperature, °C

p = Pressure, *bar*

h_f = Specific enthalpy of saturated liquid, kJ / kg

h_{fg} = Specific enthalpy of evaporation (latent heat), kJ / kg

h_g = Specific enthalpy of saturated vapour, kJ / kg

s_f = Specific entropy of saturated liquid, kJ / kg K

s_{fg} = Specific entropy of evaporation, kJ / kg K

s_g = Specific entropy of saturated vapour, kJ / kg K

v_f = Specific volume of saturated liquid, m³ / kg

v_g = Specific volume of saturated steam, m³ / kg

Temp. (°C)	Absolute pressure (bar)	Specific enthalpy (kJ/kg)			Specific entropy (kJ/kg K)			Specific volume (m³/kg)	
t	P	h_f	h_{fg}	h_g	s_f	s_{fg}	s_g	v_f	v_g
0	0.0061	– 0.02	2501.4	2501.3	– 0.0001	9.1566	9.1565	0.0010002	206.3
0.01	0.0061	0.01	2501.3	2501.4	0.000	9.156	9.156	0.0010002	206.2
1	0.0065	4.2	2499.0	2503.2	0.015	9.115	9.130	0.0010002	192.6
2	0.0070	8.4	2496.7	2505.0	0.031	9.073	9.104	0.0010001	179.9
3	0.0076	12.6	2494.3	2506.9	0.046	9.032	9.077	0.0010001	168.1
4	0.0081	16.8	2491.9	2508.7	0.061	8.990	9.051	0.0010001	157.2
5	0.0087	21.0	2489.6	2510.6	0.076	8.950	9.026	0.0010001	147.1
6	0.0093	25.2	2487.2	2512.4	0.091	8.909	9.000	0.0010001	137.7
7	0.0100	29.4	2484.8	2514.2	0.106	8.869	8.975	0.0010002	129.0
8	0.0107	33.6	2482.5	2516.1	0.121	8.829	8.950	0.0010002	120.9
9	0.0115	37.8	2480.1	2517.9	0.136	8.789	8.925	0.0010003	113.4
10	0.0123	42.0	2477.7	2519.7	0.151	8.750	8.901	0.0010004	106.4
11	0.0131	46.2	2475.4	2521.6	0.166	8.711	8.877	0.0010004	99.86
12	0.0140	50.4	2473.0	2523.4	0.181	8.672	8.852	0.0010005	93.78
13	0.0150	54.6	2470.7	2525.3	0.195	8.632	8.828	0.0010007	88.12
14	0.0160	58.8	2468.3	2527.1	0.210	8.595	8.805	0.0010008	82.85
15	0.0170	63.0	2465.9	2528.9	0.224	8.557	8.781	0.0010009	77.93
16	0.0182	67.2	2463.6	2530.8	0.239	8.519	8.758	0.001001	73.33
17	0.0194	71.4	2461.2	2532.6	0.253	8.482	8.735	0.001001	69.04
18	0.0206	75.6	2458.8	2534.4	0.268	8.444	8.712	0.001001	65.04
19	0.0220	79.8	2456.5	2536.3	0.282	8.407	8.690	0.001002	61.29
20	0.0234	84.0	2454.1	2538.1	0.297	8.371	8.667	0.001002	57.79
21	0.0249	88.1	2451.8	2539.9	0.311	8.334	8.645	0.001002	54.51
22	0.0264	92.3	2449.4	2541.7	0.325	8.298	8.623	0.001002	51.45
23	0.0281	96.5	2447.0	2543.5	0.339	8.262	8.601	0.001002	48.57
24	0.0298	100.7	2444.7	2545.4	0.353	8.226	8.579	0.001003	45.88
25	0.0317	104.9	2442.3	2547.2	0.367	8.191	8.558	0.001003	43.36
26	0.0336	109.1	2439.9	2549.0	0.382	8.155	8.537	0.001003	40.99
27	0.0357	113.2	2437.6	2550.8	0.396	8.120	8.516	0.001004	38.77
28	0.0378	117.4	2435.2	2552.6	0.409	8.086	8.495	0.001004	36.69
29	0.0401	121.6	2432.8	2554.5	0.423	8.051	8.474	0.001004	34.73
30	0.0425	125.8	2430.5	2556.3	0.437	8.016	8.453	0.001004	32.89
31	0.0450	130.0	2428.1	2558.1	0.451	7.982	8.433	0.001005	31.17
32	0.0476	134.2	2425.7	2559.9	0.464	7.948	8.413	0.001005	29.54

TABLE 15.1 Saturated Water and Steam (Temperature) Tables

(*Continued*)

Temp. (°C)	Absolute pressure (bar)	Specific enthalpy (kJ/kg)			Specific entropy (kJ/kg K)			Specific volume (m³/kg)	
t	P	$'h_f$	h_{fg}	h_g	s_f	S_{fg}	s_g	v_f	v_g
33	0.0503	138.3	2423.4	2561.7	0.478	7.915	8.393	0.001005	28.01
34	0.0532	142.5	2421.0	2563.5	0.492	7.881	8.373	0.001006	26.57
35	0.0563	146.7	2418.6	2565.3	0.505	7.848	8.353	0.001006	25.22
36	0.0595	150.9	2416.2	2567.1	0.519	7.815	8.334	0.001006	23.94
37	0.0628	155.0	2413.9	2568.9	0.532	7.782	8.314	0.001007	22.74
38	0.0663	159.2	2411.5	2570.7	0.546	7.749	8.295	0.001007	21.60
39	0.0700	163.4	2409.1	2572.5	0.559	7.717	8.276	0.001007	20.53
40	0.0738	167.6	2406.7	2574.3	0.573	7.685	8.257	0.001008	19.52
41	0.0779	171.7	2404.3	2576.0	0.586	7.652	8.238	0.001008	18.57
42	0.0821	175.9	2401.9	2577.8	0.599	7.621	8.220	0.001009	17.67
43	0.0865	180.1	2399.5	2579.6	0.612	7.589	8.201	0.001009	16.82
44	0.0911	184.3	2397.2	2581.5	0.626	7.557	8.183	0.001010	16.02
45	0.0959	188.4	2394.8	2583.2	0.639	7.526	8.165	0.001010	15.26
46	0.1010	192.6	2392.4	2585.0	0.652	7.495	8.147	0.001010	14.54
47	0.1062	196.8	2390.0	2586.8	0.665	7.464	8.129	0.001011	13.86
48	0.1118	201.0	2387.6	2588.6	0.678	7.433	8.111	0.001011	13.22
49	0.1175	205.1	2385.2	2590.3	0.691	7.403	8.094	0.001012	12.61
50	0.1235	209.3	2382.7	2592.1	0.704	7.372	8.076	0.001012	12.03
52	0.1363	217.7	2377.9	2595.6	0.730	7.312	8.042	0.001013	10.97
54	0.1502	226.0	2373.1	2599.1	0.755	7.253	8.008	0.001014	10.01
56	0.1653	234.4	2368.2	2602.6	0.781	7.194	7.975	0.001015	9.149
58	0.1817	242.8	2363.4	2606.2	0.806	7.136	7.942	0.001016	8.372
60	0.1994	251.1	2358.5	2609.6	0.831	7.078	7.909	0.001017	7.671
62	0.2186	259.5	2353.6	2613.1	0.856	7.022	7.878	0.001018	7.037
64	0.2393	267.9	2348.7	2616.5	0.881	6.965	7.846	0.001019	6.463
66	0.2617	276.2	2343.7	2619.9	0.906	6.910	7.816	0.001020	5.943
68	0.2859	284.6	2338.8	2623.4	0.930	6.855	7.785	0.001022	5.471
70	0.3119	293.0	2333.8	2626.8	0.955	6.800	7.755	0.001023	5.042
75	0.3858	313.9	2321.4	2635.3	1.015	6.667	7.682	0.001026	4.131
80	0.4739	334.9	2308.8	2643.7	1.075	6.537	7.612	0.001029	3.407
85	0.5783	355.9	2296.0	2651.9	1.134	6.410	7.544	0.001033	2.828
90	0.7014	376.9	2283.2	2660.1	1.192	6.287	7.479	0.001036	2.361
95	0.8455	397.9	2270.2	2668.1	1.250	6.166	7.416	0.001040	1.982
100	1.0135	419.0	2257.0	2676.0	1.307	6.048	7.355	0.001044	1.673

TABLE 15.1 Saturated Water and Steam (Temperature) Tables

Absolute pressure (bar) P	Temp. (°C) t_s	Specific enthalpy (kJ/kg)			Specific entropy (kJ/kg K)			Specific volume (m³/kg)	
		h_f	h_{fg}	h_g	s_f	S_{fg}	s_g	v_f	v_g
0.006113	0.01	0.01	2 501.3	2 501.4	0.000	9.156	9.156	0.0010002	206.14
0.010	7.0	29.3	2 484.9	2 514.2	0.106	8.870	8.976	0.0010000	129.21
0.015	13.0	54.7	2 470.6	2 525.3	0.196	8.632	8.828	0.0010007	87.98
0.020	17.0	73.5	2 460.0	2 533.5	0.261	8.463	8.724	0.001001	67.00
0.025	21.1	88.5	2 451.6	2 540.1	0.312	8.331	8.643	0.001002	54.25
0.030	24.1	101.0	2 444.5	2 545.5	0.355	8.223	8.578	0.001003	45.67
0.035	26.7	111.9	2 438.4	2 550.3	0.391	8.132	8.523	0.001003	39.50
0.040	29.0	121.5	2 432.9	2 554.4	0.423	8.052	8.475	0.001004	34.80
0.045	31.0	130.0	2 428.2	2 558.2	0.451	7.982	8.433	0.001005	31.13
0.050	32.9	137.8	2 423.7	2 561.5	0.476	7.919	8.395	0.001005	28.19
0.055	34.6	144.9	2 419.6	2 565.5	0.500	7.861	8.361	0.001006	25.77
0.060	36.2	151.5	2 415.9	2 567.4	0.521	7.809	8.330	0.001006	23.74
0.065	37.6	157.7	2 412.4	2 570.1	0.541	7.761	8.302	0.001007	22.01
0.070	39.0	163.4	2 409.1	2 572.5	0.559	7.717	8.276	0.001007	20.53
0.075	40.3	168.8	2 406.0	2 574.8	0.576	7.675	8.251	0.001008	19.24
0.080	41.5	173.9	2 403.1	2 577.0	0.593	7.636	8.229	0.001008	18.10
0.085	42.7	178.7	2 400.3	2 579.0	0.608	7.599	8.207	0.001009	17.10
0.090	43.8	183.3	2 397.7	2 581.0	0.622	7.565	8.187	0.001009	16.20
0.095	44.8	187.7	2 395.2	2 582.9	0.636	7.532	8.168	0.001010	15.40
0.10	45.8	191.8	2 392.8	2 584.7	0.649	7.501	8.150	0.001010	14.67
0.11	47.7	199.7	2 388.3	2 588.0	0.674	7.453	8.117	0.001011	13.42
0.12	49.4	206.9	2 384.2	2 591.1	0.696	7.390	8.086	0.001012	12.36
0.13	51.0	213.7	2 380.2	2 593.9	0.717	7.341	8.058	0.001013	11.47
0.14	52.6	220.0	2 376.6	2 596.6	0.737	7.296	8.033	0.001013	10.69
0.15	54.0	226.0	2 373.2	2 599.2	0.754 9	7.254 4	8.009 3	0.001014	10.022
0.16	55.3	231.6	2 370.0	2 601.6	0.772 1	7.214 8	7.986 9	0.001015	9.433
0.17	56.6	236.9	2 366.9	2 603.8	0.788 3	7.177 5	7.965 8	0.001015	8.911
0.18	57.8	242.0	2 363.9	2 605.9	0.803 6	7.142 4	7.945 9	0.001016	8.445
0.19	59.0	246.8	2 361.1	2 607.9	0.818 2	7.109 0	7.927 2	0.001017	8.027
0.20	60.1	251.5	2 358.4	2 609.9	0.832 1	7.077 3	7.909 4	0.001017	7.650
0.21	61.1	255.9	2 355.8	2 611.7	0.845 3	7.047 2	7.892 5	0.001018	7.307
0.22	62.2	260.1	2 353.3	2 613.5	0.858 1	7.018 4	7.876 4	0.001018	6.995
0.23	63.1	264.2	2 350.9	2 615.2	0.870 2	6.990 8	7.861 1	0.001019	6.709
0.24	64.1	268.2	2 348.6	2 616.8	0.882 0	6.964 4	7.846 4	0.001019	6.447

TABLE 15.2 Saturated Water and Steam (Pressure) Tables

(Continued)

Absolute pressure (bar) P	Temp. (°C) t_s	Specific enthalpy (kJ/kg)			Specific entropy (kJ/kg K)			Specific volume (m³/kg)	
		h_f	h_{fg}	h_g	s_f	S_{fg}	s_g	v_f	v_g
0.25	65.0	272.0	2 346.4	2 618.3	0.893 2	6.939 1	7.832 3	0.001020	6.205
0.26	65.9	275.7	2 344.2	2 619.9	0.904 1	6.914 7	7.818 8	0.001020	5.980
0.27	66.7	279.2	2 342.1	2 621.3	0.914 6	6.891 2	7.805 8	0.001021	5.772
0.28	67.5	282.7	2 340.0	2 622.7	0.924 8	6.868 5	7.793 3	0.001021	5.579
0.29	68.3	286.0	2 338.1	2 624.1	0.934 6	6.846 6	7.781 2	0.001022	5.398
0.30	69.1	289.3	2 336.1	2 625.4	0.944 1	6.825 4	7.769 5	0.001022	5.229
0.32	70.6	295.5	2 332.4	2 628.0	0.962 3	6.785 0	7.747 4	0.001023	4.922
0.34	72.0	301.5	2 328.9	2 630.4	0.979 5	6.747 0	7.726 5	0.001024	4.650
0.36	73.4	307.1	2 325.5	2 632.6	0.995 8	6.711 1	7.707 0	0.001025	4.408
0.38	74.7	312.5	2 322.3	2 634.8	1.011 3	6.677 1	7.688 4	0.001026	4.190
0.40	75.9	317.7	2 319.2	2 636.9	1.026 1	6.644 8	7.670 9	0.001026	3.993
0.42	77.1	322.6	2 316.3	2 638.9	1.040 2	6.614 0	7.654 2	0.001027	3.815
0.44	78.2	327.3	2 313.4	2 640.7	1.053 7	6.584 6	7.638 3	0.001028	3.652
0.46	79.3	331.9	2 310.7	2 642.6	1.066 7	6.556 4	7.623 1	0.001029	3.503
0.48	80.3	336.3	2 308.0	2 644.3	1.079 2	6.529 4	7.608 6	0.001029	3.367
0.50	81.3	340.6	2 305.4	2 646.0	1.091 2	6.503 5	7.594 7	0.001030	3.240
0.55	83.7	350.6	2 299.3	2 649.9	1.119 4	6.442 8	7.562 3	0.001032	2.964
0.60	86.0	359.9	2 293.6	2 653.6	1.145 4	6.387 3	7.532 7	0.001033	2.732
0.65	88.0	368.6	2 288.3	2 656.9	1.169 6	6.336 0	7.505 5	0.001035	2.535
0.70	90.0	376.8	2 283.3	2 660.1	1.192 1	6.288 3	7.480 4	0.001036	2.369
0.75	92.0	384.5	2 278.6	2 663.0	1.213 1	6.243 9	7.457 0	0.001037	2.217
0.80	93.5	391.7	2 274.0	2 665.8	1.233 0	6.202 2	7.435 2	0.001039	2.087
0.85	95.1	398.6	2 269.8	2 668.4	1.251 8	3.162 9	7.414 7	0.001040	1.972
0.90	96.7	405.2	2 265.6	2 670.9	1.269 6	6.125 8	7.395 4	0.001041	1.869
0.95	98.2	411.5	2 261.7	2 673.2	1.286 5	6.090 6	7.377 1	0.001042	1.777
1.0	99.6	417.5	2 257.9	2 675.4	1.302 7	6.057 1	7.359 8	0.001043	1.694
1.1	102.3	428.8	2 250.8	2 679.6	1.333 0	5.994 7	7.327 7	0.001046	1.549
1.2	104.8	439.4	2 244.1	2 683.4	1.360 9	5.937 5	7.298 4	0.001048	1.428
1.3	107.1	449.2	2 237.8	2 687.0	1.386 8	5.884 7	7.271 5	0.001050	1.325
1.4	109.3	458.4	2 231.9	2 690.3	1.410 9	5.835 6	7.246 5	0.001051	1.236
1.5	111.3	467.1	2 226.2	2 693.4	1.433 6	5.789 8	7.233 4	0.001053	1.159
1.6	113.3	475.4	2 220.9	2 696.2	1.455 0	5.746 7	7.201 7	0.001055	1.091
1.7	115.2	483.2	2 215.7	2 699.0	1.475 2	5.706 1	7.181 3	0.001056	1.031
1.8	116.9	490.7	2 210.8	2 701.5	1.494 4	5.667 8	7.162 2	0.001058	0.977
1.9	118.6	497.8	2 206.1	2 704.0	1.512 7	5.631 4	7.144 0	0.001060	0.929

TABLE 15.2 Saturated Water and Steam (Pressure) Tables

(Continued)

Absolute pressure (bar) P	Temp. (°C) t_s	Specific enthalpy (kJ/kg)			Specific entropy (kJ/kg K)			Specific volume (m³/kg)	
		h_f	h_{fg}	h_g	s_f	S_{fg}	s_g	v_f	v_g
2.0	120.2	504.7	2 201.6	2 706.3	1.530 1	5.596 7	7.126 8	0.001061	0.885
2.1	121.8	511.3	2 197.2	2 708.5	1.546 8	5.563 7	7.110 5	0.001062	0.846
2.2	123.3	517.6	2 193.0	2 710.6	1.562 7	5.532 1	7.094 9	0.001064	0.810
2.3	124.7	523.7	2 188.9	2 712.6	1.578 1	5.501 9	7.080 0	0.001065	0.777
2.4	126.1	529.6	2 184.9	2 714.5	1.592 9	5.472 8	7.065 7	0.001066	0.746
2.5	127.4	535.3	2 181.0	2 716.4	1.607 1	5.444 9	7.052 0	0.001068	0.718
2.6	128.7	540.9	2 177.3	2 718.2	1.620 9	5.418 0	7.038 9	0.001069	0.693
2.7	129.9	546.2	2 173.6	2 719.9	1.634 2	5.392 0	7.026 2	0.001070	0.668
2.8	131.2	551.4	2 170.1	2 721.5	1.647 1	5.367 0	7.014 0	0.001071	0.646
2.9	132.4	556.5	2 166.6	2 723.1	1.659 5	5.342 7	7.002 3	0.001072	0.625
3.0	133.5	561.4	2 163.2	2 724.7	1.671 6	5.319 3	6.990 9	0.001074	0.606
3.1	134.6	566.2	2 159.9	2 726.1	1.683 4	5.296 5	6.979 9	0.001075	0.587
3.2	135.7	570.9	2 156.7	2 727.6	1.694 8	5.274 4	6.969 2	0.001076	0.570
3.3	136.8	575.5	2 153.5	2 729.0	1.705 9	5.253 0	6.958 9	0.001077	0.554
3.4	137.8	579.9	2 150.4	2 730.3	1.716 8	5.232 2	6.948 9	0.001078	0.538
3.5	138.8	584.3	2 147.4	2 731.6	1.727 3	5.211 9	6.939 2	0.001079	0.524
3.6	139.8	588.5	2 144.4	2 732.9	1.737 6	5.192 1	6.929 7	0.001080	0.510
3.7	140.8	592.7	2 141.4	2 734.1	1.747 6	5.172 9	6.920 5	0.001081	0.497
3.8	141.8	596.8	2 138.6	2 735.3	1.757 4	5.154 1	6.911 6	0.001082	0.486
3.9	142.7	600.8	2 135.7	2 736.5	1.767 0	5.135 8	6.902 8	0.001083	0.473
4.0	143.6	604.7	2 133.0	2 737.6	1.776 4	5.117 9	6.894 3	0.001084	0.462
4.2	145.4	612.3	2 127.5	2 739.8	1.794 5	5.083 4	6.877 9	0.001086	0.441
4.4	147.1	619.6	2 122.3	2 741.9	1.812 0	5.050 3	6.862 3	0.001088	0.423
4.6	148.7	626.7	2 117.2	2 743.9	1.828 7	5.018 6	6.847 3	0.001089	0.405
4.8	150.3	633.5	2 112.2	2 745.7	1.844 8	4.988 1	6.832 9	0.001091	0.390
5.0	151.8	640.1	2 107.4	2 747.5	1.860 4	4.958 8	6.819 2	0.001093	0.375
5.2	153.3	646.5	2 102.7	2 749.3	1.875 4	4.930 6	6.805 9	0.001094	0.361
5.4	154.7	652.8	2 098.1	2 750.9	1.889 9	4.903 3	6.793 2	0.001096	0.348
5.6	156.2	658.8	2 093.7	2 752.5	1.904 0	4.876 9	6.780 9	0.001098	0.337
5.8	157.5	664.7	2 089.3	2 754.0	1.917 6	4.851 4	6.769 0	0.001099	0.326
6.0	158.8	670.4	2 085.0	2 755.5	1.930 8	4.826 7	6.757 5	0.001101	0.315
6.2	160.1	676.0	2 080.9	2 756.9	1.943 7	4.802 7	6.746 4	0.001102	0.306
6.4	161.4	681.5	2 076.8	2 758.2	1.956 2	4.779 4	6.735 6	0.001104	0.297
6.6	162.6	686.8	2 072.7	2 759.5	1.968 4	4.756 8	6.725 2	0.001105	0.288
6.8	163.8	692.0	2 068.8	2 760.8	1.980 2	4.734 8	6.715 0	0.001107	0.280

TABLE 15.2 Saturated Water and Steam (Pressure) Tables

(Continued)

Absolute pressure (bar) P	Temp. (°C) t_s	Specific enthalpy (kJ/kg)			Specific entropy (kJ/kg K)			Specific volume (m³/kg)	
		h_f	h_{fg}	h_g	s_f	S_{fg}	s_g	v_f	v_g
7.0	165.0	697.1	2 064.9	2 762.0	1.991 8	4.713 4	6.705 2	0.001108	0.273
7.2	166.1	702.0	2 061.1	2 763.2	2.003 1	4.692 5	6.695 6	0.001110	0.265
7.4	167.2	706.9	2 057.4	2 764.3	2.014 1	4.672 1	6.686 2	0.001111	0.258
7.6	168.3	711.7	2 053.7	2 765.4	2.024 9	4.652 2	6.677 1	0.001112	0.252
7.8	169.4	716.3	2 050.1	2 766.4	2.035 4	4.632 8	6.668 3	0.001114	0.246
8.0	170.4	720.9	2 046.5	2 767.5	2.045 7	4.613 9	6.659 6	0.001115	0.240
8.2	171.4	725.4	2 043.0	2 768.5	2.055 8	4.595 3	6.651 1	0.001116	0.235
8.4	172.4	729.9	2 039.6	2 769.4	2.065 7	4.577 2	6.642 9	0.001118	0.229
8.6	173.4	734.2	2 036.2	2 770.4	2.075 3	4.559 4	6.634 8	0.001119	0.224
8.8	174.4	738.5	2 032.8	2 771.3	2.084 8	4.542 1	6.626 9	0.001120	0.219
9.0	175.4	742.6	2 029.5	2 772.1	2.094 1	4.525 0	6.619 2	0.001121	0.215
9.2	176.3	746.8	2 026.2	2 773.0	2.103 3	4.508 3	6.611 6	0.001123	0.210
9.4	177.2	750.8	2 023.0	2 773.8	2.112 2	4.492 0	6.604 2	0.001124	0.206
9.6	178.1	754.8	2 019.8	2 774.6	2.121 0	4.475 9	6.596 9	0.001125	0.202
9.8	179.0	758.7	2 016.7	2 775.4	2.129 7	4.460 1	6.589 8	0.001126	0.198
10.0	179.9	762.6	2 013.6	2 776.2	2.138 2	4.444 6	6.582 8	0.001127	0.194
10.5	182.0	772.0	2 005.9	2 778.0	2.158 8	4.407 1	6.565 9	0.001130	0.185
11.0	184.1	781.1	1 998.5	2 779.7	2.178 6	4.371 1	6.549 7	0.001133	0.177
11.5	186.0	789.9	1 991.3	2 781.3	2.197 7	4.336 6	6.534 2	0.001136	0.170
12.0	188.0	798.4	1 984.3	2 782.7	2.216 1	4.303 3	6.519 4	0.001139	0.163
12.5	189.8	806.7	1 977.4	2 784.1	2.233 8	4.271 2	6.505 0	0.001141	0.157
13.0	191.6	814.7	1 970.7	2 785.4	2.251 0	4.240 3	6.491 3	0.001144	0.151
13.5	193.3	822.5	1 964.2	2 786.6	2.267 6	4.210 4	6.477 9	0.001146	0.146
14.0	195.0	830.1	1 957.7	2 787.8	2.283 7	4.181 4	6.465 1	0.001149	0.141
14.5	196.7	837.5	1 951.4	2 788.9	2.299 3	4.153 3	6.452 6	0.001151	0.136
15.0	198.3	844.7	1 945.2	2 789.9	2.314 5	4.126 1	6.440 6	0.001154	0.132
15.5	199.8	851.7	1 939.2	2 790.8	2.329 2	4.099 6	6.428 9	0.001156	0.128
16.0	201.4	858.6	1 933.2	2 791.7	2.343 6	4.073 9	6.417 5	0.001159	0.124
16.5	202.8	865.3	1 927.3	2 792.6	2.357 6	4.048 9	6.406 5	0.001161	0.120
17.0	204.3	871.8	1 921.5	2 793.4	2.371 3	4.024 5	6.395 7	0.001163	0.117
17.5	205.7	878.3	1 915.9	2 794.1	2.384 6	4.000 7	6.385 3	0.001166	0.113
18.0	207.1	884.6	1 910.3	2 794.8	2.397 6	3.977 5	6.375 1	0.001168	0.110
18.5	208.4	890.7	1 904.7	2 795.5	2.410 3	3.954 8	6.365 1	0.001170	0.107
19.0	209.8	896.8	1 899.3	2 796.1	2.422 8	3.932 6	6.355 4	0.001172	0.105
19.5	211.1	902.8	1 893.9	2 796.7	2.434 9	3.911 0	6.345 9	0.001174	0.102

TABLE 15.2 Saturated Water and Steam (Pressure) Tables

(*Continued*)

Absolute pressure (bar) P	Temp. (°C) t_s	Specific enthalpy (kJ/kg)			Specific entropy (kJ/kg K)			Specific volume (m³/kg)	
		h_f	h_{fg}	h_g	s_f	S_{fg}	s_g	v_f	v_g
20.0	212.4	908.6	1 888.6	2 797.2	2.446 9	3.889 8	6.336 6	0.001177	0.0995
20.5	213.6	914.3	1 883.4	2 797.7	2.458 5	3.869 0	6.327 6	0.001179	0.0971
21.0	214.8	920.0	1 878.2	2 798.2	2.470 0	3.848 7	6.318 7	0.001181	0.0949
21.5	216.1	925.5	1 873.1	2 798.6	2.481 2	3.828 8	6.310 0	0.001183	0.0927
22.0	217.2	931.0	1 868.1	2 799.1	2.492 2	3.809 3	6.301 5	0.001185	0.0907
22.5	218.4	936.3	1 863.1	2 799.4	2.503 0	3.790 1	6.293 1	0.001187	0.0887
23.0	219.5	941.6	1 858.2	2 799.8	2.513 6	3.771 3	6.284 9	0.001189	0.0868
23.5	220.7	946.8	1 853.3	2 800.1	2.524 1	3.752 8	6.276 9	0.001191	0.0849
24.0	221.8	951.9	1 848.5	2 800.4	2.534 3	3.734 7	6.269 0	0.001193	0.0832
24.5	222.9	957.0	1 843.7	2 800.7	2.544 4	3.716 8	6.261 2	0.001195	0.0815
25.0	223.9	962.0	1 839.0	2 800.9	2.554 3	3.699 3	6.253 6	0.001197	0.0799
25.5	225.0	966.9	1 834.3	2 801.2	2.564 0	3.682 1	6.246 1	0.001199	0.0783
26.0	226.0	971.7	1 829.6	2 801.4	2.573 6	3.665 1	6.238 7	0.001201	0.0769
26.5	227.1	976.5	1 825.1	2 801.6	2.583 1	3.648 4	6.231 5	0.001203	0.0754
27.0	228.1	981.2	1 820.5	2 801.7	2.592 4	3.632 0	6.224 4	0.001205	0.0740
27.5	229.1	985.9	1 816.0	2 801.9	2.601 6	3.615 8	6.217 3	0.001207	0.0727
28.0	230.0	990.5	1 811.5	2 802.0	2.610 6	3.599 8	6.210 4	0.001209	0.0714
28.5	231.0	995.0	1 807.1	2 802.1	2.619 5	3.584 1	6.203 6	0.001211	0.0701
29.0	232.0	999.5	1 802.6	2 802.2	2.628 3	3.568 6	6.196 9	0.001213	0.0689
29.5	233.0	1 004.0	1 798.3	2 802.2	2.637 0	3.553 3	6.190 2	0.001214	0.0677
30.0	233.8	1 008.4	1 793.9	2 802.3	2.645 5	3.538 2	6.183 7	0.001216	0.0666
30.5	234.7	1 012.7	1 789.6	2 802.3	2.653 9	3.523 3	6.177 2	0.001218	0.0655
31.0	235.6	1 017.0	1 785.4	2 802.3	2.662 3	3.508 7	6.170 9	0.001220	0.0645
31.5	236.5	1 021.2	1 781.1	2 802.3	2.670 5	3.494 2	6.164 7	0.001222	0.0634
32.0	237.4	1 025.4	1 776.9	2 802.3	2.678 6	3.479 9	6.158 5	0.001224	0.0624
32.5	238.3	1 029.6	1 772.7	2 802.3	2.686 6	3.465 7	6.152 3	0.001225	0.0615
33.0	239.2	1 033.7	1 768.6	2 802.3	2.694 5	3.451 8	6.146 3	0.001227	0.0605
33.5	240.0	1 037.8	1 764.4	2 802.2	2.702 3	3.438 0	6.140 3	0.001229	0.0596
34.0	240.9	1 041.8	1 760.3	2 802.1	2.710 1	3.424 4	6.134 4	0.001231	0.0587
34.5	241.7	1 045.8	1 756.3	2 802.1	2.717 7	3.410 9	6.128 6	0.001233	0.0579
35.0	242.5	1 049.8	1 752.2	2 802.0	2.725 3	3.397 6	6.122 8	0.001234	0.0570
35.5	243.3	1 053.7	1 748.2	2 801.8	2.732 7	3.384 4	6.117 1	0.001236	0.0562
36.0	244.2	1 057.6	1 744.2	2 801.7	2.740 1	3.371 4	6.111 5	0.001238	0.0554
36.5	245.0	1 061.4	1 740.2	2 801.6	2.747 4	3.358 5	6.105 9	0.001239	0.0546
37.0	245.7	1 065.2	1 736.2	2 801.4	2.754 7	3.345 8	6.100 4	0.001242	0.0539

TABLE 15.2 Saturated Water and Steam (Pressure) Tables

REFERENCES

[1]   E. N. Hiebert, "Historical Roots of the Principle of Conservation of Energy," *British Journal for the Philosophy of Science,* vol. 14, no. 54, pp. 160-166, 1963.

[2]   C. Iltis, "The History of Science Society," *Chicago Journals, The University of Chicago Press,* vol. 62, no. 1, pp. 21-35, 1971.

[3]   W.C. Reynolds and H.C. Perkins, Engineering Thermodynamics.

[4]   Chang. L. Tien and John H. Lienhard, Statistical Thermodynamics, Hemisphere Publishing Corp., 1979.

[5]   Herbert B. Callen., Thermodynamics and An Introduction to Thermostatics, Second Edition, John Wiley & Sons, 1985.

[6]   R. Fitzpatrick, "Thermodynamics and Statistical Mechnaics," Lulu Enterprises, Inc., 2006.

[7]   M. Kardar, Statistical Physics of Particles, Cambridge: Cambridge University Press, 2007.

[8]   J. Gibbs, Elementary principles in statistical mechanics, New York: University Press, John Wilson and Son, 1902.

[9]   P. Dhar, Engineering Thermodyanmcis-A Generalized Approach, Delhi: Elsevier, 2013. .

[10]   J. Murrell, "A Very Brief History of Thermodynamics," [Online]. Available: https://www.sussex.ac.uk/webteam/gateway/file.php?name=a-thermodynamicshistory.pdf&site=35. [Accessed 2017].

[11]   [Online]. Available: http://www.thefamouspeople.com/profiles/sadi-carnot-550.php.

[12]   S. Carnot, Réflexions sur la puissance motrice du feu et sur les machines propres à développer cette puissance. (Reflections on the motive power of fire and the clean machinery to develop this power)., 1824.

[13]   E. Clapeyron, "Mémoire sur la Puissance Motrice de la Chaleur (Memoir on the Motive Power of Heat)," *Journal de l'Ecole Royale Polytechnique,* vol. 23, pp. 153-190, 1834.

[14]   R. Clausius, The Mechanical Theory of Heat – with its Applications to the Steam Engine and to Physical Properties of Bodies, London: John van Voorst, Retrieved 19 June 2012. Contains English translations of many of his other works., 1867.

[15]   R.C. Miller and P. Kusch, "Velocity Distribution in Potasium and Thallium Atomic Beams," *Physics Review,* vol. 99, no. 4, pp. 1314-1321, 1955.

[16]   L. Boltzmann, Lectures on Gas Theory, Berkley, CA: Dover Books, 1964, reprint.

[17]   P. Atkins and J. Paula, Physical Chemistry:Thermodynamics, Structure, and Change, W H Freeman & Co., 2014, 1oth ed..

[18]   G. Ruppeiner, "Riemannian geometry in thermodynamic fluctuation theory," *Rev. Mod. Phys.,* vol. 68, p. 313, 1996.

[19]   Y. A Cengel and M. A Boles, Thermodynamics-An Engineering Approach, 8th edition, MacGraw-Hill, 2014.

[20]   I. P. Dilip Kondepudi, Modern Thermodynamics: From Heat Engines to Dissipative Structures, 2nd Edition, Wiley, 2014.

[21]   R. Clausius, The Mechanical Theory of Heat, London: MacMillan and Co., 1879.

[22]   Lars Onsager, "Reciprocal Relations in Irreversible Processes," *Physical Review,* vol. 37, pp. 405-426, 1931.

[23]   M.W. Zemansky and H.C. Van Ness, Basic Engineering Thermodynamics, New York: McGraw-Hill, 1966.

[24]   W. F. John Tyndall, Scientific Memoirs, Selected from the Transactions of Foreign Academies of Science, and from Foreign Journals. Natural Philosophy, Volume 1, Taylor and Francis, 1853, Digitized 26 Aug 2008 (Google Books).

[25]   A. Shulman, "The Discovery of Entropy," [Online]. Available: https://www.youtube.com/watch?v=glrwlXRhNsg. [Accessed 2017].

[26]   G. Porter, "An Introduction to Chemical Change and Thermodynamics," [Online]. Available: https://www.youtube.com/watch?v=lj5tqM5GZnQ. [Accessed 2017].

[27]   D. Roller, The Early Development of the Concepts of Temperature and Heat-the Rise and Decline of the Caloric Theory, Harvard University Press, 1950.

[28]   J. S. John, "A memoir to explain the principles of the methodical nomenclature," in Method of Chemical Nomenclature, London, Kearsley, 1788, pp. 19-50.

[29]   B. C. O. Rumford, "An inquiry concerning the source of the heat which is excited by friction," Philosophical Transactions of the Royal Society of London, vol. 88, pp. 80-102, 1798.

[30]   M. Tribus, Thermostatics and Thermodynamics, Princeton, New Jersey: Van Nostrand, 1961.

[31]   R. Clausius, "Ueber verschiedene für die Anwendung bequeme Formen der Hauptgleichungen der mechanischen Wärmetheorie," Ann. Phys. Chem., vol. 125, pp. 353400, 1865.

[32]   Z. S. Spakovszky, "Thermodynamics and Propulsion," MIT, 1996. [Online]. Available: http://web.mit.edu/16.unified/www/FALL/thermodynamics/notes/node5.html. [Accessed 2017].

[33]   A. Bejan, Entropy Generation Minimization: The Method of Thermodynamic Optimization of Finite-size Systems and Finite-time Processes, CRC Press, Boca Raton, 1996.

[34]   J. C. Maxwell, Theory of Heat, London: Longmans, Green and Co., 1902.

[35]   H. S. a. A. F. R. e. Leff, Maxwell's Demon 2: Entropy, Classical and Quantum Information, Computing, CRC Press, 2002.

[36]   C. L. T. a. J. H. Lienhard, Statistical Thermodynamics, Hemisphere Publishing Corp., 1979.

[37]   N. M. Laurendeau, Statistical Thermodynamics: Fundamental and Applications, Cambridge: Cambridge University Press, 2005.

[38]   E. Jaynes, "Information Theory and Statistical Mechnaics," Phys. Rev. , vol. 106, no. 4, pp. 620-630, 1957.

[39]   E. Jaynes, "Information Theory and Statistical Mechnaics. II," Phys. Rev., vol. 108, no. 2, pp. 171-190, 1957.

[40]   "Physics," Stack Exchange, [Online]. Available: http://physics.stackexchange.com/questions/47593/helmholtz-free-energy-partitionfunction. [Accessed 2017].

[41]   Thomas Engel and Philip Reid, Thermodynamics, Statistical Thermodynamics & Kinetics, second edition, Prentice Hall, 2010.

[42]   A. R. Cunha, "Understanding the ergodic hypothesis via analogies," Physicæ , vol. 10, pp. 9-12, 2011.

[43]   John M.Prausnitz, et. al., Molecular Thermodynamics of Fluid-Phase Equilibria, 3rd edition, Prentice Hall, 1998.

[44]   Courant, Richard and Hilbert, David, Methods of Methematical Physics 2, John Wiley & Sons, 2008.

[45]   Zia, R.K., Redish, E.F. and McKay, S.R., "Making sense of the Legendre transform," *American Journal of Physics,* vol. 77, no. 7, p. 614, 2009.

[46]   W. T. (Kelvin), "On a Universal Tendency in Nature to the Dissipation of Mechanical Energy," *Proceedings of the Royal Society of Edinburgh,* vol. i, no. art. 59, p. 511, 1852.

[47]   J. Joule, "On the Mechanical Equivalent of Heat," *J Philosophical Transactions of the Royal Society of London,* vol. 140, pp. 61-82, 1850.

[48]   U. Wagner, "CERN Accelator School," [Online]. Available: https://cds.cern.ch/record/808372/files/p295.pdf. [Accessed 2017].

[49]   G. Perini , G. Vandoni, T. Niinikoski,, "Introduction to Cryogenic Engineering-CERN," [Online]. Available: http://www.slac.stanford.edu/econf/C0605091/present/CERN.PDF.

[50]   R.H. Perry and D.W. Green, Perry's Chemical Engineering Handbook, McGraw-Hill, 1984.

[51]   A. M. Mathai, Jacobians of Matrix Transformations and Functions of Matrix Arguement, World Scientific Publishing Company, 1997.

[52]   E. Kreyszig, Advanced Engineering Mathematics, 10th Edition, Wiley, 2011.

[53]   "Wolfram Alpha LLC," [Online]. Available: http://www.wolframalpha.com/widgets/gallery/?query=jacobian. [Accessed 2017].

[54]   A. N. Shaw, "The Derivation of Thermodynamical Relations for a Simple System," *Phil. Tran. A, Roy. Soc. Lon.,* vol. 234, no. 740, pp. 299-328, 1935.

[55]   J. H. H. Perk, " A Simple Method to Reduce Thermodynamic Derivatives by Computer," *arXiv.org,* pp. 1-4, 2014.

[56]   G. H. Bryan, "Thermodynamics. An introductory treatise dealing mainly with first principles and their direct applications," in *ch. III,* B.G. Teubner, Leipzig, , 1907, pp. 20-26.

[57]   P. W. Bridgman, A Condensed Collection of Thermodynamic Formulas,, Cambridge, Mass.: Harvard Univ. Press., 1925.

[58]   J. H. H. Perk, "A Simple Method to Reduce Thermodynamic Derivatives by Computer," *arXiv: 1401.7728,* vol. 1, 2014.

[59]   M. Mikhailov, "Derivation of Thermodynamic Derivatives Using Jacobians," from the Wolfram Demonstrations Project, [Online]. Available: http://demonstrations.wolfram.com/DerivationOfThermodynamicDerivativesUsingJacobians. [Accessed 2017]. [60] R. Boyle, "A defence of the doctrine touching the spring and weight of the air," F. G. for Thomas Robinson, London, 1662.

[61]   J.-L. Gay-Lussac, "The Expansion of Gases by Heat," *Annales de Chimie,* vol. 43, no. 137, p. http://web.lemoyne.edu/~giunta/gaygas.html#foot3, 1802.

[62]   Thomas Engel and Philip Reid, Thermodynamics, Statistical Thermodynamics & Kinetics, 3rd edition, Pearson, 2012.

[63]   J. D. v. d. Waals, "The law of corresponding states for different substances",," *Proceedings of the Koninklijke Nederlandse Akademie van Wetenschappen,* vol. 15, no. II , pp. 971-981, 1913.

[64]   G. J. Su, "Modified Law of Corresponding States for Real Gases," *Ind. Eng. Chem.,* vol. 38, no. 8, pp. 803-806, 1946.

[65]   C. Reid, John M. Prausnitz and Thomas K. Sherwood, The properties of gases and liquids, Third Edition, McGraw-Hill, 1978.

[66]   John R. Howell and Richard O. Buckius, Fundamentals of Engineering Thermodynamics, SI version, New York: McGraw-Hill, 1987. P. 558, fig. C.2 and p. 561, fig. C.5..

[67]   B. H. Mahan, Elementary Chemical Thermodynamics, Berkley, CA: W. A. Benjamin, 1963.

[68]   Sanford Klein, Gregory Nellis, Thermodynamics, 1st Edition, Cambridge University Press, 2011.

[69]   Hirschfelder, J.O, Curtis, C.F. and Bird, R.B., The Molecular Theory of Gases and Liquids, New York: Wiley, 1954.

[70]   J. Wisniak, "Heike Kamerlingh-The Equation of State," *Indian Journal of Chemical Technology,* vol. 10, pp. 564-572, 2003.

[71]   H. Onnes, *Proc. Kon. Akad. Wetenschhappen Amsterdam,* vol. 17, 1901.

[72]   Geoffrey Maitland (Author), M. Rigby, E. Brian Smith, W. A. Wakeham, Intermolecular Forces: Their Origin and Determination, Oxford University Press, 1987.

[73]   F. Schreiber, F. Zanini, F. Roosen-Runge, "Virial Expansion – A Brief Introduction," 2011. [Online]. Available: http://www.soft-matter.unituebingen.de/teaching/Tutorial_Virial_Expansion.pdf. [Accessed 2017].

[74]   Robert J. Silbey, Robert A. Alberty, Moungi G. Bawendi, Physical Chemistry, 4th Edition, Wiley, 2005.

[75]   K. Pitzer, Thermodynamics, 3rd ed., McGraw-Hill, 1995.

[76]   J. van der Waals, *Over de Continuiteit van den Gas-en Vloeistoftoestand (on the continuity of the gas and liquid state). Ph.D. thesis,* Leiden, The Netherlands, 1873.

[77]   David R. Lide, CRC Handbook of Chemistry and Physics, 86th Edition, Boca Raton, FL.: Taylor&Francis Group, 2005.

[78]   T. Andrews, "On the continuity of the gaseous and liquid states of matter," *Philos. Trans. R. Soc. London,* vol. 159, pp. 575-590, 1869.

[79]   J.C. Holste, et. al., "Experimenmtal (p,Vm,T) for pure CO2 between 220 and 450K," *J. Chem. Thermodynamics,* vol. 19, pp. 1233-1250, 1987.

[80]   A. Bejan, Advanced Engineering Thermodynamics-Third Edition, Wiley, 2006.

[81]   J. C. Maxwell, "On the dynamical evidence of the molecular constitution of bodies," *Nature (London),* vol. 11, pp. 357-359 , 374-377, 1875.

[82]   S. Blinder, "van der Waals Isotherms," 11 February 2008. [Online]. Available: http://demonstrations.wolfram.com/VanDerWaalsIsotherms/. [Accessed 28 April 2017].

[83]   P. Abbott, "Trick of the Trade: Maxwell COnstruct," *The Mathematica Journal,* vol. 8, no. 1, 2001.

[84]   C. Dieterici, " Ann. Phys. Chem. Wiedemanns Ann.," vol. 69, p. 685, (1899).

[85]   Negi, A.S. and Anand, S.C., A Textbook of Physical Chemistry, New Delhi: New Age International Limited, 1985.

[86]   O. Redlich and J. N. S. Kwong, "On the Thermodynamics of Solutions. V. An Equation of State. Fugacities of Gaseous Solutions," *Chem. Rev.,* vol. 44, no. 1, pp. 233-244, 1949.

[87]   C. Tsonopoulos. and J.L. Heidman, "From Redlich-Kwong to the present," *Fluid Phase Equilinria,* vol. 24, no. 1-2, pp. 1-23, 1985.

[88] J. Albert, "thermod06.pdf," [Online]. Available: http://astrowww.phys.uvic.ca/~tatum/thermod/thermod06.pdf. [Accessed 2017].

[89] D. Berthelot, "A Few Remarks on the Characteristic Equation of Fluids,," *Arch. Netherl. Sci. Natural Exacts, [II],* vol. 5 , pp. 417-446 , 1900.

[90] D. Berthelot, " Travaux et Mémoires du Bureau international des Poids et Mesures – Tome XIII," Paris, Gauthier-Villars, 1907.

[91] D. Y. Peng and D. B. Robinson, "A New Two-Constant Equation of State," *Ind. Eng. Chem. Fundam,* vol. 15, no. 1, pp. 59-64, 1976.

[92] James A. Beattie and Oxcar C. Bridgemann, "A New Equation of State for Fluids," *Journal of the American Chemical Society,* vol. 49, pp. 1665-1667, July 1927.

[93] G.J. Van Wylen and R.E. Sonntag, Fundamentals of Classical Thermodynamics, SI version, 3rd ed., New York: John Wiley & Sons, 1986.

[94] Benedict, Manson; Webb, George B.; Rubin, Louis C, "An Empirical Equation for Thermodynamic Properties of Light Hydrocarbons and Their Mixtures: I. Methane, Ethane, Propane, and n-Butane," *Journal of Chemical Physics,* vol. 8, no. 4, p. 334–345, 1940.

[95] D. Miller, "Joule-Thomson Inversion Curve, Corresponding States, and Simpler Equations of State," *Ind. Eng. Chem. Fundam.,* vol. 9, no. 4, pp. 585-589, 1970.

[96] Garr W. Dilay and Robert A. Heldemann, "Calculation of Joule-Thomson Inversion Curves from Equations of State," *Ind. Eng. CHem. Fundam.,* vol. 25, pp. 152-158, 1986.

[97] K. Juris and L.A. Wenzel, "A study of Inversion Curves," *AICHE Journal,* vol. 18, no. 4, pp. 684-688, 1972.

[98] Gunn, R. D., Chueh, P. L. and Prausnitz, J. M, "Inversion Temperatures and Pressures for Cryogenic Gases and Their Mixtures," *Cryogenics,* vol. 6, pp. 324-329, 1966.

[99] F. Din, Thermodynamic functions of gases, London: Butterworths Scientific Publications, 1956.

[100] H.W. Cooper and J.C. Goldfrank, "Hydrocarbon Processing," vol. 46, no. 12, p. 141, 1967.

[101] A. Newman, 1996. [Online]. Available: http://newmanator.net/projects/vle.pdf. [Accessed August 2016].

[102] H.Orbey and S.Sandler, "Analysis of excess free energy based equation of state models," *AIchE Journal,* vol. 42, no. 8, pp. 2327-2334, 1996.

[103] T.Y.Kwak and G.A.Mansoori, "Van der Waals Mixing Rules for Cubic Equations of State, Applications for Supercritical Fluid Extraction Modeling," *Chaemical Engineering Science,* vol. 41, no. 5, pp. 1303-1309, 1986.

[104] W.B.Kay, "Gases and Vapors At High Temperature and Pressure - Density of Hydrocarbon," *Ind. Eng. Chem.,* vol. 28, pp. 1014-1019, 1936.

[105] J.W.Gibbs, The Collected Works of J. Willard Gibbs, Vol. I, New York: Longmans, Green, 1928.

[106] G. Hawkins, Thermodynamics, New York: John Wiley & Sons, Inc. , 1946.

[107] D.E.Winterbone and A.Turan, Advanced Thermodynamics for Engineers, 2nd ed., Elsevier, 2015.

[108] G. Job and F. Herrmann, "Chemical potential—a quantity in search of recognition," *Eur. J. Phys.,* vol. 27, p. 353–371, 2006.

[109] R. Baierlein, "The elusive chemical potential," *Am. J. Phys.,* vol. 69, no. 4, pp. 423-434, 2001.

[110] H. Devoe, Thermodynamics and Chemistry, Prentice-Hall, 2001.

[111] J. Powers, "Lecture Notes on Intermediate Thermodynamics," Unviversity of Notre Dame, Notre Dame, 2012.

[112] G. Job, "EDUARD-JOB-FOUNDATION FOR THERMO- AND MATTERDYNAMICS," Eduard-Job-Stiftung c/o NTU, [Online]. Available: http://www.jobstiftung.de/index.php?id=54,0,0,1,0,0. [Accessed 2017].

[113] G. Lewis, "The law of physico-chemical change," *Proceedings of American Academy of Arts and Sciences,* vol. 37, no. 4, pp. 49-69, 1901.

[114] A. Avogadro, "Essay on a Manner of Determining the Relative Masses of the Elementary Molecules of Bodies, and the Proportions in Which They Enter into These Compounds," *Journal de Physique,* vol. 73, no. 4, pp. 58-76, 1811.

[115] R. Mollier, "Ein neues diagram für dampfluftgemische (New Graphs for steam mixtures)," *ZVDI,* vol. 67, no. 9, 1923.

[116] K. J. H., Steam tables and Mollier diagram, The American society of mechanical engineers, 1930, Digitized 2007.

[117] K. Kolmetz and A. Jaya, "KLM Technology Group," Feb. 2015. [Online]. Available: http://kolmetz.com/pdf/EGD2/ENGINEERING_DESIGN_GUIDELINES_steam_turbine_ systems_rev_web.pdf. [Accessed 2017].

[118] E. Myer Kutz, "Steam Turbines," in *Mechanical Engineers' Handbook: Energy and Power, Volume 4, Third Edition.*, John Wiley & Sons Inc., 2006, pp. 844-885.

[119] "LearnEengineering.org," Learn Engineering, [Online]. Available: https://www.youtube.com/watch?v=MulWTBx3szc. [Accessed 2017].

[120] W. B. JACHENS, "STEAM TURBINES - THEIR CONSTRUCTION,SELECTION AND OPERATION," Proceedings of The South African Sugar Technologists' Association-, March 1966.

[121] C. B. Meher-Homji, "The Historical Evolution of Turbomachinery," [122] Proceedings of the 29th Turbomachinery Symposium, 2000.

[122] T. Croft, Steam-turbine principles and practice, Nabu Press , 2010.

[123] R.S., Andersen, B. Salamon, P. Berry, , "Thermodynamics in finite time," *Phys. Today,* vol. 37, no. 9, pp. 62-70, 1984.

[124] K. Hoffmann, "Recent Development in Finite Time Thermodynamics," *Technische Mechanik,* vol. 22, no. 1, pp. 14-25, 2002.

[125] A. Vaudrey, F. Lanzetta, M. Feidt, "H. B. Reitlinger and the origins of the Efficiency at Maximum Power formula for Heat Engines," arXiv.org, 2014.

[126] H. Rubin, "Optimal configuration of a class of irreversible heat engines," *I. Phys. Rev.,* vol. 19, no. 3, pp. 1272-1276, 1979.

[127] Hoffmann,K.H. , Burzler, J., Schubert, S., "Endoreversible systems," *J. Non-Equil. Therm.,* vol. 22, no. 4, pp. 311-355, 1977.

[128]   K. H. HOFFMANN, "AAPP," 1 February 2008. [Online]. Available: http://cab.unime.it/journals/index.php/AAPP/article/view/C1S0801011/294. [Accessed September 2016].

[129]   A. D. Vos, Thermodynamics of Solar Energy Conversion, Wiley-VCH, 2008.

[130]   Curzon, F.L. and Ahlborn, B., "Efficiency of a Carnot engine at maximum power output," *Am. J. Phys.,* vol. 43, pp. 22-24, 1975.

[131]   H.B.Reitlinger, On the use of heat in steam engines, Liege, Belgium: Vaillant-Carmanne, 1929, in Frech.

[132]   M. Esposito, R. Kawai, K. Lindenberg, and C. Van den Broeck,, "Efficiency at maximum power of low-dissipation Carnot engines," *Phys. Rev. Lett.,* vol. 105, no. 15, 2010.

[133]   Spalding, D.B. and Cole, E.H., Engineering Thermodynamics, 2nd ed., London: Edward Arnold, 1966.

[134]   G. Griffiths, *Phys. Can.,* vol. 30, no. 2, 1974.

[135]   A. Chierici, "Planning of a Geothermal Power PLant: Technical and Economic Principles," in *UN Conference on New Source of Energy*, New York, 1964.

[136]   I. I. Novikov, "The Efficiency of Atomic Power Stations," *Journal Nuclear Energy II,* vol. 7, pp. 125–128 (translated from Atomnaya Energiya, volume 3 (1957), pp. 409-412, 1958.

[137]   J.M. Gordon and M. Huleihil, "General performance characteristics of real heat engines," *J. App. Phys.,* vol. 72, no. 3, pp. 829-837, 1992.

[138]   Yalin Ge, Lingen Chen and Fengrui Sun, "Progress in Finite Time Thermodynamic Studies for Internal Combustion Engine Cycles," *Entropy,* vol. 18, no. 139, pp. 1-44, 2016.

[139]   A. Sharma, S.K. Shukla, and A.K. Raj, "Finite time thermodynamic analysis and optimization of solar-dish Stirling heat engine with Regenerative losses," *Thermal Scince,* vol. 15, no. 4, pp. 995-1009, 2011.

[140]   H.Leff, "Thermal efficiencies at maximum work output: new results for old heat engines," *American Journal of Physics,* vol. 55, pp. 602-610, 1987.

[141]   T. Markvart, "From steam engine to solar cells: can thermodynamics guide the development of future generations of photovoltaics?," *Wiley Interdisciplinary Reviews: Energy and Environment, Wiley Periodicals, Inc.,* vol. 5, no. 5, pp. 543-569, 2016.

[142]   Angulo-Brown,F. Paez-Hernandez,R., "Endoreversible thermal cycle with a non-linear heat transfer law," *J. Appl. Phys.,* vol. 74, pp. 2216-2219, 1993.

[143]   S. Xia, L. Cehn and F. Sun, "Optimal paths for minimizing lost available work during heat transfer process with a generalized heat transfer law," *Brazilian Journal of Physics,* vol. 39, no. 1, pp. 99-106, 2009.

[144]   Theodore L. Bergma, et. al., Introduction to Heat Transfer, 6th Edition, Wiley, 2011.

[145]   M. Tabatabaian, "COMSOL Blog," 29 Feb. 2016. [Online]. Available: https://www.comsol.com/blogs/studying-transient-heat-transfer-in-a-fin-design-with-anapp/. [Accessed 20 Jan. 2017].

[146]   P. Wormer, 2007. [Online]. Available: http://en.citizendium.org/wiki/File:Vanderwaals_equation_of_state.png. [Accessed 2017].

EVEN-NUMBERED ANSWERS TO SELECTED EXERCISES

Chapter 7

2. [**Ans.** (i) 0.4058 kg, (ii) 6252.9 kJ]
4. [**Ans.** (i) 4390.3 kJ, (ii) 614.5 kJ]
6. [**Ans.** (i) 400 ^{0}C, (ii) 0.01722 m^3/kg, (iii) 2760.82 kJ/kg]
8. [**Ans.** (i) 0.897, (ii) 0.2449 m^3/kg, (iii) 2379.67 kJ/kg]
10. [**Ans.** 0.946]

Chapter 8

Multiple choice questions:

ANSWERS

2. (a)   **4.** (b)   **6.** (b)   **8.** (b)   **10.** (a).

22. [**Ans.** 0.296]
24. [**Ans.** (a) 52.94%, (b) 59.35 × 10^4 kg/h, (c) 28.49 MW, (d) 46.2%]

Chapter 9

Multiple choice questions:

ANSWERS

2. (b)   **4.** (c)   **6.** (c)   **8.** (d)   **10.** (c).

18. [**Ans.** (i) 4.55 bar; (ii) 110.5 kW; (iii) 1018 kg/h]
20. [**Ans.** $D_{L.P.}$ = 455 mm; $D_{H.P.}$ = 312 mm]

24. [**Ans.** (*a*) 233 mm; (*b*) 516 mm; (*c*) 516 mm; (*d*) 1.029]

26. [**Ans.** (*a*) 2.34 bar. 2.87 bar; (*b*) D.F. = 0.815; 142.5 kW]

28. [**Ans.** (*a*) 5.23 bar; (*b*) 270 kW; (*c*) 284 mm]

30. $\left[\textbf{Ans.}\ (i)\ \left(p_m\right)_{\text{H.P.}} = 5.58 \text{ bar},\ \left(p_m\right)_{\text{I.P.}} = 2.1 \text{ bar}, \right.$
$\left. \left(p_m\right)_{\text{L.P.}} = 0.99 \text{ bar},\ (ii)\ loss\ in\ power = 6.7\% \right]$

Chapter 10

Multiple choice questions:

ANSWERS

2. (*c*)	**4.** (*a*)	**6.** (*d*)	**8.** (*d*)
10. (*a*)	**12.** (*c*)	**14.** (*c*)	**16.** (*d*)
18. (*c*)	**20.** (*a*).		

40. [**Ans.** 879 kW]

42. [**Ans.** (*i*) 39°, (*ii*) 1.135 kN, (*iii*) 113.5 kW]

44. [**Ans.** (*i*) 36.3°, (*ii*) 144 kW, (*iii*) 0.762]

46. [**Ans.** (*i*) 4580 r.p.m., (*ii*) 225 m/s, (*iii*) 290.5 Nm]

48. [**Ans.** (*i*) 46°, 49°; (*ii*) 243 m/s; (*iii*) 19.8 kW]

50. [**Ans.** 36°, 0.317 kg/s]

52. [**Ans.** (*i*) 64 mm, (*ii*) 218 kW, (*iii*) 19.1 kJ/kg]

54. [**Ans.** (*i*) 61.5°, (*ii*) 82.2%, (*iii*) 30.47%]

56. [**Ans.** 40.4 kJ/kg, 70.3%, 57 mm]

Chapter 11

Multiple choice questions:

ANSWERS

2. (*b*)	**4.** (*d*)	**6.** (*b*).

14. [**Ans.** 32.55%]

16. [**Ans.** 4]

18. [**Ans.** (*i*) 56.47%; (*ii*) 14.24 bar]
20. [**Ans.** 56.1%; 14.6%]
22. [**Ans.** (*i*) 61%; (*ii*) 13.58 bar]
24. [**Ans.** 2.1%]
26. [**Ans.** 61.7%]
28. [**Ans.** (*i*) 61.92%; (*ii*) 9.847 bar]
30. [**Ans.** (*i*) 66.5%; (*ii*) 4.76 bar]

Chapter 12

Multiple choice questions:

ANSWERS

2. (*a*)   **4.** (*c*)   **6.** (*d*)

8. (*a*)   **10.** (*a*)   **12.** (*a*)

24. [**Ans.** 2.393 , 37.03%]
26. [**Ans.** 920 kW]
28. [**Ans.** 655 kW, 12%, 0.168]
30. [**Ans.** 32.75%, 0.3852, 152 kJ/kg]
32. [**Ans.** 22.76%]
34. [**Ans.** 16.17%, 0.2215, 69.33 kg/s of air]
36. [**Ans.** (*i*) 613.5 m/s; (*ii*) 0.7267 kg/s; (*iii*) 4.3×10^{-5} kg / N of thrust /s; (*iv*) 44%; (*v*) 8318 kw; (*vi*) 60.6%; (*vii*) 16.58%]

Chapter 13

Fill-the-blank questions:

ANSWERS

2. Refrigerant   **4.** Relative C.O.P.   **6.** $\dfrac{T_1}{T_1 - T_2}$

8. Less   **10.** Vapor compression   **12.** Condenser

14. Evaporator   **16.** Dry compression   **18.** Sub-cooling

46. [**Ans.** 27.83 kW]

48. [**Ans**. 21.44 kW]

50. [**Ans.** (*i*) 9.54 ; (*ii*) 9.54 kW ; (*iii*) 10.54 kW]

52. [**Ans.** 69.6°C ; 88.42 kJ/kg]

54. [**Ans.** 0.502 kg/min ; 9.04 kW ; 2.73 kW]

56. [**Ans.** 21.59 kW]

58. [**Ans.** (*i*) 129.17 kJ/kg ; (*ii*) 19.355 kg/min ; (*iii*) 1.518 m³/min ; (*iv*) 7.2 kW ; (*v*) 2931 kJ/min ; (*vi*) 91 mm, 118 mm]

60. [**Ans.** 4.95, 682 kg/h, 2.2 m³/min]

Index

F

Feed pump, 261–262

Finite-time thermodynamics, 804

1st law of thermodynamics, 6–7, 12–15, 48

Fluorocarbons, 788

Flying stovepipe, 703

Freezing point, 787

Freon-12, 792–793

Freon-22, 793

Fuel oil consumption, 278

Fugacity, concept of, 186–187

Fusion curve, 218

G

Gaseous fuels, 634

Gas mixtures, 165

    algebraic mixing rules, 172–173

    Amagat's mixing rule, 167–172, 174

    chemical potential. *See* Chemical

        potential, for gas mixtures

    chemical reactions equilibrium,

        190–191

    Dalton's mixing rule, 166–167,

        170–172, 174–177

    at different temperatures, 179–182

    heterogeneou mixture, 162

    homogeneous mixture, 162

    ideal. *See* Ideal gas mixture

    Kay's mixing rules, 169–171, 173

    non-reactive mixture, 162

    reactive mixture, 162, 182

    real, 169

    total mass of, 163

    van der Waals mixing rules, 172–177

    Z-factor, 170–171

Gas power cycles

    air standard efficiency, 509–510

    definition of, 509

    overview, 509

Gas turbines, 591–592, 731

    air-fuel ratio in, 599

    application of, 614

    characteristic features of, 613

    classification of, 614

    compressor in, 619

    constant pressure combustion. *See*

        Constant pressure combustion

        gas turbines

    constant volume combustion, 614, 633

    demerits of, 616

    efficiency of, 614

    examples, 635–678

    gaseous fuels, 634

    liquid fuels, 634–635

    merits of, 615–616

    power plant, 586

    reheating, 621–622

    solid fuels, 635

    thermal efficiency of, 614

    uses of, 634

Generalized compressibility chart,

    106–109

    Lee-Kesler charts, 109

    Nelson-Obert charts, 109

Generalized departure charts, 118

Gibbs-Dalton law, 177

Gibbs energy functions, 37, 50–52

    chemical potential and, 185–186

    departure functions and factors for, 117

    of ideal gas mixture, 188–190

Gibbs phase rule, 102

Gibbs statistical thermodynamics, 43–46

Governing

    by-pass, 495–496

    of compound engines, 345–347

    method of, 383

    nozzle, 494–495

    throttle, 493

Grand canonical ensemble, 44

H

Halocarbon

    compounds, 784–785

    refrigerant, 784–785

www.ingramcontent.com/pod-product-compliance
Lightning Source LLC
Chambersburg PA
CBHW061922190326
41458CB00009B/2630